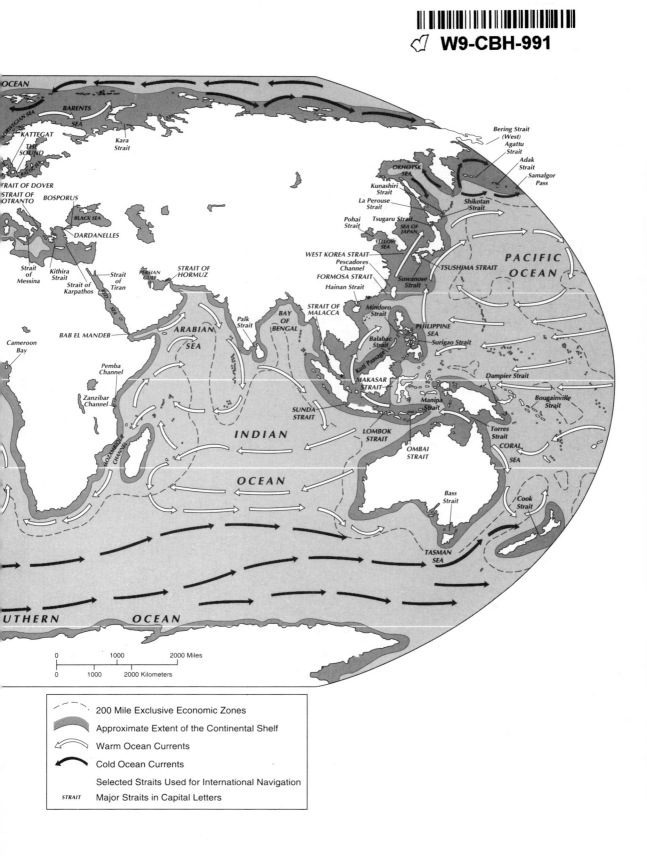

W9-CBH-991

OCEAN

NORWEGIAN SEA

BARENTS SEA

KATTEGAT
THE SOUND
BALTIC SEA

STRAIT OF DOVER
STRAIT OF OTRANTO
BOSPORUS
BLACK SEA
DARDANELLES

Strait of Messina
Kithira Strait
Strait of Karpathos

Kara Strait

Strait of Tiran
PERSIAN GULF
STRAIT OF HORMUZ

BAB EL MANDEB

RED SEA

Cameroon Bay

Pemba Channel
Zanzibar Channel

MOZAMBIQUE CHANNEL

ARABIAN SEA

Palk Strait

BAY OF BENGAL

INDIAN

OCEAN

STRAIT OF MALACCA

SUNDA STRAIT

LOMBOK STRAIT

OMBAI STRAIT

Bering Strait (West)
Agattu Strait
Adak Strait
Samalgor Pass

OKHOTSK SEA

Kunashiri Strait
La Perouse Strait

Shikotan Strait

Pohai Strait
Tsugaru Strait
SEA OF JAPAN
YELLOW SEA

WEST KOREA STRAIT
Pescadores Channel
FORMOSA STRAIT
Hainan Strait

Suwanose Strait

TSUSHIMA STRAIT

PACIFIC OCEAN

Mindoro Strait

PHILIPPINE SEA

Balabac Strait
Kati Passage

MAKASAR STRAIT

Manipa Strait

Surigao Strait

Dampier Strait

Bougainville Strait

Torres Strait

CORAL SEA

Bass Strait

Cook Strait

TASMAN SEA

SOUTHERN          OCEAN

| 0 | 1000 | 2000 Miles |

| 0 | 1000 | 2000 Kilometers |

- - - - 200 Mile Exclusive Economic Zones

Approximate Extent of the Continental Shelf

Warm Ocean Currents

Cold Ocean Currents

Selected Straits Used for International Navigation

STRAIT  Major Straits in Capital Letters

# Political Geography

---

## Second Edition

## Martin Ira Glassner

### John Wiley & Sons, Inc.

NEW YORK   CHICHESTER   BRISBANE   TORONTO   SINGAPORE

| ACQUISITIONS EDITORS | Frank Lyman/Christopher Rogers |
|---|---|
| MARKETING MANAGER | Catherine Faduska |
| PRODUCTION EDITOR | Sandra Russell |
| DESIGNER | Ann Marie Renzi |
| MANUFACTURING MANAGER | Mark Cirillo |
| PHOTO EDITOR | Lisa Passmore |
| ILLUSTRATION EDITOR | Edward Starr |
| MAPS/ILLUSTRATIONS/GRAPHICS | Alice Thiede |
| COVER DESIGN | Laura Ierardi |
| PHOTOGRAPH | © Earth Satellite Corporation |

PART OPENER CREDITS    **Part I:** Courtesy New York Public Library. **Part II:** A.C.T. Tourism Commission. **Part III:** Michael Massey/Gamma Liaison. **Part IV:** Bettmann Archive. **Part V:** Department of Defense/Still Media Record Center. **Part VI:** UN Photo. **Part VII:** Reuters/Bettmann. **Part VIII:** Mekong River Commission. **Part IX:** UN Photo

This book was set in ITC Garamond Light by ATLIS Graphics & Design, Inc. and printed and bound by Hamilton Printing Company.  The cover was printed by New England Book Components.

Recognizing the importance of preserving what has been written, it is a policy of John Wiley & Sons, Inc. to have books of enduring value published in the United States printed on acid-free paper, and we exert our best efforts to that end.

The paper in this book was manufactured by a mill whose forest management programs include sustained yield harvesting of its timberlands. Sustained yield harvesting principles ensure that the number of trees cut each year does not exceed the amount of new growth.

***Library of Congress Cataloging-in-Publication Data***
Glassner, Martin Ira, 1932 –
     Political geography / Martin Ira Glassner. – 2nd ed.
        p.   cm.
     Includes index.
     ISBN 0-471-11496-0  (cloth: alk. paper)
     1. Political geography.    I. Title
JC319.G588    1995                              95-32362
320. 1′2–dc20                                    CIP

Printed in the United States of America

10  9  8  7  6  5  4  3

*To Renée,*
*Karen, Aleta and Cindy*

# *Preface to the First Edition*

## About This Book

This book has an interesting and honorable history. In 1967 Harm de Blij published a fine, innovative introductory text called *Systematic Political Geography*. The second, improved, edition appeared in 1973. In 1978 he and Wiley asked me to prepare the third edition. I changed the format radically, with many short chapters arranged in a logical but not rigid sequence so as to allow for maximum flexibility in assigning and reading the subject matter. I also added substantial new text and many new maps, photographs, and other illustrations. *SPG III* appeared in 1980. Late in 1986 I was asked to revise it again, and again I made many improvements, as well as expanding the book even more. The result was the fourth edition of *Systematic Political Geography (SPG IV)*, which was published in 1989. Since then the numerous historic changes in the world have necessitated an accelerated revision schedule. A new title, *Political Geography,* was chosen for this edition, reflecting my sole authorship.

The present work contains a number of new features. Naturally, it has been thoroughly updated. Chapters 1, 18, and 19 have been extensively rewritten and there is a new Chapter 17 on indigenous peoples. Several new sections appear, notably those on international peacekeeping operations, the Arctic, contemporary State practice in marine affairs, the proposed New World Information and Communication Order and New International Economic Order, and the United States outer continental shelf. There are 15 new maps and numerous new graphs, photographs, and other illustrations. Finally, the references have been thoroughly revised. I have deleted hundreds of the older, less useful and less accessible items and more than replaced them with newer and better items. I have retained, however, a number of older "classic" and near-classic materials and have followed the same guidelines used for *SPG IV*. Only books, monographs, and articles in English are included. Regretfully excluded for purely pragmatic reasons are immensely valuable materials in other languages, atlases, and unpublished items such as dissertations and papers read at professional meetings. Articles appearing in anthologies or other edited volumes are not listed separately if the books themselves are listed. Nor are articles appearing in specialized journals listed; only the journals themselves are, as each is a treasure trove of material on the topic it covers. General texts in political geography, whether edited or single-author volumes, are listed with the references for Part One; studies of individual countries and regions are listed with Part Nine. There are some exceptions to most of these principles.

Placement of many references was difficult—some cover many topics and some topics could logically fall under many of the book's headings. No item is intentionally listed more than once. Readers are advised to scan *all* of the references and footnotes that might have any relation to a topic of interest and to use these references and footnote citations only as a springboard for research in the materials omitted. Researchers are especially urged to utilize the documents and publications of the United Nations, its various organs, and its specialized agencies; those of other intergovernmental organizations and of individual governments; and those of specialized, nonprofit, nongovernmental organizations. These sources can provide immensely valuable material on virtually every topic covered here. Although there is inadequate space to list all useful materials here, the collection of references and citations herein constitutes the largest bibliography of political geography in the world.[1]

## Acknowledgments

Again I express my gratitude to Harm de Blij for his kindness in giving me the opportunity to combine my study, field work, experience, and thinking to create *Political Geography*. Several other people provided assistance of various kinds and I would like to thank them publicly. Among them are:

Rich Allan, Municipality of Metropolitan Toronto

Peter Doyle, Delegation of the Commission of the European Communities, Washington

Geoffrey Martin, SCSU, for comments on Chapter 1

Bruce McDowell, Advisory Commission on Intergovernmental Relations, Washington, for comments and materials in interstate relations in the United States

Ethan Nadelmann, Princeton University, for material on drug trafficking

Geoffrey Parker, University of Birmingham, England, for suggestions and materials on geopolitics

M.A. Robinson, Fisheries Division, Food and Agriculture Organization of the United Nations, Rome

Imre Sutton, Emeritus, California State University, Fullerton, for material on American Indians

Betsy Wheeler and Don White, Tennessee Valley Authority

Mike Williams, *Congressional Quarterly*, Washington

William Wood, The Geographer, U.S. Department of State

Many other people in embassies, government agencies, research institutes, universities, nongovernmental organizations, and international organizations for bits of information, maps, papers, documents, and comments.

The following people offered suggestions for the book in response to my circular letter to members of the Political Geography Specialty Group of the Association of American Geographers:

Bruce Davis, Jackson State University, Mississippi

Michael Doran, Southern Literary Agency, Houston

Hurlburt Field, Troy State University, Florida

James Hunter, Emeritus, Georgetown University

David Knight, Carleton University, Ottawa

David Newman, Ben-Gurion University of the Negev, Israel

Henk Rietsma, University of Amsterdam

Michael Roskin, Lycoming College, Pennsylvania

Guido Weigend, Emeritus, Arizona State University

Christine Drake of Old Dominion University, Mark Bassin of University of Wisconsin at Madison, Anthony Williams of Pennsylvania State University, and four of his students reviewed *SPG IV* carefully and made a number of suggestions for improvements to be incorporated into this book. The staff of Wiley did their usual professional job of preparing and producing a complex book very rapidly and very well. They were very helpful, if at times merciless in holding me to deadlines.

Jessie Jones of SCSU, she of great skill, good humor, and infinite patience, typed all the references and much of the text. Finally, my lonely "book widow" wife, Renée, deserves accolades for sharing her very spare time and the dining room table with the manuscript over several months.

## About the Author

Most political geography textbooks are written by scholars, some of whom have done a bit of field work as well as burrowing in libraries. I have done my share of both, and my more relevant publications are listed in the references. What makes this text rather different from most—and I hope more interesting and useful—is that much of it has also emerged from my own experience.

Briefly, I was born and raised in New

---

[1] I would like to call attention to *World Eagle*, a marvelous monthly publication that has been very helpful to me. It is full of authoritative facts, statistics, and illustrations on a great many topics of current interest. It is published at 111 King Street, Littleton, MA 01460.

Jersey, where I became a lifelong history fanatic and collector of books, stamps, and maps. I received my BA in Geography from Syracuse University in 1953 and went on to do graduate work at the University of Wisconsin in Madison; in Pennsylvania, Chicago, and Mexico City; at California State University at Fullerton (MA in Geography and Political Science, 1964), and the Claremont Graduate School, California (Ph.D. in International Relations, 1968). I served in the army at Fort Belvoir, Virginia, and in the Foreign Service of the United States in Washington, D.C.; Kingston, Jamaica; and Antofagasta, Chile. I have taught geography, history, and political science at three colleges in California (1965–67), political science at the University of Puget Sound in Tacoma, Washington (1967–68), and geography and political science in New Haven since 1968. In addition, I have taught at universities in Haifa, Israel; Copiapó, Chile; and Zhengzhou, China, and have lectured in many places in the United States and in Canada, Mexico, and Yugoslavia.

I have read papers at numerous professional meetings in the United States, Canada, Mexico, Costa Rica, Trinidad, and Senegal; have had a study tour of North Africa on a federal government grant (1971); and have done field work (with the aid of research grants) in South America (1983 and 1985); Central and Eastern Europe (1988); South and Southeast Asia (1989); and West Africa and Western Europe (1991). I have been a Visiting Scholar at the Yale Law School (1974–75), represented the International Law Association (ILA, headquarters in London) and served as advisor to the delegation of Nepal at the Third United Nations Conference on the Law of the Sea (1974–82); and currently represent the ILA at the United Nations.

I have served as advisor to His Majesty's Government of Nepal in Kathmandu on negotiating transit treaties with India (1976 and 1989), and have served as a consultant to the United Nations Development Programme in South and Southeast Asia (1979) and Southern Africa (1984 and 1990), and to the United Nations Conference on Trade and Development in Geneva (1984).

Altogether, to date I have visited 96 countries—and look forward to learning about many more, first hand. This, then, helps to explain why this book tends to be pragmatic and rooted in real life, why theory is included but not emphasized, and why its perspective, both temporal and spatial, is broader than it is deep. I have written it with its intended audience—students in English-speaking countries—constantly in mind, though I would be delighted if others found it useful and enjoyable as well. I invite comments on it from all readers.

New Haven, Connecticut

February 1992

# *Preface*

Since preparing the first edition of this book, I have had the great good fortune to be able to do field work—with the aid of research grants from the Connecticut State University—in Belarus, Finland, Moldova, Russia and Ukraine, and in Kenya, Uganda, and Israel. I have also visited Saint Lucia and participated in conferences in several American cities and in London and Malta. Like my previous experiences, these have both broadened and deepened my understanding of the world. In addition to specific bits of information and illustrations that I brought back from each trip, I have tried to incorporate into this edition some of this enhanced understanding. Although I have now visited 107 countries and territories, I am still looking forward to visiting more and learning more. But the more I learn, the more I realize how little I know, and I am impelled to learn even more, in a never-ending process. I hope that in this book I have conveyed just a tiny bit of the excitement of learning.

## *Acknowledgments*

Dozens of people have helped me with this book, most of them civil servants and diplomats in numerous United States government and United Nations agencies and in embassies, consulates, and UN missions of many countries. I express my gratitude to all of them. I would also like to thank the following people for somewhat more substantial contributions:

Amy Eisenberg for material on the original people of Taiwan.

Ron Crocombe of the University of the South Pacific and Elizabeth Wright of the Ministry of Foreign Affairs of the Cook Islands for information on the Cook Islands.

Beth Marks and Charles Webb of the Antarctica Project, Jack Child of The American University, and Robert Smith of the U.S. Department of State for comments on the Antarctica chapter.

Shlomo Shalmon of Kibbutz Gesher, Israel, for the photograph and the raw material for the map of the Jordan Valley.

Imre Sutton, emeritus of the California State University at Fullerton, for comments on my treatment of U.S. Indian reservations.

Joseph Zimmerman of the State University of New York at Albany for comments on Part Three.

Four people provided detailed comments on the first edition at the request of the publisher: Simon Dalby of Carleton University, Wayne McKim of Towson State University, Howard G. Salisbury of Northern Arizona University, and one person who chose to remain anonymous. I thank them all for their helpful suggestions. Three other professors graciously provided comments on the whole book at my request: Max Barlow of Concordia University, Fillmore C.F. Earney of Northern Michigan University, and Michael Roskin of Lycoming College. They, too, deserve my gratitude. Finally, my own political geography students deserve thanks for filling out a questionnaire I devised for the evaluation of their text; their comments were both helpful and gratifying.[1]

Jessie Jones of the Public Affairs Office and Joan Boughton, secretary to the Department of Geography at Southern Connecticut State University, did a splendid job of typing the new material. The staff of Wiley also deserve kudos for their professionalism and

[1]Again I must express my gratitude to the staff of that most unusual and helpful publication *World Eagle*, from which many of the graphs and statistics in this book were obtained. The address is 111 King Street, Littleton, MA 01460.

their forbearance with an author who did not always follow the rules. And, of course, my wife Renée deserves garlands of roses for enduring months of only a part-time husband—again.

## *How to Use This Book*

This book is designed as a textbook, but it can also be useful as a reference book since it contains a vast amount of factual material without being encyclopedic, and it still contains the world's largest bibliography in political geography. Scattered throughout the book are numerous suggestions of topics that need further investigation, even in student research papers; many of the footnotes also suggest good topics. The preface to the first edition can be helpful in deriving full benefit from this edition. Finally, some people may find the book simply a good read, something with which to curl up before the fireplace on a long winter evening or to idle away a summer vacation on a sunny beach. I'd like to hear from anyone who enjoys it in this way!!

New Haven, Connecticut

March 1995

# Contents

# Maps

# PROLOGUE

## *Into the Twenty-first Century*

*There is a tide in the affairs of men*
*Which, taken at the flood, leads on to fortune;*
*Omitted, all the voyage of their life*
*Is bound in shallows and in miseries.*
*On such a full sea are we now afloat;*
*And we must take the current when it serve*
*Or lose our ventures.*
William Shakespeare, *Julius Caesar, IV, 3*

Fifteen years ago I wrote,

> The nineteenth century, for all practical purposes, ended in the fire and blood of the First World War. The twentieth century began with the hope of Versailles in 1919. No one now can foretell how the twentieth century will end or the twenty-first begin. We can see clearly, however, that the century in which we live is a time of transition, of readjustment from a world that evolved but slowly for thousands of years to a world which will be very different indeed from anything in the human experience. Of all the momentous events and processes we are witnessing in the twentieth century, five are crucial. Population . . . Science and Technology . . . Decolonization . . . The Rich–Poor Gap . . . Environment. . . .*

It is possible that many years from now historians will look back and judge that the twenty-first century (by the Gregorian calendar) began during the period in which we

are now living, roughly from the mid-1980s to the mid-1990s. Seldom in human history have so many momentous developments come cascading down upon us in such a short space of time. Let's review some of them to see why they may add up to a watershed, a revolutionary change from one way of life to another, beginning with the five I identified in 1978 and then moving on to two new areas, social conditions and politics.

*Population* During the 1980s the rate of world population growth fell and then leveled off for the first time since the early stages of the Industrial Revolution. Indications are that the rate of growth will resume its fall beginning in the mid-1990s. This could result in a reduction of the rate of environmental degradation in rural areas, especially in developing countries, and have many other beneficial effects in all areas of human life.

*Science and Technology* The world, led by the United States, has entered the long-anticipated post-industrial society. The ad-

*Martin Ira Glassner, *A Special Issue on The Law of the Sea. Focus* (American Geographical Society) 28, 4 (March–April 1978), 1–2.

1

justment process, as is usual in revolutionary situations, is difficult, even painful. It will continue to be so as waves of industrialization and then a service economy spread out from North America and Europe to the rest of the world. This period has also seen us thrusting far out into outer space, laying the foundations for dramatic new uses of our last frontier that we can scarcely imagine now. Biotechnology and especially genetic engineering have come of age and may soon be commonplace, changing the world in unforeseeable ways. We now have worldwide, instant communications, thanks to orbiting satellites and fiber optics. A glimmer of what this will mean was evident during the Gulf War in early 1991, when both national and coalition decisions were made on the basis not only of military satellite communications, but also the commercial transmissions of the Cable News Network. National stock exchanges (such as those in London, Frankfurt, Hong Kong, Tokyo, and New York) are already linked by satellite and we may soon have one global, round-the-clock stock exchange. And finally, we seem to be on the brink of mastering fusion energy. When it is fully usable, it may render obsolete most other types of electricity generation, with incalculable effects on every person on the planet.

*Decolonization*  We have now just about completed the decolonization of the world, with the breakup of the Soviet Union, the world's last big colonial empire. Of the traditional overseas empires, only fragments remain, along with a number of relatively small unsettled sovereignty questions. Peace has come to Angola, Chad, and Mozambique; Namibia is independent. How truly historic! Colonial peoples, free at last!

*The Rich–Poor Gap*  This is still, regrettably, a major feature of the world scene. But at last there are signs that the gap may not grow much wider and may begin to shrink before long. The Uruguay Round of trade negotiations under the General Agreement on Tariffs and Trade (GATT) is crucial: If it is successful in reducing agricultural subsidies in the rich countries and freeing trade in agricultural products, and including trade in services and intellectual property in a further liberalizing of the world trading system, the chain reaction is likely to bring substantial benefits to the poor countries. The debt burden acquired by many poor countries in the 1980s, resulting in a net outflow of their resources to the rich countries, seems to be about to begin shrinking, to some extent through jiggery-pokery but basically because of some fundamental restructuring that is getting under way. This includes the first reduction in a generation of the proportion of the developing world's gross domestic product being devoted to military expenditures. Economic integration efforts being undertaken now are more solidly based and more likely to bring more benefits to more people in developing countries than their predecessors. It will undoubtedly be a long time before the gap is narrowed to an acceptable size, and perhaps it never will, but it does appear that at last the process has begun.

*Environment*  This is the period in which serious concern for widespread and life-threatening degradation of our planet's physical environment spread from the rich countries to the "socialist" and the developing countries. There has been a spate of worldwide environment protection legislation: The International Whaling Commission's moratorium on nearly all whaling, the 1987 Montreal accords on phasing out use of CFCs, banning of driftnetting on the high seas, a 50-year moratorium on minerals exploitation in Antarctica and the Southern Ocean, agreements to limit trade in endangered species of animals and products derived from them, all these and many more are signs of a growing consensus on at least the reality of a threat to our global environment. Concern about depletion of the ozone layer in the atmosphere, global climate change, oceanic pollution, disappearing species of plants and animals, desertification, deforestation, a growing shortage of pure water, and other threats to the human life-support system may be a catalyst to bring together all the peoples of the world to pursue a common goal for community benefit, regardless of their differences in other areas.

*Social Conditions*   The past few years have seen the maturation and general acceptance of human rights as a genuine matter of concern in the world community and as a legitimate reason for limitations on State sovereignty. The breakdown of family life in most of the world prompted the United Nations General Assembly to proclaim 1994 as the International Year of the Family, and in 1995 the UN will sponsor the Fourth World Conference on Women, two events that, given the changed political climate in the world, could well generate concrete action to alleviate the dismal condition of women, children, and families that has characterized the past half century, if not a much longer period. The new United Nations International Drug Control Programme may provide the basic mechanism for a long-needed and long-avoided cooperative, coordinated, worldwide attack on a major cancer eating away our global society, and provide a model for a similar attack on other types of crime, some of which have gone international on a grand scale.

*Politics*   Most spectacular, though not necessarily most important, have been the absolutely astounding political developments of this period. Only a list of them is required to call to mind recent headlines and television images. The decay and disintegration of the Soviet empire, from the outer perimeter to its very heart, is certainly outstanding. It has led to the end of the Cold War; the beginning of true disarmament, even of nuclear, chemical, and biological weapons; the spread of greater democracy in the vast formerly Soviet-controlled territory. It also made possible the reunification of Germany, movement toward peace in Cambodia; and the grand coalition of some 26 disparate countries that liberated Kuwait from Iraqi rule. That, in turn, led to the beginning of the long-lasting and peripatetic negotiations on peace in the Middle East, surely one of the most sensitive, difficult, and enduring trouble spots in the world. A freely negotiated and fair peace arrangement in this re-

gion alone would end an era and begin a new one. The radical but *relatively* peaceful change in South Africa is almost as important, and itself is truly historic. Not only is that country now on the way toward a just and progressive multiracial society, but it has stopped trying to "destabilize" its neighbors, has withdrawn its forces from Angola and Namibia, and has ceased aiding the ruthless bandit/rebels in Mozambique. We can now foresee a possible unification of all of southern Africa into a powerful and reasonably prosperous entity.

In other areas, the completion of the Common Market in Europe and its subsequent broadening and deepening are also historic; Europe seems destined again to be a major world power. Democracy is sweeping across Africa as well as Europe, though without the blare of publicity about such events on the northern continent. There is only one traditional dictatorship left in Latin America, and the Philippines has the first democratic government in a generation. Communism has joined fascism in the scrap pile of discarded ideologies that no longer inspire fanatical loyalty and lead to oppression, imperialism, aggression, and mass murder.

We are witnessing the beginning of the long-term, voluntary erosion of State sovereignty, not only through regional integration efforts, but also through the United Nations and other intergovernmental organizations. It began with human rights and especially the anti-apartheid campaign, which brought the world community emphatically into the internal affairs of South Africa. But just in the past few years there have also been the UN establishment of a security zone in northern Iraq to protect the Kurds from the wrath of the Iraqi government, followed by the humiliation of having UN teams scouring that country to discover and destroy surviving weapons of mass destruction. Other examples include UN supervision of elections in Haiti and Nicaragua, and UN intervention in the British/American attempt to extradite two alleged terrorists from Libya.

One of the most important and perhaps most durable of the major developments taking place right now is the rejuvenation of the United Nations and increasing use of its organs and affiliates (especially the World Court) to resolve both international and domestic problems. In its first 43 years of life, for example, the UN was permitted to mount only 13 peace-keeping operations, but in just 1988–90 nine more were undertaken. A new Secretary-General took office on 1 January 1992 and immediately began a major restructuring of the organization that should, if it is adequately financed by its members, enable the UN to carry out its mandates with even greater efficiency and success than in the past. In February 1992 the Security Council held its first meeting ever in which all 15 members were represented by their heads of State or government. The Secretary-General now has political offices in Kabul and Islamabad, Tehran and Baghdad, to monitor potential conflict situations and render mediation and other services as needed. There is much more, of course, but this is enough to illustrate the point.

There are a number of observations to be made about the foregoing list of recent developments:

1.  This is far from a complete list; most readers could add others.

2.  None of these developments has occurred in isolation; each is related to many others, and even the classifications are somewhat arbitrary.

3.  None of them has occurred suddenly; nothing in human affairs ever happens suddenly, but people suddenly become conscious of changes that may have been gestating quietly for a long time.

4.  There are no predictions in this list, nor in the commentary to follow; there are only projections of readily observable trends. Actual events are almost impossible to predict; it is amusing to hear the "experts" pronounce events "inevitable" only *after* they have occurred.

5.  This list of positive developments should not lead one into undue optimism. We must not succumb to the glib hyperbole of some politicians who speak of our

being poised at the edge of "a new world order." Nonsense! Such shallow and misleading rhetoric does us a great disservice. Not only does it conjure up horrifying images of Adolf Hitler's "New Order," but it is simply untrue. The next few decades are likely to be quite *disorderly*, and the United States is unlikely to remain dominant throughout this period. Although we will enter a new millennium on 1 January 2001, as well as a new century and a new decade (again, by the Gregorian calendar), we are still a very long way from The Millennium.

Let us now go through our classes of developments again, this time without the proverbial rose-colored glasses.

*Population*    Although the population growth rate is not as high as it has been in the recent past, much of the drop can be attributed to China's draconian population control policies, and these policies could easily be reversed when government coercion is relaxed. Furthermore, even if the worldwide *rate* of increase declines even more, the effects will not be felt for another generation because the *absolute* increase will continue to climb for at least that long. Also, the world's population is becoming increasingly urbanized, placing great stress on available resources. Finally, although we have managed to increase food production enough to keep up with the growth in population, the effort has been very costly and very uneven and may not be sustainable. Already over a billion people in the world are living in absolute poverty—poverty by any definition—and we have experienced one famine after another in Africa, with no end in sight.

*Science and Technology*    We can all marvel at the new products of our laboratories and our factories, and some of them have undoubedly helped make life better for many people. But as good as we are with gadgetry, we have not yet developed the wisdom to use our gadgets properly—or even to determine whether they should be produced in the first place. There is, moreover, serious doubt about whether the expanding materialism that characterizes modern society is really worth its cost in the erosion of the life of the mind and the spirit. And there's a question that's been troubling me for a long time: In the glorious high-tech world of the twenty-first century, what will happen to low-tech people? Our worship of science and technology is only one facet of our notion of *progress*, a product of seventeenth-century European philosophy, a notion that must now be examined very carefully. Perhaps this is the time to reverse the spread of this concept around the world and bring science and technology once more into the service of man, rather than the reverse, and then return man to the service of nature as the steward of the earth.

*Decolonization*  There are still some difficult, unresolved problems of decolonization. The status of Mayotte, East Timor, Western Sahara, the former Palestine Mandate, the Falkland (Malvinas) Islands, New Caledonia, and a number of other places are still in dispute. And many former colonies are experiencing serious internal problems, some attributable directly to the colonial experience. Examples include Cyprus, Somalia, the Philippines, Myanmar (formerly Burma), the Cabinda exclave of Angola, and others. Some of these problems will remain well into the twenty-first century. In most of the former colonies, the post-revolutionary generation, the first to mature since independence, is taking control, but there is no assurance they will be more honest, skillful or democratic than the founders.

*The Rich–Poor Gap*  If the Uruguay Round of negotiations fails, it could mean the end of GATT as the principal regulator of world trade. That could lead to the formation of exclusive and competing world trading blocs: North America, Europe, and East Asia. If that happens, it will not be good for the poor countries. As an old African proverb expresses it. "When the elephants fight, it's the grass that gets trampled." Even if the Uruguay Round is successful, there is no assurance that protectionism will not resurge anyway, perhaps disguised in various ways. In addition, the end of the Cold War has reduced the incentive for rich and middle-class countries to assist the poor to become a little less poor. Realistically, we cannot expect the gap to close very much very soon.

*Environment*  The 1992 United Nations Conference on Environment and Development may well be as critical to our future as the Uruguay Round or the Middle East peace talks. If the scores of heads of State and government and their staffs gathered there cannot agree on a rational, comprehensive, long-range plan to reverse the deterioration of our planet's ecology, and *then commit themselves to making it work*, all the progress we have made recently may be obliterated. There are simply too many countervailing forces at work in the world to expect a miraculous transformation of a cesspool into a lovely, fragrant lake in even one generation.

*Social Conditions*  We now have more refugees, asylum seekers, and other displaced persons in the world than at any time since the Second World War, and perhaps more than ever before in history. Unless population, human rights, economic, and other fundamental problems are resolved, the refugee problem will not be resolved either. The spread of AIDS is not only accelerating, but it is gradually reaching into every corner of the earth and every sector of society. It is already epidemic in parts of Africa and could become a pandemic. The devastation this could wreak on *Homo sapiens* is unimaginable; it could be far worse than the Black Death of medieval Europe. The problems of drugs, terrorism, and other threats to civilization are far from solution. Massive numbers of people around the world are prohibited from reaching their potential as individuals and from contributing fully to the improvement of society because of their race, gender, nationality, caste, religion, physical or mental condition, sexual orientation, tribe, parentage, and other artificial distinctions that have nothing to do with their ability, motivation, or character. If the level of civilization can be measured by how people treat one another (and I think that's the *only* valid criterion), we have not made much progress in the past century and may not in the next.

*Politics*   Although we are entitled to feel some relief that fascism and communism are no longer powerful forces in the world, we must recognize that neither is dead. Both are still active in scattered places around the world, reviving in others, and dormant in still more, awaiting opportunities to rally the discontented, dispirited, and despairing to assault civilized society once again. More important, they have been replaced for the present by two other malign and waxing forces, nationalism and militant religious fundamentalism. We have experienced both before, but now, in the absence of competing ideologies and with better technology at their disposal, they menace all the social, economic, and political progress that has recently been made. We can hope that they are merely novas, flaring spectacularly before they burn out and die, but we cannot be certain of it. Yes, democracy is spreading around the world, but its spread is *not* inevitable or irresistible or irreversible. Tad Szulc, highly respected correspondent of *The New York Times*, published a book a generation ago called *Twilight of the Tyrants* (New York: Holt, 1959). In it he detailed and celebrated the virtual disappearance of the *caudillos*, the *juntas*, and assorted other brutal and primitive rulers from Latin America. The book had hardly dropped from the best-seller lists when a new wave of coups took place around the region, and before long most countries there once more had dictators. That could easily happen again, in Africa, in the former Soviet republics, nearly anywhere.

Much more could be said on all of these topics. The point, though, requires no further illustration: Although we are entering a new century in which the world will be very different from that of just a few years ago, it will not *necessarily* be a better one. The new world will be influenced, of course, by factors beyond human control, but very largely the new world will be made by the immediate descendants of those who made the old one; that is, by the people reading this book.

To use a maritime metaphor, we are sailing into uncharted waters, where no one has been before, with few aids to navigation. We will certainly be buffeted by storms, battered by rocks and shoals, perhaps run aground. But we have a chance to sail clear and pull into a snug harbor. The choices will be made by people, individuals, working alone and, perhaps more effectively, in groups. Will we take the tide at the flood, take the current when it serves, or will we lose our ventures?

## A Political Geographer's Perspective

This prologue is designed as a highly personalized overview of the contemporary world, from the perspective of one who has spent a lifetime in the study and practice of geography and politics. Not all of the topics mentioned here are examined in the body of the book, however, because some of them do not—at least not yet—fall into the area of political geography. The world is changing so rapidly and profoundly, and the changes are so complex and intertwined, that a major act of discipline, of self-restraint, is needed to confine oneself to those topics that are truly both geographic and political.

Geographers are concerned with the *spatial* aspects of both the physical and human environments of our planet, and especially with their interrelationships; *political* geographers are concerned with the *political* aspects of these interrelationships. Therefore, as examples, I do not discuss such important topics mentioned in this prologue as the debt burdens of some developing countries, the details of environmental degradation, the recent shifts in many countries from socialism to capitalism or from dictatorship to democracy, or the spread of acquired immune deficiency syndrome (AIDS). These topics in themselves are either not geographic or not political, although all of them may have politicogeographic consequences. More on the perspectives of political geography may be found in Chapter 1.[*]

[*]The two foregoing paragraphs have been inserted at the request of reviewers into the Prologue from the first edition, written in 1991, which otherwise is reprinted *verbatim*.

This book is not designed as a chronicle or an almanac. It makes no attempt to be up to date on everything. It does, however, call attention to many of the underlying forces and quiet developments in the world that will suddenly burst into the headlines and blare from the television screens tomorrow. As I said earlier, we have few aids to navigation into the future. None of the extant theories, including those summarized in this book, is likely to survive the transition unscathed, and no new grand theories are offered here. Instead, I offer information, ideas, principles, patterns, interpretations, suggestions, trends, examples. There is a lot in this book, perhaps too much to absorb in a short time, yet too little to do justice to its field or its readers. Nevertheless, I hope that it will serve as a useful guide into the twenty-first century.

*"Idealists are the salt of the earth; without them to move us, society would soon stagnate and civilization fade."*
Sir Halford Mackinder, *Democratic Ideals and Reality, 1919*

# PART ONE

# Introduction to
# Political Geography

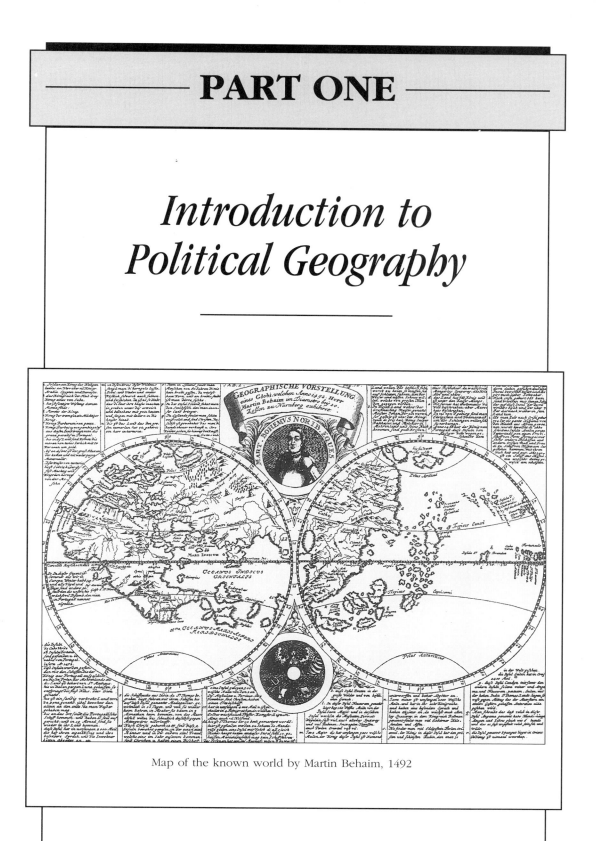

Map of the known world by Martin Behaim, 1492

# 1

# *The Field of Political Geography*

Political geography is a varied and wide-ranging field of learning and research, exciting and endlessly fascinating to the student and useful to the practitioner in many fields. Its roots go back to Aristotle's model of an ideal or perfect State. It developed in spurts over the next two thousand years and is now experiencing a new efflorescence. Interest in it has waxed and waned throughout the history of political geography, and each period of development featured a particular emphasis without discarding ideas that had emerged earlier. Despite this constant change, there is also continuity. In the words of the old gospel hymn *When the Saints Go Marching In*, "We are traveling in the footsteps of those who've gone before."

As a preface to our brief review of the development of and current trends in political geography, we offer a random selection of definitions of the field suggested by some of our more prominent contemporary political geographers.

. . . the study of the variation of political phenomena from place to place in interconnection with variations in other features of the earth as the home of man. *Hartshorne*

. . . Political Geography, a subdivision of human geography, is concerned with a particular aspect of earth–man relationships and with a special kind of emphasis . . . the relationship between geographical factors and political entities. *Weigert*

. . . the study of political regions or features of the earth's surface. *Alexander*

. . . the geographical nature, the policy, and the power of the State. *Pounds*

. . . the study of political phenomena in their areal context. *Jackson*

. . . the study of the interaction of geographical area and political process. *Ad Hoc Committee on Geography, Association of American Geographers*

. . . the study of the spatial and areal structures and interactions between political process and systems, or simply, the spatial analysis of political phenomena. *Kasperson and Minghi*

. . . political geography is concerned with the spatial attributes of political process. *Cohen and Rosenthal*

. . . political geographers are concerned with the geographical consequences of political decisions and actions, the geographical factors which were considered during the making of any decisions, and the role of any geographical factors which influenced the outcome of political actions. *Pacione*

. . . humanistic political geography is concerned with uncovering the dynamic social processes whereby the spatial dimensions of the natural and social world are organized and reorganized into geographically delimited and symbolically meaningful provinces by national and transnational groups. *Brunn and Yanarella*

These few quotations demonstrate that there is no generally accepted definition of the field. Although many critics, both within and outside geography, assail this vagueness as a fault, in reality it might be considered a virtue. The lack of a rigidly defined focus, of precise boundaries, has enabled political geographers to investigate various phenomena that exhibit both political and spatial, or geographic, characteristics without being concerned about straying from the central theme of the field or about intruding into someone else's territory. (We will have more to say about territory in the next chapter.) It has also enabled political scientists, sociologists, and other social scientists to "do" political geography—indeed, to make valuable contributions to the field—without incurring the displeasure of the "insiders." The backgrounds of political geographers, in fact, are as varied as the subjects they study, and their ranks are frequently enriched by converts from other fields. All this contributes to the variety, fascination, and utility of political geography.

## *Highlights of the History of Political Geography*

The definitions just cited, different as they may seem, really fall into only two categories. The first five emphasize the study of political areas, two focusing specifically on the State, while the last five, the most recent, introduce the study of political processes, a reflection of the behavioral approach so popular in the social sciences in the 1960s and 1970s. Even in a period of only some 40 years, we have seen shifts in the orientation of many teachers and students of political geography; the changes over 2000 years have been much greater.

The single long-term interest underlying the field of political geography from its origins to the present is the interrelationships between politics and the physical environment. When Aristotle fashioned his model State some 2300 years ago, he considered the ratio between population and territory—and the qualities of both—as funda-

mental. He examined the requirements for the capital city, the composition of the army and navy, the boundaries, and other factors, the determining one of which was the nature of the physical environment, especially climate. Although his deterministic approach has little support today (we discuss environmental determinism in more detail later), many of his ideas have been incorporated into basic concepts in political geography and are currently very much alive.

Strabo, a Greco-Roman geographer, wrote a 17-volume description of the entire known world more than three centuries after Aristotle, following this tradition but basing his work in large part on his own field observations. Although his work was designed for practical use by soldiers, administrators, and travelers, he did consider the factors necessary for the functioning of a political unit the size of the Roman Empire. Like Aristotle, he tended to be quite ethnocentric, believing that his homeland had the "ideal" climate, the best governing techniques, and so on.

After the decline of Rome and the rise of Christianity, science and scholarship were nearly eclipsed in Europe. But geography flowered anew in the Muslim world, particularly among Arab merchants, historians, travelers, and philosophers. One of the most widely traveled, observant, and prolific of these writers was Abd-al Rahman Ibn Khaldun. He wrote a detailed autobiography late in the fourteenth century in which he proposed theses about the most powerful political units of his time, the tribe and the city, and related them to the occupations determined by the physical environments of their regions. His investigations into the nature of political units led to theories about political integration and disintegration and ultimately to the concept of a cyclical nature of a State's existence.

French writers of the sixteenth and seventeenth centuries followed in this tradition. Montesquieu, for example, emphasized the role of the landscape and climate in determining systems of government. He observed many complex relationships between agriculture and population, political systems and occupations and land area, soil fertility, and government. In the mid-seventeenth century William Petty published works discussing the territorial and demographic elements in the power of States, the concept of the sphere of influence, the role of capital cities, and a host of other politicogeographical issues. Petty was a physician, teacher of anatomy at Oxford, cartographer, political scientist, and economist; his major works were *Political Anatomy of Ireland* and *Political Arithmetic.* Carl Ritter in Berlin, writing in the late nineteenth century, developed a cycle theory of State growth similar to Ibn Khaldun's, but based more on analogy with organisms. Ritter was more scientific than his predecessors but still somewhat ethnocentric and deterministic. His *General Comparative Geography* was, in a sense, the first real attempt to create models in political geography, and he is considered one of the founders of modern geography.

A new phase in the evolution of political geography began with Friedrich Ratzel, commonly referred to as the "father" of our specialization because of his innovations in both concept and methodology. Much of his work derived from that of Ritter and other

Friedrich Ratzel, 1844–1904. (Courtesy of Geoffrey J. Martin)

intellectual forebears, and their tradition continues to the present. The political geography of the late nineteenth century and the first half of the twentieth was dominated by geopolitics, rooted in some of Ratzel's voluminous work. We devote four chapters to Ratzel and four geopolitics, not only because of their influence on both thought and political action during the period, but also because some elements of the geopolitical approach can still be useful today.

Late in the nineteenth century political geography emerged in the United States, stimulated first by such practical matters as perceived European (chiefly British but also Spanish) threats to American hegemony in the Western Hemisphere under the Monroe Doctrine and the very real need for an interoceanic canal across the Central America–Panama isthmus. Daniel Coit Gilman, professor of geography at Yale, was engaged by the U.S. government in 1895 to take charge of preparing the historical and geographical background of the Venezuela–U.K. dispute over the western portion of British Guiana (now Guyana), a dispute still unsettled a century later. C. W. Hayes, Emory Johnson, and others did feasibility studies of various canal routes, and Johnson's work led directly to his pioneering paper "Political Geography as a University Subject," read to the Association of American Geographers in 1905.

The work of Ratzel was introduced to the United States by his student Ellen Churchill Semple, though she tended to draw more from his anthropogeography than from his political geography. The geopolitical work of the British geographer and statesman Sir Halford Mackinder also crossed the Atlantic and attracted considerable attention, especially as it complemented the writings of the American admiral Alfred Thayer Mahan (both discussed in Chapter 21).

The First World War, and especially the construction of the peace afterward, provided a dramatic stimulus to the development of the field, not yet as an academic speciality but primarily as an approach to resolving real and difficult problems. This was particularly true in Britain, France, Poland, Hungary, and the United States. Isaiah Bowman, a member of the American delegation to the Paris Peace Conference and already a distinguished geographer, went on to become one of the century's greatest political geographers. Like most of the prominent figures in the field until late in the twentieth century, he concentrated on practical problems at the State, regional, and global levels. He continued to serve on U.S. delegations to important international conferences, including the 1945 San Francisco conference that adopted the Charter of the United Nations.

During the interwar period Derwent Whittlesey, Richard Hartshorne, and Stephen B. Jones in the United States and Jean Brunhes in France continued the tradition of concentrating on the study of political areas, especially the State, while in Germany political geography was being distorted into a new type of geopolitics that was very different from true political geography. Whittlesey adopted a historical approach, viewing the State genetically. Hartshorne and Jones developed two of the basic theories of modern political geography, both of which we discuss later. By the outbreak of the Second

World War, political geography had become a serious academic subject.

There was little academic development of the field during World War II; most political geographers were serving their governments in civilian or military roles, trying to wrestle with the enormous politicogeographical problems that led to the war and that were generated by it. Not only was German distortion of geopolitics into a nationalistic justification for aggression rejected in Western Europe and North America, but geopolitics as a serious academic subject fell into disrepute—and, for some years after the war, so did the broader and older field of political geography. The International Geographical Union, in fact, did not permit any sessions on political geography at its quadrennial congresses until 1964.

## *Contemporary Political Geography*

Beginning in the early 1960s, several new trends emerged in political geography. These contained diverse elements and overlapped considerably, making categorization difficult. Some classification is helpful, however, and so we attempt it here by grouping the contemporary emphases in the field into four broad areas.

First, there was a notable change in the *scale* of subjects studied. Traditionally, political geographers and their kin in other disciplines have tended to work at the meso- and macroscales, essentially concentrating on the State and on interstate (international) or global problems. Recently, there has been a higher proportion of work being done at the microscale; that is, at levels below that of the State. The study of administrative districts and systems of all kinds within States spawned a new subspecialization—electoral geography—which uses the electoral district as its field. Other studies have covered urban areas, cities and sections of cities, and the supply of municipal services. This seems to have been a reflection of a general turning inward, a concern with ever-smaller social as well as political units, and a withdrawal from the universalist perspective of

Isaiah Bowman, 1878–1950. (Courtesy of Geoffrey J. Martin)

the period of World War II and the 15 or 20 years following it.

Second, there was a shift in the *subject matter* itself, with increasing emphasis on spatial interactional factors (flow of goods, services, and ideas over space). Associated with this have been studies of the effects of spatial structural elements on this spatial interaction. Geographers have also been analyzing political power, conflict resolution, value systems, territoriality, and public policy—all subjects that were previously the domain of the political scientist, the sociologist, and the psychologist—but applying geographic attitudes, approaches, and techniques.

The new emphasis on such subjects both encouraged and was stimulated by the adoption of new *methodologies.* Prominent among them were quantification and behavioralism. In both cases, political geography was among the last of the major branches of geography to adopt them, just as geography lagged behind economics, psychology, and political science in adopting them. This is merely an observation, not a criticism, for there is no particular virtue in jumping on a bandwagon merely because it happens to be passing by and the music is enticing.

The advent of computers encouraged people to find ways to use them, and geography, including political geography, offered some opportunities. New discoveries in psychology led to analyses of decision making, ideologies, and perception—all with applications in political geography. Similarly, the systems approach, which combines mathematical and behavioral techniques, has been used with increasing frequency and sophistication, but has not yet been generally accepted.

One of the chronic criticisms of political geography, even in comparison with other branches of geography, has been that it is long on facts and short on ideas. We seem to be on the way to rectifying that weakness, if indeed it is a weakness, by developing *new theories.* A number of theories have been proposed over the centuries, particularly in this century, and some of them, concerning primarily the State and geopolitics, have achieved some currency, but the theoretical underpinnings of our field do need shoring up. Attempts are being made in this direction, but so far they have resulted in more themes than theories, and certainly no one theory is generally accepted by political geographers.

Some of the major themes that may lead to new theories are areal or spatial association, which tries to explain the distribution patterns of various phenomena occupying an area; attempts to link generating processes with the forms of spatial distribution so that the type of pattern that will be generated in an area at a particular time can be inferred if not predicted; spatial interaction, which provides the basis for many hypotheses about flows of phenomena; distance decay, which proposes that the effect of an event or activity tends to decrease with increasing distance from the site of its occurrence; territoriality and perception; a number of concepts from economics, such as friction of space, localized costs and benefits, externalities, and *"homo economicus"*; and, of course, the well-established concept of the political region.

A newer version of the geopolitical concept of "heartland-rimland"' (discussed in Chapter 21) is that of *core (center)-periphery,* derived from dependency theory, which visualizes the postcolonial world as one still characterized by a core of rich and dominant former colonial States surrounded by a periphery of poor and dependent former colonies that share but little in the wealth of the center. Many varieties of Marxist (or radical or socialist) theory have been applied in political geography in the past three decades, and some of them are slowly gaining respectability, if not wide acceptance. *Antipode,* the journal of radical geography, is now well established and almost mainstream; it has come far from its origins as a distinctly marginal project of a very few British and American geographers.

There are many signs of a revival of interest around the world in political geography besides these new approaches being advanced in various quarters. The Association of American Geographers (AAG) established a political geography specialty group in 1979, and it has become one of the larger and more active of the some 40 speciality groups in the AAG. The journal *Political Geography Quarterly* was founded in 1982 and is similarly flourishing. In 1992, in fact, it became a bimonthly and dropped the *Quarterly* from its title and in 1995 went to 8 issues per year. The Institute of British Geographers now has a political geography working party, and the International Geographical Union Study Group on the World Political Map has become a full-fledged and very active commission. The Office of the Geographer in the U.S. Department of State has greatly expanded both its size and the scope of its operations and serves many government agencies.

New centers of political geography have developed outside of the traditional ones in Western Europe and North America. The most notable of these is in Israel, where there was not even one course being taught in the subject anywhere in the country in 1970. Now it is taught in several of the universities and is especially strong at the University of Haifa. Moshe Brawer, Yehuda Gradus, Nurit Kliot, David Newman, Arnon Soffer, and Stanley Waterman are among the Israeli political geographers who have already earned international reputations.

Following this new flowering of an old field is what may come to be regarded as the beginning of the fifth phase in its evolution. Since about the mid-1980s; that is, from about the time that the fourth edition of *Systematic Political Geography* was written, both the mathematical and behavioral approaches seem to have faded away, not completely, of course, but significantly. They are being replaced by a return to more pragmatic analyses of contemporary problems, considering them in their historical and cultural contexts. The computer has finally been recognized as just one more tool in the geographer's toolbox and seems to be used more selectively (and perhaps more effectively) now. The map and other graphic representations of both facts and ideas are again prominent in the literature.

There seems to have been a major shift not only in methodology but in subject matter as well. Although there is still considerable and growing interest in problems at the microscale—elections, social conflicts, factory and school closings, and so on—there has been a return to interest in national, regional, and global concerns. These include such topics as environmental degradation, the Arab-Israeli problem, international organizations, transnational corporations, the breakup of the Soviet empire, refugees, economic restructuring at all levels, nationalism, and so on.

Most dramatic of all has been the return to respectability of geopolitics as an important academic subject. This is covered in greater detail in Part V, but it deserves emphasis here. Perhaps both as professionals and as individuals, we are beginning to emerge from a period of introspection and even parochialism into a period of greater awareness of our place in the great sweep of history worldwide and of our obligation to help one another understand it.

Currently, in addition to geopolitics and electoral geography, discussed in detail later in this book, two relatively new themes have become prominent in publications devoted to or identified as political geography. One is an attempt to relate geography to changes in social theory. Much of this is a continuation of one or another of the many efforts to apply Marxist theory to geography, though with different emphases and interpretations. The other is the introduction of "gender issues" or "feminist viewpoints" into the ongoing evolution of our field. It remains to be seen whether either of these trends has enough substance and staying power to survive the inevitable assaults of reality upon them. Some of the ideas may survive in other disciplines such as sociology or political theory, but only those based on both politics *and* geography are likely to be incorporated into future mainstream political geography.

In Part Nine we present briefly some of the topics of current interest in political geography that are not discussed in detail elsewhere in the text. We also suggest other topics that need to be studied but that are currently being neglected. The bibliographies at the end of each part of the book give some indication of the vast literature available to any seeker of additional material, including much that is new.

But we wish to reiterate that the current phase of the evolution of political geography is not the first one, but the fifth, and it will very likely not be the last. There is, moreover, little in this phase that is truly new or truly unique. Most of it is derived from the work done in previous phases, though with fresh approaches and techniques, and much of it is common to geography and the other social sciences.

As the current scholars and practitioners in the field continue to develop their own interests, ideas, and methodology, and as today's students join the ranks and make their own contributions, the field will continue to evolve, perhaps into a new phase we cannot yet perceive. Ours is a dynamic field and a useful one, with unlimited opportunities for newcomers to express themselves and to attain recognition from their peers. We hope that many students will accept the challenge and join us.

### Our Approach

In this book we make no attempt to treat every aspect of political geography equally or to incorporate all of the very latest ideas and materials. Essentially what we try to do is to develop the theme of our opening sentence: "Political geography is a varied and wide-ranging field of learning and research, exciting and endlessly fascinating to the student and useful to the practitioner in many fields."

The organization of the book is orderly and rational, but arbitrary. One of the most distressing trends in modern scholarship and education is the increasing compartmentalization of human knowledge and experience. But our perceptual universe is expanding far more rapidly than our mental abilities, and so some division and classification is necessary. Although we must accept reality and discuss one topic after another, we constantly remind the reader in various ways that everything in the book is related to everything else, not only in the book but also in human knowledge and experience. We cannot emphasize too strongly the unity of our field and its interconnections.

Another theme running through the book is one emphasized in this chapter; the continuity and incompleteness of our field. Every change, every development, every advance raises more questions than answers. There is far more to be done in political geography than has already been done, and we ask questions throughout and suggest approaches and topics for future investigation as examples of the work that still needs doing.

We hope this book will serve not only as an introduction to the field of political geography, but also as a guide through it and an inducement to enter it.

# 2

# *Personal Space and Territoriality*

The subject of territoriality could well be discussed under the heading of perception, but both topics have been developed so much in recent decades that each deserves a separate chapter. Of the two, territoriality is the more controversial. We have already emphasized the interdisciplinary nature of political geography, and nowhere is this more evident than in considering territoriality, based as it is primarily on research in ethology (the study of animal behavior) and psychology. From this growing body of empirical and theoretical work, we can extract some ideas that may be helpful in understanding human political behavior. We must be extremely careful in doing so, however, for political parties and States do not necessarily behave in the same manner as animals or individual people.

## *Personal Space*

One promising area of research ought to be familiar to all of us. It relates to the way in which we spread ourselves out in classrooms and theaters, restaurants and libraries. If we do not *have* to be crowded, we like to put a little distance between ourselves and the next person. This is *personal space*—an envelope of territory we carry about with us as an extension of ourselves. People as well as animals carry such envelopes around, but the shape and size of the envelope vary from culture to culture. In terms of shape,

we can tolerate greater proximity in front of us than beside us, and still less behind us. Thus our "portable territory" is not symmetrical around our body. And, as Robert Sommer remarks, Englishmen keep farther apart than Frenchmen or South Americans. This subject was first touched on by Edward T. Hall in *The Silent Language* and later more fully developed in *The Hidden Dimension*. In the second book Hall introduced the term *proxemics* for the study of the ways in which people perceive space and use it in various cultures.

How can we interpret the territorial behavior of societies and cultures from what we know about personal space and small-group proxemics? Psychologists and geographers have had occasion to refer to ranking and hierarchies in human society. Walter Christaller made this a central issue in his work on urban centers; in Allan K. Philbrick's book *This Human World*, it is a dominant theme. In Sommer's 1969 book the idea that human territorial organization relates to rank and hierarchy is explored in some detail.

In a study of the city of Mombasa, for example, where more than a dozen of Kenya's tribal peoples were represented by substantial numbers in the population of about 200,000, it was noted that the people of the "stronger" tribal groups generally possessed better residential locations than those of "weaker" tribal groups. The strengths of several of the major tribal peoples in Kenya may be a matter for debate, but few would argue that the Kikuyu and the Luo hold positions of power compared to, say, the Nandi and the Gyriama. Now, this indefinite power hierarchy is somehow reflected by the urban residential pattern in Mombasa—the spatial adjustment those peoples have adopted within that city. This is an example of *dominance behavior*, and it may apply not just to tribal peoples living in urban centers, but to States and even power blocs as well. We shall return to this point shortly.

Although the minimum acceptable personal space around an individual is chiefly determined by culture, reactions to invasions of this "forbidden zone" vary with the circumstances and the individuals involved. They range from "cocooning," or withdrawing into oneself, through displays of discomfort to flight, leaving the place to the intruder, the most extreme nonviolent reaction. Regardless of the degree of tolerance for closeness of physical presence of others in various cultures and individuals, there is a limit for everyone. In some animals behavior is grossly distorted whenever there is extreme crowding of members of the same species. Similar reactions have been noted among people, but social adaptations sometimes work to prevent violent reactions among them. A great deal of research remains to be done on the effects of crowding on people, but we already know enough about their seriousness to make us add crowding to our list of adverse effects of excessive population growth.

## *Animal Territoriality*

Territoriality has been investigated since the seventeenth century and was first described in detail in 1920. Many vertebrates exhibit it in varying degrees, including cats, some birds, some fish, and monkeys. Robert Ardrey first brought the subject forcefully to public attention in his book *The Territorial Imperative.* Ardrey makes some statements political geographers can hardly ignore. "A territory," he writes, "is an area of space, whether of water or earth or air, which an animal or group of animals defends as an exclusive preserve. The word is also used to describe the inward compulsion in animate beings to possess and defend such a space." He concludes that man "is as much a territorial animal as a mockingbird singing in the clear California night" and that "the territorial nature of man is genetic and ineradicable."

We do have detailed information to supplement and support many of Ardrey's observations. We know, for example, that territorial animals mark out their territories in various ways, commonly by a glandular secretion or urination. Territories are rarely rigidly bounded and exclusive; they generally overlap and sometimes overlap almost completely. They have gradations of sensitivity so that the animal will react differently according to species, the type of intruder, and the distance the intruder penetrates into the animal's territory. Territory is not immutable; it tends to get larger when food is scarce and shrink when food is plentiful. Finally, territory serves the animals that mark and defend it in several ways. It functions, for example, to regulate population by limiting breeding, reduce frictions among members of the species, protect feeding and nesting sites, reduce the rate of spread of diseases and parasites, and afford some security for weak or subordinate animals.

If we define the concept of territoriality as a pattern of behavior whereby living space is fragmented into more or less well-defined territories whose limits are viewed as inviolable by their occupants, it becomes clear that political geographers have for some time recognized this phenomenon. Any of these territories, after all, will acquire certain particular characteristics. Gottmann called it *iconography*, and we discuss this idea later. The Ratzel–Kjellén–Haushofer school of geopoliticians was in the process of defining a form of territoriality when they conceptualized the State as an organic being. Hartshorne and Jones urged that the State be viewed as an entity whose characteristics could be linked, ultimately, to the behavior of the individuals constituting it.

Therein lies one of the problems with the concept of territoriality. It seems inappropriate to make inferences about aggregate or group behavior from research dealing mainly with the behavior of individuals; it is also questionable to what extent human behavior may safely be equated with animal behavior.

## *Human Territoriality*

By combining the concepts of personal space, including dominance behavior, and animal territory, we can interpret many aspects of human behavior as indicating some form of territoriality. We begin with very small units of territory and move to progressively larger ones, at each stage making some observations and reserving our analysis for the end.*

Obviously, individuals exhibit territorial behavior in those small, confined places where they spend most of their time—home, office, factory floor, automobile—whether or not they own these places legally. The home, in its interior furnishings and decor especially, expresses an individual's (or family's) personality. It is more than just a residence; it is a refuge, a fortress shielding the individual from the problems of the world outside. There is certainly a strong desire to defend this home from intruders and even to discourage intruders from a favorite armchair or "my" room or "my" place at the dining table. People who have grown up in the United Kingdom or North America tend to idealize the single-family detached house on its own lot, perhaps with a fence or shrubbery around it. Some observers speculate that this dream and its realization respond to territorial needs. Many people, however, prefer apartments instead of houses, and many prefer to rent rather than buy; perhaps other needs are more strongly felt than territoriality.

Bureaucrats (in government, business, or university) have similar feelings about "their" offices, parking places, and similarly assigned or adopted spaces. Many of the perquisites of office, the coveted status symbols, have little intrinsic or functional value but do enhance the space occupied by the individual or simply symbolize his or her right to use it. The elegant desk, the carpet or rug on the floor, the key to the executive washroom, the view from the window all go with the position and are defended by the possessor. In our culture and in many others, there is a distinct correlation between status and size of territory. Dominant people tend to have more and larger territories. "Important" people generally are expected to—and desire to—have big homes, big lots, big offices, big cars, and often several homes, offices, cars. These symbols of status also reinforce status itself, so someone with a big home, office, or car must be "important."

Even in temporary situations, in the diplomatic world, for example, status is symbolized by the space assigned to or occupied by a person. It really does make a difference whether one sits at a formal dinner to the left or the right of the host, or above or below the salt. In formal business sessions, the presiding officer sits in a prominent place with the rapporteur and other officials nearby in a particular order. These and many other sensitive matters relating to rank and precedence have been painfully worked out over centuries and are summarized in the term *protocol*, a prerequisite for conducting diplomatic business effectively.** In

*Much of the human behavior that we consider under the headings of personal space and human territoriality—those gestures and signals that, though differing from one culture to another, communicate messages to other people about ownership of space—is included in the studies known variously as psycholinguistics, paralinguistics, and nonverbal communication, which have been popularized as "body language."

**Who sits where at international conferences can be an important substantive problem to be resolved before negotiations actually begin because the seating symbolizes the relationships among the participants' countries or bears on the issue itself. Recent examples include the U.S.–North Vietnamese peace talks in Paris in the early 1970s and the Arab–Israel peace talks in Madrid, Washington, and elsewhere 20 years later. In both cases, even the shape of the negotiating table was an issue. At UN meetings, States are normally seated in English alphabetical order to avoid status disputes, and generally seating is rotated regularly so no country is permanently at the front or rear of the hall. This system deliberately diminishes the value of each territory (seat) so it is not worth fighting for.

most cultures there is a less formal but equally effective protocol that regulates the temporary possession of territory, such as in libraries, cafeterias, and classrooms. A seat is known to "belong" to a particular person by prescription, even if not assigned, or a seat may be territorially claimed in the "owner's" absence by markers, such as books, coats, or purses.

On a larger scale, territoriality is manifested in neighborhoods. Street gangs often claim particular "turf" as their own, sometimes marking it out with graffiti or other symbols. Ethnic groups tend to congregate in particular areas within cities, frequently creating a replica (or a pale imitation) of a neighborhood in "the old country." If groups are forced by law or custom into segregated areas, the territorial feeling is unlikely to be as strong, but such enforced segregation, as occurred in the Jewish ghettos of Europe and most Muslim countries, may help to preseve a culture that would tend to change along with the general culture if place of residence were unrestricted. A new kind of segregated neighborhood is widespread in our society: the specialized housing area designed for particular groups of people. Examples would be the retirement communities for "senior citizens," summer beach colonies, and garden apartments for "swinging singles," all similar in some ways to the old "company towns," but more likely to inspire territorial feelings.

Depending on where people live and their length of residence as well as personal factors, they may extend the territoriality they exhibit toward their homes and neighborhoods to their towns, counties, and states. They may take pride in these places and advertise their virtues everywhere they go. This kind of "boosterism" seems to be fading with increasing mobility and standardized entertainment, however. Even regional loyalties have lost much of their importance in the United States; one seldom hears heated arguments any more over the respective merits and vices of New England, the South, the Midwest, the Far West, or other regions. The Civil War is no longer a topic to inspire passions as it was when the

author was in high school. Yet people still do identify themselves as being from a particular town, county, state, or region with a touch of pride.

People generally feel more comfortable and perform better at home, among familiar surroundings. Sports teams are well aware of the "home court (or field or ice or arena) advantage," even if the advantage is only psychological. Performers put out extra effort for the hometown folks. A person will be more confident in a confrontation with another in his own home, office, or other territory than in the adversary's territory. This principle does not seem to be applicable, however, to individuals or groups that either have no territory or habitually operate outside their territory. In international affairs this could include transnational corporations, international terrorist groups, international political movements, and perhaps religious movements as well. It does apply to farmland one owns, and land tenure is an important factor in economic development, even in a communist country such as the former Soviet Union, where agricultural production was higher on the small private plots of land than on State-owned farms (*sovkhozi*) or collectively owned farms (*kolkhozi*). But it does not seem to apply to continents; people rarely have territorial feelings toward South America or Asia or even Africa.

This background helps us to examine the aspect of territoriality that most interests political geographers: political territoriality, particularly as expressed in the State. As we see in greater detail in subsequent chapters, the modern nation-state based on nationalism is quite a recent phenomenon; in its most important form, it dates back not more than about four centuries. It emerged out of feudalism in Europe as the concept of *regnum*, or personal sovereignty, was gradually replaced by that of *dominium*, or national sovereignty. That is, people gradually transferred their allegiance from an individual sovereign (king, duke, prince) to an intangible but territorial political entity, the State. It is no longer "Mary, Queen of Scots" but Elizabeth II, Queen "of the United Kingdom of Great Britain and Northern Ireland."*

Nationalism is probably the strongest political force in the world today, and nationalism is territorially based. People identify not only with others of their group but with a particular portion of the earth's surface. A person is French not only because he or she speaks French and identifies with other Frenchmen, but also because France is home. A person of French descent living abroad may still speak French and feel a kind of kinship toward Frenchmen, yet owe loyalty and feel warmly toward the country in which he or she was born and raised. In the modern world the ideal political ties are those of place, not descent (although, as we see later, there are many places where this ideal has not yet been attained.) Most countries confer nationality on the basis of *jus soli*, or place of birth, instead of *jus sanguinis*, or parentage (although the United States and many other countries recognize both).

Sometimes a people or a nation will be sustained in its travail by a mystical concept of territory, a dream of a land of their own once possessed but now lost or of a land never possessed but longed for. This mystical territoriality is best exemplified by the Jewish people, who endured 40 years in the wilderness until they reached the Promised Land, who longed for Eretz Yisrael during the Babylonian exile, who through 2000 years of dispersion in nearly every country in the world repeat every single year in their Passover prayers, "Next year in Jerusalem." Zionism, the national liberation movement of the Jewish people, developed in the nineteenth century to convert the spiritual dream into political reality. The Jewish nation was to return home, to Zion. They rejected offers of homelands in Uganda, British Guiana (now Guyana), Sinai, Cyprus, Angola, and other countries. Finally, in 1948, they reestablished the Jewish State on its original territory and called it Israel.

We will examine other examples of homeless nationalism, but none can compare with this one which inspired countless generations of people to dream of dying in the ancient homeland or at least of being buried there, inspired them to walk from Russian Poland to Turkish Palestine and reclaim the wilderness there with their hands, to live and work and die on the land.

Was this an expression of some animal instinct? Do the other examples of human territoriality mean that it is instinctive, or has it been acquired and modified by learning—through cultural evolution? In other words, Is it really an "imperative"? And the concomitant, human aggression: Is it similarly universal? Is it genetic in its origins? Or is it neither?

Edward Soja writes:

> Only when human society began to increase significantly in scale and complexity did territoriality reassert itself as a powerful behavioral and organizational phenomenon. But this was a cultural and symbolic territoriality, not the primitive territoriality of the primates and other animals. . . . Thus, although "cultural" territoriality fundamentally begins with the origins of the cultured primate, man, it achieves a central prominence in society only with the emergence of the state. And it probably attains its fullest flowering as an organizational basis for society in the formally structured, rigidly compartmentalized, and fiercely defended nation-state of the present day.**

Soja is suggesting that territoriality as it is expressed by our complex societies is not the same phenomenon as that which commanded our distant ancestors—or our animal contemporaries. Nor, according to some anthropologists, is man the "naked ape," driven by savagery and killer instincts. Alexan-

---

*There are still some technical survivors *of regnum*. Albert II, for example, is still "King of the Belgians," and a citizen of England or Scotland is a British subject. But these anachronisms are purely symbolic, part of the national culture. Even the very few remaining absolute monarchs in the world demand loyalty not as individuals but as Chiefs of State. The Chief of State is still recognized in international law as the sovereign, regardless of domestic title, but the sovereign is now the personification of the State.

**Edward W. Soja, *The Political Organization of Space*, Association of American Geographers Resource Paper No. 8, Washington, D.C., 1971, p. 30.

der Alland, Jr. argues forcefully that aggressiveness and territoriality are not universal human "imperatives" at all; in fact, he goes so far as to say that territoriality is born with and nurtured by culture. Viewing some "primitive" peoples like those still living a hunting-and-gathering existence, Alland notes that these have the weakest of territorial imperatives; supposedly they are the ones who should have the strongest instinct of this sort if territoriality is indeed a biological urge. And Alland is not so convinced either about the vigor with which nation-states are inevitably defended; he views draft laws, anthem singing, pledge reciting, and flag waving as evidence that States do not find defenders so easy to come by.

Torsten Malmberg, however, a Swedish biologist and geographer, accepts territoriality as innate in humans, and says, in fact, "Territoriality probably reaches its highest development in the human species." He reviews much of what we have just presented but from a different perspective and in much greater detail. He warns strongly about the very serious consequences for the human species of unchecked population growth. Then he summarizes:

> Behavioural territoriality has an important place in the man-environment systems of evolutionary adaptation. What evolved from environmental demands on our ancestors is now required by the individual from the milieu. Thus the only relevant criterion by which we can consider the natural adaptedness of any particular part of present-day-man's behavioural equipment is the degree to which and the way in which it can contribute to population survival in his primeval environment.*

Even if we accept Malmberg's argument, however, must we also accept the argument that countries go to war over resources strategic positions, of *Lebensraum* ? Not

even Malmberg claims that! More recently, Robert Sack has developed a rather elaborate theory about territoriality based on two basic forms of spatial relation: action by contact and territoriality. Essentially, he conceives of human territoriality as only a strategy for access and control. While this concept can be applied to political actions of States, it leaves unanswered the question—which Sack deliberately avoids—of whether territoriality in human beings is a biological drive or an instinct.

More recent detailed studies, however, have been unequivocal on the subject. Michael Rosenberg, for example, writing in the *American Anthropologist*, concludes that "among modern human hunter-gatherers, territories, or similar such allocation systems, are essentially sociocultural phenomena. . . . As such, they are largely social arrangements, based as much on mutual recognition by neighbors of each other's rights as on defense." And Ralph B. Taylor, in a lengthy analysis of individual and small-group territoriality, concludes: "In sum, both because of flaws in argumentation, and misconceptualizations regarding the evolutionary foundations of territorial functioning, Ardrey's instinct-based view of human territoriality, and Malmberg's similar view, can be rejected."

It would seem that people desire territory in general or a particular piece of territory only if and when they perceive that territory to have some real or potential value for them—economic, strategic, or psychological. Remote and barren areas of the earth, even when well-known, were seldom claimed as anyone's owned territory and remained *res nullius* until nationalism (a cultural phenomenon) arose so recently in the western peninsula of Asia and led some people there to imperialism.

The debate over human territoriality will continue to rage in the professional literature and provide diversion at cocktail parties, but meanwhile we would be wise to be skeptical about it. The reality, however, may be less important than the perception, and it is to perception that we now turn our attention.

---

*Torsten Malmberg, *Human Territoriality*, The Hague, Mouton, 1980, pp. 319–320.

# 3

# *Perceptions of the Political World*

Decisions emanate from a complex set of processes. They are made because the person or group of persons making them desires a state of affairs different from the one that prevails. They are based on information, but in order to assess the process of decision making in a given context, we must know the amount of information the decision makers had and the degree of accuracy of that information. Furthermore, there are different *kinds* of decisions; they may be made as a matter of habit, they may be made subconsciously, or they may be controlled—that is, made in response to a perceived choice of alternatives.

Fortunately, geographers other than political geographers have also been interested in research on decision making, and political geography has benefited from this work. Economic geographers have focused on the element of rationality as part of a decision-making model. This is the concept of the "intendedly rational man," who may have an incomplete and imperfect information set but who makes rational decisions when confronted with alternative courses of action. Julian Wolpert incorporated this factor into his search for the behavioral elements involved in people's decisions to move elsewhere—the decision to migrate. In several articles of great relevance to political geography, Wolpert stresses a number of dimensions of decision making. *Place utility* has to do with a decision maker's relationship with the locale where the decision is being made. The *decision environment* is constituted by the decision maker's perceptions of alternatives, information, cues, and stresses. This brings up another related item: *threat*. Often decisions must be made in haste, in a crisis atmosphere. This tends to reduce not only the time involved, but also the range of choices that might have been considered. This has obvious relevance to strategic decision making. In the face of armed hostilities or during actual conflict, decisions often must be made rapidly. This leads to another dimension of decision making: *risk and uncertainty*. Voters, too, face uncertainty when they select among presidential candidates; they cannot be sure platform promises will be kept.

## Mental Maps

One of the most important factors influencing us as decision makers is our images of places, our *mental maps* of our immediate surroundings, our country, the world. When we are asked to draw our "mental map" of an area, we will distort it in a way that may significantly reflect our misconceptions. In other words, we hold in our mind a "model" of the spatial environment involving notions of distance, direction, shape, accessibility, and so forth—and on the basis of this model, which is at variance with reality, we operate.

So, too, do politicians. Complicated issues are simplified and thus distorted, and then decisions are made (and opinions swayed) accordingly. A good example of this is the comment made by the experienced William A. O'Neill, then Governor of Connecticut, at the annual conference of New England governors and eastern Canadian premiers in June 1990. When asked about the controversy over the status of Quebec, which dominated the meeting, he admitted his ignorance of it, and then said, "What you're talking about here is approximately one third of the country, *geographically located in the center of the country*, removing itself." (emphasis added)*

Spatial perception, of course, is only one dimension of a complex of images we hold of the world around us (including ourselves), a totality that has been called the *perceptual field*. This perceptual field is affected by numerous conditions: our cultural conditioning and the values attached to it, our attitudes, motivations, goals. In the spatial context, at least within cultures, it may be possible to discover broad patterns of common behavior. John Sonnenfeld addresses himself to these matters. He recognizes geographical, operational, perceptual, and behavior environments.

These generalizations deserve some elaboration.

In Chapter 2 we point out that home is the most important place to people. It is, in fact, central in their world. This *centrality* of home—the most familiar, the most important place in the world—manifests itself clearly in children's maps of their neighborhoods or hometowns, in which their homes, schools, and other special places are shown as both larger and more central than they really are according to the scale of the drawing. It may be observed in the familiar humorous, but quite serious, maps of the United States as seen by Texans, New Yorkers, Bostonians, or others, in which home is enormous and fairly detailed, with the rest of the country fading vaguely into the distance. This is also seen in the

all-too-common maps showing Alabama or Jordan or Podunk surrounded by concentric circles proclaiming its centrality in the universe, and hence its ideal location for a vacation or for new industry. A view with profound political implications is the traditional Chinese perception of the world with China as "The Middle Kingdom" surrounded by barbarians.

In fact, every point on the globe is central, and concentric rings around it could demonstrate this clearly. A place is central only because individuals or groups perceive it that way, based on subjective criteria that are subject to change. Read "strategic" for "central" and one is thinking geopolitically. More on these points later.

*Mental*, or *cognitive*, *maps* are based on individual perceptions of the world, which are nearly always influenced strongly by location and culture. Mountain dwellers and valley dwellers have different surroundings, see the world differently, and hence have different attitudes in environmental, social, and political affairs, among others. Similarly, seafarers and landlubbers have different perspectives; so do desert nomads and city residents. Perceptions of space are conditioned by occupation, by technology, and possibly by numerous other factors not fully understood.

*Distance* is also a matter of perception. It is a cliché by now to observe that the world is shrinking. It isn't actually, it just seems that way because circulation (transportation and communications) is so much faster, safer, and more efficient now and places are closer together in time than formerly. Americans, in fact, have become accustomed to measuring distance in time—minutes', hours', or days' travel time—or in money—bus zones.

But our perceptions of distance still vary from one individual to another. Is a place 200 miles away far away? That depends. People in Rhode Island and Texas would very likely have quite different answers. Imagine the difference in attitude between a citizen of a country in which it is physically or politically impossible to travel overland more than a few hours in any direction

---

*New Haven Register*, 20 June 1990.

## Comparison of Spatial Perceptions

A good way to understand how spatial perspectives are created, reinforced, and expressed is to compare atlases produced in different parts of the world. Here is a summary of two student atlases produced in the United States and Australia, with the maps listed in the order in which they appear.

| *Goode's World Atlas, 18th ed.* *Chicago: Rand McNally, 1990* | | *The New Jacaranda Atlas,* *3rd ed. Milton, Queensland:* *Jacaranda Wiley, 1987* | |
|---|---|---|---|
| Area Covered | Pages of Maps | Area Covered | Pages of Maps |
| World | 53 | World | 12 |
| Northern Lands and Seas | 1 | Australia | 38 |
| United States and Canada | 53 | Papua New Guinea | 4 |
| Middle America | 9 | South Pacific | 10 |
| South America | 8 | New Zealand | 4 |
| Europe | 29 | Asia | 2 |
| Former USSR | 11 | South and Southeast Asia | 4 |
| Asia | 9 | East Asia | 8 |
| Middle East and Southwest Asia | 4 | India | 4 |
| South and East Asia | 12 | Middle East | 2 |
| Pacific Ocean | 2 | Europe | 13 |
| Australia and New Zealand | 7 | Former USSR | 3 |
| Africa | 13 | Africa | 10 |
| Antarctica | 1 | United States and Canada | 6 |
| Ocean Floor | 7 | Middle America | 2 |
| Metropolitan Areas | 18 | South America | 6 |
| (5 US, 13 foreign) | | Arctic | 2 |
| | | Antarctica | 2 |
| Total | 237 | Total | 132 |

within the State's boundaries and one who lives in a country as huge as the United States or Russia, or one who lives in Micronesia and customarily sails in an open canoe 150 kilometers just to get a particular brand of tobacco.

Cognitive distance affects the orientation of peoples. Until the 1970s, Australians felt closer to England than to the Philippines, and Jamaicans still feel closer to England or Canada than to Antigua only 1600 kilometers (1000 miles) away. These perceptions help explain why Australia has only weak links with the Philippines and why the West Indies Federation was doomed to failure.

*Direction* is a matter of cognition also, conditioned largely by culture and in particular by frequent exposure to maps ori-

ented in a particular direction and by the kind of ethnocentrism already discussed in connection with centrality. Very few Americans really understand, for example, that their nearest noncontiguous neighbor is not Cuba but Russia. Our geographic vocabulary is full of expressions such as Near, Middle, and Far East; Down Under; Top of the World; and Subsaharan Africa. In the United States, the four cardinal points of the compass are not north, south, east, and west, but "Up North," "Down South," "Out West," and "Back East." We commonly speak of going up to Canada or down to Baja, of above the equator or below the Rio Grande.

None of these terms has any meaning whatever on a globe or on the surface of the

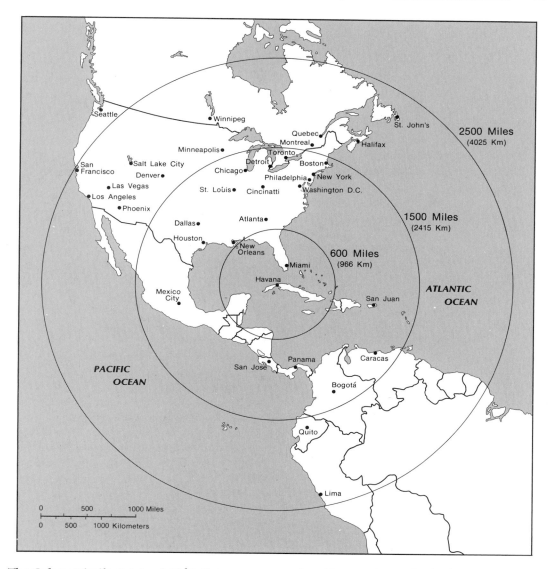

**The Cuban Missile Crisis of 1962.** This map, centered on Havana, shows clearly the potential threat to Latin America as well as North America posed by Soviet offensive weapons based in Cuba. The inner circle represents the approximate range of the smaller missiles, the second that of the nuclear weapon-carrying bombers and the medium-range ballistic missiles, while the outer circle shows the approximate range of the intermediate-range missiles.

earth itself. They are all based on the perspective of the viewers who coined the phrases and helped diffuse them around the world to distort other peoples' perceptions. Indonesia may be the Far East to a Britisher, but to an Australian it's the Near North. This was brought forcefully to the attention of Australians when Japanese troops occupied Indonesia during World War II, invaded eastern New Guinea, and threatened their homeland. Australia had to reorient its attention away from Britain to the west and toward Japan to the north and the United States to the northeast.

Cognitive direction is also linked with centrality; that is, a single situation viewed from two perspectives can look very different. At the beginning of the cold war in the 1950s, Americans were told that their country was surrounded by a chain of their own military bases protecting them from attack by the Soviet Union. But a map appearing in a popular weekly news magazine was centered on the Soviet Union rather than the United States and showed unmistakably that it was the Soviet Union that was surrounded—and by hostile, not friendly, forces. Then, in October 1962, when the United States revealed that the USSR had placed intermediate-range ballistic missles and long-range bombers in Cuba that were capable of delivering nuclear warheads to nearly every major city in North America, many Latin Americans chuckled about a small Latin American country pulling the tail feathers of the Yanqui eagle. The smiles faded, however, when a wire service map appeared in local newspapers showing concentric rings around Havana representing the ranges of the weapons. To their surprise and dismay, they saw that those very same nuclear warheads could be delivered anywhere in Latin America as far south as Lima. Fidel Castro lost many friends and admirers in Latin America that day. A final example illustrates the changing role of Mexico in the world as it develops economically and invests heavily in Central America. Mexicans used to refer to the United States as "The Colossus of the North." They seldom use that sobriquet any more since they no longer feel so inferior to the United States, but now Central Americans are using it to refer to Mexico!*

Another type of cognitive map is largely *imaginary*, an idealized or stereotyped vision of a place that bears little resemblance to the real place but nevertheless influences decisions and actions. Soldiers go off to war to fight and die for the green fields of England or the Volga steppe or Main Street,

USA. In truth, these images have come to them only through folklore, and the soldiers may really live in the industrial slums of Birmingham or Kiev or Pittsburgh. But, be it ever so humble, a slum tenement is hardly worth dying for 20,000 kilometers away, and governments foster images of picket fences and pastures of plenty as morale boosters.

Boosterism has a long and honored history in the United States, and it has helped to shape the iconography that inspires loyalty. We discuss iconography in more detail in Chapter 19, but here's an example designed to create a mental image of the United States. It's a song from the late nineteenth century, typical of the poems, advertisements, and songs created to entice immigrants to settle the new lands and build industries.

### *Bounding the US*

Of all the mighty nations in the East and in the West,
Why, the glorious Yankee nation is the wisest
and the best.
There is room for all creation, and the banner is
unfurled,
It's a general invitation to the people of the world.

Chorus:

Come along, come along, make no delay;
Come from every nation, come from every way,
Our land it is broad enough, and don't you be alarmed,
Uncle Sam is rich enough, he'll give you all a farm.

The St. Lawrence bounds our Northern line,
the crystal waters flow,
And the Rio Grande, the southern bound,
way down in Mexico.
From the great Atlantic Ocean where the sun
begins to dawn,
It peaks the Rocky Mountains, clear away in Oregon.

Chorus

In the South, they raise the cotton, in the West,
the corn and pork,
While New England's manufacture, they do
the finer work,
And the little creeks and waterfalls, that force

---

*Then there is the classic story of the headline not long ago in *The Times* (London) that symbolizes why Britain was so reluctant to join Europe: "Terrible Gale in the Channel—Continent Isolated."

along our hills,
Are just the thing for washing sheep and
driving cotton mills.

Chorus


This magnificent vision of America as a Garden of Eden was replaced not long after by a less attractive one as settlers began to move out of the prairies of the Old Northwest and the Mississippi Valley, out onto the Great Plains, where rainfall was scantier and the sod would not yield to the simple plows of the period. Then developed the legend of the Great American Desert blocking the way to California and Oregon, a region fit only for grazing, not for large-scale settlement. This legend, like the song, influenced migration patterns profoundly.


## *Map Distortions*

All maps are distortions of reality. They must be, for no flat map can accurately portray even a small portion of the earth's curved surface, to say nothing of the entire earth. The problem for the cartographer is how to manipulate the distortion so as to minimize it in those characteristics that are most important for the purpose of the map. The three essential qualities of every map are *projection, scale,* and *symbolization.*

In order to control distortion, the cartographer chooses the projection most appropriate for the purpose: an equal-area projection if the map is to show comparisons of areas in different parts of it or an equidistant projection if distance is the critical factor. The wrong projection can convey the wrong impression. The scale must be large enough for the amount of detail in the map to be seen clearly, but not so large that the map wastes space or becomes unwieldy. Symbols must fit the map scale and be easy for the reader to interpret, with the aid of a legend or key if necessary. A responsible cartographer will select and combine these qualities in various ways in

order to transmit information about a part or all of the earth's surface. But even a map produced with the best of intentions is not always perfect; it is subject to errors in the original data on which it is based, the inexperience or subconscious biases of the cartographer, and the interpretations of the reader.

Even accurate maps can be misleading. The projection with which we are most familiar is Mercator's. It is a splendid projection for navigation, which is what Mercator designed it for in the sixteenth century, and is widely used today by aerial and maritime navigators all over the world. Unfortunately, however, it is all too commonly used for general-purpose or reference maps, and, especially on small-scale maps, it presents a grossly distorted picture of the world. Yet this is the map of the world most of us have indelibly printed on our minds. Wrong as it is, it still shapes our thinking. The original maps in this book have been designed and executed with the greatest care, each one individually, so as to be as useful as possible within the technical limitations imposed by the book itself. Nevertheless, to derive the fullest value from them, the reader must study each map carefully and try to understand its message.

Our Mercator view of the world prior to World War II led us to exaggerate the size of areas in the high latitudes and ignore the polar areas entirely, except when reading or hearing accounts of the exploits of polar explorers. But the global nature of the war plus the extensive use of aircraft made polar projection maps and other azimuthal (or zenithal) maps popular. American insularity has been reinforced by such historical and cultural factors as the central role of the Monroe Doctrine in our foreign policy, the location of the Pan American Union headquarters in Washington, Roosevelt's "Good Neighbor" policy toward Latin America, and the desire of many European immigrants to turn away from the misery and oppression they had left behind and look into the interior of North America for hope and opportunity. All this was changed by the war

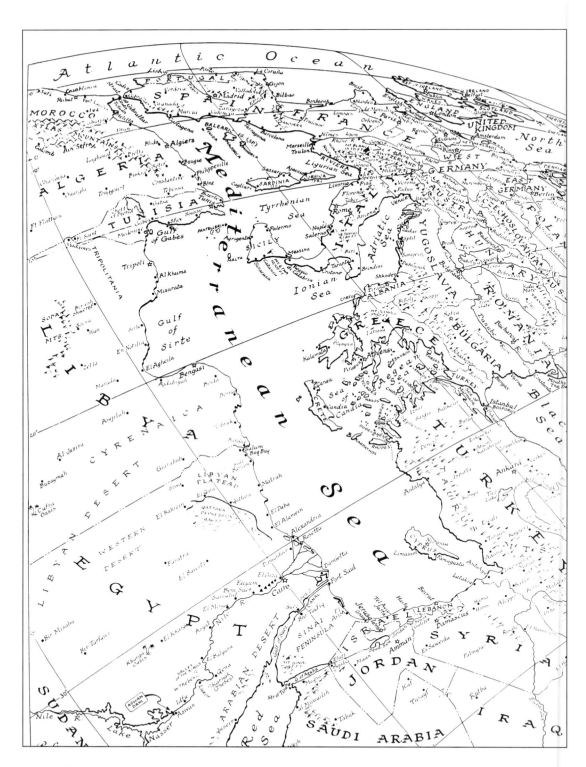

**North Africa as seen from the Mideast.** One of a series of 10 maps published in *The Christian Science Monitor* in late 1968 designed to show how people in various parts of the world perceive nearby areas that look very different when viewed from other perspectives. An understanding of the other person's perspective of the world can help us to appreciate his or her attitudes and decision making. (© 1968 Russel H. Lenz. All rights reserved.)

when Asia, the Pacific islands, North Africa, and the Soviet Union suddenly became our neighbors.

While we have tended to lapse back into our more traditional view of the world, we are now more accustomed to seeing maps not drawn to cylindrical projections and not oriented with north at the top.* During the war, Richard Edes Harrison produced a number of such unconventional maps, and J. Paul Goode's homolosine projection became popular. Both helped us to develop mental maps that are somewhat more realistic than a Mercator view. In the 1960s *The Christian Science Monitor* ran a series of full-page maps that showed portions of the earth as viewed from different points above it. The series, called "Global Perspectives," was an attempt to help us understand the mental maps of the world carried around by people living in Canada, the Soviet Union, China, Africa, and other places. More of these maps more widely used would help us develop a more accurate mental image of the world.

Cartographers use graphic images resembling maps but deliberately drawn out of scale and not even drawn to a projection, but emphasizing size or shape as the primary symbol. Pictograms, cartograms, and block diagrams are the most common. We use a cartogram in Chapter 19 to emphasize the gross disparity in per capita incomes around the world. Maps and other graphic devices can be used to inform, as in most of the cases cited so far, or to persuade. Maps designed deliberately to channel the reader's thinking along certain lines are called *propaganda maps*, and are far more common than we realize.

## Propaganda Maps

Some propaganda maps are relatively benign, such as the maps mentioned earlier that show the United States or the world revolving around Alabama or Jordan or Podunk. Others are found widely in commercial advertising: they depict how *our* bank blankets the state with its branch offices, or that *our* restaurant chain has 7479 locations in 59 states and therefore it must be better (or at least nearer) than the competition, or that *our* hotel is located precisely halfway between here and there and is therefore the best possible place to spend the night. We are concerned here, however, with maps that carry political messages.

Political propaganda maps, often in cartoon form, are not new. The late nineteenth and early twentieth centuries saw a spate of them, particularly in Europe, depicting countries in shapes or caricatures purporting to picture the national character. Prussian, Russian, Turk and Englishman were the favorite cartoon-map characters portraying stereotypes of the respective States. With changing tastes and circumstances and growing sophistication among propagandists and cartographers, propaganda maps became more serious and perhaps more persuasive. During the 1930s, the German "geopoliticians" made propaganda maps one of their principal weapons in generating support among intellectuals and the general public at home and abroad for the Nazi policy of German expansion.

The Germans clearly used projection, scale, and symbol to distort reality, always retaining enough truth in their maps to lend them credibility, and appealed to the emotions rather than the intellect of their readers. It is difficult to gauge their effectiveness, even with hindsight, but they did leave us a legacy of greater understanding of the utility of maps in persuasion as well as in education. Even during World War II, the Allies used maps to dramatize the threat posed by the enemy or to justify strategies such as attacking the "soft underbelly of Europe" and invading one island in the Pacific after another in stepping-stone fashion toward Japan.

*There is no particular reason why north should be placed at the top of the map, except perhaps that most of the world's land area, people, States, and economic activity are located in the Northern Hemisphere. Medieval European maps were oriented (from orient—east) with east at the top, the general direction of Jerusalem, and some even had Jerusalem in the center with the rest of the world around it, and all were perfectly well understood by their readers.

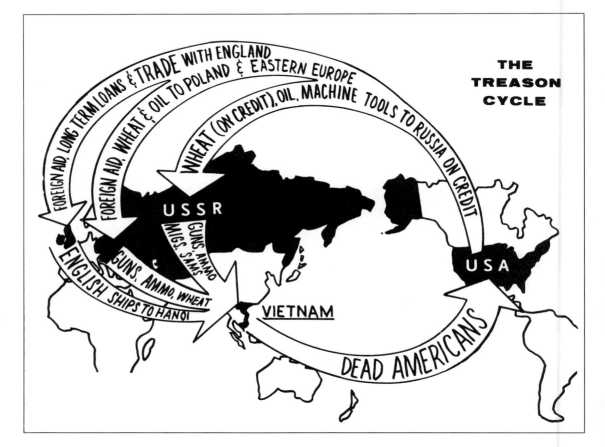

**A modern propaganda map.** This map was distributed in Connecticut (and probably elsewhere) by the John Birch Society at the height of the Vietnam War in 1968. It uses a number of standard propaganda techniques to convey the message that dealing with those who engage in any way with our enemy of the moment is treason. An older example of a propaganda map may be found in Chapter 21.

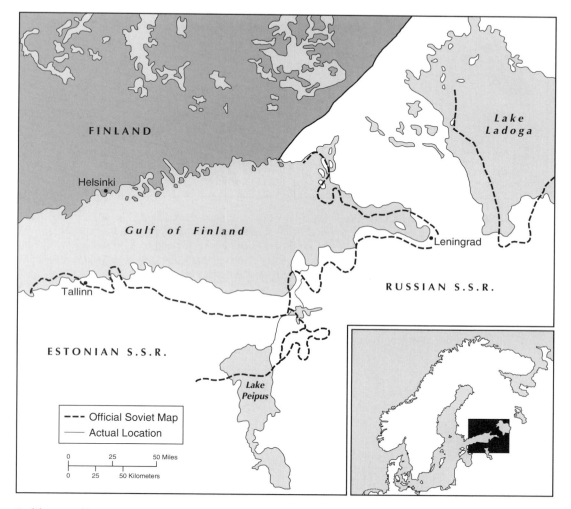

**Deliberate distortion of a Soviet map.** In the late 1960s the Soviet Union, which had a reputation for cartographic excellence, began producing maps such as this one (redrawn for clarity), which distort many areas of the country. Apparently, two techniques were used: creation of a pseudoprojection on which to draw the map, and deliberate shifting of features from their actual positions. Such distortions are relatively easy to detect by comparison with older, accurate maps and through remote sensing from aircraft or space vehicles.

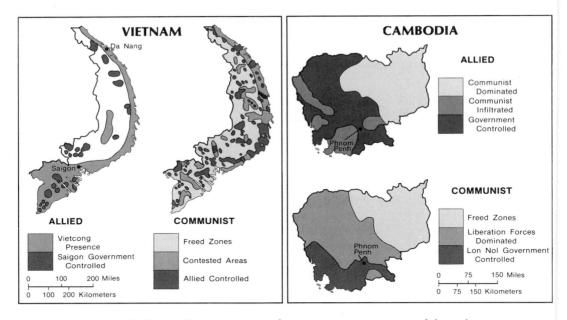

**The map war in Indochina.** These two pairs of maps represent versions of the military situation in South Vietnam and Cambodia toward the end of 1970, produced by the United States and its allies on one hand and the Communist-led rebels on the other. Each side was trying to convince the world that it controlled the most territory, in order to win support and establish credibility. Similar map wars to influence public opinion have been waged in Portuguese Guinea and elsewhere.

Both sides, but particularly the Allies, deliberately falsified maps that the enemy was allowed to "capture." The maps misled the enemy about planned military operations, part of the intelligence/counterintelligence battles that are so much a part of modern warfare. Even well after the war, the Soviet Union produced falsified maps of its own territory in apparent attempts to mislead NATO strategic planners about the locations of cities and natural features. These are not really propaganda maps, but specialized types of military maps.

Maps are also used extensively in psychological warfare. During the Cold War in the 1950s and 1960s, both the United States and the USSR used maps to impress their peoples with the danger of imminent attack by the other, across Europe or over the north polar region. We have mentioned the maps of U.S. bases surrounding the USSR; there were many more. An example may be found in Chapter 23.

Maps continue to play an important role in attempts to mold public opinion on political issues. The Vietnam War, the Panama Canal treaties, the Egyptian blockade of the Strait of Tiran, "wars of national liberation," the Argentine-Chilean dispute over the Beagle Channel, and many other battles have been fought in maps as well as on the ground. There is no reason to believe that maps will be used in this way any less in the future. Examples are scattered throughout this book, and so are other instances of perception in political decision making. The concept could well be kept in mind as we explore some elements of political geography.

# BIBLIOGRAPHY FOR PART ONE

## *Books and Monographs*

**A**

Agnew, John A., *Place & Politics*. London: Allen & Unwin, 1987.

Alexander, Lewis M., *World Political Patterns*. Chicago: Rand McNally, 1963.

Alland, A., *The Human Imperative*. New York: Columbia Univ. Press, 1972.

Amadeo, Douglas and Reginald Golledge, *Introduction to Scientific Reasoning in Geography*. New York: John Wiley, 1975.

Anderson, Ewan, *The Atlas of World Political Flash Points*. London: Belhaven, 1992.

Ardrey, Robert, *The Territorial Imperative*. New York: Atheneum, 1966.

**B**

Berry, Brian J.L., *The Nature of Change in Geographical Ideas*. De Kalb: Northern Illinois Univ. Press, 1979.

———, *Long-Wave Rhythms in Economic Development and Political Behavior*. Baltimore: Johns Hopkins Univ. Press, 1991.

Bowman, Isaiah, *The New World: Problems in Political Geography*. Yonkers, NY: World Book Co., 1921.

Boyd, Andrew, *An Atlas of World Affairs*. 9th ed. New York: Routledge, 1992.

Brigham, Albert Perry, *Geographic Influences in American History*. Boston: Ginn, 1903.

Burnett, Alan D. and Peter J. Taylor (eds.), *Political Studies from Spatial Perspectives*. New York: John Wiley, 1981.

**C**

Canters, Frank and Hugo Decleir, *The World in Perspective: A Directory of World Map Projections*. New York: John Wiley, 1989.

Corbridge, Stuart, *Capitalist World Development; A Critique of Radical Development Geography*. Totowa, NJ: Rowman & Littlefield, 1986.

Cox, Kevin and others, *Locational Approaches to Power and Conflict*. New York: Halsted, 1974.

**D**

Dear, Michael J. and Jennifer R. Wolch (eds.), *Territory & Reproduction*. Beverly Hills, CA: Sage, 1988.

Demko, George J. and William B. Wood, *Reordering the World: Geopolitical Perspectives on the Twenty-First Century*. Boulder, CO: Westview, 1994.

Dikshit, Ramesh K., *Political Geography: A Contemporary Perspective*. New York: McGraw-Hill, 1982.

Downs, Roger M. and David Stea, *Maps in Minds; Reflections on Cognitive Mapping*. New York: Harper & Row, 1977.

Dunn, Ross E., *The Adventures of Ibn Battuta, A Muslim Traveller of the Fourteenth Century*. London and Sydney: Croom Helm, 1986.

**E**

East, W. Gordon and A.E. Moodie (eds.), *The Changing World: Studies in Political Geography*. London: Methuen, 1968.

East, W. Gordon and J.R.V. Prescott, *Our Fragmented World*. New York: Macmillan, 1975.

**F**

Fisher, Charles A. (ed.), *Essays in Political Geography*. London: Methuen, 1968.

Fitzgerald, Charles P., *The Chinese View of Their Place in the World*. Oxford University Press, 1969.

Freeman, T. W., "Political Geography," in *A Hundred Years of Geography*. London: Duckworth, 1961.

**G**

Glacken, Clarence, *Traces on the Rhodian Shore*. Berkeley: Univ. of California Press, 1967.

Gladwin, Thomas, *East Is a Big Bird*. Cambridge, MA: Harvard Univ. Press, 1970.

Goblet, Yann Morvan, *Political Geography and the World Map*. New York: Praeger, 1955.

Gottmann, Jean, The *Significance of Territory*. Charlottesville: Univ. Press of Virginia, 1973.

Gould, Peter and Rodney White, *Mental Maps*. 2nd ed. Boston: Allen & Unwin, 1986.

**H**

Hall, E. T., The *Silent Language*, Garden City, NY: Doubleday, 1959.

———, *The Hidden Dimension*. Garden City, NY: Doubleday, 1966.

Harrison, Richard Edes, *Look at the World: The Fortune Atlas for World Strategy*. New York: Knopf, 1944.

Hartshorne, Richard, "Political Geography," in Clarence F. Jones and Preston James (eds.), *American Geography: Inventory and Prospect*. Syracuse, NY: Syracuse Univ. Press, 1954, 167–225.

Harvey, David, *Explanation in Geography.* New York: St. Martin's, 1969.

Hoggart, Keith and Eleanore Kofman (eds.), *Political Geography and Social Stratification.* London: Croom Helm, 1986.

Hooson, David (ed.), *Geography and National Identity.* Oxford: Blackwell, 1994.

Huff, Darrell, *How to Lie with Statistics.* New York: Norton, 1954.

**J**

Jackson, W.A. Douglas (ed.), *Politics and Geographic Relationships.* 2nd ed. Englewood Cliffs, NJ: Prentice-Hall, 1971.

Jackson, W.A. Douglas and Edward F. Bergman, *A Geography of Politics.* Dubuque, IA: Brown, 1973.

Johnston, Ronald J., *Geography and the State: An Essay in Political Geography.* New York: Macmillan, 1982.

Johnston, Ronald J. and Peter J. Taylor (eds.), *A World in Crisis?: Geographical Perspectives.* 2nd ed. Oxford: Blackwell, 1989.

Jones, Stephen B. and Marion F. Murphy, *Geography and World Affairs.* Chicago: Rand McNally, 1971.

Jufte, Edward R., *Envisioning Information.* Cheshire, CT: Graphics Press, 1990.

**K**

Kasperson, Roger E., *Frontiers of Political Geography.* Englewood Cliffs, NJ: Prentice-Hall, 1973.

Kasperson, Roger E. and Julian Minghi (eds.), *The Structure of Political Geography.* Chicago: Aldine, 1969.

Kidron, Michael and Ronald Segal, *The New State of the World Atlas.* London: Pluto Press, 1987.

Kirby, Andrew, *The Politics of Location; An Introduction.* London: Methuen, 1982.

**L**

Lewis, D., *We, the Navigators.* Honolulu: Univ. of Hawaii Press, 1972.

Lorenz, Konrad, *On Aggression.* New York: Harcourt, Brace, 1966.

Lowenthal, David and Martyn J. Bowden (eds.), *Geographies of the Mind.* Oxford: Oxford University Press, 1976.

**M**

Martin, Geoffrey J., *The Life and Thought of Isaiah Bowman.* Hamden, CT: Shoe String Press, 1980.

Mellor, Roy E.H., *Nation, State, and Territory.* Routledge, 1989.

Monmonier, Mark, *How to Lie with Maps.* Chicago: Univ. of Chicago Press, 1991.

Montagu, M.F.A., *Man and Aggression.* Oxford: Oxford University Press, 1968.

Moodie, A. E., *Geography Behind Politics.* London: Hutchinson's Univ. Library, 1949.

Muir, Richard, *Modern Political Geography.* New York: Halsted, 1975.

Muir, Richard and Ronan Paddison, *Politics, Geography and Behaviour.* London and New York: Methuen, 1981.

**N**

National Academy of Sciences, *Studies in Political Geography.* Washington, DC: Committee on Geography, 1965.

**P**

Pacione, Michael (ed.), *Progress in Political Geography.* London: Croom Helm, 1985.

Pearcy, George Etzel and others, *World Political Geography.* 2nd ed. New York: Crowell, 1957.

Peet, Richard, *Radical Geography.* London: Methuen, 1978.

Poulsen, Thomas M., *Nations and States; A Geographic Background to World Affairs.* Englewood Cliffs, NJ: Prentice Hall, 1995.

Pounds, Norman J.G., *Political Geography.* 2nd ed. New York: McGraw-Hill, 1972.

**R**

Rościszewski, Marcin (ed.), *The State, Modes of Production and World Political Map;* Seminar of the IGU Study Group on the World Political Map. Warsaw: Institute of Geography and Spatial Organization, Polish Academy of Sciences, 1989.

Roskin, Michael, *Countries and Concepts: An Introduction to Comparative Politics.* 3rd ed. Englewood Cliffs, NJ: Prentice Hall, 1992.

**S**

Saarinen, Thomas F., *Perception of Environment,* Washington, DC, AAG, Resource Paper No. 5, 1969.

———, *Environmental Planning; Perception and Behavior.* Prospect Heights, IL: Waveland Press, 1976.

Sack, Robert David, *Human Territoriality.* Cambridge University Press, 1986.

Sanderson, Marie, *Griffith Taylor; Antarctic Scientist and Pioneer.* Ottawa: Carleton Press, 1988.

Short, John R., *An Introduction to Political Geography.* 2nd ed. New York: Routledge, 1989.

Sommer, R., *Personal Space: The Behavioral Basis of Design.* Englewood Cliffs, NJ: Prentice-Hall, 1969.

Sonnenfeld, J. "Geography, Perception, and the Behavioral Environment," in Paul W. English and R. C. Mayfield (eds.), *Man, Space, and*

*Environment.* Oxford University Press, 1972, 244-251.

Stoddart, David R. (ed.), *Geography, Ideology and Social Concern.* Oxford: Blackwell, 1981.

**T**

Taylor, Peter J. (ed.), *Political Geography of the Twentieth Century; A Global Analysis.* London: Belhaven, 1994.

———, *Political Geography; World-Economy, Nation-State and Locality.* 3rd ed. London: Longman, 1994.

Taylor, Peter J. and John W. House (eds.), *Political Geography; Recent Advances and Future Directions.* London; Croom Helm, 1984.

Taylor, Ralph B., *Human Territorial Functioning.* Cambridge, Cambridge University Press, 1988.

**V**

Van Valkenburg, Samuel and Carl L. Stotz, *Elements of Political Geography.* Englewood Cliffs, NJ: Prentice-Hall, 1954.

**W**

Wanklyn, Harriet G., *Friedrich Ratzel: A Bio-graphical Memoire and Bibliography.* Cambridge University Press, 1961.

Weigert, Hans and Richard E. Harrison (eds.), *New Compass of the World: A Symposium on Political Geography.* New York: Macmillan, 1949.

Weigert, Hans and Vilhjalmur Stefansson (eds.), *Compass of the World: A Symposium on Political Geography.* New York: Macmillan, 1945.

Weigert, Hans and others, *Principles of Political Geography.* New York: Appleton-Century-Crofts, 1957.

Whittlesey, Derwent, *The Earth and the State.* Madison, WI: Published for the U.S. Armed Forces Institute by Holt, 1944.

Wolch, Jennifer and Michael Dear (eds.), *The Power of Geography; How Territory Shapes Social Life.* New York: Routledge, 1988.

Wood, Denis, *The Power of Maps.* New York: Guilford, 1992.

**Z**

Zukin, Sharon, *Landscapes of Power; From Detroit to Disney World.* Berkeley: Univ. of California

## *Periodicals*

**A**

Agnew, John A., "Socializing the Geographical Imagination: Spatial Concepts in the World-System Perspective," *Political Geography Quarterly,* 1, 2 (April 1982), 159–166.

———, "An Excess of 'National Exceptionalism': Towards a New Political Geography of American Foreign Policy," *Political Geography Quarterly,* 2, 2 (April 1983), 151–166.

———, "The Return of Time and the Need for a New Materialism," *Political Geography,* 12, 1 (Jan. 1993), 84–86.

Alwin, John A., "North American Geographers and the Pacific Rim: Leaders or Laggards," *Professional Geographer,* 44, 4 (Nov. 1992), 369–376.

*Antipode; A Radical Journal of Geography.* Published quarterly since 1969. Oxford: Blackwell.

**B**

Boggs, S. Whittemore, "Cartohypnosis," *Scientific Monthly,* 64 (1947), 469–476.

———, "Geographic and Other Scientific Techniques for Political Science," *American Political Science Review,* 42 (1948), 223–238.

Bosque-Maurel, Joaquin and others, "Academic Geography in Spain and Franco's Regime, 1936–55," *Political Geography,* 11, 6 (Nov. 1992), 550–562.

Brunn, Stanley D. and Ernie Yanarella, "Towards a Humanistic Political Geography," *Studies in Comparative International Development,* 22, 2 (Summer 1987), 3–86.

Buleon, Pascal, "The State of Political Geography in France in the 1970s and 1980s," *Progress in Human Geography,* 16, 1 (March 1992), 24–40.

Bunge, William W., "Geography of Human Survival," *Annals, AAG,* 63, 3 (Sept. 1973), 275–295.

Burghardt, Andrew, "The Dimensions of Political Geography: Some Recent Texts," *Canadian Geographer,* 9 (1965), 229–233.

———, "The Core Concept in Political Geography: A Definition of Terms," *Canadian Geographer,* 13, 4 (1969), 349–353.

**C**

Chappell, John E., Jr., "Marxism and Geography," *Problems of Communism,* 14 (1965), 12–22.

Cox, Kevin R., "Comment: The Politics of Globalization: A Sceptic's View," *Political Geography,* 11, 5 (Sept. 1992), 427–429.

———, Review Essay—"The Division of Labor,

the State and Local Politics," *Political Geography*, 12, 4 (July 1993), 382–385.

**D**

Dalby, Simon, "Reading Peet, (re)Reading Fukuyama: Political Geography at 'The End of History'," *Political Geography*, 12, 1 (Jan. 1993), 87–90.

de Blij, Harm, "Political Geography of the Post Cold War World," *Professional Geographer*, 44, 1 (Feb. 1992), 16–19.

DeBres, Karen, "George Renner and the Great Map Scandal of 1942," *Political Geography Quarterly*, 5, 4 (Oct. 1986), 385–394.

Demko, George J. and W. Hezlep, "USSR: Mapping the Blank Spots," *Focus*, (Spring 1989), 20–21.

Dikshit, Ramesh Dutta, "Comment: Another False Dawn?," *Political Geography*, 11, 6 (Nov. 1992), 523–525.

Dodds, Klaus-John, "Eugenics, Fantasies of Empire and Inverted Whiggism" (An essay on the political geography of Vaughan Cornish), *Political Geography*, 13, 1 (Jan. 1994), 85–99.

Duncan, C. J. and W. R. Epps, "Comment: GIS and the Role of the State 'Down Under'," *Political Geography*, 12, 1 (Jan. 1993), 3–7.

**E**

East, W. Gordon, "The Nature of Political Geography," *Politica*, 2 (1937), 259–263.

**F**

Fischer, Eric, "German Geographical Literature, 1940–1945," *Geographical Review*, 36, 1 (Jan. 1946), 92–100.

**G**

*Geographic and Global Issues*. Published quarterly by the Office of the Geographer and Global Issues, U.S. Department of State.

Golledge, Reginald G., "Behavioral Approaches in Geography," *Australian Geographer*, 12 (1972), 59–79.

Goodchild, Michael F., "Comment: Just the Facts," *Political Geography Quarterly*, 10, 4 (Oct. 1991), 335–337.

Gould, Peter and N. Lafond, "Mental Maps and Information Surfaces in Quebec and Ontario," *Cahiers de Géographie de Québec*, 23, 60 (1979), 371–398.

**H**

Hall, E.T., "Proxemics," *Current Anthropology*, 9 (1968), 83–108.

Hall, Peter, "The New Political Geography," *Transactions, Institute of British Geographers*, 63 (1974), 48–52.

Harrison, Richard Edes, "The War of the Maps," *Saturday Review of Literature*, 26 (7 August, 1943), 24–27.

———, "The Face of the World: Five Perspectives for an Understanding of the Air Age," *Saturday Review of Literature*, 27 (1944), 5–6.

Harrison, Richard Edes and Robert Strauz-Hupe, "Maps, Strategy, and World Politics," *Smithsonian Institution Report* (1943–44), 253–258.

Hartshorne, Richard, "Recent Developments in Political Geography," *American Political Science Review*, 29 (1935), 785–804, 943–966.

———, "The Politico-Geographic Pattern of the World," *Annals, American Academy of Political and Social Science*, 218 (1941), 45–57.

———, "Political Geography in the Modern World," *Journal of Conflict Resolution*, 4, 1 (March 1960), 52–66.

Heffernan, Michael, "On Geography and Progress: Turgot's Plan d'un Ouvrage sur la Géographie Politique (1751) and the Origins of Modern Progressive Thought," *Political Geography*, 13, 4 (July 1994), 328–343.

Henrickson, Alan K., "The Map as an 'Idea': The Role of Cartographic Images During the Second World War," *American Cartographer*, 2, 1 (1975), 19–53.

Herb, Henrik G., "Persuasive Cartography in Geopolitik and National Socialism," *Political Geography Quarterly*, 8, 3 (July 1989), 289–303.

Hershkovitz, Linda, "Tiananmen Square and the Politics of Place," *Political Geography*, 12, 5 (Sept. 1993), 395–420.

Heske, Henning, "German Geographic Research in the Nazi Period—A Content Analysis of the Major Geography Journals 1925–1945," *Political Geography Quarterly*, 5, 3 (July 1986), 267–281.

Holdar, Sven, "The Ideal State and the Power of Geography: The Life-work of Rudolf Kjellen," *Political Geography*, 11, 3 (May 1992), 307–324.

Holdich, Thomas H., "Some Aspects of Political Geography," *Geographical Journal*, 34, 6 (Dec. 1909), 593–607.

**J**

Jackson, W.A. Douglas, "Whither Political Geography?" *Annals, AAG*, 48, 2 (June 1958), 178–183.

Johnson, Douglas Wilson, "A Geographer at the Front and at the Peace Conference," *Natural History*, 19 (1919), 511–521.

Johnston, Ronald J., "Texts, Actors and Higher Managers: Judges, Bureaucrats and the Political Organization of Space," *Political Geography Quarterly*, 2, 1 (Jan. 1983), 3–19.

————, "On the Practical Relevance of a Realist Approach to Political Geography," *Progress in Human Geography* 9 (1985), 601–604.

————, "One World, Millions of Places: The End of History and the Ascendancy of Geography," *Political Geography*, 13, 2 (March 1994), 111–121.

Johnston, Ronald J. and Mickey Lauria, Review Essay: "Political Geography: Improving the Theoretical Base," *Political Geography Quarterly*, 4, 3 (July 1985), 251–257.

Jones, Stephen B., "Field Geography and Postwar Political Problems," *Geographical Review*, 33, 4 (Oct. 1943), 446–456.

————, "Views of the Political World," *Geographical Review*, 45, 3 (July 1955), 309–326.

**K**

Kirby, Andrew, "Pseudo-random Thoughts on Space, Scale, and Ideology in Political Geography," *Political Geography Quarterly*, 4, 1 (Jan. 1985), 5–18.

————, "Editorial Comment: Publishing Deca(ye)de," *Political Geography*, 11, 3 (May 1992), 235–238.

Kliot, Nurit, "Contemporary Trends in Political Geography: A Critical Review," *Monadnock*, 50 (1976), 54–70.

————, "Contemporary Trends in Political Geography," *Tijdschrift voor Economische en Sociale Geografie*, 73 (1982), 270–279.

Knight, David B., "The International Geographical Union Study Group on the World Political Map," *Political Geography Quarterly*, 8, 1 (Jan. 1989), 87–94.

Kofman, Eleanore and Linda Peake, "Editorial Comment on the Special Issue," *Political Geography Quarterly*, 9, 4 (Oct. 1990), 311–312.

————, "Into the 1990s: A Gendered Agenda for Political Geography," *Political Geography Quarterly*, 9, 4 (Oct. 1990), 313–336.

Kolosov, V.A., "Political Geography in the USSR," *Soviet Geography*, 30 635-650.

Kriesel, Karl Marcus, "Montesquieu: Possibilistic Political Geographer," *Annals, AAG*, 58, 3 (Sept. 1968), 557–574.

Kristof, Ladis K.D., "Review Essay on Political Geography," *American Political Science Review*, 79 (1985), 1178–1179.

**L**

Ley, David and Roman Cybriewsky, "Urban Graffiti as Territorial Markers," *Annals, AAG*, 64, 4 (Dec. 1974), 491–505.

Lloyd, Robert, "Cognitive Maps: Encoding and Decoding Information," *Annals, AAG*, 79, 1 (March 1989), 101–124.

Lundén, Thomas, "Swedish Contributions to Political Geography," *Political Geography Quarterly*, 5, 2 (Oct. 1986), 181–186.

**M**

Macdonald, Gerald M. and John T. O'Hara, "Samuel Van Valkenburg: Politics and Regional Geography," *Political Geography Quarterly*, 7, 3 (July 1988), 288-290.

Mashbits, Ya. G., "The Nature of Political Geography and Ways of Increasing Its Geographical Content," *Soviet Geography*, 30 (1989), 650–657.

Mayer, Tom, "Reflections on 'The End of History'," *Political Geography*, 12, 1 (Jan. 1993), 79–83.

McColl, Robert W., "Political Geography as Political Ecology," *Professional Geographer*, 18, 3 (May 1966), 143–145.

Mercer, John, Review Essay: "Diversity as Strength," *Political Geography Quarterly*, 2, 1 (Jan. 1983), 81–87.

Mitchell, Don and Neil Smith, "Comment: The Courtesy of Political Geography: Introductory Textbooks and the War Against Iraq," *Political Geography Quarterly*, 10, 4 (Oct. 1991), 338–341.

Moran, Warren, "Rural Space as Intellectual Property," *Political Geography*, 12, 3 (May 1993), 263–277.

**N**

Newman, David, "Comment: On Writing 'Involved' Political Geography," *Political Geography Quarterly*, 10, 3 (July 1991), 195–199.

Nientied, P., "A 'New' Political Geography: On What Basis?," *Progress in Human Geography*, 9 (1985), 597–600.

**O**

Obenbrugge, Jürgen, "Political Geography Around the World: West Germany," *Political Geography Quarterly*, 2, 1 (Jan. 1983), 71–80.

O'Loughlin, John, "Political Geography: Bringing the Context Back," *Progress in Human Geography*, 12, 1 (March 1988), 121–137.

————, "Political Geography: Coping with Global Restructuring," *Progress in Human Geography*, 13, 3 (Sept. 1989), 412–426.

————, "Political Geography: Attempting to Understand a Changing World Order," *Progress in Human Geography*, 14, 3 (1990), 420–437.

————, "Political Geography: Returning to Basic Conceptions," *Progress in Human Geography*, 15, 3 (1991), 322–339.

————, "Ten Scenarios for a 'New World Order'," *Professional Geographer*, 44, 1 (Feb. 1992), 22–28.

Ormeling, F. J., Jr., "Cartographic Consequences of a Planned Economy—50 Years of Soviet Cartography," *American Cartographer*, 1, 1 (1974), 39–50.

Otok, Stanislaw, "Political Geography Around the World: Poland," *Political Geography Quarterly*, 4, 4 (Oct. 1985), 321–327.

O'Tuathail, Gearóid, "Beyond Empiricist Political Geography: A Comment on van der Wusten and O'Loughlin," *Professional Geographer*, 39, 2 (May 1987), 196–197.

———, "Putting Mackinder in His Place: Material Transformations and Myth," *Political Geography*, 11, 1 (Jan. 1992), 100–118.

**P**

Paasi, Anssi, "The Institutionalization of Regions: A Theoretical Framework for Understanding the Emergence of Regions and the Constitution of Regional Identity," Fennia, 164 (1986), 105–146.

———, "Deconstructing Regions: Notes on the Scales of Spatial Life," *Environment and Planning A*, 23 (1991), 239–256.

———, "Reading Fukuyama: Politics at the End of History," *Political Geography*, 12, 1 (Jan. 1993), 64–78.

Peet, Richard, "The End of Prehistory and the First Human," *Political Geography*, 12, 1 (Jan. 1993), 91–95.

Penrose, Jan, "The Great Category of Nation and Specific Nations," *Political Geography*, 13, 2 (March 1994), 195–203.

Pike, Steve, "'A Load of Bloody Idiots': Somerset Dairy Farmers' View of Their Political World," *Political Geography Quarterly*, 10, 4 (Oct. 1991), 405–421.

*Politics and Geography.* Special Issue of *International Political Science Review* 1, 4 (1980).

Prescott, John Robert Victor and J. G. Hajdu, "A Review of Some German Post-War Contributions to Political Geography," *Australian Geographical Studies*, 2, 1 (1964), 35–46.

**Q**

Quam, Louis O., "The Use of Maps in Propaganda," *Journal of Geography*, 42, 1 (Jan. 1943), 21–32.

**R**

Reynolds, David R., "Political Geography: Thinking Globally and Locally," *Progress in Human Geography*, 16, 3 (Sept. 1992), 393–405.

———, "Political Geography: Closer Encounters with the State, Contemporary Political Economy, and Social Theory," *Progress in Human Geography*, 17, 3 (Sept. 1993), 389–403.

———, "Political Geography: The Power of Place and the Spatiality of Politics," *Progress in Human Geography*, 18, 2 (June 1994), 234–247.

Rosenberg, Michael, "The Mother of Invention: Evolutionary Theory, Territoriality, and the Origins of Agriculture," *American Anthropologist*, 92, 2 (June 1990), 399–415.

Rumley, Dennis, "Conflict and Compromise: Political Geography at the IGU Congress in Sydney, Australia, August 1988," *Political Geography Quarterly*, 8, 2 (April 1989), 197–200.

———, "The Political Organisation of Space: A Reformist Conception," *Australian Geographical Studies*, 29, 2 (Oct. 1992).

Rumley, Dennis and others, "The Content of Ratzel's *Politische Geographie*," *Professional Geographer*, 25, 3 (1973), 271–276.

**S**

Sack, Robert D., "The Power of Place and Space," *Geographical Review*, 83 (1993), 326–329.

Sandner, Gerhard, "The *Germania Triumphans* Syndrome and Passarge's *Erdkundliche Weltanschauung*: The Roots and Effects of German Political Geography Beyond Geopolitik," *Political Geography Quarterly*, 8, 4 (Oct. 1989), 341–352.

———, "Introduction to Special Issue," *Political Geography Quarterly*, 8, 4 (Oct. 1989), 311–314.

Sanguin, André-Louis, "André Siegfried, An Unconventional French Political Geographer," *Political Geography Quarterly*, 4, 1 (Jan. 1985), 79–83.

Sauer, Carl O., "The Formative Years of Ratzel in the United States," *Annals, AAG*, 61, 2 (June 1971), 245–254.

Sayer, Andrew, "Radical Geography and Marxist Political Economy: Towards a Re-evaluation," *Progress in Human Geography*, 16, 3 (Sept. 1992), 343–360.

———, "Comment: Realism and Space: A Reply to Ron Johnston," *Political Geography*, 13, 2 (March 1994), 107–109.

Schultz, Hans-Dietrich, "Fantasies of Mitte: Mittellage and Mitteleuropa in German Geographical Discussion in the 19th and 20th Centuries," *Political Geography Quarterly*, 8, 4 (Oct. 1989), 315–340.

Sevrin, Robert, "Research Themes in Political Geography—A French Perspective," *Political Geography Quarterly*, 4, 1 (June 1985), 67–78.

Sharp, Joanne P., "Publishing American Identity:

Popular Geopolitics, Myth and the Reader's Digest," *Political Geography*, 12, 6 (Nov. 1993), 491–503.

Smith, Neil, "Isaiah Bowman: Political Geography and Geopolitics," *Political Geography Quarterly*, 3, 1 (Jan. 1984), 69–76.

———, "Bowman's New World and the Council on Foreign Relations," *Geographical Review*, 76 (1986), 438–460.

Spate, O.H.K., "Toynbee and Huntington: A Study in Determinism," *Geographical Journal*, 118, 4 (Dec. 1952), 406–428.

Sprout, Harold H. and Margaret T. Sprout, "Geography and International Politics in an Era of Revolutionary Change," *Journal of Conflict Resolution*, 4, 1 (March 1960), 145–161.

Stea, David, "Space, Territoriality and Human Movements," *Landscape*, 15 (1965), 13–16.

Stevenson, Ian, "Comment: Don't Shoot the Coachman: A Rejoinder to Andrew Kirby," *Political Geography*, 12, 2 (March 1993), 99–102.

Stoddart, David R., "Geography and War: The 'New' Geography and the 'New Army' in England, 1899–1914," *Political Geography*, 11, 1 (Jan. 1992), 87–99.

**T**

Taylor, Peter J., "Editorial Comment: 1688 and All That," *Political Geography Quarterly*, 7, 3 (July 1988), 207–208.

———, "Editorial Comment: In Praise of Lord Rennell of Rodd," *Political Geography Quarterly*, 7, 4 (Oct. 1988), 305–306.

———, "Editorial Comment: Children and Politics," *Political Geography Quarterly*, 8, 1 (Jan. 1989), 5–6.

———, "Editorial Comment: Geographical Dialogue," *Political Geography Quarterly*, 8, 2 (April 1989), 103–106.

———, "Editorial Comment: Oneworldism," *Political Geography Quarterly*, 8, 3 (July 1989), 211–214.

———, "Editorial Comment: GKS," *Political Geography Quarterly*, 9, 3 (July 1990), 211–212.

———, "Tribulations of Transition," *Professional Geographer*, 44, 1 (Feb. 1992), 10–12.

———, "Editorial Comment: Politics in Maps, Maps in Politics: A Tribute to Brian Harley," *Political Geography*, 11, 2 (March 1992), 127–129.

———, "*Contra* Political Geography," *Tijdschrift voor Ecomische en Sociale Geografie*, 84 (1993), 82–90.

———, "Review Essay: Modern, Postmodern and Post Postmodern," *Political Geography*, 13, 3 (May 1994), 279–289.

Terlouw, C.P., "World-System Theory and Regional Geography," *Tijdschrift voor Economische en Sociale Geografie*, 80 (1989), 206–221.

Thrift, Nigel and Dean Forbes, Review Essay: "A Landscape with Figures: Political Geography with Human Conflict," *Political Geography Quarterly*, 2, 3 (July 1983), 247–263.

Troll, Carl, "Geographic Science in Germany During the Period 1933–1945," *Annals, AAG*, 39, 2 (June 1949), 99–137.

**V**

van der Wusten, Herman and John O'Loughlin, "Back to the Future of Political Geography: A Rejoinder to O'Tuathail," *Professional Geographer*, 39, 2 (May 1987), 198–199.

Veness, April R., "Neither Homed Nor Homeless: Contested Definitions and the Personal Worlds of the Poor," *Political Geography*, 12, 4 (July 1993), 319–340.

**W**

Wallerstein, Immanuel, "The Rise and Future Demise of the World Capitalist System: Concepts for Comparative Analysis," *Comparative Studies in Society and History*, 16 (1974), 387–415.

Warf, Barney, "Can the Region Survive Post-Modernism?" *Urban Geography*, 11 (1990), 586–593.

Whittlesey, Derwent, "The Horizon of Geography," *Annals, AAG*, 35, 1 (March 1945), 1–36.

Wilson, David, "Excavating the Dialectic of Blindness and Insight: Anthony Giddens' Structural Theory," *Political Geography*, 14, 3 (April 1995), 309–318.

Wright, John Kirtland, "Training for Research in Political Geography," *Annals, AAG*, 34, 4 (Dec. 1944), 190–201.

# PART TWO

# *The State*

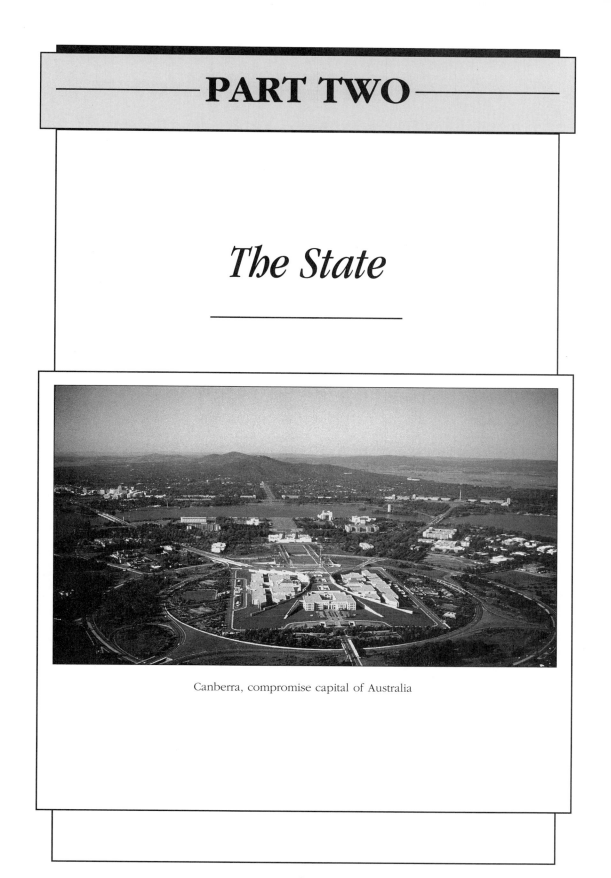

Canberra, compromise capital of Australia

# 4

# *State, Nation, and Nation-State*

In the history of political geography, more attention has been devoted to studying the State than any other topic. The State has been described, classified, analyzed, discussed, and argued about from the time of Aristotle and Plato—if not earlier. Today it is still the focus of our attention, and despite many other important and alluring subjects to investigate, some of them relatively new, this should not be surprising. In the past two centuries the State has become the dominant form of political organization in the world and is likely to remain so for some time to come.

In much of the discussion, not only among nonspecialists but even among scholars in many disciplines, the term *State* is confused with *state* or *nation*. Often State or state or nation is used interchangeably with *nation-state*, and sometimes the hyphenated word is used rather pompously in place of a simpler one. Before we proceed further, we should clarify the proper meanings of these terms.

## State

A State is a place. It is also a concept represented by certain symbols and demanding (though not always obtaining) the loyalty of people. Its intangible qualities are discussed later, but here we can examine some of its more measurable qualities.

In order for a place to be considered a State in the strictest sense, it must possess to a reasonable degree certain characteristics:

1.  **Land territory.** A State must occupy a definite portion of the earth's land surface and should have more or less generally recognized limits, even if some of its boundaries are undefined or disputed.

2.  **Permanent resident population.** An area devoid of people altogether, no matter how large, cannot be a State. An area only traversed by nomads or occupied seasonally by hunters or scientists similarly cannot be a State. A State is a human institution created by people to serve some of their particular needs.

3.  **Government.** The people living within a territory must have some sort of administrative system to perform functions needed or desired by the people. Without political organization, there can be no State.

4.  **Organized economy.** While every society has some form of economic system, a State invariably has responsibility for many economic activities, even if they include little more than the issuance and supervision of money and the regulation of foreign trade, and even if economic activities are managed badly.

5.  **Circulation system.** In order for a State to function, there must be some organized means of transmitting goods, peo-

ple, and ideas from one part of the territory to another. All forms of transportation and communication are included within the term *circulation*, but a modern State must have more sophisticated forms available to it than runners or the "bush telegraph."

These requirements for a State are all geographic, and political geographers have long been studying them. Two other requirements, however, have traditionally been in the realms of political science and international law. If we are to be realistic in any attempt to classify the political units of the world, we must give careful consideration to them. In many cases the geographic criteria may be quite clear, but only the political criteria can be decisive in determining whether or not a political unit is a State. These criteria are

1. **"Sovereignty."** There is general agreement that a State is "sovereign" or "possesses sovereignty," but no agreement on precisely what constitutes sovereignty. In a general sense it means power over the people of an area unrestrained by laws originating outside the area, or independence completely free of direct external control. If we try to get more specific than this, however, we encounter serious problems. There are so many qualifications and exceptions to the general concept that it would seem wise to use quotation marks around "sovereign" or "sovereignty" in most cases. However, that is often awkward and we may omit them as long as we understand that the term is imprecise and often used too loosely.

2. **Recognition**. For a political unit to be accepted as a State with "an international personality" of its own, it must be recognized as such by a significant portion of the international community— the existing States. In effect, it must be voted into the club by the existing members. Specialists in international law disagree about whether such recognition is "declarative" or "constitutive"; that is, whether the formal act of recognition simply puts the label of "State" on a political unit that has already demonstrated that it possesses most or all of the five characteristics and criteria already mentioned, or whether it is the act of recognition itself that brings the State into being for the first time. There is also disagreement over what proportion of the club's members must extend such a welcome and even over whether numbers alone have any value or whether affirmative acts by most of the "great powers" are necessary. These details need not concern us here, but we cannot forget that recognition is important, perhaps decisive, in determining the proper application of the word "State."

Among all the political units in the world, many are clearly States. The United States, for example, and Austria and Thailand, Venezuela and Tunisia. Many others are not: Hong Kong, Puerto Rico, Kalaalit Nunaat (Greenland), and French Polynesia, for example, are all dependent territories, even though they possess many of the attributes of statehood, including varying degrees of self-government. Still others are questionable. Is Taiwan a State? Or the Turkish Republic of Northern Cyprus? Or Vatican City? Or were a number of other entities whose status has been clarified, such as Biafra, the Baltic States, Rhodesia, Andorra, or Sikkim? We need not answer any of these questions. For our purposes, it is enough to understand that the term *State* has a fairly precise meaning and should be used with great care.

Nearly all States have civil divisions; that is, administrative districts within them that perform governmental functions and even, in the case of federal States, possess some measure of sovereignty. These civil divisions are frequently ranked in a hierarchy of size and/or authority, from first-order civil divisions on down. States give many names to their first-order civil divisions. Department, province, and county are among the most common.* In some States, all of which have at least nominally federal systems, the first-order civil divisions are called states. These include Mexico, Brazil, India, and, of course, the United States. In this book we follow the United Nations practice of capitalizing State when we mean an independent country and using a small "s" when we mean a first-order civil division, such as Nuevo León, Mato Grosso, West Bengal, or South Carolina.†

## *Nation*

Unlike a State, a nation is a group of people. Although definitions of the term vary considerably, political scientists and political geographers generally use it to mean a reasonably large group of people with a common culture, sharing one or more important culture traits, such as religion, language, political institutions, values, and historical experience. They tend to identify with one another, feel closer to one another than to outsiders, and believe that they belong together. They are clearly distinguishable from others who do not share their culture.

This may seem rather vague, but it may help a bit if we point out that the term "people" when used as a singular noun is roughly equivalent to "nation," just as "country" is roughly equivalent to "State." Thus, we may properly speak of the Polish nation or the Navajo nation or the Ewe nation be-

*Others include krai, prefecture, division, parish, governorate, commune, circonscription, republic, region, wilaya, and district.

†It should be noted that the word is customarily capitalized when giving the official name of the division, as in State of New Hampshire or State of New South Wales. On the other hand, some countries use the word in their official long-form names: some are the State of Bahrain, State of Israel, State of Kuwait, Independent State of Western Samoa, and the new (and probably temporary) Commonwealth of Independent States.

cause each is a people distinguished from all others by its distinctive culture.

How small a group may be considered a nation? There is no answer to this question. Certainly the group has to be larger than a family, a clan, or even a tribe, but how much larger no one can say. This is one of the dilemmas in the application of the principle of "national self-determination" or "self-determination of peoples," which we discuss later. We have in the world today some very large nations and some very small ones. Compiling a list would be impossible. Nevertheless, the concept of the nation as a cultural entity is very old, and it has been very important in the development of the modern political world. Perhaps the critical factor is whether the group in question considers itself to be a nation.

The feeling among a people that they constitute a nation may be carried further, into the political realm. They may begin to feel that they ought to have a State of their own. Or an individual may identify with the State in which he or she resides—or with another State. This linkage of emotion with a political entity called the State is termed, for lack of a better word, "*nationalism.*" *Nationalism* is an old sentiment, apparently dating from our earliest civilizations, but we are primarily concerned here with its modern manifestations and its effects on the political organization of space.

Nationalism can be a very healthy emotion. It can encourage people to identify with a group larger than their family, clan, or tribe. It can get them interested, even involved, in the affairs of modern society. It can stimulate them to create new ideas and institutions that will advance society as a whole, in partnership with other people and other nations. But if the emotion becomes more powerful it can lead to a demand for self-determination among ever-smaller groups and to the fragmentation of society instead of its unification. Carried further, the emotion can lead to *irredentism*, the desire to incorporate within the State all areas that had once been part of it or in which live people who belong to the nation but not to the State. A still more extreme form of nationalism is *chauvinism*,* which may be defined as "superpatriotism," excessive and bellicose feelings of superiority over all other peoples and countries. Clearly this is unhealthy; it has frequently led to wars of aggression and to imperialism.

## Nation-State

Just as the State is the dominant form of political organization today, the nation-state is the ideal form to which most nations and States aspire. That this ideal is very difficult to attain becomes evident when we try to count the number of true nation-states in the world today. Put in the simplest terms, a nation-state is a nation with a State wrapped around it. That is, it is a nation with its own State, a State in which there is no significant group that is not part of the nation. This does not mean simply a minority ethnic group, but a nationalistic group that either wants its own State or wants to be part of another State or wants at least a large measure of autonomy within the State in which it lives.

Despite being composed of a large number of minorities, the United States is arguably a nation-state. Japan certainly is. Sweden, Uruguay, Egypt, and New Zealand are all nation-states, though they have very different histories and demographic characteristics. The Soviet Union, on the other hand, was an empire, composed of many "nationalities," most of which retained much of their cultural heritage in their ancestral homelands. Canada is frequently referred to as "two nations in one State." Belgium, South Africa, Afghanistan, and China are other examples of older countries that are not nation-states, and very few countries that have become independent since World War II can be considered nation-states. As usual, other States fall in the question-mark column; it would take precise definitions, adequate data, and careful analysis to determine whether Israel, Jordan, Bolivia, or Liberia, for example, can truly be called nation-states.

In recent times we have seen a number of nations that are split among two or more States attempt to break away and form States of their own. Usually those in one country are more enthusiastic about independence than their compatriots in adjacent countries. In no case yet has the group succeeded in assembling its politically separated parts and creating from them a new nation-state. Some examples are demands for independent States of Macedonians living in Bulgaria, Greece, Yugoslavia, and perhaps Albania; Basques living in Spain and France; and Kurds of Turkey, Iraq, Iran, Syria, and possibly the Caucasus.

In the first case the issue had been raised from time to time primarily by Bulgaria, and only recently has there been any enthusiasm for it among those who may be considered Macedonian. The great majority of Spanish Basques are probably content with a large measure of autonomy within Spain, which most of them received in 1977. Only a small minority is actively fighting for independence, while across the border, the French Basques seem to be content with the cultural autonomy they have long enjoyed. The Kurdish rebellion against Iraq has been prolonged and frequently intense, while the Kurds in Turkey and Iran have also rebelled at times, demanding autonomy within these countries—at the least. These examples indicate the very great difficulty today of detaching a portion of a State to form a new one on ethnic grounds. The creation of Bangladesh is a special case in which ethnic considerations were important but not decisive.

The other method of trying to create nation-states from existing States is the secession of areas within a State that are inhabited largely or exclusively by minority ethnic groups. We have seen numerous examples of this in the past few decades alone, and we are likely to see more in the future. Indonesia, for example, has had serious rebellions in the Riau Archipelago, in Southern Sulawesi (Celebes), in Aceh, Irian Jaya, East Timor, and the Southern Maluku (Molucca)

*From Nicolas Chauvin, a French soldier wounded and decorated many times during the Revolutionary and Napoleonic wars but who retained a simple-minded devotion to Napoleon. He came to typify the cult of military glory and ultranationalism.

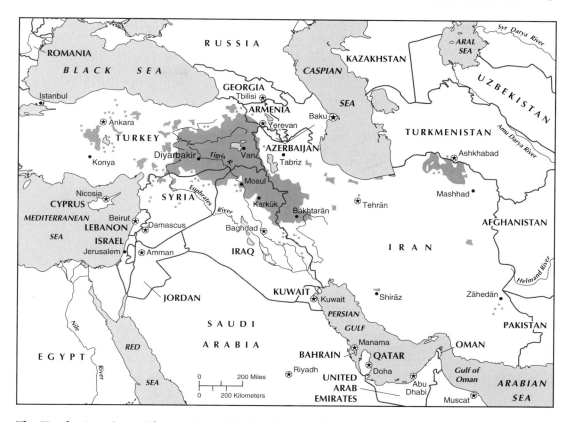

**The Kurds: A nation without a State.** The Kurdish people are scattered through Southwest Asia, but their heartland is a contiguous area divided among Turkey, Syria, Iran, Iraq and Armenia, a region they have been striving for generations to establish as a sovereign State of Kurdistan. This effort has always been thwarted by outside powers, local sovereigns and their own internal divisions. Iraqi Kurdistan is a major oil producing region. The projected capital of an independent Kurdistan is Diyarbakir, Turkey.

Islands. Minority ethnic, or national, groups have engaged in full-scale rebellion or at least terrorism in order to obtain independence or autonomy in such countries as Chad (Muslim Toubou tribesmen), India (Mizos, Nagas), France (Bretons, Corsicans), Myanmar (Burma) (Karens, Kachins, Shans), Ethiopia (Eritreans, Somalis), Nigeria (Ibos), China (Uigurs, Tibetans, others), Canada (Québécois), Vietnam ("Montagnards"), Philippines (Moros), Sudan (southern blacks), Sri Lanka (Tamils), and many others. Among these rebellions, only that of Eritrea has been successful, and only after long and bitter fighting. None of the other rebellions has been any more successful in creating new States than the attempts to create new States from parts of several others.

## Irredentism

Another manifestation of nationalism, mentioned earlier, is irredentism. The term derives from an area in northern Italy that was still part of Austria in 1871 after the rest of Italy had been unified into a nation-state. Young Italian nationalists referred to it as *Italia irredenta*, or unredeemed Italy. Today there are a number of irredentist movements around the world, although it is often difficult to distinguish them from relatively routine border disputes, ideological conflicts, or simple aggression. There are also many cases in which people who speak the language or share other cultural attributes with those of a particular State live across the border of a neighboring State but are not in-

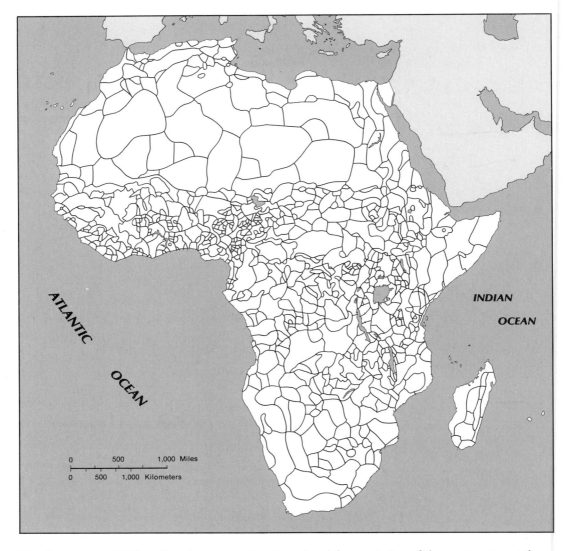

**Ethnic groups of Africa**. The almost unbelievable cultural fragmentation of the continent contributes greatly to its political instability. It is nearly impossible to draw political boundaries along strictly ethnic lines that would make sense in other respects. Clearly, the concepts of nation-state and self-determination have their limitations in the real world. This problem is discussed in more detail in Parts IV and VIII.

volved in irredentist claims. They live in peace as citizens of their State, and the neighboring States are on good terms.

On occasion, however, such circumstances lead to political problems. Sometimes it is felt that members of a nation living in an adjoining State are being mistreated and the State embodying their nation tries to defend them. This may lead to more vigorous action on their behalf, claims to the territory in which they live, and ultimately to military action designed to incorporate the territory into its own. Strong factions in Greece and Cyprus, for example, have long campaigned for *enosis*, or union of the two countries. Turkey has tended in the same direction to protect the Turkish Cypriot minority and in 1974 actually invaded Cyprus, ultimately occupying some 37 percent of the island. While they have not annexed the occupied area, they have sponsored therein a "Turkish Republic of Northern Cyprus."

**One version of the proposed State of Pushtunistan.** This crude map claims as the territory for a new Pushtu (Pakhtun) State nearly all of Pakistan west of the Indus River, including all of Baluchistan, but none of the adjacent area of Afghanistan occupied by Pushtu-speaking peoples. Note the absence of such major cities as Peshawar and Sukkur, whereas many quite unimportant places are shown. (Source of original map: Rahman Pazhwak, *Pakhtunistan; The Khyber Pass as The Focus of the New State of Pakhtunistan*, London: Afghan Information Bureau, [1951].)

Indonesia in the 1950s waged a vigorous campaign for the incorporation into its national territory of Dutch-held western New Guinea. It was finally successful in 1962 in forcing out the Dutch, and in May 1963 it obtained control after a nine-month period of United Nations administration. Indonesia then turned its attention increasingly to its "confrontation" with the new State of Malaysia, demanding all the former British territories on the island of Borneo. In this endeavor, it received no support from the United Nations and temporarily withdrew from that body in January 1965. A change in government in late 1965 and the forced retirement of President Sukarno led to the abandonment of the irredentist (or imperialist) campaign in August 1966.

The longest running, most violent, and most clear-cut irredentist campaign in our time has been the Somali drive for a "Greater Somalia." The Somali Republic, itself the product of a merger of the former British Somaliland and Italian Somaliland, has since independence in 1960 been waging a propaganda and guerrilla war against Ethiopia and less violent, but occasionally intense, battles with Kenya and Djibouti for control of the Somali-inhabited portions of those countries. The campaign reached a crescendo with a full-scale Somali invasion of the Ogaden region of Ethiopia in the summer of 1977. By October, the Somalis had occupied most of the Ogaden, including all but one of its important cities, and cut the vital Ethiopian railway to the port of Djibouti. This aggression received little support from abroad, and within a few months the Ethiopians, with considerable Cuban help, had driven the Somali regulars and most of the guerrillas back across the border. Since then the Ogaden has been relatively peaceful, but Somali irredentism has not been extinguished; it has only been submerged by problems of drought, famine, civil war, and foreign intervention.

A different and rather strange case is that of Pushtunistan (also called Pakhtoonistan and other varieties of the name). Afghanistan was not happy about the creation of Pakistan in 1947, and late that year made its first demand of the new State for the formation of an independent entity in its Northwest Frontier Province. The people (called Pathans in Pakistan) on both sides of the border are closely related, but the Afghans have never suggested either adding the Pathan-inhabited area of Pakistan to their own State or ceding the Pushtu-inhabited area of Afghanistan to a new State carved out of Pakistan. Thus, there is no attempt to reunite a people divided by a colonial boundary. It seems instead to be an artificial political issue created by Afghanistan for its own purposes, waxing and waning according to political conditions in the two countries and relations with the former Soviet Union, Iran, Arab States, and others. It may also be related to the Afghan desire for a seacoast, as evidenced by a map of the proposed new country produced by an Afghan diplomat. It has resulted in border closings, broken relations between Afghanistan and Pakistan, and even warfare in 1947, but the issue has not been raised publicly since 1979.

# 5

# *The Emergence of States*

The State is such a familiar phenomenon that we tend to forget that only recently has all of the earth's land surface except Antarctica come under the jurisdiction and control of States. Today there are nearly 200 States in the world. In the past century and a half, the Europeanization of the world has spread not only revolutionary techniques in agricultural, industrial, financial, medical, scientific, and other technical areas, but also European philosophies and techniques of political organization. But in many non-European areas of the world—in Middle and South America, in black Africa, in North Africa and Southwest Asia, and in East and Southeast Asia—States had been developing for thousands of years. Some of these ancient States collapsed almost immediately in the face of the European onslaught, but others, such as Ashanti in West Africa, held out until early in the twentieth century.

We do not truly understand all the forces and stimuli that contributed to the development of the first true States. Explanations usually involve factors of soil productivity, opportune location, natural protection against enemies, advantageous climate, and similar assets supposedly possessed by those ancient human agglomerations that managed to forge States out of their habitats. But what is not explained is why, in other places where conditions were the same or even more attractive, States failed to emerge.*

## Clan and Tribe

One way to approach the whole question of man's territorial organization is to look for the least complicated example that can be found. Early humans found that the number of people that could associate closely was limited by mode of life and environmental circumstance. In southern Africa live a people, the !Kung San (Bushmen), who face just such limitations. Having been driven into dry, inhospitable country by their enemies, the !Kung San continue to subsist on hunting and gathering. They do so mainly in groups, or clans, of about 60 individuals. Life for them revolves around the tenuous supply of the water hole, which must draw the animals the people trap and kill while providing them with this essence of life in the Kalahari Desert and Steppe. But,

> Although Bushmen are a roaming people and therefore seem to be homeless and vague about their country, each group of them has a very specific territory which that

*One interesting hypothesis, which rejects most traditional theories of State origins and emphasizes the role of warfare over circumscribed environments, is found in Robert L. Carneiro's "A Theory of the Origin of the State," *Science*, *169*, 3947 (21 August 1970), 733–738.

group alone may use, and they respect their boundaries rigidly. Each group also knows its own territory well; although it may be several hundred square miles in area, the people who live there know every bush and stone, every convolution of the ground.*

Like the !Kung San, the Australian aborigines were not organized to resist a challenge from more advanced competitors, and except in their reserve areas, nothing remains today of such political institutions as existed when the Europeans arrived. Had Australia experienced the evolution of strong and well-organized tribal States, the sequence of events that led to the formation of the modern Australian federation would have been very different. What might have happened is well illustrated in Africa, where the San suffered a fate similar to that of their Australian contemporaries, but many of the more advanced tribal States in other parts of that continent resisted the European advance for some time. Indeed, although these States were eventually overthrown, their political and economic organization not only withstood the European impact, but to some extent channeled it. And since the termination of the colonial period, many characteristics of the precolonial African tribal States are reflected in the modern black African States.

The pastoral Hottentots, who themselves were to fall victim to the onslaught of still more powerful African peoples, were members of a more complex form of politicoterritorial organization, namely, the *tribe*. It is difficult to define this concept comprehensively; there are very large tribes and very small ones. But comparing it to the San's clans reveals several important differences. The power of the tribal leader is greater. Personally given laws emerge. Territorial limits are even more jealously guarded. Hundreds of thousands of people may pay allegiance to the tribal chief or king. And it is especially important to note that the concept and reality of the *central place* exist at this level.

*Elizabeth Marshall Thomas, *The Harmless People*, New York, Knopf, 1959, p. 10.

Thus the tribal habitat will center on a headquarters (possibly with several subsidiary, minor foci)—a seat of government in the real sense.

Such a central place may be positioned according to some of the same rules that governed the beginnings of some towns that grew into capital cities of world importance: a defensible site, a place at the center of an area rich in resources, a location especially favorable as a market, a point of historical or religious significance. Long before the coming of whites to Africa and many other parts of the world, important capital cities had developed in tribally organized areas. Many a European traveler described with awe the characteristics of those early cities.

## Ancient States

### The Middle East

For thousands of years progress was made in the Middle East that overshadowed all else on the globe. In the Fertile Crescent, people learned to domesticate certain plants. By 5000 B.C. the small farming villages on the lower slopes of the hills were showing the effects of improving farming techniques and higher yields; they were growing larger than they had ever been. Irrigation, metal working, writing, planning, organization— innovations multiplied in numerous spheres and were transmitted to the Nile region, to Mesopotamia, and to the lowland of the Indus River. By 3000 B.C. urban life had developed in the Middle East, and unprecedented political consolidation was taking place.

Despite the fact that Mesopotamia probably was ahead of Egypt and the Indus region in the earliest times, ancient Egypt was destined to have the most lasting impact on the politicogeographical world. The political philosophies and practices of Pharaonic Egypt were diffused to West Africa's savanna kingdoms and to many of black Africa's tribal States; to this day there are tribal ceremonies as far away as southern Africa that bear the unmistakable imprint of Egyptian origins.

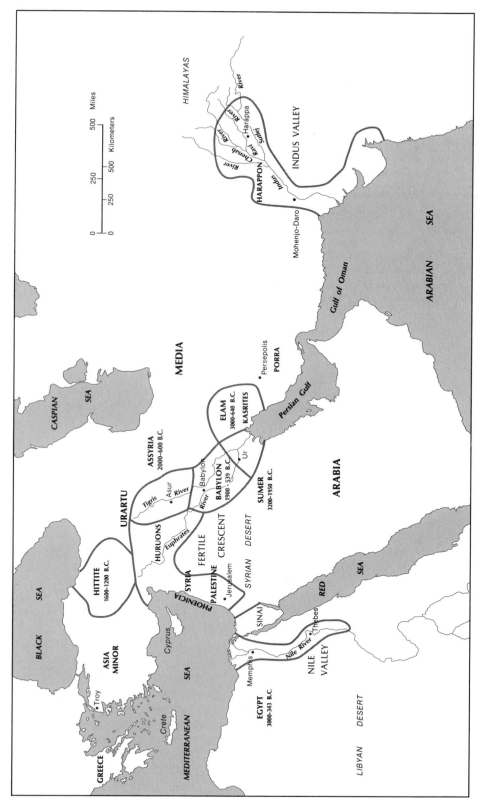

**Three culture hearths.** The Lower Nile Valley, the Fertile Crescent, and the Indus Valley are three of the world's eight major culture hearths. In these areas agriculture, cities, organized religion, States, and other attributes of civilization first appeared. These areas are still scenes of discord and even bitter warfare among competing cultures and nationalities.

**Angkor**. One of the world's eight great culture hearths developed in Southeast Asia more than a thousand years ago. It was centered on Angkor, near Siem Reap just north of Tonle Sap (Great Lake) in what is now Cambodia. The site was first settled about A.D. 819 and soon became the focus of a large agricultural region with a very sophisticated irrigation system. The magnificent temple dedicated to Vishnu (called Angkor Wat) was built early in the twelfth century. For some 600 years Angkor was a powerful State but was challenged by Siam late in the fourteenth century. It was abandoned about A.D. 1431 and was soon covered by jungle. The ruins were discovered in 1861. Restoration began in the 1920s.

What were the particular attributes of ancient Egypt, other than its pivotal location, felicitous protection, and productive capacity? The State's longevity is remarkable. Historians often identify the period of "true" empire to have extended from 2650 to 1100 B.C., but there were many centuries of consolidation prior to 2650 and, although subjected to invasion and defeats, the State hardly ceased to exist during the last millenium B.C. At a time when territorial organization was just beginning (in a world context), Egypt rose and survived longer than any other political entity ever had. In the process, the State changed from a theocratic to a militaristic one, and, in common with some other States we know, it developed an incredibly complex bureaucracy. The roots of "Western" civilization thus lie in Egypt and Mesopotamia, and among those are the roots of statecraft.

### East Asia

Middle Eastern contributions may have reached China as early as 2000 B.C., per-haps even before that. From Mesopotamia and the Indus area, it is thought, innovations in agriculture, techniques of irrigation, metal working, and even writing were diffused to the valley of the Hwang He (Yellow River) in the vicinity of its confluence with the Wei River. Whatever the sources of ancient China's stimuli, there is a record of Chinese civilization around five thousand years in length, and from 2000 B.C. on, that record is continuous, quite reliable, and positive proof of China's cultural individuality.

Out of those beginnings in northern China, in the lowland of the Hwang–Wei Rivers, rose a series of dynasties. The oldest of these about which there is substantial knowledge is the Shang (Yin) Dynasty, which lasted from about 1900 B.C. until 1050 B.C. Centered on the ancient capital of Anyang, this dynasty secured the consolidation of China as a true State. The Bronze Age reached China during this period; walled towns developed. Coupled with this were innovations in political organization and control, administration, and the use of military forces. The original core was ready to expand and absorb adjacent frontiers.

This expansion was both easiest and most profitable in a southward direction, and during the Ch'in Dynasty the lands of the Yangtze Jiang were integrated into the Chinese State. During the Han Dynasty, ancient China experienced a crucial period of growth and further consolidation. Between 202 B.C. and A.D. 220 the Han rulers brought unity and stability to their State, and they also extended the Chinese sphere of influence to include Korea, Manchuria, and Mongolia as well as distant Xinjiang and Annam.

The Han period was a critical time in the life of the Chinese State. Not only was there military might and great territorial expansion, but the country's internal order changed decisively. The old feudal system of land ownership broke down, and individual property was recognized. Internal circulation was unprecedentedly effective. Towns grew into cities; internal trade intensified. The silk trade grew into China's first regular external commerce. Along the silk route across Asia came ideas and innovations—China was being transformed. To this day the people of China, recognizing the germinal character of the Han period, call themselves the People of Han.

Like Egypt, Mesopotamia, the Indus area, and so many other great ancient States and cultures, China faced a conquering enemy. But there was a difference. In A.D. 1280 the Mongols captured control of China and made the State part of a vast empire that stretched all the way across Asia to Eastern Europe. It was not the Mongols who imposed a permanent new order upon China, however, but the Chinese whose culture was imprinted upon the Mongols. The Mongol period ended in 1368, and the Mongol conquest was only an interlude in the continuity of Chinese life. Egypt and Mesopotamia fell, but China managed to shake off its severest challenge. And so today, with China still focused on its ancient northern heartland, it is the only major ancient State whose lineage traces back—spatially as well as temporally—directly to its origins.

## West Africa

The savannas of West Africa lie between the desert to the north and the tropical rain forest to the south. For many centuries the peoples of the desert have traded goods with the peoples of the forests. West Africa's economic mainstays were gold and salt, and the West African savanna peoples found themselves positioned very advantageously between the traders. From the Nile Valley these fortunate savanna dwellers received some political ideas about divine kingship, about the power inherent in a monopoly over the use of iron, and about the use of military forces to exact tribute and maintain control. And so the cities of West Africa grew into great market centers and generated wide realms of control. Here was a form of the *city-state*. But West Africa was also a place of invention and innovation. There are anthropologists who argue that West Africa, along with Southwest Asia, Southeast Asia, and Middle America, was one of the four major agricultural complexes of the ancient world, where crops were domesticated and farming techniques invented and refined.

In West Africa a series of States of impressive strength and durability arose. The oldest of these about which much is known was ancient Ghana, located in parts of present-day Mali and Mauritania. For perhaps more than 1000 years (the first millennium A.D.), ancient Ghana's rulers managed to weld many diverse groups of people into a stable State. Ghana eventually fell to its northern competitors, and it was succeeded in the savanna belt by the State of Mali, centered on Timbuktu on the Niger River. Timbuktu in its heyday was a university city, a place of learning of international significance.

Mali, too, eventually collapsed. Its initial successor was the State of Songhai, slightly to the east; still later, the center of power shifted farther east again, to the region that is today northern Nigeria and western Chad. But by then the European impact was being felt along the West African coast, and the era of the slave trade was soon to upset the pat-

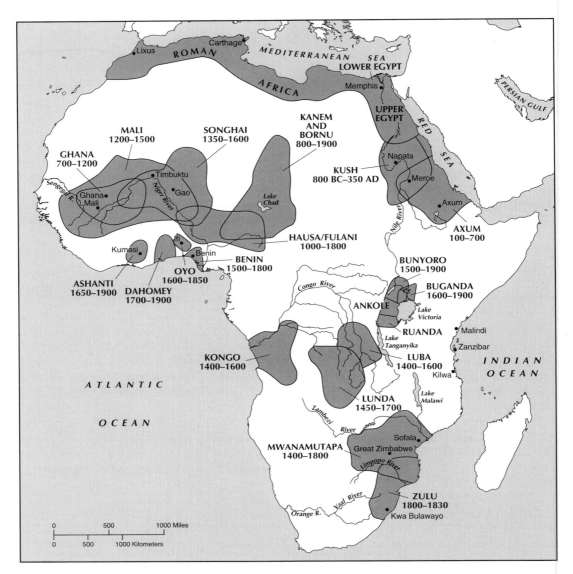

**Early African States.** Long before Europeans arrived south of the Sahara, Africans had developed sophisticated and durable States, complete with armies, schools, royal courts, and even tax collectors. Most striking were the succession of States in the Sudan-Sahel region of West Africa, between the desert and the forests. Many of the others were also outstanding.

terns of development of nearly two millennia in this part of the world.

### Middle and South America

In the Americas, prior to the European invasion, there were also large, stable, and well-organized States. Toward the end of the fifteenth century, various sectors of the American Indian realm were at different stages of politicoterritorial organization. In the southern part of South America lived peoples who subsisted by hunting and gathering. This economy prevailed also in parts of what is today Canada and the western half of the United States (though not the Southwest). In the Amazon Basin rain forests, in southeastern North America, and elsewhere in the hemisphere, some agriculture accompanied hunting and gathering,

**Ancient Zimbabwe.** While Mali, Ghana, and Songhai flourished in West Africa, the Shona peoples of South Central Africa between the Zambezi and Limpopo rivers also had a number of States. Of their remnants, those of Zimbabwe, near Masvingo, Zimbabwe are the most imposing. Apparently, the site was occupied by at least three successive peoples. This "Temple" or "Elliptical Building" is located on the plain below the "Acropolis" atop a granite hill. It may date from the twelfth century. (Photo courtesy of the Zimbabwe Tourist Office)

but the practice of agriculture was not always accompanied by more advanced organization.

In other parts of the Americas, tribal organization was achieved by many Indian peoples, who established permanent villages, practiced more intensive agriculture, and developed considerable technological skill. The idea of a form of federation also came to the leaders of these peoples, and many multicommunity tribal States developed through voluntary alliance and through conquest, although such alliances were often short-lived.* North of the Rio Grande, only the Iroquois achieved a confederacy prior to the European invasion. Nevertheless, these

*It is important to distinguish between indigenously conceived "federal" associations and those that developed after the European impact was felt. The second type were the direct result of common enmity toward the European invader and conceived of as a better means of defense.

tribal political entities represented a major advance in the quest for the State.

### Aztec State and Inca Empire

The culmination of indigenous political evolution in the Americas, so closely bound up with the development of organized agriculture, was achieved in Middle America and the Central Andes. Although the Mayan civilization, which flourished in Yucatan and adjacent areas between A.D. 300 and 900, was a forerunner of the significant events to take place later in what is today Mexico, important differences existed between the Mayan societies and those of the Toltecs and Aztecs. The Mayan societies were theocratic, controlled by priest-rulers; Toltec and Aztec society was militaristic and structurally much more complex. The main centers of Mayan culture lay in the forested sections of Yucatan, to the north of the mountain slopes of Guatemala and El Salvador. The central places, however, were not central places in the usual sense. They were ceremonial centers and served as markets, but did not contain large permanent populations.

One ceremonial site, Mayapan, did ultimately become a true central place, but only after the decline of Mayan power and organization had begun. Mayapan was a walled city, the location of a despotic authority, the home of thousands of residents, and the heart of all Yucatan. But by this time the area that had been unified by Mayan civilization was breaking into warring tribal States, with a corresponding decline in the arts and other fields of achievement.

Elsewhere in Middle America the strands of progress remained intact, ultimately to lead to the formation of one of the most impressive States to develop anywhere in the world at any time prior to the rise of modern Europe. In fact, the Aztec State was achieved by a relatively small number of Indians who were acquainted with Toltec technology, scattered by the disastrous breakup of Toltec society, and finally regrouped in a very fortuitous location. The choice of headquarters proved a permanent one: the site of the Aztec capital, Tenochtitlán, was also to become the capital of an empire, the seat of government for New Spain, and finally the focus for the modern State of Mexico.

A highly complex bureaucracy developed in the capital, and there were systems of tax collection, law enforcement, mail delivery, rapid messenger communication, and a provincial administration with governors and district commissioners. In the Valley of Mexico, the Aztecs had found the pivotal geographic feature of Middle America; agricultural productivity was high, and technology, including irrigation practices, had been developing for centuries. The population was relatively dense: the region at that time contained perhaps over 2 million inhabitants. Central places grew rapidly and to sizes that were remarkable for any part of the world: Tenochtitlán had well over a hundred thousand inhabitants, as did a number of other centers of trade and administration.

The American region marked by the culmination of indigenous politicogeographic organization is the Central Andes. Here, over a distance of 2500 miles from Ecuador to central Chile, the Inca Empire evolved. The Inca Empire was itself preceded by earlier civilizations, notably that centered on Tiahuanaco in the area of Lake Titicaca. Better integrated, more stable, less violently militaristic, and generally more benevolent than the Aztec sphere of influence, the Inca domain in many ways was the only real empire to develop in the Americas, Aztec achievements notwithstanding. The Inca Empire was centered on Cuzco, a city located 11,000 feet above sea level, 200 miles from the ocean, and with a population of over a quarter of a million. Here the Inca rulers built the mighty fortress of Sacsahuaman, the Temple of the Sun, and other huge megalithic structures.

Cuzco and its surrounding region formed the core area for a vast State with far-flung possessions and an intricate organization that may with justification be called its major achievement. At the height of its power the empire contained perhaps as many as 11 million subjects, each of whom had an assigned task of supplying a certain tax or tribute to the central authority, the Inca. In view of the contrasts in terrain within the empire, its stability is admirable, and it could not have been accomplished without a system of central places and efficient communication lines.

A system of roads connected the central places, and along these roads were a number of supply stations located at regular intervals and staffed by professional messengers; in relay fashion, an order issued in Cuzco would reach outlying districts in the shortest possible time. When European visitors first arrived in this area, they wondered how such broad and sturdy roads could have been constructed in such difficult terrain. In this respect, as in others, the Inca Empire resembled the Roman Empire, but without the use of wheels or large animals.

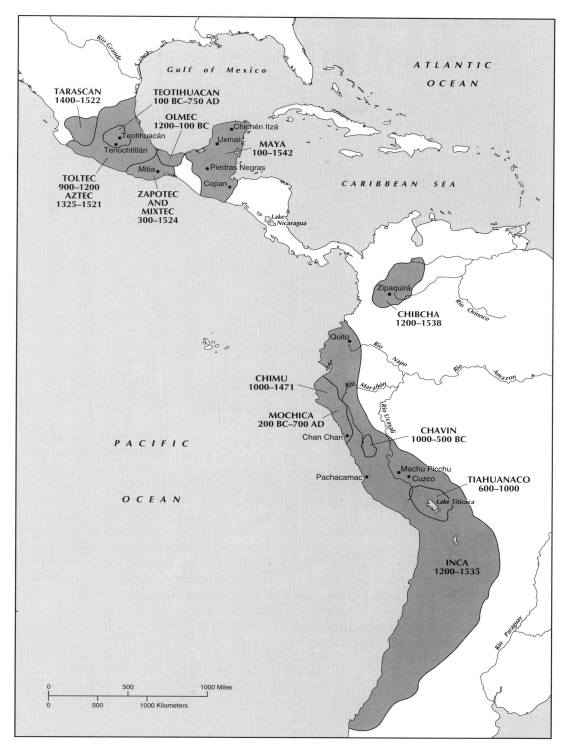

**Early States of the Americas.** While the Americas south of the Tropic of Cancer were dominated for some 500 years until the Spanish Conquest by the Toltec, Aztec (or Mexica), Maya, and Inca empires, many other people had established States in the region. The Chibchas alone had five States, two of them well organized and quite prosperous for the period; they won the admiration of the first Spanish arrivals who saw them in full flower. Other cultures had stone cities, elaborate legal and religious systems, and extensive international commerce.

## *The State in Europe*

The sequence of events that led to the emergence of the European State system began in pre-Christian times. A major factor in the transmission of early stimuli from the Middle East to Southeast Europe lay in the growing need for raw materials in the advanced societies of the Nile Valley and the lowland of the Tigris and Euphrates rivers. These societies wanted metals, textiles, and oils, and their search for these and other commodities brought contact with southeast Europe. Greek traders were active in Egyptian towns, and Phoenician ships linked many Mediterranean ports. The traffic was by no means confined to copper and tin; the political ideas of Middle Eastern States filtered into Europe as the trade grew.

### *Ancient Greece*

During the first millennium B.C. the Greeks created what in many respects is the foundation of European civilization. The Hellenic people occupied much of southeast Europe, but they had to adjust to the landscapes of the region, which are characteristically rugged, peninsular, and insular. Hellenic settlements were founded in river valleys and on estuaries, on the narrow littorals of peninsulas, and on islands. They were in contact with one another, but mostly by sea; water often provided a better avenue than the land. Thus the Hellenic civilization was extensive—but it was not contiguous.

This situation produced another version of the city-state. Individuality and independence were strong features of the ancient Greeks, and in their comparative isolation they experimented with government and politics. At times several of the city-states would join together in a *league* to promote common objectives, but the participants would never give up their autonomy altogether. And just as the Greek city-states competed with each other individually, so there were rival leagues; Athens, Sparta, and Thebes each headed powerful and competing leagues.

During Greek times political philosophy and public administration became sciences, pursued by the greatest of practitioners. Plato (428 to 347 B.C.) and Aristotle (384 to 322 B.C.) are two of the most famous philosophers of this period, but there were a host of other contributors to the greatness of ancient Greece.

### *The Roman Empire*

The Romans succeeded where the Greeks had failed. Theirs was not a nation dispersed through a string of city-states, but a contiguous and unified *empire* in the true sense of the word, strongly centralized in one unrivaled city, Rome. But Rome itself did not achieve supremacy overnight. Around 400 B.C. (about the time of Greece's greatest glory), Italy was a territory of many tribes and nations. Well-fortified hilltop towns dotted the area; these were the centers of power and protection for local groups. As in Greece, but on a smaller scale, towns entered leagues of mutual defense. As time went on places whose natural protections were less than adequate were abandoned as their inhabitants moved to larger and safer centers. The regional influence of Rome gradually increased. Its great achievements were based on successful absorption of immigrants, stability of political instutitions, and an effective campaign of Roman colonization in outlying districts. The great empire that was to come had its local predecessor: by the middle of the third century B.C. there was a true Roman Federation extending from near the Po Valley in the north to the Greek-occupied south.

Having achieved unification on its peninsula, Rome now was a major force in the entire Mediterranean area. The defeat of Carthage removed the major obstacle to westward expansion; to the east the Greeks were forced to recognize Roman superiority. But Greek culture had a major impact on the Roman civilization, and Greek identity survived despite the strength of the Roman invasion. It is more accurate to speak of a Roman-Hellenic civilization than of a Roman civilization alone, for the Romans provided the vehicle to disseminate Greek cultural achievements throughout Mediterranean and Western Europe.

Greek culture was a major component of Roman civilization, but the Romans made their own essential contributions. The Greeks never achieved national organization on the scale accomplished by the Romans. In such fields as land communications, military organization, law, and public administration, the Romans made unprecedented progress. Comparative stability and peace marked their vast realm, and for centuries these conditions promoted social and economic advances.

Only in Asia did Rome have contact with a State of any significance. Thus the empire could organize internally without interference. But its internal diversity was such that isolation of this kind did not lead to stagnation. The Roman Empire was the first truly interregional political unit in Europe. In many ways the Roman Empire was centuries ahead of its time. It was also the first true empire by modern criteria, with outlying colonies and diverse racial groups under its control.

Whatever the impetus for its development, the Roman Empire stood out from all its predecessors and contemporaries, especially in the area of organization. Almost everything else—its power, its size, its longevity, its internal variety—was a function of the Romans' dedication to organizing their land and possessions. One aspect of this organization was the effort to build a network of communications. The concept was totally new, and the scale of its execution was grand. Naturally, the network was not dense by modern standards, but it was carefully planned to meet the needs of the State.

Although highly centralized, the Roman Empire's political structure contained elements of the federal as well as the unitary State of today. Peace—and a certain amount of autonomy—was granted to areas west of the Rhine and east of the lower Danube. A true body politic evolved and permitted the disparate peoples of the empire to develop their areas and their abilities in peace.

## Feudal Europe

*Feudalism* Europe during the period from about the middle of the eighth century to the end of the twelfth was dominated by feudal institutions. Roman judicial processes and other institutions that protected individual rights had weakened. The new kings ruled by private law, personal power, influence, and wealth and through the allegiance of counts, barons, dukes, and other representatives. Although some kings were successful in forging and maintaining States of considerable size, the framework of government was not sufficiently strong, and there was not enough contact and communication with the country to withstand disorder and disorganization. Although he managed to hold his dominions together for a long time, even Charlemagne was compelled to grant increasing powers to local lords. In this way feudalism served the kings. To secure their allegiance, feudal rights were granted to counts and dukes in outlying parts of the State.*

When a large political unit disintegrated through the death of a king, a weak succession, or an invasion, the feudal counts and barons were quick to assert complete hegemony over their local domains, including all land and people within them. Thus, during the feudal period much of Europe was organized into dozens of jigsaw-like feudal units, each with its nobility and vassals and its more or less tenuous connections with higher authority. The connectivity that Rome had brought to Europe was now minimal and reforms were needed to rekindle economic as well as political progress. In England the Norman invasion (1066) destroyed the Anglo-Saxon nobility and replaced it with one drawn from the new immigrants. William the Conqueror strengthened the feudal organization of England to such a degree that he became the most powerful ruler in Western Europe; at the same time he introduced changes in the system of government

*These old units can still be seen in the locations and names of the political subdivisions of certain countries of the Old World, such as the *county*, the *duchy*, and the *march*, or *mark*.

organization that were to outlast the feudal period.

Despite oppression and excesses perpetrated on the population under the divine rights of kings, the roots of modern European nationalism lie in this period. There was a slow but noticeable improvement in circulation, especially in Western Europe, and a feeling of belonging together within a State framework developed among the peoples. Eventually they began to perceive certain physiographic features of Europe—a river or mountain range—as "their" country's borders. Because the kings had amalgamated diverse groups of people into their empires, the peoples within those borders often were quite varied in terms of language and religion. But if the State or king provided a strong common interest, a nation was in the making.

## The Development of Nationalism

The emergence of modern Europe may be said to date from the second half of the fifteenth century. Western Europe's monarchies, as we have seen, began to represent something more than mere authority; increasingly, they became centers of an emerging national consciousness and pride. At the same time the trend toward fragmentation, which had dominated the feudal period, was reversed as various monarchies, through marriage, alliance, or both, combined to promote territorial unity. Feudal privileges were being recaptured by the central authority, and there was progress in the parliamentary representation of the general population. Renewed interest was shown in Greek and Roman philosophies of government and administration. Europe was ready for change.

Change did come to Europe, and in many forms. The emerging States engaged in intense commercial competition, and *mercan-*

*tilism* was viewed as the correct economic policy to serve the general interest. The search for precious metals (the standard of wealth) sent the ships of several countries traversing the oceans to lands that lay open to discovery—and appropriation. Kings and nobles sent many of those ships on their way, but in the growing cities, an influential new class was on the rise: the merchants. Before long they were demanding a greater voice in the politics of their countries.

A landmark in the evolution of the State was the Peace of Westphalia, which in 1648 brought to an end the Thirty Years' War between Protestants and Catholics. It was a complex settlement among a number of civil and ecclesiastical authorities, but by reducing the power of the Holy Roman Empire and strengthening the emerging States, it made the territorial State, rather than the individual sovereign, the cornerstone of our modern political system. This began a radical reduction in the number of States in Europe. Before Westphalia there were around 900 German States, for example. The settlement reduced them to 355. Napoleon I eliminated more than 200 of these, and by the time the Germanic Confederation was formed in 1815, only 36 were left to join it.

The unification process continued, with the survivors growing stronger, until one German Empire finally emerged in 1871 with the unification of the remaining 24 German States. Similar processes were going on elsewhere in Europe, with Italy (1870) and Germany being the last to consolidate into national States based on a common culture, primarily language. In the course of consolidation, people transferred their loyalties from the sovereign or the Pope to the State, a State they often helped to create by participating in a revolution that swept away the old systems of tribalism, feudalism, and absolutism. This revolution was the development of nationalism.

# 6

# *Modern Theories About States*

An academic discipline is often judged not so much by its empirical studies, its discoveries, and its inventory of "facts," but by its theories. In Part One we review some of the theories in the body of political geographic thought. Now we turn our attention to some specific theories about the central object of study by political geographers: the State. We do not attempt to discuss or even mention all the theories that have been propounded about the State by political scientists, historians, anthropologists, political geographers, and others, but select only a few for discussion.

No responsible scholar would suggest that any theory describes exactly the real world of territorial political behavior. Neither do any combination of them nor all of them together provide answers to our innumerable questions about the State and State behavior. They are, after all, more like impressionistic paintings than like photographs. This is not to imply that their only value is to provide the student with material for momentary diversion, coffee-shop debates, or mental gymnastics. These theories have been formulated carefully and scientifically by responsible, experienced, and idealistic individuals who are groping, like the rest of us, toward definitive explanations and solutions we know we may never reach.

Some theories have been rendered obsolete by new information, changing circumstances, or new patterns of thought. Others may not have survived the frequent questioning, analysis, and testing that any good theory

provokes. And many others are still undergoing this process and their validity is uncertain. Nevertheless, they all provide compact descriptions, clues to explanations, and tools for more and better work. They should inspire us to challenge them, extract their valid and useful elements, apply them to real and important problems, and use them as the nuclei of new and better theories.

## *Environmental Determinism*

One of the most important books ever published was *On the Origin of Species* by Charles Darwin, which appeared in 1859. It set off a revolution in many branches of science and led to the popularization of science. Out of the ferment of thought, discussion, and speculation generated by Darwin came attempts to apply "science"—particularly concepts of natural selection—to many areas of human endeavor and study. Henry Adams and Frederick Jackson Turner in the field of history, for example, Sir Henry Maine in political science, Thorstein Veblen in economics, Oliver Wendell Holmes in law, Lewis Henry Morgan in anthropology, and William Graham Sumner in sociology were only a few of the people who applied Darwin's concepts to human society. This application of Darwinian ideas about the natural world to society is loosely called Social Darwinism. One outgrowth of Social Darwinism in geography is the organic State theory, whose earliest exponent was Friedrich Ratzel. Because of its complexity and importance, we reserve our discussion of this theory for Part Five, which is on geopolitics. Another derivative of Social Darwinism was the concept of *environmental determinism.*

We have all been exposed at one time or another to such notions as: "Because Japan is very mountainous and has little good agricultural land, its people were forced to turn to the sea for a living." Or, "Because Britain had many good natural harbors and fine oak forests, it was able to build a navy that could control the seas." Or, "The searing deserts and rain-soaked forests to the south and the cold, dreary lands to the north keep their inhabitants in a state of savagery, or at best barbarism; civilization can develop only in the equable climate of the Mediterranean Basin, as in Greece." This concept—that one or another element of the physical environment determines the type and level of civilization a society can attain—is called environmental determinism.

Many variations of this theory, from extreme determinism to more modest formulations, were expounded by eminent geographers from the 1880s to the 1960s. Among them were the German Friedrich Ratzel, the Americans Ellen Churchill Semple and Ellsworth Huntington, and the British-Australian-Canadian Griffith Taylor. By the late 1930s the theory was coming under increasing attack; by the late 1950s "determinist" had become almost a pejorative term. The inadequacy of the theory was revealed in the light of newer concepts backed by more experience and solid evidence. Preston James, for example, demonstrated the importance of culture in determining how a particular people will use a particular environment, contrasting the very different societies that have developed in the very similar physical environments of Iowa and Uruguay. He summed up the importance of culture by stressing the importance of the "attitudes, objectives, and technical abilities" of a people. Derwent Whittlesey contributed a temporal perspective, demonstrating how, as in California, different cultural groups that occupy a particu-

lar region in sequence over time frequently utilize the physical environment in very different ways. Newer studies of the influence of people on their environments and of environmental perception demonstrate even more emphatically that neither individuals, societies, nor States are controlled by any element of their physical environment.

Nevertheless, we cannot ignore the environment altogether. We have not yet learned to control nature, nor should we. The physical environment still offers opportunities to, and imposes limitations on, the people resident in every State. The French geographers Vidal de la Blache and Jean Brunhes, among others, developed concepts of probabilism and possibilism, which recognized these opportunities and limitations as alternatives to determinism. Neither people nor States can do as they please regardless of the environment. It seems highly unlikely, for example, that Bhutan will ever be a great naval power or Tonga a great industrial power. We examine the significance of the physical environment again in our discussion of the power inventory; for the present we can say that the physical environment is important in the development of States, though not necessarily determinative.*

## *The Functional Approach*

A very different view of the State was taken by Richard Hartshorne in his presidential address delivered before the Association of American Geographers in 1950. It was titled "The Functional Approach in Political Geography" and was an elaboration of a theme he had introduced a decade earlier in an ar-

---

*Another manifestation of Social Darwinism was the development in 1939 of a cycle theory of the development of States by Samuel van Valkenburg. It was based on the cyclic theory of landscape evolution propounded in 1899 by another American, William Morris Davis, himself strongly influenced by Darwin. Van Valkenburg theorized that States develop through identifiable stages of youth, adolescence, maturity, and old age, each with its own characteristics of State behavior. He also allowed for the possibility of rejuvenation of a State after it has begun to decline and for interruptions of the cycle at any time, bringing the State back to a former stage.

ticle on the *raison d'être* and maturity of States. In his address he proposed that the study of the functioning of the State was the central task of political geography, and that political geographers should view the State (and other politically organized areas) in terms of structure and functions. The State is a politically organized space that functions effectively. How does it succeed? By overcoming the *centrifugal forces*—forces that tend to break a State apart—with the cohesion provided by prevalent *centripetal forces*, which bind the State together. Centrifugal forces exist in every State; in some States they are so powerful that they disrupt the State system completely. In Nigeria the Biafran secession movement of the late 1960s was barely overcome; in the early 1970s East Pakistan successfully seceded from Pakistan and became Bangladesh. Most recently, the centrifugal forces in Yugoslavia, Czechoslovakia, and the Soviet Union, principally nationalism, overcame the binding forces and the States fell apart.

**Richard Hartshorne, 1899–1992.** (Courtesy of Geoffrey J. Martin)

The centripetal, binding forces must prevail if the State is to have a *raison d'être*, a reason for existing. Hartshorne suggests that "the greatest single weakness in our thinking in political geography" has been our preoccupation with the disruptive, divisive forces affecting politically organized areas, without considering the forces that manage to hold the State together. This set of centripetal forces is the *functioning* State, and, Hartshorne argued a half-century ago, this process should be studied.

Hartshorne's discourse touched on several additional topics, including the concepts of nation and core area and the internal and external relations of States—all within the context of the functioning of the political unit involved. It still warrants a careful reading today.

## Patterns of Integration and Disintegration

Karl Deutsch has contributed much to our understanding of the formation and growth of States; his analyses combine geographic and behavioral approaches. In a classic 1953 article he summarized his findings of uniformities in the development of nations, or "recurrent patterns of integration."* They include

1. The shift to exchange economies from subsistence agriculture.
2. Appearance of core areas.
3. Growth of towns.
4. Development of basic communication grids.
5. Concentration of capital.
6. Growth of individual self-awareness and of group interests.
7. Awakening of ethnic awareness.
8. Merging of this ethnic awareness with political compulsion and sometimes the social stratification of society.

*Karl W. Deutsch, "The Growth of Nations: Some Recurrent Patterns of Political and Social Integration," *World Politics*, 5, 2 (January 1953), 168–195.

As a nation develops and forms a political system in this manner, it begins to grow and integrate its territory in a number of ways. It perpetuates itself through the generations, it increases the social and political divisions of labor, it accumulates more national symbols, and it acquires the ability to utilize its resources for its own development. Thus, the nation-state "represents a more effective organization than the supranational but largely layer-cake society or the feudal or tribal localisms that preceded it." By leading to great differences in living standards among States, however, this process tends to turn nationalism to imperialism. But since the process that led one State to become dominant over others is also functioning in the subordinate States, they tend to grow strong enough to weaken or destroy the dominant State.

### The Unified Field Theory

Stephen B. Jones in 1954 integrated some of the ideas of Whittlesey, Hartshorne, and Jean Gottmann, added some of his own, and organized the whole into a coherent explanation of how States develop. He published his concept in an important paper entitled "A Unified Field Theory of Political Geography." He viewed movement as a process involving the flow of ideas as well as goods and other tangibles in which movement becomes linked with several other processes. The unified field theory holds that there are, in the process of spatial-political organization, at least five clearly recognizable "hubs" of activity, each related to and connected to the other. The first of these is the generation of an *idea*.

A political idea—say, an ideology—may well have spatial ramifications. Colonialism was such an idea, as geopolitics was in interwar Germany and Zionism is today. Ideas may lead to *decisions* to occupy territory, to announce claims, to make promises. All this can generate *movement*, of people, goods, ideas, money, and more. While the process proceeds, the "links" in the chain interact with each other. Movement can influence the decision and alter it; the original idea

can be changed, too. The whole development takes place in a *field* that, like the movement phase of the model, has tangible (spatial) as well as intangible characteristics. And at the end of the "chain" lies the politically organized *area*.

The unified field theory came to be known as the idea-area chain, suggesting a one-way sequence of development from political idea to politically organized area. But Jones himself cautioned against misinterpretation by emphasizing the underlying principle of two-way interaction; the links in the chain have reverse as well as forward effect.

### The Territorial State

We stress in Chapter 4 that a State is different from a nation in that it has territory; without territory a State cannot exist. John H. Herz in 1957 made this fact the central theme in his article "Rise and Demise of the Territorial State." In fact, it is his basic thesis that the territorial State assumed its modern importance only after military technology, particularly gunpowder and large mercenary armies, had rendered obsolete the castles and walled cities of feudal rulers. Security— the primary consideration in the formation of political regions—dictated the expansion of the defensive perimeter to the boundaries of the emerging territorial State. The rise of nationalism personalized the new political units as self-determining national groups that were thus entitled to the same protection as a sovereign. Even the system of collective security (or defensive alliances among sovereign States) has not diminished the significance of the territorial State, but rather is an attempt to maintain and reinforce the "impermeability" of each individual State.

Just as artillery made stone walls obsolete, Herz argues, recent developments in airborne propaganda and thermonuclear weapons have enabled belligerents "to overleap or by-pass the traditional hard-shell defense of States." Even industrialization has weakened the territorial State by making it dependent on foreign sources of commodities and foreign markets for its manufactured goods.

Therefore, the fundamental need of humans for protection can no longer be served by the territorial State, and it is bound to vanish. Rationally, its replacement would have to be some form of "universalism" that would afford protection to all humanity conceiving of itself as a unit.

Events of the next decade stimulated Herz to rethink both his thesis and his conclusion. In his 1968 article "The Territorial State Revisited—Reflections on the Future of the Nation-State," he explained how some of his ideas had changed. He maintained that his analysis both of "classical" territoriality and of the factors threatening its survival was still valid, but he was no longer certain that "universalism" would replace it. Indeed, area or territory seem to be assuming a new importance to States; there is a trend toward a "new territoriality." He pointed to the survival of Israel and Vietnam, both of which in 1967 faced extinction by States possessing nuclear weapons. Decolonization, then in its most frenetic period, meant that all remaining colonies were bound to be added to a global "mosaic of nation-states." The threat of nuclear (or thermonuclear) annihilation has stabilized the world through nuclear stalemate rather than forcing States to seek new means of protection, and nationalism "of the self-determining and self-limiting variety (in contrast to expansionist imperialism)" will strengthen the foundations of the individual States. He elaborated on the themes of legitimacy, foreign intervention, and nonalignment and detected diminished antagonisms among States.

In his conclusion, Herz rejected the notion of instinctive human territoriality, but he reiterated the centrality of the need for protection—for security against deprivation of scarce resources as well as physical attack. This protection can be provided at least in part by States, but only if the world becomes "modernized" through science and technology and thus freed from scarcity. The "new-old nation-state" may be "the polity of the last decades of this century," but only if four conditions are met. First is "the spread of political, economic, and attitudinal modernity" throughout the world. Territorial dis-

putes must be settled in such a manner that nationalities, or peoples, are satisfied. Systems based on world-revolutionary doctrines must be deradicalized, and States must not interfere in the internal affairs of others. Finally, international violence must be reserved only for self-defense in cases of direct attack or invasion and for resistance of the population of an occupied area against the aggressor. Under these conditions, the "neo-territorial world of nations" can not only provide the essentials of group identity, protection, and welfare, but also preserve the diversity of life and culture, traditions, and civilizations that are threatened by our accelerating rush to "the technological conformity of a synthetic planetary environment."

## Political Systems Analysis

Early in 1971 an article appeared whose authors, Saul Cohen and L. D. Rosenthal, sought to build further on the methodological heritage of Whittlesey, Hartshorne, Gottmann, and Jones.* "What is needed," say the authors, "is a methodology to link effectively political process and its spatial attributes—a methodology that, having identified a specific political process, can pinpoint spatially significant phenomena that relate to this process, can observe the impact of these phenomena within some control context, and can connect one process or parts thereof to another." Thus, they argue here that political geographers should focus on political processes in their research and on the spatial expression of these processes.

Their point of departure for analysis of political process is the *political system*, "a set of related political objects (parts) and their attributes (properties) in an environment. . . . political system refers to a functional entity composed of interacting, interdependent parts." It "can be viewed as the end product of the processes by which man organizes

*Saul B. Cohen and L. D. Rosenthal, "A Geographical Model of Political Systems Analysis," *Geographical Review*, *61*, 1 (January 1971), 5–31.

himself politically in his particular social and physical environment and in response to outside political systems with their unique environments." Other elements in their analysis are the *locational perspective* and the *open/closed political system*. We dicussed aspects of locational perspective in Chapter 3; the degree of openness of a political system is almost as important in understanding decision making within it. Cohen and Rosenthal also consider the societal forces and legal systems to be especially useful in analyzing the broader relationship between political process and spatial attributes.

They construct a model that attempts "to show man in his political role in society and the relation of that role (as expressed in political ideology, political structure, and political transaction) to the land (places, areas, and the general landscape), and to indicate the consequences of that relation in the formation of the political system."

The authors conclude their paper with a case study of Venezuela, where two dominant forces—nationalism and social democracy—provide vantage points for the study of the impact of process on geographical space. Aspects of Venezuela's spatial organization are examined in the context of laws relating to petroleum exploitation and immigration. The conclusion is that the Venezuelan case confirms the law-landscape thread as the key to the analysis of the political system.

The six theories presented here were developed during more than a century and differ widely in their approaches. As we proceed, it would be well to bear in mind the essentials of these theories and recall them as we discuss many political aspects of our world. Perhaps some of them can help us understand this amazingly complex world. One of its most complicated elements is territory, to which we now turn our attention.

# 7

# *The Territory of the State*

All modern theories about States agree on at least one thing: a State must have territory. Neither in law, custom, nor current practice, however, are there any guidelines about the territorial characteristics necessary either for formal recognition of a State or for its survival. In this chapter we examine some essential characteristics of a State's territory—acquisition, size, and shape. In subsequent chapters we consider other characteristics.

## *Acquisition of Territory*

A State must have territory, but only a State can acquire territory under international law (with a few minor exceptions). This would seem to be a proverbial vicious circle. In fact, however, international law was and is being created by States, so there can be no formal rules for the acquisition of territory until States make the rules. Furthermore, States came into existence gradually, as we have seen, through the actions of peoples or sovereigns. Therefore, we begin with a survey of how existing States add to their national territory. In Part Four we discuss how new States come into existence out of the territory of old ones.

### *Occupation*

Originally, States became identifiable as such largely by common consent or tacit agreement. European discovery of new lands in the Americas and Africa, however, presented the problem of the legal grounds on which a European State could take possession of land elsewhere. At first, discovery alone was given some status as a basis for a claim, but it was frequently challenged and seldom sustained. By the eighteenth century discovery alone was no longer adequate; it had to be followed by effective occupation. There have been a great many disputes over definitions of "effective occupation" and its importance vis-à-vis other claims to territory, and some of these disputes survive today. But since there is probably no undiscovered land remaining in the world and little unclaimed land, this basis for claims is of historical and legal interest only.

### *Prescription*

If an area claimed by a State is occupied by another for many years without serious objection by the original claimant, the title, whether or not clear and recognized, may be considered abandoned and may pass to the occupying State. The rules for this are similar to those for occupation. The rise of nationalism, however, has virtually eliminated prescription as a means of transferring territory, except for a few small and remote islands. Some contemporary territorial disputes, however, still involve surviving claims to prescriptive rights.

## Conquest and Annexation

Historically, territory has changed hands through conquest at least as often as by any other means. Conquest alone, however, does not confer title. Conquerors must take steps to annex the new territory and incorporate it into their own, extending their laws over it, giving it representation in the national legislatures, appointing administrators, or otherwise making the annexation effective. This annexation is generally recognized in a peace treaty or other legal instrument of cession. This method of acquiring territory is no longer considered acceptable in polite society, however, and is rarely used. Territory is still being conquered, or course, but seldom annexed. Even when it is, the annexation is not normally recognized by the international community. A good example is the portion of Palestine conquered by Transjordan in 1948. Its annexation in 1950 was formally recognized only by Pakistan and the United Kingdom. Even Jordan's Arab allies never accepted the annexation, and in 1988 Jordan renounced its claim to the territory. More recently, when Iraq conquered and annexed Kuwait in August 1990, the international community mobilized immediately and Iraq was driven out early in 1991.

There are cases of assimilation of a territory under coercion short of conquest. Korea was annexed by Japan in 1910, Austria by Germany in 1938, and the Baltic States by the Soviet Union in 1940, all without war and without effective opposition from elsewhere. More recently, Western Sahara (formerly Spanish Sahara or Río Muñi) was divided between Morocco and Mauritania with the passive agreement of Spain, but the annexations were not recognized by the international community, and Mauritania withdrew, leaving Morocco in charge of the entire territory.

## Voluntary Cession

Formerly, it was quite common for territory (with its inhabitants) to pass from one country to another simply by agreement, with or without cash or other compensation in-volved. Much of the United States was acquired in this way, for example. Today voluntary cession is rare. Remote islands are still being transferred, as when Christmas Island in the Indian Ocean was transferred from Britain to Australia in 1950. Minor boundary adjustments, generally involving exchanges of territory, have been quite common in Europe since World War II. Some disputed or loosely held areas are being divided by agreement involving transfer of territory, as in the northern Arabian settlements of 1965–84. Generally, however, States just do not like to give up territory, no matter how remote, sparsely settled, or useless it may seem to be.

## Accretion

Accretion is the addition of land to a State by natural processes. This most commonly results from a gradual shift in the bed of a river that has been adopted as an international boundary. If the river changes course suddenly, as the result of a flood or earthquake, for example, the process is called *avulsion*, and the boundary customarily remains in place. Land is also accreted in deltas, along emerging coastlines, as islands built up by rivers and ocean currents, and so on. The amount of territory involved is generally not very great, but sometimes bitter disputes break out over the ownership of accreted lands.

## Acquisition of Rights

Often the use of territory is granted by one State to another without title or sovereignty actually changing hands. Such transfers of rights take the form of leases and servitudes. Russia, Germany, the United Kingdom, and France all leased territories from China in the late nineteenth century, for example, and the Russian and German leases eventually passed to Japan without even the assent of China. The United Kingdom still leases from China the New Territories, adjacent to its crown colony of Hong Kong. The United States in 1903 acquired from Panama rights to the "use, occupation, and control" of a

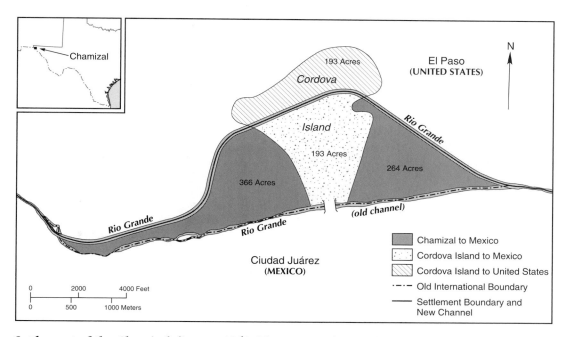

**Settlement of the Chamizal dispute, 1963.** The Rio Grande is a classic example of the inadequacy of rivers as political boundaries. This stream meanders over much of its length, constantly shifting channels and rearranging the land by accretion and avulsion. Meander cutoffs in the Rio Grande Valley are called *bancos.* There were several hundred of them, most disputed between Mexico and the United States, until they were eliminated by agreement between 1905 and 1970. One, however, known as El Chamizal, resisted settlement because of its location in a densely populated area. The eventual agreement provided for partition of the disputed tract and construction of a new, permanent concrete channel for the river, through which the international boundary now runs.

zone for the construction, operation, and defense of an interoceanic canal, which it is now gradually giving up. Similarly, the Soviet Union in 1962 granted Finland a 50-year lease over a 4-km strip of land along the Saimaa ship canal, permitting Finland to renovate, improve, and operate the canal built in 1856 to connect the Saimaa Lakes with the Gulf of Finland at Viipuri (now Vyborg).* And many military bases around the world have since 1940 been leased from host countries, primarily by the United States but also by the Soviet Union (Porkkala, Finland, and Saseno, Albania, for example) and other countries.

A *servitude* in international law is roughly comparable to an easement in domestic law.

*In 1971 the United States gave up its 1914 lease of the Corn Islands in the Caribbean, although Nicaragua had been administering them with the acquiescence of the United States.

Generally, it is a restriction on the sovereignty of a State over its own territory in the form of an obligation to permit a certain use to be made of it by another State or States. There are several forms of servitude, ranging from rights to free navigation on some international rivers and in the territorial waters of States to special servitudes concerning land territory. These were formerly quite common, but since the consolidation of the nation-state system after the Congress of Vienna in 1815, they have become rare. Rights of way are still in use. Panama had a right of way across the former Canal Zone, for example, and Peru has the right to enter Chilean territory at will to maintain or enlarge the Mauri and Uchusuma irrigation canals, parts of which lie in territory acquired from Peru by Chile in the War of the Pacific. Finally, Norway is required by a 1920 treaty to permit other countries to en-

**Helgoland.** This small island strategically located in the North Sea (1 mile [1.6 km] long, one-third mile [.5 km] wide), 64 kilometers north of the mouth of the Elbe River, was first settled in ancient times by the Frisians. The island was occupied by the Netherlands from 1402 to 1714, then by Denmark and the United Kingdom before Germany acquired Helgoland in 1890. It was heavily fortified by the German empire, which made it a major naval base. After World War I the fortifications were demolished, and the island was formally demilitarized. Hitler fortified it again and after World War II his submarine base, air base, and fortifications were destroyed by the British. The island is again formally demilitarized, and its approximately 2400 citizens live primarily by fishing and tourism. (Photo courtesy of the German Information Center)

gage in commercial activities in Svalbard (Spitsbergen) under their own laws.* There are also negative servitudes, the most important of which are demilitarized zones. Even since World War II, many of these have been established. Among them have been a number of border areas in southern Europe, some Mediterranean islands, Trieste, places along the 1949 armistice lines around Israel, and strips along both sides of the military demarcation lines separating North and South Vietnam and North and South Korea.

Here is a fertile field of investigation for political geographers. There has not yet been a good geographic study of leased territories or demilitarized zones and other

*Only the former Soviet Union has so far taken very much advantage of this, however, primarily to mine coal there and to maintain weather, navigation, and intelligence-gathering facilities.

servitudes. Perhaps more information about them would indicate their practicality in resolving many contemporary boundary and territorial disputes.

## Size

We all know that States vary greatly in size. They range from half a dozen giant States to half a hundred that are very small indeed. But does it really matter how big a country is? The simple answer is that the advantages and disadvantages attributable to size alone seem to be distributed quite randomly over large and small alike. A large country may not necessarily be endowed with resources commensurate with its size, and many of those it has may remain untapped because of the difficulty and expense of utilizing them. It may be easier and cheaper for a small State to import its primary require-ments than for a large State to develop its own. The location, physiography, and shape of a State often enhance or diminish the value of large or small size. Defensive depth may be nullified by difficulties of adminis-tration and circulation. Population may be large or small, evenly or unevenly distrib-uted, ethnically homogeneous or variegated regardless of the measurements of the terri-tory. This is not to imply that size is irrele-vant; it is sufficiently important to warrant further discussion.

A very large State that is sparsely popu-lated may experience internal division, espe-cially if the areas intervening between the populated regions are both difficult to cross and unproductive. Australia's central desert, Siberia, and the Canadian Shield all exem-plify the barrier effect of vastness, although in each greater political unity exists than in many smaller States that do not have size problems. Nevertheless, most very large States attempt to diminish the "empty" as-pect of their sparsely populated regions by encouraging settlement in those areas or by practicing population policies aimed at rapid growth. In Sudan the problem of size is di-rectly related to the centrifugal forces that have confronted that country almost con-stantly since independence. Sudan is so large that it extends from Arab Africa into black Africa. Thus it contains an Arab popu-lation in the north, focusing on the capital of Khartoum, while the southern provinces are occupied almost exclusively by peoples who are more closely related to black Africa—racially, culturally, and historically. As it hap-pens, the two "heartlands" of the State are separated by desert and swamp, so that communication has always presented prob-lems. The State has had to expend much of its energy in controlling divisive forces, with distance complicating virtually every aspect of the matter.

The size of a State is related in many ways to its effective national territory, or *ec-umene*. Many of the States that evolved in various parts of the world ultimately broke up because their frontiers extended too far outward to be sufficiently integrated with the central area of the State. Continued growth meant growing strength—up to a certain point, after which it meant increas-ing vulnerability. This was one of the rea-sons for the collapse of the Aztec Empire, ancient Ghana, and the Roman Empire. It also has been a major factor in the breakup of more recent colonial empires, and in such States as India and Sudan, it is a criti-cal matter today.

On the other hand, it is obvious that size can present advantages. If some attention is also paid to location (relative location, with reference to environmental regions, mineral-ized belts, trade routes), a generalization re-garding size might even be possible. After all, the United States fits comfortably within the Sahara, and its size there would be meaningless. But the United States lies in middle latitudes, in a world zone of many transitions (in terms of soils and climate, to name two), and fronting two oceans. De-pending on location, then, size and environ-mental diversification are indeed related.

We may consider the matter in this fash-ion: the total land area of the world is lim-ited, and it contains the bulk of the re-sources on which progress is based. A State that has a larger area than another obviously has a chance to find a greater percentage of such resources within its borders. But these known resources themselves are not evenly distributed; they are scattered in patches across the globe. The soils of the lowland tropics are often incapable of sustaining sedentary agriculture; those of the highest latitudes are too shallow, acidic, or frozen. The climates of the polar regions are not conducive to agriculture; those of the low-latitude regions are often excessively dry or moist. A series of maps of mineral resources indicate the remarkable concentrations of significant deposits in rather well-defined belts. Taking these world distributions into consideration when evaluating the effects of size on the wealth and self-sufficiency of the State, we see why some of the largest States are comparatively poor while others are in-comparably rich.

Generally, States exceeding 2.5 million square kilometers (1 million square miles) are described as very large, while those under 25,000 square kilometers (10,000 square miles) are referred to as very small. Small States range from 25,000 to 150,000 square kilometers (60,000 square miles), and medium-size States from 150,000 to 350,000 square kilometers (135,000 square miles). Those over 350,000 but under 2.5 million square kilometers are referred to as large. Thus the following would apply:

| | | |
|---|---|---|
| *Very small* | Burundi | Lebanon |
| *Small* | Netherlands | Liberia |
| *Medium* | United Kingdom | Poland |
| *Large* | France | Mexico |
| *Very large* | Russia | Canada |

One of the remarkable aspects of the group of very large States is the clustering of several of these States around the 8 million square kilometer mark:

| | km$^2$ | mi$^2$ |
|---|---|---|
| Canada | 9,974,382 | 3,851,113 |
| China | 9,758,475 | 3,767,751 |
| USA | 9,363,394 | 3,615,210 |
| Brazil | 8,511,631 | 3,286,344 |
| Australia | 7,704,165 | 2,974,581 |

At the lower end of the size range there is another cluster of very small States:

| | km$^2$ | mi$^2$ |
|---|---|---|
| Liechtenstein | 160 | 62 |
| San Marino | 62 | 24 |
| Tuvalu | 26 | 10 |
| Nauru | 21 | 8 |
| Monaco | 243 hectares | (600 acres) |

With such an enormous range in the territorial size of States, it is useful to have some terms to identify States within certain size categories. For example, the very smallest States, such as Liechtenstein and Monaco, are referred to as *microstates*. States somewhat larger than these smallest units, such as Brunei (5765 square kilometers, 2226 square miles), The Gambia (11,292 square kilometers, 4360 square miles), and Cyprus (9251 square kilometers, 3572 square miles) often are called *ministates*. But neither term has any precise definition.

Since the mid-1960s small States and territories have not been mere curiosities or sub-

jects for trivia quizzes (in which geographers and stamp collectors have a natural advantage), but have become a major international concern. As more and more small colonies became independent and applied for membership in the United Nations, it became evident that before long they would constitute a significant proportion of the international community. Even the superpowers, despite their public protestations of support for the principles of self-determination and independence, quietly expressed doubts about the desirability of having a large number of UN members representing collectively only a tiny fraction of the world's land area, population, wealth, and military strength casting votes that they probably would not be able to control. Many other questions were raised about the viability of such small States or their potential for sparking conflicts far out of proportion to their size. Numerous studies were produced by the United Nations, governments, and scholars who examined all the questions in great detail.

The result of all the studies, conclusions, and recommendations was that there has been no change at all. The trend toward miniaturization in the international community continues unabated. Nearly all of the new small countries have joined or will join the United Nations. But there should be no more concern about small States than there was not long ago about large ones. Size is no indicator of wisdom, talent, or virtue among States any more than among individual human beings. Nor are small States, old or new, necessarily beggars at the tables of the large ones; some, such as Liechtenstein, Singapore, and Brunei, are, in fact, quite prosperous.

There is also nothing immutable about the size of States. States throughout history, as we have seen, expand and contract, appear and disappear. Although Ratzel and his followers envisioned something inexorable about State growth and modern territorialists foster belief in a biological urge of each society, no matter how small, to have its own sovereignty, the fact remains that the State is only one of a number of institutions created by people to serve their needs.

There seems to be, in fact, a tendency in our time toward equilibrium in the size of States. Even since the consolidation of the modern State system, we have seen countries shrink as well as grow. Thailand, Pakistan, Bolivia, and Germany, for example, are smaller than they used to be. The fragmentation of the Soviet Union, Czechoslovakia, Ethiopia, and Yugoslavia has set even more midget actors on the world stage. At the same time movements emerge for closer association among small States, new and old, leading toward varying degrees of economic and eventually political integration into larger units. The two Yemens, the two Germanies, and the two Vietnams have reunited, and the two Koreas have begun the same process. The evidence belies all theories about the size of States: large or small size is neither good nor bad; what matters are how well the area meets the needs of the people of a State and how they react to that size.

### Shape

Size is only one of the morphological characteristics of a State that influence its functioning and its international behavior. Another is shape. Again, a quick glance at a map reveals countries of widely varying shapes, some comfortably geometric (roughly) and some disturbingly erratic. The reactions of the observer to the shapes on the map, however, are not nearly as significant as the effects of those shapes on the people who live in them and on their neighbors. Since no two States have the same shape, it may be helpful to classify them into a few categories and discuss each in turn.

An *elongated* or *attenuated* State may, on the basis of the Chilean example, be defined as one that is at least six times as long as its average width. Thus, Norway, Sweden, Togo, The Gambia, Italy, Panama, and Malawi are among the States in this category. Depending to some extent on the

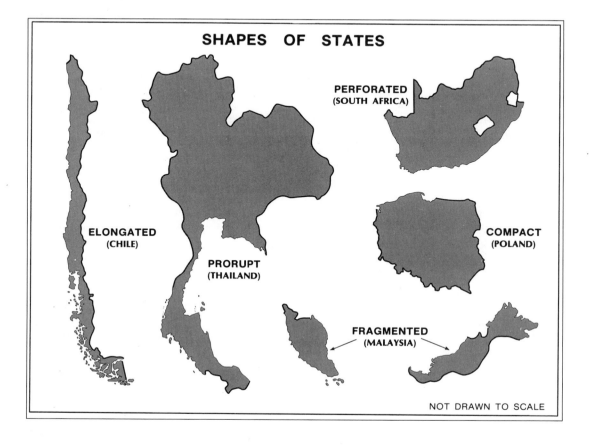

State's location with reference to the world's cultural areas, elongation may involve internal division. The north-south division of Italy is one that permeates life in that country and is related to the different exposures of the two regions to European mainstreams of change. The Norwegian administration of the Laplanders (Saami) is a regional matter, involving the northern parts of that State. Furthermore, the physiographic contrasts within the elongated State may accentuate other divisions. Chile, for example, possesses at least three distinct environmental regions. The central region is Mediterranean in nature, the south is under Marine West Coast conditions, and the north is desert. The capital is in the central (Mediterranean) part, and the effects of distance and remoteness are very evident on the Peruvian-Bolivian borders in the north and on Tierra del Fuego in the south. On the other hand, the internal diversification of the State resulting from its straddling of several environmental or cultural zones may be advantageous. Again, as in the case of size, much depends on location. Physiographically, location also plays its role. Chile lies astride the Tropic of Capricorn; Norway is bisected by the Arctic Circle. The resulting differences are obvious.

Many States appear to lie spread about their central area and are nearly round or rectangular in shape. Again, some generalizations are possible. Theoretically, since all points of the boundary of such a *compact* State lie at about the same distance from the geometrical center of the State, there are many advantages. First, the boundary is the shortest possible in view of the area enclosed. Second, since there are no peninsulas, islands, or other protruding parts, the establishment of effective communications to all parts of the country should be easier here than under any other shape conditions (unless there are severe physiographic barriers). Third, and consequent to the second point, effective control is theoretically more easily maintained here than in other countries. Of course, additional factors should be considered. In many compact States, the capital city is located along the periphery rather than at or near the geometric center of the State. Often the area of greatest productive capacity lies in one of the quadrants of the State rather than at the center. Finally, a compact State, no less than an elongated one, may be located in a zone of transition—cultural, physiographic, or both. Thus, internal divisions may still occur. Uruguay, Belgium, Poland, Sudan, and Cambodia are a few compact States.

Certain States are nearly compact, but they possess an extension of territory in the form of a peninsula, or "corridor," leading away from the main body of territory. Such *prorupt* States and territories often face serious internal difficulties, for the proruption frequently is either the most important part of the political entity or is a distant problem of administration. Perhaps the best example is that of Zaïre, which consists of a huge, compact area with two proruptions, both of which are vital to the country and are in many ways its most important areas. The capital city itself lies on the western proruption, which also forms a corridor to the ocean via the Zaïre port of Matadi. The most important area of revenue production, on the other hand, is Shaba Province (formerly Katanga), itself a proruption in the far southeast. Separating the two areas lies the vast Congo Basin, in many ways more a liability than an asset to the State. Another African example is Namibia, whose proruption extends to the Zambezi River. Long a useless area, this corridor (the Caprivi Strip) has recently attained great strategic importance. In Asia, Myanmar (Burma) and Thailand have large territories, fairly compact in shape, but they share a section of the Malayan Peninsula along a boundary that runs almost through the middle of this narrow strip of land.

Other States consist of two or more individual parts, separated by land or international water, and are therefore *fragmented*. Such fragmentation brings with it obvious consequences. Contact between the various population sectors is more difficult than in a contiguous State, and the sense of unity so necessary in the forging of a nation may be slow to develop. Because any State must have a capital, it will be located in one of the fragments, the choice of which may itself be a source of friction. Governmental control can be rendered ineffective by distance, as was proved repeatedly by the case of Indonesia. In contrast, the fragmented State may lie entirely on land. Pakistan exemplified the continental type. Its two "wings," the West and the East, were united by a common religious faith but divided by numerous cultural contrasts. As happens frequently with fragmented States, one part of the State (in this case the East), felt itself the victim of political discrimination. Charges of "domestic colonialism" abounded, and in the end the State broke up as East Pakistan—now Bangladesh—fought for independence, aided by neighboring India. A third type of fragmented State is that which lies partly on one or more islands. Malaysia consists of a mainland section on the Malayan Peninsula, a host of smaller intervening islands, and a sizable portion of the large Indonesian island of Borneo (Kalimantan). Technically, Italy, with its island territories of Sicily and Sardinia, belongs in this category, as do many other States with island possessions.

Finally, there are a few States that completely enclose other States. Such States are *perforated*, and it is impossible to reach the perforating State without crossing the territory or air space of the perforated State. This means, obviously, that the perforated State is in a strong position with reference to the land-locked perforator. These terms usually refer to tiny enclaves, such as San Marino (which perforates Italy, hence Italy is the perforated State), which have little if any political significance. But in southern Africa, the case of the Republic of South Africa is in a different class. South Africa is perforated by Lesotho, a State of 30,344 square kilometers (11,716 square miles) with a population of about 1 million. About the size of Belgium or Costa Rica, Lesotho is a substantial impediment to the territorial integrity of South Africa. And most of the successor States of the Soviet Union contain ethnic enclaves that might themselves become independent.

## Exclave and Enclave

One or two additional aspects of territorial morphology are relevant. Certain States might appear to be compact on a small-scale map. Close scrutiny of the boundaries, however, reveals that these States also have small pockets of land lying outside the main territory, as islands within the territory of neighboring States. These tiny areas are far too small to render the State fragmented. They may be smaller than dozen square kilometers in area, and their populations may number a few hundred. Nevertheless, these *exclaves* are of some importance in political geography, for they may depend for their survival on their connections with the "homeland." Hence, their boundaries may be under great stress, and from the study of exclaves may come a greater understanding of the nature and functions of boundaries elsewhere.

Exclaves are not always small, and neither are they always unimportant in terms of area or population. West Berlin, as an exclave of West Germany, was one of Europe's most important urban centers and was considered a vital part of the German Federal Republic.* Curiously, West Berlin had its own exclaves embedded within East Germany. All were very small, however, and only one, the hamlet of Steinstücken (population 190 in 1970), is permanently inhabited.

Exclaves can be classified according to their degree of separation from the "homeland." *Normal* exclaves are parts of certain States completely and effectively surrounded by the territory of other States. Such normal exclaves are usually small. Those in Europe are all under 26 square kilometers (10 square miles) in area and have populations under 2000. There are small Belgian ex-

---

*Sometimes the terms *exclave* and *enclave* are confused. The correct usage depends on the point of reference. When an outlying, surrounded area is considered in connection with the homeland of which it forms a part, it is described as an exclave. In this sense, West Berlin was an exclave of West Germany. On the other hand, a State that has such a small territory within its borders (though not necessarily surrounding it) would view the entry as an enclave. West Berlin, then, was an enclave in East Germany.

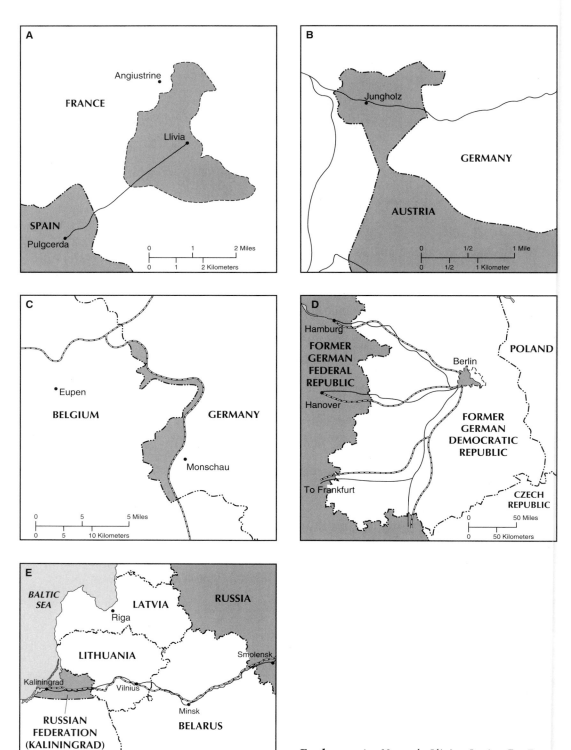

**Exclaves.** A—Normal: Llivia, Spain. B—Pene: Jungholz, Tyrol. C—Quasi: Raeren—Weywertz Railway Zones, Germany. D—Temporary: West Berlin. (All after G.W.S. Robinson) E—Coastal: Kaliningrad.

**Exclaves.** The top photo shows Point Roberts, an American territory at the tip of a peninsula jutting south of the 49° border with Canada, surrounded on three sides by the Strait of Georgia and Boundary Bay. It is a popular resort area for residents of Vancouver, some 35 km (21 miles) to the north, and much of the land is owned by Canadians. (Photo courtesy of the Point Roberts Chamber of Commerce) The bottom photo shows Campione, an Italian exclave on the east shore of Lake Lugano, Switzerland. Its gambling casino is a great attraction and its only "industry." It is separated from the main part of Italy by a high ridge, seen at the right. (Martin Glassner)

claves within the Netherlands, and there is Dutch territory within Belgium. *Pene*-exclaves are "parts of the territory of one country that can be approached conveniently—in particular by wheeled traffic—only through the territory of another country." In other words, these are proruptions, barely connected to the main territory of the State, with the connecting links so narrow or difficult that the only transport lines lie through neighboring territory. *Quasi*-exclaves are those that are technically separated from the motherland, but in reality they are so completely connected with it that they do not function as exclaves. *Virtual* exclaves are areas treated as the exclaves of a country of which they are not legally an integral part. Finally, *temporary* exclaves result from the fragmentation of a State through an armistice; occupation zones or demilitarized areas may create temporary exclaves.* One need only look at a map of the former Soviet Union to find examples of nearly every size and shape of state described here, and of various types of exclaves as well. Many of these were created for reasons no longer valid and may serve as excuses for conflict in the future.

Size and shape are two important characteristics of a State. We turn our attention now to the limits that enclose them.

*The material in this paragraph is taken from G.W.S. Robinson, "Exclaves," *Annals of the Association of American Geographers*, 49, 3, Part I (September 1959), 283–295.

# 8
# *Frontiers and Boundaries*

Many studies in political geography have dealt with frontiers and boundaries. Boundaries, on the map and on the ground, mark the limit of the State's jurisdiction and sovereignty. Along boundary lines States make physical contact with their neighbors. Boundaries have frequently been a source of friction between States, and the areas through which they lie are often profoundly affected by their presence. In many ways boundaries are the most obvious politico-geographical features that exist, for we are constantly reminded of them—when we travel, when we read a newspaper map or an atlas, and in many other ways.

Discussions of boundary problems sometimes treat the word "frontier" as though it were synonymous with "boundary." We read of the boundary between Germany and France in one paragraph, and of the French-German frontier in the next. Both references are to the same phenomenon, represented on the map by a line separating these two States. But the terms really represent different concepts.

## The Frontier

We have already described briefly the expansion of States from their heartlands or core areas. Such States as the Aztecs' Mexican Empire or the Empire of the Incas were able to expand into areas that were unable to resist their power. At times in history,

several States grew to local power and prominence simultaneously but never made effective contact. Separating them were natural impediments to communication: lakes, swamps, dense forests, mountain ranges. These States, with very few exceptions, possessed no boundaries in the modern sense of the word, but they were nevertheless separated from their neighbors. Whatever the separating agent—perhaps sheer distance—it functioned effectively to prevent contact.

The modern map showing all States bounded by thin lines that can be precisely represented on maps is, in the politicogeographical world, a very new phenomenon. Although some of the old States, such as the Roman Empire, attempted to establish real boundaries by building stone lines across the countryside and natural features, such as rivers, served as trespass lines, the present almost total framework of boundaries is a recent development. Maps representing the situation one, two, three, or more centuries ago show vast areas that are either unclaimed, unsurveyed, or merely spheres of influence.

Thus the States and embryonic States of the past were separated not by lines, but by areas. Still, they were separated, and they were either not in contact or only sporadically and ineffectively so. This intervening area functioned to prevent contact. Today's boundaries do not prevent contact; along them States *make* physical contact!

*Frontier = politicogeographical - area lying beyond the integrated region of the political unit + into which expansion could take place*

The frontier can thus be described as a politicogeographical area lying beyond the integrated region of the political unit and into which expansion could take place. This is the same principle we use when we speak of the "frontiers of science": we mean an ill-defined outer belt, vague and unknown, but into which we are penetrating. So it was with the geographical frontier. In southern Africa, the European-settled core area forming after 1652 around Cape Town grew stronger, unaware of the powerful Zulu State that was developing in Natal. The frontier functioned to separate these two political units until penetration of the intervening belt began in the 1830s. Then, finally, a confrontation occurred, and the frontier was replaced by uneasy and temporary truce lines.

Through frontiers, boundaries were often drawn. Expanding States or spheres met; sometimes they fought over the area involved, and sometimes they settled the disputed area by boundary treaty. The colonial invasion of Africa is replete with examples of boundary treaties. The frontier, in the original sense of the word, is no more, for the space into which States had been expanding has been fully occupied. Hence the boundary, a manifestation of integration, is inner-oriented. (The sea, however, is still a political frontier, and coastal States are now expanding into it. More on this later.)

It is often still possible to recognize frontier characteristics in an area where a boundary does exist. Among the functions of boundaries, as we shall see, is that of division, separation—not physically, but in other ways, such as in economic terms. On either side of an international boundary, therefore, there may be a discernible zone suffering from the interruptive effects of the boundary. This zone obviously has spatial characteristics and, hence, may be a frontier.

There are other kinds of frontiers besides natural or politicogeographical ones. Despite the expansion of *Homo sapiens* into nearly every nook of the earth in which the species can survive, there are still settlement frontiers. These areas of relatively sparse population but relatively abundant resources

capable of supporting larger populations today are found primarily in the more extreme climatic regions of the world: the high latitudes, the humid tropics, and the arid zone. Many States located in these regions, such as Russia, Colombia, and Egypt, are actively trying to settle and develop these frontiers because they are viewed as real, though still potential, economic assets. Other States are encouraging their nationals to settle in frontier areas near their boundaries, even if they are not very inviting physically. This is largely a defensive measure to discourage or render more costly invasion or even peaceful encroachment by their neighbors. China, Bolivia, and Israel are among the countries applying this policy adopted in the ancient empires more than 2000 years ago.

Frontiers are frequently zones of transition, and few transitions are more significant on the map or on the landscape than those between one culture and another. Cultural frontiers are often traversed by political boundaries that sometimes obscure the transitional nature of the region. People living along an international boundary tend to be more concerned with their local affairs, which in many cases straddle the boundary, than with affairs in their own capitals, which may be very distant in many ways. Links with neighbors across the border may be stronger than those with compatriots some distance away from the border. The trend, however, since the rise of nationalism and the solidification of States, has been toward hardening of boundaries and complete integration of a country's nationals right up to the borders. Nevertheless, in regions in which boundaries were drawn without regard to ethnic considerations and in regions of considerable international migration, cultures frequently overlap the borders and fade gradually with increasing distance from them. Often as one approaches, passes through, and leaves such a cultural frontier, the observable changes are so gradual it is difficult to know where the international boundary is unless it is clearly marked. Nevertheless, it is there, and it affects the lives of everyone on both sides of it.

## The Boundary

Boundaries appear on maps as thin lines marking the limit of State sovereignty. In fact, a boundary is not a line but a plane, a vertical plane that cuts through the airspace, the soil, and the subsoil of adjacent States. This plane appears on the surface of the earth as a line because it intersects the surface and is marked where it does so. But boundaries can be effective underground, where they mark the limit of adjacent States' mining operations in an ore deposit they may share, and they can be effective above the ground, for most countries jealously guard their airspace. The matter of airspace arises again in Chapter 34.

### Boundary Making

The ideal sequence of events in establishing a boundary is as follows. The first stage involves the description of the boundary and the terrain through which it runs. This description identifies, as exactly as possible, the location of the boundary being established. It may refer to hilltops, crestlines, rivers, and even to cultural features, such as farm fences and roads. The more detailed the description, the less likely it is that subsequent friction will occur. As will be seen later, even the most prominent physical features in the landscape have given rise to serious disputes when used as political boundary lines. This first stage, often formalized in treaties, is referred to as the *definition* of the boundary.

When the treaty makers have completed their definition of the boundary in question, their work is placed before cartographers who, using large-scale maps and air photographs, plot the boundary as exactly as possible. The period of time separating this stage of *delimitation* from the initial stage of definition may amount to decades; for example, several African States whose boundaries were defined toward the end of the last century are only now in the process of exactly delimiting their borders. Some are discovering that the original work of definition was done rather crudely and imprecisely, causing many problems today.

Then there is the task of marking the boundary on the ground. For this purpose, both the actual treaty and the cartographic material are employed. Boundary *demarcation*, as this process is called, has by no means taken place along every boundary defined and delimited. When a boundary is demarcated, a wide variety of methods may be employed. A mere line of poles or stones may suffice. Cement markers may be set up, so that from any one of them the adjacent ones on either side will be visible. Fences have been built in certain delicate areas where exact demarcation is required, and on rare occasions walls have been built. Boundary demarcation is an expensive process, and when States do not face problems along their boundaries that absolutely require demarcation, they often delay this stage indefinitely.

The final stage in boundary making is *administration*; that is, establishing some regular procedure for maintaining the boundary markers, settling minor local disputes over the boundary and its effects, regulating the use of water and waterways in the border area, and attending to other "housekeeping" matters. Sometimes special commissions, such as the International Joint Commission of the United States and Canada and the International Boundary Commission of the United States and Mexico, will be established to perform these functions. The tasks are usually assigned to an office in the foreign ministry or some other government agency. For many boundary segments, however, there is no regular administration; special ad hoc commissions are appointed as needed, or a particular problem is assigned to an official to handle.

We must stress that this is an ideal pattern of boundary-making. It appears in practice most commonly in Europe and North America, rather regularly in Latin America, less commonly in Asia, and rarely in Africa.

### Criteria for Boundaries

Political geographers, among others, have searched for the "ideal" criteria for boundary definition in hopes of reducing international tensions created by boundary disputes. This

**A demarcated boundary.** Variations of this type of pillar are found on many boundaries around the world. This one is on the U.S.—Mexican border west of Tijuana. The fence is not at all common, however. Erected by the United States and offensive to Mexico, it runs for many kilometers east and west of Tijuana. It is largely symbolic, however, as it is easily penetrated or surmounted and ends on the beach several meters from the Pacific Ocean, seen in the background. (Martin Glassner)

activity was especially common during the interwar period and led to intense debate regarding the merits of "artificial" boundaries as opposed to "natural" (physical) ones. However, a boundary is one of the parts of the State system, and we are concerned not only with criteria for its establishment, but also with the effect the use of these criteria will have; that is, the function of the border. For example, some political geographers have felt that ethnic criteria may be the most appropriate for the definition of international boundaries. In other words, boundaries should be drawn so as to separate peoples who are culturally uniform so that a minimum of stress will be placed on them.

However, world population is too heterogeneous and interdigitated to permit the definition of boundaries that completely and exactly separate peoples of different character. The map of African ethnic groups in Chapter 4 clearly illustrates the impossibility of using ethnic criteria alone. Occasionally, instead of altering boundaries, States have attempted to exchange such groups: after

World War I, some hundreds of thousands of Turks were moved into Anatolia, from where Greeks were repatriated. Among the solutions proposed for the dilemma of Cyprus have been repatriation of Turkish Cypriots to Turkey and partition of the island into Greek and Turkish States. The partition has, in fact, already been achieved by a haphazard exchange of populations since the Turkish invasion in 1974, but it has not noticeably reduced Greek-Turkish friction.

Language might also be proposed as a basis for boundary definition. But a map of the world's languages shows a patchwork of great complexity that would immeasurably compound the boundary framework existing today. Many States are multilingual and would be fragmented in any such effort. This brings up a question that kept political geographers at odds for some time—namely, whether a boundary should be a barrier or a bond between adjacent States. If a boundary separates people who speak different languages, they are not likely to understand each other well, with the result that relations may remain hostile across their mu-

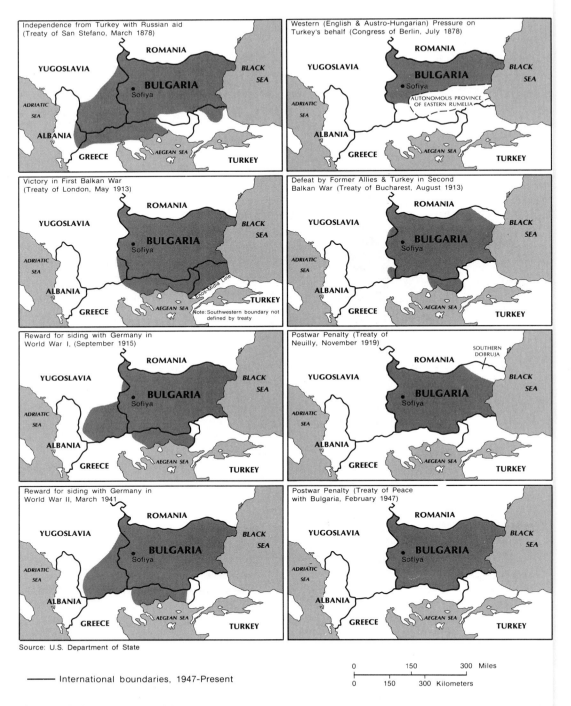

Source: U.S. Department of State

—————— International boundaries, 1947–Present

0        150        300  Miles

0        150        300  Kilometers

**The instability of national boundaries** through history is particularly evident in the world's shatter belts. The classic example is Bulgaria. Since the formation of the United Nations, however, with its emphasis on the "sovereignty and territorial integrity" of States, boundary changes have been much less frequent.

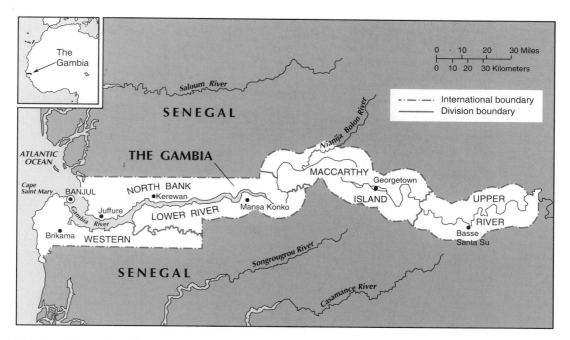

**The Gambia** is the only country in the world whose boundaries consist largely of arcs of circles. One of the world's smallest and poorest states, this former British colony consists of just the lower portion of the Gambia River and the land a short distance back from both banks of the river. Tourism has replaced peanuts as The Gambia's chief foreign exchange earner, with many Western Europeans attracted to the beaches around Cape Saint Mary and some American blacks to Juffure, the ancestral village of Alex Haley, the American author of *Roots*.

tual boundary. On the other hand, a boundary running through a region of linguistic homogeneity would ensure that people on either side would at least have a language in common, and as a result could communicate more easily. This common language across the border would then act as a bond between the two States involved. The point can also be put differently, in terms of physical criteria: a mountain range may serve as a barrier, with divisive consequences; a river, with its two banks close together, would form a bond.

Religion, like language, is an important component of ethnicity and even of nationhood, as explained in Chapter 4. Peoples of widely varied races and tongues have accepted the same faith, and peoples speaking the same language have adopted different religions. Nevertheless, in areas where religion has been a strong source of internal friction, it has been a major basis for bound-

ary definition. A good example is the partition of the Indian subcontinent into (mainly Hindu) India and (mainly Muslim) Pakistan. The latter, as a result, became a fragmented State and has demonstrated the weakness of religion as a centripetal force.

Inspection of the world framework of boundaries will indicate that many political boundaries lie along prominent physical features in the landscape. Such boundaries have become known as "physiographic political boundaries," a term not to be confused with the physiographic boundary used in physical geography. In political geography a physiographic boundary refers to any prominent physical feature paralleled by a political boundary: a river, mountain range, or escarpment. These would seem to be especially acceptable criteria, since pronounced physical features often also separate culturally different areas. But in practice such boundaries have also produced major prob-

**Two contrasting boundaries.** At the top is a fine example of a relict boundary, the one formerly separating the French and Spanish protectorates in Morocco. This impressive ruin is on the main highway near the Atlantic coast. (Martin Glassner) On the bottom is a segment of the Berlin Wall, erected in 1961 and torn down in 1989. While it stood, it represented the world's best example of a superimposed boundary. (Photo courtesy of German Information Center)

lems. No State corresponds exactly to a physiographic province, and very homogeneous physiographic regions are sometimes divided by boundaries based on some insignificant feature, such as a small stream or a low divide. In the early days of boundary establishment, physiographic features were useful because they were generally known and could be recognized as trespass lines. But this function has ceased now, and the most divisive and obvious physical features have created major difficulties between States.

Any mountain range has recognizable crestlines, but they rarely coincide with the watershed. As a result, water flowing from one side of the State boundary (which coincides with a physiographic, but not with a hydrographic feature) feeds the streams of the State on the other side of the boundary. If the State that possesses the source areas decides to dam those waters, it may impede the water supply of the other State and friction may result. There are many other examples of problems arising along international boundaries in mountains: transhumance, the use of passes, the need for tunnels.

Rivers, too, seem to be obvious and useful boundary features, but again many problems attend their use as boundaries. Use of the water by the riparian States is one major issue for debate.* Furthermore, rivers tend to shift their course, producing new circumstances that require a redefinition of the boundary, which may lead to friction. In addition, rivers have breadth as well as length and depth, and countless disputes have erupted over whether the boundary should be along the left bank, the right bank, the *thalweg* (main navigable channel), the middle, or somewhere else.

Thus physical features are at times as problematic as boundaries based on the other criteria enumerated. We might also observe that arguments on behalf of "natural" boundaries are usually advanced by representatives or academics of States that wish to expand their territories. Can anyone imagine a State offering to move its boundary *back* to a "natural" line deep inside its own territory?

Since *all* political boundaries are by definition artificial, a *morphological* classification of them is not particularly useful. It hardly matters when one is at a border and looking across or flying over whether or not it is geometric. How the border functions, how it is reflected on the landscape, and how the people on both sides feel about it are much more important. Thus other classifications have been devised that can be

quite useful in analyzing boundaries and boundary problems.

A *functional* classification might reflect, for example, whether the boundary was originally or is still designed primarily for defensive purposes, as a separator of cultures, according to economic factors, simply for legal or administrative purposes (as in Spanish America and French Africa), or on ideological bases (communist or noncommunist areas, Catholic or Protestant areas, black- or white-controlled areas in Africa). A *genetic* classification would be based on when the boundary was laid out. A *pioneer* boundary is drawn through essentially unoccupied territory. Or a boundary may be *antecedent* to intensive settlement and land use, *subsequent* to the establishment of peoples with different cultures in the region and taking them into account, *superimposed* on an existing cultural pattern, or *relict*, a former boundary that no longer functions as such.

A *legal* classification could consider those boundaries that are settled and recognized in international law; those recognized only by the adjacent countries and some others; *de facto* boundaries, which are disputed by one adjacent State; and *fictitious* boundaries, which exist on maps but not in the real world, usually relict boundaries or the limits dreamed of by separatist or irredentist groups.

Although all of these concepts are useful for analyzing and understanding boundaries, we must remember that no type of boundary is necessarily better than any other. The best boundary may be the one that performs the fewest functions; certainly, it is the one falling between good neighbors.

## Functions

The functions of boundaries change over time. Until quite recently, for example, it was conceivable for a State to attempt to fortify its boundary to such an extent that it would be invincible. French defensive strategy until 1940 was pinned on the Maginot Line, a line of fortifications constructed along its northeastern boundary. The idea is as old as the Great Wall of China, and the principle

*Riparians are those States through or along which a river flows. The term is also used as a synonym for the adjective "littoral" to indicate a State that abuts a lake or the sea.

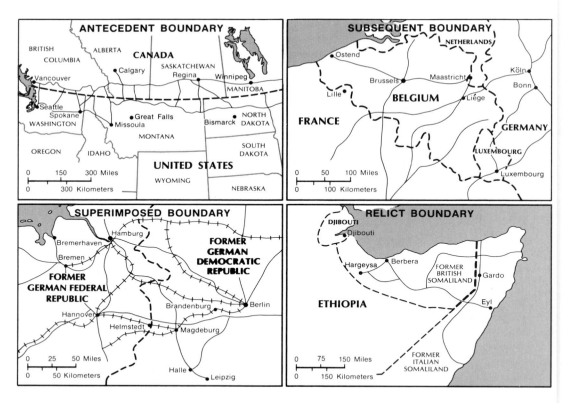

**Genetic classification of boundaries.**

is the same. Plateaus with sheer escarpments have afforded protection to societies that used these natural barriers to their advantage and considered the scarps to mark the limit of their domain. Lesotho and Ethiopia are good examples.

But advancing technology has diminished the importance of the defensive function of boundaries, and States no longer rely on fortified borders for their security. In some parts of the world, where guerrilla activities short of open warfare occur, a river or mountain range may still present strategic advantages. To the major powers of the world, however, and to those States possessing modern military equipment, the naturally or artificially fortified boundary is no longer an asset. Consequently, many States are presently demarcating their boundaries without intending or attempting to fortify them. Rather, the aim is to mark the limit of

State sovereignty, which may have become necessary as a result of emerging friction.*

The boundary has an impact on many forms of organization within the State and, in turn, on the way in which the State's various arms of organization affect the nature of

*Nevertheless, boundary walls and fences continue to be built. The most famous, of course, was "the Iron Curtain," purely metaphoric, in fact, but symbolized on the borders of East Germany and Czechoslovakia with the West by fences, minefields, guard towers, and so on. The similar installations along Israel's borders with Lebanon and Syria, however, perform a very different function: to keep armies and terrorists out, not to keep citizens in. Morocco in 1985 completed a 4000-kilometer (1550-mile) wall of stone and sand in Western Sahara to protect the most valuable portion of the territory from Algerian-backed rebels seeking independence. Thailand has built an antismuggler and antiterrorist fence along the Malaysian border. South Africa erected an electrified fence along parts of its border with Zimbabwe. And there are other examples.

**Two very different border towns.** At the top is Jogbani, which is on the border between the Indian state of Bihar and the southeastern corner of Nepal, just south of Biratnagar. Nepal and India have an open border, and there are no restrictions on the movement of their citizens across it. There is a tiny customs-immigration post on the Indian side to service the occasional foreign traveler; on the Nepali side, beyond the striped pole, there is no post at all. At the bottom is Le Perthus, a Catalan town straddling the boundary between France and Spain in the Pyrenees south of Perpignan, France. This is the main highway between the French and Spanish Rivieras, and the town is generally crowded with tourists. A small stone pillar to the left of the car marks the actual boundary. The customs and immigration posts for each country are some distance down the mountain slopes where there is some level land. (Martin Glassner)

the boundary. There is, for example, the commercial function of the boundary. The government can erect tariff walls against outside competition for its market and thus assist internal industries. However, the price differential on either side of the boundary will affect the location of outlets for the products of the various industries affected by tariff provisions, and while the industry may prosper, the area under the shadow of the boundary may not and smuggling will be encouraged.

Some boundaries, because of the close and positive relationship between the societies they separate, do not mark great changes in modes of government, prosperity, or other phenomena. The U.S.-Canada border is an excellent example. Some other borders, on the contrary, mark lines across which contrasts are severe and prominent. The societies they separate may have different institutions, different levels of development, and different ideologies. Such a boundary may be very interruptive in every way. The partition of Hispañola by the boundary between Haiti and the Dominican Republic is a case in point.

The boundary, of course, also has a legal function. National law prevails to this line. Taxes must be paid to the government by anyone legally subject to taxation, whether the person resides 1 or 100 kilometers from the border. Even though residents living within sight of the border may have closer linguistic, historical, and religious ties with the people on the other side, they are subject to the regulations prevailing on their side of the boundary. Furthermore, the government usually attempts to control emigration and immigration at points along the border.

A number of contemporary trends are combining to reduce the significance of boundaries, even as many countries are energetically seeking to fix and demarcate theirs. For one thing, as peoples are brought closer together by vastly better communications and transportation, they are more and more being forced or enticed to look beyond their own national boundaries and become more familiar with the people out there. As trade, travel, and tourism increase, boundary formalities are being recognized as unnecessary and very harmful impediments and are being removed. Many groups of States around the world are integrating their economies in varying degress, and fundamental to all integration is free movement across borders. Boundaries are utterly meaningless to missiles and television broadcasts alike; not only terror but understanding can readily pass through a political boundary, no matter how well defended. Boundaries are becoming more permeable and less hostile. This is not to say, however, that boundaries are disappearing or even that States no longer care much about them; indeed, quite the contrary is true. There are still a great many boundary and territorial disputes around the world. Some of them are dor-

---

**Disputed islands in the East and South China Seas.** As competition grows for marine resources, ever smaller islands are becoming objects of dispute. The seas off East and Southeast Asia provide many examples. The Senkaku islands are currently controlled by Japan but are also claimed by China and Taiwan; the Paracels are occupied by China, which displaced the Vietnamese claimants in a brief battle in 1974; various Spratly islets are controlled by China, Taiwan, Vietnam, Malaysia, and the Philippines, all of which have claims on some or all 200 of them, as does Brunei. All these islets, reefs, rocks, and cays are uninhabited except, in some cases, where they have military garrisons. At a conference in Bandung, Indonesia in July 1991, the six claimants offered to resolve the problem peacefully, to proceed cautiously with development in the interim, and to cooperate in maintaining the islands' fragile environment. Vietnam and Cambodia have not agreed on their maritime boundary or on ownership of the islands in the disputed area. The value of all of these islands lies in the fisheries nearby, the potentially large petroleum deposits in their continental shelves, and, as shown in the outset map, their proximity to some of the world's most strategic sea lanes. Japan and Korea, unable to agree on their southern continental shelf boundary, established the disputed area as a zone to be developed jointly for at least 50 years. China has also claimed a portion of this shelf.

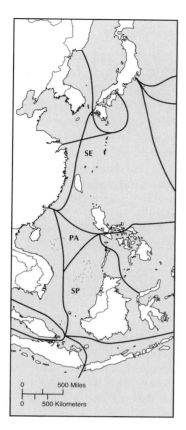

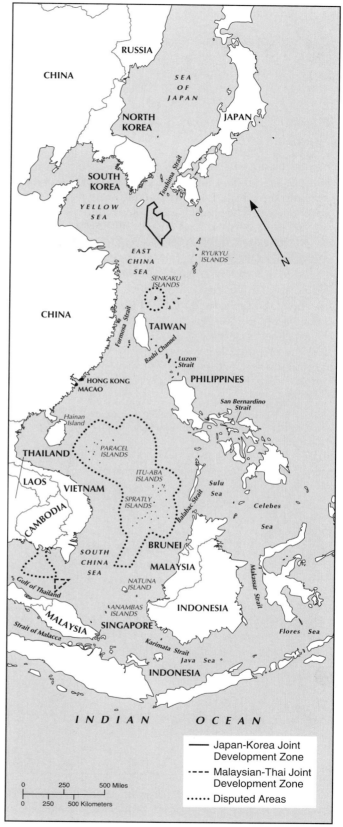

SE

PA

SP

0    500 Miles

0    500 Kilometers

RUSSIA

CHINA

*SEA OF JAPAN*

NORTH KOREA

JAPAN

SOUTH KOREA

*Tsushima Strait*

*YELLOW SEA*

N

*EAST CHINA SEA*

*RYUKYU ISLANDS*

*SENKAKU ISLANDS*

CHINA

*Formosa Strait*

TAIWAN

*Bashi Channel*

*Luzon Strait*

HONG KONG
MACAO

PHILIPPINES

*San Bernardino Strait*

*Hainan Island*

THAILAND

*PARACEL ISLANDS*

*ITU-ABA ISLANDS*

*Sulu Sea*

LAOS

VIETNAM

*SPRATLY ISLANDS*

*Balabac Strait*

*Celebes Sea*

CAMBODIA

*SOUTH CHINA SEA*

BRUNEI

MALAYSIA

*Makassar Strait*

*Gulf of Thailand*

*NATUNA ISLAND*

*ANAMBAS ISLANDS*

INDONESIA

*Flores Sea*

MALAYSIA

SINGAPORE

*Strait of Malacca*

*Karimata Strait*

*Java Sea*

INDONESIA

I N D I A N     O C E A N

0    250    500 Miles

0    250    500 Kilometers

—— Japan-Korea Joint Development Zone

---- Malaysian-Thai Joint Development Zone

····· Disputed Areas

mant, some very much alive, and some of uncertain status. All, however, are interesting and deserve study.*

## Boundary and Territorial Disputes

It is impossible to count precisely all contemporary boundary and territorial disputes. Certainly the number is near 100, far more if we count every disputed island separately. And these are only land areas; maritime boundary disputes constitute a huge category of their own, although they are closely related to disputes over land, and we discuss them later. All we can do now is to survey a few outstanding problems to illustrate the point that there is still a great deal of work to be done by political geographers—and students of political geography—on boundary and territorial disputes.

*Many of the points made here about boundaries are illustrated in the case study at the end of this chapter, excerpts from the Israel-Jordan peace treaty of 1994.

As we turn more and more to the sea for its resources as well as its transport and strategic value, and as States expand their national jurisdiction farther and farther out into the sea frontier, more and more islands are being disputed, often ones scarcely known until recently and that may have no intrinsic value. What makes them worth disputing is the 200-nautical-mile exclusive economic zone to which each island is entitled under the Law of the Sea. The United States alone disputed 18 islands in the Pacific with the United Kingdom and 7 with New Zealand. Although these disputes have been resolved as the United States relinquished its claims, other island disputes are getting more intense.

One type of island dispute is represented by the Falkland Islands, administered since the eighteenth century by the United Kingdom (with an interval of abandonment) but claimed by Argentina. This is more of a traditional territorial dispute, similar to those between Guatemala and Belize or between Guyana and Venezuela. In each case the problems are complex, resulting from com-

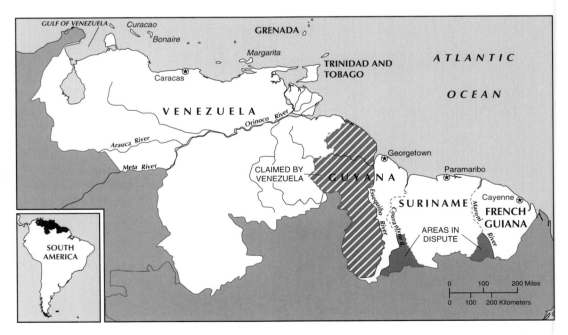

**Traditional territorial disputes in northern South America.** All three of these areas are in dispute because of varying interpretations of boundary definitions and conflicting maps made at a time when the region was largely unexplored and scarcely known to Spain, France, the Netherlands, and the United Kingdom. At the moment all of these disputes are quiet, but none has been resolved.

**Propaganda sign at La Quiaca, Argentina on the Bolivian border.** "The Malvinas (Falklands) are Argentine." The people are Bolivians crossing the border to buy cheap food in Argentina to take back to La Paz for sale. (Martin Glassner)

peting claims among colonial powers to little-known territories based on conflicting principles of international law and disputed facts. In 1982, the Falklands dispute broke into open warfare as Argentina occupied them briefly before being displaced by a sizable military force sent from Britain. There are still no signs of a peaceful resolution of this problem.

There are other boundary and territorial disputes in Latin America, but considering the casual way those boundaries were de-

termined in the first place, it is remarkable more don't occur. This is in part attributable to the adoption by the leaders of the struggle for independence from Spain early in the nineteenth century of the ancient principle of Roman law, *uti possidetis juris*. This means that whoever possesses a thing has a right to it. The Latin American leaders called their version "The *Uti Possidetis* of 1810," meaning that the boundaries of the new States that would emerge there would be the administrative limits in effect when the rev-

**Propaganda mural in Szczecin, Poland (formerly Stettin, Germany).** The slogan, superimposed on a stylized Polish eagle, reads, "West Pomerania Always Polish." This photograph was taken in 1988; less than three years later, Poland had overthrown its Communist government, the two Germanies had reunited and Germany had formally undertaken to respect the post–World War II Polish boundary, including the transfer of West Pomerania to Poland. (Martin Glassner)

**Boundary and Territorial disputes on stamps.** Top row, left to right: "Belize is Ours" on a map of Guatemala; Colombia and Nicaragua both claim San Andrés and Providencia Islands in the Caribbean; and the claims of Ecuador and Peru to the same territory in the Amazon basin. Second row: Chilean stamp celebrating settlement of the Beagle Channel dispute with Argentina through papal arbitration; Honduras marks settlement of her territorial dispute with Nicaragua; Argentina marks her brief occupation of the Falkland Islands in 1982. Third row: German stamp celebrating conquest of Danzig in 1939; stamp issued under the temporary French administration of the Saarland after WWII; Polish map stamp showing territories acquired from Germany in 1945; 1990 stamp celebrating German reunification; and Venezuela's claim to two-thirds of Guyana. Fourth row: Danish and German stamps showing settlement of their mutual boundary and minority communities left on both sides; American Military Government overprints on Italian stamps for temporary use in the disputed areas of Venezia-Giulia and Trieste. Fifth row: United Nations stamp honoring its Temporary Executive Authority in the former Dutch New Guinea and an Indonesian stamp marking the country's acquisition of this territory; Greek stamp mourning Turkish occupation of northern Cyprus and stamp issued by the Turkish Federated State of Cyprus in the occupied area. Bottom row: Stamps issued by Egypt showing Egyptian troops invading Palestine in 1948, Morocco asserting the Arab claim to all of the former Palestine mandate, and Iran urging the "liberation" of Jerusalem; Pakistani map stamp showing that country's claim to all of Jammu and Kashmir. (Martin Glassner)

olutions broke out in 1810. The Organization of African Unity adopted the same principle in 1963 in an effort to prevent innumerable conflicts over imposed colonial boundaries of the emerging States of Africa. As in Latin America, the effort has not been wholly successful: witness the wars between Algeria and Morocco and between Somalia and Ethiopia over desert areas. Nevertheless, the policy did prevent major wars that might have spread from the Nigerian civil war and the repeated attempts by Shaba to secede from Zaïre. It is likely that more dormant or latent boundary and territorial disputes will develop in Africa, but they stand a good chance of being resolved peacefully. There is very strong resistance to a "Balkanization of Africa."

While Spain has been trying to wrest Gibraltar from the United Kingdom, Morocco would like to retrieve from Spain five presidencies—small islands and enclaves along the Mediterranean coast—that Spain has held as part of its sovereign territory for centuries. This poses a problem for Spain: Does it believe in self-determination? Or decolonization? In one case but not the other? No resolution is in sight. To complicate matters further, an irredentist group in Portugal has claimed Olivenza, a 1500-square-kilometer (600-square-mile) wedge of territory just south of the Spanish town of Badajoz, which was part of Portugal from 1228 to 1801. But the Portuguese government could hardly press the claim while stoutly refusing to give up its colonial possessions in Africa and Asia. Boundary disputes are seldom conducted in a vacuum; they are generally linked with many other factors in each of the disputing countries and even considerations halfway around the world.

The irredentist claims of China to huge areas of the former Soviet Union have been well publicized, as have the several wars fought between India and Pakistan over Kashmir and the Rann of Kutch. Unlike the examples given earlier, these cases (and a number of others in Asia) involve areas of considerable size, population, natural resources, and/or strategic importance. They could well lead to more serious fighting before the disputes are settled. In Asia there is no regional organization comparable to the Organization of American States or the Organization of Africa Unity to moderate and help resolve disputes in the region. And no *uti possidetis* principle operates there either.

That the principle does not always work even in the region that first adopted it in modern times is illustrated by the chronic boundary disputes between Argentina and Chile. Their mainland boundary in the Andes has been largely settled one segment at a time in a series of agreements extending over more than a century, the final one concerning the Laguna del Desierto being signed in October 1994. Despite the inscription on the giant statue in Uspallata Pass, erected in 1902 to commemorate settlement of an earlier boundary dispute, the two countries are anything but friendly. The inscription reads, "These mountains will crumble into dust before the peoples of Argentina and Chile break the peace sworn at the feet of Christ the Redeemer." Yet the two countries came perilously close to war in 1978 over their boundary in the Beagle Channel at the southern tip of the continent, and the dispute was not resolved until 1985, when they finally accepted an arbitral decision of Pope John Paul II. The mutual suspicion remains, however. Speaking of the Christ of the Andes, the Argentines will say that he faces Argentina with his hand raised in benediction over them because he favors them over the Chileans, on whom he has turned his back. The Chileans will respond that Christ is only facing the Argentines because he has to watch them carefully; he doesn't dare turn his back to them!

It may be true, as Robert Frost said, that "Good fences make good neighbors." It is much more likely, though, that it takes good neighbors to make good fences.*

---

*A splendid source of information about boundaries and related matters is the International Boundaries Research Unit, Suite 3, Mountjoy Research Centre, University of Durham, Durham DH1 3UR, U.K.

## Case Study: The Israel-Jordan Boundary

After nearly half a century of tension and two wars between them, Israel and Jordan signed a treaty of peace on 26 October 1994. In reality, many provisions of the treaty merely formalized the many informal arrangements long in effect between the parties, which had permitted quasinormal relations since about 1969. Of special interest here, however, is Article 3 of the treaty, which reads as follows.

### Article 3: International Boundary

1. The international boundary between Jordan and Israel is delimited with reference to the boundary definition under the Mandate, as is shown in Annex I(a), on the mapping materials attached thereto and coordinates specified therein.

2. The boundary, as set out in Annex I(a), is the permanent, secure and recognized international boundary between Jordan and Israel, without prejudice to the status of any territories that came under Israeli military government control in 1967.

3. The Parties recognize the international boundary, as well as each other's territory, territorial waters and airspace as inviolable, and will respect and comply with them.

4. The demarcation of the boundary will take place as set forth in Appendix (I) to Annex I and will be concluded not later than 9 months after the signing of the Treaty.

5. It is agreed that where the boundary follows a river, in the event of natural changes in the course of the flow of the river as described in Annex I(a), the boundary shall follow the new course of the flow. In the event of any other changes the boundary shall not be affected unless otherwise agreed.

6. Immediately upon the exchange of the instruments of ratification of this Treaty, each Party will deploy on its side of the international boundary as defined in Annex I(a).

7. The Parties shall, upon the signature of the Treaty, enter into negotiations to conclude, within 9 months, an agreement on the delimitation of their maritime boundary in the Gulf of Aqaba.

8. Taking into account the special circumstances of the Baqura/Naharayim area, which is under Jordanian sovereignty, with Israeli private ownership rights, the Parties agree to apply the provisions set out in Annex I(b).

9. With respect to the Al-Ghamr/Zofar area, the provisions set out in Annex I(c) will apply.

As is typical in many treaties, details are set forth in appended annexes, protocols, memoranda, or other instruments. Annex I(a) of this treaty reads:

### Jordan-Israel International Boundary
### Delimitation And Demarcation

1. It is agreed that, in accordance with Article 3 of the Treaty, the international boundary between the two States consists of the following sectors:
   A. The Jordan and Yarmouk Rivers.
   B. The Dead Sea.
   C. The Wadi Araba/Emek Ha'arava.
   D. The Gulf of Aqaba.
2. The boundary is delimited as follows:
   A. JORDAN AND YARMOUK RIVERS
      1. The boundary line shall follow the middle of the main course of the flow of the Jordan and Yarmouk Rivers.
      2. The boundary line shall follow natural changes (accretion or erosion) in the course of the rivers unless otherwise agreed. Artificial changes in or of the course of the

rivers shall not affect the location of the boundary unless otherwise agreed. No artificial changes may be made except by agreement between both Parties.

3. In the event of a future sudden natural change in or of the course of the rivers (avulsion or cutting of a new bed) the Joint Boundary Commission (Article 3 below) shall meet as soon as possible, to decide on necessary measures, which may include physical restoration of the prior location of the river course.

4. The boundary line in the two rivers is shown on the 1/10,000 orthophoto map dated 1994 (Appendix III attached to this Annex).

5. Adjustment to the boundary line in any of the rivers due to natural changes (accretion or erosion) shall be carried out whenever it is deemed necessary by the Boundary Commission or once every five years.

6. The lines defining the special Baqura/Naharayim area are shown on the 1:10,000 orthophoto map (Appendix IV attached to this Annex).

7. The orthophoto maps and image maps showing the line separating Jordan from the territory that came under Israeli military government control in 1967 shall have that line indicated in a different presentation and the legend shall carry on it the following disclaimer:

"This line is the administrative boundary between Jordan and the territory which came under Israeli military government control in 1967. Any treatment of this line shall be without prejudice to the status of that territory."

**B.** DEAD SEA AND SALT PANS

The boundary line is shown on the 1:50,000 image maps (2 sheets, Appendix II attached to the Annex). The list of geographic and Universal Transverse Mercator (UTM) coordinates of this boundary line shall be based on Israel Jordan Boundary Datum (IJBD 1994) and, when completed and agreed upon by both Parties, this list of coordinates shall be binding and take precedence over the maps as to the location of the boundary line in the Dead Sea and the salt pans.

**C.** WADI ARABA/EMEK HA'ARAVA

1. The boundary line is shown on the 1:20,000 orthophoto maps (10 sheets, Appendix I attached to this Annex).

2. The land boundary shall be demarcated, under a joint boundary demarcation procedure, by boundary pillars which will be jointly located, erected, measured and documented on the basis of the boundary shown in the 1/20,000 orthophoto maps referred to in Article 2-C-(1) above. Between each two adjacent boundary pillars the boundary line shall follow a straight line.

3. The boundary pillars shall be defined in a list of geographic and UTM coordinates based on a joint boundary datum (IJBD 94) to be agreed upon by the Joint Team of Experts appointed by the two parties (hereinafter the JTE) using joint Global Positioning System (GPS) measurements. The list of coordinates shall be prepared, signed and approved by both Parties as soon as possible and no later than 9 months after this Treaty enters into force and shall become part of this Annex. This list of geographic and UTM coordinates when completed and agreed upon by both Parties shall be binding and shall take precedence over the maps as to the location of the boundary line of this sector.

4. The boundary pillars shall be maintained by both Parties in accordance with a procedure to be agreed upon. The coordinates in Article 2-C-(3) above shall be used to reconstruct boundary pillars in case they are damaged, destroyed or displaced.

5. The line defining the Al-Ghamr/Zofar area is shown on the 1/20,000 Wadi Araba/Emek Ha'arava orthophoto map (Appendix V attached to this Annex).

**D.** THE GULF OF AQABA

The Parties shall act in accordance with Article 3.7 of the Treaty.

3. Joint Boundary Commission

**A.** For the purpose of the implementation of this Annex, the Parties will establish a Joint Boundary Commission comprised of three members from each country.

    **B.**    The Commission will, with the approval of the respective governments, specify its work procedures, the frequency of its meetings, and the details of its scope of work. The Commission may invite experts and/or advisors as may be required.

    **C.**    The Commission may form, as it deems necessary, specialized teams or committees and assign to them technical tasks.

Annexes I(b) and I(c) are special provisions for small areas along the border where the 1949 armistice line between Israel and Transjordan did not follow the boundary established by the United Kingdom in 1922 when it first partitioned its Palestine mandate. These two areas, totaling 341 km², between the armistice line and the international boundary, have been occupied and cultivated by neighboring Israeli *kibbutzim* (collective settlements). The provisions of the annexes are nearly identical; Annex I (b) is given below and illustrated by the map on page 102.

*Annex I (b)*
*The Baqura/Naharayim Area*

**1.**    The two Parties agree that a special regime will apply to the Baqura/Naharayim area ("the area") on a temporary basis, as set out in this Annex. For the purpose of this Annex the area is detailed in Appendix IV.

**2.**    Recognizing that in the area which is under Jordan's sovereignty with Israeli private land ownership rights and property interests ("land-owners") in the land comprising the area ("the land") Jordan undertakes:

    **a.**    to grant without charge unimpeded freedom of entry to, exit from, land usage and movement within the area to the land-owners and to their invitees or employees and to allow the land-owners freely to dispose of their land in accordance with applicable Jordanian law;

    **b.**    not to apply its customs or immigration legislation to land-owners, their invitees or employees crossing from Israel directly to the area for the purpose of gaining access to the land for agricultural, touristic or any agreed purpose;

    **c.**    not to impose discriminatory taxes or charges with regard to the land or activities within the area;

    **d.**    to take all necessary measures to protect and prevent harassment of or harm to any person entering the area under this Annex;

    **e.**    to permit with the minimum of formality, uniformed officers of the Israeli police force access to the area for the purpose of investigating crime or dealing with other incidents solely involving the land-owners, their invitees or employees.

**3.**    Recognizing Jordanian sovereignty over the area, Israel undertakes:

    **a.**    not to carry out or allow to be carried out in the area activities prejudicial to the peace or security of Jordan;

    **b.**    not to allow any person entering the area under this Annex (other than the uniformed officers referred to in paragraph 2(e) of this Annex) to carry weapons of any kind in the area; unless authorized by the licensing authorities in Jordan after being processed by the liaison committee refered to in Article 8 of this Annex.

    **c.**    not to allow the dumping of wastes from outside the area into the area.

**4.**  **a.**    Subject to this Annex, Jordanian law will apply to this area.

    **b.**    Israeli law applying to the extra-territorial activities of Israelis may be applied to Israelis and their activities in the area, and Israel may take measures in the area to enforce such laws.

    **c.**    Having regard to this Annex, Jordan will not apply its criminal laws to activities in the area which involve only Israeli nationals.

**5.**    In the event of any joint projects to be agreed and developed by the parties in the area, the terms of this Annex may be altered for the purpose of the joint project by agreement between the Parties at any time. One of the options to be discussed in the context of the joint projects would be the establishment of a Free-Trade Zone.

6. Without prejudice to private rights of ownership of land within the area, this Annex will remain in force for 25 years, and shall be reviewed automatically for the same periods, unless one year prior notice of termination is given by either Party, in which case, at the request of either Party, consultations shall be entered into.

7. In addition to the requirement referred to in Article 4(a) of this Annex, the acquisition of land in the area by persons who are not Israeli citizens shall take place only with the prior approval of Jordan.

8. A Jordanian-Israeli Liaison Committee is hereby established in order to deal with all matters arising under this Annex.

Other treaty articles of interest to political geographers deal with security, water, economic relations, refugees and displaced persons, places of historical and religious significance and interfaith relations, combating crime and drugs, transportation and roads, freedom of navigation and access to ports, civil aviation, environment, energy, Rift Valley development, and Aqaba and Eilat. One of the possible joint development projects in the Rift Valley (which extends from Syria to Mozambique) is resurrection of the branch of the Hijaz Railway (built by the Turks to connect Damascus and Medina in Saudi Arabia) that ran down the Yarmouk Valley and across Palestine to Haifa before it was largely destroyed during Israel's War of Independence in 1948. Part of it is shown in the photograph on page 103.

This peace treaty, with its imaginative solutions to difficult problems, is an example of what can be achieved by enemies who decide to make peace instead of war.

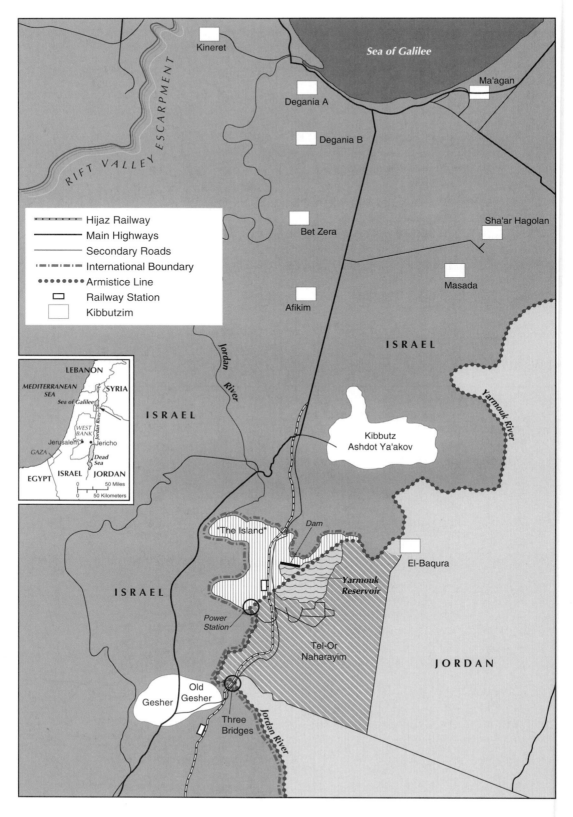

**Boundary accommodation in the Jordan Valley**. The 1994 Israel-Jordan peace treaty incorporates imaginative and pragmatic provisions to deal with some minor but knotty boundary problems. This map, based largely on materials provided by Shlomo Shalmon of Gesher, illustrates the most important of these accommodations. The Naharayim power station on the Yarmouk River was built (1924-1928), operated and maintained by Palestine Jews on land leased by Transjordan to the Palestine Electric Corporation of London. Many workers lived in a Jewish settlement in Transjordan called Tel-Or; others commuted from nearby Jewish settlements in Palestine on the Haifa branch of the Hijaz Railway. The power station was destroyed in May 1948 and the Palestinian Jews withdrew to the new State of Israel. The power station is too small to be worth restoring now, but the Hijaz Railway may be rebuilt and operated profitably. The Gesher railway station was the lowest on earth (267 meters below sea level) and the nearby customs post was also the lowest on earth. The 1949 armistice line drawn between Israel and Transjordan left a portion of the emirate under Israeli control. It has been farmed since then by members of Israeli kibbutzim to the north. The 1994 treaty recognizes Jordanian sovereignty over "The Island" (as the Israelis call this area) but permits continued Israeli use of it. This may be considered a resumption of the Jewish-Arab cooperation in the Naharayim area between 1924 and 1948.

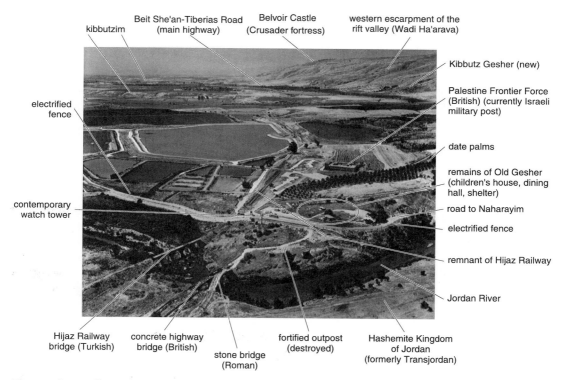

**The Jordan Valley.** This aerial photograph overlaps the accompanying map on the south. It shows a portion of the great rift valley system that extends from Syria to Mozambique, looking roughly southwest from a point above Jordan, across the Jordan River, the Israeli kibbutz (collective settlement) of Gesher (Hebrew for bridge) and other kibbutzim to the south. The strategic location of the kibbutz is as clear as that of the Crusader fortress overlooking the valley. Gesher was founded here in 1939 precisely because of its location. Almost from the moment of Israel's independence on 14 May 1948, the tiny settlement was besieged by troops of Transjordan and Iraq for three weeks. After the war the members of the kibbutz abandoned the ruined settlement and moved west to higher and safer ground, where the kibbutz of some 500 members now prospers. Old Gesher is now a museum. (Photograph taken by OFEK Aerial Photography of Israel in May 1994 and provided by Omri Shalmon of Gesher.)

# 9

# Core Areas and Capitals

We noted earlier that the original States of Europe, Asia, Africa, and Latin America developed around core areas of relatively dense population, generally focused on at least one city, served by a good circulation system, and supported by a firm agricultural base. During the past few hundred years, however, this pattern has been joined by others, and there are examples of core areas that do not fit into any particular pattern. Every adequately functioning State system, however, has a nucleus, a central, essential, enduring heart. It was probably the German geographer Friedrich Ratzel who first tried to define this reality in politicogeographical terms. States, he said, tended to begin as "territorial cells," which would then become larger through the addition of land and people, and eventually evolve into States or even empires. Derwent Whittlesey elaborated on this theme, emphasizing the role of the core as the area in or about which a State originates. Normally, the capital city is situated in this core area. These generalizations are all incontestable; problems arise, however, in applying them to the real world of the twentieth century.

For one thing, there is the problem of quantifying the definition. How dense should the population be? Or the transportation network? What is the relative importance of agriculture, industry, ethnic homogeneity in delimiting the core area? It would seem that it is both impossible and unnecessary to quantify the definition, for in fact the criteria would necessarily differ from one country to another and even in different periods of history. Thus, when we study core areas, we must study among other things the level of political, social, and economic development of each State system individually rather than try to establish universally applicable criteria. As we do, we find different types of core areas.

## Types of Core Areas

Core areas may be classified in several ways. The Canadian geographer Andrew Burghardt identified three types based on historical development. The case in which a small territory grows into a larger State, perhaps over a period of centuries, as described by Ratzel and Whittlesey, he calls a nuclear core. In some States the original core was always the area of greatest importance within an already larger framework. Finally, the contemporary core is the area within the State with the greatest current economic and/or political importance, although it may have superseded one or more earlier cores. Norman J.G. Pounds and Sue Simons Ball analyzed European core areas in some detail and concluded that they can be classified on the basis of function. Thus they recognized States with distinct core areas, such as France, Czechoslovakia, and Russia; those with peripheral core areas, including Yugoslavia and Portugal; and those without distinct core areas, such as Albania and Belgium. Another possible classification is based on spatial considerations. In France and South Africa, for example, the core area is *centrally* located. In Brazil, Argentina, and many other States, the core area is *marginal* in the national territory. It is even noted that in some States that have experienced territorial division or shifting boundaries, the core area may currently be *external*.

Spatial considerations immediately lead to another problematic characteristic of States and core areas: Certain States possess more than one focus. Thus, we might recognize *multicore*, *single-core*, and *no-core* States. Nigeria, for example, has three core areas: one in the southwestern part of the country, one in the southeast, and a third in the north. Ecuador has two core areas: one centered on Guayaquil on the coast and another on Quito in the highland interior. Thailand has a single-core area; Mauritania and Chad might well classify as no-core units.

In addition, some States have quite distinct core areas *and* incipient core areas emerging elsewhere. Are such States multicore States or are they of a different variety? Consider, for example, the United States. Although there might be argument about the exact boundary of the core area of this country, there is little doubt that it lies in the northeastern corner. But core area characteristics are developing to the west of this core, notably on the West Coast. Thus we suggest that the major core area, that of the Northeast, be designated as the *primary* core, while the western (and other emerging areas) be identified as *secondary* core areas. Burghardt suggests that this criterion of scale can be carried further, that there are *continental* and *world* core areas as well. In a sense, the United States-Canadian core in eastern North America is such a continental core area, and in Europe a developing continental core area can be recognized as well.

## Core Areas Around the World

The types of core areas just described can be illustrated by a survey of different parts of the world. In Europe most core areas are of the type described by Ratzel, Whittlesey, and Pounds, perhaps because they drew their conclusions primarily from analyses of those very core areas. Most of them—nuclear, original, or contemporary—include one or more urban centers. The Paris Basin is the core area of France, and Paris is the focus of the Paris Basin. Normally in Europe, these cities are national capitals as well as the largest cities. In the United States and Canada, the core area is located in the eastern portions of the countries. In each case the core area contains roughly half the total population of the country and nearly three quarters of the industrial employment. It is also the cultural and political heart of the State, the area in which the State idea originated, from which the westward movement began, where the capital cities emerged. Both countries are developing subsidiary core areas to the west, but the eastern core area, essentially one core area shared by two States, remains unrivaled.

In Latin America most States similarly have distinct core areas. This results largely from the original settlement pattern. When Europeans first arrived, they found dense clusters of Indian population in highland basins from Mexico to Chile. Many of these clusters, or nodes, of settlement became centers for Spanish administration and economic activity and eventually became the core areas of States that emerged from the revolutions of the period 1810 to 1825. Two other types of settlements were founded by the Spaniards and Portuguese: mining centers and seaports. Each became the focus of a lively and productive region, but many declined or disappeared entirely when, because of exhaustion of minerals, changing trade patterns, or other factors, they lost their *raison d'être*. Some, though, survived to become nuclei of core areas and later of States; Panamá, Tegucigalpa, and Asunción are examples. Chile represents still a different pattern, evident primarily in the Southern Cone of South America. Here the core area is the Central Valley, which contains neither a dense Indian population nor a mining center nor a seaport. It is almost entirely an agriculture-based core.

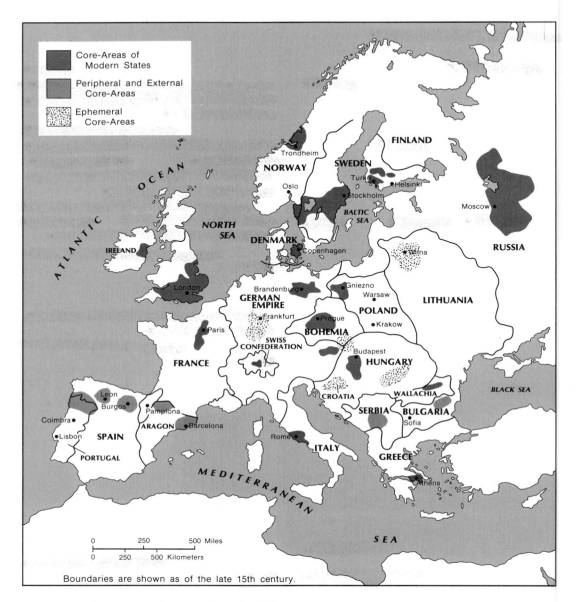

**Core areas of Europe.** (After Pounds and Ball.)

African core areas are still developing. Without a doubt the most significant, in terms of productivity, degree of development, variety of activities, and intensity of urbanization, among other factors, is the core of South Africa. This consists of an east-west axis that extends from the coal-mining town of Springs in the east to the gold-mining town of Klerksdorp in the west, along the Witwatersrand, centered on Johannes-burg. There is also a north–south axis to this core area and a number of secondary core areas elsewhere in the country, notably at Cape Town. Most of the countries south of the Sahara have single cores, and most of the capitals are located in these cores. An important exception is Zambia, whose core area is the Copperbelt along the Zaïre border. Most of the West African coastal States are dominated by their capital cities, which are nearly all important seaports. This is no accident, as the ports were developed and used by Europeans for penetration into the interior and remained their primary links with home. Along the North African coast, the core area of Egypt is one of the best defined in the world. Westward, the core areas that have developed are on or near the coast.

A number of African countries possess more than one core area, and since several of these States are also among the most important and populous of the continent, they deserve special mention. Zaïre has two major core areas, which we have already described. Tanzania has no traditional core, but Dar es Salaam, the capital and largest seaport, constitutes a relatively new one. The area around Arusha, near the Kenyan border, has developed spectacularly in recent years and, together with Moshi and the port of Tanga, constitutes a second core. A third is on the shores of Lake Victoria, cen-

tered on the port of Mwanza. Smaller ones are found in the southeast and southwest. In Ethiopia the area around the capital, Addis Ababa, is clearly the dominant region of the country, but a second core is developing around Diredawa.

A number of States in Africa are so sparsely populated or such recent creations that they have no true cores at all. Most of them are in former French West and Equatorial Africa. In each case a core is developing around the capital, a process quite the reverse of that observed in Europe.

The core of Russia is a rather large area centered on Moscow. It is difficult to say whether this core is simply the heart of a larger but more diffuse core that includes St. Petersburg, Ukraine, and the Volga-Urals region, or whether each of these constitutes a separate secondary core. In any case the areas around Novosibirsk and Irkutsk constitute secondary cores, but they have become so only in the past half-century.

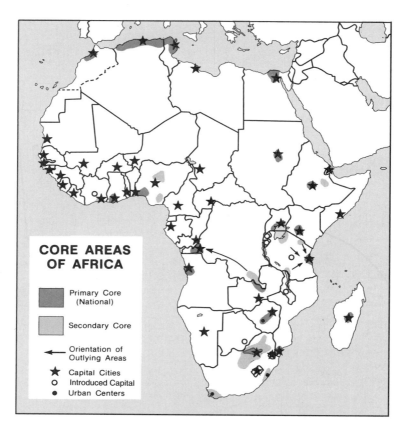

**CORE AREAS OF AFRICA**

Core areas in Asia show a great deal of variety. China's core, like those of Mexico, Egypt, Peru, and Iraq, is one of the world's great culture hearths. Here in the northeast, in the North China Plain, are such cities as Shanghai, Nanjing, Jinan, Tianjin, and Beijing, as well as most of the country's manufacturing and much of its productive agricultural land. India, on the other hand, has several cores. For a long time the major one has extended from Delhi along the Ganges to the Bangladesh border and from there south to Calcutta. Other cores focus on Madras and Bombay, and there may be still others. Japan is unusual in that its core area, stretching from just north of Tokyo along the north shore of the Inland Sea to Kitakyushu on Kyushu Island, occupies a considerable percentage of the national territory. Tokyo-Yokohama dominates the region—and the country—more completely than any other large-country capital region in Asia.

Vietnam, since reunification in 1975, again has two distinct cores; the lower reaches of the Red River in the north and the Mekong in the south, dominated, respectively, by Hanoi and Ho Chi Minh City (formerly Saigon). Should Korea be reunified, it too would have two distinct cores, since Pyongyang is unlikely to lose completely the importance it has attained as the capital and focal point of the Democratic People's Republic of Korea. Most of the other countries of mainland Asia also have a major core area and one or more secondary ones. All the island and archipelago States—and Taiwan—have their core areas on the most productive and densely populated island.

The Australian core area is located in the southeastern part of the country, with the two ports of Melbourne and Sydney forming the focal points of the region. Beyond its margins are several nuclei that are developing into subsidiary areas. Each is the area of a state capital and each is located on tide-

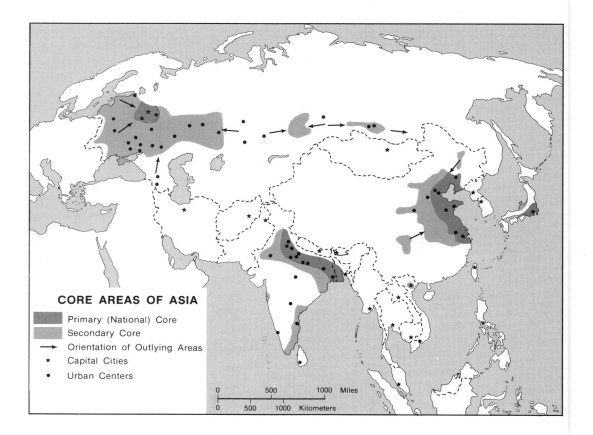

**CORE AREAS OF ASIA**

■ Primary (National) Core
■ Secondary Core
→ Orientation of Outlying Areas
★ Capital Cities
● Urban Centers

0   500   1000  Miles
0   500  1000  Kilometers

water, a reflection of the pattern of population distribution of the country as a whole. New Zealand's less well-defined core area, around Auckland on North Island, is similarly on the periphery of the State.

## Significance of the Core Area

Only a few core areas of the States of the world have been briefly outlined here, but it is clear that core areas must be viewed in two ways: first on a world scale and second on a State scale. We have seen that we cannot apply the same criteria to define the core areas of the United States and Niger. But we can use the same criteria in determining the core areas of the United States, Argentina, China, the United Kingdom, Japan, and Germany. Comparisons among these States, therefore, are possible. When considering the core areas of individual States, our attention is drawn to the contrasts between conditions within the core and those in the remaining parts of the country. Here we must vary our judgments, although the significant factors remain the same (population density, productivity, and communications, for example).

The most important question regarding core areas, and the most difficult one to answer, is whether the core functions satisfactorily as the binding agent for the State. This depends on a number of factors, including the adequacy of the capital city, its position with reference to the population distribution within the State, its ethnic content and the values held, the efficacy of various core-centered organizations in areas outside the core, and many more.

The core area may also be viewed as performing two major functions within a State, both related to scale. First, it may be considered as the nucleus of a State's ecumene. Ecumene derives from the Greek word oik-oumene, which originally meant the inhabited world. Now we define it in several different ways. It can mean that part of the total territory of a State that actually contributes to its economic viability, in which the inhabitants participate in the national economic system, in which the economic

system functions effectively. It can be defined politically also: It would be that part of a State's territory in which the government functions effectively, in which the people participate somehow in the political life of the country (if there is any). Preston James summed up the concept by defining the ecumene as the *effective* national territory, as distinguished from the *total* territory. In many countries even today there are significant areas outside the effective national territory in which the government does not exercise effective control, and/or where the people are oriented toward a neighboring country rather than toward their own capital. Among these countries are Colombia, Brazil, Chad, Myanmar, and Papua New Guinea. We mention this factor again when we discuss the power inventory in Chapter 20.

We have observed that a country's capital is frequently located in its core area. Even if the capital is outside the core, a city within the core probably was the capital at some time in the past. Generally, the country's largest city is located in the core, whether or not it is the capital. We may then view the core area as the hinterland of the capital, largest city, or cluster of large cities. This interpretation assigns to the core the dynamic role of supporting a significant urban area instead of simply the passive role of being the area in which certain things happen or have happened in the past. Good core area studies testing this hypothesis would be most welcome.

## Capitals

Many States, as we have seen, originally grew around urban centers that possessed nodality and attained strength and permanence. Many were market centers for large tributary areas; others were fortifications to which the population retreated every night after farming the surrounding lands. As the influence of those cities expanded, far-flung territories came under the control of the political authority located there. Rivalries between "city-states" occurred, in which the largest and best organized center had the

**A planned capital: Brasilia, Brazil.** It is centrally located, replacing a peripheral capital established in colonial times. The city is located in open, level land and has considerable open space between modernistic buildings, most uncharacteristic of other cities in Latin America. Still not very popular with officials, Brasilia is achieving one objective by attracting some pioneer settlers to the areas around it. (Martin Glassner)

greatest chance of survival and of absorbing its competitors.

Capitals have often evolved as centers of trade and government because of their fortuitous situation. But not all capitals can claim such a past. In simpler as well as more complex societies in history, rulers and their entourages moved about within their realm not by preference, but because supplies could be stored for them at various places visited at regular intervals; it was simpler to take the rulers to their food than food to the rulers. A ruler may have expressed a liking for one of his temporary abodes over others, and ultimately arrangements were made to permit the Chief of State to occupy such a place permanently.

Whatever the origins of the capital, it is destined to embody and exemplify the nature of the core area of the State and be a reflection of its wealth, organization, and power. And indeed, it sometimes is more than that. Some States have poured money into their capital cities to create there an image of the State as it will be in the future, a goal for the people's aspirations, and a

source of national pride. Anyone who has seen the shining capitals of some Latin American countries and the abject poverty of areas within sight of the rooftops of the skyscrapers will attest to this phenomenon. Addis Ababa, capital of Ethiopia, has been described as "a mask, behind which the rest of the country is hidden."[*]

Some countries have invested heavily in the creation of wholly new capitals that are constructed with specific aims. Brasilia was built at tremendous expense in large part to draw Brazilian attention toward the interior. Here the functions of the capital, long performed by Rio de Janeiro, were taken out of the core area and placed elsewhere. Other motives underlay Pakistan's decision to replace Karachi as its capital with Islamabad, in what may still be described as a northern frontier. Islamabad lies close to areas disputed with India and Afghanistan, and thus the government is in a critical position to determine policy.

[*]David Mathew, *Ethiopia*, London; Longmans, Green, 1947, p. 5.

Formerly colonial States have at times decided to replace their European-developed capital with another town having either traditional importance or a more favorable location from the point of view of the government. Hence, Malaŵi decided in 1964 to move its capital from the town of Zomba to more centrally located Lilongwe; Tanzania in 1973 announced it would move its capital from coastal Dar es Salaam to more central Dodoma; Nigeria has developed a new capital district near Abuja; Côte d'Ivoire plans to move its capital from Abidjan to Yamoussoukro (the president's hometown); even Argentina plans to replace Buenos Aires as the country's capital with the small Patagonian town of Viedma, and Sri Lanka began in 1988 to transfer its capital from Colombo to Sri-Jayewardenepura-Kotte.

## Functions and Types

Capital cities perform certain distinct functions. Some are obvious: this is the place for legislative gatherings and the residence of the Chief of State. It is a prime place for the State's reception of external influences, for embassies and international trade organization offices are located there and the turnover of foreign visitors is likely to be greater than anywhere else. In most States the capital city is also the most "cosmopolitan" city.

Capital cities must also act as binding agents, for example, in a federation such as Brazil or Nigeria. In a federal State of great internal diversity, the capital city may be the only place to which all the people can look for guidance. Often in such a State, a territory is separated from the rest of the country and made into "federal territory," so that none of the other entities in the federation can claim bias in the location of the national capital. In the United States the District of Columbia serves this function.

Capital cities must also be a source of power and authority, either to ensure control over outlying and loosely tied districts of the State or to defend the State against undesirable external influences. This function is diminishing, but the capital is most frequently located in the economic heart of the country, from which much of the image of strength of the State emanates.

Functions of capital cities have changed much over time. London provides a good example. After the Roman conquest in the first century A.D., it was the trade center of the Roman Province of Britain, but not the capital, which was located about 20 miles away at Verulamium. Under Norman influence, Londonium became the capital, and after the conquest of Wales and Scotland those territories fell under its jurisdiction also. The city's salutary location became an even greater asset during the years of increasing overseas trade and empire building. Thus, once a local center, then a regional capital, London became the capital of a State and then the headquarters of an empire.

### Types

There are several distinct types of capitals, so classification may be possible. But caution is needed here. Some geographers, for example, have argued that there are "natural" (that is, *evolved*) capitals and "artificial" capitals. This categorization would suggest that certain capitals have emerged and developed as the State system grew increasingly complex, whereas others have been simply the result of arbitrary decision. O.H.K. Spate attacked this whole concept, arguing that any decision leading to the establishment of an "artificial" capital is itself the result of pressures created within and by the system.

It was perfectly "natural," for example, for Belize in 1970 to inaugurate a new, planned capital in almost the exact center of the country, leaving Belize City to continue serving as the colony's chief port. Like Brasilia, Islamabad, and other new capitals, Belmopan was created for a variety of reasons. The trigger for the move was Hurricane Hattie in 1961, which destroyed 80 percent of Belize City, but the city of more than 40,000 people was already in a deplorable condition, with no feasible means of improving it. Establishment of a second major urban center and the need for development of the interior were also important motives. There was nothing arbitrary about the move.

## The Primate City ✳

Another approach might be based on Mark Jefferson's "The Law of the Primate City": "A country's leading city is always disproportionately large and exceptionally expressive of national capacity and feeling."* In many countries the largest city is the capital, but this need not be so. In States that have selected new capital sites, for example, such as Pakistan, the primate city did not lose its primacy when it lost its major political function. Certain federal States, such as Canada, Australia, and Nigeria, similarly have cities larger and culturally more expressive of their nations than their capitals. In some other States, such as Spain and Ecuador, no city, not even the capital, dominates the country sufficiently to be designated as primate. Although the status of national capital automatically confers political power on a city, other cities, by virtue of their size, wealth, and concentration of influential people, can also be politically powerful.

## A Morphological Approach

Perhaps the most productive approach is a morphological one. Here we view capital cities in relation to their position with reference to the State territory and the core area of the State. This results in three classes of capital cities:

1. **Permanent Capitals.** Permanent capitals might also be called historic capitals. They have functioned as the leading economic and cultural center for their State over a period of several centuries. Obvious examples are Rome, London, Paris, and Athens, the leading cities in their respective States for many centuries and through numerous stages of history. The Japanese capital, Tokyo, however, is not in this category. It has been the Japanese headquarters for little

more than a century—the century that spanned the emergence of the modern nation-state of Japan. A whole new cycle began in the life of the Japanese State in the late 1860s, and with it the old capital of Kyoto gave up its primacy. Tokyo symbolized the new Japan, and although the city seems destined to become the permanent Japanese capital, it does not yet rank with Rome and Paris (or Beijing) as a headquarters of continuity.

2. **Introduced Capitals.** Tokyo, in fact, was introduced to become the focal point for Japan when the revolutionary events referred to as the Meiji Restoration occurred. Recent history has seen similar choices made in other countries, but while Tokyo (then called Edo or Eastern City) was already a substantial urban center, other capitals were created, literally, from scratch. They replaced older capitals in order to perform new functions, functions perhaps in addition to those normally expected of the seat of government. Several examples of this type have already been mentioned.

Introduced capitals have also come about by less lofty actions. Intense interstate rivalries among Australia's individual states made it impossible to select one of that country's several large cities as the permanent national capital, and a compromise had to be reached. That compromise was the new capital of Canberra, built in federal territory carved out of the State of New South Wales.

One other factor must be taken into account in connection with introduced capitals. Despite the general absence of planning for a time when the colonial city in Africa would serve as a national capital, the vast majority of former colonial States have retained the former European headquarters as the national capital. Some of these capitals are still rather embryonic, such as Mbabane (Swaziland), Libreville (Gabon), and Kigali (Rwanda), but others have been converted into modern symbols of the nation's aspirations. In a most unusual case, when the British protectorate of

---

*Mark Jefferson, "The Law of the Primate City," *Geographical Review*, 29, 2 (1939), p. 226.

Bechuanaland attained independence, it had—incredibly—no capital, since the territory had for 80 years been administered from a town called Mafeking, located *outside* its boundaries. Thus Gaborone was chosen as a new, internal (and introduced) capital of Botswana, and construction of the necessary facilities began.

3. **Divided Capitals.** In certain States the functions of government are not concentrated in one city, but are divided among two or even more. Such a situation suggests—and often reflects—compromise rather than convenience. In the Netherlands (a kingdom) the parliament sits in The Hague (the legislative capital), but the royal palace is in Amsterdam (the "official" capital). In Bolivia, intense rivalry between the cities of La Paz and Sucre produced the arrangements existing today whereby the two cities share the functions of government. In South Africa, following the war between Boer and Briton, a union was established in which the Boer capital, Pretoria, retained administrative functions, while the British headquarters, Cape Town, became the legislative headquarters. As a further compromise, the judiciary functions in Bloemfontein, capital of one of the old Boer republics that fought in the Boer War.

## *The Search for Conceptual Utility*

Political geographers have described many capital cities, they have made attempts at classification, and they have explained why certain cities became capitals while others did not. But as in the case of the core-area concept, it has not been easy to define exactly *how* the capital functions in the integration of the total State system. Among the first geographers to consider this whole subject was Vaughan Cornish. More than a half-century ago he published a book entitled *The Great Capitals*, and it is still worth a careful reading. He discusses the origin and

evolution of capital cities in many parts of the world, and suggests that

> they fall into three categories of *Natural Storehouses, Crossways,* and *Strongholds.* The first is the original and fundamental character, the second and third dependent upon it, for easy movement and natural barriers are only important if there be inducement to develop or obstruct traffic. It is therefore the world's Storehouses of natural wealth which sooner or later determine the importance of Crossways and Strongholds.[*]

Following an elaboration of this theme, Cornish adds still another category: the *forward capital,* one whose strategic position vis-à-vis frontier regions or potential enemies gives it special importance. Some governments today use capital cities for purposes over and above those having to do with administration. As national foci they enjoy the attention of the nation, an attention that can be manipulated through the manipulation of the capital itself. The frontier-like position of Islamabad, Pakistan's capital, underscores that State's determination to confirm its presence in nearby, still-contested areas.

Spate, too, struggles with the problem of defining the functional qualities of the capital in the State system. He proposes the idea of the *headlink* function, whereby the capital is viewed as the primary link in a chain representing the organizing direction of the State. He then analyzes the cases of Belgrade and Prague, the capitals of Yugoslavia and Czechoslovakia, two States that, at the time his article was written, were still quite young. Belgrade became a "headlink" capital for at least two reasons: its preemptive position (it was already a regional capital) and the role played by Serbia, the "active organizing principle in the creation of the new state." In Czechoslovakia, Prague was the regional center of Bohemia (one of the regions of the new State) and, as Spate points out, the city was "the very flower of Slav civiliza-

*Vaughan Cornish, *The Great Capitals*, London, Methuen, and New York, Doran, 1923, p. vii. Italics added.

tion, enriched by western contacts." Although "the geographical center of Czechoslovakia would have been somewhere near Brno . . . there could be only one political center, and that was eccentric Prague; centrally focal to Bohemia, to Czechoslovakia as a whole it was the link with western cultural and social influences, a 'headlink' capital."

The short-lived Federation of the West Indies presents a fascinating study of both the importance and the difficulty of proper selection of a national capital. The federation broke apart before it could become independent for many reasons. There was little enthusiasm for it in the larger territories, Jamaica and Trinidad and Tobago, and a State idea never really developed. The ten participants disagreed on many vital matters that had to be settled before a State could be created. Among them were representation in the federal parliament, control over economic development, freedom of movement among the islands, and—most emotional of all—the location of the federal capital.

What we note about the West Indies applies to much of the politicogeographical world, although the parochialism and individualism of the several competitors for the capital's functions were made worse by the insular character of the federal State that was being created. The three major aspirants were Jamaica, Trinidad, and Barbados, and for a while there was the possibility of a small-island compromise: St. George's, Grenada. Both Barbados and Trinidad rejected this proposal, however, and the compromise was negated. That negation was confirmed by a special commission that argued that certain minimal service and living conditions were to be met by the town or city chosen as capital. This eliminated all but three of the proposed Federation's cities: Bridgetown, Bardados; Port of Spain, Trinidad; and Kingston, Jamaica. The commission chose Barbados.

All this emanated from a London—and still colonial—commission, of course. But in the West Indies, it was received with anger almost everywhere, except Barbados. When, finally, the Standing Federation Committee

met to consider the Barbados proposal, it rejected it outright. In the subsequent balloting, Jamaica was first eliminated, and in the second ballot Trinidad was selected by 11 votes to 5 over Barbados.

Finally, the parties were able to agree on a site: Chaguaramas, near Port of Spain. It was far from centrally located, but it did have a number of advantages, chief among which may have been that the parties were too exhausted and dispirited after years of haggling and bitterness to object to it. There was just one small problem with the site chosen, however: it happened to be the location of a U.S. naval base!

The federal government had to send a delegation to negotiate with the United States for the land for their new capital. By the time the Americans finally agreed to relinquish a portion of the base area, it was too late; Jamaica was voting to leave the federation, signaling its inevitable and ignominious demise. The Federation of the West Indies, born with much fanfare in 1958, died almost unnoticed in 1963—and it never even had a capital.

The West Indies case is certainly an extreme, at least in modern times, but it could be repeated elsewhere as small countries gradually move toward closer association and perhaps eventual political union. And such problems will surface not only among the newer countries of the world. We need only recall the intense competition in the 1960s and 1970s among Luxembourg, Strasbourg, and Brussels for selection as the headquarters of the merged European Communities (the European Coal and Steel Community, Economic Community, and Atomic Energy Community) to realize that these problems can arise anywhere. The economic, political, cultural, strategic, and—perhaps most important—psychological and emotional importance of a national capital cannot be overstressed.

We have discussed some of the features of capitals of both federal and unitary States; now we will see what these classes of States are like.

# 10

# *Unitary, Federal, and Regional States*

Political geographers are naturally interested in the ways in which States are organized in spatial-political terms. Such organization is the result of lengthy processes of experimentation and modification. State systems are continually being altered, sometimes through deliberation and consultation and at other times because the system cannot withstand certain centrifugal pressures or forces. We are all aware that the federal system in the United States is very different from what it was 100 or even 25 years ago. Napoleon laid certain foundations for modern France, but the France inherited by Charles de Gaulle was also very different. Here we review the major forms of territorial organization of States.

## *The Unitary State*

The word *unitary* derives from the Latin *unitas* (unity), which, in turn, comes from *unus* (one). It thus emphasizes the *oneness* of the State and implies a high degree of internal homogeneity and cohesiveness. The term *federal*, on the other hand, has its origins in the Latin *foederis*, meaning league. It implies alliance, contract, and coexistence of the State's internal, diverse regions and peoples. The unitary State, therefore, theoretically has one strong focus.

All States are divided for political purposes into administrative units. Each of these units has a local government to deal with local matters. In the case of the unitary State, the central authority controls these local governments and determines how much power they will have. The central authority may, under certain circumstances, temporarily take over the functions of a local government. It can impose its decisions on all local governments, regardless of their unpopularity in certain parts of the country. In a national emergency, the central authority can assume greater powers to meet the crisis, while in times of stability it may grant increased responsibilities to local governments.

An ideal unitary State should not be in the "large" or "very large" categories of State territory. The larger a State, the more likely it is to straddle more than one cultural region, and the greater may be the physiographic impediments to effective communications and transportation. Anything that would intensify the centrifugal forces present in any State reduces the efficacy of a single, central authority.

Second, it is compact in shape. A fragmented or prorupt territory may present obstacles to unity and cohesion and require a measure of autonomy for various individual regions. Racial, religious, and linguistic diversity are likely in a fragmented State, and only a federal political framework may function to the satisfaction of the majority of the people.

Third, the unitary State should be relatively densely populated and effectively inhabited. There should be no vast territory with separate concentrations of population interspersed with empty and unproductive areas. This leads to isolation and regionalism.

Finally, the unitary State should have only one core area. Multicore States reflect strong regionalism, an undesirable condition in unitary States. Theoretically, the most suitable location for the single core area of the unitary State is central to its compact territory. This brings all peripheral areas within the shortest distance of the capital city and makes the presence of the core area and capital strongly felt in all parts of the State.

### Evolution and Present Distribution

Few of the unitary States in existence today conform to the model just described. Several examples that show a close approximation to the ideal occur in Western Europe. The old European States fostered a strong central authority, and it is here that the unitary State as it is known today emerged. But this system of politicoterritorial organization has been copied in many other parts of the world—and the copies are often far removed from the original. Nevertheless, in 1995 there were fewer than two dozen federal States and more than 150 unitary States on the world map.

France is often cited as the best example of the unitary State. Though large by European standards (544,000 square kilometers), this country, apart from Corsica, is compact in shape, has a core area with a lengthy history, and at its heart is a capital city of undoubted eminence as well as a large, politically conscious population with much historical momentum and strong traditions. Modern France was forged, in effect, by Napoleon, who swept away the old system of loosely tied divisions (Artois, Picardy, etc.) and replaced them with 90 separate "departments," based on rough equality of size. Each of these departments had the same relationship to the central political authority as did the next, and each sent representatives to Paris. Napoleon also developed an entirely new system of communications, focusing very strongly on Paris, to act as a unifying agent. Until the days of Napoleon,

allegiances in France had been to individual divisions rather than to France, despite the forces of revolution and the overthrow of the monarchy. France today may be considered as nearly completely a nation-state.

Most of the formerly colonial territories of the world have adopted a unitary form of organization, especially those in which the indigenous population has taken control. In those areas where Europeans remain dominant (Australia, Canada, Brazil), the federal arrangement has sometimes been employed. In Latin America, including the Caribbean, all countries except Venezuela, Brazil, Mexico, and Argentina function as unitary States. In Africa and the Middle East, all Arab countries are unitary States, and the majority of the black African States have also chosen this form of organization. Africa south of the Sahara affords some excellent examples of recent experimentation with European concepts of government. The former French territories have become unitary States, but some of the former British dependencies have adopted a federal system. In Asia, only India and Malaysia are not unitary States.

As we noted above, comparatively few unitary States approach the ideal. A number are territorially fragmented, including Japan, the Philippines, Indonesia, and New Zealand, and several are in size categories that contradict the required internal homogeneity and unity, such as China and Sudan.

### Types of Unitary States

Whatever the mode of government in the unitary State—whether a monarchy, a dictatorship, or a democracy—certain adjustments are made in the politicoterritorial system that reflect the internal conditions of the State. As we pointed out earlier, some unitary States have emerged simply because of the almost total assent and satisfaction on the part of the population with regard to the existing authority. On the other hand, certain other unitary States have seen an increase in the power vested in the central authority, whether a person, a group of persons, or the representatives of a minority ethnic group. As we shall see, changes of

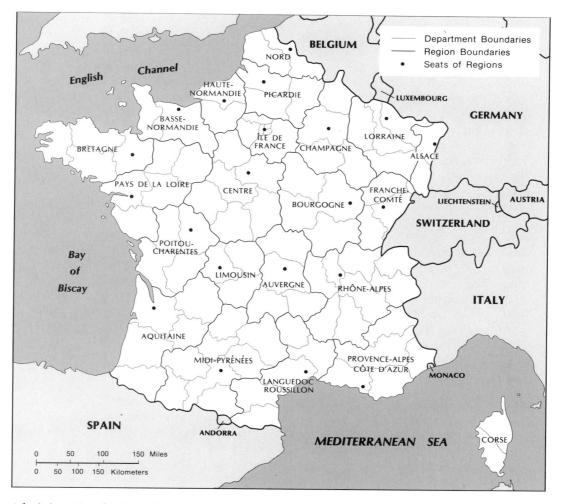

**Administrative districts of France.** Since Napoleon's reforms, France has been divided into *départements*, the first order civil divisions. At present there are 96 *départements*, each headed by a prefect representing the government. All these *départements* have been regrouped for economic and administrative purposes into 22 regions, each headed by a regional prefect who coordinates activities in the *départements* composing his region. Many of the regions' names are the names of pre-revolutionary provinces.

this kind—whether in the direction of greater centralization of power or the reverse—often reflect geographical conditions within the State. In some States, new and old, tribalism remains a force that obstructs the evolution of real nations; in many the governing authority has taken over more and more of the powers of the local chiefs. The following types may be recognized:

**1. Centralized.** This is the "average" unitary State, true to the basic rule of cen-

tralization of governmental authority but without excesses either in the direction of totalitarianism or in the direction of devolution of power. Normally, in such a State stability has been achieved by virtue of the homogeneity of the population and the binding elements of culture and traditions.

States in this category usually possess only one core area. They are generally older States, in the mature stage of the Van Valkenburg classification.

Most examples are found in Europe, such as Denmark, Sweden, and The Netherlands. These three monarchies, by their very retention of this form of government, reflect the satisfaction of the majority of the population with the status quo. In centralized unitary States, the population participates in government through the democratic election of representatives. No single ethnic minority or political party has sole claim to leadership, which doesn't mean that there is no ethnic diversity within centralized unitary States and no regionalism. Despite its small size, the Netherlands has at times been mildly aware of regionalism in Friesland, with Frisians demanding that their language be taught in Dutch schools if Frisians must learn Dutch. The overriding factors of proximity, interdigitation, interdependence, and historical association produce the centripetal forces that bind the State together.

2.  **Highly Centralized.** In this type of unitary State, internal diversity or dissension, ethnic heterogeneity, tribalism, or regionalism that threaten to disrupt the State system are countered by tight and omnipotent control. The leader or leaders often are the representatives of a minority group within the country or of the only political party that may operate within the State. Three major groups of States may be recognized within this category: unitary States within the former communist sphere, one-party States in Africa and other parts of the decolonized world, and dictatorships elsewhere.

    Cuba in Latin America, Equatorial Guinea and Libya in Africa, and Saudi Arabia and Indonesia in Asia are some current highly centralized unitary States, and it is not difficult to list others. Some of these States have recently emerged from several generations of colonial rule, and although some were endowed with a multiparty system, many have become what is described as "one-party democracies."

In other emergent States, the shift toward highly centralized control on the part of a single party, a small group of individuals, or a single individual simply involved the transfer of power from the European colonists to a local, acculturated, and privileged elite. This occurred most commonly in the French-influenced States in Africa and Asia.

## The Federal State

In the unitary State, the central government exercises its power equally over all parts of the State. The federal framework, on the other hand, permits a central government to represent the various entities within the State where they have common interests—such as defense, foreign affairs, and communications—yet allows these various entities to retain their own identities and to have their own laws, policies, and customs in certain fields. Thus, each entity (such as a state, province, or region) has its own capital city, its own governor or premier, and its own internal budget. And each entity is, in turn, represented in the federal capital, so that it has a voice in matters concerning the entire federation.

Federal States, like unitary States, evolve and change over time. In North America, when the 13 original states found themselves with more common interests to unite them than conflicting ideas to divide them, a federation emerged that was, in many ways, a far cry from the United States of today. The number of functions of government has grown tremendously, and the federal government must involve itself more and more in the affairs of the states, affairs that at one time would have been considered domestic to these states. Thus, just as some unitary States show signs of adjustment in the direction of decentralization, many federations are shifting toward greater centralization of authority. Some federal States are so highly centralized that they function, in effect, as unitary States.

Theoretically, the federal framework is especially suitable for States in the large and very large categories. Poor communications and ineffective occupation of large areas within the State still affect for example, Brazil and Zaïre. These impediments to contact and control might disrupt a unitary State, whereas a federal framework can withstand such centrifugal forces. In terms of shape it is obvious that fragmented States and States with pronounced and important proruptions may be best served by a federal system. An elongated State possessing more than one core area also might turn to a federal arrangement.

Federal States can adjust to the presence of more than one core area (or a number of subsidiary cores) more easily than unitary States. Several contemporary federal States are multicore States, and in the case of Canada, Nigeria, and Australia, it is probable that only a federal constitution could have bound the diverse regions together. Thus, the federal State often has a number of individual population agglomerations separated by large areas that are sparsely populated and relatively unproductive. Canada and Australia, two of the largest federations in terms of area and two of the smallest in terms of population, illustrate this principle.

Ideally, the government of the federal State functions in a capital city located in an area of federal territory set off within the State for the specific purpose of administration. This forestalls any friction over the choice of an existing major city as the capital and prevents regional favoritism from occurring, as it has in some federal States. In Nigeria, an existing city (Lagos) was separated from the Western Region and made the capital for two reasons: first, it was a long-term colonial capital and housed most government records and existing facilities, and second, it happens to be the leading port of the country, thus guaranteeing the landlocked north an exit through a federal rather than a regional port. Even the new capital, Abuja, is located in a federal district.

The federal arrangement is also a political solution for those territories occupied by peoples of widely different ethnic origins, languages, religions, or cultures. This is especially true in cases where these differences have regional expression, where various peoples see individual parts of their

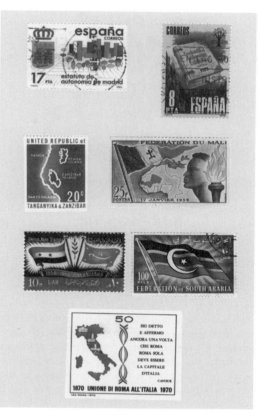

**Territorial organization of States shown on stamps.** Two Spanish stamps marking the grant of autonomy to the Madrid and Basque regions; stamps honoring the federation of Tanganyika and Zanzibar, which still survives, and the federation of Senegal and French Soudan, which broke up in August 1960, only two months after independence; the United Arab Republic (Syria and Egypt) welcomes Yemen into the federation in 1958, and the short-lived federation based in Aden; and Italy celebrates the 1870 union of Rome with Italy, completing the unification of the country as a unitary state. (Martin Glassner)

country as a homeland. The promise of an autonomous Macedonian Republic, for example, to be part of a Balkan Federation, rallied Macedonians around Tito's war effort. Although the Macedonian Republic (which was to include parts of Greece and Bulgaria as well as Yugoslavia) did not materialize, as we noted in Chapter 4, Macedonians did obtain a small national territory within the Yugoslav federal State. For them it was the only

**MACEDONIA**

**UNITED MACEDONIANS**

**МАКЕДОНИЈА
Е МОЕТО СРЦЕ**

**"Macedonia of My Heart".** This slogan and a silhouette map of a greater Macedonia comprise a label or sticker prepared and distributed in The Former Yugoslav Republic of Macedonia. It is a graphic representation of the dreams of some chauvinist Macedonians and the fears of some Greeks. At present there is no credible evidence that this represents the view of any but a tiny minority of Macedonians, but the sentiment could grow and spread to neighboring countries. If it does become a case of a divided people seeking a State of their own created out of territory of three or four additional States, it would resemble the case of the Kurds (discussed in Chapter 4) and reinforce the characterization of the Balkans as the original shatterbelt (discussed in Chapter 21) and as "The tinderbox of Europe." (Courtesy of Leon Yacher).

alternative to absorption, and it constituted a closer approximation of their original goal than they had enjoyed in centuries. In unitary Bulgaria and Greece, such status would have been impossible. Secession from Yugoslavia in the early 1990s was relatively easy, but detaching Macedonian-occupied territories from two or three neighboring unitary States would be much more difficult.

By their flexibility, federal frameworks have been able to accommodate expanding territories. Provisions are often made for areas not yet incorporated. The United States, India, Mexico, and Brazil have promoted several of their territories into the ranks of states, equal to the others already in these federations. The reunification of Germany was facilitated by a clause of the West German federal constitution that provided for the admission into the federation of any of the former states then in centralized East Germany.

Another advantage offered by the federal arrangement is its encouragement of individual and local enterprise. Economic development in the United States took place as fast as it did largely for this reason; the westward push of Brazil, as exemplified by the relocation of the capital, is an effort to stimulate a similar chain of events in its hinterlands. Of course, this accommodation of individualism has not prevented the emergence of conflicts of interest. Some, like the Civil War in the United States, helped bring about revisions in the federal framework; others resulted in the collapse of the State itself.

### Types of Federal States

"A federation," wrote Kenneth Robinson in an article about Australian federalism, "is the most geographically expressive of all political systems. It is based on the existence of regional differences, and recognizes the claims of the component areas to perpetuate their individual characters. . . . Federation does not create unity out of diversity; rather, it enables the two to coexist."* In the creation of the State system, internal variety and diversity can be treated in two ways. One way, as we noted earlier in this chapter, is to render the State the great equalizer, the eliminator of internal differences, even to the point of making certain expressions of regional individuality illegal. The other way permits the perpetuation of internal contrasts and provides the possibility of coexistence within one State of peoples with varied backgrounds and interests. Thus, the advantages of participation in the State must, for all these peoples, outweigh the liabilities that inevitably are involved. To put it another way, the centripetal forces must outweigh the centrifugal forces present in any federal State.

*Kenneth Robinson, "Sixty Years of Federation in Australia," *Geographical Review, 51,* 1 (January 1961), p. 2.

There are several different *types* of federal States. In some present-day federal States, such as Australia and Argentina, internal variety and diversity seem so insignificant (compared to that existing in other countries) that a unitary arrangement might be just as effective. Other federal States, including Canada and Nigeria, incorporate such diversity that a certain amount of give and take was, and remains, essential for the well-being of the State. In still other federal States, the geographical obstacles to any unitary system rendered a federal arrangement imperative. It is important to realize that these types are not mutually exclusive, for the elements of two or more kinds of federalism may be recognized in some federal States today. Furthermore, a federal State established for one set of reasons may continue to exist today for an entirely different set.

Another category of federations that must be considered is those that have *failed*. The world is littered with the wreckage of federations that have been proposed but never consummated, that have been created only to fragment relatively quickly, and that have survived but only after conversion into unitary States. Most resulted from the breakup of empires, and federation was seen as a way of managing, if not solving, many of the different problems engendered by decolonization.

In the early 1950s there was much talk of a united, independent Maghreb, to be composed of Morocco, Tangier, Algeria, Tunisia, and perhaps Libya. Organizations were formed to support the idea and at least the intellectuals in each country favored it. Despite the many real differences among them, all except Libya had a great deal in common. France granted independence to Morocco and Tunisia in 1956, and Tangier and the Spanish protectorates soon joined Morocco. But France fought a long and bloody war to keep Algeria French. By the time Algeria was granted independence in 1962, the other two potential partners had gone their separate ways, and the chance for federation (if indeed there ever was one) was lost. A similar situation developed in East Africa. The British territories of Kenya, Uganda, and Tanganyika had been cooper-

ating in increasingly closer association since the 1920s, and the movement toward formal federation after independence (possibly to include Zanzibar) was growing stronger. But there was no great enthusiasm for federation among the ordinary people and little among the black African leaders. Moreover, while Tanganyika received independence in 1961 and Uganda in 1962, Kenya's was delayed, at least in part by the Mau Mau rebellion, and it did not become a State until December 1963. By that time it was certainly too late. Interest in close cooperation among them, however, revived in the early 1990s.

Early in the nineteenth century, when most of the Spanish colonies in the Americas won their independence, two federations emerged. Within a short time, however, both broke up. Central America split into Guatemala, Honduras, El Salvador, Nicaragua, and Costa Rica; Gran Colombia splintered into Venezuela, Colombia (including Panama), and Ecuador.

A century and a half later, another decolonization period spawned a number of federations. The British promoted the federation as, it was thought, an honorable and practical device for cutting loose a large number of small dependent territories while giving them a chance to survive in the highly competitive world of sovereign States. They tried the device in the West Indies, Central Africa, South Arabia, and Malaysia. The first three were failures and the last only a limited success. The West Indies and Central African federations broke up before independence, and each unit went its own way, most toward independence. A civil war broke out in Aden and South Arabia before the British could escape, and the winning faction almost immediately formed a unitary State centered in Aden. Malaysia remains a federation (with a federation as its major constituent), but without Singapore, which was expelled in 1965 after only eight years of togetherness.

But federations failed elsewhere besides in the British Empire. French Soudan and Senegal joined in January 1959 to form the federation of Mali, became independent in June 1960, and split just two months later

**Twilight of the Federation of the West Indies.**
Fourteen months after this photograph was taken at the Queen's Birthday Parade at Up Park Camp in Kingston, Jamaica, the colony was an independent State and the Federation was dead. The flags, from top to bottom, are the Union Jack—the official flag of Jamaica as well as of the UK at the time; the Royal Standard, flown only when the sovereign is present, on this occasion in the person of the Governor-General; and the flag of the Federation, rarely seen anywhere in the Federation except on public occasions. (Martin Glassner)

into Senegal and Mali. In a move unconnected with decolonization, Egypt and Syria joined in 1958 to form the United Arab Republic, but after an unhappy marriage were divorced in 1961. A bewildering number of other federations involving various combinations of Libya, Yemen, Algeria, Egypt, Sudan, Iraq, Syria, and Jordan have been proposed, announced, or attempted—but none has ever become effective.

Then there are the countries that have been federations in recent years but have become unitary States. Eritrea was federated

with Ethiopia by the United Nations in 1952, but just a decade later lost its autonomy and, after a long and dreadful civil war, became independent in 1993. French Cameroon became independent in 1960 and was joined in 1961 by the southern portion of British Cameroon. The Federal Republic of Cameroon, after a referendum in 1972, became the United Republic of Cameroon. Uganda at independence was a federation of four kingdoms and one territory, plus another territory and ten administrative districts (with the King of Buganda as president of the country). Only five years later, however, the kingdoms were abolished, and the country became a unitary State. In 1993 the kingdoms were reconstituted by a democratic government, but their jurisdiction was limited to cultural matters. Uganda remains, for the present at least, a unitary State. Indonesia's federal period was the shortest of all, lasting less than eight months after independence and ending in August 1950.

How can we explain this dreary succession of failures? Isn't the federation "the most geographically expressive of all political systems," as Robinson, Dikshit, and others have proclaimed? If so, why are there only about a dozen true federations in the whole world, and why are even some of these dubiously federal? There are many answers, of course, at least as many answers as failures. Because a federation is a compromise form of territorial organization, it requires very special conditions for initiation and a great deal of hard work and dedication for survival. A common factor in all the failures in the first two categories—those that were never born and those that broke up—was the lack of a *raison d'être*, or a State idea. Typically, the federation was justified in terms of certain practical advantages that could be quickly realized. In such a case, if the advantages do materialize, the federation may no longer be needed; if they do not, the federation is exposed as bankrupt. Without a real will for a federation on the part of the people, failure is inevitable. Since both a balance of forces within a federation and its practical, day-to-day management are so difficult, the tendency toward centralization is strong indeed,

**The headquarters of the Permanent Secretariat of the former Senegambian Confederation.** Since independence in 1965, The Gambia has cooperated closely with Senegal in a number of fields. In 1982 the relationship was strengthened and formalized by a confederation of the two countries. There was a common parliament, council of ministers and other organs; a customs union, confederation army and security service. The Secretariat coordinated these activities and performed other functions. Neither country gave up any sovereignty, however, and no effort was made for closer union. Instead, cracks appeared in the structure very soon and widened as time went on. The confederation was dissolved peacefully in 1990 and replaced by a treaty of friendship. This is one more example of the difficulties of achieving even limited federation, and the faded and forlorn sign in front of the former secondary confederation headquarters in Banjul, The Gambia symbolizes this. (Martin Glassner)

as evidenced by the third category of failure. Federation in the classic sense may not survive even in those countries where it is best developed.

## Regional States

In territorial organization, the same force we have observed in the size of States may well be operating: a tendency toward equilibrium. As federal States become more centralized and unitary States grant more au-

tonomy to regions within them, we face increasing difficulty in applying the old labels to new situations. A solution to the dilemma, and a possible label for States approaching a midway area between federalism and unitarianism, is suggested by the title of a book by the Spanish scholar Juan Ferrando. This is the term *regional State.*\*

Into this category we may place those unitary States in which considerable autonomy has been granted to regions within them, generally regions of ethnic distinctiveness or remoteness from the core area. Most of the really good examples at the moment are in Europe. Perhaps that is not so strange, however, for both unitary and federal States, after all, began there.

The United Kingdom is perhaps a good example of a regional State. Though now experiencing a period of direct rule, Northern Ireland long enjoyed a large measure of home rule. Scotland has had some autonomy and is gradually receiving more, and even Wales may receive some before long as the "devolution" of power from the center to the peripheries continues.† In Italy, Sardinia, Sicily, Trentino-Alto Adige, Valle d'Aosta, and Friuli-Venezia Giulia have for some time been functioning under special statutes, while another 15 regions elected their first regional councils (parliaments) in 1970. In Finland, the Aaland Islands constitute an autonomous county, demilitarized and maintaining their own traditions. The Faeroe Islands and Kalaallit Nunaat (Greenland) are largely—but not completely—independent of Denmark. Spain granted regional autonomy to Cataluña in 1977 and to three of the four Basque provinces in 1978, completing the process in 1983 by granting autonomy to Extremadura, Castilla-León, the Balearic Is-

---

\*Juan Ferrando Badia, *El Estado Unitario, El Federal y El Estado Regional*, Madrid, 1978.

†Despite the rejection by the people of Scotland and Wales of proposals for their own assemblies that would give them a measure of home rule, in referenda on 1 March 1979, it still seems likely that the United Kingdom, while remaining a unitary State technically, will continue to devolve some responsibilities to these regions, especially to Scotland.

**Administrative districts of Italy.** Like France, Italy has conceded to regional feelings to some extent by grouping its 94 provinces into 20 regions with some autonomy. Those labeled with capital letters were the original five regions and still exercise somewhat greater authority than the other 15, which have been functioning only since 1970. Considering the long and difficult struggle to unite Italy, its existence as a strictly unitary State for less than a century is instructive.

lands, and Madrid.* The Azores were granted autonomy by Portugal in 1975. In 1970, Belgium adopted a form of "federal-

*In 1990 Spain's administrative structure was further reorganized. The number of first-order civil divisions was reduced from 55 to 22, and the provinces became second-order divisions. Seventeen autonomous communities were created, of which ten consist of two or more provinces and seven are uniprovincial. Such reorganization would be unthinkable in a truly federal system.

ization without federalism" by creating four linguistic regions (Dutch [or Flemish], French, German, and Brussels [French and Flemish]) that have autonomy in many cultural matters.

A special case that might fit into this category is Cyprus, in which a *de facto* Turkish federated State was established by the Turkish Cypriots in the northern part of the island, which was occupied by troops from Turkey in 1974. Other candidates for consideration as regional States are Myanmar (formerly Burma), Iraq, and China, which are all nominally unitary States with regions exercising uncertain degrees of self-government. Perhaps the Netherlands Antilles' status would allow the Netherlands to be included. France may move in this direction by giving some autonomy to Brittany, Corsica, and the overseas departments.

The trend toward devolution of powers and/or decentralization of activity from the center to the constituent civil divisions is spreading around the world, reaching even some of the newer States of the South Pacific. The best examples at present are Papua New Guinea, Vanuatu, and Solomon Islands.

Other regional States are those with federal constitutions in which federalism either was never very real, has gradually given way to centralization, or alternates with unitarianism in law or practice. Here we might consider the former Union of Soviet Socialist Republics, whose constitution provided that each constituent republic could conduct its own foreign relations and secede at will. In practice, however, no republic conducted its own foreign affairs, not even the Ukraine and Byelorussia, which were charter members of the United Nations. And no republic would have dared to try to secede, not even one in which local nationalism was very strong. The state was, in fact, highly centralized and kept under control by the national army, the Communist Party, and the practice of settling Great Russians in all nationality areas, frequently giving them the best jobs. This rock-solid edifice, however, was shattered during 1989–91 under the impact of

economic collapse, rising nationalism among many ethnic groups, the defeat in Afghanistan, and President Mikhail Gorbachev's policies of *glasnost* and *perestroika*. Now most of the successor States face the same problem of determining the degree of centralization appropriate to their situations. Yugoslavia similarly splintered at the same time and for essentially the same reasons.

Mexico is essentially a "one-party democracy" in which the party and the central government manipulate state elections and in which from time to time the president will oust an elected state governor and appoint his own person to bring the deviant state back into line. Tanzania has been whittling away at the autonomy of Zanzibar almost since federation, and the trend is apparently toward total incorporation. In Nigeria, Brazil, Argentina, and India, there are different kinds of limitations on federalism in practice that raise some questions about how they should actually be classified. We could provide other examples, but the point is clear.

This notion of a regional State is quite new and still untested. Even the countries we have listed and the categories in which we have placed them are offered tentatively. There is both ample scope and real need for investigations into the utility of home rule or autonomy as a device for governing areas inhabited by minority ethnic groups, those separated from the core area of a country, or those with greatly different economic, physiographic, or political conditions. Even the suggestion offered here of a tendency toward equilibrium is worth testing. Perhaps it offers hope for resolution of some problems of government in this restless world.

# 11

# *Anomalous Political Units*

Although the European State system has spread around the world during the past several hundred years, resulting in the creation of nearly 200 States, by no means has all the earth's land area been organized in such a neat and seemingly definitive manner. We have already discussed leased areas as an example of territory legally under the sovereignty of one State but functioning as part of another, and we later discuss territories possessing little or no sovereignty, but rather functioning merely as colonies of a State. Here we are concerned with types of territorial organization that fall between the two extremes—or outside them altogether. Some of these anomalies are remnants of the period before the organization of nation-states; others are simply ad hoc arrangements resulting from war or decolonization, many of which seem to be taking on the appearance of permanence. The following review, by no means a complete inventory of such anomalies, attempts to indicate some alternatives to either complete sovereignty or none whatsoever.

## *Military Occupation*

We have already discussed how territory is acquired through conquest and annexation and through assimilation involving coercion short of overt violence. States also acquire control over territory, though not sovereignty, through conquest without annexa-

tion. We need not go back further than World War II to find many examples. Even since that disastrous war, portions of sovereign States and disputed territories have been occupied by invading forces during smaller wars. Examples of such areas that have taken on a degree of semipermanence are northern Cyprus, occupied by Turkey since 1974, and "the West Bank" of the Jordan River.

After the Turkish invasion of Cyprus in the summer of 1974, Turkish troops occupied the northern 37 percent of the island, including some areas that were inhabited almost exclusively by Greek Cypriots. During and especially after the fighting, Greek Cypriots living in the north were uprooted and fled to the south, held by the Cypriot national guard, while Turkish Cypriots fled northward across the cease-fire line. In 1975 the Turkish Cypriots proclaimed the "Turkish Federated State of Cyprus", backed by the Turkish army, and in 1983 adopted the name "Turkish Republic of Northern Cyprus". This territory is still occupied by Turkish troops and is recognized as a State only by Turkey. Even its passports are accepted only by the U.S. and the U.K.; elsewhere its citizens must use Turkish passports. Sporadic efforts were continuing in the mid-1990s to unify the island again, with the United Nations providing its good offices and conciliation.

Cyprus harbors another anomaly, another type of military occupation. When the coun-

**Northern Cyprus.** Here the customs/immigration post of the "Turkish Republic of Northern Cyprus" is seen from the neutral zone manned by United Nations peacekeeping forces that bisects the capital city of Cyprus and the entire island. The northern sector of the city is called Lefkosa by the Turkish Cypriots, who have made it their capital. The southern portion is still called Nicosia. (Photograph by Leon Yacher).

try became independent of the United Kingdom in 1960, one of the terms of the independence agreement was British retention of two large military complexes on the south coast. Known as Sovereign Base Areas, these are completely separate British enclaves within Cyprus. Within these enclaves, however, are a number of enclaves of Greek Cypriots, another example of this type of territory discussed in Chapter 7.

In June 1967, during the Six-Day War, Israeli forces dislodged the Jordanian army from that portion of Palestine (known as "The West Bank") that it had been occupying since 1948. Jordan renounced its claim to the territory in 1983, and Israel continues to rule it under military administration. Its legal status is still unclear. In May 1994, however, Israel began what is likely to be a lengthy and difficult process of withdrawing from the West Bank and the Gaza Strip, granting the Palestine Liberation Organization (PLO) limited jurisdiction, or autonomy, in the Gaza Strip and the Jericho area of the West Bank. These autonomous areas, no longer under Israeli military occupation, are likely to be forerunners of an eventual State of Palestine, but meanwhile, they are curious but pragmatic anomalies.

## Territories of Intermediate Status

On occasion, partition of a dependent territory has been used as a method of dealing with otherwise intractable problems resulting from cultural heterogeneity in the territory. Examples are Ireland in 1922, Palestine in the same year and again in 1948, and India in 1947. (Partition has also been suggested for Cyprus and Lebanon.) In each case, the status of the separated pieces is quite clear; each is a sovereign State or is part of one (except for the West Bank and Gaza). In other cases, however, the status is not so clear. Since 1945, Korea, Germany, Vietnam, China, and Kashmir have been divided into two or more segments by the fortunes of war. Although the circumstances in each case are different, only Vietnam and Germany have been reunited, and the rest may remain divided for some time.

In this intermediate status also have been such *quasi-states* as Andorra, Vatican City, Taiwan, and the South African Bantustans of Transkei, Bophuthatswana, Venda, and Ciskei. Andorra was governed by foreign co-princes, the president of France and the bishop of Urgel in Spain, and they con-

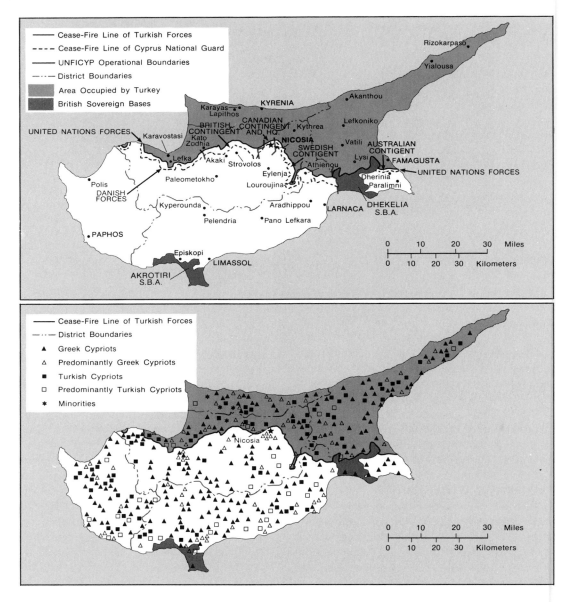

**Cyprus.** The top map shows the present politicogeographical situation. UN peacekeeping forces have been separating contending Greek and Turkish Cypriot communities pending a political settlement to the communal dispute. The bottom map is a simplified depiction of the distribution of the Cypriot population in 1960 by ethnic group. It is evident that no partition plan based on ethnic criteria as of 1960 was feasible; since 1974, however, there has been an ad hoc exchange of populations, and now the area occupied by Turkey is predominately Turkish Cyrpiot. All the large cities had mixed but predominately Greek Cypriot populations. The "minorities" are primarily Armenians and Maronites.

ducted its foreign affairs, defense, and judicial system. But it was not a colony, has had its own international personality since 1278, and was a member of UNESCO. Its admission to the United Nations in July 1993 has probably confirmed its new status as a State.

The State of Vatican City is the tiny remnant of the former Papal States, annexed by Italy in 1870. After that act, the number of States having diplomatic relations with the Holy See fell to four, but now there are over 150. It is still questionable, however, whether a

**Entrance to the autonomous area of Jericho.** This ceremonial arch marks the southern limit of the autonomous area on the main north-south highway through the Jordan Valley. Just to the south of here is a checkpoint manned jointly by Israeli and Palestinian police. Note the Palestinian flags at the top of the arch. (Martin Glassner)

territory of 44 hectares (109 acres), whose residents are all citizens of other States, can properly be called a "State" under our definition. The South African government for years pursued a policy of "separate development" of the "races" in the country, as it defined these terms. The logical outcome of this policy was the subtraction of portions of the national territory and their conversion into independent States. The government claimed to have done this, but since the Bantustans were not created by the peoples of the territories and had very greatly circumscribed "sovereignty," few in South Africa and fewer still outside the country took them seriously. They were reincorporated into the new South Africa in 1994.

Taiwan (with a number of smaller islands) is the only province of China not taken over by the Communists during the civil war that ended in 1949. The Nationalist government fled to Taiwan and ever since has claimed that it is the legitimate government of all of China. So does the government in Beijing. Technically, then, Taiwan (officially still called the Republic of China) is merely a province of China, yet it functions in every way as a State. It simply lacks recognition as a State except by a few small countries and, of course, it does not claim to be a State.

Another group of countries in this intermediate status are a number of colonies, protectorates, and other units of present or former colonial empires that have been or have recently become autonomous or semi-independent. They include such territories as Puerto Rico, Niue, the Cook Islands, the Faeroe Islands, Kalaallit Nunaat (Greenland), the Northern Marianas, and a number

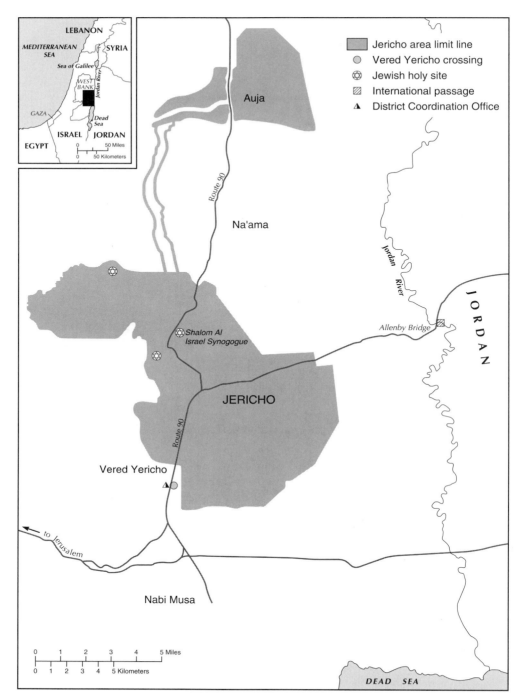

**The autonomous area of Jericho.** Along with the Gaza Strip, this is the first sector of Israeli-occupied territory to be granted some degree of self-government under the Palestinian Authority. Autonomy is to be gradually extended to other areas of the West Bank and to be broadened to include jurisdiction over more and more matters. The long, narrow strips connect outlying Arab settlements and refugee camps with the town of Jericho. The Agreement on the Gaza Strip and the Jericho Area of 4 May 1994 between Israel and the Palestine Liberation Organization (PLO), the source of this slightly modified map, spells out in a number of protocols, six maps and other attachments the details of the arrangement.

**The border of Bophuthatswana with Botswana.** The Tswana people of Botswana (formerly Bechuanaland) have been separated politically from their brethren in South Africa and these were further separated from the rest of South Africa and one another by South Africa's creation of a badly fragmented "Bantustan," recognized by no one except South Africa and therefore not a State. Nevertheless, it maintained border posts as a symbol of its "independence" until it was reincorporated into South Africa in 1994. (Martin Glassner)

of others. In each case, while a metropolitan country handles its foreign relations, currency, and other matters, it enjoys such a substantial degree of self-government that it cannot be called a colony without some qualification. Some, in fact, are teetering on the brink of independence, but others, particularly the smaller ones, may opt to continue indefinitely their semi-independent status.

Undoubtedly, the most curious anomaly is the Sovereign Military Order of Malta, commonly known as the Knights of Malta and also known as the Knights of St. John Hospitaler. The order was founded in 1099 and was a power in the Mediterranean for 700 years, but now is sovereign over only an estate of 1.2 hectares (three acres) in Rome. Nevertheless, it operates hospitals and leper asylums around the world; issues its own coins, stamps, and passports; and has diplomatic relations with some 60 countries, including Spain, Italy, Brazil, Austria, Togo, and Mauritania. In August 1994 it became an observer at the United Nations, has been invited to several UN conferences and appears

to be moving toward full membership in the United Nations. Observer status, was, however, opposed by the U.S. and the U.K. on the grounds that it is neither a State nor an intergovernmental organization but only an NGO (non-governmental organization). It may be some time before its status is clarified and generally accepted.

### Territories of Uncertain Status

The decolonization process has not been easy. Not all former colonies, protectorates, mandates, and trusteeships have become independent States or legally parts of independent States. One example is Western Sahara, formerly Spanish Sahara. Spain wanted to withdraw from the territory in 1975, but it was claimed by Morocco, Mauritania, and Algeria, and no one knew what the people of the territory, mostly nomads, wanted. As Spain withdrew, Morocco and Mauritania divided it between them, and later Mauritania withdrew. This action has been protested by Algeria, which has been supporting the local fighters for independence, the POLISARIO

**Frente POLISARIO supporters in Western Sahara.** These people in Ausred, which is located in the southern part of the former Spanish Sahara, are demonstrating their support for the Algerian-backed rebels against the Moroccan occupation of their country. (Courtesy of the United Nations. Photo by Y. Nagata)

Front. In early 1995 the United Nations was still working with Morocco to arrange a plebiscite on the territory's future.

Slovenia and Croatia declared independence from Yugoslavia on 25 June 1991 and were followed in this action by Macedonia and Bosnia-Herzogovina. For some months their status was uncertain, but now all are recognized as States and belong to the United Nations.*

Another region of uncertain status is Antarctica. Part of the continent has never been formally claimed by any State, and the United States and the former Soviet Union (which have been most active in Antarctic research and exploration) have never made any territorial claims. Most of the continent and its outlying islands have, though, been claimed by several countries. In the area of the Antarctic Peninsula (also called Graham Land and the Palmer Peninsula), the claims of Chile, Argentina, and the United Kingdom overlap. All claims, however, are frozen by the 1959 Antarctic Treaty. (No pun intended; there's just no better way of describing the

situation.) Some countries assert that their claims are part of the national territory, others that they are dependent territories. But none is recognized by the United States or the former Soviet Union and each claim, as well as the unclaimed area, must be considered as legally *res nullius*, or belonging to no one. Antarctica is discussed in detail in Chapter 33.

## Insurgent States and Nonstate Nations

The current turmoil of decolonization and intensifying nationalism among peoples in multinational or plural States has spawned a large number of revolutions, rebellions, guerrilla wars, terrorist activities, and other acts of violence by a wide assortment of "national liberation movements" and many political groups that have no particular national distinctiveness. These activities, in turn, have spawned a large number of studies. Some of the basic work has been done by the geographer Robert McColl and the political scientist Judy Bertelsen, on insurgent states and nonstate nations, respectively. Since neither term is widely accepted, we may take some liberties with them (perhaps doing them an injustice) and consider them as separate stages of the same process. We might consider a *nonstate nation* as being a people living as a minority in one or more States who want a State of their own in territory

---

*Because of Greece's objection to Macedonia's use of that name, the country is officially known in the United Nations as The Former Yugoslav Republic of Macedonia and listed under "T" in alphabetical lists. Serbia and Montenegro, the remaining Yugoslav republics, together claim to be the continuation of Yugoslavia. This claim is recognized by the United Nations but not by the United States.

currently included in one or more States. It may engage in a variety of activities to win its point, ranging from polite petitions to intense propaganda and economic sanctions to widespread terrorism. It does not, however, effectively control any significant territory within the area it claims as a national homeland. It may, in exceptional cases, perform certain activities normally performed by States, but only with the permission of, and under the rules laid down by, the government. An *insurgent state* might develop within a country if the rebellious group is actually able to win and retain control of a territorial base of operations. Within this territory, the insurgents operate a state within a State in which only they and not the regularly constituted authorities perform all governmental functions.

According to these criteria, we might consider among the nonstate nations of the postwar era such groups as the republicans of Ulster, the Greek Cypriots demanding *enosis*, the Puerto Rican nationalists in New York, the South Moluccans in The Netherlands demanding independence from Indonesia, the Palestine Liberation Organization demanding an independent Palestinian Arab State, and other nationalist groups that use both peaceful and violent tactics but without a territorial base of their own.

Other groups in this same period, however, have had territorial bases in which they have established something resembling a State, what McColl has labeled an "insurgent state." This has occurred in Greece, China, South Vietnam, the Philippines, Malaya, Indonesia, Cuba, Portuguese Guinea, Angola, Mozambique, Algeria, Nigeria, Eritrea, Thailand, Myanmar (Burma), Iraq, and elsewhere. In most cases a nationalist group was fighting for independence from colonial rulers; in some cases national minorities were fighting for autonomy within or secession from an independent State; and in others the movement was political, generally opposing a corrupt or tyrannical government or favoring establishment of a communist government. Regardless of motive, however, the patterns of operation have been similar. McColl and others have tried to apply the same analysis to "urban guerillas" and even

the university campus, but with less justification and less success. The basic theses of the nonstate nation and insurgent state concepts will undoubtedly be tested repeatedly in the coming years, for we are not yet free of our "times of troubles."

## *Binational Territories*

A number of small areas around the world are administered by two States, neither of which has exclusive sovereignty over them. Although they fall into two different classes of entities, we may group them together as binational territories. They are *condominiums* and *neutral zones*.

Undoubtedly, the largest and most important condominium of modern times was the Anglo-Egyptian Sudan. This vast territory, the largest in Africa, was governed jointly by Britain and Egypt from 1899 through 1955. Britain was definitely the senior partner, and the Sudan was generally accepted as being part of the British Empire, but in various ways Egypt did participate in the country's operation until it abrogated the condominium agreements in 1951. Sudan became independent on 1 January 1956.

The last remaining condominium of any importance was New Hebrides, a group of 12 main islands and nearly 70 smaller ones in the South Pacific, administered jointly by France and Britain from 1906. Its government was known as the Joint Administration, of which the joint and equal heads were British and French officials. Some services were provided by the British and French National Services and some by the Joint Service. This arrangement ended with the granting of independence to the territory in 1980, when it became Vanuatu. Another condominium was also in the South Pacific, uninhabited Canton and Enderbury Islands, claimed by both the United States and the United Kingdom and administered jointly by them until the United States vacated its claims. Andorra, as mentioned earlier, was technically governed by Spanish and French coprinces, but in 1993, after a referendum on the question, the country became a republic and then was admitted to the United Nations on 28 July of that year. Finally, em-

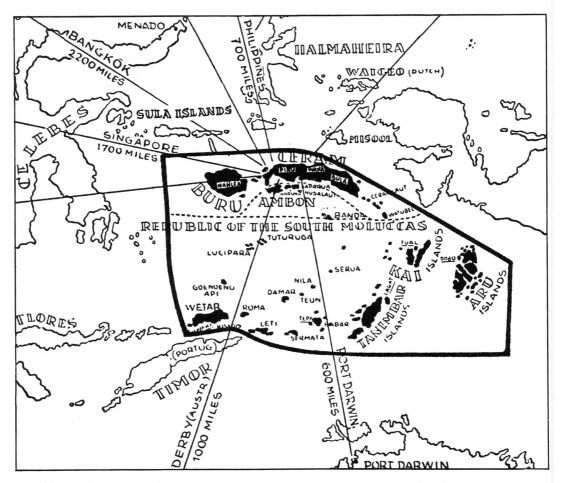

**Republic of the South Moluccas.** This map is a slightly enhanced and modified version of one found in the pamphlet *The Forgotten War: An Appeal from the Republic of the South Moluccas* (New York: Information Office of South Moluccas Republic). This is a good example of a "nonstate nation." The "republic" does not exist; its advocates live mostly in the Netherlands, control no territory in Indonesia, and try to win sympathy for their cause chiefly through propaganda and terrorism.

bedded within and adjacent to the United Arab Emirates (UAE) are a number of enclaves administered jointly by Oman and Ajman, by Oman and Sharjah, and by Sharjah and Fujairah. Oman is not a member of the federation.*

Neutral zones also used to be rather common around the world in frontier regions or separating contending States. Today there are few left. The largest and most important, the one between Kuwait and Saudi Arabia, was partitioned between them in 1969. Iraq and Saudi Arabia jointly administered a similar neutral zone that lay between them. They agreed in 1975 to divide the zone and began in 1981 to produce maps showing the new boundary. The British colony of Gibraltar, at the tip of a peninsula jutting south from Spain, is separated from Spain by a mile-wide neutral zone called La Linea running across the peninsula. Within the UAE is

---

*There is still at least one condominium left in Europe. The Ile des Faisans lies at the mouth of the Bidassea River, which for seven miles is the boundary between Spain and France. Under 1856 and 1901 treaties, the two countries police the island in turn for six month periods.

another neutral zone, 1 kilometer (0.6 mile) wide and 18 kilometers (11.2 miles) long, between Dubai and Abu Dhabi. A map of the UAE, in fact, resembles a map of seventeenth-century Europe or nineteenth-century India, with one large component (Abu Dhabi), six more semiautonomous states linked together with it in a loose confederation (Ajman, Dubai, Fujairah, Ras al Khaima, Sharjah, and Umm al Qawain), and 12 enclaves besides the neutral zone and three condominiums already mentioned. And the UAE itself is the survivor of the former Gulf Federation, which from 1968 until 1971 included Bahrain and Qatar. Like the West Indies Federation, it could not agree on terms for unity before independence, and all three became independent separately in 1971.

In addition, there are a number of maritime joint development zones (discussed in more detail in Chapter 32, dozens of parks straddling national boundaries around the world, and several buffer zones separating hostile forces and occupied by United Nations peace-keeping forces (discussed in Chapter 28).

## International Territories

Finally, a number of territories have been placed under international supervision and/or control in the twentieth century. By far the largest and most important class in this category is composed of the mandated territories of the League of Nations and the trust territories of the United Nations. We discuss *mandates* and *trusteeships* in greater detail in Chapter 18; here we mention only the most unusual ones.

German South-West Africa was assigned by the League of Nations to the Union of South Africa after World War I. Under the terms of the mandate agreement, South Africa was to guide the territory for the benefit of its people. After World War II, when the United Nations succeeded the League of Nations, all the mandates were either terminated (those in the Middle East) or replaced by UN trusteeships—except South-West Africa. South Africa instead extended its apartheid system and other laws to the terri-

tory and moved toward its annexation. After futile efforts to convince South Africa to fulfill its mandatory obligations, sign a trusteeship agreement, or grant the territory independence, the United Nations voted in 1966 to assume control of the territory itself. It established a council to administer it, but South Africa would not recognize the council or permit it to enter the territory (now called Namibia). Instead, South Africa proceeded with plans to grant autonomy to Bantustans which it had created within the territory and otherwise to modify but strengthen its own position there. Under enormous pressure and with considerable assistance from the United Nations, South Africa finally granted independence to Namibia in March 1990.

Quite a different situation prevailed in the Pacific Ocean. There, Japan was forced after World War II to relinquish its mandate over the former German (originally Spanish) colonies of the Marshall, Caroline, and Mariana Islands. All these islands (except Guam, which has been a U.S. colony since 1898) were grouped together into the Trust Territory of the Pacific Islands, with the United States as trustee.* Since it was designated a strategic trust, with fortifications permitted, it was placed under the supervision of the UN Security Council rather than the General Assembly—a unique situation. The United Nations continued to supervise Palau, the last remaining fragment of the Trust Territory until it became independent in December 1994. The Northern Marianas were separated from it in 1976 after a plebiscite in which the people voted to become a commonwealth of the United States, much like Puerto Rico. The rest of the territory became independent in 1990 as the Federated States of Micronesia and the Marshall Islands.

The United Nations has also had experience with *direct administration* of territory. Its own headquarters in New York is international territory, owned and administered

---

*This territory was also known, somewhat inaccurately, as Micronesia, which properly describes a cultural area only partially overlapping the former trust territory.

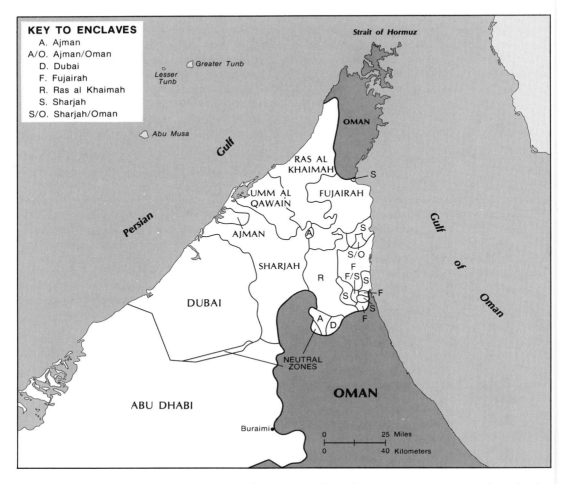

**KEY TO ENCLAVES**
A. Ajman
A/O. Ajman/Oman
D. Dubai
F. Fujairah
R. Ras al Khaimah
S. Sharjah
S/O. Sharjah/Oman

Strait of Hormuz

Greater Tunb

Lesser Tunb

Abu Musa

Gulf

Persian

OMAN

RAS AL KHAIMAH

UMM AL QAWAIN

FUJAIRAH

AJMAN

SHARJAH

DUBAI

Gulf

of

Oman

NEUTRAL ZONES

OMAN

ABU DHABI

Buraimi

0    25 Miles
0    40 Kilometers

**The United Arab Emirates.** In this one small peninsula of Southwest Asia, we see a number of politi-coterritorial phenomena very common in Europe in the seventeenth century but rare anywhere today. Oman is a fragmented State and the federation itself contains not only seven members, but also a neutral zone, 3 condominiums, and 12 enclaves. The Buraimi Oasis was long fought over by Oman (formerly Muscat and Oman), Abu Dhabi, and Saudi Arabia. The Tunb Islands and Abu Musa in the Persian Gulf were occupied and annexed by Iran when the British withdrew in late 1971, despite claims to them by several of the emirates (formerly the Trucial States).

by the United Nations under an agreement with the United States. More important as a precedent, however, was the United Nations Temporary Executive Authority. Between 10 October 1962, and 1 May 1963, this agency directly administered the western half of New Guinea. It served as a temporal rather than a spatial buffer, separating the departing Dutch and the arriving Indonesians, as control was passed from one to the other after a long and bitter struggle. It performed

all the functions of a government, except for conducting foreign relations. Should the United Nations actually assume control of any other territory in dispute, even on an interim basis, it would have a useful and successful precedent on which to draw.

Another class of international territory is the *free city*. Three times in this century disputes over ownership of strategically located cities have resulted in the establishment of separate regimes in these cities and their immediate hinterlands, creating in effect modern city-states, but ones lacking sovereignty.*

*A proposal to make Fiume (now Rijeka, Croatia) a free state run by an Italo-Fiuman-Yugoslav consortium was approved by the city's electorate in 1921 but was never implemented.

The first was Danzig, governed by a largely German administration from its creation in 1919 to its reincorporation into Germany after Hitler's invasion of Poland in 1939. Tangier was governed by an international administration, largely French and British, from 1923 until it was incorporated into Mo-

---

### Case Study: The Cook Islands

The Cook Islands consist of two groups of islands in the South Pacific Ocean. The southern group, or Lower Cooks, consists of four volcanic islands, three low coral islands, and one atoll. The largest and most important of all the Cook Islands is Rarotonga, a high volcanic island; the capital, Avarua, is located here. The Northern Cooks are seven low coral atolls, the largest of which is Penrhyn. The total land area of the Cook Islands is 93 square miles, and the 1991 population was about 18,000. Polynesians settled the islands in successive migrations from about 800 A.D. Of the many Europeans who visited these islands from the sixteenth century onward, the most important were Captain James Cook (1773), Lieutenant William Bligh (1789), and Reverend John Williams (1823).

Williams led the London Missionary Society in converting the native Polynesians and providing a revised framework of self-government for them. The United Kingdom established a protectorate over the islands in 1881, and they were annexed to New Zealand in 1901. After the Second World War, Cook Islanders started migrating to New Zealand in growing numbers. Currently, more than twice as many Cook Islanders live in New Zealand as in the islands, and about 10,000 live in Australia. The territory became self-governing in 1965. The largest exports today are cultured pearls for the Japanese market and fresh fruits and vegetables flown to the U.S.A. and New Zealand. Imports are mainly manufactured goods.

The 1965 self-government agreement with New Zealand has been superseded not by subsequent formal documents, but by evolving practice. The "free association" status is written into the constitution and can only be repealed by a two-thirds vote of the islands' parliament. The country is democratic, with multiple political parties and free elections.

Many theorists consider an independent foreign policy to be the touchstone of "sovereignty." By this standard, the Cook Islands are more independent than many allegedly "sovereign" States. They have had their own Ministry of Foreign Affairs since the 1970s and since then have been making their own foreign and defense policies. When New Zealand banned nuclear-powered and nuclear-armed vessels from its waters in the early 1980s, for example, the Cook Islands welcomed them. On 10 December 1982, the Cook Islands signed the United Nations Convention on the Law of the Sea in its own right, the only non-"sovereign State" permitted to do so (with the exception of the United Nations Council for Namibia).

The Cook Islands belong to several United Nations specialized agencies and to the Asian Development Bank, as well as a number of South Pacific regional organizations. They have negotiated aid agreements with a score of United Nations and Commonwealth bodies and participate actively in world affairs generally. Yet they have chosen to retain their advantageous links with New Zealand: free trade in both directions, full New Zealand citizenship, representation by New Zealand for routine diplomatic and consular matters, and so on.

Not only the government but also the people of the Cook Islands seem content with this arrangement, and there is no "independence" movement. The United Nations does not include the country in its list of non-self-governing territories. Perhaps a status of this type would be appropriate not only for some of the remaining "dependencies and areas of special sovereignty" (for which see Chapter 19), but even for many nominally independent States for which independence has proven both chimerical and burdensome, a difficult situation indeed!

**Anomalous political units on stamps.** The top four were produced by South Africa for the "independent" Bantustans. The next row has stamps of Andorra, the Aaland Islands, Kalaallit Nunaat (formerly Greenland), and the Faeroe Islands, all of which are self-governing but not sovereign. In the third row are stamps from the Isle of Man and three of the Channel Islands, neither colonies nor parts of the United Kingdom, but dependencies of the Crown. In the next row are stamps issued by rebel regimes of Biafra (in Nigeria), Katanga (in Zaïre), the South Moluccas (in Indonesia), Viet Minh (in French Indochina), and Viet Cong (in South Vietnam). Next row: stamps of the former condominium of New Hebrides, now Vanuatu; Taiwan, which still claims to be part of the "Republic of China," the former international city of Tangier, showing a map of the city and Rhodesia, which managed to retain its independence (though not statehood) from 1965 to 1980, when it became Zimbabwe; and in the last row are a stamp issued by "The Turkish Republic of Northern Cyprus," and a *se tenant* set of stamps from the Knights of Malta, which claims to be the world's smallest "state" but which consists only of a mansion and a tiny hillside enclave in Rome and a few buildings scattered around other countries. (Martin Glassner).

rocco in 1956. The Free Territory of Trieste was a little different for several reasons: (1) it was a much larger area than either Danzig or Tangier; (2) it was only a temporary arrangement pending settlement of the dispute over its disposition between Italy and Yugoslavia; and (3) the territory was governed by military authorities of the United States and Britain in Zone A (Trieste city and a strip of territory to the northwest) and of Yugoslavia in Zone B (the remainder of the territory to the east and south). It was partitioned in 1954, with Italy acquiring Zone A plus about 5 square miles and Yugoslavia receiving the rest. None of these international city administrations was notably successful. If experience is any guide, such an administration for Jerusalem, recommended by the United Nations in 1947 but rejected by the Palestinian Arabs and the Arab States at the time, would be no more successful. This proposal is still raised in varying forms from time to time but is unlikely to be adopted.

## *The Importance of Anomalies*

It is possible that the current *fin de siècle* will, in addition to the criteria discussed in the Prologue, be understood in the future as marking the zenith of the European State system introduced in the first three chapters of this part, and only three and a half centuries after Westphalia. The imminent demise of the State has been predicted for more than a century. Karl Marx proclaimed the inevitable "withering away of the State," but his disciples Lenin and Stalin reinforced the power of the Soviet State and had no intention of letting it wither away. Immediately after the Second World War, many people advocated—even expected—replacement of the old "balance of power" by collective security and of a world of States by a world government, but very soon came the Cold War with new alliances of States, and decolonization, bringing with it the near quadrupling of the number of States.

Now the Soviet Union is gone, the Cold War is behind us, and the momentum of decolonization is nearly spent. Yes, there are still nationalist movements among national minorities in many countries trying to attain statehood, but it is unlikely that many of them will be successful. Existing States, old and new, rich and poor, homogeneous and diverse, core and peripheral, are all experiencing the effects of many current trends that are gradually but noticeably eroding their "sovereignty" and their "power." We cannot discuss them all in detail here, but they do deserve at least a listing and a brief analysis, and some are discussed elsewhere in this book. Virtually all of them are interrelated or even interdependent, and cause and effect are often difficult to determine. Finally, this list is not definitive, only suggestive, and nearly anyone could reorganize it or add to it.

### *The Erosion of the State*

1. **Internationalization of illegal activities** Crime of all kinds, including narcotrafficking, terrorism, arms trafficking, piracy, and money laundering, is increasingly controlled by international gangs who are able to evade or even ignore the authorities of States. The same is true to a lesser but growing extent as regards theft and smuggling of art works, archaeological treasures, endangered species of plants and animals, and intellectual property, including trade secrets. This is leading to the strengthening of Interpol, greater cooperation among States in exchanging information and in extradition, and the creation of an International Criminal Court.

2. **Global health problems** Acquired Immunodeficiency Syndrome (AIDS), tuberculosis, plague, and other communicable diseases are spreading at an increased rate because of the increased mobility of people, the breakdown of national health programs, and other causes. AIDS is particularly pernicious and is already decimating the young people in many developing countries, particularly in Africa, who already are or are about to become middle-level

managers and civil servants, thereby weakening governments and inviting international cooperation.

3. **The global economy** Many transnational corporations have grown larger, richer, and more powerful than many governments and make decisions independent of government policies. We will shortly have global stock markets and a relatively free flow of investment capital, which is also likely to weaken governments' control over their own economies. The new World Trade Organization aims to free international trade as much as possible and has the authority to discipline violators of its rules.

4. **Devolution of internal power** Many countries are decentralizing their government powers and granting more authority to local governmental units in response to the agitation of ethnic minorities, pressure from growing business interests, and the sheer burden of managing States in an increasingly complex and costly world. Some ethnic minorities no longer have faith in national governments to protect them and guarantee them equal opportunities and are seeking to break loose in various ways.

5. **Cultural globalism** The rapid penetration of Western cultural elements into nearly every crevice of every State and territory is degrading or smothering local cultures and reducing the differences among them. Massive migrations among countries, voluntary and involuntary, are reducing both cultural distinctiveness and nationalistic fervor, reducing support for many governments and challenging the *raison d'être* of many States.

6. **Environmental Degradation** No longer can ecology be a strictly local concern. Decay of the ozone layer and potential global warming can profoundly alter the conditions for life on the entire planet. Already transborder pollution of all kinds and international movements of wastes of all kinds have become serious problems beyond the ability of individual States to manage. A growing body of international environmental law is not only encouraging cooperation but also imposing sanctions on violators.

7. **International intervention in the internal affairs of States** This began with arms controls imposed on the countries that began the Second World War, spread to attacking apartheid in South Africa and subsequently to other human rights violations, and now includes international sanctions—including on-the-ground inspections—imposed on States such as Iraq that violate international norms of behavior.

8. **Science and technology** Vastly improved transportation and communications have enabled individuals and groups to bypass governments and form separate associations that ignore international boundaries. Utilization of the sea and outer space have become so complex and costly that no State can any longer conduct sophisticated activities alone. The computer may soon become more powerful than the bureaucrat or even the elected official.

9. **The growth of NGOs and IGOs** The past two decades have seen a proliferation of nongovernmental organizations (NGOs) with increasing wealth, sophistication, and dedication that have begun to influence national and international policies and developments, thereby reducing somewhat government power. At the same time, most States have found it necessary to assign some decision-making and functions to *inter*governmental organizations (IGOs), especially at the regional and subregional levels.

10. **The spread of democracy** Democracies are generally less inclined toward bellicosity than dictatorships and are more willing to cooperate with other States and even to surrender some of their "sovereignty." This is leading to the strengthing of both international law and international organizations,

and to greater regard for human rights, including those of indigenous peoples and minorities.

### Alternatives to the State

In this chapter we have introduced some forms of spatial political organization other than the State that have served peoples and individuals very well for a very long time. Other alternatives to the State can be revived or devised, especially for countries, independent or dependent, that are too small or poor to bear the many and heavy costs of statehood today. They can opt out of sovereignty, perhaps contracting with States or intergovernmental organizations to carry out their foreign relations and defense. The United Nations Trusteeship Council, since Palau's independence, no longer has any trusteeships to supervise, but it is still being retained with a skeleton staff and could usefully be employed to supervise "territories voluntarily placed under the [trusteeship] system by States responsible for their administration" (Article 77(1)c of the United Nations Charter). Alternatively, the United Nations itself could administer territories directly, drawing on its experience in Dutch New Guinea, the Congo, and Cambodia. Condominiums, joint development zones, and the type of sovereignty typified by the Cook Islands could serve as models for new arrangements. The State, as pointed out in Chapter 5, is only the latest form of political territorial organization to dominate the world. There is no reason that it should be the last.

# BIBLIOGRAPHY FOR PART TWO

## *Books and Monographs*

**A**

Abu-Dawood, Abdul-Razzak S. and others, International *Boundaries of Saudi Arabia*. New Delhi: Galaxy, 1990.

Alexander, David G. and others, *Atlantic Canada and Federation*. Toronto: Univ. of Toronto Press, 1983.

Aliriza, Bulent, *The Cyprus Problem Revisited*. Washington, DC: Carnegie Endowment, 1992.

Almog, Shmuel, *Nationalism and Antisemitism in Modern Europe 1915-1945*. Oxford: Pergamon, 1990.

Al-Nageeb, Khaldoun H., *Society and State in the Gulf and Arab Peninsula*. New York: Routledge, 1991.

Alter, Peter, *Nationalism*. Frankfurt am Main: Hodder & Stoughton, 1985.

Anderson, James (ed.), *The Rise of the Modern State*. Brighton: Harvester Press, 1986.

Anderson, Thomas P., *The War of the Dispossessed: Honduras and El Salvador, 1969*. Lincoln: Univ. of Nebraska Press, 1981.

Arnade, Charles W., *The Emergence of the Republic of Bolivia*. Gainesville: Univ. of Florida Press, 1957.

Arshi, Ziba and Khosro Zabihi, *Kurdistan*. London: Kegan Paul, 1991.

Asiwaju, A. I. (ed.), *Partitioned Africans; Ethnic Relations Across Africa's International Boundaries 1884–1984*. New York: St. Martin's, 1985.

Asiwaju, A. I. and B. M. Barkindo (eds.), *The Nigeria-Niger Transborder Cooperation*. Lagos: Malthouse Press for National Boundary Commission, 1993.

**B**

Banac, Ivo, The National Question in Yugoslavia; Origins, History, Politics. Ithaca, NY: Cornell Univ. Press, 1994.

Barros, James, *The Aland Islands Question; Its Settlement by the League of Nations*. New Haven, CT: Yale Univ. Press, 1968.

Barton, Brian and Patrick J. Roche (eds.), *The Northern Ireland Question: Myth and Reality*. Hanover, NH: Dartmouth, 1991.

Basch, Linda and others (eds.), *Nations Unbound; Transnational Projects, Postcolonial Predicaments, and Deterritorialized Nation-States*. U.K.: Gordon and Breach, 1994.

Bashevkin, Sylvia, *True Patriot Love; The Politics of Canadian Nationalism*. Oxford Univ. Press, 1991.

Bavkis, Herman and William Chandler (eds.), *Federalism and the Role of the State*. Toronto: Univ. of Toronto Press, 1987.

Bender, Barbara and John Gledhill (eds.), *State and Society: The Emergence and Development of Social Hierarchy and Political Centralization*. New York: Routledge, 1988.

Bennett, R. J. (ed.), *Territory and Administration in Europe*. London: Pinter, 1990.

Benvenisti, Meron, *The West Bank Data Project: A Survey of Israel's Policies*. Lanham, MD: Univ. Press of America, 1984.

Bernier, Luc, *From Paris to Washington: Quebec's Foreign Policy in a Changing World*. Washington, DC: CSIS, 1994.

Bertelsen, Judy S. (ed.), *The Non-State Nation in International Politics*. New York: Praeger, 1977.

Bethlehem, Daniel and others, *The 'Yugoslav' Crisis in International Law*. Cambridge: Grotius, 1993.

Blaut, James M., *The Nationalist Question: Decolonising the Theory of Nationalism*. London: Zed Books, 1987.

Boal, Frederick W. and J. Neville H. Douglas (eds.), *Integration and Division: Geographical Perspectives on the Northern Ireland Problem*. London and New York: Academic Press, 1982.

Boggs, S. Whittemore, *International Boundaries: A Study of Boundary Functions and Problems*. New York: Columbia Univ. Press, 1940.

Bose, Sumantra, *States, Nations, Sovereignty: Sri Lanka, India and the Tamil Zalam Movement*. Newbury Park, CA: Sage, 1993.

*Boundary Briefings, Territory Briefings, Newsletters* and other publications are produced aperiodically by the International Boundaries Research Unit of the Univ. of Durham, England.

Brass, Paul, *Ethnicity and Nationalism; Theory and Comparison*. Newbury Park, CA: Sage, 1992.

Braveboy-Wagner, Jacqueline Anne, *The Venezuela-Guyana Border Dispute; Britain's Colonial Legacy in Latin America*. Boulder, CO: Westview, 1984.

Breuilly, John, *Nationalism and the State*. 2nd ed. Chicago: Univ. of Chicago Press, 1994.

Briand, Michael (ed.), *A Way Out: Federalist Options for South Africa*. San Francisco: ICS Press, 1987.

Brown, C. M., *Boundary Control and Legal Principles*. New York: Wiley, 1986.

Brown, Peter G. and Henry Shue (eds.), *Boundaries: National Autonomy and Its Limits*. Totowa, NJ: Rowman & Littlefield, 1981.

Brownlie, Ian, *African Boundaries: A Legal and Diplomatic Encyclopedia*. London: Hurst, 1979.

Burgess, Michael (ed.), *Canadian Federalism; Past, Present and Future*. Leicester, U.K.: Leicester Univ. Press, 1990.

Burns, Ronald M., and others, *Political and Administrative Federalism*. Canberra: Australian National Univ. Press, 1976.

Butler, Jeffrey and others, *The Black Homelands of South Africa*. Berkeley: Univ. of California Press, 1977.

**C**

Cairns, Alan C., *Charter Versus Federalism*. Montreal: McGill-Queens, 1992.

Calvert, Peter, *The Falklands Crisis: The Rights and Wrongs*. London: Pinter, 1982.

Campbell, David, *Politics Without Principle*. Boulder, CO: Lynne Rienner, 1993.

Carney, John and others (eds.), *Regions in Crisis; New Perspectives in Europe Regional Theory*. London: Croom Helm, 1982.

Carter, F. W., *Dubrovnik (Ragusa): A Classic City-State*. New York: Seminar Press, 1972.

Carter, Stephen K., *Russian Nationalism*. New York: St. Martin's, 1991.

Catudal, Honoré M., *The Enclave Problem of Western Europe*. University: Univ. of Alabama Press, 1979.

Chang, Luke T., *China's Boundary Treaties and Frontier Disputes*. New York: Oceana, 1982.

Chay, John, and Thomas E. Ross (eds.), *Buffer States in World Politics*. Boulder, CO: Westview, 1986.

Chazan, Naomi (ed.), *Irredentism and International Politics*. Boulder, CO: Lynne Rienner, 1991.

Coakley, John (ed.), *The Social Origins of Nationalist Movements: The Contemporary West European Experience*. Beverly Hills, CA: Sage, 1992.

———, *The Territorial Management of Ethnic Conflict*. London: Frank Cass, 1993.

Cohen, Leonard J., *Broken Bonds; The Disintegration of Yugoslavia*. 2nd ed. Boulder, CO: Westview, 1993.

Cohen, Saul B., *The Geopolitics of Israel's Border Question*. Boulder, CO: Westview, 1987.

Connah, Graham, *African Civilizations; Precolonial Cities and States in Tropical Africa: An Archaeological Perspective*. Cambridge Univ. Press, 1987.

Connor, Walker, *Ethnonationalism; The Quest for Understanding*. Princeton, NJ: Princeton Univ. Press, 1994.

Cook, Ramsay, *Canada, Quebec and the Uses of Nationalism*. Ontario: McClelland & Stewart, 1986.

Copper, John F., *Taiwan: Nation-State or Province?* Boulder, CO: Westview, 1989.

Cowgill, George L. and Norman Yoffee (eds.), *The Collapse of Ancient States and Civilizations*. Tucson: Univ. of Arizona Press, 1988.

Crawford, James, *The Creation of States in International Law*. Oxford Univ. Press, 1979.

Crepeau, Paul André and C.B. Macpherson (eds.), *The Future of Canadian Federalism*. Toronto: Univ. of Toronto Press, 1965.

Currie, David P. (ed.), *Federalism in the New Nations*. Univ. of Chicago Press, 1964.

Curzon, Lord, Frontiers. Oxford Univ. Press, 1908.

**D**

Daly, Martin and Ahmad Alawad Sikainga, *Civil War in the Sudan*. London: Tauris, 1992.

Day, Alan J. (ed.), *Border and Territorial Disputes*. Detroit: Gale Research Co., 1982.

Denktash, Rauf R., *The Cyprus Triangle*. Rev ed. New York: Office of the Turkish Republic of Northern Cyprus, 1988.

Dennis, William Jefferson, *Documentary History of the Tacna-Arica Dispute*. Port Washington, NY: Kennikat Press, 1971.

Diamond, Larry and Marc F. Plattner (eds.), *Nationalism, Ethnic Conflict, and Democracy*. Baltimore: Johns Hopkins Univ. Press, 1994.

Dima, Nicholas, *From Moldavia to Moldova; The Soviet-Romanian Territorial Dispute*. New York: Columbia Univ. Press, 1991.

Duchacek, Ivo D., *Comparative Federalism: The Territorial Dimension of Politics*. Lanham, MD: Univ. Press of America, 1987.

Dugard, John, *Recognition and the United Nations*. Cambridge: Grotius, 1987.

Duncan, W. Raymond and G. Paul Holman, Jr., *Ethnic Nationalism and Regional Conflict: The Former Soviet Union and Yugoslavia*. Boulder, CO: Westview, 1994.

Dyson, K.H.F., *The State Tradition in Western Europe: A Study of an Idea and Institution*. Oxford: Martin Robertson, 1980.

**E**

East, W. Gordon, *An Historical Geography of Europe*. London: Methuen, 1935.

———, *The Political Division of Europe*. London: Univ. of London Press, 1948.

East, W. Gordon and O.H.K. Spate, *The Changing Map of Asia*. London, 1961.

Eberstadt, Nicholas, *Korea Approaches Reunification*. Armonk, NY: Sharpe, 1995.

Egerö, Bertil, *South Africa's Bantustans; From Dumping Grounds to Battlefronts*. Uppsala: Scandinavian Institute of African Studies, 1991.

Elazar, Daniel J., *Judea, Samaria, and Gaza: Views on the Present and Future*. Lanham, MD: Univ. Press of America, 1982.

———, *American Federalism: A View from the States*. New York: Harper & Row, 1984.

———, *Federalism and Political Integration*. Lanham, MD: Univ. Press of America, 1985.

———, *Two Peoples . . . One Land: Federal Solutions for Israel, the Palestinians, and Jordan*. Lanham, MD: Univ. Press of America, 1991.

Elkins, T. H. and B. Hoffmeister, *Berlin: The Spatial Structure of a Divided City*. London: Methuen, 1988.

Eriksen, Thomas Hylland, *Ethnicity and Nationalism; Anthropological Perspectives*. London: Pluto Press, 1993.

**F**

*Falklands/Malvinas: Whose Crisis?* London: Latin America Bureau, 1982.

Farah, Tawfic E. (ed.), *Pan-Arabism and Arab Nationalism*. Boulder, CO: Westview, 1987.

Farnan, Russell F. (ed.), *Nationalism, Ethnicity, and Identity*. New Brunswick, NJ: Transaction, 1994.

Fawcett, Charles Bungay, *Frontiers: A Study in Political Geography*. New York: Oxford Univ. Press, 1918.

Ferrer Vieyra, Enrique, *An Annotated Legal Chronology of the Malvinas (Falkland) Islands Controversy*. Córdoba, Argentina: 1985.

Fine, John V.A., *Bosnia and Hercegovina; A Tradition Betrayed*. New York: Columbia Univ. Press, 1994.

Finnie, David H., *Shifting Lines in the Sand: Kuwait's Elusive Frontier with Iraq*. Cambridge, MA: Harvard Univ. Press, 1992.

Fishel, Wesley R., *The End of Extraterritoriality in China*. Berkeley: Univ. of California Press, 1952.

Fox, Hazel and others, *Joint Development of Offshore Oil and Gas*. London: British Institute of International and Comparative Law, 1989.

Franck, Thomas Meral, *Why Federations Fail*. New York: New York Univ. Press, 1968.

Franklin, H., *Unholy Wedlock: The Failure of the Central African Federation*. London: Allen & Unwin, 1964.

Fry, Earl H., *Canada's Unity Crisis; Implications for U.S.-Canada Relations*. New York: Twentieth Century Fund, 1992.

**G**

Gallusser, Werner A. (ed.), *Political Boundaries and Coexistence; Proceedings of the IGU Symposium, Basle, Switzerland*. Peter Lang, 1994.

Galnoor, Itzhak, *The Partition of Palestine; Decision Crossroads in the Zionist Movement*. Ithaca, NY: SUNY, 1994.

Gamba, Virginia, *The Falklands/Malvinas War*. London: Allen & Unwin, 1986.

Geldenhuys, Deon, *Isolated States; A Comparative Analysis*. Cambridge Univ. Press, 1991.

Gerner, Deborah J., *One Land, Two People: The Conflict of Palestine*. 2nd ed. Boulder, CO: Westview, 1994.

Giliomee, Hermann and Lawrence Schlemmer, *From Apartheid to Nation Building*. Oxford Univ. Press, 1993.

Girot, Pascal O. (ed.), *World Boundaries: The Americas*. New York: Routledge, 1994.

Goertz, Gary and Paul Diehl, *Territorial Changes and International Conflict*. New York: Routledge, 1992.

Gottlieb, Gidon, *Nation Against State; A New Approach to Ethnic Conflicts and the Decline of Sovereignty*. New York: Council on Foreign Relations, 1993.

Gottmann, Jean, *Megalopolis*. New York: Twentieth Century Fund, 1961.

———, *The Significance of Territory*. Charlottesville: Univ. Press of Virginia, 1973.

Grundy-Warr, Carl (ed.), *World Boundaries: Eurasia*. New York: Routledge, 1994.

Gunter, Michael M., *The Kurds in Turkey; A Political Dilemma*. Boulder, CO: Westview, 1990.

Gustafson, Lowell S., *The Sovereignty Dispute Over the Falkland (Malvinas) Islands*. Oxford Univ. Press, 1988.

**H**

Haas, Jonathan and others, *The Origins and Development of the Andean State*. Cambridge Univ. Press, 1987.

Hancock, M. Donald and Helga A. Welsh (eds.), *German Unification; Process and Outcomes*. Boulder, CO: Westview, 1993.

Hanna, Willard A., *The Republic of the South Moluccas*. AUFS Reports, Southeast Asia Series, Vol. 23, No. 2. 1975.

Hasan, Mushirul (ed.), *India's Partition Process, Strategy, and Mobilization*. Oxford Univ. Press, 1993.

Heiberg, Marianne, The *Making of the Basque Nation*. Cambridge Univ. Press, 1989.

Hendry, T. D. and M. C. Wood, *The Legal Status of Berlin*. Cambridge: Grotius, 1987.

Heraclides, Alexis, *The Self-Determination of*

*Minorities in International Politics.* London: Frank Cass, 1990.

Hertslet, Sir Edward, *The Map of Europe by Treaty 1875-1891.* London: Butterworths, 1891.

———, *The Map of Africa by Treaty.* London: Frank Cass, 1967.

Hitti, Philip K., *Capital Cities of Arab Islam.* Minneapolis: Univ. of Minnesota Press, 1973.

Hobsbawm, E. J., *Nations and Nationalism Since 1780; Programme, Myth, Reality.* Cambridge Univ. Press, 1990.

Hof, Frederic C., *Galilee Divided: The Israel-Lebanon Frontier, 1916–1984.* Boulder, CO: Westview, 1985.

Holdich, Thomas H., *Political Frontiers and Boundary Making.* New York: Macmillan, 1916.

**I**

Iivonen, Jyrki (ed.), *The Future of the Nation State in Europe.* Helsinki: Edward Elgar, 1993.

*International Boundary Studies.* U.S. Department of State. Office of the Geographer.

Ireland, Gordon, *Boundaries, Possessions and Conflicts in South America.* Cambridge, MA: Harvard Univ. Press, 1938; reprinted, New York: Octagon, 1971.

———, *Boundaries, Possessions, and Conflicts in Central and North America and the Caribbean.* Cambridge, MA: Harvard Univ. Press, 1941; reprinted, New York: Octagon, 1971.

Izady, Mehrdad, *The Kurds; A Concise Handbook.* Crane Russak, 1992.

**J**

Jackson, Robert H., *Quasi-States: Sovereignty, International Relations and the Third World.* Cambridge Univ. Press, 1990.

Jackson, Robert H. and Alan James (eds.), *States in a Changing World; A Contemporary Analysis.* Oxford Univ. Press, 1993.

James, Alan, *Sovereign Statehood: The Basis of International Society.* London: Allen & Unwin, 1986.

Jennings, R. Y., *The Acquisition of Territory in International Law.* Manchester, U.K.: Manchester Univ. Press, 1963.

Johnston, Ronald J., David B. Knight and Eleanore Kofman, *Nationalism, Self-Determination and Political Geography.* London: Croom Helm, 1988.

Jones, Stephen B., *Boundary Making.* Washington, DC: Carnegie Endowment, 1945.

**K**

Kacowicz, Arie Marcelo, *Peaceful Territorial Change.* Columbia: Univ. of South Carolina Press, 1994.

Kaikobad, Kaiyan Homi, The *Shatt-al-Arab Boundary Question.* Oxford Univ. Press, 1988.

Kalia, Ravi, *Chandigarh.* Carbondale: Southern Illinois Univ. Press, 1987.

Kellerman, Aharon, *Society and Settlement: Jewish Land of Israel in the Twentieth Century.* Albany, NY: SUNY Press, 1993.

Kelly, Ian, *Hong Kong; A Political-Geographic Analysis.* Honolulu: Univ. of Hawaii Press, 1989.

Kelly, J. B., *Eastern Arabian Frontiers.* London: Faber & Faber, 1964.

Kennan, George F., *The Other Balkan Wars; A 1913 Carnegie Endowment Inquiry in Retrospect, with a New Introduction and Reflections on the Present Conflict.* Washington, DC: Brookings, 1993.

Khalidi, Rashid and others, *The Origins of Arab Nationalism.* New York: Columbia Univ. Press, 1993.

Khalidi, Walid, *Palestine Reborn.* London: Tauris, 1992.

Khoury, Philip S. and Joseph Kostiner, *Tribes and State Formation in the Middle East.* London: Tauris, 1991.

Kiang, Ying Cheng, *China's Boundaries.* Lincolnwood, IL: Institute of China Studies, 1984.

Kimmerling, Baruch (ed.), *Zionism and Territory —The Socio-territorial Dimension of Zionist Politics.* Berkeley, CA: Institute of International Studies, 1983.

———, *The Israeli State and Society; Boundaries and Frontiers.* Ithaca, NY: SUNY, 1994.

Kimmich, Christoph M., *The Free City; Danzig and German Policy, 1919–1934.* New Haven, CT: Yale Univ. Press, 1968.

Kliot, Nurit, and Stanley Waterman (eds.), *Pluralism and Political Geography.* New York: St. Martin's, 1983.

Knight, David B. (ed.), *Our Geographic Mosaic.* Ottawa: Carleton Univ. Press, 1985.

———, *Choosing Canada's Capital; Conflict Resolution in a Parliamentary System.* Ottawa: Carleton Univ. Press, 1991.

Knop, Karen and others, *Rethinking Federalism, Citizens, Markets, and Governments in a Changing World.* Vancouver: UBC Press, 1994.

Kofman, Eleanore and Colin H. Williams (eds.), *Community Conflict, Partition and Nationalism.* New York: Routledge, 1989.

Kohn, Hans, *Nationalism: Its Meaning and History.* Melbourne, FL: Krieger, 1982.

Kraemer, Joel L. (ed.), *Jerusalem; Problems and Prospects.* New York: Praeger, 1980.

Kratochwil, Friedrich and others, *Peace and Disputed Sovereignty: Reflections on Conflict over Territory.* Lanham, MD: Univ. Press of America, 1985.

Krenz, Frank E., *International Enclaves and Rights of Passage*. Geneva and Paris: Droz & Minard, 1961.

Kreyenbroek, Philip G. and Stefan Sperl (eds.), *The Kurds; A Contemporary Overview*. New York: Routledge, 1991.

**L**

Landau, Jacob M., *Pan-Turkism; From Irredentism to Cooperation*. Bloomington: Indiana Univ. Press, 1995.

Lane, Kevin P., *Sovereignty and the Status Quo: The Historical Roots of China's Hong Kong Policy*. Boulder, CO: Westview, 1990.

Lapidoth, Ruth and Moshe Hirsch (eds.), *The Jerusalem Question and Its Resolution: Selected Documents*. Norwell, MA: Kluwer, 1994.

Lataruski, Paul (ed.), *The Reconstruction of Poland, 1914–23*. New York: St. Martin's, 1992.

Latouche, Daniel, *Canada and Quebec, Past and Future: An Essay*. Toronto: Univ. of Toronto Press, 1987.

Lattimore, Owen, *Inner Frontiers of Asia*. American Geographical Society, Research Series, No. 21. New York, 1940.

———, *Inner Asian Frontiers of China*. Boston: Beacon Press, 1962.

———, *Studies in Frontier History: Collected Papers 1928-1958*. Paris: Mouton, 1962.

Lee Yong Leng, *Razor's Edge: Boundaries & Boundary Disputes in Southeast Asia*. Vermont: Gower, 1980.

Leslie, Peter, *Federal State, National Economy*. Toronto: Univ. of Toronto Press, 1987.

Levie, Howard S., *The Status of Gibraltar*. Boulder, CO: Westview, 1983.

Little, David, *Sri Lanka; The Invention of Enmity*. Arlington, VA: USIP Press, 1993.

Liu, Xucheng, *The Sino-Indian Border Dispute and Sino-Indian Relations*. Lanham, MD: Univ. Press of America, 1994.

Lo, Chi-Kin, *China's Policy Towards Territorial Disputes; The Case of the South China Sea Islands*. New York: Routledge, 1989.

Luciani, Giacomo (ed.), *The Arab State*. Berkeley: Univ. of California Press, 1990.

Lyons, Gene M. and Michael Mastanduno (eds.), *Beyond Westphalia? State Sovereignty and International Intervention*. Baltimore: Johns Hopkins Univ. Press, 1995.

**M**

Maddy-Weitzman, Bruce, *The Crystallization of the Arab State System: Inter-Arab Politics, 1945–1954*. Syracuse, NY: Syracuse Univ. Press, 1992.

Magas, Branka, *The Destruction of Yugoslavia*. New York: Routledge, 1993.

McRoberts, Kenneth (ed.), *Beyond Quebec, Taking Stock of Canada*. Montreal: McGill-Queens, 1995.

Minghi, Julian V. and Dennis Rumley (eds.), *The Geography of Border Landscapes*. Routledge, 1991.

Miyoshi, Masahiro, *Considerations of Equity in the Settlement of Territorial and Boundary Disputes*. Norwell, MA: Kluwer, 1993.

Moodie, A. E., *The Italo-Yugoslav Boundary; A Study in Political Geography*. London: George Philip, 1945.

Morton, W. L., *The Canadian Identity*. Madison: Univ. of Wisconsin Press, 1961.

Müller, Wolfgang and Vincent Wright (eds.), *The State in Western Europe; Retreat or Redefinition?* London: Frank Cass, 1994.

Munck, R., *The Difficult Dialogue: Marxism and Nationalism*. London: Zed Books, 1986.

Muni, S.D., *The Pangs of Proximity: India and Sri Lanka's Ethnic Crisis*. Beverly Hills, CA: Sage, 1993.

**N**

Necatigil, Zaim M., *The Turkish Republic of Northern Cyprus in Perspective*. Nicosia: Cyprus, 1985.

———, *The Cyprus Question and the Turkish Position in International Law*. Oxford Univ. Press, 1993.

Nelson, Brian and others (eds.), *The Idea of Europe; Problems of National and Transnational Identity*. Herndon, VA: Berg, 1992.

Newbigin, Marion I., *Geographical Aspects of the Balkan Problems in Their Relation to the Great European War*. New York: Putnam, 1915.

Newman, David (ed.), *The Impact of the Gush Emunim: Politics and Settlement on the West Bank*. London: Croom Helm, 1985.

———, *Population, Settlement and Conflicts: Israel and the West Bank*, Cambridge Univ. Press, 1991.

Norbu, Dawa, *Culture and the Politics of Third World Nationalism*. London: Routledge, 1992.

Norrie, K., and others, *Federalism and Economic Union in Canada*. Toronto: Univ. of Toronto Press, 1987.

**O**

O'Halloran, Clare, *Partition and the Limits of Irish Nationalism*. Atlantic Highlands, NJ: Humanities, 1987.

Ojeda, Mario, *Mexico: The Northern Border as a National Concern*. El Paso: Univ. of Texas Press, 1983.

Olson, Robert, *The Emergence of Kurdish Nationalism, 1800–1925.* Austin: Univ. of Texas Press, 1989.

Onuf, Peter S., *Statehood and Union; A History of the Northwest Ordinance.* Bloomington: Indiana Univ. Press, 1992.

Orgill, Andrew, *The Falklands War; Background, Conflict, Aftermath.* Massell, 1992.

**P**

Pankhurst, E. Sylvia and Richard K. Pankhurst, *Ethiopia and Eritrea; The Last Phase of the Reunion Struggle, 1941–1952.* Essex, U.K.: Lalibela House, 1953.

Patrick, Richard A., *Political Geography and the Cyprus Conflict 1963–1971.* Univ. of Waterloo (Ontario), Dept. of Geography, 1976.

Playfair, Emma (ed.), *International Law and the Administration of Occupied Territories: Two Decades of Israeli Occupation of the West Bank and Gaza Strip.* Oxford Univ. Press, 1992.

Plischke, Elmer, *A Geographical Study of the Dutch-German Border.* Münster in Westfalen: Selbstverlag der Geographischen Kommission, 1958.

———, *Microstates in World Affairs: Policy Problems and Options.* Washington, DC: American Enterprise Institute, 1977.

Prescott, John Robert Victor, *Map of Mainland Asia by Treaty.* Melbourne: Melbourne Univ. Press, 1975.

———, *Frontiers of Asia and Southeast Asia.* Melbourne: Melbourne Univ. Press, 1977.

Prescott, John Robert Victor and others, *Political Frontiers and Boundaries.* London: Allen & Unwin, 1987.

**R**

Rabel, Roberto Giorgio, *Between East and West: Trieste, the United States, and the Cold War, 1941–1954.* Durham, NC: Duke Univ. Press, 1988.

Ramet, Sabrina P., *Nationalism and Federalism in Yugoslavia, 1962–1991*, 2nd ed. Bloomington: Indiana Univ. Press, 1992.

Reese, David, *The Soviet Seizure of the Kuriles.* New York: Praeger, 1985.

Reeves, T. Zane, *The U.S.-Mexico Border Commissions: An Overview and Agenda for Further Research.* El Paso: Univ. of Texas Press, 1984.

Regier, Philippe, *Singapore, City-State in Southeast Asia.* Honolulu: Univ. of Hawaii Press, 1992.

Reinharz, Jehuda and George L. Mosse (eds.), *The Impact of Western Nationalisms.* Beverly Hills, CA: Sage, 1992.

Roberts, Kenneth, *Quebec; Social Change and Political Crisis.* London: UCL Press, 1992.

Roff, William R., *The Origins of Malay Nationalism.* Oxford Univ. Press, 1994.

Romann, Michael and Alex Weingrod, *Living Together Separately, Arabs and Jews in Contemporary Jerusalem.* Princeton, NJ: Princeton Univ. Press, 1991.

**S**

Sahlins, Peter, *Boundaries; The Making of France and Spain in the Pyrenees.* Berkeley: Univ. of California Press, 1989.

Sakamoto, Yoshikazu, *Global Transformation: Challenges to the State System.* Tokyo: United Nations Univ. Press, 1994.

Schofield, Clive H. (ed.), *World Boundaries: Global Boundaries.* New York: Routledge, 1994.

Schofield, Clive H. and Richard Schofield (eds.), *World Boundaries: Middle East and North Africa.* New York: Routledge, 1994.

Schofield, Richard, *Kuwait and Iraq: Historical Claims and Territorial Disputes.* London: Royal Institute of International Affairs, 1991.

Segulja, Kristina and Leonard Unger, *The Trieste Negotiations.* Lanham, MD: Univ. Press of America, 1990.

Shain, Yossi (ed.), *Governments-in-Exile in Contemporary World Politics.* New York: Routledge, 1991.

Shaw, R. Paul and Yuwa Wong, *Genetic Seeds of Warfare; Evolution, Nationalism and Patriotism.* New York: Routledge, 1989.

Shemesh, Moshe, *The Palestinian Entity 1959–1974; Arab Politics and the PLO.* London: Frank Cass, 1988.

Shingleton, Bradley and others, *Dimensions of German Unification: Economic, Racial, and Legal Analysis.* Boulder, CO: Westview, 1995.

Shlaim, Avi, *The Politics of Partition; King Abdullah, The Zionists, and Partition, 1921–1951.* New York: Columbia Univ. Press, 1990.

Sicker, Martin, *Judaism, Nationalism, and the Land of Israel.* Boulder, CO: Westview, 1992.

Sickerman, Harvey, *Palestinian Self-Government.* Boulder, CO: Westview, 1992.

Singh, Elen C., *The Spitsbergen (Svalbard) Question: United States Foreign Policy, 1907–1935.* Oslo: Univ. of Oslo Press, 1980.

Smith, Anthony D., *The Ethnic Origins of Nations.* Oxford: Blackwell, 1987.

Smith, Barbara J., *The Roots of Separation in Palestine.* Syracuse, NY: Syracuse Univ. Press, 1992.

Smith, David M., *Apartheid in South Africa.* Cambridge Univ. Press, 1985.

Smith, Wayne S. (ed.), *Toward Resolution?: The*

*Falklands/Malvinas Dispute.* Boulder, CO: Lynne Rienner, 1991.

Snyder, Louis L., *Contemporary Nationalism: Intensity and Persistence.* Melbourne, FL: Krieger, 1992.

Southall, Roger, *South Arica's Transkei; The Political Economy of an "Independent" Bantustan.* New York: Monthly Review Press, 1983.

Spencer, Robert and others (eds.), *The International Joint Commission Seventy Years On.* Toronto: Univ. of Toronto, Centre for International Studies, 1980.

Sugar, Peter F. (ed.), *Eastern European Nationalism in the Twentieth Century.* Lanham, MD: Univ. Press of America, 1994.

Suliman, Hassan S., *The Nationalist Movements in the Maghrib.* Uppsala: Scandinavian Institute of African Studies, 1985.

Suny, Ronald Grigor, *The Making of the Georgian Nation.* London: Tauris, 1988.

**T**

Taylor, John and others (eds.), *Capital Cities/Les Capitales, Perspectives Internationales/International Perspectives.* Ottawa: Carleton Univ. Press, 1992.

Tekle, Amare (ed.), *Eritrea and Ethiopia: From Conflict to Cooperation.* Lawrenceville, NJ: Red Sea Press, 1994.

Thom, Derrick J., *The Niger-Nigeria Boundary 1890–1906: A Study of Ethnic Frontiers and a Colonial Boundary.* Papers in International Studies, Africa Series No. 23. Athens: Ohio Univ. Center for International Studies, 1975.

Thomas, Raju G.C. (ed.), *Perspectives on Kashmir.* Boulder, CO: Westview, 1992.

Tilly, Charles and others (eds.), *Cities and the Rise of States in Europe, A.D. 1000 to 1800.* Boulder, CO: Westview, 1994.

Touval, Saadia, *Somali Nationalism.* Cambridge, MA: Harvard Univ. Press, 1963.

———, *The Boundary Politics of Independent Africa.* Cambridge, MA: Harvard Univ. Press, 1972.

**V**

van Walt van Pragg, Michael C., *The Status of Tibet.* Boulder, CO: Westview, 1987.

Volovici, Leon, *Nationalist Ideology and Antisemitism.* Oxford: Pergamon, 1991.

**W**

Walker, Comor, *Ethnicity, Ethnocentrism and Ethnonationalism.* Melbourne, FL: Krieger, 1993.

Wambaugh, Sarah, *The Doctrine of Self-Determination.* Oxford Univ. Press, 1919.

———, *A Monograph on Plebiscites.* Oxford Univ. Press, 1920.

———, *Plebiscites Since the World War.* New York: Carnegie Endowment, 1936.

Watkins, Susan Cotts, From *Provinces into Nations.* Princeton, NJ: Princeton Univ. Press, 1990.

Weaver, R. Kent (ed.), *The Collapse of Canada.* Washington, DC: Brookings, 1992.

Weitzer, Ronald, *Transforming Settler States; Communal Conflict and Internal Security in Northern Ireland and Zimbabwe.* Berkeley: Univ. of California Press, 1990.

Whalley, John, *Regional Aspects of Confederation.* Toronto: Univ. of Toronto Press, 1987.

Wilkinson, John C., *Arabia's Frontiers: The Story of Britain's Boundary Drawing in the Desert.* London: Tauris, 1991.

Williams, Colin and Eleanore Kofman (eds.), *Community Conflict, Partition and Nationalism.* London: Croom Helm, 1987.

Williams, Frederick D. (ed.), *The Northwest Ordinance.* Michigan State Univ. Press, 1989.

Wilson, Constance M. and Lucien M. Hanks, *The Burma-Thailand Frontier over Sixteen Decades.* Athens: Ohio Univ. Press, 1984.

Wood, Bryce, *Aggression and History; The Case of Ecuador and Peru.* Ann Arbor, MI: Univ. Microfilms, 1978.

Wu, Hsin-Hsing, *Bridging the Strait; Taiwan, China, and the Prospects for Reunification.* Oxford, 1994.

**Y**

Yesilada, Birol Ali, *Social Progress in Northern Cyprus.* UFSI Reports Europe Series 1989/90 No. 3.

Yohannes, Okbazghi, *Eritrea, a Pawn in World Politics.* Gainesville: Univ. Press of Florida, 1991.

Young, Robert, *The Secession of Quebec and the Future of Canada.* Montreal: McGill-Queens, 1995.

**Z**

Zahlan, Rosemarie Said, *The Making of the Modern Gulf States.* London: Unwin Hyman, 1989.

Zelinsky, Wilbur, *Nation into State; the Shifting Symbolic Foundations of American Nationalism.* Chapel Hill: Univ. of North Carolina Press, 1988.

# *Periodicals*

## A

Agnew, John A., "Place and Political Behaviour: The Geography of Scottish Nationalism," *Political Geography Quarterly,* 3, 3 (July 1984), 191–206.

Ahmad, Feroz, "Arab Nationalism, Radicalism and the Specter of Neocolonialism," *Monthly Review,* 42 (Feb. 1991), 30–37.

Austin, D. G., "The Uncertain Frontier: Ghana-Togo," *Journal of Modern African Studies,* 1 (1963), 139–146.

## B

Baram, Amatzia, "Territorial Nationalism in the Middle East," *Middle East Studies,* 26 (Oct. 1990), 425–448.

Bar-Gal, Yoram, "Boundaries as a Topic in Geography Education; The Case of Israel," *Political Geography,* 12, 5 (Sept. 1993), 421–435.

Barlow, I. Max, "Political Geography and Canada's National Unity Problem," *Journal of Geography,* 79, 7 (Dec. 1980), 259–263.

Barnard, W.S., "The Political Geography of an Exclave: Walvis Bay," *South African Geographer,* 15 (1988), 85–99.

Berry, Brian J.L., "Comparative Geography of the Global Economy: Cultures, Corporations and the Nation-State," *Economic Geography,* 65 (1989), 1–18.

Bird, J., "The Foundation of Australian Seaport Capitals," *Economic Geography,* 43, 3 (Oct. 1965), 283–289.

Bond, A. R. and M. K. Sagers (eds.), "Panel on Nationalism in the USSR: Environmental and Territorial Aspects," *Soviet Geography,* 30 (1989), 441–509.

Bouchez, Leo J., "The Fixing of Boundaries in International Boundary Rivers," *International and Comparative Law Quarterly,* 12, 31 (July 1963), 789–817.

Brigham, Albert Perry, "Principles in the Determination of Boundaries," *Geographical Review,* 7 (1919), 201–219.

Burghardt, Andrew F., "Nation, State and Territorial Unity: A Trans-Outaouais View," *Cahiers de Géographie du Québec,* 24 (1980), 123–134.

Byrnes, Andrew and Hilary Charlesworth, "Federalism and the International Legal Order: Recent Developments in Australia," *American Journal of International Law,* 79, 3 (July 1985), 622–640.

## C

*Canadian Review of Studies in Nationalism.* Published annually since 1973 by the Univ. of Prince Edward Island, Charlottetown.

*Capital Cities.* Special Issue of *Ekistics,* 50, 299 (March–April 1983).

Catudal, Honoré M., "Steinstucken: The Politics of a Berlin Enclave," *World Affairs,* 134, 9 (Summer 1971), 51–62.

Chao, John K.T., "South China Sea: Boundary Problems Relating to the Nansha and Hsisha Islands," *Chinese Yearbook of International Law and Affairs,* 9 (1989–90), 66–157.

Charney, Jonathan I., "The Delimitation of Lateral Seaward Boundaries Between States in a Domestic Context," *American Journal of International Law,* 75, 1 (Jan. 1981), 28–68.

Chiu, Hungdah, "The International Legal Status of the Republic of China," *Chinese Yearbook of International Law and Affairs,* 8 (1988–89), 1–19.

Chou, David S., "The ROC's Membership Problems in International Organizations," *Outlook* (Taipei), 26 (May 1991), 18–26.

Christopher, A. J., "Continuity and Change of African Capitals," *Geographical Review,* 75, 1 (1985), 44–57.

———, "South Africa: The Case of a Failed State Partition," *Political Geography,* 13, 2 (March 1994), 123–136.

Clove, Ronald I., "Who Owns Wrangel Island?," *Marine Policy Reports,* 1, 3 (1989), 197–206.

Cohen, Saul B. and Nurit Kliot, "Place-Names in Israel's Ideological Struggle over the Administered Territories," *Annals, AAG,* 82, 4 (Dec. 1992), 653–680.

Connell, John, "The India-China Frontier Dispute," *Royal Canadian Asian Journal,* 47, 3–4 (1960), 270–285.

Cox, Kevin R., "Comment; Redefining 'Territory,'" *Political Geography Quarterly,* 10, 1 (Jan. 1991), 5–7.

Crone, G. R., "The Turkish-Iranian Boundary," *Geographical Journal,* 91 (1938), 57–59; 92 (1938), 149–150.

## D

Dale, Edmund H., "The West Indies: A Federation in Search of a Capital," *Canadian Geographer,* 5, 2 (Summer 1961), 44–52.

———, "The State-Idea: Missing Prop of the West Indies Federation," *Scottish Geographical Magazine,* 78, 3 (Dec. 1962), 166–176.

Darby, H. C., "The Medieval Sea-State," *Scottish Geographical Magazine*, 47 (1932), 136–149.

Deutsch, Herman J., "The Evolution of the International Boundary in the Inland Empire of the Pacific Northwest," *Pacific Northwest Quarterly*, 51, 2 (1960), 63–79.

———, "A Contemporary Report on the 49° Boundary Survey, Pacific Northwest," *Pacific Northwest Quarterly*, 53, 1 (Jan. 1962), 17–33.

Dikshit, Ramesh D., "The Failure of Federalism in Central Africa; a Politico-Geographical Postmortem," *Professional Geographer*, 23, 3 (July 1971), 224–228.

Dore, Isaak I., "Recognition of Rhodesia and Traditional International Law: Some Conceptual Problems," *Vanderbilt Journal of Transnational Law*, 13, 1 (Winter 1980), 25–41.

Douglas, Neville, "Amorphous Peoples Will Not Succeed: A Lesson for 'The North'," *Political Geography*, 12, 2 (March 1993), 156–160.

Driver, Felix, "Political Geography and State Formation: Disputed Territory," *Progress in Human Geography*, 15, 3 (1991), 268–280.

Dupont, Louise, "My Nation Is Not a Country; My Country Has No Nation," *Political Geography*, 13, 2 (March 1994), 188–191.

**E**

Edwards, K. C., "Luxembourg: How Small Can a Nation Be?" *Northern Geography* (1966), 256–267.

Ennals, David, "Tibetan Question; Report on International Consultation on Tibet," *Journal of Asian and African Affairs*, 2 (Dec. 1990), 157–167.

Erasmus, Gerhard and Debbie Mannan, "Where Is the Orange River Mouth? The Demarcation of the South African/Namibian Maritime Boundary," *South African Yearbook of International Law*, 13 (1987–88), 49–71.

Eyre, John D., "Japanese-Soviet Territorial Issues in the Southern Kurile Islands," *Professional Geographer*, 20, 1 (Jan. 1968), 11–15.

**F**

Fifer, J. Valerie, "Washington, D.C.; The Political Geography of a Federal Capital," *Journal of American Studies*, 15 (1981), 5–26.

Fischer, Eric, "On Boundaries," *World Politics*, 1, 2 (Jan. 1949), 196–222.

Flint, Colin, "Back to Front? The Existence and Threat of Extremism in English Nationalism," *Political Geography*, 12, 2 (March 1993), 180–184.

Floyd, Barry N., "Pre-European Political Patterns in Sub-Saharan Africa," *Bulletin, Ghana Geographical Association*, 8, 2 (1963), 3–11.

Franck, Dorothea Seelye, "Pakhtunistan Dispute—Disposition of a Tribal Land," *Middle East Journal*, 6, 1 (1952), 49–68.

**G**

Galnoor, I, "Territorial Partition of Palestine: The 1937 Decision," *Political Geography Quarterly*, 10, 4 (Oct. 1991), 382–404.

Gibson, James R., "Russia on the Pacific: The Role of the Amur," *Canadian Geographer*, 12, 1 (1968), 15–27.

Griffiths, Ieuan, "The Scramble for Africa: Inherited Political Boundaries," *Geographical Journal*, 152, 2 (1986), 204–216.

Griffiths, Ieuan L. and D. C. Funnell, "The Abortive Swazi Land Deal," *African Affairs*, 90, 358 (Jan. 1991), 51–64.

Grossman, David and Zeev Safrai, "Satellite Settlements in Western Samaria," *Geographical Review*, 70, 4 (Oct. 1980), 446–461.

———, "The Expansion of the Settlement Frontier of Hebron's Western and Southern Fringes," *Geographical Research Forum*, 5 (1982), 57–73.

Gruffudd, Pyrs, "Remaking Wales: Nationbuilding and the Geographical Imagination, 1925–50," *Political Geography*, 14, 3 (April 1995), 219–239.

**H**

Haas, M., "Paradigms of Political Integration and Unification," *Journal of Peace Research*, 21 (1984), 47–60.

Hamdan, G., "Capitals of the New Africa," *Economic Geography*, 40 (July 1964), 239–253.

Harris, Chauncey D., "Unification of Germany in 1990," *Geographical Review*, 81 (April 1991), 170–182.

Hartshorne, Richard, "Geographic and Political Boundaries in Upper Silesia," *Annals, AAG*, 23, 4 (Dec. 1933), 195–228.

———, "Suggestions on the Terminology of Political Boundaries," *Annals, AAG*, 26, 1 (March 1936), 56–57.

———, "The Concept of Raison *d'Être* and Maturity of States," *Annals, AAG*, 30, 1 (1940), 59–60.

———, "The Functional Approach in Political Geography," *Annals, AAG*, 40, 2 (June 1950), 95–130.

Haupert, John, "Political Geography of the Israeli-Syrian Boundary Dispute, 1949–1967," *Professional Geographer*, 21, 3 (May 1969), 163–171.

———, "Jerusalem: Aspects of Reunification and Integration," *Professional Geographer*, 23, 4 (Oct. 1971), 312–319.

Helin, Ronald A., "Finland Regains an Outlet to the Sea: The Saimaa Canal," *Geographical Review*, 58, 2 (April 1968), 167–194.

Hennayake, Shantha K., "Interactive Ethno-nationalism: An Alternative Explanation of Minority Ethnonationalism," *Political Geography*, 11, 6 (Nov. 1992), 526–549.

Herz, John H., "The Rise and Demise of the Territorial State," *World Politics*, 9, 4 (July 1957), 473–493.

———, "The Territorial State Revisited; Reflections on the Future of the Nation-State," *Polity*, 1, 1 (Fall 1968), 11–34.

Hill, James E., Jr., "El Chamizal: A Century-Old Boundary Dispute," *Geographical Review*, 55, 4 (Oct. 1965), 510–522.

———, "El Horcon: A United States—Mexican-Boundary Anomaly," *Rocky Mountain Social Science Journal*, 4, 1 (April 1967), 49–61.

Hodgkiss, A. G. and Robert W. Steel, "The Shatter-belt in Relation to the East-West Conflict," *Journal of Geography*, 51, 7 (Oct. 1952), 265–275.

Holdich, Thomas H., "Political Boundaries," *Scottish Geographical Magazine*, 32 (1916), 497.

———, "Geographical Problems in Boundary Making," *Geographical Journal*, 47, 6 (June 1916), 421–439.

Horvath, R. J., "The Wandering Capitals of Ethiopia," *Journal of African History*, 10, 2 (1969), 205–219.

Howe, Geoffrey, "Sovereignty and Interdependence; Britain's Place in the World," *International Affairs*, 66 (Oct. 1990), 675–695.

Huaraka, Tunguru, "Walvis Bay and International Law," *Indian Journal of International Law*, 18, 2 (1978), 160–174.

Hyde, Charles Cheney, "Notes on Rivers as Boundaries," *American Journal of International Law*, 6 (1912), 901–909.

———, "Maps as Evidence in International Boundary Disputes," *American Journal of International Law*, 27, 2 (April 1933), 311–316.

**I**

Innis, H.A. and Jan O. Broek, "Geography and Nationalism: A Discussion," *Geographical Review*, 35, 2 (1945), 301–311.

**J**

James, Alan, "Sovereignty in Eastern Europe," *Millennium*, 20 (Spring 1991), 81–89.

Jefferson, Mark, "The Problem of the Ecumene: The Case of Canada," *Geografiska Annaler*, 16 (1934), 146–158.

Johnston, Ronald J., "Marxist Political Economy; The State and Political Geography," *Progress in Human Geography*, 8 (1984), 473–492.

Jones, Stephen B., "The Forty-Ninth Parallel in the Great Plains: The Historical Geography of a Boundary," *Journal of Geography*, 31, 9 (Dec. 1932), 357–368.

———, "The Cordilleran Section of the Canada-United States Borderland," *Geographical Journal*, 89, 5 (May 1937), 439–450.

———, "The Description of International Boundaries," *Annals, AAG*, 33 (1943), 99–117.

———, "A Unified Field Theory of Political Geography," *Annals, AAG*, 44, 2 (June 1954), 111–123.

———, "Boundary Concepts in the Setting of Place and Time," *Annals, AAG*, 49, 3 Pt. 1 (June 1959), 241–255.

**K**

Karan, Pradyumna P., "Dividing the Waters: A Problem in Political Geography," *Professional Geographer*, 13, 1 (Jan. 1961), 6–10.

———, "The Sino-Soviet Boundary Dispute," *Journal of Geography* (1964), 216–222.

———, "The India-Pakistan Enclave Problem," *Professional Geographer* 18, 1 (Jan. 1966), 23–25.

———, "The Indochinese Boundary Dispute," *Journal of Geography*, 59, 1 (Jan. 1969), 16–21.

Katz, Yossi, "The Re-emergence of Jerusalem: New Zionist Approaches in Attaining Political Goals Prior to the First World War," *Political Geography*, 14, 3 (April 1995), 279–293.

Kearns, Kevin C., "Belmopan: Perspective on a New Capital," *Geographical Review*, 63, 2 (April 1973), 147–169.

Khadduri, Majid, "Iraq's Claim to the Sovereignty of Kuwait," *New York Journal of International Law and Politics*, 23 (Fall 1990), 5–34.

Klemencic, Vladimir and Milan Bufon, "Cultural Elements of Integration and Transformation of Border Regions (The Case of Slovenia)," *Political Geography*, 13, 1 (Jan. 1994), 73–83.

Knight, David B., "Gaberones: A Viable Proposition," *Professional Geographer*, 17, 6 (Nov. 1965), 38–39.

———, "Impress of Authority and Ideology on Landscape: A Review of Some Unanswered Questions," *Tijdschrift voor Economische en Sociale Geografie*, 63 (1971), 383–387.

———, "Regionalisms and Nationalisms and the Canadian State," *Journal of Geography*, 83 (1984), 212–220.

———, "Statehood: A Politico-Geographic and Legal Perspective," *GeoJournal*, 28 (1992), 311–318.

———, "Skating on Thin Ice: Comments on 'Mon Pays ce N'est Pas un Pays' Full Stop," *Political Geography*, 13, 2 (March 1994), 182–187.

Kovács, Zoltán, "Border Changes and Their Effect on the Structure of Hungarian Society,"

*Political Geography Quarterly*, 8, 1 (Jan. 1989), 79–86.

Kratochwil, Friedrich, "Of Systems, Boundaries and Territoriality: An Inquiry into the Formation of the State System," *World Politics*, 39 (1987), 27–52.

Kristof, Ladis K.D., "The Nature of Frontiers and Boundaries," *Annals, AAG*, 49, 3 Pt. 1 (June 1959), 269–282.

———, "The State Idea, the National Idea and the Image of the Fatherland," *Orbis*, 11 (1967), 238–255.

**L**

La Ponce, J. A., "On Three Nationalist Options," *Political Geography*, 13, 2 (March 1994), 192–194.

Lattimore, Owen, "The New Political Geography of Inner Asia," *Geographical Journal*, 119, 1 (March 1953), 17–32.

Lauterpacht, E., "River Boundaries: Legal Aspects of the Shatt-Al-Arab Frontier," *International and Comparative Law Quarterly*, 9 (1960), 206–236.

Lewis, John P., "Some Consequences of Giantism: The Case of India," *World Politics*, 43 (April 1991), 367–389.

Lowenthal, David, "The West Indies Chooses a Capital," *Geographical Review*, 48, 3 (July 1958), 336–364.

**M**

Magnusson, W. and R.B.J. Walker, "Decentering the State: Political Theory and Canadian Political Economy," *Studies in Political Economy*, 26 (1988), 37–71.

May, R. J., "Ethnic Separatism in Southeast Asia," *Pacific Viewpoint*, 31 (Oct. 1990), 28–59.

McCune, Shannon, "Physical Bases for Korean Boundaries," *Far Eastern Quarterly*, (1946), 272–288.

———, "The Thirty-eighth Parallel in Korea," *World Politics*, 1, 2 (Jan. 1949), 223–232.

McEwen, Alec C., "The Establishment of the Nigeria/Benin Boundary, 1889–1989," *Geographical Journal*, 157 (March 1991), 62–70.

McKay, J. Ross, "The Interactance Hypothesis and Boundaries in Canada: A Preliminary Study," *Canadian Geographer*, 11 (1958), 1–8.

McKee, J. O., "The Rio Grande: The Political Geography of a River Boundary," *Southern Quarterly*, 4, 1 (1965), 29–40.

———, "An Application of the Jones Field Theory to Rhodesia and the Concept of the Historical Time Period," *Virginia Geographer*, 6, 1 (Spring–Summer 1974), 11–14.

McManis, Donald R., "The Core of Italy: The Case for Lombardy–Piedmont," *Professional Geographer*, 19, 5 (Sept. 1967), 251–257.

Melamid, Alexander, "The Economic Geography of Neutral Territories," *Geographical Review*, 45, 3 (July 1955), 359–374.

———, "Political Boundaries and Nomadic Grazing," *Geographical Review*, 55, 2 (April 1965), 287–290.

Mercer, John, "A Comment from the Real North to the Southerner, P.J. Taylor," *Political Geography*, 12, 2 (March 1993), 169–173.

Mikesell, Marvin W., "The Myth of the Nation State," *Journal of Geography*, 82 (1983), 257–260.

Miles, William F.S. and David A. Rochefort, "Nationalism Versus Ethnic Identity in Sub-Saharan Africa," *American Political Science Review*, 85 (June 1991), 393–403.

Minghi, Julian Vincent, "Point Roberts, Washington: The Problem of an American Exclave," *Yearbook, Association of Pacific Coast Geographers*, 24 (1962), 29–34.

———, "Boundary Studies in Political Geography," *Annals, AAG*, 53, 3 (Sept. 1963), 407–428.

Modelski, George, "The Long Cycle of Global Politics and the Nation-State," *Comparative Studies in Society and History*, 20 (1978), 214–235.

Morris, Michael A., "The 1984 Argentine-Chilean Pact of Peace and Friendship," *Oceanus*, 28, 2 (1985), 93–96.

Mumme, Stephen P., "Regional Power in National Diplomacy: The Case of the U.S. Section of the International Boundary and Water Commission," *Publius*, 14, 4 (Fall 1984), 115–136.

Murphey, Rhoads, "New Capitals of Asia," *Economic Development and Cultural Change*, 5, 3 (April 1957), 216–243.

Murphy, Alexander B., "Historical Justifications for Territorial Claims," *Annals, AAG*, 80, 4 (Dec. 1990), 531–548.

**N**

"Namibia, South Africa, and the Walvis Bay Dispute," *Yale Law Journal*, 89, 5 (April 1980), 903–922.

*National and International Boundaries, Thesaurus Acroasium* (Thessaloniki), 14 (1985), whole volume.

Newman, David, "The Evolution of a Political Landscape: Geographical and Territorial Implications of Jewish Colonization in the West

Bank," *Middle Eastern Studies*, 21, 2 (April 1985), 192–205.

**O**

*On the State*. Special issue of *International Social Science Journal*, 32, 4 (1980).

Orridge, Andrew W., and Colin H. Williams, "Autonomous Nationalism," *Political Geography Quarterly*, 1, 1 (Jan. 1982), 19–39.

O'Sullivan, Patrick, "Further Reply to Taylor's 'The Meaning of the North': The Displacement of Identity: On Being Irish," *Political Geography*, 13, 3 (May 1994), 270–278.

**P**

Paddison, Ronan, "Scotland, the Other and the British State," *Political Geography*, 12, 2 (March 1993), 165–168.

Pearcy, G. Etzel, "Boundary Types," *Journal of Geography*, 64, 7 (1965), 300–303.

———, "Boundary Functions," *Journal of Geography*, 64, 8 (1965), 346–349.

Penrose, Jan, "Mon Pays ce N'est Pas un Pays' Full Stop: The Concept of Nation as a Challenge to the Nationalist Aspirations of the Parti Québécois," *Political Geography*, 13, 2 (March 1994), 161–181.

Pfaff, D., "The Capital Cities of Africa with Special Reference to New Capitals Planned for the Continent," *Africa Insight*, 18, 4 (1988), 187–196.

Pickles, John and Jeff Woods, "South Africa's Homelands in the Age of Reform: The Case of Qwa Qwa," *Annals, AAG*, 82, 4 (Dec. 1992), 629–652.

Platt, Rutherford H., "Berlin: Island of Detente," *Journal of Geography*, 86 (1987), 263–268.

———, *Pluralism and Federalism*. Special issue of *International Political Science Review*, 5, 4 (1984).

Portugali, Juval and Michael Sonis, "Palestinian National Identity and the Israeli Labor Market: Q-Analyses," *Professional Geographer*, 43, 3 (August 1991), 265–279.

Potts, Deborah, "Capital Relocation in Africa: The Case of Lilongwe in Malawi," *Geographical Journal*, 151, 2 (July 1985), 182–196.

Pounds, Norman J.G., "The Origin of the Idea of Natural Frontiers in France," *Annals, AAG*, 41, 2 (June 1951), 146–157.

———, "France and 'Les Limites Naturelles' from the Seventeenth to the Twentieth Centuries," *Annals, AAG*, 44, 1 (March 1954), 51–62.

———, "History and Geography: A Perspective of Partition," *Journal of International Affairs*, 18, 2 (1964), 161–172.

Pounds, Norman J.G. and Sue Simons Ball, "Core

Areas and the Development of the European States System," *Annals, AAG*, 54, 1 (March 1964), 24–40.

Pringle, D.G., "An Irish Perspective on the English Question," *Political Geography*, 12, 2 (March 1993), 161–164.

**Q**

Qureshi, S.M.M., "Pakhtunistan: The Frontier Dispute Between Afghanistan and Pakistan," *Pacific Affairs*, 39, 1–2 (Spring–Summer 1966), 99–114.

**R**

Ratner, Steven R., "The Cambodia Settlement Agreements," *American Journal of International Law*, 87, 1 (Jan. 1993), 1–41.

Ratzel, Friedrich, "The Territorial Growth of States." *Scottish Geographical Magazine*, 12 (1896), 351–361.

Redner, Harry, "Beyond Marx-Weber: A Diversified and International Approach to the State," *Political Studies*, 38 (Dec. 1990), 638–653.

*Regional Politics and Policy*. Published thrice a year since 1991 by Frank Cass, London.

Reich, Robert B., "What Is a Nation?" *Political Science Quarterly*, 106 (Summer 1991), 193–209.

Reitsma, Hendrik A., "Agricultural Changes in the American-Canadian Border Zone, 1954–1978," *Political Geography Quarterly*, 7, 1 (Jan. 1988), 23–38.

Richards, J. Howard, "Changing Canadian Frontiers," *Canadian Geographer*, 5, 4 (Winter 1961), 23–29.

Richardson, Elliot L., "Jan Mayen in Perspective," *American Journal of International Law*, 82, 3 (July 1988), 443–458.

Ritter, G. and G. J. Hajdu, "The East-West German Boundary," *Geographical Review*, 79 (1988), 326–344.

Roberts, Geoffrey K., "'Emigrants in Their Own Country'; German Reunification and Its Political Consequences," *Parliamentary Affairs*, 44 (July 1991), 373–388.

Robinson, G.W.S., "West Berlin: The Geography of an Exclave," *Geographical Review*, 43, 4 (Oct. 1953), 540–557.

Roeder, Philip G., "Soviet Federalism and Ethnic Mobilization," *World Politics*, 43 (Jan. 1991), 196–232.

Ruggie, J. G., "Territoriality and Beyond: Problematizing Modernity in International Relations," *International Organization*, 47, 1 (1993), 139–174.

Rumble, Gary A., "Federalism, External Affairs and Treaties: Recent Developments in Australia," *Case Western Reserve Journal of International Law*, 17, 1 (Winter 1985), 1–42.

**S**

Salisbury, Howard G., III, "The Israeli-Syrian Demilitarized Zone: An Example of Unresolved Conflict," *Journal of Geography*, 71, 2 (Feb. 1972), 109–116.

Sanguin, André-Louis, "Territorial Aspects of Federalism: A Geography Yet to Be Made," *Scottish Geographical Magazine*, 99, 2 (Sept. 1983), 66–75.

Schwartz, Lee, "USSR Nationality Redistribution by Republic, 1979–1989; From Published Results of the 1989 All-Union Census," *Soviet Geography*, 32 (April 1992), 209–248.

Semple, Ellen Churchill, "Geographical Boundaries," *Bulletin, American Geographical Society*, 39 (1907), 385–397, 449–463.

Seyersted, Finn, "Federated and Other Partly Self-Governing States and Mini-States in Foreign Affairs and in International Organizations," *Nordic Journal of International Law*, 57, 3 (1988), 369–375.

Slowe, Peter M., "The Geography of the Borderlands; The Case of the Quebec-US Borderlands," *Geographical Journal*, 157 (July 1991), 191–198.

Smith, Fiona M., "Politics, Place and German Reunification: A Realignment Approach," *Political Geography*, 13, 3 (May 1994), 228–244.

Snyder, D. E., "Alternate Perspectives on Brasilia," *Economic Geography*, 40, 1 (Jan. 1964), 34–45.

Sorenson, John, "Discourses on Eritrean Nationalism and Identity," *Journal of Modern African Studies*, 29 (June 1991), 301–317.

Spate, O.H.K., "Factors in the Development of Capital Cities," *Geographical Review*, 32 (1942), 622–631.

———, "The Partition of the Punjab and of Bengal," *Geographical Journal*, 110, 4–6 (Oct. 1947), 201–222.

———, "The Partition of India and the Prospects for Pakistan," *Geographical Review*, 38, 1 (Jan. 1948), 5–29.

———, "Two Federal Capitals: New Delhi and Canberra," *Geography Outlook*, 1, 1 (Jan. 1956), 1–8.

Steinberg, Philip E., "Comment: Territory, Territoriality and the New Industrial Geography," *Political Geography*, 13, 1 (Jan. 1994), 3–5.

*Studies in Zionism*, published twice a year since 1979 by Frank Cass in London.

**T**

Tamaskar, B.G., "The Concept of Boundaries in India Through the Ages," *Indian Political Science Review*, 12, 1 (Jan. 1978), 59–72.

Tamir, Yael, "The Right to National Self-determination," *Social Research*, 58 (Fall 1991), 565–590.

Taylor, Peter J., "Disunited Kingdom," *Political Geography*, 12, 2 (March 1993), 185–189.

———, "The Meaning of the North: England's 'Foreign Country' Within?," *Political Geography*, 12, 2 (March 1993), 136–155.

———, "The State as Container: Territoriality in the Modern World-System," *Progress in Human Geography*, 18, 2 (March 1994), 151-162.

Tietze, Wolf, "What Is Germany—What Is Central Europe (Mitteleuropa)?" *Geojournal*, 19 (1989), 173–176.

Turnock, David, "Bucharest: The Selection and Development of the Romanian Capital," *Scottish Geographical Magazine*, 86, 1 (April 1970), 53–68.

**V**

van der Vyver, J. D., "Statehood in International Law," *Emory International Law Review*, 5, 1 (Spring 1991).

Vining, Daniel R., Jr., "The Growth of Core Regions in the Third World," *Scientific American*, 252 (April 1985), 45.

**W**

Waterman, Stanley, "Partitioned States," *Political Geography Quarterly*, 6 (1981), 90–106.

Weber, C., "Reconsidering Statehood: Examining the Sovereignty/Intervention Boundary." *Review of International Studies*, 18 (1992), 199–216.

Weissberg, Guenter, "Maps as Evidence in International Boundary Disputes: A Reappraisal," *American Journal of International Law*, 57 (1963), 781–803.

Whitney, Robert A., "New Apartheid: International Law and the Transfer of South African Territory to the Kingdom of Swaziland," *Boston Univ. International Law Journal*, 2, 3 (Summer 1984), 417–448.

Whittam, Daphne E., "The Sino-Burmese Boundary Treaty," *Pacific Affairs*, 34, 2 (1961), 174–183.

Whittlesey, Derwent, "Trans-Pyrenean Spain: The Val d'Aran," *Scottish Geographical Magazine*, 49 (1933), 217–228.

———, "The Impress of Effective Central Authority Upon the Landscape," *Annals, AAG*, 25, 2 (June 1935), 85–97.

Williams, Colin H., "On England's Beleaguered North," *Political Geography*, 12, 2 (March 1993), 174–179.

Williams, Colin H. and A. D. Smith, "Conceived in

Bondage—Called into Liberty: Reflections on Nationalism," *Progress in Human Geography,* 9, 3 (Sept. 1985), 331–355.

Wurtel, D., "Okinawa: Irredenta on the Pacific," *Pacific Affairs,* 35, 4 (1962), 353–374.

**Y**

Yu, Peter Kien-hong, "Issues on the South China Sea: A Case Study," *Chinese Yearbook of International Law and Affairs,* 11 (1991–92), 138–200.

Yu, Steven Kuan-Tsyh, "Who Owns the Paracels and Spratlys? An Evaluation of the Nature and Legal Basis of the Conflicting Territorial Claims," *Chinese Yearbook of International Law and Affairs,* 9 (1989–90), 1–29.

# PART THREE

# *Political Geography Within the State*

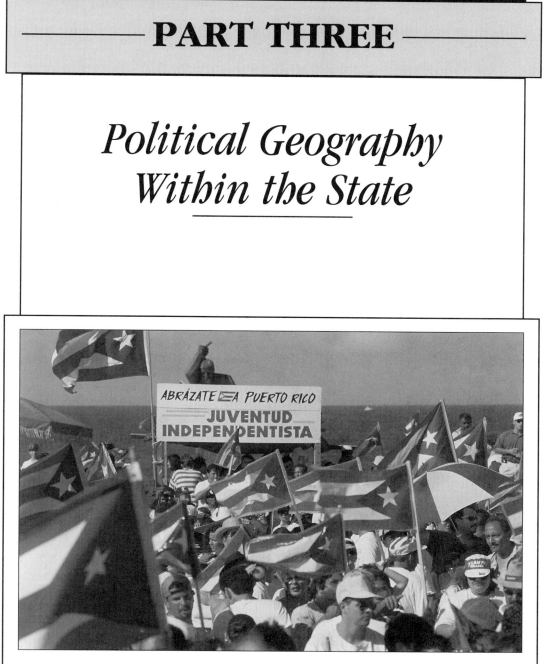

Campaign rally preceding the landmark Puerto Rico plebiscite in November 1993. Continuation of commonwealth status won narrowly over statehood; independence received just over 4 percent of the votes

# 12

# *First-Order Civil Divisions*

In Part Two we described many characteristics of States, including origin, growth, morphology and organization. One characteristic, however, deserves special attention: the internal administrative organization of the State. All States, even the smallest, have internal divisions for administrative purposes.* The largest general-purpose administrative or governmental units within a State are called *first-order civil divisions.* In most States these have subdivisions within them, ranging downward in some cases through second-, third-, and fourth-order civil divisions. Local names for all these divisions vary considerably and can be very confusing if we try to generalize too much about them. The commune, for example, is a first-order civil division in Liechtenstein but fourth order in France. The important thing for us, however, is not their names but their functions, and we consider organization and function in this chapter and the next.

By definition, the first-order civil divisions in federal States have more authority and usually more functions than those in unitary States. In many States, federal and unitary, the capital city is located in a separate district in which its functions are performed by the national government and the municipality. In a few States, notably the United States, the first-order civil divisions may form regional consultative groups and even formal regional authorities to perform specific functions. All are governmental in character, whether or not they have independent taxing power (probably the single most important measure of "sovereignty" at the sub-State level), and even if they are simply administrative arms of the central government. In addition, many States have established special-purpose districts, which we discuss in Chapter 14.

## *Unitary States*

In unitary States, any civil divisions are created by the central government. The central government, whether by constitutional or other procedures, determines their number, boundaries, authority, and operations. Generally, the central government finances them, largely or completely, and appoints their chief officials. In many unitary States, even the first-order civil divisions have no legislative or judicial functions and little decision-making power. Thus they function as administrative districts, designed to make the work of the central government easier or more effective. When conditions in the country change, or the needs of the government, or even the philosophy of the party or

---

*San Marino, for example, has 9 castellos, Burundi has 8 provinces, The Gambia has 5 divisions, Andorra has 6 parroquias (parishes), Liechtenstein has 11 Gemeinden (communes), and Nauru has 14 districts. Even the British colony of Hong Kong has 2 departments divided into 17 districts, and the tiny British Caribbean colony of Montserrat has 3 parishes.

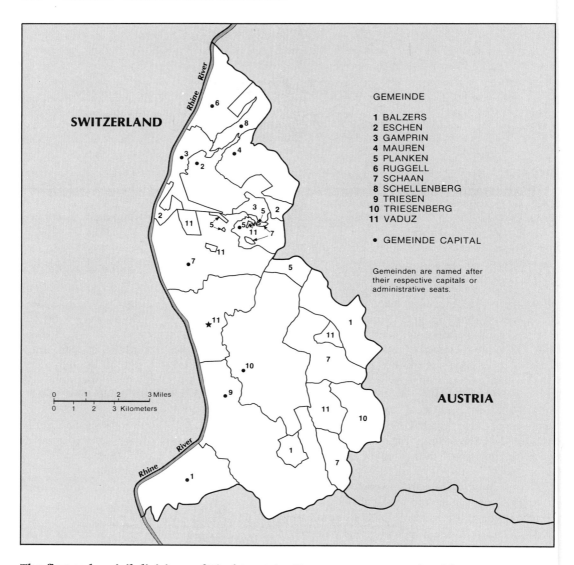

**The first-order civil divisions of Liechtenstein.** Here we see an example of fragmentation of administrative districts within a very small country. Note the numerous exclaves and the oddly shaped *Gemeinden*. Something other than administrative convenience must account for them.

faction in power, then modifications are made in the civil divisions.

We are not saying that civil divisions in unitary States have no personalities of their own or that in practice they are mere appendages of the central government. This is largely true with the lower order civil divisions but does not necessarily describe those of the first order, particularly in old countries or those in which the civil divisions are based on linguistic, tribal, ecclesiastical, historical, or other cultural associations. In these cases the central government must exercise great care in making changes, lest the people involved become angered and even rebellious.

Central governments also face the classic dilemma of the number of levels of civil divisions: if there are too many levels, power tends to gravitate toward the lower levels since the central government finds it difficult to keep track of who is doing what at which level; on the other hand, if there are too few levels, the central government can be overwhelmed with the detail passed upward from a large number of units at each level. In some cases the government decides to create districts on the basis of some cultural, economic, or physiographic criteria so as to aid rational planning and encourage the cooperation of the people involved. In other cases it decides to make the divisions arbitrarily, deliberately breaking up and rearranging traditional or "natural" units so as to weaken or destroy traditional loyalties and foster loyalty to the State.

There is nothing necessarily undemocratic about unitary government or even about repeated reshuffling of civil divisions. In both unitary and federal democratic States, these changes are usually made with the feelings and needs of the people concerned in mind. In those States that are not democratic, whether unitary or federal, the process of creating, changing, or abolishing civil divisions is much easier than in their democratic counterparts. A few examples will help to illustrate the general observations made thus far.

## The United Kingdom

The United Kingdom of Great Britain and Northern Ireland is composed of three ancient kingdoms—England, Scotland, and Wales—and part of a fourth—Northern Ireland. Over a period of several centuries they were welded together into a single unitary State headquartered in London. (The nearby Isle of Man and the Channel Islands are dependencies of the Crown, having a great deal of local autonomy, but they are not part of the United Kingdom.) Each of the four constituent units of the United Kingdom is divided into civil divisions that, despite reorganizations, retain many characteristics of, and links with, political units of ancient times, at least as far back as the Saxon period. Numerous studies, both official and academic, had shown that the traditional hierarchy of governmental units was anachronistic—long on sentiment, short on efficiency. Repeated tinkering with the system

met with considerable opposition from traditionalists, could only compromise on piecemeal changes, and generally resulted in more complexity but no greater efficiency. Finally, Parliament passed the Local Government Acts of 1972 (for England, Wales, and Northern Ireland) and 1973 (for Scotland).

The general effect of the reorganization has been a simplification of the administrative system. Prior to reorganization, there had been 48 counties and 78 county boroughs in England (the latter chiefly cities that had been detached from their counties after 1888 and given equal status); after 1 April 1974 there were only 47 counties and 6 metropolitan counties. In Wales 13 counties and 4 county boroughs were reduced to 8 counties. In each case the lower order civil divisions were also simplified. The county and metropolitan boroughs and the rural and urban districts were abolished. The counties were divided into districts and these into parishes.

By 1991 there had been another reorganization. Now there are 39 counties incorporating 296 districts within which there are more than 10,200 parishes. In addition, there are 32 London boroughs and the City of London, 36 metropolitan districts, and the Isles of Scilly. In 1994 even this new structure was under review, with perhaps more changes coming.

For Scotland, which still retains some autonomy from its period as a kingdom, special legislation has often been passed by Parliament taking into account the special conditions there. Thus, the former 33 counties and 4 counties of a city were reduced on 16 May 1975 to 9 regions and 3 island areas (Shetland, Orkney, Western Isles). Each region has from 3 to 19 districts. In 1994, however, elections were being planned for May 1995 for 28 unitary authorities, including three for the islands, to replace the current two-tier system. Similarly, in Wales, the plan is to replace the current 8 counties, 37 districts within them, and over 800 communities at the lowest level with a single tier of 21 local authorities, to be called, variously, counties, county boroughs, or cities (Cardiff and Swansea). Northern Ireland has only one level of local administration; now there are 26 districts instead of the former 6 counties and 2 county boroughs, and no changes are planned. The Northern Ireland Parliament, however, suspended since 1969, is likely to be revived once a settlement is reached between the British government and the rebel Irish Republican Army.

All of these reorganizations have been carried out by the central government, of course, but only after lengthy consultations with the citizens involved and with cultural factors as well as administrative efficiency in mind. Furthermore, the differing types of civil divisions in each of the four provinces contribute to the rationale for considering the United Kingdom to be already a regional rather than a unitary State.

### France

One of the many changes wrought in France by the Revolution of 1789 was the reorganization of its administrative structure. The major objective was to create a highly centralized State to which all citizens would transfer their allegiance from the provinces, remnants of medieval and earlier independent States. One device for achieving this centralization was the fragmentation of the historic provinces into 83 *départements,* which have since been increased to 96 shown on the map in Chapter 10. (In addition, there are now 5 overseas *départements:* French Guiana, Guadeloupe, Martinique, Reunion, and St. Pierre and Miquelon.) Rather than retaining any of the traditional names, the revolutionary government assigned to the new units names of physical features in the area. The old provinces did not die, however. The *département* boundaries often follow old provincial boundaries, people still identify with the provinces, and the 22 new regions of France (groupings of *départements* for economic and administrative purposes) generally bear the names of old provinces, such as Bretagne, Picardie, Lorraine, Alsace, Aquitaine, Languedoc, and Auvergne.

The *départements* are divided into *arrondissements,* these into *cantons,* and these

into *communes.* All, however, serve as administrative arms of the central government in Paris. The creation of regions in the early 1970s was in part a response to the evident need for larger units for planning and carrying out economic development and environmental management programs, and in part a concession to the durable loyalties of the French people to their traditional provinces. We may see more changes in France in the near future, perhaps the secession of the overseas *départements,* the abolition of the largely nonfunctional *arrondissements* and *cantons,* and the consolidation of many mainland *départements.*

### China

China is the oldest State in the world, with a continuous history of some 5000 years. During that time its territory has expanded and contracted; it has been split, invaded, and occupied; its rulers have frequently changed the administrative structure. Throughout, however, China has remained either in theory or practice a unitary State. During times of weakness and disorder at the center, peripheral portions of the Empire detached and maintained themselves as autonomous States, often still recognizing the nominal suzerainty of the emperor. Even within what used to be called China Proper, local warlords often controlled large regions with little deference paid to the central government. This was the situation that prevailed in China during the waning days of the Manchu (Ch'ing) Dynasty and throughout the Nationalist period (1911–49). When the communists took over the mainland in 1949, they did as many emperors had done before them: they established a strong government over nearly all of China.

After the new government had brought under its control all of what had nominally been China at the end of the Manchu Dynasty (except for Mongolia and Tannu Tuva), it began a series of reorganizations of its territory. At times a number of provinces were grouped together into large economic and political units that were subsequently broken up. Today the first-order civil divi-

**The first-order civil divisions of China.**

sions consist of 23 provinces (24 if Taiwan is included, as the government insists), 5 autonomous regions, and the centrally administered metropolitan areas of Beijing, Tianjin, and Shanghai. The autonomous regions are all on the peripheries and are the homes of most of the non-Han citizens of China. They are Nei Mongol, Xinjiang Uygur, Xizang (Tibet), Ningxia Huizu, and in the south Guangxi Zhuangzu, formerly a province. These autonomous regions constitute 42 percent of the territory of China but contain only some 6 percent of the population. Even some of the lower order civil divisions are based on ethnic groups, but in no case does their autonomy extend to more than cultural affairs; China's government is highly centralized. It uses a variety of technical and administrative devices to maintain control, one

of which is the commune, the smallest minor civil division.

## Nepal

Nepal was united into a kingdom in the latter half of the eighteenth century by King Prithvi Narayan Shah of Gorkha out of a large conglomeration of petty kingdoms. At first the kingdoms retained varying degrees of local autonomy within the unified State, depending on how vigorously they had resisted Prithvi. After they consolidated power, however, the Shah Dynasty began reducing the prerogatives of the local rajas and centralizing the administration. By 1950 the powers of the local rulers had been reduced to limited judicial functions, revenue collection, and the right to exact forced labor. By

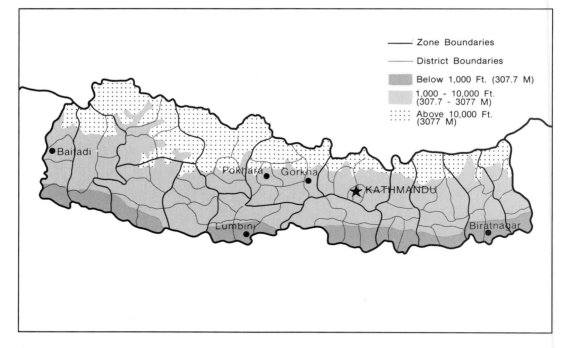

**The districts and zones of Nepal.** Note how most of the zones included districts at different elevations, thus bringing together for planning purposes and administration not only a variety of natural resources, but also different ethnic groups, while separating related ethnic groups that tend to occupy the same types of physical regions.

1961 even these had been abolished. All this time local administration had been developing in those areas controlled by the king. By 1950 there were 32 districts headed by chief administrative officers and three units in the Kathmandu Valley headed by magistrates, all appointed by the central government. The districts were reorganized partially in 1956 and completely in 1961. The 1961 reorganization divided the country into 14 zones, most of which were arranged vertically, including the lowland Terai in the south, the central belt of "hills," and the Himalayas in the north.

The major objectives of this arrangement were stated to be administrative uniformity and better implementation of development projects. Unstated, but very likely of considerable importance, was the objective of breaking with the past and orienting the diverse peoples of Nepal (some 30 ethnic groups) toward the national development program. The zones are subdivided into 75 districts. Within the districts, villages or groups of villages with a population of 2000 or more are organized into village assemblies, while a town of 10,000 or more may have a town council. These lower order divisions sent representatives to the district assemblies, which in turn sent members to the national legislature. This system took some time to put completely into effect, but it functioned as an administrative system of the king, in whom the real power resided until popular demonstrations forced democratic reforms in 1990. The king's power is now greatly reduced, and the new constitution makes no reference to zones.

*Norway*

While Norway is a unitary State, it is also an old one and a democratic one. Its administrative structure is similar to that of the United Kingdom in some ways, since similar forces are at work in both countries. The first-order civil division is the county municipality, of which there are 19 on the mainland, plus Svalbard and Jan Mayen. The counties (except for Oslo) are subdivided into 45 urban and 403 rural municipalities, down from a total of 680 in 1950. The counties are primarily responsible for social services, utilities, firefighting, education, health care, and roads, and they play a large role in development planning. Norway devolved more responsibility from the State to the county and from the county to the municipality. This was not a shift toward either federalism or regionalism, however, but simply a reordering of administrative tasks.

The central government retains the real authority and assigns tasks to the other levels of government as it deems appropriate according to a general formula. The Local Government Act of 1992 did not change the number, boundaries, or functions of the municipalities or the county municipalities, but it did strengthen their autonomy by expanding their freedom to organize their activities and their role as popularly elected local government bodies. The central government continues to supervise both the counties and the municipalities, but they now have less central control than formerly.

## Federal States

Although there are only a relative handful of federal States today, they exhibit as much variety in their territorial organization as the unitary States. We have already discussed the nature of federalism and have attempted to classify federal States on the basis of certain criteria. Now we can look briefly at a few federal States to see how they function. No paper description, however, can give an accurate picture of federalism in action. One of the most bewildering aspects of the United States to a visitor or an immigrant from Sweden or Egypt or Honduras, for example, is just this federal system—the differences from state to state in marriage and divorce laws, motor vehicle codes, professional certification, educational systems, elections, and a thousand other aspects of everyday life. All we can do here is describe

certain salient features of a few federal States and try to derive some patterns from them.

### The Russian Federation

Although the USSR is now history, its former territorial organization is still important to understand, if only because it may reappear in one or more of the successor States.

Its complexity was due not so much to its size as to its ethnic diversity and centrally planned economy. The Tsarist Empire had a relatively simple administrative structure because it made few concessions to its non-Russian subjects and because, while it was autocratic and highly centralized, it engaged in little detailed economic planning. When the Bolsheviks came to power in 1917, their first task was to win the support of the masses, including, if possible, the non-Russian masses.

The Soviet Union was organized in 1922 as a federation of Ukraine, White Russia (Belarus), and, largest of all and itself a federation, the Russian Soviet Federated Socialist Republic (RSFSR). The first-order civil divisions, called union republics, were organized on an ethnic basis, the first experiment of its kind, at least in modern times. Through persuasion and coercion, the outlying nationalities were gradually brought into the federation. The Turkmen and Uzbek Soviet Socialist Republics joined in 1925; the Tadjik SSR in 1929; Kazakhstan, Armenia, Azerbaijan, and Georgia in 1936; and Moldavia (formerly Bessarabia, now Moldova), Estonia, Latvia, and Lithuania in 1940. The Karelo-Finnish SSR was created in 1940 out of territory acquired from Finland, but lost its union republic status in 1956. The 15 union republics had, according to the Soviet constitution, a great deal of autonomy; as pointed out earlier, however, this autonomy was, in fact, fictitious. The republics also were co-equal in theory; in practice, however, the Russian republic was far more equal than all the others.

The lower order civil divisions were organized in two parallel tiers, one based on nationality, the other on essentially economic criteria. In the first tier were the autonomous

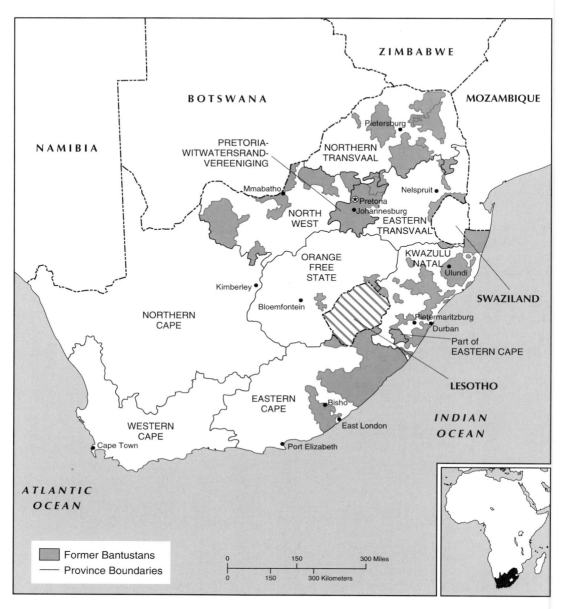

**The new South Africa.** Before the transition to democracy in 1994, South Africa consisted of four provinces and four "Bantustans" (or "homelands") that had been excised from the provinces, which were governed directly from Pretoria. Now the Bantustans have been reincorporated into the nine new provinces, which enjoy some degree of autonomy, making South Africa now arguably a federal State. The Western Cape and PWV (Pretoria-Witwatersrand-Vereeniging) provinces remain the core areas of the country, but Kwazulu Natal (the most populous province and the only one with a monarchy specifically provided for in the national constitution) seems destined to rival them before long.

soviet socialist republics (20 of them, of which 16 were within the RSFSR), then the autonomous *okrugs* (areas). All of the autonomous *okrugs* and most of the autonomous *oblasts* were within the RSFSR.

The second tier consisted of *oblasts* and *krays* (territories). The *rayon* was a still lower order civil division, and the municipality was below that. All these units varied greatly in size and population, and all were subject to change at the will of the central government that created them. In practice, however, the nationality areas remained relatively stable, with only a few up- or down-gradings, while the economic areas were changed more frequently.

The Russian Federation has simplified somewhat this cumbersome administrative structure, but it is still rather complex and fluid. On 31 March 1992, most of Russia's autonomous republics, *oblasts*, *krays*, and autonomous regions signed the Russian Federation Treaty. Four of the old *oblasts* became republics (the new name for the old autonomous republics), but two of the older autonomous republics, Tatarstan and Checheno-Ingushetia, refused to sign, insisting on full independence. Checheno-Ingushetia was divided into Chechen and Ingush republics, but still refused to join the federation. In late 1994 the central government invaded Chechnya but by mid-1995 had still failed to suppress the rebellion. Meanwhile, two regions declared themselves republics: the Urals, with its capital in Yekaterinburg, and the Maritime, with Vladivostok as its capital. By mid-1995 these had still not been recognized as full republics. On 15 February 1994, however, Tatarstan signed an agreement with the Russian Federation that reads almost like a treaty between two sovereign States. The result of all this activity is that Russia today, with 21 republics, 11 autonomous regions, and a total of 90 civil divisions, can probably be listed as a true federation rather than a nominal one as was the USSR.

## *India*

The Indian federation has required a great deal of compromise and adjustment on the part of many people. With partition of the Indian subcontinent and the independence of India and Pakistan in 1947, the Indian State consisted of a patchwork of more than two dozen provinces and some 562 "princely states," the largest of which was nearly the size of France, and the smallest no larger than a large American city. These traditional states had survived the period of British rule, but in a modern federation they clearly had

no place. Thus the internal boundaries of India's provinces were altered to include these principalities, while some "unions" were created out of a large number of these political entities. One of the greater subdivisions thus created—Saurashtra—included more than 200 of these small states. The process of partition and consolidation was not always readily accepted, and there were riots in which thousands were killed. Finally, the Union of India emerged with 27 states and a number of federal territories covering the capital city, several islands, recently acquired colonial enclaves, and some of the northern frontier areas.

Almost as soon as this framework was established, pressures on the government came from several of the states for a revision of boundaries according to the distribution of the more than 200 languages and dialects spoken in India. Thus the state of Bombay was divided into a Gujarati-speaking north and a Marathi-speaking south (Gujarat and Maharashtra, respectively), a move that led to street demonstrations and violence in Bombay. Eventually, the 27 states were reduced to 14, but riots, guerrilla warfare, and other pressures forced renewed reorganization, so that by 1974 there were 21 states and 9 union territories, but not the original 9, some of which had become states. Then in 1975 India annexed the former protectorate of Sikkim, bringing the total number of states to 22. In 1987 the union territories of Arunachal Pradesh and Mizoram in the far northeast were granted statehood, as was Goa, but not the three other former Portuguese territories in India.

Although the basis for the first-order civil divisions in India is similar to that in the former Soviet Union, in this case generally language, the states have been created and changed not by the acts of a few political leaders in the capital but through the democratic process in response to the will of the people (sometimes expressed quite violently). The central government has this power under Article 3 of the Indian constitution. The process has never been easy, since any state created on linguistic grounds is bound to include speakers of other languages and exclude some speakers of the

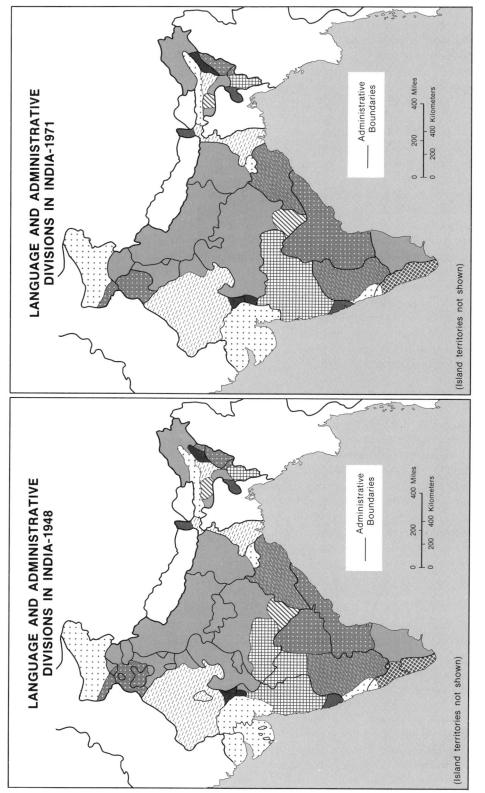

LANGUAGE AND ADMINISTRATIVE
DIVISIONS IN INDIA–1971

— Administrative
Boundaries

0        200        400 Miles

0        200        400 Kilometers

(Island territories not shown)

LANGUAGE AND ADMINISTRATIVE
DIVISIONS IN INDIA–1948

— Administrative
Boundaries

0        200        400 Miles

0        200        400 Kilometers

(Island territories not shown)

Note how the current state and territorial boundaries nearly coincide with the distribution of the major languages. This experiment in ethnic federalism will be interesting to watch over the next generation or two. Since 1971 a number of territories have become states and several other changes in politicoterritorial patterns have taken place, but few of them have been based on language. The "linguapolitical" situation appears stable for the present.

dominant language. In addition, India has many tribal groups and a number of religious minorities whose interests are sometimes given inadequate attention in the process of creating or abolishing states.

The most recent partition of a state, for example (other than the states created out of former districts of Assam), occurred in 1966 when Punjab (which had already been partitioned with Pakistan in 1947) was reduced by the creation of the Hindi-speaking state of Haryana and the transfer of other areas to the union territory of Himachal Pradesh. This left the remainder of Punjab largely Punjabi-speaking, but also largely Sikh rather than Hindu. A problem arose in this partition that is rather interesting. The beautiful new planned city of Chandigarh, capital of Punjab, was on the linguistic boundary, and the two states vied for possession of it as their capital. Eventually they agreed to share it, so today one city serves as the capital of two states.

Although this division of India on a linguistic basis is still experimental, it does appear to be working. Its flexibility has enabled the country to avoid—or postpone—many problems, and there do not appear to be significant centrifugal forces operating within the country today, except for the Sikh extremists.

## Nigeria

Nigeria constitutes another compromise federation. Its initial federal framework, developed during discussions between the (British) colonial rulers and Nigerian political leaders, was established before independence, in 1954. Nigeria has a population approximating 120 million and three distinct and separate core areas. These core areas are the three dominant homelands of nations in the country: the Yoruba in the southwest, the Ibo in the southeast, and the Hausa in the north. When the State achieved independence in 1960, it consisted of three major federal regions and a federal district of Lagos, the federal capital. Because Nigeria contains literally hundreds of tribal peoples and more than 200 languages are spoken today, the

measure of compromise was considerable.

Some Nigerian politicians immediately warned against the inadequacies of the system, insisting that the State would not survive unless a greater number of regions were established to satisfy the demands of minority peoples. The Northern Region, by far the most populous, was in a position to dominate the State. Counterbalancing this was the north's land-locked situation, which rendered it dependent on southern ports for its external trade. Exacerbating the situation was the great contrast between Nigeria's northern and southern regions. Indeed, the north is a world apart from the south, with the strength of its Islamic traditions deriving in large measure from British methods of indirect colonial rule.

Northern fears were aggravated by the prevalence of southerners (especially Ibos) in key positions and jobs in the north. Following independence, outbreaks of violence occurred against Ibo people living in the north, and the survivors returned to the south. In May 1967, in an attempt to placate further many ethnic groups and avoid civil war, the government reorganized the country into 12 states. The Ibo-controlled government of the Eastern region rejected the plan and seceded, declaring itself the sovereign State of Biafra. This touched off a civil war that ended in the surrender of the Ibos and the dissolution of Biafra on 12 January 1970. In 1976 the country was again reorganized, this time into 19 states and a Federal Capital Territory, to which the capital has been moved. Another reorganization in 1991 brought the number of states to 30.

Nigeria is an example of a country that inherited a federal system upon attaining independence, but one clearly not suited to its needs. Tribal unrest, internal frictions and jealousies, economic dislocations, repeated violent changes of government, and experiments with territorial organization have kept the country from achieving what its size, population, location, and resources (including petroleum) had promised for it at independence.

Few generalizations can usefully be made about first-order civil divisions in federal States, and examinations of them in such true federations as Canada, Australia, Switzerland, Germany, and the United States will reveal as much diversity as appears in the three nominal federations just described. Since the United States was the first modern State to be organized as a federation, however, and is still one of the most successful, we devote the entire next chapter to a review of its civil divisions, with some references to other federal States.

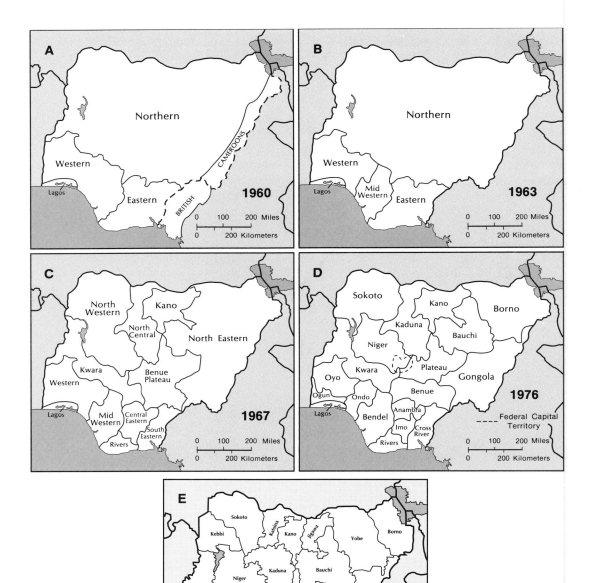

**Nigeria: Changes in first-order civil divisions.** *Map A* In 1954 the territory was reorganized into three regions; the federal district of Lagos; and the Southern Cameroons, part of a British trust territory administered through Nigeria. This was the situation at independence in 1960. *Map B* In a United Nations-sponsored plebiscite in 1961, the Northern Cameroons voted to join Nigeria, and Southern Cameroons voted to join French Cameroun to create the Federal Republic of Cameroun. In 1963, the Western Region was partitioned to create a Midwestern Region. *Map C* In May 1967 the four regions were reorganized into 12 states, but this move led to the secession of the Eastern Region (which called itself Biafra) and to civil war. *Map D* Biafra was defeated in 1970 but ethnic rivalries continued, so the 12 states were redivided into 19 in 1976, and a new Federal Capital Territory was designated at Abuja in a sparsely populated, centrally located area not dominated by any major ethnic group. *Map E* In 1991 the states were reorganized again, this time into 30 states plus Abuja. Most of the newly divided states are in the north and southeast, perhaps because these have tended to be the regions least satisfied with the central government over the years.

# 13

# *Civil Divisions of the United States*

The territorial organization of the United States of America may be considered representative of the class of true federations. The State was created by the Constitution, which in turn was created by representatives of the people acting through their state governments under the Articles of Confederation. Federation was chosen as the method of organization partly because of theoretical advantages, but largely as a compromise, an expedient dictated by the objective conditions of the time and the perceptions of the decision makers. Unity of the 13 former colonies was clearly necessary for survival. Confederation had proved unworkable, and a unitary State was both impossible and undesirable; therefore the only way to achieve unity without destroying diversity and enforcing uniformity was through federalism.

The framers of the Constitution drew ideas from British and French political philosophers, from the experience of Switzerland and the United Provinces of the Netherlands, from the practice of the Iroquois Confederacy, from their own experience under the Articles of Confederation, and from the conditions in their individual states. The result was "a bundle of compromises" so complex and so subtle that generations of scholars, officials, and judges have made careers of analyzing and interpreting it. It has, moreover, been so transformed by amendment, interpretation, and practice that the federal

system of today works quite differently from that envisaged by the founders over 200 years ago. Yet its elements have remained unchanged, and the founders would have little difficulty in recognizing the descendant of their creation, a unique experiment in a form of government unknown in the world at the time. Small wonder that most foreigners—and many Americans—are bewildered by the system!

## The States

The first-order civil division of the United States is the state. There are presently 50 states plus the Commonwealth of Puerto Rico and the semiautonomous federal district in which the capital is located, the District of Columbia. Thirty-seven states have joined the union so far "on an equal footing with the original states in all respects whatever," in the words of the Northwest Ordinance of 1787. This act, one of the last of the Confederation, proved to be among the sturdiest foundations of the federal system, a device for expanding the national territory without creating a permanent colonial empire. It was similar to the Iroquois system, under which the Tuscorora nation was first admitted to the Confederacy in about 1714 as a kind of junior partner (much like the later American territories) and then, in about 1789, as a full member with an equal vote.*

The Constitution granted a number of executive, legislative, and judicial powers (called the enumerated powers) to the national government; others are implied in the wording of the Constitution, and still others, particularly in the area of foreign affairs, are inherent in the existence of a national government. The Tenth Amendment provides that "The powers not delegated to the United States by the Constitution, nor prohibited by it to the States, are reserved to the States respectively, or to the people."† The states thus retain some measure of sovereignty; within their sphere, they are free to do as they please, provided they do not violate the federal Constitution, federal laws, or treaties.

The Constitution provides also for an elaborate system of checks and balances regulating not only the separation of powers of the branches of the federal government, but also the division of powers between the federal government and the state governments. The federal judiciary acts as the major interpreter of all these provisions and leavens the whole system. An extraconstitutional "fourth branch of government," the independent regulatory agencies, has also developed since the late nineteenth century. Their members are appointed by the President, confirmed by the Senate and subject to the jurisdiction of the federal courts, but otherwise operate independently. The system is enormously complex, firm yet flexible, and constantly changing. In our brief survey we

can only scan some aspects of the system that are particularly geographic.

Since the states are in some measure sovereign, they have exclusive jurisdiction over their territory, except for federal land. Thus, even before the Constitution was adopted, the United States could not govern the newly acquired lands west of the Appalachians that were claimed by seven states until these states had ceded the lands to the federal government (1781–1802). In the subsequent territorial acquisitions west of the Mississippi, all the land became part of the public domain except for the state of Texas.*

There have been scores of boundary disputes, not only between the United States and its neighbors, but also among the several states. This is because the territory of the United States was acquired over a long period of time through royal grants, purchases, conquest, annexation, and other means; because the states have much sovereignty in territorial matters; and because much of the land over which they claimed sovereignty was sparsely populated and scarcely known, and surveying methods were primitive, especially under difficult conditions. Most of these disputes have been settled by negotiation, arbitration, or Supreme Court decisions, but some persist today. Most involve rivers whose channels are meandering; some involve maritime boundaries.

In a unitary State, such problems either would not arise at all or would be very quickly resolved by a decision made in the national capital. In a federal state, however, these problems can result in protracted, costly, and bitter struggles.† Geographers

*The Land Ordinance of 1785 had provided for the surveying of the Old Northwest and the laying out of townships 6 miles square (93 sq. km), each consisting of 36 sections of 640 acres (259 hectares) with one section reserved for education. This township and range system was later extended to much of the western land acquired by the United States and was another instrument for the peaceful and orderly integration of new territory. It is also vividly impressed on the map and on the landscape even today in the form of political boundaries, property lines, and road and field patterns.

†In Canada the reverse is true; powers not specifically delegated to the provinces are reserved for the central government.

*As an independent republic before voluntarily joining the United States as a state without first being a territory, Texas was able to negotiate unusual terms for admission, including retention of public lands, its 9-mile territorial sea, its navy, its land survey system, and many other features that still distinguish Texas from other states.

†The United States is not alone in this; even in Brazil, where the states have less sovereignty, a boundary dispute between Espirito Santo and Minas Gerais was not settled until 1963, when the two governors signed a treaty incorporating the suggestions of a joint commission.

have been employed as experts in a number of these interstate boundary disputes, and there is every likelihood that more will be so employed in the future.

As with boundaries, so with state capitals; the states are responsible for their locations. We have already discussed the importance of national capitals. State capitals, while generally not arousing the emotions that national capitals do, may have much greater importance in the everyday life of the average citizen of a federal State—and in the economy and politics of the individual state. Consequently, state capital locations have often engendered much negotiation, propaganda, and even improper conduct as rival factions sought to locate the capital here or there.

In some cases existing cities competed for the prize; in others a new central location was desired for the convenience of the citizens or to thwart the designs of existing claimants; in still others the location was based on existing or projected population concentration, transportation facilities, or economic activity—or simply on the contemporary balance of political forces in the state. Decisions have been made by state legislatures, by special commissions, and by popular vote. Some states have changed their capitals one or more times. Pennsylvania, for example, moved its capital from Philadelphia to Harrisburg in 1812; Ohio's first capital was Chillicothe, then Zanesville for two years, then back to Chillicothe, and finally Columbus in 1816; Michigan followed a similar pattern; Illinois has had three capitals, Oklahoma two, and North Carolina eight. Connecticut even had two capitals at the same time, New Haven and Hartford, until 1873.

Most of these changes took place during the early years of statehood, since capitals, like boundaries, tend to become fixed with time. Alaska, however, which became a state in 1959, decided in a referendum in 1976 to move its capital from Juneau in the far south to a more central site near Willow, north of Anchorage. Thereafter, doubts arose about whether the site was far enough north of Anchorage, if the original cost estimates were realistic, and whether the target date of 1982

for transfer could be met. In 1978 the voters rejected a proposed bond issue to pay for the move but approved an initiative requiring voter approval of all "bondable costs" before any funds could be spent on the move. In 1982 the voters apparently killed the move completely, at least for the foreseeable future.

Over the years many proposals have been made for changes in state boundaries, frequently by disgruntled citizens of a peripheral area of a state demanding that the area be detached and added to an adjoining state or made into a new state. These secession movements were successful a number of times in the nineteenth century. Vermont, Maine, Kentucky, Tennessee, and West Virginia were all parts of, or claimed by, other states before they entered the union. After becoming states, Maryland and Virginia ceded land to the federal government for the creation of the new federal capital district, Missouri added territory in its northwest corner in 1836, Nebraska received small tracts from Dakota Territory in 1870 and 1882 and exchanged some small tracts with Iowa in 1943, New York annexed a part of Massachusetts in 1853, Pennsylvania bought an area in its northwest corner from the federal government in 1792, West Virginia annexed two counties of Virginia in 1866, and a number of states gained or lost small bits of territory after settling boundary disputes.*

Even in recent years there have been proposals, generally short-lived, to transfer southeast Oregon to Nevada; to attach the part of Inyo County, California lying east of Death Valley National Monument, to Nevada; to make a separate state of the Upper Peninsula of Michigan; and, in 1977, to detach Nantucket and Martha's Vineyard islands from Massachusetts (whereupon then-Governor Ella Grasso of Connecticut graciously offered to take them in!). And as regularly as the summer fogs of San Francisco or the Santa Ana winds of Los Angeles, someone proposes the partition of California

---

*In 1846 the federal government returned to Virginia all ceded land south of the Potomac River, a move it has long since regretted. This land is now Arlington County and the City of Alexandria.

into northern and southern states.* No secession movement has been successful, however, since the exceptional period of the Civil War, and it seems unlikely that any state boundaries will be substantially changed in the future. Even imaginative proposals to reorganize the country into more rational first-order civil divisions arouse little enthusiasm in a conservative population satisfied with the present arrangement.†

## Interstate Relations

One virtue of the American federal system is flexibility. Without built-in flexibility, it is doubtful if the system could have survived the changes of the past 200-plus years, changes that would have been almost inconceivable to the system's creators. Enormous demographic and territorial growth; the development of party politics; new technology; involvement in world wars and world affairs generally; immigration of masses of people of almost every existing race and culture; the greatest migration in history, from the countryside to the city; great increases in crime and poverty; environmental degradation; the shift of the population to the South and West—all and more have created problems too big for individual states to handle and that frequently also fall outside the province of the federal government. The states have therefore been forced to cooperate with one another, formally and informally, on a bilateral, regional, and national basis, despite their historic propensity to compete instead. At times this cooperation has had to be encouraged by the federal government.

The states are bound together by an intricate network of conferences, information exchanges, informal agreements, and formal compacts. Many agreements grow out of governors' conferences and the Council of State Governments. The National Governors' Association meets frequently; there are also conferences of Democratic and Republican governors. Most important in practice, however, are the regional conferences of governors, often held several times a year in the South, New England, Middle Atlantic states, the Midwest, and the West.* Other state officials and legislators meet from time to time with their out-of-state counterparts. Both the conferences and the agreements they spawn sometimes have permanent offices and staffs. Boston, for example, contains several regional offices, including those of the New England Governors' Conference and the New England Board of Higher Education.

The most formal agreement among states is the interstate compact. Some compacts require the consent of Congress before they can become effective. There are currently more than 160 of these compacts in operation. Some that are of interest to geographers are the Atlantic States Marine Fisheries Commission, the Southern States Forest Fire Compact, the Ohio River Valley Water Sanitation Commission, the Colorado River Compact, and the Pacific Ocean Resources Compact, in which British Columbia is eligible for associate membership.

One of the more interesting is the Western States Water Council. Organized in 1965 at the initiative of the Western Governors' Conference, it provides a forum for the free discussion and consideration of many vital water policies. It serves in an expert advisory capacity to the governors of the 18 member states. Full members include most of the states from Alaska to South Dakota to Texas; Hawaii, Minnesota, and Oklahoma

---

*Recently, some wags have been proposing a division of the state into *three* parts, to be called, respectively, from south to north, Smog, Fog, and Log!

†In 1973, for example, G. Etzel Pearcy proposed a 38-state arrangement, and in his 1974 book, *Geography and Politics in America,* Stanley Brunn suggested 16 states. In both cases all the proposed states would have completely new boundaries.

*In addition, the New England governors meet annually with the premiers of the eastern provinces of Canada. Out of these meetings have come agreements on energy, joint tourism promotion, and other regional matters. And in late 1989 the governors of Maine, New Hampshire, and Massachusetts and the premiers of Nova Scotia and New Brunswick signed an Agreement on Conservation of the Marine Environment of the Gulf of Maine Between the Governments of the Bordering States and Provinces.

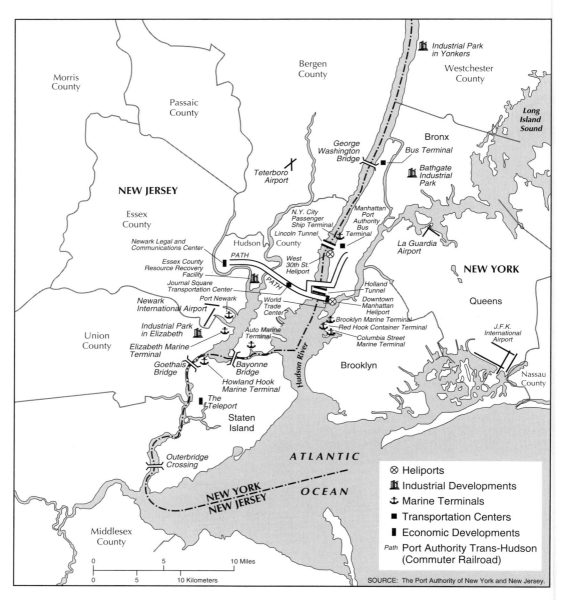

**Facilities operated by the Port Authority of New York and New Jersey.** The Port Authority also has Trade Development Offices in Chicago, Cleveland, New York, Tokyo, London, and Zurich.

are associate members.* The Council conducts quarterly meetings, organizes conferences and symposia, carries out scientific studies, produces many valuable publications, and issues formal resolutions and position statements. Its offices are in Midvale,

Utah. In its own words, "The Council strives to protect western states' water interests, while at the same time serving to coordinate and facilitate efforts to improve western water planning and management." It has played an important role in the long-running struggle over uses of the limited water resources of the West. We shall return to this issue more than once later in this book.

The Appalachian Regional Commission

*In 1993, however, because of severe budget cuts, Montana requested and the Council approved temporary associate membership status, leaving 17 full members.

was established by the Congress in 1965 to plan and coordinate economic development programs in the 13-state area on a joint federal-state basis. Similar commissions for regional economic development in 11 other regions were established between 1966 and 1979 under Title V legislation, but all of the commissions, except for the Appalachian, were abolished by the Reagan administration during the 1980s. In addition, seven commissions were established during the same period for river basin planning under Title II. None of these interstate and regional organizations has received sufficient attention from geographers; even the information about the better known ones is scattered throughout the political science literature. In this book we can only examine the principal features of three of the best known interstate and regional agencies.

*The Port Authority of New York and New Jersey* (originally the Port of New York Authority) was organized in 1921 through an interstate compact that supplemented an 1834 harbor treaty between the two states. It operates under the direction of six unsalaried commissioners from each state, and its programs are subject to the veto of the governor of each state. It has the power to purchase, construct, lease, and operate any terminal or transportation facility within the port district, but it may not levy taxes, and it receives no governmental financial support. It does, however, have the power of eminent domain. At present it owns and operates four major commercial airports, four bridges, two vehicular tunnels under the Hudson River, two heliports, two interstate bus terminals, Port Newark, several marine terminals, most of the docks in New York City, a grain terminal in Brooklyn, a commuter railroad between New York and Newark, the World Trade Center, and a number of other facilities. In addition, it engages in a wide range of activities designed to promote and protect commerce using the Port of New York. Despite criticism for its fiscal policies, it has been remarkably successful in managing a difficult assignment.

A different type of interstate compact is the Delaware River Basin Compact, which created the *Delaware River Basin Commis-*

*sion.* Its unique feature was that the federal government joined New York, New Jersey, Pennsylvania, and Delaware in both the compact and the commission, thus creating a unified federal-state agency for "the planning, conservation, utilization, development, management and control of the water resources of the basin." In order to carry out this mandate, it has been given powers similar to those of the New York Port Authority. It was formed because of a series of disputes over uses of Delaware River water, most immediately the fear of Philadelphia that it might be hurt by New York City's extraction of water from the upper river for its own needs.

The commission now has authority to deal with additional aspects of river use, such as pollution control, watershed management, recreation, and hydroelectric power. While the commission was still being formed in 1961 it was put to its first test—a prolonged drought that brought interstate relations to a water-sharing crisis by 1965. The commission managed to work out a solution that helped the states survive the drought until it ended in 1967. For some years after, the commission largely confined itself to technical matters: gathering data, making analyses, investigating complaints, warning of pollution spills and impending floods, and similar management functions. More recently, it has become active in such areas as wetlands conservation, preparation of an annual Water Resources Program, studies of waste treatment and water salinity, flood stage mapping, an ice jam project, and preparation of recreation maps.

*The Tennessee Valley Authority* is yet another type of regional organization, this one a wholly federal corporation. Its genesis was in facilities the federal government had built at Muscle Shoals on the Tennessee River in Alabama to manufacture explosives during World War I and fertilizer from the nitrogen after the war. There was considerable disagreement, however, about what to do with the facility after the war. The debate dragged on in the Congress until finally the vision and dynamism of Senator George Norris of Nebraska prevailed in the early days of the Roosevelt administration. In May 1933 Con-

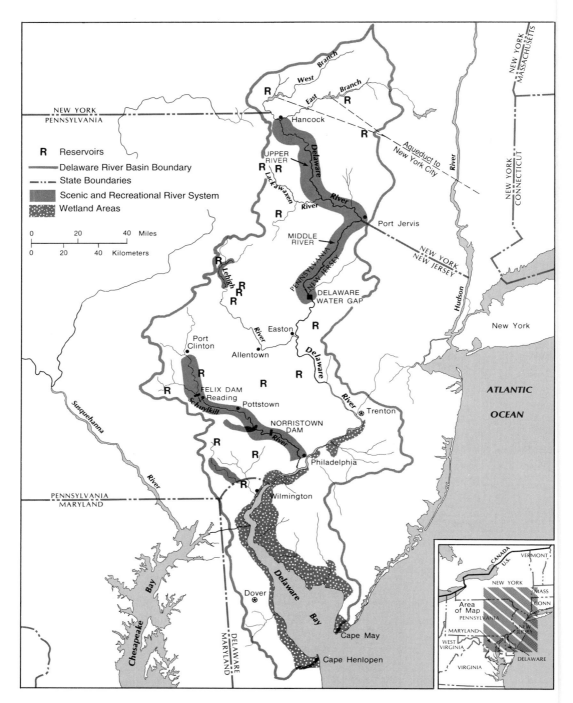

**The Delaware River Basin Commission.** The major physical operations of the Commission are shown on this map. In 1978 the federal government designated both the Upper and Middle River stretches of the Delaware as part of the National Wild and Scenic Rivers system. The other scenic rivers have been so designated by the Commonwealth of Pennsylvania. The Delaware Water Gap is a National Recreation Area under the National Park system.

gress passed a comprehensive act establishing the TVA "to improve the navigability and to provide for the flood control of the Tennessee River; to provide for reforestation and the proper use of marginal lands in the Tennessee Valley; to provide for the national defense." Many specific powers were enumerated in the act, enabling the TVA to be instrumental in the rebirth of a large region of the United States that had been chroni-

cally depressed and poverty-ridden since the Civil War.

TVA operates in a 106,000-square-kilometer (41,000-square-mile) basin covering parts of seven states, sells electric power in an area more than twice this size, and has influence throughout the United States. It has pioneered in the comprehensive, integrated development of an entire river drainage basin, and, despite incessant criti-

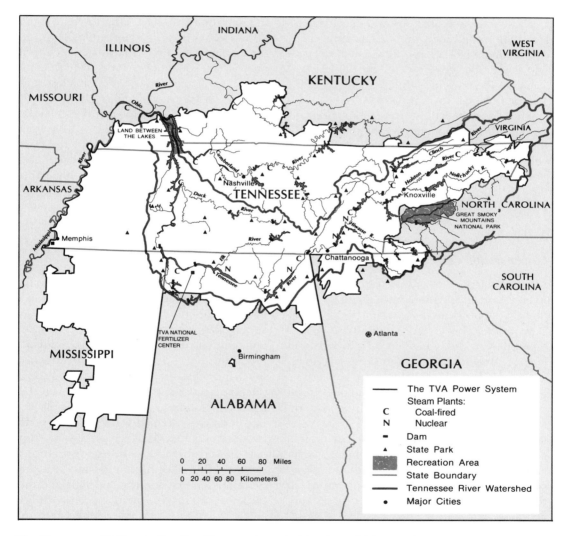

**The Tennessee Valley Authority.** The area bounded by the solid line is the area within which TVA is authorized to sell electric power. (Note that it does not coincide with the river basin itself.) The 11 coal-burning plants produce 68 percent of TVA's electricity, and the four nuclear power plants another 13 percent. Clearly, the region's abundant hydroelectricity cannot meet the needs of its people and their rapidly expanding economy. The dams are multipurpose, providing for hydropower, flood control, navigation, fishing, irrigation and other controlled water uses, and recreation.

cism from special interests and ideologues, has been a conspicuous success. Its organization and operations have been reproduced, with suitable local variations, in many countries around the world—but not in the United States. Despite its remarkable achievements, it is likely to remain singular, largely because of the preemption of its work by other federal agencies in other river basins. Yet it demonstrated the value of unified regional development, and the lessons learned there have been applied elsewhere.

These interstate relations and many others like them, combined with imaginative and ever-changing relations between the states and the central government, are the kind of "creative federalism" found wanting in Latin America, Africa, and most of Asia. All successful federations must make adaptations of this kind. In the Commonwealth of Australia, for example, we find a similar pattern, developed in response to comparable conditions on the basis of a broadly similar, though much newer, federal structure (dating from 1901).

### *Lower Order Civil Divisions*

All the states of the United States except Alaska are subdivided into counties, and these are subdivided into minor civil divisions.* These lower order civil divisions are "creatures of the states" in which they are located, and for this reason they vary considerably. The older states have tended to retain the civil divisions they inherited from colonial times, while in general the pattern for the newer states was set while they were territories. In the early settlement period the county was an important governmental unit, less so in New England than elsewhere. Fascinating stories abound in the literature of the formation, division, combination, and abolition of counties, especially in the West.

Representation was by county in some state legislatures, and that factor alone made them important. Land values, accessibility to government services, law enforcement, and regional pride all contributed to their importance and generated many fierce political battles. The location of the county seat aroused much emotion; the stories of our county seats could alone fill several volumes.

Today there are more than 3000 counties in the United States, ranging in number from three in Delaware to 254 in Texas; and in area from Arlington, Virginia (62.2 square kilometers or 24 square miles)* to San Bernardino County, California, which, with over 52,000 square kilometers (20,000 square miles), is bigger than the four smallest states plus the District of Columbia. Thirty counties have more than 1 million people, including four in New York City, while 24 have fewer than 1000. In 1992 74 percent of the counties had fewer than 50,000 inhabitants each, and Loving County, Texas, had only 107! At the other extreme, Los Angeles County had more than 8 million people. Virginia had 41 independent cities in 1992, with charters similar to those of both cities and counties.

In most states, the functions of the county are assigned to it by the state. In general, counties have been concerned primarily with law enforcement, the administration of justice, building and maintaining highways, supervising county property, assessing and collecting taxes, recording vital statistics, recording legal documents, directing poor relief (now usually called welfare), and conducting elections. Other functions were performed by some counties, others only during certain periods. By the middle of the twentieth century, however, many functions had passed upward to the states or downward to the mushrooming cities. As their functions shrank, so did their staffs and their budgets. Many of them became hollow shells, little more than lines on the map, and

*Connecticut and Rhode Island have deprived their counties of all governmental functions; they survive as statistical and judicial units only. In Louisiana the counties are called parishes. The borough governments in Alaska resemble county governments in other states.

*Arlington County is a Washington suburb that has no incorporated places within it; it functions much like a city. Technically, New York County, 57 square kilometers (22 square miles) is smaller, but it is coterminus with the Borough of Manhattan and is entirely within the City of New York.

there was much talk of abolishing them altogether. Two states did abolish their county governments, but the others retained them, often only for sentimental or political reasons. The series of Supreme Court decisions in the postwar period that required representation in state legislatures by population rather than by county further weakened the county. Now, about 10 percent of the U.S. population is not served by a county government.

At the same time, however, the same societal changes that forced interstate cooperation—population growth, urban sprawl, rise in crime rates, pollution, demands of citizens for more amenities, and so on—also forced many states to assign new functions to their counties. While rural areas continue to lose population, the suburbs are still growing, and it is suburban counties and those containing smaller urban areas that have seen the greatest revival in the past three decades. Many suburban counties, in fact, have become virtually urban governments themselves in terms of both functions and organization. They now provide and maintain recreation areas; deal with solid waste, water supply, flood control, and sewage disposal; and provide library, airport, bus, police, fire, and health and hospital services, among other things. They have become big business, with more than 2 million employees spending over $50 billion in 1977. Many counties have outgrown the old unsalaried board of three or four commissioners or supervisors; they now have full-time, well-paid chief executives and legislatures, and employ computers and full-time planners and managers. Although a case can still be made for the abolition of counties in rural and large urban areas, it is either too soon or too late to consider elimination of the county entirely as a governmental unit. It appears to be here to stay.

In New England the original unit of settlement was the *town* rather than the county, and it performed many of the functions performed by counties elsewhere. As population grew, the towns tended to be divided instead of having their powers increased. As they became more urbanized, they tended to lose most of their functions to the municipalities. Under the same pressures, the famed New England town meeting has tended to be modified into a representative town meeting or to be replaced by more orthodox forms of government. The town system functioned well for a century and a half. It spread, in fact, southward through the Middle Atlantic states, then westward through the Midwest and even to parts of Washington State, until today there are nearly 17,000. Outside New England, New York, and Wisconsin, however, it is called a *township.**

Three factors probably explain the confinement of this tier of government to these areas: the plantation system in the South, which like *latifundia*† everywhere, involved a large measure of political as well as economic and social function; the sparseness of population in the West, which made the county much more significant than it had been in the Northeast; and the migrations of New Englanders westward through the northern tier of states, carrying the town concept in their cultural baggage. Today, in the 20 states in which it exists, there is great variation in the scope of governmental powers and activities of the township, but only about 27 percent of them own and operate even one public service facility, and in 1992 only 6.5 percent of them had as many as 10,000 inhabitants, while 55 percent had fewer than 1000. Many are atrophying and may eventually disappear, though others are likely to endure for some time.

The smallest general-purpose administrative unit in the United States is the *municipality.* Today there are more than 19,000 of them, and they range in population from a few score people to over 7 million in New York City, and in area from a few hectares to the 2142 square kilometers (827 square

---

*Some are called "plantations" in Maine and "locations" in New Hampshire.

†A *latifundium* is a large, privately owned estate worked by slaves or hired workers, chiefly for the benefit of the landowner and his family. For more details, see the section on land reform in Chapter 38.

## LOCAL GOVERNMENTS IN THE UNITED STATES

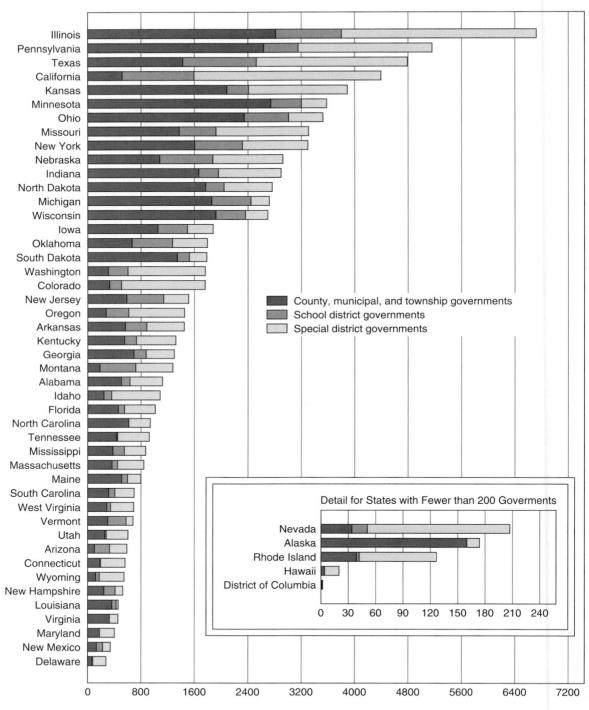

*Source:* Compiled from U.S. Bureau of the Census, *1992 Census of Governments,* Vol. 1, Appendix A. Note particularly the high proportions of special districts in many states. These are discussed in detail in the next chapter.

miles) of Jacksonville, Florida (nearly the size of Rhode Island!).* Since municipalities are creatures of the state and since the country's population is rapidly urbanizing, the trend is toward the creation of ever more municipalities. This contrasts sharply with the other units of government, whose numbers have remained essentially stable in the twentieth century. In 1992 Illinois had the most municipalities (1282), as well as the most governments of all kinds (6622).

There is almost as much variety in the organization and functions of municipalities as there is in their size, and there is not necessarily any correlation among size, organization, and functions. State law (and in some cases the state constitution) determines all these factors, except for population, though some states specify minimum and/or maximum populations for different types of incorporation. In California, for example, a community may incorporate as a "general law" city with as few as 500 people, or as a "charter" city with a minimum of 3500 residents—and these are the only types of municipalities. In Oregon only 150 people are needed for incorporation as a city or a town, which have the same status and functions. In Connecticut there are only towns (townships) and cities. New Jersey, however, charters municipalities as cities, towns, boroughs, and villages, all without regard to size; unincorporated areas are governed by the townships and, as in many southern and western states, by the county.

Functions are also assigned in some states without regard to the size of the municipality, so that in a small one a big-city function may simply be performed by fewer people on a smaller scale for fewer residents than in a big city. In other states, however, formulas determine which classes of municipalities provide which services. Similarly, the state determines which municipalities may raise money in which ways, and what kind and

what level of financial support the state will provide. All of this variety has the advantage of flexibility, of course, but it is increasingly being considered a nuisance at best and frequently a serious impediment to good government at the local level. The relations among municipalities (particularly adjacent ones) and among municipalities, townships, counties, and state are becoming so complex that many states have embarked on reform programs—or at least are discussing them. At the risk of overgeneralization, we can say that the overall trend at present is for functions to be passed upward through the hierarchy of civil divisions and for financial aid to be passed downward, although in the early 1990s it seemed possible that this pattern was being reversed. Both movements include the federal government. There is considerable experimentation with methods of dealing with the manifold problems of urban areas, many involving the political organization of space.

### Dealing with Urban Problems

One way to deal with urban problems is to avoid them altogether by building and governing *new towns* with the best available methods. Although there were scattered precedents elsewhere, the new-town movement began in Britain and Scandinavia just after World War II and has spread over much of the world. In the United States the first two in the postwar era, and perhaps still the best known, are Reston, Virginia, and Columbia, Maryland. During that period, 15 others were started with federal assistance (until the program was discontinued in 1975) in Minnesota, Maryland, Illinois, Texas, Arkansas, New York, North Carolina, South Carolina, Georgia, and Ohio. The success of these communities is debatable (probably Columbia is the most successful so far), but even if it were possible to avoid problems in new towns, they cannot help solve problems in old ones.

Another approach is to *restrict the growth of communities* deliberately. This has long been done through zoning—for instance, requiring minimum building lot sizes of 1 acre or more. Since the Supreme Court has out-

---

*Three municipalities in Alaska (Sitka, Juneau, and Anchorage) have areas exceeding 4400 square kilometers (1700 square miles), but these are special cases, and they all now have consolidated city–borough governments. Another special case is the City and County of Honolulu, which includes the entire island of Oahu.

lawed such "exclusionary zoning," in some places used to keep out "undesirables," other devices have been tried. These include a simple ban on more people moving in through a refusal to register new deeds or issue building permits. Another, more constructive, approach is to purchase vacant land around the community to maintain as publicly owned greenbelts.

Oregon in the 1970s adopted a comprehensive system of statewide growth man-agement, including urban growth bound-aries (UGBs) to promote compact urban de-velopment and to discourage sprawl. Since then, Maine, Vermont, Rhode Island, Mary-land, Georgia, Florida, New Jersey, and Washington have adopted statewide growth control plans, but so far the results have been mixed and inconclusive.

The most common technique for provid-ing urban services to people in unincorpo-rated areas surrounding a municipality has

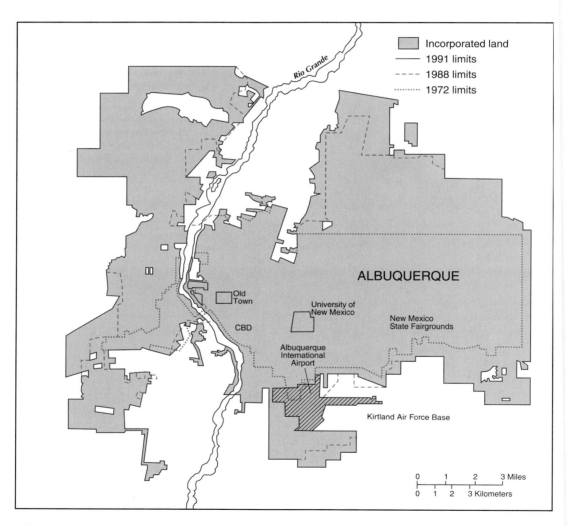

**Albuquerque, New Mexico:** A city that has grown by annexation. Settled in 1706 at a ford on the Rio Grande, the town began expanding eastward when a railroad was laid down a mile to the east of the center (now called Old Town) in 1880. Since then its expansion has been primarily eastward and most recently northward and northwestward. Further expansion is limited by the Sandia Mountains to the east and northeast, the Air Force base and Sandia Scientific Laboratories to the south, and Indian lands to the southwest. Additional annexations are planned in the west and northwest.

been *annexation*. Some cities, notably Albuquerque and Los Angeles, have sprawled out over enormous areas in this way. Others include San Diego, Phoenix, Houston, and San Antonio. There have been two major surges of municipal annexations in the United States. The first occurred roughly from 1850 to 1920, chiefly in the East. Such cities as Philadelphia, Boston, Chicago, Cleveland, Indianapolis, and Denver expanded rapidly. Annexations virtually ceased during the Great Depression and the Second World War. The second big surge has occurred since the war, primarily in the South and West. In the period 1970–77, for example, there were 48,000 annexations by municipalities with populations of over 2,500; the annexed areas had over 2.5 million.

Annexations are generally very small in area, and most of the large ones have been in the South. Chattanooga and Jackson, for example, have more than doubled in area through annexation, and Kansas City, Kansas has almost doubled. This technique has always been very rare in New England because there the towns (townships) are municipal corporations with traditions and powers that render them almost immune to annexation. Other factors that influence the rate of annexations are state laws and policies, the ability of suburban areas to provide urban services, the growth of special-purpose districts (discussed in some detail in the next chapter), socioeconomic differences between cities and their surroundings, the federal Voting Rights Act of 1964, and the declining attractiveness of the city to the suburb or rural area because of at least the perception of rampant crime, corruption, pollution, congestion, and other evils in the city. Furthermore, annexation does not always solve the problems of either the city or the surrounding area.

In some cases annexation solved some major problems, but in other cases it simply postponed the day of reckoning, and, when it came, the problems were bigger because the cities were bigger. More important, however, annexation is of little or no value to cities hemmed in by physiographic features or by other incorporated places. Many unincorporated communities, in fact, have incorporated at considerable expense solely to avoid being "swallowed up" by a neighboring city. Nevertheless, annexations continue at the rate of about 5000 per year, most commonly by cities in the 25,000 to 500,000 population range.

The opposite of annexation is *secession,* or in Census Bureau parlance, detachment. Earlier in this chapter we discussed secession movements in a number of states; they also occur for a variety of reasons at the local level. During the 1980s there were 1635 "detachments" involving 597 square miles but only 60,000 people. More than half of these were in the South. In the early 1990s the East Shore section of New Haven, Connecticut (a city of 130,000 people), containing the wards of Morris Cove, The Annex, and Fair Haven Heights, was the scene of a lengthy secession battle. Some residents complained about high city taxes, poor services, and declining property values. Since these wards are almost entirely inhabited by people of European origin and abut wards with large proportions of darker skinned people, there may have been other motives not publicly stated. Secession would deprive New Haven of several city-owned facilities, including its sewage treatment plant, a large beachfront park, and the airport.

During the same period, there was a secession movement in Staten Island, the only predominantly white borough of New York City, closer to New Jersey than to Manhattan, largely middle-class, conservative, and suburban (even still rural in places), voicing similar complaints. Neither movement seemed likely to be successful. A possible precedent, though in reverse, occurred in 1986 when several predominantly black sections of Boston threatened to secede and form a new city called Mandela. The proposal was rejected in a referendum.

Other methods of dealing with urban problems include authorization by states of *extraterritorial* powers for municipalities (i.e., they may engage in certain activities outside their corporate limits) and *transfer of*

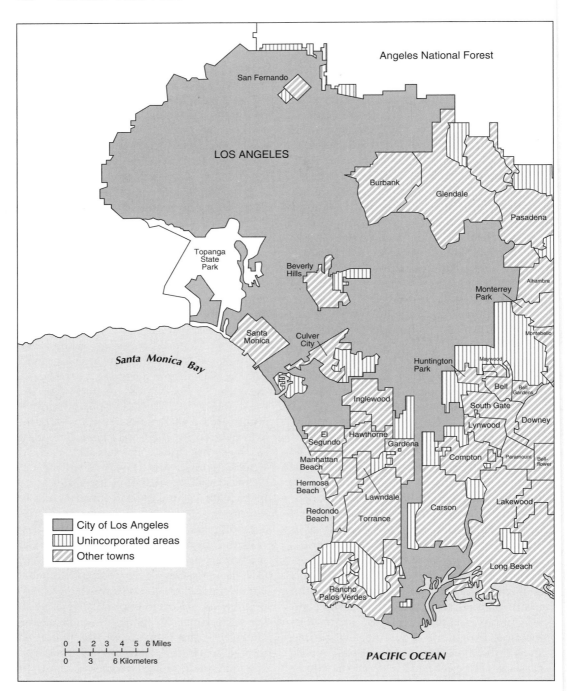

**Los Angeles: A patchwork city.** Founded in 1781 and incorporated in 1850, the town did not begin to grow rapidly until connected by rail with San Francisco in 1876 and the East in 1885. Another milestone was the opening of the "Free Harbor" of San Pedro to the south in 1914 and its connection with Los Angeles through the "shoestring strip," annexed in 1906. Expansion since World War I has been rapid, but numerous communities have resisted annexation and survive as independent municipalities surrounded by Los Angeles. The county of Los Angeles controls the unincorporated areas, which creates many problems, including some related to law enforcement.

**Secession movement in New Haven.** Many signs of this type were prepared by the East Shore Secession Organization and displayed in the early 1990s on the lawns of private homes along Townsend Avenue, the major street in the area, and nearby streets. Like so many other secession movements, this one failed, but we may expect more such attempts in the future. (Martin Glassner)

*functions* to county government. There are also *councils of governments* or other voluntary associations of small and large governments within a metropolitan area. Most are devoted to research, communications among the governments, coordination, and recommendation, but some have modest governmental powers as well. *Intergovernmental contractual arrangements* are most popular in California but have been adopted nationwide. Under such an arrangement a county will perform certain functions for a city on a fee basis, a city will provide services for its suburbs, or one suburb will even provide another with services, all on a contract basis.

None of the methods mentioned so far involves disturbing existing governmental structures. While some are imaginative, all are moderate in their approach to metropolitan area problems. Many observers feel, however, that the governmental structure itself is one of the major causes of contemporary urban problems and that the best way to resolve the problems is to reorganize urban area government in some way. If we consider all the special-purpose districts, including school districts, as governmental units, there are 85,000 governments in the United States. (The total has been reduced from more than 90,000 in 1962, largely because of school district consolidation.) Although this large number of governmental units is not entirely bad and in fact has some very real advantages, there is no question that it has caused overlapping, duplication, "empire building," jurisdictional disputes, petty jealousies, waste, corruption, dangerous gaps, and a host of other ills. Four approaches to reform of this patchwork arrangement have been tried.

*City-county separation* was adopted in Baltimore, San Francisco, and St. Louis in the nineteenth century, in Denver in 1912, and later in Carson City, Nevada. In some cases, however, the experiment was abandoned after a while. Today only in Virginia are cities commonly separated from their counties, when they reach a population of 10,000 and request separation. Separation (except in Virginia) would very likely cause more problems than it would solve today, and it is not seriously considered.

*Consolidation of similar government units* is more attractive to many people, but it also arouses much more opposition, some of which tends to be on emotional rather than practical grounds. It is often based on fear of higher taxes—and often the fear is justified. The opposition has been overcome and some consolidation effected in a few places, generally where it seemed to be the best way to deal with a real crisis situation.

Very few counties have merged, mostly before World War II, but a number of municipalities have. In 1958 Warwick and Newport News, Virginia, consolidated; in 1961 the town of Winchester and the city of Winsted, Connecticut were merged, and Tampa and Port Tampa, Florida, consolidated; and in New Mexico in 1970 Las Vegas city and Las Vegas town merged. One of the largest consolidations in the United States involved North Sacramento and Sacramento, California, which in 1965 formed a united city of over a quarter of a million people. Nevertheless, consolidation efforts have failed more often (as in Pittsburgh, Cleveland, and St. Louis) than they have succeeded recently, and it does not appear that the success rate will increase notably in the near future.*

*City-county merger*, on the other hand, has met with somewhat more acceptance. Usually it involves one major city in a rapidly urbanizing county in a moderate-sized metropolitan area in the population range of about 150,000–800,000. Normally, the smaller municipalities have the option of joining or remaining outside the merged unit. Consolidation by referendum, however, is largely a postwar phenomenon. Mergers in the nineteenth century and the first decades of the twentieth were generally accomplished by state legislation. The first was New Orleans-Orleans Parish in 1805, and others included Honolulu, Denver, Boston, New York, Philadelphia, and San Francisco. Since World War II the only city-county mergers attempted or accomplished solely by legislation have been Indianapolis-Marion County in 1969, Las Vegas-Clark County in 1975, and several in Virginia.

Even those accomplished by referendum have been relatively few. Although scores of such proposals have been made by reform-minded citizens all over the country, only 57 different communities have gotten beyond the proposal stage and actually held referenda. Of the 77 consolidation referenda held between 1947 and 1979, only 17 were approved by the voters, and only four communities approved after having rejected consolidation at least once. Three have rejected it three successive times and Macon, Georgia, four times. Most referenda and most successful consolidation attempts have occurred in the South and more recently the West.†

The distribution of blacks and whites be-

tween the central city and the surrounding county has certainly been an issue in the South, as in other cases outside the South, but its effect is unpredictable: an equal number of consolidation efforts in the South have failed.* Consolidation efforts will undoubtedly continue around the country, since consolidation does result in more efficient government at lower cost. In order to be successful, however, these efforts will have to include carefully planned and executed campaigns to educate the voters on the advantages, counter the often emotional appeals of the opposition, and turn out the vote.

Finally, governmental reform in urbanized areas is being tried through what has been called a "two-tier" approach, a kind of federalism at the local level. In one version of this system the entire metropolitan area, city and suburbs, is placed under a single government that performs a variety of areawide functions, while the preexisting municipalities retain their identity and responsibility for some local functions. This is known as *metropolitan government*.

The prototype for this system is the Municipality of Metropolitan Toronto, created by the Ontario government in 1954. Subsequent provincial action in 1967 consolidated the region's 13 member governments into six units. This was the result of a petition to the Ontario government by the city of Toronto for consolidation and a petition from the suburb of Mimico for some type of areawide administration for the provision of urban services. Public hearings were held, studies were conducted, recommendations made and debated, and finally the legislature acted.

Metro Toronto provides major services for the whole region, including water supply,

*During the period 1970–75, for example, there were only 24 mergers of this type in the whole country, nearly all of them involving very small communities.

†The successful ones included five in Virginia plus Nashville-Davidson County, Tennessee (1962); Jacksonville-Duval County, Florida (1967); Columbus-Muscogee County, Georgia (1970); and Lexington-Fayette County, Kentucky (1974).

*Of the 24 cases so far in which some form of city-county merger has taken place, 8 involve state capitals: Baton Rouge, Boston, Carson City, Denver, Honolulu, Indianapolis, Nashville, and Juneau, which merged with Greater Juneau Borough in 1970. Richmond is an independent city. The most recent consolidation was Athens-Clarke County, Georgia, in 1991. Smaller mergers occurred in 1994 in Missouri and Louisiana with one in South Carolina scheduled to take effect in 1996.

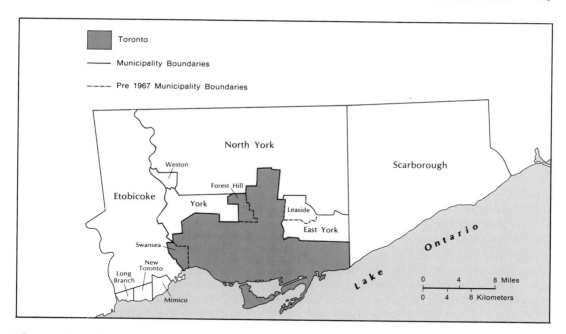

**The Municipality of Metropolitan Toronto.** Created in 1954 with the city of Toronto as its core, Metro Toronto was reorganized in 1967 to bring its component units from 13 to 6. The entire area within the outer boundary is governed as one unit, but the cities of Etobicoke, York, North York, and Scarborough and the borough of East York have some degree of local autonomy. This is the prototype for metropolitan government elsewhere.

sewage system, policing, public transit, and trunk roads. It also funds and provides day care, financially assisted housing, ambulance services, solid waste management, homes for the aged, hotel accommodations, and business licensing. It also owns and operates four major cultural facilities. Despite the usual federal problems of distribution of powers and representation on the governing council, and complaints that the area covered is not large enough, the experiment has been quite successful, and the residents are generally enthusiastic about it.

Not only has Metro Toronto been widely praised, it has also been used as a model for federative governments in Ottawa, Sudbury, and other Ontario cities, and in Edmonton, Alberta and Winnipeg, Manitoba. In the United States this approach has so far been adopted only in Miami-Dade County, Florida, in 1957. It differs from the Toronto model, however, because in Ontario the county was abolished, whereas in Florida home rule was granted in 1956 to the county, which then

adopted a new charter reorganizing county government and transferring many city functions to the county. It was not a consolidation, however, for the 26 municipalities in the county still exist and retain jurisdiction over certain local matters. Nevertheless, the county's authority is very broad, and it has been expanding its jurisdiction. The success of this experiment is by no means certain; it has barely survived a variety of crises and may have to be revised again in the near future. The current trend is toward centralization.

Variations on metro government are being applied in Alaska and the Twin Cities. The Alaska legislature provided in 1961 for the creation of boroughs (essentially regional governments incorporating at least one metropolitan area) on a local option basis, and in 1963 boroughs were made mandatory under certain conditions. Two classes of boroughs were established, based not on population but on powers, and the borough chooses its class. The newest one is Denali

Borough, created early in the 1990s. Yakutat, in the southeastern panhandle, was tentatively approved for borough status in 1992. The remainder of the state constitutes a single unorganized borough. All functions normally performed by lower order civil divisions are divided between the borough and municipal governments.

In the Minneapolis-St. Paul area of Minnesota, the Metropolitan Council, created by the state legislature in 1967, was given jurisdiction over areawide matters, as in Toronto and Dade County, but it was strictly a policy-making body and had no executive powers. It relied on the preexisting metropolitan commissions to implement its policies. Since establishment, however, it has been granted taxing power and the power to execute its proposals in matters of zoning, solid waste disposal, pollution, and noise abatement. In the late 1980s, though, it seemed to be declining as an effective instrument of local government. In 1995, however, the council was given jurisdiction over sewer systems and transportation and made accountable to the governor, thereby strengthening it.

Another form of federative government is the multifunctional *metropolitan service district* (MSD). Few metropolitan areas have established MSDs, and most are in the West. Its range of functions is usually limited. One example is the Municipality of Metropolitan Seattle (commonly known simply as Metro). It was established in 1958 and expanded in 1971—after a considerable struggle in the state legislature—to include all of King County. But King County has its own expanded and modernized government, and both the county and the city are financially viable and administratively effective. Metro, moreover, still does not include Snohomish County, the other county in the metropolitan area, and it performs few functions.

Much younger, and off to a better start, is the MSD of Portland, Oregon, the first *elected* regional government in the country. It covers the urbanized area of the three-county Portland metropolitan area and has jurisdiction over sewerage, water supply, en-

vironmental quality, mass transit, regional parks, cultural and sports facilities, libraries, and correctional facilities. It has a 12-member board elected on a nonpartisan basis and a full-time chief elected officer. Yet it is still not well established and lacks a stable funding source, and its full potential may never be realized.

With so many advantages to be expected from metro and other forms of federative government, why has it not been adopted more widely, especially in the United States? Suburbs often resist metro governments for the same reasons they resist annexation and consolidation. But recently, as this resistance has begun to weaken, new opposition has arisen from within the central cities themselves. Because of changing demographic patterns (primarily the exodus of whites to the suburbs and the inflow of blacks into the cities) and sweeping political and social changes, blacks have at last attained some measure of political power in many cities, and they are unwilling to have this power diluted or possibly obliterated in a governmental unit that would include a number of white-dominated suburbs. Another reason for the greater success of metro government in Canada than in the United States is that Canadian provinces have greater power to create such units than American states do.

Metropolitan reorganization is another field in which geographers can make major contributions, but as yet few have done so. The best effort so far has been by the Canadian geographer Max Barlow. In a recent book he examines various types of metropolitan governments in Australia, Britain, Canada, and the United States.* But there is a great deal more to be done on the subject. It is enormously complex and calls for a thoroughly interdisciplinary approach. One of the chief complicating factors is the multiplicity of special-purpose districts, to which we now turn our attention.

---

*I. M. Barlow, *Metropolitan Government*, London and New York, Routledge, 1991.

# 14

# *Special Purpose Districts*

As numerous as the general purpose governmental units are in the United States, they are more than equaled by special purpose districts. Although other countries have them also, and for similar reasons, nowhere else are they so prolific and so varied. They have become so important partly because of the nature of the American federal system; the high mobility of Americans (changing residence on average every five years); the emphasis on private enterprise and individual decision making regarding land use; the almost total lack of planning in most American cities until recently and its continuing ineffectiveness in many places; and the postwar suburbanization or urban sprawl that has so transformed the American landscape. Historically, the special district (aside from the school district) was created in rural areas to administer drainage, irrigation, water conservation, flood control, fire protection, soil conservation, hydroelectric power, river transportation, and similar functions. Since World War II, however, with cities and suburbs exploding (spatially), the special district has experienced a phenomenal rate of growth within or overlapping metropolitan areas, especially those included in metropolitan statistical areas (MSAs).*

The traditional special districts continue to function in metropolitan areas, though often their organization and services are adjusted to meet changing needs, and new ones are being created to meet increased demands for their services. Newer types of districts, however, associated with largely urban needs, are being formed in even greater numbers. In addition to those shown in the bar graph, they include transportation (bus, rail, seaport, bridge, tunnel), law enforcement, courts, utilities, air and water pollution control, and solid waste disposal districts. They are usually operated by commissions or boards. Some districts perform multiple but related functions. Some districts that straddle state boundaries were discussed in the previous chapter. The only essential difference between them and *intrastate* districts is that interstate districts must generally be created by interstate compacts.

In 1987 there were more than 29,500 special districts, not including school districts. As may be recalled from the graph in Chapter 13, these districts are not distributed evenly throughout the country. Sixty-one percent of them were functioning in only 12 states, of which all but New York and Pennsylvania are in the Midwest and West. Illi-

*A *metropolitan statistical area* (MSA) is defined by the federal Office of Management and Budget as "an integrated economic and social unit with a recognized large population nucleus." The boundaries are normally county boundaries and thus often enclose rural as well as urban areas. Two important MSAs that have no cen-

tral city as such are located in suburban Long Island, New York and Orange County, California. Besides the basic MSA, of which there were 249 in 1989, the Bureau of the Census has identified 54 larger Primary MSAs and 17 still-larger Consolidated MSAs, containing one million or more people.

191

nois and California had the most, with about 2800 each. Alaska and Hawaii had the fewest with 14 and 16, respectively.

How can we account for the popularity of special purpose districts? Why can't minor civil divisions (counties, townships, municipalities) perform these functions? At least four sets of circumstances related to local general purpose governments seem to encourage the creation of special districts.

1.  They frequently cover inadequate areas and/or lack the resources to do the job.

2.  Their expansion is restricted by statutory and even constitutional provisions.

3.  Their quality of service or performance record may be poor.

4.  Special interest groups want independent units established that will obtain the best possible results from their point of view.

The districts themselves can operate within areas delimited for their specific purposes, disregarding existing political boundaries. (About 90 percent of them, however, are located wholly within a single county.) They can hire experts to concentrate on particular problems; they are generally nonpolitical and do not threaten existing political interests; and, since most have independent power to tax, borrow, and levy service charges, they can be financed in a more efficient and flexible manner than frequently is possible under the numerous restrictions and limitations imposed on minor civil divisions. There is not necessarily a conflict either between a special and general government unit. In Maryland, California, and other states, the county serves as the tax-collection agent for a number of special districts; its regular tax bill includes an itemized list of these special taxes.

In addition, they are often able to take advantage of economies of scale, especially in capital-intensive infrastructure provision and maintenance. Many special districts are staffed entirely by volunteer or part-time appointed or elected officials; over 70 percent of them have no full-time employees. Especially in rural areas, where these volunteer or part-time employee-staffed districts pre-dominate, they are a convenient, inexpensive, and collaborative way of obtaining needed services. In fact, the major advantage of special districts is that they are often the only mechanism for providing government services.

Special districts are not without their problems and detractors, however. After a decision has been made to establish such a district (which itself may have been a complex process), a major problem and one of special interest to geographers is how to draw its boundaries. Management and staffing methods are sometimes questionable; responsibility to the voters and taxpayers is sometimes obscured or forgotten; single purpose sometimes becomes single minded, with the district pursuing its own plans without balancing them against other needs; the existence of a multiplicity of special districts that overlap one another and the local government units leads to jurisdictional disputes and duplication of effort; and individual citizens are likely to be ignorant, confused, or dismayed (or all three) about the dozen or more governmental agencies that tax them and regulate their activities.* These and other problems have led to the demands for reform of local government organization described in the last chapter. They have not, however, noticeably slowed the proliferation of special purpose districts. Instead, these are likely to continue increasing in the future—although the greater growth will probably be in metropolitan rather than rural areas.

While political scientists, historians, and economists have studied various kinds of special purpose districts, geographers have tended to ignore them. Here we single out two of the most interesting types of districts for special attention: school districts and public lands.

---

*To cite only one small example, East Haven, a town of 54,000 people adjacent to New Haven, Connecticut, has *three* fire districts within it, each with its own career and volunteer firemen, very well-paid commissioners, union contracts, taxing power and jurisdiction. The three chiefs are paid $184,000 to supervise 114 firefighters. The annual budgets total some $13 million. *New Haven Register*, 27 February 1994.

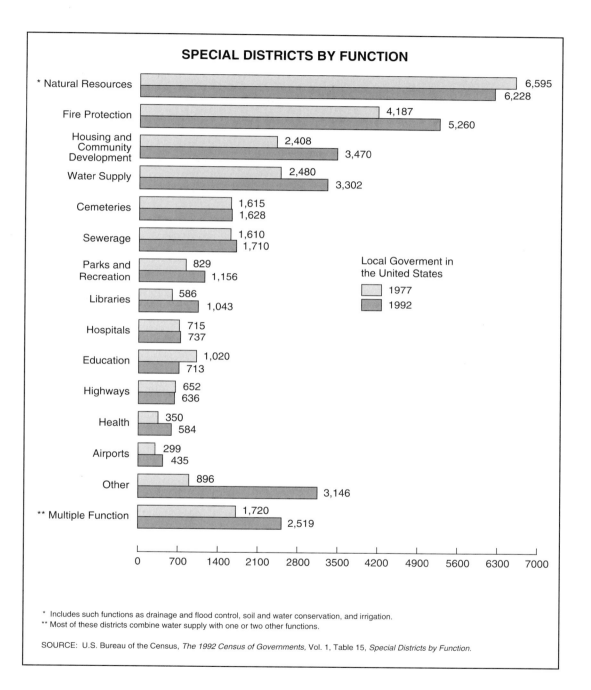

## SPECIAL DISTRICTS BY FUNCTION

| Function | 1977 | 1992 |
|---|---|---|
| * Natural Resources | 6,595 | 6,228 |
| Fire Protection | 4,187 | 5,260 |
| Housing and Community Development | 2,408 | 3,470 |
| Water Supply | 2,480 | 3,302 |
| Cemeteries | 1,615 | 1,628 |
| Sewerage | 1,610 | 1,710 |
| Parks and Recreation | 829 | 1,156 |
| Libraries | 586 | 1,043 |
| Hospitals | 715 | 737 |
| Education | 1,020 | 713 |
| Highways | 652 | 636 |
| Health | 350 | 584 |
| Airports | 299 | 435 |
| Other | 896 | 3,146 |
| ** Multiple Function | 1,720 | 2,519 |

Local Goverment in the United States

☐ 1977
▨ 1992

0   700   1400   2100   2800   3500   4200   4900   5600   6300   7000

* Includes such functions as drainage and flood control, soil and water conservation, and irrigation.
** Most of these districts combine water supply with one or two other functions.

SOURCE: U.S. Bureau of the Census, *The 1992 Census of Governments*, Vol. 1, Table 15, *Special Districts by Function*.

## School Districts*

Education has had a high priority in American society since colonial times, and the United States was one of the first countries to encourage publicly financed and operated educational systems. Only two years after independence was secured by a peace treaty with Britain, the Land Ordinance of 1785 required that one section of each township in the new lands of the West was to be reserved for public education. The land grant college system, established by the Morrill Act in the midst of the Civil War in 1862, was also a demonstration of this commitment. Education itself, however, was a matter left by the Constitution to the states.

Although the federal government has come to the aid of education since the 1950s by supplying financial and technical assistance and setting some standards and guidelines, the states still control public education. In only five states (Alaska, Hawaii, Maryland, North Carolina, and Virginia) and the District of Columbia are *all* local school systems dependent on some state or local governmental unit; in Tennessee and five New England states *most* school systems are dependent in this way. Elsewhere all public school systems are operated through independent school districts, in order, the theory goes, to keep them free from local politics.†

In practice, the local school districts, whether dependent or independent, have been the locale of some of the most momentous political and social battles of modern times: church/state separation, division of powers between state and federal governments, local versus state control of education, racial segregation and desegregation, censorship, sex education, equal opportunity for homosexuals and handicapped people, the fight against drugs (including tobacco and alcohol), AIDS, Darwinian concepts of evolution—all have been fought out in both the educational and political systems. Geographers, however, are concerned primarily with the spatial aspects of education, and at the local level, these focus on the independent school district.

In the 1930s there were more than 120,000 independent school districts in the United States. By 1992 there were only 14,422 (plus 1412 dependent systems). Of these, over 3300 operated only one school, 6200 operated three to nine schools, and only 455 operated 20 or more. Despite the cost reductions resulting from this consolidation, education still accounts for nearly half the expenditures of local government and 60 percent of all state aid to local governmental units.* Whether the schools are referred to as consolidated, unified, central, or regional, they still cost money, and school financing is currently the major issue in education, to which even the desire to improve the quality of education seems to be subordinated.

School districts typically are endowed by state governments with considerable administrative, curricular, and fiscal independence. Most are administered by unsalaried elected boards that hire professional superintendents and other administrators and are, in turn, supervised by state boards or departments of education. Many are large, wealthy, and dynamic units that completely overshadow some of the local general purpose governments. Many, at the other extreme, particularly in the Midwest, are "nonoperating" districts that own no property and provide no instruction but only furnish tuition and transportation for resident students at-

---

*School districts are not considered special districts by the federal Office of Management and Budget or by the Bureau of the Census for various reasons. Since they are organized for similar reasons and function in much the same way, however, we follow common practice and include both school and special districts under the heading of special purpose districts.

†A unique bistate school district is that encompassing Norwich, Vermont, and Dresden (Hanover), New Hampshire, which took an Act of Congress to make possible. The community of interest focuses on Dartmouth College and Mary Hitchcock Hospital in Hanover.

---

*By 1990, the proportion of total state budgets devoted to education had increased (frequently under court order) to about 50 percent. Federal financial aid to education is still very small.

tending schools in other districts. Most of these result from consolidation.

Consolidation has become not only desirable, but both essential and feasible as the country and society have changed. It became increasingly difficult to attract and retain good teachers and administrators for rural districts, and the growing complexity and mobility of society increased the need for expansion, improvement, and standardization of curricula. The most important development impelling school district consolidation, however, has been the great increases in the costs of construction, maintenance, instruction, and administration. Most of the districts abolished were unable to finance needed expansion and improvements or even their current operating budgets. In addition, there are economies of scale to be realized; it costs less to operate one high school for 1000 students than five for 200 each. Consolidation makes feasible the purchase and operation of expensive equipment for transportation, science instruction, athletics, dramatics, vocational training, and so on. Adult education, programs for the handicapped and bedridden, joint programs with universities, exchange programs with other districts and even other countries, sabbaticals for teachers, and many other valuable contributions to quality education for a democratic society become feasible in larger districts.

Two major problems arise with consolidation, however: how to delineate the new districts and how to work out relations between the new districts and other units of government. When consolidation is proposed, frequently by the state, a number of questions arise. Among them, the most important from our perspective are: Which districts should be included in the plan? How should the boundaries be drawn? Where should the new central schools be located? What procedures and routes should be arranged for transportation? What would the social and economic effects be on both the communities losing schools and those gaining them?

Opposition to consolidation comes from those who philosophically oppose surrendering "local control" of schools, even though local control does not necessarily mean good education; from rural families who do not want their children infected with an urban viewpoint or "radical" ideas; from parents who do not want their children mixing with others of a different race, religion, or social background; from local school board members and administrators who fear losing prestige and power, if not jobs; and from a variety of other groups with other objections. These objections have slowed but not halted consolidation.

School districts can, in fact, get too big, so big that economies of scale pass the point of diminishing returns; that parents and taxpayers lose contact with their educational system; that the new education establishment becomes a powerful political force itself, sometimes challenging political officials and institutions over many issues; that community identities and values are sacrificed to the cause of efficiency. For these reasons, New York City decentralized its school system in 1969, creating 31 (later 32) neighborhood school districts that have their own school boards and considerable autonomy. The central board retained control over high schools and certain financing and contractual matters. The results of this decentralization so far have not been particularly encouraging. It has not noticeably increased local participation in decision making, produced innovative programs, or saved money. A few have provided superior service, most are barely adequate, and some are scandal-ridden.

Consolidation of districts, especially in metropolitan areas, often brings the new districts into conflict with established general or special purpose districts. Problems arise regarding police and fire protection, traffic control and pedestrian safety, use of library and recreational facilities, adult education and vocational training, and health services. These problems have to be resolved by formal or informal arrangements among the various jurisdictions, or neither the educational process, the children, the parents, or the taxpayers will realize the benefits of either consolidation or local control of education through school districts.

## Public Lands

Another type of area is not normally considered a special purpose district, but should be mentioned here. This is land in the public domain that is managed under special provisions of law for particular purposes. All land acquired by the United States from other countries and from individual states since independence (except for properties privately owned under laws of the former sovereigns) automatically entered the public domain. This totaled 728 million hectares (1.8 billion acres) between 1781 and 1867 (exclusive of Pacific and Caribbean territories), of which 445 million hectares (1.1 billion acres) had been disposed of by 1993. The remaining public domain plus another 27 million hectares (68 million acres) acquired since then (usually by purchase or gift from citizens) is managed by a number of governmental agencies, most of them within the U.S. departments of Interior, Agriculture, and Defense, and the Nuclear Regulatory Commission. In 1993 about 29 percent of all the land in the United States was owned by the federal government and about 7 percent by state and local governments.* See Tables 14-1 and 14-2 for details.

These lands are scattered widely throughout the United States, but nearly 94 percent of all federally owned land is in the West; 64 percent of the land in the 13 western states is federally owned, including 85 percent of Nevada and over 86 percent of Alaska. The 1971 Native Claims Settlement Act granted outright title to 40 million acres of federal land in Alaska to native Indians, Aleuts, and Eskimos, and also authorized the Secretary of the Interior to withdraw from the public domain 32 million hectares (80 million acres) for possible inclusion in national parks, national forests, wildlife refuges, and wild and

**Table 14-1**   *U.S. Federal Public Lands (in acres), 1992*

| | |
|---|---|
| Department of the Interior | |
| Bureau of Land Management | 272,029,418 |
| Fish and Wildlife Service | 91,318,691 |
| National Park Service | 72,842,856 |
| Bureau of Reclamation | 5,503,093 |
| Bureau of Indian Affairs | 2,747,438 |
| Department of Agriculture | |
| Forest Service | 189,380,078 |
| Department of Defense | |
| Department of the Army | 9,683,074 |
| Corps of Engineers | 5,474,171 |
| Department of the Air Force | 8,113,532 |
| Department of the Navy | 2,361,756 |
| Department of Energy | |
| Nuclear Regulatory Commission | 2,166,341 |
| Tennessee Valley Authority | 1,040,231 |
| Total (Including agencies not shown) | 724,066,171 |

*Source:* *World Almanac and Book of Facts,* 1995, p. 502.

scenic rivers. The Statehood Act of 1959 had allocated another 42 million hectares (104 million acres) to the new state of Alaska. These three allocations total in area nearly the size of Texas, and Texas-sized controversies over their precise distribution and use are still raging.

These public lands represent zoning on a state and national scale since their uses are controlled by statute and regulation. Even land once owned by the federal government but granted or sold cheaply to homesteaders, railroads, irrigators, and states, amounting to over 400 million hectares (1 billion acres), carried with it requirements for, or limitations on, its use. Through a variety of programs, the federal government encourages the states to support soil, water, forest, and wildlife conservation, in part by restrictions on land use, required reclamation of strip-mined land, and other sound management practices.

Federal land ownership and management, particularly the environmental and natural resource regulation involved, have long been resented in the West. This smoldering resentment broke into the open in 1979 when the State of Nevada filed a lawsuit claiming the land within the state that was under the jurisdiction of the federal Bureau

*In western Oregon more than 2.6 million acres comprise the Oregon and California (O&C) revested lands. Originally public land, this area was granted to private concerns for the construction of the Oregon and California Railroad and the Coos Bay Military Wagon Road, but it was reconveyed to and revested in the federal government. Most of the revested lands are administered by the Bureau of Land Management.

*Table 14-2   Some Examples of Land Owned by the Federal Government in the United States (in million acres)*

| States | Federally Owned | Total Land in States | Percentage Federally Owned |
|--------|-----------------|----------------------|----------------------------|
| Alaska | 248.0 | 365.5 | 67.8 |
| Nevada | 58.3 | 70.3 | 82.9 |
| Utah | 33.7 | 52.7 | 63.9 |
| Michigan | 4.6 | 36.5 | 12.6 |
| Hawaii | 0.6 | 4.1 | 15.5 |
| West Virginia | 1.0 | 15.4 | 6.7 |
| Connecticut | 0.06 | 3.1 | 0.2 |
| New York | 0.2 | 30.7 | 0.7 |
| Kansas | 0.4 | 52.5 | 0.8 |

*Source:*   U.S. Department of the Interior, Bureau of Land Management, *Public Land Statistics 1993,* p. 5.

of Land Management. This triggered the "Sagebrush Rebellion," which was soon joined by six other western states. In 1981 the new Reagan administration gave considerable support to the rebellion, but only by seeking to loosen the regulations, not by giving away the land. But support in the West faded when the costs and benefits of acquiring huge tracts of federally owned land were carefully calculated. The Reagan administration, meanwhile, did sell off some public land, but only a small fraction of it; relaxed enforcement of land-use regulations; accelerated timber and minerals leasing; froze acquisition of parkland and wildlife refuges; slowed down wilderness designations; and encouraged private rather than public conservation efforts. Short-term economic considerations took precedence over long-term environmental ones. Thus, the federal government defused the rebellion by making many concessions; Nevada lost its lawsuit in 1981, and the Sagebrush Rebellion wound down.

Though the rebellion ended, the controversy lingers on. It flared again in 1986 when the Bureau of Land Management (BLM) and the Forest Service proposed a trade of 14 million hectares (35 million acres) of lands under their jurisdiction in the West, and again in 1987 over the addition of public lands to the National Wilderness Preservation System.

In the early 1990s another rebellion gathered force: some 20 western states and hundreds of counties passed resolutions demanding federal lands be turned over to them. Whether the devolution of power favored by the new Republican Congress will include transfers of this kind remains to be seen.

Three aspects of the public lands controversy deserve special mention: water, usage rights, and revenue sharing. Of these, water is by far the most important.

Generally speaking, the western part of the United States—"beyond the hundredth meridian"—is dry. Not uniformly so, of course, but on the whole drier than the East. The problem has grown increasingly acute over time, as population, manufacturing, and agriculture have all increased in the West while the water supply has not. Pressure on the water supply has been particularly heavy, of course, in Southern California.

Los Angeles has for more than three quarters of a century been drawing water from Northern California and Nevada, leaving some source areas parched and dry. When its latest spurt of growth threatened the swimming pools and car washes of the basin, Los Angeles proposed a gigantic project to bring water from the Columbia and Snake rivers in the Pacific Northwest. The governors of those states immediately and emphatically rejected the proposal, saying they needed to reserve their water for their own future growth. The resident geniuses in Southern California thereupon, in true Hollywood fashion, proposed an even more

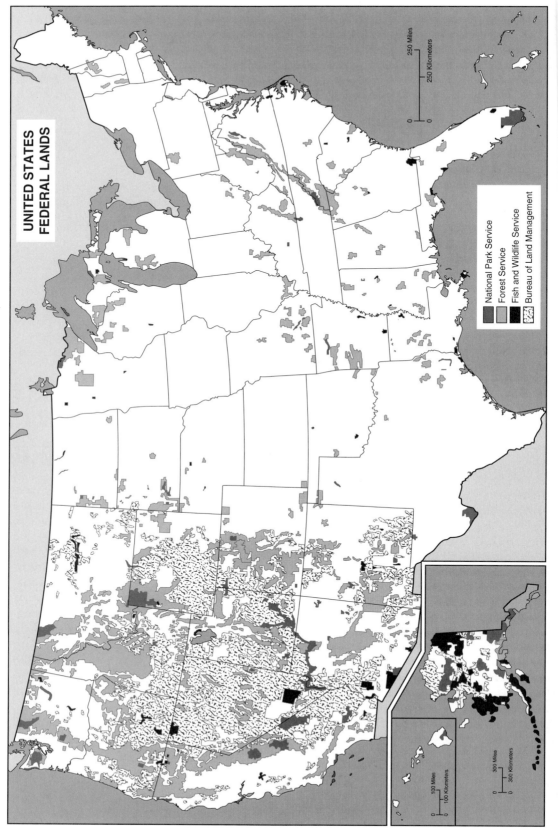

UNITED STATES
FEDERAL LANDS

National Park Service
Forest Service
Fish and Wildlife Service
Bureau of Land Management

250 Miles
250 Kilometers

100 Miles
100 Kilometers

300 Miles
300 Kilometers

Federal lands in the United States.

grandiose project, the North American Water and Power Alliance (NAWPA), which would bring not only water to them from the Pacific Northwest, but also water and hydro-electricity from the Fraser and Mackenzie rivers in western Canada!

Fortunately, the NAWPA was never adopted. However, the massive Central Arizona Project was, and it now brings water from a huge catchment area into the Phoenix region so retired easterners can enjoy golf courses and air-conditioned homes—at the taxpayers' expense. At some point the American people will have to decide whether this is the best way for the water of the West, most of which comes from public lands, to be used; whether the long-term cost of irrigated avocados and strawberries is not just a bit too high, whether it is not appropriate to begin charging fair market prices for water from public lands and adopt other measures to keep population growth more in line with the carrying capacity of the land.

The situation is similar regarding the use of public lands for ranching, mining, and forestry. Since in most cases public lands in the West are unsuitable for agriculture, these activities often represent the "highest, best use" of the land. Theoretically, then, users of the land should be charged accordingly for their grazing, mining, and timber rights. In fact, however, for many decades the taxpayers have been subsidizing these activities, which are usually carried out by large corporations, not small families, for which leases and usage rights are granted for nominal fees. Many times bills have been introduced in Congress to rectify this situation, to reduce this form of "welfare for the rich" and reduce the federal deficit by raising fees for the use of public land to something approaching market rates. Yet each time industry lobbyists have been able to defeat the proposals (most recently in 1994), and the subsidies continue.

Federal lands are also used as communications sites for broadcasting and transmitting radio, television and other electronic signals. The Forest Service alone has issued about 6300 such permits and the BLM about 3200. Their lease fees are substantially below market rates, another taxpayer subsidy to wealthy corporations. Furthermore, the BLM estimates that about 60 percent of the houses on public land do not qualify as exempt homes under the 1872 mining law but are simply illegal vacation homes. The bureau has been forced to burn down a number of them. This is still another cost borne by the taxpayer.

Finally, it seems appropriate to note that some of the revenues received by the federal government for the use of public lands is returned to the states in which the revenues originate as compensation for the states' inability to tax land owned by the federal government. In 1984, for example, the 13 western states alone received nearly $900 million under this program, more than half of it from oil and gas leases. As might be expected, this federal money is not distributed evenly. In 1984 the payments ranged from under $11 million to North Dakota to over $214 million to Wyoming. Virtually everything about this program is controversial. In fact, its history is long and complex, rooted as it is in changing philosophies and policies regarding the use and value of land. It is likely to remain controversial until the people of the United States are able to reach an informed and rational consensus on this and related matters.

---

**United States Outer Continental Shelf.** This map shows the areas considered by the U.S. Geological Survey to have the greatest potential for commercial oil and gas production, and the planning areas of the Minerals Management Service's Five-Year Leasing Program (1987–1992). Vigorous campaigns by environmental groups and others have delayed indefinitely proposed exploratory drilling in the Georges Bank and Baltimore Canyon areas. Part of the Navarin Basin area is disputed with Russia, but leasing of exploratory tracts has been conducted anyway. The chief offshore production areas continue to be the Beaufort Basin, Cook Inlet, Southern California and the Central Gulf. The 200-meter isobath approximates the outer edge of the continental shelf. *Sources:* U.S. Minerals Management Service. *Federal Offshore Statistics: 1993* and *U.S. Geological Survey Yearbook (Fiscal Year 1979).*

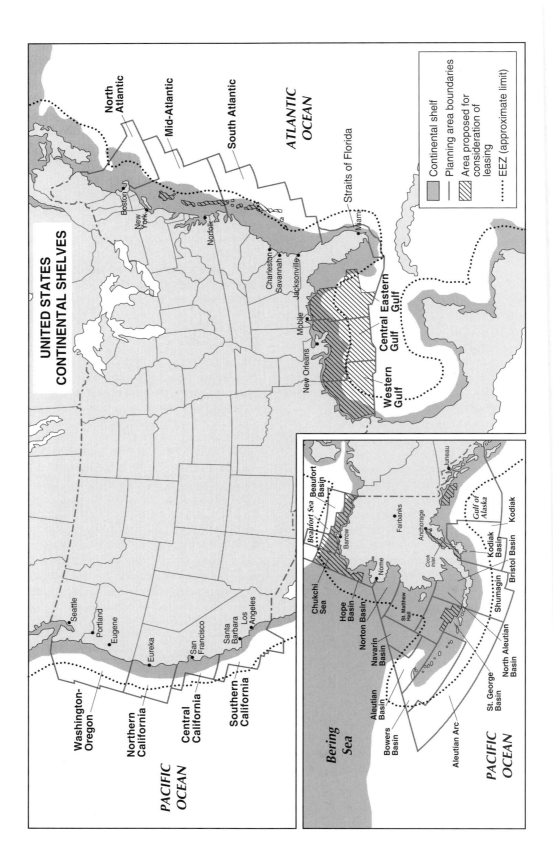

UNITED STATES
CONTINENTAL SHELVES

**Legend:**
- Continental shelf
- Planning area boundaries
- Area proposed for consideration of leasing
- EEZ (approximate limit)

ATLANTIC OCEAN

North Atlantic
Mid-Atlantic
South Atlantic

Boston
New York
Norfolk
Charleston
Savannah
Jacksonville
Mobile
New Orleans

Straits of Florida

Western Gulf
Central Gulf
Eastern Gulf

PACIFIC OCEAN

Seattle
Portland
Eugene
Eureka
San Francisco
Santa Barbara
Los Angeles

Washington-Oregon
Northern California
Central California
Southern California

Bering Sea

Beaufort Sea
Beaufort Basin
Chukchi Sea
Hope Basin
Norton Basin
Navarin Basin
Aleutian Basin
Bowers Basin
Aleutian Arc
St. George Basin
North Aleutian Basin
Shumagin
Kodiak Basin
Kodiak
Bristol Basin
St. Matthew Hall

Barrow
Nome
Fairbanks
Anchorage
Juneau

Gulf of Alaska
Cook Inlet

PACIFIC OCEAN

Even in a huge, rich, and relatively sparsely populated country such as the United States, land for grazing, mining, and forestry is highly prized, no less now than in pioneer days. The controversies over land and water use in the West are as bitter as those over land ownership or land use in the crowded Northeast. These continuing controversies underscore the absence of, and need for, a national land-use policy, one that will help us make wisest use of our most precious natural resources.

## The Outer Continental Shelf

Few Americans realize that on 10 March 1983 the territory of the United States more than doubled. On that day President Reagan issued a proclamation establishing an exclusive economic zone (EEZ) for the United States, more or less in accordance with new provisions of the Law of the Sea. We discuss the background of this move in Chapters 31 and 32, but here we are concerned with some of its effects.

The EEZ extends out to sea 200 nautical miles from the baseline from which the U.S. territorial sea is measured, around all the coasts of the mainland and the island territories. Thus the United States has added some 1.58 billion hectares (3.9 billion acres) of underwater land to the .93 billion hectares (2.3 billion acres) of exposed land. The United States does not have sovereignty over this land, however. In a fashion roughly similar to the leased lands described in Chapter 7, the United States has acquired, in the words of the proclamation, "sovereign resources, both living and non-living, of the seabed and subsoil and the superjacent waters."

Within this new zone of federal jurisdiction lies the continental shelf over which President Truman claimed jurisdiction for the United States on 28 September 1945. In a series of cases decided in the 1950s, the Supreme Court determined that the individual coastal states would have jurisdiction over the submerged lands underlying the territorial sea and obtain revenues directly from them; the shelf beyond the territorial sea, however, would be under the jurisdiction of the federal government. This area is

known in the United States as the outer continental shelf (OCS). On 27 December 1988 President Reagan extended the territorial sea from the traditional 3 nautical miles to the now nearly universal 12 nautical miles. Nevertheless, the OCS continues to extend from the old 3-mile limit outward to the edge of the continental shelf. This submerged land may be considered a special form of public land.

At present, the only commercially exploited resources in this area are petroleum and natural gas. In the future, other potential mineral resources, such as phosphates, polymetallic nodules, polymetallic sulfides, manganese crusts, and metaliferous muds may become commercially valuable, and we may find uses for other elements of this still largely unexplored area, but for at least the next several decades, oil and gas will be the important ones. They are extracted by private corporations operating under leases issued and supervised by the Minerals Management Service of the Department of the Interior. As of 30 September 1990, 1.5 million of the .57 billion hectares (37 million of the 1.4 billion acres) of the OCS were under lease, broken down roughly as shown in Table 14-3.

This is a greatly simplified picture. There are many factors that already complicate the production, transportation, processing, and use of resources from the outer continental shelf. They include costs of production, hazards to people and their equipment in this unfamiliar environment, pollution resulting from commercial activities, competition with land-based resources, rights of in-

***Table 14-3***   *Federal Offshore Leases Under Supervision of the Department of the Interior*

| OCS Region | Leases | Acres Under Lease |
|---|---|---|
| Alaska | 346 | 1,849,455 |
| Atlantic | 53 | 301,739 |
| Gulf of Mexico | 5,174 | 26,349,931 |
| Pacific | 92 | 465,126 |
| Totals as of 31 Dec. 1993 | 5,665 | 28,966,251 |

*Source:* Minerals Management Service, U.S. Department of the Interior, *Federal Offshore Statistics: 1993,* p. 25.

digenous peoples (especially off Alaska and Washington State), maritime boundary disputes with Canada, Russia, and Mexico, and conflicts in uses of the EEZ (which includes nearly all of the continental shelf), including fishing (both sport and commercial), navigation, marine parks and sanctuaries, waste disposal, and military activities. More complications are likely to develop in the future as our uses of the sea proliferate and intensify.

It is not unreasonable to anticipate that in the intermediate future, as part of a comprehensive national land-use plan, the management of public lands, both exposed and below the sea, will be removed from the jurisdiction of the existing federal bureaucracy and placed under the jurisdiction of newly created special-purpose districts. The notion is at least worth considering, not only in the United States and other federal States, but in unitary and regional States as well.

# 15

# The Geography of Elections

In 1913 *André Siegfreid* published a seminal study of elections in western France and their relationship to geographic and socio-economic factors. In 1949 he produced another landmark study of elections in the *département* of Ardèche, on the west bank of the river Rhône, from 1871 to 1940. During the intervening period, not only did Siegfried develop the technique of comparing maps of electoral results with maps of geographic and other factors, but other French scholars, including geographers, did similar work. So did a few people elsewhere, mostly sociologists and political scientists. Also in 1949, the noted American political scientist *V. O. Key, Jr.,* published his classic study of voting in the American South. After that more Americans became involved in the study of elections. It was not until the late 1960s, however, after the advent of the so-called quantitative revolution, that American, British, and other political geographers became seriously interested in the subject. Since then we have been inundated with studies of all kinds on elections. The flood has become so great, in fact, that, on the basis of sheer magnitude alone (if for no other reason), electoral geography has come to be recognized as a subdivision of political geography.

Examining the literature on voting produced since 1913 in North America, Western Europe, Australia, New Zealand, India, Israel, and the very few other places where it has attracted professional interest, we find some (but not nearly enough) good studies on referenda and plebiscites; voting in the United Nations, the European Community, and other intergovernmental organizations; and voting in national legislatures.* Overwhelmingly, attention has been fixed on contested elections for seats in legislatures at the national and subnational level, and elections for president in States with a presidential system. We may wonder whether this fixation is based on the true significance of these elections or on the ready availability of enormous quantities of statistical material and the machines to play with them. Certainly, playing with election returns is a marvelous exercise for graduate students; whether professional political geographers can make a really useful contribution by doing so, however, is not entirely clear. In any case, some interesting patterns and theories are beginning to emerge about voting at the national and subnational level, and for this reason electoral geography is being presented here along with subdivisions of the State.

## Themes in Electoral Geography

*Peter Taylor* and *Ronald Johnston,* in their numerous publications on the subject, have identified "three main foci of geographical interest in electoral studies." First is *the ge-*

*Some of these we discuss in Chapter 18, on decolonization and Chapter 29, on regional and subregional organizations.

*ography of voting.* Generally, studies in this genre try to explain the patterns of voting after a particular election or group of elections. The emphasis is on statistical methods and, while the work follows the tradition established by Siegfried, maps have been largely replaced by statistics and formulas to illustrate the results.* It is questionable whether this represents a true advance toward a clearer understanding of elections and their results. Are numbers really better than maps?

The second major theme, or focus, is *the geographic influences on voting*. There are four aspects of voting that can be explained in part by examining the geographic background of an election. They are voting on issues, voting for candidates, the effects of election campaigns, and—most geographic of all—"the neighborhood effect." This is the relationship between election results and the hometowns or home districts of the candidates.

A study of either of these two main themes requires an understanding of the electoral system in use at the time. There are many such systems, and many combinations and variations. To name only the major ones: proportional representation, winner-take-all, single-member constituency, multiple-member constituency, majority, plurality, weighted plurality, representation by party, representation by socioeconomic group, etc. Then there are the methods of selecting candidates: primary elections, party conventions, nomination by party chiefs, selection by traditional social/political leaders, and so on. The more one delves into the intricacies of the political process, the farther one goes from geography and the less geography has to contribute to an understanding of it.

Undoubtedly, the most geographic theme in electoral geography is that of *the geography of representation*. In those countries in

which there are elections to the legislature (or legislatures if the civil divisions also have elected legislatures) based on constituencies or districts, the number of districts and their boundaries can have profound influences on the composition of the legislature independent of the actual total votes for candidates and/or parties. This often, even in the most "democratic" countries, results in a disparity between the number of votes won by a party and the number of seats it wins in the legislature. This electoral bias is most evident in those countries, such as the United States, United Kingdom, Canada, and New Zealand, in which the plurality-majority system is used. Here the winner of a seat is the candidate with a plurality of votes, even if it is not close to a majority, and the party that organizes the legislature (and in a parliamentary system forms the government) is the one that can command a majority of seats.* Because of their importance and their geographic nature, we turn our spotlight now on electoral districts, using the United States to illustrate our main points. It should be borne in mind, however, that similar situations exist in other countries. Recent studies of electoral districts in the United Kingdom, Norway, France, and Australia show this graphically.

## Electoral Districts

Most Americans live within a number of electoral districts, typically a congressional district and separate districts for the two houses of the state legislature. Many, especially in metropolitan areas, live within additional districts (wards, zones) for local or metropolitan legislative bodies. Thus, not only does the individual citizen cast votes in as many as six or seven districts within a few seconds on election day, but the thousands of electoral districts generate enough statistics to enable us to make some useful generalizations. Each district is in itself a study, for its size, shape, population numbers and

---

*In Taylor and Johnston's 1979 book, *Geography of Elections,* for example, of 466 pages of text, only 22 are devoted to maps, many of which are of poor quality. The proportion is slightly higher in Johnston, Shelley, and Taylor's collection of essays, *Developments in Electoral Geography,* 1990: Of 270 pages of text, maps appear on 13, but occupy only five of them fully.

---

*In the United States, several states and many local governments have multi-member legislative districts.

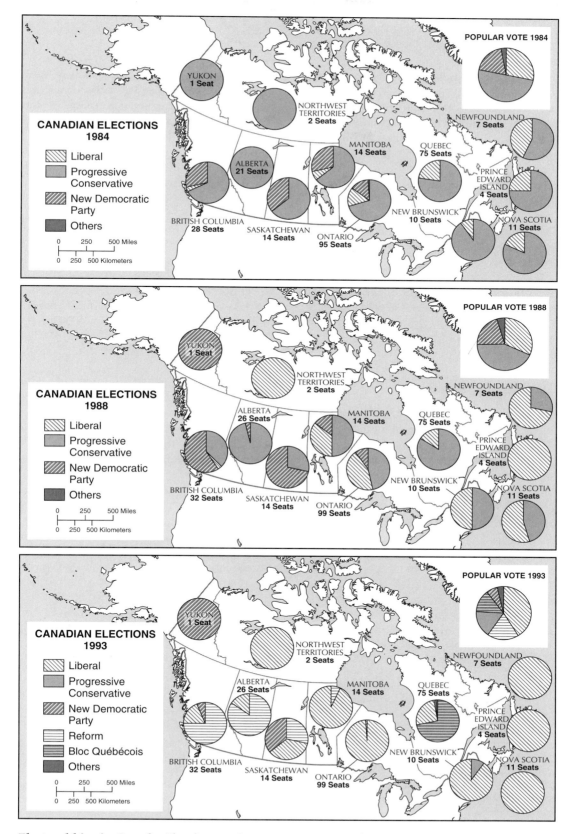

**Electoral bias in Canada.** The disparity between votes received by the various parties in successive elections and the seats won in the national parliament is clearly evident in these maps. This is a common problem in many countries that do not have proportional representation. However, proportional representation itself frequently generates other and more serious problems.

characteristics, land use, and so on. Combining districts in various ways reveals patterns and raises questions only hinted at by study of the individual district. Studies of the same information for a number of successive elections or selected elections over a longer period of time reveal patterns of change and raise additional questions. Studies of this type must be based, however, on a clear understanding both of the voting system used and the method of application of the system. In the United States this is relatively simple, but in other countries it can trap the unwary. A basic principle in studying election data is that the most meaningful patterns are revealed by the smallest possible voting units. This requires manipulation of more figures and makes mapping more difficult, but it is frequently worth the extra trouble.

The aspect of the electoral process in the United States that most readily lends itself to geographical analysis is the voting district. Because the number of districts in legislative elections is limited by the number of seats to be filled, while the electorate itself changes (sometimes dramatically) in both number and distribution, the size (population) and shape of the districts can be important influences on the outcome of the elections.

The federal Constitution mandates that a decennial census of the population be conducted to determine representation in the House of Representatives. Since the number of members of the House is fixed by law (currently 435), increases in population (even through the admission of new states) cannot be accommodated by simply adding more House seats. States that are growing relatively slowly or are actually losing population must therefore surrender seats to the more rapidly growing states. Each census, therefore, results in a *reapportionment* of the 435 seats among the several states. In the states that neither gain nor lose seats, the congressional districts normally remain unchanged. In the rest, *redistricting* is necessary; that is, the district boundaries must be redrawn to allow for the addition or loss of seats. Since this process has been mandated by the Constitution and enforced by the courts, it has generally been accomplished on schedule.

This regularity does not necessarily apply to *state* legislative districts, however, which in the first instance are governed by state constitutions and statutes. By 1960, a number of states had not redistricted in half a century or more, sometimes in defiance of their own constitutions and statutes. Idaho had not redistricted since 1911, for example, Louisiana since 1912, or Tennessee since 1901. Others had not redistricted since 1931 (South Dakota, Colorado, Rhode Island, Connecticut, Georgia, and Alabama). In all of them (and others), major demographic changes had taken place, so that the districts were grossly malapportioned. To cite only a few examples, lower house districts ranged in population from 236,000 to 635,000 in Alabama, 257,000 to 410,000 in Idaho, 319,000 to 690,000 in Connecticut, and 216,000 to 952,000 in Texas.

The effects of this malapportionment were many. For one thing, rural districts were heavily overrepresented while rapidly growing urban districts could scarcely be heard in the legislatures. The workload for the legislators varied widely, conservatives were perpetuated in office, committee chairmanships and other positions based on seniority were held by conservative rural representatives, and, most glaring of all, one voter's vote was worth two or three times another's.

It was this last point that formed the basis for the landmark Supreme Court decision of 1962 in the case of *Baker v. Carr*. The Court ruled that the apportionment of the Tennessee General Assembly by means of its 1901 statute debased the votes of the plaintiffs and denied them equal protection of the laws under the Fourteenth Amendment. Subsequent decisions in 1964 ruled (in *Reynolds v. Sims*) that state senates as well as lower houses had to be apportioned on the basis of population, not counties or other units ("Legislators represent people, not trees or acres."), and that (*Wesberry v. Sanders*) in congressional districts as well, as nearly as practicable, one person's vote in an election is to be worth as much as another's. These and similar decisions precipitated a wave of reapportionments throughout the country based on the 1960 and 1970 censuses. They did not, however, solve all the problems connected with apportionment and districting.

For one thing, the reapportionment came too late to help many urban areas, for the great rural-to-urban migration had already largely given way to an urban-to-suburban migration, and many of the seats lost by rural voters went straight to the suburbs, bypassing the cities entirely. Second, a number of states either did not redistrict at all or did so in such a manner that there was still a fairly wide discrepancy between the districts with the largest and smallest populations. Third, while the Court established broad guidelines regarding the shape of districts, saying only that they should be contiguous and compact, ample scope remained for the state legislatures in drawing new boundaries to continue the time-dishonored practice of gerrymandering. Subsequent Court decisions and federal and state legislation have mitigated some of these problems, but gerrymandering survives.

*Gerrymandering* is a device to give an advantage to a particular party or group by drawing district boundaries in advantageous shapes. Blacks and whites, Republicans and Democrats, rural and urban and suburban voters have all been helped or hurt by this tactic in nearly every state for nearly two centuries. There are several types of gerrymandering. One of them—simply failing to redistrict as population changes—is no longer very common, but three others are. Perhaps the most common is the practice of drawing the boundaries in such a way that one group (e.g., Chicanos, Republicans, blue-collar workers) is concentrated in the fewest possible districts so that they win overwhelmingly there but their influence is not felt in other districts where they might have had a chance of winning. This is called the *excess vote* technique. Its counterpart is the *wasted vote* technique, in which the lines are drawn so as to break up a concentration of voters (e.g., urbanites, Democrats, blacks) into a number of districts so that their votes are wasted through dispersion and they are unable to elect anyone. A third method is to draw circuitous boundaries circumscribing grotesque shapes to enclose pockets of strength of the group in power or to avoid

**The original gerrymander.** The term (though not the practice) originated in 1812 when Governor Elbridge Gerry of Massachusetts signed into law an oddly shaped district in Essex County. Painter Gilbert Stuart saw a map of the new district and penciled in a head, wings, and claws, saying, "That will do for a salamander!" Editor Benjamin Russell replied, "Better say a Gerrymander."

areas of weakness. This is the *stacked* type of gerrymander. Regardless of the type, the practice does infringe the "one person one vote" concept and is inherently undemocratic. It is also exceedingly difficult to purge from the political scene.

## Reforming Electoral Districts

Nearly everyone can agree in theory that gerrymandering is "wrong." It makes one person's vote (for a candidate or in the legislature) worth more than another's, and this is quite undemocratic, though in some cases it might be more efficient and practical than strict equality of voting strength. If reform is desired, geographers can contribute to it by analyzing the problem and offering some solutions. There seem to be, for example, three nonpolitical problems that lead to drawing of unfair electoral districts, even inadvertently. First, even though we are being drowned in statistics and our census is becoming ever more sophisticated, we still do not have sufficient accurate data on population numbers, characteristics, and distribution. Second, there is little or no correlation between census units and traditional political precincts, though results are tabulated by precincts within districts, while seats are allocated on the basis of census tracts. Third, while we can produce photographs of the earth from orbiting satellites so large and sharp that we can read the license plates on the cars parked in the Kremlin, we still do not have adequate large-scale maps suitable for use as base maps in drawing boundaries of electoral districts.

The solutions to these problems seem simple: use computers to compile adequate databases, including all relevant population factors; use computers to generate adequate base maps; and have the census people and local politicians get together to create statistical units that will serve the needs of both. Ignoring the practical obstacles to each of these steps and assuming that they have been accomplished, we can then proceed to utilize our computers once again either to adjust the existing districts by shifting their boundaries slightly until all districts have roughly the same population, or create entirely new idealized districts and then make the necessary minor adjustments.

Even assuming that we have overcome the problems inherent in all of these procedures, we will still be faced with difficulties. If, for example, some independent, apolitical body has produced strictly neutral district boundaries, the new districts will inevitably favor the majority party in whatever area (such as a state or county) is being redistricted. An alternative technique is to have the redistricting done by a bipartisan (or multipartisan) commission. The inevitable horse-trading and jockeying for power that goes on in such bodies will inevitably produce a number of safe seats for both (or all) parties, so that the parties are satisfied but the voters will really have little influence on the outcome of the elections. In either case, democracy will be sacrificed on the altar of statistical purity. There must be a better way to reform the process. Until one is found, we may expect the courts to be forced to intervene more into what should be solely a politicogeographical problem.*

Redistricting after the 1990 census did, in fact, utilize computers extensively as well as the latest theories and models. Comparison of the sample post–1980 and post–1990 districts shown in the accompanying maps reveals, however, more art than science. Gerrymandering, it seems, not only survives but thrives, using the computer to produce results more bizarre than ever. At least part of the explanation lies in a series of laws and court decisions between the two censuses designed to produce more equitable representation in the House for "minorities," principally blacks and "Hispanics." The resulting convoluted districts are anything but compact, or even contiguous, but they are, in most cases, "majority minority" districts, if only just barely. We may know by the next census if the laws, court orders, and computers have truly brought to the "minorities," or to the American people as a whole, greater benefits than previous systems did.

## Themes and Theories in Electoral Geography

Besides the three themes just discussed, others have attracted some interest among political geographers in recent years, but they really deserve more attention than they have been getting. Perhaps when the fascination with numbers and computers has run its course, scholars will again turn their attention to serious questions that are not amenable to quantification. Some of these themes are: political parties themselves and their bases of support; representation and voting patterns in legislatures; the relationships among voting, geography, and power (or influence); and the role of ideology in producing electoral results and the geographic variations in this role.

---

*The political scientist Mark Rush, in a study published in 1993, concludes that analyses of gerrymandering have been based on false assumptions about voting behavior and that redistricting may not really be as important as generally assumed in determining the outcome of elections.

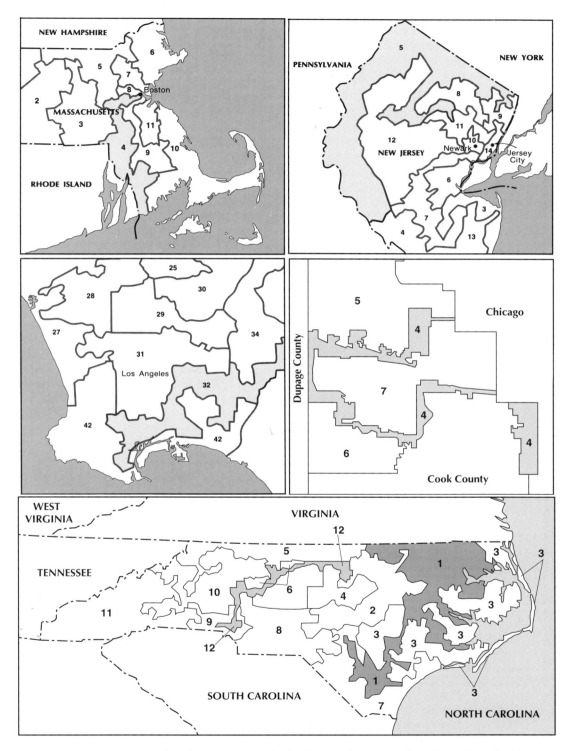

**Contemporary gerrymandered congressional districts.** Redistricting after the 1980 and 1990 censuses left a number of congressional districts throughout the conterminus United States in what appear to be flagrant violations of the concept of compact and contiguous districts. Here are a few examples of the more egregious cases; there are many more in all parts of the country. Courts have already ordered redrawing of gerrymandered district boundaries in several states, not only for Congress, but even for city councils.

Recently, a few political geographers have begun to go beyond merely counting votes to try to develop a geography of elections that will help to explain more general political patterns. This includes studying the relationships between geography and elections in (1) the process of forming a government, (2) the functioning of a State, and (3) the world economy. The last has probably been best developed so far. In the rather elaborate body of theory emerging from its study, mostly with a Marxist perspective, political parties are viewed as a part of the overall political development of the world economy. Although much of this theory is quite abstruse and perhaps even far-fetched, we must remember that this whole field is quite new and give credit to those who are trying to learn something useful from huge masses of raw election statistics not all of which are accurate.

This effort was advanced considerably by a conference on electoral geography and social theory in Los Angeles in 1988, sponsored by the Commission on the World Political Map of the International Geographical Union. The papers read there reported on elections in New Zealand, the United Kingdom, the United States, and Belgium, and also tried to derive some useful generalizations from the empirical studies. In addition, four dealt primarily with the field of electoral geography itself and its future.

Perhaps the theme of this conference is best expressed by the editors of the published papers, in their introductory essay. In summarizing the paper read by John Agnew, they say:

> Agnew briefly reviews research that has been undertaken from what he terms "modernization-nationalization" and "social welfare" approaches. Both of these approaches represent the philosophical underpinnings of the paradigm of liberal pluralism. Indeed, the modernization-nationalization perspective is a statement of the relationships between liberal pluralism and contemporary regional development theory. . . . More directly relevant from the perspective of social theory are two alternative approaches examined by Agnew:

the perspective of uneven development inspired by the world-systems analyses of Wallerstein and that of place context, which examines electoral geography in terms of daily experience within social contexts associated with particular places.

Most of the remaining papers illustrate some of Agnew's arguments and those of David Reynolds, who concentrated on local concerns within the same general framework. We shall have occasion to utilize these concepts of modernization—nationalization and social welfare in various contexts later in this book, but for the present they offer useful guides for understanding the outcomes of many elections, at least those in the rich Western and westernized countries. Peter Taylor closed the conference with a discussion of the very different conditions prevailing in most developing countries. "Thus," say the editors, "Taylor concludes that the task ahead for political geographers is to analyze the causal mechanisms relevant to the success and failure of liberal social democracy over space and through time."

This will not be easy to do, however. One of the basic problems in trying to detect and explain worldwide patterns in election statistics is that neither the statistics in many places nor the maps necessary for plotting them are adequate, and electoral systems and statistics are not compatible from one country to another.* Furthermore, elections themselves are important in choosing governments in relatively few countries in the

---

*Indeed, after all the statistics from the 1990 U.S. Census were digested, the Secretary of Commerce, in whose jurisdiction the Census Bureau lies, had to admit publicly that the population had been undercounted by a minimum of 5 million people, most of them in urban areas in urbanized states, yet he refused to allow the figures to be adjusted to conform more closely to reality. This undercounting has a serious impact on representation in the House of Representatives, distribution of federal and state financial assistance, and many other aspects of daily life. It also raises the question of the reliability, hence the utility, of census figures elsewhere and statistics generally.

world; perhaps three dozen countries have governments that are freely elected according to democratic, constitutional procedures on a more or less regular basis. Most of these countries are rich, industrialized, heavily engaged in international trade, and members of the Organization for Economic Cooperation and Development (OECD, discussed in Chapter 27). There are a few others: Israel, India, Costa Rica, and most of the former British colonies in the Caribbean and the Pacific, for example. But elsewhere governments are chosen through heredity, councils of chiefs, the Party, the military, or some other method. Elections, if held at all, have largely symbolic value, and their statistics are meaningless and useless for analytical purposes.

Furthermore, the geography of elections will continue to have only limited and local value until more cross-cultural studies are included, until techniques are more standardized, and until there are more free elections around the world to begin with. Geographers can help to bring about all three conditions. If they do not, electoral geography will continue to be little more than an intellectual exercise of little practical value, and the really useful electoral studies will continue to be done by sociologists and political scientists. One step toward accomplishing this goal would be a return to mapping election results. Anyone can play with computers and statistics, but only geographers can map the data in such a way as to make them meaningful to ordinary mortals.

# 16

# *Indigenous Peoples*

In the last chapter of the fourth edition of *Systematic Political Geography* (1988), we suggested that

> political geographers have simply not done careful studies of the number, distribution, and characteristics of indigenous or aboriginal peoples and their relations with the dominant ethnic group, their roles in the political and economic life of the States which have grown up around them, and their relations with their fellows across national boundaries. . . . We may be certain that by the end of this century we will be hearing much more about indigenous peoples around the world through our media of mass communication. . . .

Anthropologists, lawyers, and human rights advocates have long been advancing the cause of these peoples, and their efforts are now bearing fruit. It has been moving up on the priority list of matters of international concern and reached the top on 18 December 1990. On that day the United Nations General Assembly proclaimed 1993 as the International Year for the World's Indigenous Peoples. One of its outcomes was the designation of 1995–2004 as The International Decade of the World's Indigenous Peoples. Even some political geographers are now working on the subject. Imre Sutton in California and David Knight in Ontario have evinced a serious and abiding interest in it. And at the 1991 meeting of the Canadian Association of Geographers, a special session on Indigenous Peoples and the State

was sponsored by the Native Canadian Specialty Group of the CAG and the International Geographical Union (IGU) Commission on the World Political Map. The themes were Contemporary Perspectives, The Colonial Legacy, Indigenous Peoples and Resource Development, and Critical Evaluations of Colonialism and Environmentalism. Clearly it is time to move the subject from the back of the book to its heart. As political geographers, however, we consider only those aspects of the situation that are both spatial and political.

First, a matter of terminology. Many terms are used in the literature: native, aboriginal, autochthonous, first, and so on. Over the years, however, *indigenous* has become the most widely accepted term and has been adopted by the United Nations, and so we generally use it here. All of the terms are synonymous, however, and, in the proper context, are equally acceptable. Generally, we are referring to the original inhabitants of a place, those who were there before the arrival of peoples of different culture. Thus, the Indians of South America, the San (Bushmen) of southwestern Africa, the Ainu of Japan, the Maoris of New Zealand, the Hmong of Thailand, and the Aleuts of Alaska were all the first people to occupy these places. Among these peoples there are often many subgroups, sometimes tribes, sometimes identifying themselves in other ways. The term *tribal peoples*, then, is appropriate in many cases, but not as a synonym for indigenous.

Two more aspects of nomenclature must be mentioned. The term *indigenous people* is not generally applied to peoples who remain the dominant group within a country. Thus the Han Chinese, the Papuans of New Guinea, and the native people of Somalia would not be considered "indigenous" in this particular context. Second, the term is not a component of or substitute for "ethnic" or "minority." The indigenous peoples themselves resent these terms and insist that they are nations with distinctive cultures living on their own land, not "ethnics" or "minorities."

We must emphasize, however, that there is still no generally accepted definition of *indigenous peoples.* In an attempt to accommodate these and other problems, the special United Nations rapporteur on the subject, José Martínez Cobo, included in his lengthy, authoritative 1983 report a definition that ran to 87 words! Any figures given for these people, then, depend on definitions, as well as other problems. One reasonable guess is that there are currently some 250 million indigenous people in more than 70 countries.

In our brief survey of the topic we address only four of its many facets: (1) self-determination and self-government; (2) land and the physical environment; (3) self-government on the land (i.e., reserves); and (4) political organization at the national and global levels. We thus move gracefully from our focus on the State in this part of the book to international affairs in the next part. It may also be helpful to note that most of these factors are discussed elsewhere in this book in different contexts—a clear indication of the wide applicability of most of the principles of political geography.

## Self-Determination and Self-Government

Until very recently indigenous peoples nearly everywhere were subjected to all manner of abuse. Although commonly associated with European colonialism, this ill treatment continued long after colonies became States. They are still, for example, subjected to forced assimilation into the dominant society, often as part of the process of nation building; to population control, including involuntary sterilization; to forced migrations away from their ancestral homelands to far less desirable land to make way for new settlers; to evacuations, restrictions, and other deprivations in the name of national security, especially in areas considered unstable or insecure; to involuntary inclusion in or exclusion from national economic development programs, always to their disadvantage.

Therefore, aboriginal peoples nearly everywhere are either at the very lowest level of the dominant society or excluded from it altogether. It is not so much a matter of numbers as it is of domination by more aggressive and powerful peoples. Essentially, they are treated as colonials, even within States that were once colonies themselves. Their main objectives now are very simple: sheer physical survival, equality with the dominant groups in society, cultural survival, and economic rights, including secure legal ownership of their own land.

These objectives can best be achieved by self-determination. Again, there is no clear-cut definition of the term, but broadly it means that each group would determine for itself its relationship to the larger society. This may—and generally does—mean some degree of autonomy, of self-government, but it may not.

Self-government may take many forms, limited only by the prevailing conditions and the imaginations of the participants. It can range from very limited authority over purely cultural or tribal matters to the kind of near-independence enjoyed by the Torres Strait Islanders of Australia. The only important requirement is that it be developed with the informed consent of the people themselves.

Self-government, however it is structured, can, however, pose some very serious problems. It can conflict, for example, with the sovereignty of the State. Although some type of federal system could be worked out, or some type of servitude, we have already seen how difficult it is to create and maintain these arrangements. Also, special status with special rights for aborigines could (and frequently does) conflict with national economic development plans, or conservation regulations, or laws pertaining to public health, education, wages, working conditions, and so on. Ancient, traditional governmental systems may also be perfectly sensible and acceptable to a majority of the people involved, yet conflict with the democratic society surrounding them. Or the converse: Can a ruthless dictatorship tolerate within its territory an area in which democracy prevails?

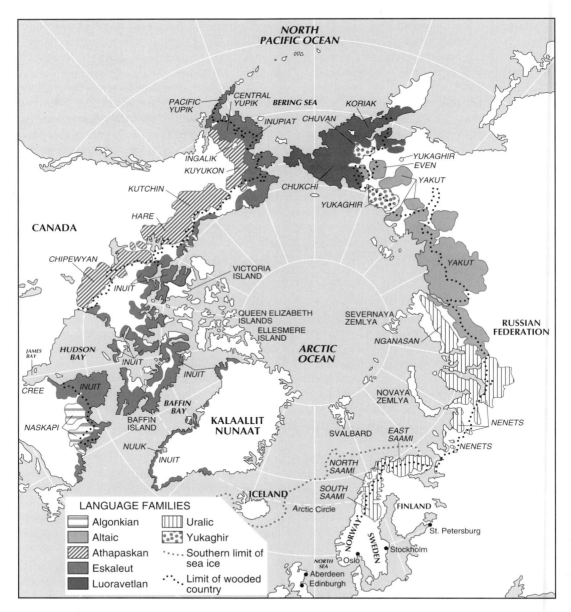

**Circumpolar Indigenous Peoples.** The native peoples of the lands surrounding the Arctic Ocean are probably at present the best organized at the international level. Although there are many groups, most of their cultures appear to have common origins and they experience many of the same problems. The Inupiat and Inuit are commonly known as Eskimos and the Saami (Sami) as Lapps.

## *Case Study: The First Nations Of Canada*

In 1975 the Northern Québec Inuit Association and the Grand Council of the Crees (of Québec) signed Canada's first comprehensive land claim agreement with the governments of Canada and Québec and three Québec crown corporations. It is known as the James Bay and Northern Québec Agreement. Under this agreement, the natives gave up all claims to the territory covered by the agreement in exchange for a number of rights, benefits, and payments. This was followed in 1978 by the Northeastern Québec Agreement, which settled the land claim of the Naskapi Indians. Together, these two agreements gave 17,000 aboriginal people property rights over 5400 square miles (14,000 square kilometers) as well as various resource rights and cash. They also have some measure of self-government.

In 1984 came the Inuvialuit Final Agreement, providing 2500 Inuvialuit people in the Western Arctic with 35,000 square miles (91,000 square kilometers) of land and similar, though not as extensive, resource rights and self-government, and cash payments. Other major land claims have been or are being settled with the 2200 Gwich'in Indians of the Mackenzie River Delta and 7000 Yukon Indians. Some 50 other major land claims and hundreds of smaller ones are in various stages of negotiation. The Canadian government is committed to resolving all outstanding land claims by the end of the century.

### Métis

In sharp contrast to the situation in the United States, people of mixed Indian/European heritage have a distinct culture, organization, and legal status. There are some 200,000 Métis (1991 est.) living chiefly in the Prairie Provinces. Descendants of French, Scots, Ojibway, and Cree, since the eighteenth century they have tended to marry among themselves, have developed their own language (Métis), and consider themselves a separate aboriginal group. Led by Louis Riel, they were instrumental in creating the Province of Manitoba in 1870, based on the Red River Settlement (Colony), in which Métis were the majority. For the next century, the Métis struggled not only for legal status but also for land rights and other concessions granted to other indigenous peoples of Canada.

Although they have gradually won some individual rights, mostly in Manitoba, history was made in November 1990 when the Alberta provincial government granted fee simple ownership of lands, rights to management of the subsurface resources, and local self-government in relation to these lands to eight Métis Settlements. This involved the transfer of 1875 square miles (4865 square kilometers), a substantial cash payment, and a package of legislation and agreements that are contributing substantially to the economic, social, and political self-sufficiency of the 4000 to 5000 people in the Settlements. Other action is being taken to improve the lives of about 60,000 Métis people living "off-settlement."

### Nunavut

The Inuit people of the Arctic (formerly known as Eskimos) have won a very different kind of land claim settlement from those just described. Most of them live in the Northwest Territories, the largest political unit in Canada. In 1976 the Inuit Tapirisat of Canada, a national organization, submitted a formal proposal to the federal government calling for the creation of a new territory out of the Northwest Territories (NWT), to be called Nunavut ("our land" in Inuktitut, the Inuit language). The territorial Legislative Assembly also expressed interest in such a partition. In 1982 a plebiscite was held in the NWT, and partition won 56.5 percent of the vote, subject to certain conditions. Negotiations among representatives of the Inuit and the territorial and federal governments continued for a decade.

A second plebiscite was held in May 1992 which approved the proposed boundaries of the new territory. This will separate the land claims of the Inuit in the central and eastern region of the NWT from those of the Dene Indians and the Métis in the west. On 30 October 1992 the Nunavut Political Accord was signed in Iqaluit (formerly Frobisher Bay), NWT, by representatives of the territory, the federal government, and the Tungavik Federation of Nunavut, representing

**The new Canadian territory of Nunavut.**

the 17,500 Inuit occupying the region to become the new territory of Nunavut. It supplements the Final Agreement of December 1991 on the settlement of Inuit land claims in the region. The remaining Inuit land claim in Canada was being negotiated in the early 1990s among the federal government, the provincial government of Newfoundland, and the Labrador Inuit Association.

The new territory of Nunavut is to be formally created not later than April 1999. Meanwhile, much preparatory work remains to be done, with the Inuit people participating fully in the planning and implementation of the new government. Among other things, this preparation involves selecting a capital, designing the government, training government workers, arranging financing and infrastructure, and planning for elections. In 1999 the new government will undertake some functions immediately and assume others gradually. This remarkable achievement of the Inuit of Canada can serve as an example for the indigenous peoples of other countries and their governments.

*The Crees and Québec*

The government of Canada is embarrassed by the secession movement in Québec for many reasons, but one of them has been deliberately omitted from the public debate on the question: the firm opposition by the Crees, the major Indian group in the province, to inclusion in an independent Québec State. The Crees claim that their territory, *Eenou Astchee* ("the People's land") has never been part of Québec because they were never consulted about the incorporation of their territory into the province in two parts, in 1898 and 1912, and never consented to it. The separatist leaders, however, have repeatedly insisted on the "territorial integrity" of the proposed State, ignoring altogether the position of the Crees. In fact, since the Cree territory, like that of other first nations of Canada, is under the jurisdiction of the federal government, not the province, it is questionable whether Québec would have any legal right to drag Eenou Astchee out of Canada should the province ultimately secede, especially without the consent of the Crees.

   The embarrassment of the Canadian government over this question is evident from their refusal to discuss it in public and from their instructions to Canadian diplomats to refer only to aboriginal "people" who, as individuals, have no right to self-determination, and not to aboriginal "peoples," who do have this right. It is for this very reason that Canadian statements made in the United Nations explain that they do not recognize their indigenous groups as "peoples." Otherwise, the Crees could have a legal claim to independence, whereas "minority" individuals do not.

   This Canadian position contrasts sharply with its practice in negotiating with the Crees and the Inuit to develop the James Bay and Northern Québec Agreement in 1975 and with other aboriginal *peoples* since then. Apparently, this Janus-like position is designed to mollify the separatists and keep Québec in the Confederation. The policy may not be successful, however, and the Crees are not waiting for Québec to secede before making their position clear in the United Nations and elsewhere.

## Land and Physical Environment

Virtually all aboriginal peoples, even nomads, have a special relationship with the land. To them, the land is not just a provider of material things. It is their mother, their spiritual sustenance, the home of their ancestors. It is the very heart and soul of their entire culture, their very existence as distinctive peoples. An individual cannot own the land; he or she can only hold it in trust for the next generation, either alone or as a partner in a communal undertaking. The land is sacred to the tribe, to the society, to each individual. Around the world, when the tie is broken, the person withers in often complete demoralization. Often the result is drunkenness, alienation, antisocial behavior, and early death.

   What happens when such a culture is dominated by a society that values land only as a commodity? Land, in this view, is meant to be used, stripped bare, built upon for the presumed benefit of the ravager. What happens when a people who had lived in harmony with their environment for generations, centuries, millennia, maintaining a stable population through sustainable economic activities well within the carrying capacity of the land, finds their habitat encroached upon or taken from them by people who are convinced that the Creator made the animals and the forests and the minerals solely for them to exploit and gave them the tools to do the job?

   The indigenous peoples of the world have dwindled rapidly since the beginning of the Age of Imperialism and the Industrial Revolution. Now only vestiges remain, perhaps 5 to 8 percent of the world's population. But this may be the saving remnant. They are the repositories of centuries of experience, of intimate understanding of the environment. And yet there is little respect for this wisdom among the more "enlightened" peoples of the "developed" world. It is instructive to note that most of the environmental problems that now concern us—destruction of

**Bedouin in Sinai.** Like nomads everywhere, often under government pressure or even coercion, Egypt's Bedouin are settling down. Nomads in many countries have been politically significant and warrant study by political geographers (Laura Zito/Photo Researchers).

forests, advancing deserts, flooding by huge dams, mining, disposal of hazardous wastes, soil erosion, testing of nuclear weapons—occur on aboriginal land, or land from which aborigines have recently been evicted. How can we reconcile exploitation with preservation? Progress with tradition? Destruction with creation?

## Aboriginal Reserves

In many places around the world the dominant societies dealt with these questions in a rather straightforward manner. Those native people who did not die or assimilate were herded into reserves, or reservations, and frequently ignored thereafter. Later in this chapter we examine in detail the Indian reservation in the United States, but at this point some generalizations can be helpful.

In many cases reserves have been created not on the ancestral land of a people, but far away from their homes and from the settlements of the intruders. Often it is land that for one reason or another is considered valueless, the territory least desired by the newcomers. But the tide of settlement of the dominant people eventually reaches the reserves. Even where the indigenous people remain on their own land, they are often invaded by farmers, ranchers, miners, loggers, seeking their land and its resources.

Sometimes reserves are actually established by governments as refuges for the native peoples, to protect them from the diseases and depredations of the outsiders. But however well-intentioned, these refuges seldom protect their residents for long. For one thing, disruption of a traditional way of life can seldom be repaired; once changed, it is gone forever. Government interaction frequently encourages (or requires) either the forced adoption of "modern" methods of agriculture, government, dress, communication, and so on, or the denial of these things to people who want them, forcing them to remain "primitive" for the benefit of local officials, anthropologists, or tourists.

Regardless of the motives for establishing reserves or the methods used to govern them, few are able to escape modernization entirely. One of the few that did, that of the Kuna (Cuna) reserve in Panama, has been adversely affected by the construction of the Bayamo Hydroelectric Complex in 1976. The best agricultural land of the Kuna was flooded, and they were forced to move to higher ground, where the soil is poorer and eroded. This has thrust upon them wage labor, concessions to logging companies, and encroachment by colonists.* Many

*To address these and other problems, all Panamanian indigenous groups met in November 1993 in a congress organized by the Coordinator of Indigenous Peoples. Similar events are occurring in many countries.

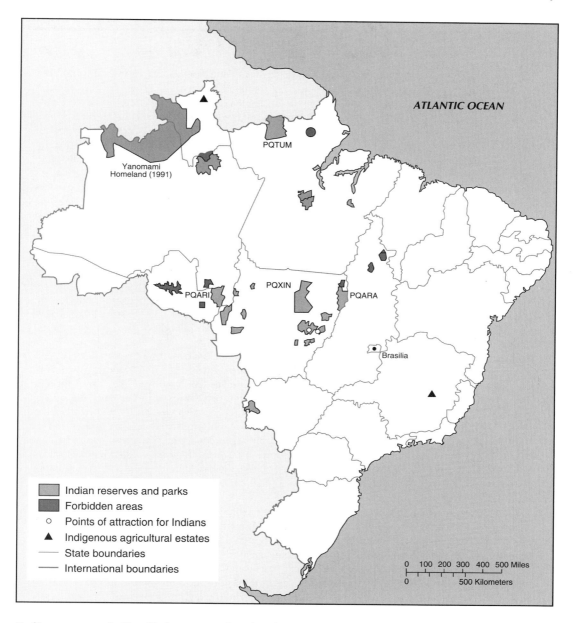

**Indian reserves in Brazil.** Over many decades, the Brazilian government has established various types of protected areas for the few large and numerous smaller groups of native peoples of the country. They have not been notably successful, however, in protecting the people from diseases, exploitation, and the destruction of their habitats by miners, loggers, ranchers, and others.

reserves in Brazil, the Philippines, and elsewhere have suffered the same or similar fate, not to mention those in North America. Does this mean that the reservation is inappropriate and should be abolished? Not at all. If the people are protected, permitted to govern themselves and make their own decisions about matters that affect them more or less directly, they can, if they wish, either preserve the old ways or gradually adopt selected features of the dominant culture.

## Case Study: The Originals of Taiwan

The first people to occupy Taiwan may have come from China or Malaya or from any number of Pacific islands, perhaps from all three places, perhaps 5000 years ago. They spread out over the eastern and western plains and for many centuries lived there undisturbed. Sporadic incursions of Chinese probably began in the third century, but the first permanent settlers, Hakka from Guangdong Province in south China, are thought to have arrived in the fourteenth century. In the sixteenth century Hoklo people from Fujian Province began arriving and gradually drove the natives out of all the plains areas except the east coast, up into the mountains of the central spine. The Hoklo also pushed the Hakka into the mountains, and they in turn pushed the aborigines higher up. Thereafter, they were largely left alone by the successive European and Chinese rulers of the island.

Japanese occupation of the island, beginning in 1895, however, brought intensive exploitation of its resources, especially its forests, most of which were located in the mountains. Some of the aborigines were assimilated into the Chinese Hoklo culture, and many moved back down to the plains. The remainder were assigned by the Japanese to reserves in the mountains. Both the Japanese and the Nationalist Chinese from the mainland, who received the island back from Japan in 1945, treated the aborigines as inferior people, and the Nationalists permitted the reduction of the mountain reserves by private sale. Since then, reserves have been set aside for the native people, but they have been steadily whittled away by private sale, transfer by the government to developers, creation of national parks out of reserve land, and seizure for "national security needs."

The policy of the Nationalist government in Taiwan, which is slowly being taken over by Taiwanese, themselves only earlier immigrants from China, has always been ambiguous toward the true natives. It insisted, for example, that they call themselves "Mountain People," even though they originally lived in the plains and were forced into the mountains, and currently the majority of them again live in the plains. (Today, these people even reject the most common appellation for them, "aborigines," in favor of "original people" or "Originals.") They are required to use Chinese names on all official documents, are arbitrarily assigned common Chinese family names, and are victimized (unofficially) by child labor, alcoholism, prostitution, assimilation, discrimination, and other evils. On the other hand, villages of Originals have been turned into theme parks for the entertainment of foreign tourists, and aboriginal dance troupes are sent touring abroad as Taiwan public relations attractions.

The result of all this mistreatment is that the Originals have been greatly reduced in numbers to about 330,000 (1990 census), some 1.7 percent of Taiwan's population, and only 9 tribes survive out of the original 20 or so. Their young people have migrated to the cities, little of their original land is left to them, and their cultures have been degraded. At the same time, however, they have been attracting outside support, much of it from Christian churches, to which about 80 percent of them belong. They have gradually raised their education levels (though they still lag far behind the Chinese), have begun to organize politically (though they belong to seven different linguistic groups), and now have some representatives in the Legislative Yuan, the provincial legislature.

The government is implementing a program of aid to the Originals, dating from 1988, much of it devoted to such basic economic development projects as improving transportation links among their villages. The government has an Aboriginal Administration Bureau (established only in July 1990) and other agencies supervising education and health services, but the Originals still suffer considerable discrimination and have no real security of land tenure.

Significantly, however, the Originals have begun using their incipient political strength to protest, publicly and dramatically, against human rights abuses—especially enforced prostitution—desecration of their cemeteries, loss of their land, neglect of their health needs, and other kinds of mistreatment. The Alliance of Taiwan Indigenous People, founded in 1984, is the more activist of the two associations of Originals, and thus far it has concentrated largely on cultural and legal matters. For all its enthusiasm, however, it numbers only some two hundred members. The native peoples of Taiwan still face an uncertain but difficult future.

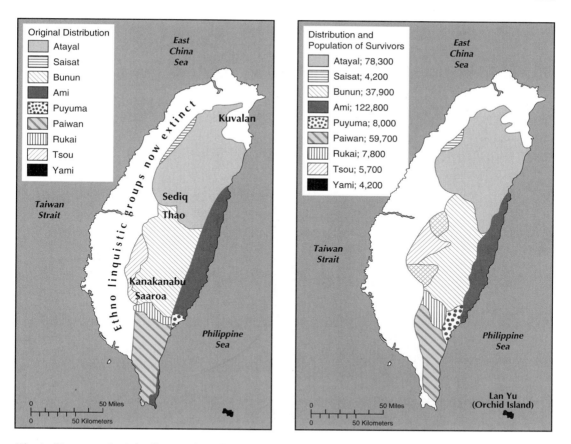

**The indigenous (original) peoples of Taiwan.** Note how the territory occupied by these peoples has shrunk from their former territory, the whole island. Those tribes not shown on the contemporary map are extinct.

## Political Organization

One of the features of Western culture that has been adopted by indigenous peoples is political action to advance their own interests. They are doing this through nongovernmental organizations (NGOs) at both the national and international levels and through intergovernmental organizations, including the United Nations. There are now scores of local native organizations in dozens of countries. There is ample evidence of their influence on government policy, particularly in Australia and Canada. They have had less success elsewhere, especially in Asia, Africa, and the former Soviet Union, where most governments refused even to admit that they had indigenous peoples, only citizens with different cultures. The in-

ternational NGOs (INGOs) have recently had considerable influence in the United Nations, and this merits some attention.

The United Nations first addressed itself formally to indigenous peoples in 1949, but the United States objected to its consideration of the matter, and it languished for years. The first major international convention relating specifically to indigenous peoples was Convention No. 107, adopted by the International Labour Organisation in 1957. This established the first binding standards on indigenous land rights. The ILO, continuing its pioneering role, adopted in June 1989 its Convention No. 169, Convention Concerning Indigenous and Tribal Peoples in Independent Countries.

As international publicity grew about the threats to isolated tribal peoples in South

反核
台湾不要死亡科技

反如工
原委會

⊕綠色小姐

**Protest demonstration by Orginals of Taiwan.**
The Yami Tribe of Orchid Island shown here in
1992 protesting a government proposal to build a
nuclear waste facility on tribal land. The banners
demand the abandonment of the project and respect for their traditions and human rights. This is
symbolic of the heightened conflicts between indigenous peoples around the world and the industrialized societies that continue to degrade their
environment, doing irreparable damage to native
cultures as well. (Poster prepared by The Green
Team of Taiwan and provided by Amy Eisenberg)

America in the late 1960s, the United Nations
sought some way to deal with the problem
that would be politically acceptable. In 1971
the UN Sub-Commission on Prevention of
Discrimination and Protection of Minorities
authorized a study, mentioned earlier, by
a Special Rapporteur on "The Problem of
Discrimination Against Indigenous Populations." In 1982, while the Rapporteur was
conducting his study (prolonged because he
was working virtually alone), the UN Economic and Social Council (ECOSOC) authorized the sub-Commission to establish a
Working Group on Indigenous Populations.

For the first time indigenous peoples now
had direct access to the United Nations
rather than having to go through governments. The Working Group has been meeting annually, taking testimony and collecting
other materials that would help it reach its
goal: preparation of a draft UN declaration
on indigenous rights. Its meetings have attracted more participants than any other UN
human rights body.

This process continues, with the aboriginal peoples becoming better organized and
more assertive, continuing to influence the
United Nations and its related entities, and
even the International Whaling Commission,
which has allowed Alaskan natives to continue their traditional subsistence hunting of
bowhead whales despite the worldwide ban
on whaling except for scientific purposes.
Another result was the 1989 UN Seminar on
the Effects of Racism and Racial Discrimination on the Social and Cultural Relations Between Indigenous Peoples and States.

The pioneer support organizations were
the International Work Group for Indigenous Affairs and Survival International,
founded in 1968 and 1969, respectively. An
umbrella organization, the World Council of
Indigenous Peoples, was founded in British
Columbia in 1975. The first INGO Conference on Discrimination Against Indigenous
Peoples of the Americas was held in 1977.
Since then, both within and outside the UN
system, work has continued on improving
the lot of the world's aborigines. ECOSOC
has so far accredited five of the leading
INGOs in the field and five local and national NGOs, including the Grand Council of
the Cree, the first individual tribal group to
receive such recognition.

Some of the NGOs have matured considerably, going far beyond their original
human rights concerns. The Inuit Circumpolar Conference, for example, has also been
concerned with peace and security issues,
working toward the establishment of an Arctic zone of peace; Arctic policy development
in a wide range of social, cultural, and economic issues affecting the native peoples of
the region; and environmental issues, including participation in preparations for the

**Declaration on Circumpolar Cooperation.** At the Arctic Leaders Summit meeting in Copenhagen on 20 June 1991, representatives of the major organizations of far northern indigenous peoples signed a cooperation agreement, a milestone in the political action of these peoples. Seated at the signing table are the presidents of (l to r) Inuit Circumpolar Conference, Association of the Indigenous Peoples of the Soviet Union, and the Nordic Saami Council. Standing are members of the ICC Council, representing (l to r) Alaska, Alaska, Greenland, Canada, USSR, Greenland, Canada, and the USSR. (Photo courtesy of the Inuit Circumpolar Conference)

1992 UN Conference on Environment and Development and in the conference itself.

### *The American Indian Reservation*

Although there are aboriginal reserves in many countries, even one as small as the Caribbean island State of Dominica, nowhere else are they as numerous and complex as in the United States. At present there are some 259 Indian reservations in the United States, excluding Alaska. This number alone may be very misleading and requires some explanation. In the first place, the very word reservation is not clearly defined.* In general, it is an area of land belonging to an Indian tribe, some of which may be held in trust by a state or the federal

government.* An examination of a reservation, however, may reveal much land within it that is owned by or leased to whites, owned by individual Indians either in trust or in fee simple, and owned or controlled by a state, county, or municipality. Further-

*The term *tribe* as applied to Indians also requires some explanation. In many cases it was simply a term of convenience applied by whites to loosely associated Indians who shared some cultural traits. Some of these groups were nations, as we have defined the term earlier. Some were simply hunting bands or independent villages of much larger cultural groups. Some were amalgamations or confederations of tribes. Some were branches or offshoots of other tribes. Most of the Indian groups in the West were not organized tribally, but generally by clan (lineage) or village. Today the term is a formal one applied to those Indian groups that organized as such under the Indian Reorganization Act of 1934 or other legislation. A number of Indian groups in the East, still lacking official tribal status, are at a disadvantage in applying for federal or state Indian benefits and in pressing land and other claims, whereas others have secured the coveted designation and are entitled to some federal services.

*See Imre Sutton, "Sovereign States and the Changing Definition of the Indian Reservation," *Geographical Review, 66*, 3 (July 1976), 281–295.

more, both Indians and whites may own land adjacent to the reservation, much of it formerly reservation land but left outside when the reservation was reduced in size.

Of the 259 reservations, 26 are state reservations, primarily in the East; the remainder are federal reservations, mostly in the West. Many of them are very small. Of the 76 federal reservations in California, for example, 17 of them are smaller than 40 hectares (100 acres), including three of 2.2, 0.5, and 0.37 hectares (5.5, 1.32, and 0.92 acres). One state reservation in Connecticut is 0.1 hectare (0.26 acre). The largest, on the other hand, is the Navajo reservation, with nearly 5.66 million hectares (14 million acres). Land quality, population density, accessibility, mineral and forest resources, and other characteristics are similarly varied. Generally, however, reservations are located in areas bypassed or unwanted by white settlers and land speculators. White society, however,

has reached many of them and engulfed some, generating many problems.

The Indian reservation is a distinctive kind of territory. It does not fit into the hierarchy of civil divisions—state, county, township, municipality—yet it performs many of the same functions and overlaps all of them. It is not a special purpose district because it was not created to serve a public purpose as an arm of the state. Yet all the jurisdictional problems we discussed in earlier chapters exist in reservation areas. They are governed in part by treaties that have little or no standing in international law and cannot be amended as international treaties can. They have some of the boundary and territorial problems of States, yet seek redress in federal court. Over the past two centuries government policy toward Indians has fluctuated from extermination to paternalism to termination to toleration. This indecision, combined with different conditions in vari-

---

### Some Unusual Indian Reservations

There are some interesting anomalies with regard to reservations. Before the mid-1970s there were 13 reservations in Alaska for Eskimos, Aleuts, and two different Indian groups. Under the Alaska Native Claims Settlement Act of 1971, all outstanding native claims against the government for lost land were settled by allowing newly formed Native Village Corporations (205 of them) and the new Regional Corporations (11) to select allocations of public land totaling 15.38 million hectares (38 million acres). The village corporations took the place of 12 of the reservations. The remaining one, the Annette Island Reserve of the Tsimshian Tribe in the southeastern part of the state, is exempted from the provisions of the Act and continues in its previous status.

In Oklahoma there are 32 tribes and tribal groups that own 43,500 hectares (107,500 acres) tribally and 607,000 hectares (1.5 million acres) in allotments, yet the state has no reservations as such.* Many tribes are organized under state or federal laws or regulations and most members live on or near their "reservations," which are federal trust areas. Most of them have assimilated into the white society.

The Red Lake Reservation of the Chippewa Tribe in Minnesota is an autonomous unit in the traditional tribal homeland that was never formally ceded to the United States. It is thus exempt from all state jurisdiction that may apply to other reservations in Minnesota, although it is under federal jurisdiction. The Warm Springs Reservation in Oregon (Warm Springs, Northern Paiute, and Wasco Confederated Tribes) enjoys a somewhat similar status.

*There is a continuing dispute over the Osage Reservation. The federal government claims it still exists, but the state denies it.

✗

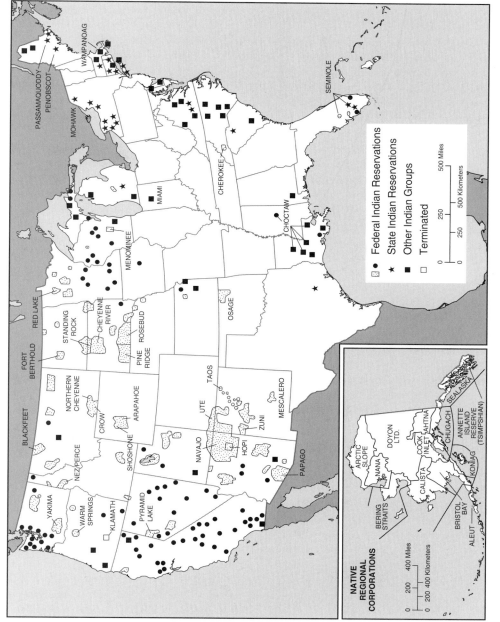

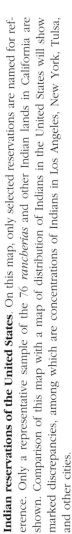

**Indian reservations of the United States.** On this map, only selected reservations are named for reference. Only a representative sample of the *76 rancherías* and other Indian lands in California are shown. Comparison of this map with a map of distribution of Indians in the United States will show marked discrepancies, among which are concentrations of Indians in Los Angeles, New York, Tulsa, and other cities.

ous parts of the country, different Indian cultures and reactions to white intrusion, and changing circumstances in American society generally, accounts for the great variety we find today in Indian reservations and their relations with other units of government and with individual Indians.

As American citizens, Indians are free to come and go as they please, and there is considerable movement to and from the reservations. The trend for the past third of a century, though, has been for increasing numbers to move away from the reservation more or less permanently, aided by government relocation programs. Today only about half of the nearly 2 million Indians (outside Alaska) are living on reservations. Off the reservation they have the same status as other citizens and are subject to the same laws. On the reservation the situation is more complex.

In general, reservation Indians are subject only to federal law and, when on the reservation, to tribal law, and the jurisdiction of each is fairly well defined. In practice, however, the federal government has gradually whittled away the immunities granted to Indians by constitutional, treaty, and statutory provisions, particularly in regard to land tenure. Indian lands were opened to white homesteaders, for example, although tribes retained their legal status. Land was allotted to individual Indians, eroding the political power of many tribes and allegedly increasing state authority over the privately owned land—and the landowners. The most serious erosion of tribal authority has occurred since 1953. In that year the federal government adopted a policy of working toward termination of federal responsibility for the tribes. It also passed P.L. 280, which permitted the states to assume civil and criminal jurisdiction over the federal reservations if their legislation was acceptable to the federal government.

Between 1954 and 1960 there were 61 tribes, groups, communities, allotments, and *rancherías* (small reservations in California) terminated by the withdrawal of federal services and protection. The largest tribes choosing termination were the Klamath in Oregon and the Menominee in Wisconsin (whose reservation became a county). The termination policy was just short of disastrous and has been abandoned.* P.L. 280 continues in force, though much modified by the Civil Rights Act of 1968 and numerous judicial decisions. Under it Alaska, California, Iowa, Kansas, and Minnesota assumed jurisdiction over the reservations within their boundaries. Wisconsin, Florida, Idaho, Nevada, and Washington assumed jurisdiction in whole or in part by state legislation under the authority of the federal law. Other states have taken the opportunity to extend certain kinds of jurisdiction over reservations. All of this has led to great confusion, much squabbling over jurisdiction, and many court battles. Some issues involved are land tenure, housing codes, environmental protection (including air and water pollution control, forest-fire prevention, restrictions on traditional fishing), and, most important, resource exploitation and commercial and industrial development on Indian reservations.

Some future damage to Indian reservations was prevented by a provision in the 1968 Civil Rights Act that requires tribal consent before a state may assume civil and criminal jurisdiction over reservations, but the current picture is still very much a muddle. Federal, tribal, and state (including county and municipal) jurisdictions coexist (and sometimes overlap) in different strengths in different states and even within states. Current laws conflict with old treaties. Different philosophies of land ownership and land use still clash. It will be a long time before all the problems are resolved.

The Indian reservation is, in Sutton's words (pp. 292–293), "a unique and semiautonomous enclave. . . . There is no obvious counterpart in our political system: a legally sanctioned island that only indigenous Americans can possess." The most significant

---

*Federal trusteeship was restored to the Menominees in 1973 and their reservation was reestablished, but a great deal of damage had already been done to tribal resources and institutions.

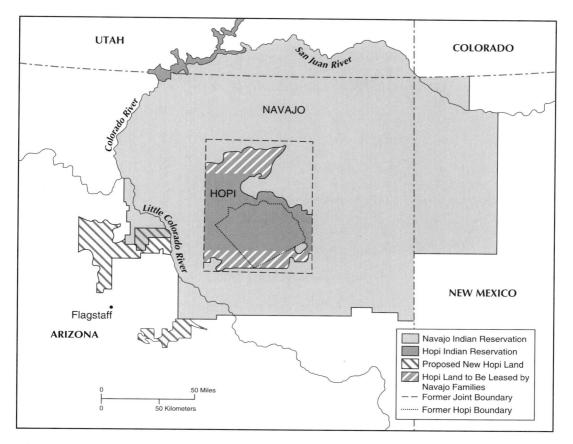

**The Navajo-Hopi territorial dispute.** This is similar in many ways to international boundary and territorial disputes. The Hopi reservation is an enclave within the Navajo reservation. For more than a century the pastoral Navajos have been occupying Hopi land without Hopi consent, and the agricultural Hopis have been trying to oust the Navajos. In 1943 the Bureau of Indian Affairs approved the Navajo request for "joint use" of three-quarters of the Hopi reservation, and this was confirmed by a 1963 court decision. But the Navajos prevented Hopi use of the land and the dispute went on. In 1976 the Hopis sued the Navajos in court. In 1977 a compromise was reached: the disputed land would be partitioned and the new boundaries would be fenced. Many Navajos living on Hopi land refused to move, however, and the compromise could not be implemented. Since then a number of new proposals have been made and new lawsuits filed. This map illustrates a proposal to compensate the Hopis for Navajo-occupied land with new land to be added to their reservation. The dispute seems likely to drag on for years.

element of this uniqueness is tribal autonomy within the reservation. But the dimensions and even the basis for this autonomy are questionable. The Supreme Court has ruled that Indian sovereignty is nonterritorial; that is, it is limited to functions within the reservation instead of applying to the territory of the reservation itself. The reservation is not a State, and its boundaries are more permeable than those of the weakest State. It is autonomous but not sovereign. It bears many similarities to a developing country (as we will see later) as it begins to assume control over its own economic development.

Recent efforts of reservations to develop economically so as to improve the living conditions of their people, and perhaps attract back home tribe members who have relocated into cities, can usefully be grouped into three categories:

**Choctaw reservation.** The reservation consists of over 6800 ha (17,000 A) of ancestral lands in east-central Mississippi. Tribal headquarters is in Pearl River, Neshoba County. Most of the tribe was removed to Oklahoma in the nineteenth century, and over 90,000 Choctaws live there today. Only about 30,000 live on this reservation. Most of the tribal income here derives from forestry, with additional funds coming both from agriculture and an arts and crafts shop in nearby Philadelphia, Mississippi. (Photo by Julie Kelsey of the Choctaw Community News.)

1. Traditional, resource-based activities, such as agriculture, ranching, forestry, fishing, and mining, especially for uranium and fossil fuels. In all of these activities, modern technology is being applied, often with the aid of outside capital, and income is also derived from leasing fees, royalties, and other transfer payments.

2. Nontraditional, land-based activities, including tourism of many kinds, manufacturing (much of which, however, is footloose and not land-based), and, most recently, providing dumping sites for solid and hazardous wastes of all kinds.

3. Non-land-based activities such as provision of services for passersby, wage labor off the reservation, profitable sales of nontaxed cigarettes and small-scale but high-stakes bingo.

High-stakes bingo, which used to be found on relatively few, small, and isolated reservations, has grown impressively in the past decade or so but has been largely eclipsed by the spectacular growth of casino gambling. This activity is one indirect result of the congressional passage of the Tribal Self-governance Demonstration Act in 1987, which allows the councils of 30 tribes to determine how and where to spend funds received from the federal government and their own economic activities, and makes them accountable for the expenditures. This is similar in some ways to the devolution of power from national governments to lower civil divisions discussed in Chapter 13. In due course, if successful, this experiment may extend to all federally supervised tribes and be made permanent.

Perhaps of less long-term significance but certainly producing more dramatic initial changes is the Indian Gaming Regulatory Act of 1988. Among other things, this law specifies three types of gambling on Indian reservations: Class I, traditional social games for prizes of minimal value or forming part of tribal ceremonies; Class II, bingo and related games, for which large cash prizes are permitted; and, most important, Class III, all other kinds of gambling. Despite many restrictions and requirements, it is the authorization of Class III gambling that has triggered an incredible growth in gambling on reservations all over the country. Within six years, scores of reservations from coast to coast had instituted high-stakes bingo and/or casinos. The amount of money now flowing into these reservations is staggering. Most impressive of all, and in a class by itself, is the Foxwoods High Stakes Bingo and Casino on the Mashantucket Pequot reservation near Ledyard, Connecticut, by 1995 the world's largest casino and still growing.

**The Foxwoods casino complex.** Located in rural Ledyard, Connecticut, this enormous hotel-gambling-entertainment-shopping-recreation-museum complex draws gamblers and other visitors from the entire northeast of the United States. The Pequots are expanding both this complex and their reservation and plan to expand their catchment area to Montreál. (Martin Glassner)

Before the arrival of Europeans, the Pequots occupied about 650 square kilometers (250 square miles) in southern Connecticut and Rhode Island and numbered about 10,000 to 15,000 people. By 1974, the remaining people in the Mashantucket branch of the tribe numbered only 55, only one of whom lived on their residual reservation of 86 hectares (214 acres). In 1976 they began a struggle for federal recognition as a tribe, finally achieving success in 1983. They then opened a bingo parlor, which brought in considerable revenue, provided jobs, attracted back many tribal members, and financed many community developments.

Foxwoods opened in February 1992. A huge complex of resort hotel, gambling facilities, shopping, and entertainment (including headline singers and professional boxing) in a rural area convenient to New York and Boston, it was profitable from the beginning. Within three years its annual revenue had reached nearly a *billion* dollars, of which the State of Connecticut was receiving nearly one-eighth. Profits were being invested in more land, bringing the reservation to over 3000 acres by 1995, and in hotels, expansion of Foxwoods, an annual Green Corn Dance and Festival (called Schemitzun) in Hartford, a consulting firm to advise other tribes around the country, a lobbying office in Washington, a $10 million gift to the National Museum of the American Indian in Washington, D.C., and millions to other charities, including the Special Olympics World Games and both major political parties. The operation has been so successful that the tribe has grown to 331 members, with more clamoring to be admitted, and other small tribes in southern New England are proposing casinos of their own.

Naturally, gambling operations also have their negative features. Twelve states have refused to negotiate the compacts with tribes that are required by the 1988 law, and many tribes have refused to be lured into the gambling business. Also, of course, as competition increases, profits from individual operations will fall, and, as the novelty wears off and new diversions draw patrons away, the number of casinos will gradually decline. Meanwhile, gambling is providing many small, poor, and isolated tribes with the means to revive their cultures and assure the futures of their members. The massive flow of funds from whites to Indians may also be viewed as the beginning of restitution of some of the assets stolen over the centuries from Indians by whites and of compensation for the physical and cultural degradation of the Indian peoples.*

*A good source of information and publications about indigenous peoples worldwide is Cultural Survival, Inc., 215 First Street, Cambridge, MA 02142.

# BIBLIOGRAPHY FOR PART THREE

## *Books and Monographs*

**A**

Abedin, Najmul, *Local Administration and Politics in Modernizing Societies: Bangladesh and Pakistan.* Oxford University Press, 1973.

Agnew, John A., *Place and Politics.* Winchester, MA: Allen & Unwin, 1987.

Allen, Howard W. and Vincent A. Lacey (eds.), *Illinois Elections, 1818–1990; Candidates and County Returns for President, Governor, Senate, and House of Representatives.* Carbondale: Southern Illinois Univ. Press, 1992.

Archer, J. Clark and Fred M. Shelley, *American Electoral Mosaics.* Washington, DC: Association of American Geographers, 1986.

Archer, J. Clark and Peter J. Taylor, *Section and Party: A Political Geography of American Presidential Elections from Andrew Jackson to Ronald Reagan.* New York: Wiley, 1981.

Arian, Asher, *The Elections of Israel—1969.* Jerusalem: Jerusalem Academic Press, 1972.

———, *The Elections of Israel 1973.* Jerusalem: Jerusalem Academic Press, 1974.

Arian, Asher and Michael Shamir (eds.), *The Elections in Israel—1988.* Boulder, CO: Westview, 1990.

———, *The Elections in Israel 1992.* Albany, NY: SUNY Press, 1994.

Arnold, Millard W. and others, *Zimbabwe: Report on the 1985 General Elections.* Washington, DC: International Human Rights Law Group, 1986.

Arnold, R., *Alaska Native Land Claims.* Anchorage: Alaska Native Foundation, 1976.

Assies, W. J. and A. J. Hoekema (eds.), *Indigenous Peoples' Experiences with Self-government.* Copenhagen: IWGIA,1994.

Aucoin, Peter, Regional Responsiveness and the National Administrative State. Toronto: Univ. of Toronto Press, 1987.

**B**

Barlow, I. Max, *Metropolitan Government.* London: Routledge, 1991.

Barnett, M. R. and others (eds.), *Electoral Politics in the Indian States.* Delhi: Monohar, 1975.

Bartlett, Richard H., *Aboriginal Water Rights in Canada: A Study of Aboriginal Title to Water and Indian Water Rights.* Calgary: Canadian Institute of Resources Law, 1988.

———, *Indian Reserves and Aboriginal Lands in Canada; A Homeland.* Saskatoon: Univ. of Saskatchewan Native Law Centre, 1990.

———, *Resource Development and Aboriginal Land Rights.* Calgary: Canadian Institute of Resources Law, 1991.

Bartolini, Stefano and Peter Mair, *Identity, Competition and Electoral Stability; The Stabilisation of European Electorates, 1885–1985.* Cambridge Univ. Press, 1990.

Bates, Sarah F. and others, *Searching Out the Headwaters; Change and Rediscovery in Western Water Policy.* Washington, DC: Island Press, 1993.

Batteau, Allen (ed.), *Appalachia and America: Autonomy and Regional Dependence.* Lexington: Univ. Press of Kentucky, 1983.

Beaglehole, John Holt, *The District: A Study in Decentralization in West Malaysia.* Oxford Univ. Press, 1976.

Beckett, Jeremy, *Torres Strait Islanders; Custom and Colonialism.* Cambridge Univ. Press, 1987.

Beigbeder, Yves, *International Monitoring of Plebiscites, Referenda and National Elections; Self-Determination and Transition to Democracy.* Kluwer, 1994.

Bennett, R. J., *Territory and Administration in Europe.* Leicester, UK: Pinter, 1989.

———, *Local Government in the New Europe.* London: Belhaven, 1993.

Berkley, George E. and Douglas M. Fox, 80,000 *Governments: The Politics of Subnational America.* Boston: Allyn & Bacon, 1978.

Berry, M. C., *The Alaska Pipeline: The Politics of Oil and Native Land Claims.* Bloomington: Indiana Univ. Press, 1975.

Beyle, Thad L. (ed.), *State Government: CQ's Guide to Current Issues and Activities 1991–92 and 1994–95.* Washington, DC: Congressional Quarterly, 1991 and 1994.

Biolsi, Thomas, *Organizing the Lakota; The Political Economy of the New Deal on the Pine Ridge and Rosebud Reservations.* Tucson: Univ. of Arizona Press, 1992.

Bolt, Christine, *American Indian Policy and American Reform; Case Studies of the Campaign to Assimilate.* New York: Routledge, 1989.

Bradshaw, Michael, *The Appalachian Regional Commission.* Lexington: Univ. of Kentucky Press, 1992.

Brice-Bennett, C. (ed.), *Our Footprints Are Everywhere; Inuit Land Use and Occupancy in Labrador.* Labrador: Inuit Association, 1977.

Brownlie, Ian [Brookfield, F. M. (ed.)], *Treaties and Indigenous Peoples.* Oxford Univ. Press, 1992.

Brubaker, Sterling (ed.), *Rethinking the Federal Lands*. Washington, DC: Resources for the Future, 1984.

Brugge, David M., *The Navajo-Hopi Land Dispute*. Albuquerque: Univ. of New Mexico Press, 1994.

Burger, Julian, *The Gaia Atlas of First Peoples; A Future for the Indigenous World*. New York: Anchor Books, 1990.

Burgess, Michael (ed.), *Canadian Federalism; Past, Present, and Future*. Leicester, UK: Pinter, 1991.

Burton, Lloyd, *American Indian Water Rights and the Limits of Law*. Lawrence: Univ. Press of Kansas, 1994.

Bushnell, E., *Impact of Reapportionment on the Thirteen Western States*. Salt Lake City: Univ. of Utah Press, 1970.

Butler, David and Austin Ranney (eds.), *Referendums Around the World; The Growing Use of Direct Democracy*. LaVergne, TN: AEI, 1994.

**C**

Cain, Bruce E., *The Reapportionment Puzzle*. Berkeley: Univ. of California Press, 1984.

Campbell, P., *French Electoral Systems and Elections Since 1789*. London: Faber & Faber, 1958.

Carter, Sarah, *Lost Harvests: Prairie Indian Reserve Farmers and Government Policy*. Montreal: McGill-Queens, 1990.

Case, D. S., *The Special Relationship of Alaska Natives to the Federal Government*. Anchorage: Alaska Native Foundation, 1978.

Castile, George Pierre and Robert L. Bee (eds.), *State and Reservation; New Perspectives on Federal Indian Policy*. Tucson: Univ. of Arizona Press, 1992.

Caves, Roger W., *Land Use Planning: The Ballot Box Revolution*. Newbury Park, CA: Sage, 1992.

Cawley, R. McGreggor, *Federal Land, Western Anger; The Sagebrush Rebellion and Environmental Politics*. Lawrence: Univ. Press of Kansas, 1994.

Chisholm, Michael and David M. Smith (eds.), *Shared Space; Divided Space; Essays on Conflict and Territorial Organization*. London: Unwin Hyman, 1990.

Clarke, Harold D. and others, *How Voters Change; The 1987 British Election Campaign in Perspective*. Oxford Univ. Press, 1990.

————, *Absent Mandate; Interpreting Change in Canadian Elections, 2nd ed.* Agincourt, Ont.: Gage, 1991.

Cohen, Abner, *Custom and Politics in Urban Africa; A Study of Hausa Migrants in Yoruba Towns*. Berkeley: Univ. of California Press, 1969.

Cohen, Fay G., *Treaties on Trial; The Continuing Controversy over Northwest Indian Fishing Rights*. Seattle: Univ. of Washington Press, 1986.

Confederation of American Indians, *The Indian Reservations; A State and Federal Handbook*. Jefferson, NC and London: McFarland, 1986.

————, *State Politics and Redistricting*, Parts I and II. Washington, DC, 1982.

Congressional Quarterly, *Congressional Districts in the 1980's*. Washington, DC, 1983.

————, *Jigsaw Politics; Shaping the House after the 1990 Census*. Washington, DC, 1990.

————, *Congressional Districts in the 1990s; A Portrait of America*. Washington, DC, 1993.

————, *Congressional Quarterly's Guide to 1990 Congressional Redistricting* (two volumes). Washington, DC, 1993.

————, *Guide to U.S. Elections*. 3rd ed. Washington, DC, 1994.

————, *Presidential Elections 1789–1992*. Washington, DC, 1995.

Council of State Governments, *The Book of the States*. Biennial.

Cox, Bruce Alden, *Native People, Native Lands; Canadian Indians, Inuit and Metis*. Ottawa: Carleton Univ. Press, 1988.

Cox, Lindsay, *Hotahitanga; The Search for Maori Political Unity*. Oxford Univ. Press, 1993.

Cullen, Richard, *Australian Federalism Offshore*. Melbourne: Univ. of Melbourne Press, 1988.

Cullingworth, J. Barry, *Urban and Regional Planning in Canada*. New Brunswick, NJ: Transaction, 1987.

**D**

Dacks, Gurston (ed.), *Devolution and Constitutional Development in the Canadian North*. Ottawa: Carleton Univ. Press, 1992.

Davidson, Chandler (ed.), *Minority Vote Dilution*. Washington, DC: Howard Univ. Press, 1994.

Davis, Shelton H., *Land Rights and Indigenous Peoples; The Role of the Inter-American Commission on Human Rights*. Cambridge, MA: Cultural Survival, 1988.

de la Garza, Rodolfo O. and others (eds.), *Barrio Ballots; Latino Politics in the 1990 Elections*. Boulder, CO: Westview, 1993.

Deloria, Vine, Jr. (ed.), *American Indian Policy in the Twentieth Century*. Norman: Univ. of Oklahoma Press, 1985.

De Vorsey, Louis, Jr., *The Georgia-South Carolina Boundary; A Problem in Historical Geography*. Athens: Univ. of Georgia Press, 1982.

Dickerson, Mark O., *Whose North? Political*

*Change, Political Development, and Self-Government in the Northwest Territories.* London: UCL Press, 1992.

Dumars, C. T. and others, *Pueblo Indian Water Rights: Struggle for a Precious Resource.* Tucson: Univ. of Arizona Press, 1984.

Dyck, Noel and James B. Waldram, *Anthropology, Public Policy, and Native Peoples in Canada.* Montreal: McGill-Queens, 1993.

**E**

Edwards, Beatrice and Gretta T. Siebentritt, *Places of Origin: The Repopulation of Rural El Salvador.* Boulder, CO: Lynne Rienner, 1991.

Elliott, Jean Leonard and Augie Fleras, *The "Nations Within;" Aboriginal-State Relations in Canada, the U.S., and New Zealand.* Oxford Univ. Press, 1992.

Emmanuel, Patrick A.M., *General Elections in the Caribbean: A Handbook.* Cave Hill, Barbados: Institute of Social and Economic Studies, Univ. of the West Indies, 1979.

**F**

Fairfax, Sally K. and Carolyn E. Yale, *Federal Lands: A Guide to Planning, Management, and State Revenues.* Washington, DC: Island Press, 1987.

Farley, Lawrence T., *Plebiscites and Sovereignty.* Boulder, CO: Westview, 1986.

Fierman, William (ed.), *Soviet Central Asia; The Failed Transformation.* Boulder, CO: Westview, 1991.

Foss, Phillip O. (ed.), *Federal Lands Policy.* Westport, CT: Greenwood, 1987.

Fournier, Pierre, *A Meech Lake Post-Mortem; Is Quebec Sovereignty Inevitable?* Montreal: McGill-Queen's, 1991.

Freemuth, John C., *Islands under Seige; National Parks and the Politics of External Threats.* Lawrence: Univ. of Kansas Press, 1991.

Frey, William H. and Alden Speare, Jr., *Regional and Metropolitan Growth and Decline in the United States.* New York: Russell Sage Foundation, 1988.

Frizzell, Alan and Anthony Westell, *The Canadian General Election of 1984: Politicians, Parties, Press and Polls.* Ottawa: Carleton Univ. Press, 1985.

Frizzell, Alan and others, *The Canadian General Election of 1988.* Ottawa: Carleton Univ. Press, 1989.

——, *The Canadian General Election of 1993.* Ottawa: Carleton Univ. Press, 1994.

Furtak, Robert K. (ed.), *Elections in Socialist States.* New York: St. Martin's, 1990.

**G**

Galligan, Brian and others (eds.), *Intergovernmental Relations and Public Policy.* New York: HarperCollins, 1991.

Galster, George C. and Edward W. Hill (eds.), *The Metropolis in Black and White; Place, Power and Polarization.* Piscataway, NJ: CUPR Books, 1992.

Garber, Larry and Eric Bjornlund (eds.), *The New Democratic Frontier: A Country by Country Report on Elections in Central and Eastern Europe.* Arlington, VA: Public Interest Pubs., 1994.

Getches, David H. and Charles Wilkinson, Cases and *Materials on Federal Indian Law.* 2nd ed. St. Paul: West, 1986.

Ghurye, G. S., *The Scheduled Tribes of India.* New Brunswick, NJ: Transaction, 1980.

Ginsberg, Benjamin and Alan Stone (eds.), *Do Elections Matter?* 3rd ed. Armonk, NY: Sharpe, 1995.

Goldman, R. M. (ed.), *Transnational Parties: Organizing the World's Precincts.* Lanham, MD: Univ. Press of America, 1983.

Goldwin, Robert A. and others, *Forging Unity Out of Diversity: The Approaches of Eight Nations.* Lanham, MD: Univ. Press of America, 1989.

Gordon, Robert J., *The Bushman Myth; The Making of a Namibian Underclass.* Boulder, CO: Westview, 1992.

Gottlieb, Robert and Margaret Fitz Simmons, *Thirst for Growth; Water Agencies as Hidden Government in California.* Tucson: Univ. of Arizona Press, 1994.

Graf, William L., *Wilderness Preservation and the Sagebrush Rebellions.* Lanham, MD: Univ. Press of America, 1990.

Grofman, Bernard (ed.), *Political Gerrymandering and the Courts.* Bronx, NY: Agothon, 1990.

Gumbert, Marc, *Neither Justice Nor Reason; a Legal and Anthropological Analysis of Aboriginal Land Rights.* St. Lucia: Univ. of Queensland Press, 1985.

Gunlicks, Arthur B., *Local Government in the German Federal System.* Durham, NC: Duke Univ. Press, 1986.

**H**

Hardy, LeRoy and others (eds.), *Reapportionment Politics: The History of Redistricting in the 50 States.* Beverly Hills, CA: Sage, 1981.

Hart, E. Richard (ed.), *Zuni and the Courts, A Struggle for Sovereign Land Rights.* Lawrence: Univ. of Kansas Press, 1994.

Hasager, Ulla and Jonathan Friedman (eds.),

*Hawaiʻi Return to Nationhood.* Copenhagen: IWGA, 1994.

Hawkes, David C. (ed.), *Aboriginal Peoples and Government Responsibility: Exploring Federal and Provincial Roles.* Ottawa: Carleton Univ. Press, 1989.

Heard, Kenneth A., *General Elections in South Africa, 1943–1970.* Oxford Univ. Press, 1974.

Heath, A. and others, *How Britain Votes.* Oxford: Pergamon, 1985.

Hoggart, Keith, *People, Power and Place; Perspectives on Anglo-American Politics.* London: Routledge, 1991.

Hoggart, Keith and David R. Green (eds.), *London: A New Metropolitan Geography.* Edward Arnold, 1992.

Hudson, R., *Wrecking a Region: State Policies, Party Politics and Regional Change in North East England.* London: Pion, 1989.


**J**

Jaimes, M. Annette, *The State of Native America.* Monroe, ME: South End Press, 1992.

Jayanntha, Dilesh, *Electoral Allegiance in Sri Lanka.* Cambridge Univ. Press, 1991.

Jeffery, Charlie and Peter Savigear (eds.), German *Federalism Today.* Leicester, UK: Leicester Univ. Press, 1991.

Johnson, Donald D., *The City and County of Honolulu; A Governmental Chronicle.* Honolulu: Univ. of Hawaii Press, 1991.

Johnston, Richard and others, *Letting the People Decide; The Dynamics of a Canadian Election.* Stanford, CA: Stanford Univ. Press, 1992.

Johnston, Ronald J., *People, Places and Votes.* Armidale, Australia: Univ. of New England Press, 1977.

———, *Political, Electoral and Spatial Systems.* Oxford Univ. Press, 1979.

Johnston, Ronald J., Fred M. Shelley and Peter J. Taylor, *Developments in Electoral Geography.* London: Routledge, 1990.


**K**

Kammer, Jerry, *The Second Long Walk: The Navajo-Hopi Land Dispute.* Albuquerque: Univ. of New Mexico Press, 1980.

Keeping, Janet, *The Inuvialuit Final Agreement.* Calgary: Canadian Institute of Resources Law, 1989.

Kennett, Steven, *Managing Interjurisdictional Waters in Canada: A Constitutional Analysis.* Calgary: Canadian Institute of Resources Law, 1991.

Knight, J., and N. Baxter-Moore, *Republic of*

*Ireland: The General Elections of 1969 and 1973.* London: Arthur MacDougall Fund, 1973.

Kugisaki, C. *Reapportionment in Hawaii.* Honolulu: Hawaiian Legislative Reference Bureau, 1978.

Kulchyski, Peter (ed.), *Unjust Relations, Aboriginal Rights in Canadian Courts.* Oxford Univ. Press, 1994.

Kux, Stephan, *Soviet Federalism; A Comparative Perspective.* Boulder, CO: Westview, 1990.


**L**

Larmour, Peter and Rapate Qalo (eds.), *Decentralisation in the South Pacific: Local, Provincial and State Government in Twenty Countries.* Suva, Fiji: Univ. of the South Pacific, 1985.

Lasater, Martin L., *A Step Toward Democracy; The December 1989 Elections in Taiwan, Republic of China.* Lanham, MD: AEI Press, 1990.

Leach, Richard H., *Interstate Relations in Australia.* Lexington: Univ. Press of Kentucky, 1965.

Little Bear, Leroy and others (eds.), *Pathways to Self-Determination—Canadian Indians and the Canadian State.* Toronto: Univ. of Toronto Press, 1984.

Lodge, Juliet, *Direct Elections to the European Parliament 1984.* New York: St. Martin's, 1986.

Lopach, James J. and others, *Tribal Government Today; Politics on Montana Indian Reservations.* Boulder, CO: Westview, 1990.


**M**

Mackie, Thomas T. and Richard Rose, *The International Almanac of Electoral History.* 3rd ed. Washington, DC: Congressional Quarterly, 1991.

Markus, Andrew, *Governing Savages.* Sydney: Allen & Unwin, 1991.

Markusen, Ann R., *Regions; The Economics and Politics of Territory.* Totowa, NJ: Rowman & Littlefield, 1987.

Markusen, Ann and others, *The Rise of the Gunbelt; The Military Remapping of Industrial America.* Oxford Univ. Press, 1991.

Masson, Jack, *Alberta's Local Governments and their Politics.* Edmonton: Univ. of Alberta Press, 1985.

McAllister, I. and R. Rose, *The Nationwide Competition for Votes; The 1983 British Election.* London: Pinter, 1984.

McCool, Daniel, *Command of the Waters; Iron Triangles, Federal Water Development and Indian Water.* Tucson: Univ. of Arizona Press, 1994.

McGillivray, Alice V., *Presidential Primaries and Caucuses: 1992, A Handbook of Election Statistics.* Washington, DC: Congressional Quarterly, 1992.

————, *Congressional and Gubernatorial Primaries 1991–1992; A Handbook of Election Statistics.* Washington, DC: Congressional Quarterly, 1993.

McGillivray, Alice V. and others, *America Votes 20; A Handbook of Contemporary Election Statistics.* Washington, DC: Congressional Quarterly, 1993.

McGillivray, Alice V. and Richard M. Scammon, *America at the Polls; A Handbook of Presidential Election Statistics.* Washington, DC: Congressional Quarterly, 1994. Two volumes.

McGuire, Thomas R. and others (eds.), *Indian Water in the New West.* Tucson: Univ. of Arizona Press, 1993.

McNeil, Kent, *Common Law Aboriginal Title.* Oxford Univ. Press, 1989.

Mehra, Ajay K., *The Politics of Urban Redevelopment; A Study of Old Delhi.* Newbury Park, CA: Sage, 1991.

Mendelsohn, Oliver and Upendra Baxi (eds.), *The Rights of Subordinated Peoples.* Oxford Univ. Press, 1994.

Michelmann, Hans J. and Danayotis Soldatos, *Federalism and International Relations; The Role of Subnational Units.* Oxford Univ. Press, 1990.

Miles, William F.S., *Elections in Nigeria: A Grassroots Perspective.* Boulder, CO: Lynne Rienner, 1988.

Miller, Warren E. and Santa A. Traugott, *American National Election Studies Data Sourcebook, 1952–1986.* Cambridge, MA: Harvard Univ. Press, 1989.

Morse, Bradford W. (ed.), *Aboriginal Peoples and the Law: Indian, Metis and Inuit Rights in Canada.* Revised First Edition. Ottawa: Carleton Univ. Press, 1989.

Murray, Christina and Catherine O'Regan, *No Place to Rest; Forced Removals and the Law in South Africa.* Oxford Univ. Press, 1990.

**N**

Nice, David C., *Federalism; The Politics of Intergovernmental Relations.* New York: St. Martin's, 1987.

Nicholson, Norman L., *The Boundaries of Canada, Its Provinces and Territories.* Ottawa: Dept. of Mines and Technical Surveys, Geographical Branch, 1954.

**O**

O'Brien, Sharon, *American Indian Tribal Governments.* Norman: Univ. of Oklahoma Press, 1989.

Olson, Paul A., *The Struggle for the Land.* Lincoln: Univ. of Nebraska Press, 1990.

O'Toole, Laurence J. Jr., *American Intergovernmental Relations,* 2nd ed. Washington, DC: Congressional Quarterly, 1993.

**P**

Page, Edward C. and Michael J. Goldsmith (eds.), *Central and Local Government Relations.* Newbury Park, CA: Sage, 1987.

Palmer, N. D., *Elections and Political Development: The South Asian Experience.* Durham, NC: Duke Univ. Press, 1975.

Parker, Linda S., *Native American Estate; The Struggle over Indian and Hawaiian Lands.* Honolulu: Univ. of Hawaii Press, 1989.

Paterson, Lindsay, *The Autonomy of Modern Scotland.* Edinburgh Univ. Press, 1994.

Paul, Sharda, *1989 General Elections in India.* New Delhi: Associated Publishing House, 1990.

Peleg, Ilan and Ofira Seliktar (eds.), *The Emergence of a Binational Israel: The Second Republic in the Making.* Boulder, CO: Westview, 1989.

Penniman, Howard and Brian Farrell (eds.), *Ireland at the Polls: 1981, 1982, & 1987: A Study of Four General Elections.* Durham, NC: Duke Univ. Press, 1987.

Peroff, Nicholas C., *Menominee Drums: Tribal Termination and Restoration, 1954–1974.* Norman: Univ. of Oklahoma Press, 1982.

Pevar, Stephen L., *The Rights of Indians and Tribes; The Basic ACLU Guide to Indian and Tribal Rights.* 2nd ed. Carbondale: Southern Illinois Univ. Press, 1992.

Ponting, J. Rick (ed.), *Arduous Journey; Canadian Indians and Decolonization.* Toronto: McClelland & Stewart, 1986.

Prucha, Francis Paul, *The Great Father: The United States Government and the American Indians.* Lincoln: Univ. of Nebraska Press, 1984.

————, *Atlas of American Indian Affairs.* Lincoln: Univ. of Nebraska Press, 1990.

————, *Documents in United States Indian Policy.* Lincoln: Univ. of Nebraska Press, 1990.

**R**

Ranney, Austin, *The American Elections of 1984.* Durham, NC: Duke Univ. Press, 1985.

Rao, Nirmada, *The Making and Unmaking of Local Self-Government.* UK: Dartmouth, 1995.

Reed, Steven R., *Japanese Prefectures and Policy-making*. Pittsburgh: Univ. of Pittsburgh Press, 1986.

Reisner, Marc and Sarah Bates, *Overtapped Oasis; Reform or Revolution for Western Water*. Washington, DC: Island Press, 1990.

Resnick, Philip, *Toward a Canada-Quebec Union*. Montreal: McGill-Queen's, 1991.

Reynolds, Henry (ed.), *Dispossession; Black Australians and White Invaders*. Sydney: Allen & Unwin, 1990.

Rogers, Peter, *America's Water; Federal Roles and Responsibilities*. Cambridge, MA: MIT, 1993.

Rousseau, Mark O. and Ralph Zariski, *Regionalism and Regional Devolution in Comparative Perspective*. New York: Praeger, 1987.

Rubenstein, James M., *The French New Towns*. Baltimore: Johns Hopkins Univ. Press, 1978.

Rule, Wilma and Joseph F. Zimmerman, *United States Electoral Systems: Their Impact on Women and Minorities*. Westport, CT: Greenwood, 1992.

———, *Electoral Systems in Comparative Perspective*. Westport, CT: Greenwood, 1994.

Rumer, Boris Z., *Soviet Central Asia; "A Tragic Experiment."* New York: Routledge, 1990.

Runte, Alfred, *National Parks: The American Experience*. 2nd ed. Lincoln: Univ. of Nebraska Press, 1987.

Rush, Mark E., Does *Redistricting Make a Difference? Partisan Representation and Electoral Behavior*. Baltimore: Johns Hopkins Univ. Press, 1991.

Rusk, David, *Cities Without Suburbs*. Baltimore: Johns Hopkins Univ. Press, 1993.

**S**

Sager, Eric W. and others, *Atlantic Canada & Confederation: Essays in Canadian Political Economy*. Toronto: Univ. of Toronto Press, 1983.

Salisbury, Richard F., *A Homeland for the Cree; Regional Development in James Bay, 1971–1981*. Montreal: McGill-Queen's, 1986.

Sauder, Robert A., *The Lost Frontier; Water Diversion in the Growth and Destruction of Owens Valley Agriculture*. Tucson: Univ. of Arizona Press, 1994.

Saunders, J. Owen, *Managing Natural Resources in a Federal State; Essays from the Second Institute Conference on Natural Resources Law, April 17–20, 1985*. Toronto: Carswell, 1985.

———, *Interjurisdictional Issues in Canadian Water Management*. Calgary: Canadian Institute of Resources Law, 1988.

Savole, Donald J., *Federal-Provincial Collaboration; the Canada-New Brunswick General Development Agreement*. Montreal: McGill-Queen's, 1987.

Schlozman, Kay L. (ed.), *Elections in America*. New York: Routledge, 1987.

Sharp, Andrew, *Justice and the Maori; Maori Claims in New Zealand*. Oxford Univ. Press, 1990.

Sharpe, L. J. (ed.), *The Government of World Cities: The Future of the Metro Model*. New York: Wiley, 1995.

Short, C. Brant, *Ronald Reagan and the Public Lands; America's Conservation Debate 1979–1984*. College Station: Texas A&M Press, 1989.

Sierra Club, *The Great Giveaway: Public Oil, Gas and Coal and the Reagan Administration*. San Francisco, 1984.

Silva, Maynard (ed.), *Ocean Resources and US Intergovernmental Relations in the 1980's*. Boulder, CO: Westview, 1986.

Simeon, Richard, *Intergovernmental Relations*. Toronto: Univ. of Toronto Press, 1987.

Simon, David J. (ed.), *Our Common Lands; Defending the National Parks*. Washington, DC: Island Press, 1988.

Singh, V. B., *Elections in India*. 2nd ed. Newbury Park, CA: Sage, 1987.

Singh, V. B. and Shankar Bose, *Elections in India; Data Handbook on Lok Sabha Elections 1952–85*. 2nd ed. New Delhi: Sage, 1986.

———, *State Elections in India; Volume I: The North (Part 1)*. Newbury Park, CA: Sage, 1987.

Smiley, Donald V. and Ronald L. Watts, *Intrastate Federalism in Canada*. Toronto: Univ. of Toronto Press, 1987.

Smith, Carol A. (ed.), *Guatemalan Indians and the State*. Austin: Univ. of Texas Press, 1990.

Smith, Michael P. (ed.), *After Modernism: Global Restructuring and the Changing Boundaries of City Life*, New Brunswick, NJ: Transaction, 1992.

Stasiulis, Daiva and Nira Yuval-Davis, *Unsettling Settler Societies*. Newbury Park, CA: Sage, 1995.

Stevenson, Garth (ed.), *Federalism in Canada; Selected Readings*. Toronto: McClelland & Stewart, 1989.

Sutton, Imre (ed.), *Irredeemable America; the Indians' Estate and Land Claims*. Albuquerque: Univ. of New Mexico Press, 1985.

**T**

Taylor, Peter J. and Ronald J. Johnston, *Geography of Elections*. London: Croom Helm, 1979.

Teaford, Jon C., *City and Suburb: The Political Fragmentation of Metropolitan America, 1850–1970*. Baltimore: Johns Hopkins Univ. Press, 1979.

Tennant, Paul, Aboriginal Peoples and Politics; *The Indian Land Question in British Columbia, 1849–1989.* Vancouver: UBC Press, 1990.

Tester, Frank and Peter Kulchyski, Tammarniit (Mistakes); Inuit Relocation in the Eastern Arctic, 1939–63. Vancouver: UBC Press, 1994.

Thorson, John E., *River of Promise, River of Peril; The Politics of Managing the Missouri River.* Lawrence: Univ. Press of Kansas, 1995.

Tsai, Wen-hui, *Toward Greater Democracy: An Analysis of the Republic of China on Taiwan's Major Elections in the 1990s.* Univ. of Maryland School of Law, Occasional Papers/Reprints Series in Contemporary Asian Studies Number 6-1994 (125).

**U**

Urban, Greg and Joel Sherzer (eds.), *Nation-States and Indians in Latin America.* Austin: Univ. of Texas Press, 1991.

Utter, Jack, *American Indians; Answers to Today's Questions.* Lake Ann, MI: National Woodlands, 1993.

**V**

van den Berg, Leo and others, *Governing Metropolitan Regions.* Brookfield, VT: Ashgate, 1993.

van der Wusten, Herman (ed.), *The Urban Political Arena.* Amsterdam: Netherlands Geographical Studies 140, 1992.

Van Horn, Carl E. (ed.), *The State of the States.* 2nd ed. Washington, DC: Congressional Quarterly, 1992.

Vasil, Raj, *What Do the Maori Want? New Maori Political Perspectives.* London: Random Century Group, 1990.

**W**

Wahl, Richard W., *Markets for Federal Water; Subsidies, Property Rights, and the Bureau of Reclamation.* Baltimore: Johns Hopkins Univ. Press, 1989.

Waldram, James B., *As Long as the Rivers Run; Hydroelectric Development and Native Communities in Western Canada.* Winnipeg: Univ. of Manitoba Press, 1988.

Warne, William E., *The Bureau of Reclamation.* Boulder, CO: Westview, 1985.

Watkins, Michael (ed.), *Dene Nation: The Colony Within.* Toronto: Univ. of Toronto Press, 1977.

Weaver, Sally M., *Making Canadian Indian Policy.* Toronto: Univ. of Toronto Press, 1980.

Weiner, Myron and Ergun Ôzbudun (eds.), *Competitive Elections in Developing Countries.* Durham, NC: Duke Univ. Press, 1987.

*West Papua: The Obliteration of a People.* London: Tapol, 1983.

Whalley, John, *Regional Aspects of Confederation.* Toronto: Univ. of Toronto Press, 1987.

Wilkinson, Charles F. and Christine L. Miklas (eds.), *Indian Tribes as Sovereign Governments.* Oakland, CA: American Indian Lawyer Training Program, 1988.

Wilmer, Franke, *The Indigenous Voice in World Politics Since Time Immemorial.* Newbury Park, CA: Sage, 1993.

Wondolleck, Julia M., *Public Lands Conflict and Resolution; Managing National Forest Disputes.* New York: Plenum, 1988.

Wood, Robert, *1400 Governments: The Political Economy of the New York Metropolitan Region.* Cambridge, MA: Harvard Univ. Press, 1961.

Wright, John B., *Rocky Mountain Divide; Selling and Saving the West.* Austin: Univ. of Texas Press, 1993.

Wunder, John R., *"Retained by the People;" A History of American Indians and the Bill of Rights.* Oxford Univ. Press, 1994.

**Y**

Young, Elspeth, *The Third World in the First; Development and Indigenous Peoples.* London: Routledge, 1995.

**Z**

Zaslowsky, Dyan and T. H. Watkins (eds.), *These American Lands; Parks, Wilderness, and the Public Lands.* Covelo, CA: Island Press, 1994.

## *Periodicals*

**A**

*Aboriginal Law Bulletin.* Published since 1981 by the Faculty of Law, Univ. of New South Wales, Kensington, NSW, Australia.

Agnew, John A., "A Place Anyone? A Comment on the McAllister and Johnston Papers," *Political Geography Quarterly,* 6, 1 (Jan. 1987), 39–40.

———, "The National Versus the Contextual: The Controversy Over Measuring Electoral Change in Italy Using Goodman Flow-of-Vote Estimates," *Political Geography,* 13, 3 (May 1994), 245–254.

Akins, Nancy, "New Direction in Sacred Land Claims: Lyng v. Northwest Indian Cemetery Protective Association." *Natural Resources Journal,* 29 (Spring 1989), 593–605.

Alfredsson, Gudmundur, "International Law,

International Organizations and Indigenous Peoples," *Journal of International Affairs*, 36, 1 (1982), 113–125.

Archer, J. Clark, "Some Geographical Aspects of the American Presidential Election of 1980," *Political Geography Quarterly*, 1, 2 (April 1982), 123–135.

———, "Macrogeographical Versus Microgeographical Cleavages in American Presidential Elections: 1940–1984," *Political Geography Quarterly*, 7, 2 (April 1988), 111–126.

Archer, J. Clark and others, "Counties, States, Sections, and Parties in the 1984 Presidential Election," *Professional Geographer*, 37, 3 (August 1985), 279–287.

———, "The Geography of U.S. Presidential Elections," *Scientific American*, 259, 1 (July 1988), 44–51.

Axford, Nicholas and Steven Pinch, "Growth Coalitions and Local Economic Development Strategy in Southern England. A Case Study of the Hampshire Development Association," *Political Geography*, 13, 4 (July 1994), 344–360.

**B**

Backstrom, Charles H., "The Practice and Effect of Redistricting," *Political Geography Quarterly*, 1, 4 (Oct. 1982), 351–359.

Barsh, Russell Lawrence, "Indigenous Peoples: An Emerging Object of International Law," *American Journal of International Law*, 80, 2 (April 1986), 369–385.

Berezkin, A. V. and others, "The Geography of the 1989 Elections of Peoples' Deputies of the USSR," *Soviet Geography*, 30 (1989), 607–634.

Berman, Howard R., "The International Labour Organization and Indigenous Peoples," *The Review, International Commission of Jurists*, No. 41 (Dec. 1988), 48–57.

Bernier, Lynne Louise, "Decentralizing the French State; Implications for Policy," *Journal of Urban Affairs*, 13, 1 (1991), 21–32.

Berry, David, "The Geographic Distribution of Governmental Powers: The Case of Regulations," *Professional Geographer*, 39, 4 (Nov. 1987), 428–437.

Blumm, Michael C., "Native Fishing Rights and Environmental Protection in North America and New Zealand: A Comparative Analysis of Profits à Prendre and Habitat Servitudes," *Wisconsin International Law Journal*, 8, 1 (Fall 1989), 1–50.

Boddy, Martin, "Central-local Government Relations Theory and Practice," *Political Geography Quarterly*, 2, 2 (April 1983), 119–138.

Bondi, Liz, "School Closures and Local Politics: The Negotiation of Primary School Ration-

alization in Manchester," *Political Geography Quarterly*, 6, 3 (July 1987), 203–224.

Bowler, Shaun, "Introduction to Special Issue. Contextual Models of Politics: The Political Impact of Friends and Neighbours," *Political Geography Quarterly*, 10, 2 (April 1991), 91–96.

Boyles, Kristen L., "Saving Sacred Sites; The 1989 Proposed Amendment to the American Indian Religious Freedom Act," *Cornell Law Review*, 76 (July 1991), 1117–1149.

Boyne, George and Martin Powell, "Territorial Justice: A Review of Theory and Evidence," *Political Geography Quarterly*, 10, 3 (July 1991), 263–281.

Brady, David W., "Research Note: Modelling the Determinants of Swing Ratio and Bias in U.S. House Elections, 1850–1980," *Political Geography Quarterly*, 10, 3 (July 1991), 254–262.

Brierly, Allen Bronson and David Moon, "Electoral Coalitions and Institutional Stability; The Case of Metropolitan Reform in Dade County, Florida," *Journal of Politics*, 53 (August 1991), 701–719.

Briggett, Marlissa S., "State Supremacy in the Federal Realm: The Interstate Compact," *Boston College Environmental Affairs Law Review*, 18 (Summer 1991), 751–772.

Brownill, Sue and Susan Halford, "Understanding Women's Involvement in Local Politics: How Useful Is a Formal/Informal Dichotomy?," *Political Geography Quarterly*, 9, 4 (Oct. 1990), 396–414.

Brunk, Gregory C., "On Estimating Distance-Determined Voting Functions," *Political Geography Quarterly*, 4, 4 (Oct. 1985), 55–65.

Brunk, Gregory C. and others, "Contagion-Based Voting in Birmingham, Alabama," *Political Geography Quarterly*, 7, 1 (Jan. 1988), 39–48.

Bunge, William, "Gerrymandering, Geography, and Grouping," *Geographical Review*, 56, 2 (April 1966), 256–263.

Burghardt, Andrew, "The Redesigning of Scotland's Administrative Areas: An Outsider's View," *Scottish Geographical Magazine*, 98, 3 (Dec. 1982), 130–142.

Busteed, M. A., "Reform of Local Government Structure in Northern Ireland," *Irish Geography*, 6 (1971).

**C**

Cant, Garth and Eric Pawson (eds.), *Indigenous Land Rights in Canada, Australia and New Zealand*. Special issue of *Applied Geography*, 12, 2 (April 1992), whole volume.

*Caribbean Quarterly*, 21, 1 (March 1981), Four articles on electoral politics.

Cassidy, Frank, "Aboriginal Governments in

Canada: An Emerging Field of Study," *Canadian Journal of Political Science*, 23, 12 (1990), 73–99.

Chadjipadelis, Theodore and Costas Zafiropoulos, "Electoral Changes in Greece 1981–90: Geographical Patterns and the Uniformity of the Vote," *Political Geography*, 13, 6 (Nov. 1994), 492–514.

Christopher, A. J., "Parliamentary Delimitation in South Africa, 1910–1980," *Political Geography Quarterly*, 2, 3 (July 1983), 205–217.

———, "Apartheid Planning in South Africa: The Case of Port Elizabeth," *Geographical Journal*, 72, (1987), 195–204.

———, "Changing Patterns of Group-Area Proclamations in South Africa, 1950–1989," *Political Geography Quarterly*, 10, 3 (July 1991), 240–253.

Clark, G. L. and K. Johnston, "The Geography of U.S. Union Elections, Parts 1–5," *Environment and Planning*, A19,(1987), 33–57, 153–172, 179–234, 289–311, and 447–469, .

Cloke, Paul and Jo Little,"The Rural State? Limits to Planning in Rural Society," *Political Geography*, 11, 2 (March 1992), 228–230.

Cohen, Saul, "A Federated Solution to Division, Conflict and the Need for Geopolitical Restructuring in New York City ," in *Eretz-Israel*. Jerusalem: Israel Exploration Society, 22 (1991), 1–11.

Comeaux, Malcolm L., "Attempts to Establish and Change a Western Boundary," *Annals, AAG*, 72, 2 (June 1982), 254–271.

Cox, Kevin Robert, "The 'Individual,' the 'Social' and Reconceptualizing Contextual Effects," *Political Geography Quarterly*, 6, 1 (Jan. 1987), 41–43.

Cox, Kevin R. and Andrew E.G. Jonas, "Urban Development, Collective Consumption and the Politics of Metropolitan Fragmentation", *Political Geography*, 12, 1 (Jan. 1993), 8–37.

*Cultural Survival Quarterly.* Published since 1976 by Cultural Survival, Inc., Cambridge, MA.

Cutter, Susan L. and others, "From Grass Roots to Partisan Politics: Nuclear Freeze Referenda in New Jersey and South Dakota," *Political Geography Quarterly*, 6, 4 (Oct. 1987), 287–300.

Cycon, Dean E., "When Worlds Collide; Law, Development and Indigenous People," *New England Law Review*, 25 (Spring 1991), 761–794.

**D**

Davies, Richard B. and Robert Crouchley, "The Determinants of Party Loyalty: A Disaggregate Analysis of Panel Data from the 1974 and 1979 Elections in England," *Political Geography Quarterly*, 4, 4 (Oct. 1985), 307–320.

Davis, Stephen, "Indigenous Maritime Claims on the North Australian Coast," *Marine Policy Reports*, 1, 2 (1989), 135–150.

Dawson, Andrew H., "Local Government and the Idea of the Region: A Comment on the Present Situation in Scotland," *Scottish Geographical Magazine*, 100, 2 (Sept. 1984), 113–122.

Diehl, Paul F. and Gary Goertz, "Interstate Conflict over Exchanges of Homeland Territory, 1816–1980," *Political Geography Quarterly*, 10, 4 (Oct. 1991), 342–355.

Dikshit, Ramesh D. and J. C. Sharma, "Electoral Performance of the Congress Party in Punjab (1952–1977): An Ecological Analysis," *Transactions, Institute of Indian Geographers*, 4 (1982), 1–15.

Dikshit, R. K., "Spatial Analysis of Electoral Participation in Haryana," *Geographical Review of India*, 50 (1988), 1–7.

Dix, R. H., "Incumbency and Electoral Turnover in Latin America," *Journal of Interamerican Studies*, 26 (1984), 435–438.

Doran, Michael F., "A Political Definition of Virginia's Nuclear Core," *Political Geography Quarterly*, 6, 4 (Oct. 1987), 301–311.

Duncan, C. J., "Ethnicity, Election and Emergency: The 1987 Fiji General Election in the Context of Contemporary Political Geography," *Political Geography Quarterly*, 10, 3 (July 1991), 221–239.

Duncan, C. J. and W. R. Epps, "The Demise of 'Country Mindedness:' New Players or Changing Values in Australian Rural Politics?," *Political Geography*, 11, 5 (Sept. 1992), 430–448.

**E**

Eagles, Munroe and Stephen Erfle, "Community Cohesion and Working-Class Politics: Workplace-Residence Separation and Labour Support, 1966–1983," *Political Geography Quarterly*, 7, 3 (July 1988), 229–250.

*Electoral Studies.* Butterworth-Heinemann, Letchworth, England. Published thrice a year since 1981.

**F**

Fincher, Ruth, "Caring for Workers' Dependents: Gender, Class and Local State Practice in Melbourne," *Political Geography Quarterly*, 10, 4 (Oct. 1991), 356–381.

Foster, Kathryn A., "Exploring the Links Between Political Structure and Metropolitan Growth," *Political Geography*, 12, 6 (Nov. 1993), 523–547.

**G**

Gale, Fay, "The Participation of Australian Aboriginal Women in a Changing Political

Environment," *Political Geography Quarterly,* 9, 4 (Oct. 1990), 381–395.

Gieman, A. and G. King, "Estimating the Electoral Consequences of Legislative Redistricting," *Journal of the American Statistical Association,* 85 (1990), 274–282.

Gilbert, Christopher P., "Religion, Neighborhood Environments and Partisan Behavior: A Contextual Analysis," *Political Geography Quarterly,* 10, 2 (April 1991), 110–131.

Glassner, Martin Ira, "Drawing Boundaries of Planning Regions: A Political Geographer's Contribution," *ITCC Review.* 1, 3 (July 1972), 35–40.

———, "The Bedouin of Southern Sinai Under Israeli Administration," *Geographical Review,* 64, 1 (Jan. 1974), 31–60.

———, "The New Mandan Migrations: From Hunting Expeditions to Relocation," *Journal of the West,* 13, 2 (April 1974), 59–74.

Glendening, Parris N. and Patricia S. Atkins, "City-County Consolidations: New Views for the Eighties," *The Municipal Year Book* (1980).

Gottman, Jean (ed.), *Urban Growth and Politics.* Symposium in *Ekistics,* 57 (Jan.-April 1990), whole volume.

Gradus, Yehuda, "The Role of Politics in Regional Inequality—the Israeli Case," *Annals, AAG,* 73, 3 (Sept. 1983), 388–403.

Greenberg, Michael R. and Samy Amer, "Self-interest and Direct Legislation: Public Support of a Hazardous Waste Bond Issue in New Jersey," *Political Geography Quarterly,* 8, 1 (Jan. 1989), 67-78.

Grofman, Bernard, "Reformers, Politicians, and the Courts: A Preliminary Look at U.S. Redistricting," *Political Geography Quarterly,* 1, 4 (Oct. 1982), 303–316.

Guest, Avery M. and others, "Industrial Affiliation and Community Culture: Voting in Seattle," *Political Geography Quarterly,* 7, 1 (Jan. 1988), 49–74.

Guillorel, Herve and Jacques Levy, "Space and Electoral System," *Political Geography,* 11, 2 (March 1992), 205–224.

**H**

Hajdú, Zoltan, "Administrative Geography and Reforms of the Administrative Areas in Hungary," *Political Geography Quarterly,* 6, 3 (July 1987), 269–278.

Hare, F. Kenneth, "Regionalism: A Development in Political Geography," *Public Affairs,* (1946), 38–39.

———, "Regionalism and Administration: North American Experiments," *Canadian Journal of Economic and Political Sciences,* 15, 3 (August 1949), 344–352.

———, "The Labrador Frontier," *Geographical Review,* 42, 3 (July 1952), 405–424.

Hart, Paxton, "The Making of Menominee County," *Wisconsin Magazine of History,* 43 (Spring 1960), 181–184.

Hasson, Shlomo and Eran Razin, "What Is Hidden Behind a Municipal Boundary Conflict?," *Political Geography Quarterly,* 9, 3 (July 1990), 267–283.

Helin, Ronald A., "The Volatile Administrative Map of Rumania," *Annals, AAG,* 57, 3 (Sept. 1967), 481–502.

Helmsing, A.H.J., "Transforming Rural Local Government: Zimbabwe's Municipal Elections," *Government and Policy,* 8 (1990), 87–110.

Hibbing, John R. and Sara L. Brandes, "State Population and the Electoral Success of U.S. Senators," *American Journal of Political Science,* 27, 4 (Nov. 1983), 808–819.

Hodge, D. and Lynn Staeheli, "Social Relations and Geographic Patterns of Urban Electoral Behavior," *Urban Geography,* 13 (1992), 307–333.

Horn, David L., and others, "Practical Application of District Compactness," *Political Geography,* 12, 2 (March 1993), 103-120.

**I**

Ikporukpo, C. O., "Politics and Regional Policies: The Issue of State Creation in Nigeria," *Political Geography Quarterly,* 5, 2 (April 1986), 127–139.

Inamete, Ufot B., "Federalism in Nigeria; The Crucial Dynamics," *Round Table* (April 1991), 191–207.

*Indigenous Affairs,* Published quarterly by the International Work Group for Indigenous Affairs in Copenhagen, Denmark.

Irving, R.E.M. and W. E. Paterson, "The 1990 German General Election," *Parliamentary Affairs,* 44 (July 1991), 353–572.

**J**

Jaensch, D., "A Functional Gerrymander—South Australia, 1944–1970," *Australian Quarterly,* 42 (1970), 96–101.

Johnston, Ronald J., "Xenophobia and Referenda: An Example of the Exploratory Use of Ecological Regression," *L'Espace Géographique,* 9 (1980), 73–80.

———, "Changing Voter Preferences, Uniform Electoral Swing, and the Geography of Voting in New Zealand," *New Zealand Geographer,* 37 (1981), 13–19.

———, "Short-term Electoral Change in England: Estimates of Its Spatial Variation," *Political Geography Quarterly,* 1, 1 (Jan. 1982), 41–55.

———, "Redistricting by Independent Commissions: A Perspective from Britain," *Annals, AAG*, 72, 4 (Dec. 1982), 457–470.

———, "Dealignment, Volatility, and Electoral Geography," *Studies in Comparative International Development*, 22, 3 (Fall 1987), 3–59.

———, "The Beginnings of Realignment? Ecological Analysis of the 1984 and 1987 New Zealand General Elections," *New Zealand Geographer*, 45 (1989), 50–57.

Johnston, Ronald J. and J. Ballantine, "Geography and the Electoral System," *Canadian Journal of Political Science*, 10 (1977), 857–866.

Johnston, Ronald J. and Rex Honey, "The 1987 General Election in New Zealand: The Demise of Electoral Cleavages?," *Political Geography Quarterly*, 7, 4 (Oct. 1988), 363–368.

Johnston, Ronald J. and C. Hughes, "Constituency Delimitation and the Unintentional Gerrymander in Brisbane," *Australian Geographical Studies*, 16 (1978), 79–110.

Johnston, Ronald J. and C. J. Pattie, "Family Background, Ascribed Characteristics, Political Attitudes and Regional Variations in Voting Within England, 1983: A Further Contribution," *Political Geography Quarterly*, 6, 4 (Oct. 1987), 347–350.

———, "A Nation Dividing: Economic Well-being, Voter Response and the Changing Electoral Geography of Great Britain," *Parliamentary Affairs*, 42 (1989), 37–57.

———, "Changing Voter Allegiances in Great Britain, 1979–87: An Exploration of Regional Patterns," *Regional Studies*, 22 (1989), 179–192.

———, "Class, Attitudes and Retrospective Voting: Exploring Regional Variations in the 1983 General Election in Great Britain." *Environment and Planning A*, 22 (1990), 893–908.

———, "Class Dealignment and the Regional Polarization of Voting Patterns in Great Britain, 1964–1987," *Political Geography*, 11, 1 (Jan. 1992), 73–86.

Johnston, Ronald J. and others, "The Changing Electoral Geography of the Netherlands: 1946–1981," *Tijdschrift voor Economische en Sociale Geografie*, 74 (1983), 185–195.

———, "The Neighborhood Effect Won't Go Away: Observations on the Electoral Geography of England in the Light of Dunleavy's Critique," *Geoforum*, 14, (1983), 161–168.

———, "Proportional Representation and Fair Representation in the European Parliament Area," *Area*, 15 (1983), 347–355.

———, "The Geography of the Working Class and the Geography of the Labour Vote in England 1983: A Prefatory Note to a Research Agenda," *Political Geography Quarterly*, 6, 1 (Jan. 1987), 7–16.

———, "What Price Place," *Political Geography Quarterly*, 6, 1 (Jan. 1987), 51–52.

———, Review Essay: "Can We Leave Electoral Reform to Politicians?" (House of Commons . . . Redistribution of Seats; Report of the Royal Commission on the Electoral System . . . *Political Geography Quarterly*, 6, 3 (July 1987), 279–282.

———, "Great Britain's Changing Electoral Geography: The Flow of the Vote and Spatial Polarization." *Tidjschrift Economische en Sociale Geografie*, 81 (1989), 189–213.

———, "The Role of Ecological Analysis in Electoral Geography: The Changing Patterns of Labour Voting in Great Britain, 1983–87." *Geografiska Annaler*, 70B (1990), 307–324.

**K**

Kaplan, David H., "Nationalism at a Micro-Scale: Educational Segregation in Montreal," *Political Geography*, 11, 3 (May 1992), 259282.

Karim, M. Bazlul, "Decentralisation of Government in the Third World: A Fad or a Panacea?," *International Studies Notes*, 16, 2 (Spring 1991), 50–53.

Kegley, Charles, W., Jr. and Steven W. Hook, "U.S. Foreign Aid and U.N. Voting: Did Reagan's Linkage Strategy Buy Deference or Defiance?," *International Studies Quarterly*, 35, 3 (Sept. 1991), 295–312.

Kenny, Christopher B., "Partisanship and Political Discussion," *Political Geography Quarterly*, 10, 2 (April 1991), 97–109.

Kincaid, John (ed.), *American Federalism; The Third Century. Annals, American Academy of Political and Social Science*, 509 (May 1990), whole volume.

Knowles, R. D., "Malapportionment in Norway's Parliamentary Elections Since 1921," *Norsk Geografisk Tidsskrift*, 35 (1981), 147–159.

Kolosov, V. A., "Socialist Federalism and Spatial Equity: Review of a Discussion," *Soviet Geography*, 30 (1989), 662–669.

———, "The Geography of Elections of USSR People's Deputies by National-Territorial Districts and the Nationalities Issue." *Soviet Geography*, 31 (Dec. 1990), 753–766.

Koulov, Boian, "Geography of Electoral Preferences: The 1990 Great National Assembly Elections in Bulgaria," *Political Geography*, 14, 3 (April 1995), 241–258.

Kramer, Daniel C., "'Those People Across the Water Just Don't Know Our Problems': An Analysis of Friends and Neighbours Voting in a Geographically-Split Legislative District," *Political Geography Quarterly*, 9, 2 (April 1990), 189–196.

## L

Lake, Robert W., "Negotiating Local Autonomy," *Political Geography*, 13, 5 (Sept. 1994), 423–442.

Laponce, J. A., "Assessing the Neighbor Effect on the Vote of Francophone Minorities in Canada," *Political Geography Quarterly*, 6, 1 (Jan. 1987), 77–78.

Lauria, Mickey, "The Transformation of Local Politics: Manufacturing Plant Closures and Governing Coalition Fragmentation," *Political Geography*, 13, 6 (Nov. 1994), 515–539.

Lawrey, Andr[aa]lee, "Contemporary Efforts to Guarantee Indigenous Rights under International Law," *Vanderbilt Journal of Transnational Law*, 23, 4 (1990), 703–706.

Laws, Glenda, "Oppression, Knowledge and the Built Environment," *Political Geography*, 13, 1 (Jan. 1994), 7–32.

Leitner, Helga, "Cities in Pursuit of Economic Growth. The Local State As Entrepreneur," *Political Geography Quarterly*, 9, 2 (April 1990), 146–170.

Leonski, Zbigniew, "Aspects of Territorial Subdivisions in European Socialist States," *International Social Science Journal*, 30, 1 (1978), 47–56.

Levy, Roger, "Finding a Place in the World-Economy; Party Strategy and Party Vote: The Regionalization of SNP and Plaid Cymru Support, 1979–92," *Political Geography*, 14, 3 (April 1995), 295–308.

Ley, David L., "Styles of the Times: Liberal and Neoconservative Landscapes in Inner Vancouver, 1968–1986," *Journal of Historical Geography*, 13 (1987), 40–56.

Logan, W. S., "The Changing Landscape Significance of the Victoria-South Australia Boundary," *Annals, AAG*, 58, 1 (March 1968), 128–154.

Long, J. Anthony, "Political Revitalization in Canadian Native Indian Societies," *Canadian Journal of Political Science,* 23 (Dec. 1990), 751–773.

Lutz, James M., "Diffusion of Nationalist Voting in Scotland and Wales: Emulation, Contagion, and Retrenchment," *Political Geography Quarterly*, 9, 3 (July 1990), 249–266.

## M

Mackay, J. Ross, "The Interactance Hypothesis and Boundaries in Canada: A Preliminary Study," *Canadian Geographer*, 11 (1958),1–8.

Mamadouh, V. D. and Herman H. van der Wusten, "The Influence of the Change of Electoral System on Political Representation: The Case of France in 1985," *Political Geography Quarterly*, 8, 2 (April 1989), 145–160.

Marando, V. C., "Inter-local Cooperation in Metropolitan Areas: Detroit," *Urban Affairs Quarterly*, 4 (1968), 185–200.

Marando, Vincent L. and Mavis Mann Reeves, "Counties as Local Governments; Research Issues and Questions." *Journal of Urban Affairs,* 13, 1 (1991), 45–53.

Martin, R., "The Political Economy of Britain's North-South Divide," *Transactions, Institute of British Geographers, NS*, 13 (1988), 389–418.

Martis, Kenneth C. and others, "The Geography of the 1990 Hungarian Parliamentary Elections," *Political Geography*, 11, 3 (May 1992), 283–306.

McAllister, Ian, "Social Context, Turnout and the Vote: Australian and British Comparisons," *Political Geography Quarterly*, 6, 1 (Jan. 1987), 17–30.

———, "Comment on Johnston," *Political Geography Quarterly*, 6, 1 (Jan. 1987), 45–49.

———, "Comment on Johnston and Pattie," *Political Geography Quarterly*, 6, 4 (Oct. 1987), 351–354.

McBurnett, Michael, "The Instability of Partisanship Due to Context," *Political Geography Quarterly*, 10, 2 (April 1991), 132–148.

McDonough, P., "Repression and Representation in Brazil," *Comparative Politics*, 14 (1982), 73–99.

McGee, T. G., "The Malayan Elections of 1959: A Study in Electoral Geography," *Journal of Tropical Geography*, 16 (1962), 72–99.

———, "The Malayan Parliamentary Elections 1964," *Pacific Viewpoint*, 6 (1965), 96–101.

McQuillan, D. Aidan, "Creation of Indian Reserves on the Canadian Prairies 1870–1885," *Geographical Review*, 70, 4 (Oct. 1980), 379–396.

Meir, Avinoam, "Nomads and the State: The Spatial Dynamics of Centrifugal and Contripetal Forces among the Israeli Negev Bedouin," *Political Geography Quarterly*, 7, 3 (July 1988), 251–270.

Mercer, David, "Terra Nullius, Aboriginal Sovereignty and Land Rights in Australia: The Debate Continues," *Political Geography*, 12, 4 (July 1993), 299–318.

Mercer, J. and John A. Agnew, "Small Worlds and Local Heroes: The 1987 General Election in Scotland," *Scottish Geographical Magazine*, 104 (1988), 138–145.

Meyer, William B. and Michael Brown, "Locational Conflict in Nineteenth-Century City," *Political Geography Quarterly*, 8, 2 (April 1989), 107–122.

Miller, Byron, "Political Empowerment, Local-

Central State Relations, and Geographically Shifting Political Opportunity Structures: Strategies of the Cambridge, Massachusetts, Peace Movement," *Political Geography*, 13, 5 (Sept. 1994), 393–406.

Mintz, Eric, "Election Campaign Tours in Canada," *Political Geography Quarterly*, 4, 1 (Jan. 1985), 47–54.

Mitchneck, Beth A., "Territoriality and Regional Economic Autonomy in the USSR," Studies in *Comparative Communism*, 24 (June 1991), 218–224.

Moore, Michael R., "Native American Water Rights: Efficiency and Fairness," *Natural Resources Journal*, 22, 3 (Summer 1989), 763–791.

Morehouse, Thomas A., "Sovereignty, Tribal Government, and the Alaska Native Claims Settlement Act Amendments of 1987," *Polar Record*, 25, 154 (1989), 197–206.

Morrill, Richard L., "Redistricting Standards and Strategies after 20 Years," *Political Geography Quarterly*, 1, 4 (Oct. 1982), 361–370.

———, "Redistricting, Region and Representation," *Political Geography Quarterly*, 6, 3 (July 1987), 241–260.

Murauskas, G. T. and others, "Metropolitan, Non-metropolitan and Sectional Variations in Voting Behavior in Recent Presidential Elections," *Western Political Quarterly*, 41 (1988), 63–84.

**N**

Nietschmann, Bernard, "Third World War: The Global Conflict over the Rights of Indigenous Nations," *Utne Reader* (Nov.-Dec. 1988), 84–91.

**O**

Ogborn, Miles, "Ordering the City: Surveillance, Public Space and the Reform of Urban Policing in England 1835–56," *Political Geography*, 12, 6 (Nov. 1993), 505–521.

Okafor, S. I., "Jurisdictional Partitioning Distribution Policies and the Spatial Structure of Health-Care Provision in Nigeria," *Political Geography Quarterly*, 6, 4 (Oct. 1987), 335–346.

O'Loughlin, John, "The Identification and Evaluation of Racial Gerrymandering," *Annals, AAG*, 72, 2 (June 1982), 165–184.

O'Loughlin, John and Anne Marie Taylor, "Choices in Redistricting and Electoral Outcomes: The Case of Mobile, Alabama," *Political Geographical Quarterly*, 1, 4 (Oct. 1982), 317–339.

O'Loughlin, John and others, "The Geography of the Nazi Vote: Context, Confession, and Class in the Reichstag Election of 1930," *Annals of AAG*, 84, 3 (Sept. 1994), 351-380.

Osei-Kwame, Peter and Peter J. Taylor, "A Politics of Failure: The Political Geography of Ghanaian Elections, 1954–1979," *Annals, AAG*, 74, 4 (Dec. 1984), 574–589.

Owen, Guillermo and Bernard Grofman, "Optimal Partisan Gerrymandering," *Political Geography Quarterly*, 7, 1 (Jan. 1988), 5–22.

**P**

Painter, Martin, "Intergovernmental Relations in Canada; An Institutional Analysis." *Canadian Journal of Political Science*, 24 (June 1991), 269–288.

Parker, A. J., "The 'Friends and Neighbors' Voting Effect in the Galway West Constituency," *Political Geography Quarterly*, 1, 3 (July 1982), 243–262.

———, "Localism and Bailiwicks: The Galway West Constituency in the 1977 General Election," *Proceedings, Royal Irish Academy*, 83c (1983), 17–36.

———, "Geography and the Irish Electoral System," *Irish Geography*, 19 (1986), 1–14.

Parysek, J. J. and others, "Regional Differences in the Results of the 1990 Election in Poland as the First Approximation to a Political Map of the Country," *Environment and Planning A*, 23 (1991), 1315–1329.

Paxman, John T., "Minority Indigenous Populations and Their Claims for Self-determination," *Case Western Reserve Journal of International Law*, 21, 2 (Summer 1989), 185–202.

Peake, Linda J., Review Essay: "How Sarlvik and Crewe Fail to Explain the Conservative Victory of 1979 and Electoral Trends in the 1970's," *Political Geography Quarterly*, 3, 2 (April 1984), 161–167.

Perkins, Charles, "Governments and Aboriginal Communities," *Australian Journal of Public Administration*, 48, 1 (March 1989), 21–28.

Phelps, Glenn, "Representation Without Taxation: Citizenship and Suffrage in Indian Country," *American Indian Quarterly*, (Spring 1985).

Picchi, Debra, "The Impact of an Industrial Agricultural Project on the Bakairi Indians of Central Brazil," *Human Organization*, 50 (Spring 1991), 26–38.

Pinch, Philip, "Ordinary Places?: The Social Relations of the Local State in Two 'M4-Corridor' Towns," *Political Geography*, 11, 5 (Sept. 1992), 485–500.

Pirie, Gordon H. and others, "Covert Power in South Africa: The Geography of the Afrikaner Broederbond," *Area*, 12 (1980), 97–104.

*Political Centralization and Decentralization in Europe and North America.* Symposium in *Policy Studies Journal,* 18 (Spring 1990).

Pommerscheim, Frank, "The Crucible of Sovereignty: Analyzing Issues of Tribal Jurisdiction," *Arizona Law Review,* 31, 2 (1989), 329–363.

Prescott, John Robert Victor, "The Function and Methods of Electoral Geography," *Annals, AAG,* 49, 3 (Sept. 1959), 296–304.

———, "The Evolution of the Anglo-French Inter-Cameroons Boundary," *Nigerian Geographical Journal,* 5, 2 (Dec. 1962), 103–120.

*Publius: The Journal of Federalism,* published quarterly since 1970 by the Center for the Study of Federalism, Temple Univ., Philadelphia.

**R**

Rantala, O., "The Political Regions of Finland," *Scandinavian Political Studies,* 2 (1967), 117–142.

*Regional Politics and Policy.* Published thrice a year since 1991 by Frank Cass in London.

Regulska, J. and others, "Women, Politics, and Place: Spatial Patterns of Representation in New Jersey," *Geoforum,* 22 (1991), 203–221.

Robinson, J., "'A Perfect System of Control': State Power and Native Locations in South Africa," *Society and Space (Environment and Planning D),* 8 (1990), 135–162.

Rule, S. P., "Emergence of a Racially-Mixed Residential Suburb in Johannesburg: Demise of the Apartheid City," *Geographical Journal,* 155 (1989), 196–03.

———, "Language, Occupation and Regionalism as Determinants of White Political Allegiances in South Africa: The 1981 and 1987 General Elections," *South African Geographical Journal,* 71 (1989), 94–101.

Rumley, Dennis, "Structural Effects in Different Contexts," *Political Geography Quarterly,* 6, 1 (Jan. 1987), 31–37.

———, "The Political Geography of Australian Federal-State Relations," *Geoforum,* 19 (1988), 367–379.

**S**

"Sacred Spaces; Beyond the Concept of Resource Exploitation," *Environmentalist,* 11 (Spring 1991), 55–61.

Sander, Robert, "Patenting an Arid Frontier: Use and Abuse of the Public Land Laws in Owens Valley, California," *Annals, AAG,* 79, 4 (Dec. 1989), 544–569.

Sanders, Douglas, "The UN Working Group on Indigenous Populations," *Human Rights Quarterly,* 11 (1989), 406–433.

Sauer, Carl O., "Geography and the Gerrymander," *American Political Science Review,* 12 (1918), 403–426.

Savage, Mike, "Understanding Political Alignments in Contemporary Britain: Do Localities Matter?" *Political Geography Quarterly,* 6, 1 (Jan. 1987), 53–76.

Seethal, Cecil, "Restructuring the Local State in South Africa: Regional Services Councils and Crisis Resolution," *Political Geography Quarterly,* 10, 1 (Jan. 1991), 8–25.

Shelley, Fred M., "A Constitutional Choice Approach to Electoral District Boundary Delineation," *Political Geography Quarterly,* 1, 4 (Oct. 1982), 341–350.

———, "Structure, Stability and Section in American Politics," *Political Geography Quarterly,* 7, 2 (April 1988), 153–160.

———, "Political Geography and the City in the 1990's," *Urban Geography,* 14 (1993), 165–176.

———, "Local Control and Financing of Education: A Perspective from the American State Judiciary," *Political Geography,* 13, 4 (July 1994), 361–376.

Shelley, Fred M. and J. Clark Archer, "Sectionalism and Presidential Politics: Illinois, Indiana and Ohio," *Journal of Interdisciplinary History,* 20 (1989), 227–256.

———, "Some Geographical Aspects of the American Presidential Election of 1992," *Political Geography,* 13, 2 (March 1994), 137–159.

Simpson, Jeffrey, "The Two Canadas," *Foreign Policy,* No. 81 (Winter 1990–91), 71–86.

Smith, Neil, "The Region Is Dead! Long Live the Region!," *Political Geography Quarterly,* 7, 2 (April 1988), 141–152.

Steinberg, Philip E., "Territorial Formation on the Margin: Urban Anti-Planning in Brooklyn," *Political Geography,* 13, 5 (Sept. 1994), 461–476.

Sutton, Imre (ed.), *The Political Geography of Indian Country.* A symposium in *American Indian Culture and Research Journal,* 15, 2 (1991), whole volume.

**T**

Tamir, Orit, "Relocation of Navajo from Hopi Partitioned Land in Pinon," *Human Organization,* 50 (Summer 1991), 173–178.

Tarlock, A. D., "One River, Three Sovereigns; Indian and Interstate Water Rights," *Land and Water Law Review,* 22, 2 (1987), 631–671.

Trevera, Donald S., "Voting Patterns in Zimbabwe's Elections of 1980 and 1985," *Geography,* 74 (1989), 162–164.

Tucker, Harvey, J., "State Legislative Apportionment: Legal Principles in Empirical Perspective," *Political Geography Quarterly*, 4, 1 (Jan. 1985), 19–28.

**U**

Upton, Graham J.G., "Displaying Election Results," *Political Geography Quarterly*, 10, 3 (July 1991), 200–220.

**V**

Valencia, Mark J. and David VanderZwaag, "Maritime Claims and Management Rights of Indigenous Peoples: Rising Tides in the Pacific and Northern Waters," *Ocean and Shoreline Management*, 12 (1989), 125–167.

**W**

Wade, Larry L., "The Influence of Sections and Periods on Economic Voting in American Presidential Elections: 1828–1984," *Political Geography Quarterly*, 8, 3 (July 1989), 271–288.

Walton, W. and Ronald J. Johnston, "The Politics of Municipal Annexation: A California State Study," *Tijdschrift voor Economische en Sociale Geografie* 80, (1989), 2–13.

Warde, Alan and others, "Class, Consumption and Voting: An Ecological Analysis of Wards and Towns in the 1980 Local Election in England," *Political Geography Quarterly*, 7, 4 (Oct. 1988), 339–352.

Warf, Barney, "The Port Authority of New York—New Jersey," *The Professional Geographer*, 40, 3 (August 1988), 288–296.

Waterman, Stanley, "The Dilemma of Electoral Districting in Israel," *Tijdschrift voor Economische en Sociale Geografie*, 71, (1980), 88–97.

———, "Electoral Reform in Israel—A Geographer's Viewpoint," *Geographical Research Forum*, 3 (1981), 16–24.

———, "The Non-Jewish Vote in Israel in 1992," *Political Geography*, 13, 6 (Nov. 1994), 540–558.

Waterman, Stanley and Eliahu Zefadia, "Israeli Electoral Reforms in Action," *Political Geography*, 11, 6 (Nov. 1992), 563–578.

Webster, Gerald R., "Partisanship in American Presidential, Senatorial and Gubernatorial Elections in Ten Western States," *Political Geography Quarterly*, 8, 2 (April 1989), 161–180.

———, "Demise of the Old South," *Geographical Review*, 82 (1992), 43–55.

———, "Congressional Redistricting and African-American Representation in the 1990s: An Example from Alabama," *Political Geography*, 12, 6 (Nov. 1993), 549–564.

Wellhofer, E. S., "Core and Periphery: Territorial Dimensions in Politics," *Urban Studies*, 26 (1989), 340–355.

Wesche, Rolf, "Ecotourism and Indigenous Peoples in the Resource Frontier of the Ecuadorian Amazon," *Yearbook, Conference of Latin Americanist Geographers*, 19 (1993), 35–45.

Whebell, C.F.J., "Core Areas in Intrastate Political Organisation," *Canadian Geographer*, 12, 2 (1968), 99–112.

Wildgen, J. K., "The Matrix Formation of Gerrymanders: The Political Interpretation of Eigenfunctions of Connectivity Matrices," *Environment and Planning B*, 17 (1990), 269–276.

Wolfe, James H., "International Law and the Regime of the Sea in Mississippi's Coastal Zone," *Mississippi College Law Review*, 2, 3 (June 1981), 239–264.

**Y**

Yiftachel, O., "Boundary Change and Institutional Conflict in the Planning of Central Perth," *New Zealand Geographer*, 45 (1990), 58–67.

**Z**

Zamora, Stephen, "Voting in International Economic Organizations," *American Journal of International Law*, 74, 3 (July 1980), 566–608.

# PART FOUR

# *Imperialism, Colonialism, and Decolonization*

New Delhi, India. Government Buildings

# 17

# *Colonial Empires*

In Chapter 4 we saw how nationalism carried to extremes can lead to imperialism. In Chapter 16 we examined the plight of indigenous peoples, the most seriously afflicted victims of imperialism. Now we come to imperialism itself.

The study of imperialism or colonialism is in part the study of power and geopolitics, which we study in the next part. Expansionism has been a characteristic of many States, and it has always been made possible by superior power and organization. Such expansionist tendencies were exhibited by States long before the most recent wave of colonialism. The Inca Empire grew by colonial acquisitions and the subjugation of outlying areas and peoples; so did the Roman Empire. We are concerned here with colonialism as a phenomenon related to the emergence of the modern nation-state, so that our primary interest is in the European drive for colonial possessions. However, it should be remembered that territorial acquisitiveness is not unique to European States, nor to the States that have existed only during the past two or three centuries.

## Modern Imperialism and Colonialism

We have seen that some terms in political geography are not well defined, and few present as many difficulties as *colonialism*. Cohen has defined colonialism as separate from imperialism: "Colonialism, as a process, involves the settlement from a mother country, generally into empty lands and bringing into these lands the previous culture and organization of the parent society. Imperialism, as distinct from colonialism, refers to rule over indigenous people, transforming their ideas, institutions, and goods."[*] Frankel has attempted to clarify the term by recognizing *primary* colonization as the occupation of the lands and the domination of indigenous peoples, and *secondary* colonization as the acquisition by a colonial power of virtually empty territory.[†] Although scholars generally agree that colonialism and imperialism are *not* interchangeable terms, the distinction between them is not very significant here.

For our purposes the modern colonial empire may be divided into three categories. First is the empire that resulted from the overland expansion of a State, conquering or otherwise acquiring territories occupied by peoples of different cultures and in some cases different races. These would include the Russian, Austro-Hungarian, Prussian, Ottoman, Chinese, and other empires. Second is the empire built up primarily of overseas territories by the States of Europe during and following the Age of Exploration beginning

---

[*]Saul B. Cohen, *Geography and Politics in a World Divided*, New York, Random House, 1963, p. 204.

[†]S.H. Frankel, *The Concept of Colonization*, London/New York, Oxford University Press (Clarendon), 1949.

in the late fifteenth century and extending into the nineteenth century. During this period the Netherlands, Portugal, Britain, France, and Spain not only built up enormous empires, but also lost significant portions of them.

After the Napoleonic Wars a third phase of imperialism began, one based largely on nationalism, geopolitics, and to a lesser extent economic and missionary motives. It resulted in the creation of new or expanded continental or overseas empires. Some of the original colonizing powers dropped out of the new competition, simply holding on to what they had; others expanded their holdings, and other countries (mostly newly unified as nation-states or just emerging from colonial status themselves) joined in the competition. This was the heyday of "the white man's burden," "manifest destiny," "the civilizing mission," and similar euphemisms for the political, economic, and cultural domination of weaker peoples. We are mainly concerned here with the empires founded, flowering, or terminating between roughly 1815 and 1919, and on the methods employed in ruling these empires. Although nearly all colonies have received independence or otherwise "decolonized," these policies are of much more than historical interest, for they have had a powerful influence on the process of decolonization and on the behavior of the newly independent States.

## Motives and Rewards

The geopolitical aspects of colonialism are self-evident. Taking the term at its narrowest meaning (i.e., referring to the German school of *Geopolitik*), the relationships between German fears of "encirclement" and British imperialist activity are clear. The geopolitician General Karl Haushofer had personally visited the Eurasian perimeter, ringed by British bases; his doctrines constituted a direct reaction to that situation. Comparisons between land and sea powers, such as those of Alfred Thayer Mahan and Sir Halford John Mackinder, which we examine in Chapter 21, served to emphasize the value of colo-

nial possessions. Britain developed a worldwide network of bases that often were located within larger colonial entities, and Mahan advocated the acquisition of similar facilities by the United States.

While the British achieved success on all continents and established the greatest colonial empire ever to exist, the French acquired a vast contiguous colonial territory in Africa and another in Southeast Asia. German colonial activity was in large part an attempt to thwart the designs of Britain and France. In West Africa the Germans succeeded in establishing themselves in Togo, thereby canceling any British plans to link their Gold Coast and Nigerian dependencies. In Kamerun the Germans pushed far northward and came close to splitting the extensive French West African domain. In East Africa Germany obtained Tanganyika, thereby frustrating British plans for a Cape-to-Cairo railway. In South-West Africa, Germany made use of a lapse in British attentiveness to gain a foothold adjacent to the richest part of the continent and came very close to uniting South-West Africa with Tanganyika. A relic of that effort is the Caprivi Strip, shown on the map of Namibia in the next chapter.

The Treaty of Tordesillas (1494) may have greatly influenced the course of colonial history, for it appeared to divide between Spain and Portugal not only the Americas but all newly discovered and undiscovered lands. It has been postulated that this may be why Portugal, which for decades had almost unobstructed opportunity to extend its sphere of influence in Africa, actually established only a few centers needed to facilitate and protect the Cape Route to the Indies. Subsequently, when Britain, France, Germany, and Belgium entered the scene, Portugal could only claim historic rights over areas that by the realities of power fell to other States. Even in South America, the British, French, and Dutch established colonial domains in the area where, according to the treaty, Portuguese and Spanish colonial realms should have met.

While European expansion overseas was proceeding, Russia was expanding eastward and the United States was expanding westward. Neither stopped at the Pacific Ocean but continued onward, with the Russians establishing colonies as far from home as San Francisco and the Americans as far away as the Philippines, for essentially the same motives and by use of the same methods as the Europeans.

Coupled with the geostrategic motives for colonization were a number of other inducements. "God, glory, and gold" motivated the Spanish in their penetration of the Americas, implying that they were there not only to gain wealth, but also to advance the spiritual well-being of their wards. This missionary zeal was not unique to the Spanish in America. Indeed, the first penetration was often made by missionaries, who established stations that sometimes became the centers from which colonial control was extended. One example is South-West Africa. The small German missionary population there found itself in trouble during an intra-African war, appealed for help—and brought German occupation to this area.

Missionary zeal was not confined to missionaries. A relatively small number of European explorers and adventurers, wishing to make a contribution to their fatherlands, had an enormous impact on the process of colonization. Cecil Rhodes had this spirit, and he almost singlehandedly penetrated the region of the Zambezi River on behalf of the British Crown. He strongly influenced the course of events in southern Africa, and the map of Africa eventually carried his name in Northern and Southern Rhodesia. F.A.E. Lüderitz, a wealthy German merchant, similarly helped initiate and sustain the German drive in South-West Africa, while Karl Peters did the same in East Africa. Pierre de Brazza represented France in the area of the lower Congo River. In nineteenth-century Europe, such activities were profusely honored.

At times the governments of Europe, as a result of the colonial scramble in which they were involved, found themselves confronted with requests for "protection" by peoples facing the threat of penetration by colonists, often from other countries. Such protection, if granted by the government, would prevent the colonists from conquering the indigenous people by force and taking their land. We might consider this type of protectorate as resulting from a particular kind of moral obligation, as distinct from that exhibited by the individual missionaries and explorers.

In many parts of the colonized world the major motive that drove the colonial power to control was the economic one. For three centuries the Netherlands East Indies produced enormous revenues for the Dutch government from the sale of coffee, rubber, petroleum, and other commodities, for example. But very often the vast wealth of the colonial world turned out to be more imagined than real, and some colonies became liabilities rather than assets. Mauritania, Chad, Somalia, Surinam, New Guinea, Laos—all produced net economic losses to the colonial power. Then some colonized territories produced richly for a few decades, but their resources were ultimately exhausted. The quest for gold and silver in the Americas initially met with great success, but the sources were rapidly depleted.

## Varieties of Imperialism

We have spoken so far of a particular kind of imperialism, one that results in a radical change in the political and military status quo by the extension of a State's sovereignty over territory that is either empty or occupied by peoples of different cultures. This *military imperialism* may be practiced in adjacent territory, as in the cases of Russia's eastward expansion and the westward expansion of the United States, or it may be practiced in noncontiguous lands, as in the British and French empires. It may or may not involve the settlement of large numbers of colonists from the imperial country. If it does, it can certainly be called colonialism; if not, it may still be colonialism, depending at least in part on the degree of self-government permitted the acquired territories.

There are, as we have seen, many motives for imperialism/colonialism. Vladimir Ilyich Ulyanov (Lenin) wrote a book titled *Imperi-*

*alism, the Highest Stage of Capitalism.* However, Lenin seems to have been wrong, for a careful study of modern imperialism shows that it was seldom practiced for exclusively economic reasons (although one of its products was "the colonial economy," which we discuss later). It would be much more accurate to refer to "imperialism, the highest stage of nationalism." Virulent nationalism has long been a driving force behind imperialism. The missionary drive has also been instrumental in extending political power over weaker peoples, to offer them—or thrust upon them—the conqueror's culture, or selected parts of it.

One variety of imperialism that does not depend on military conquest is *economic imperialism.* This is a more subtle and generally less effective process. It involves a number of techniques, all designed to tie one country to another so tightly through economic means that one becomes in practice an economic colony of the other without giving up its formal sovereignty and its trappings. The United States, for example, has long practiced "dollar diplomacy" in much of Latin America; the Central American countries are virtually economic appendages of the United States and others are tied nearly as strongly. The British have been the masters in this field, making London the world center for banking, insurance, shipping, and other "invisible exports" that we discuss further in Chapter 25. The politics of oil, in the Middle East and elsewhere, was until 1974 an aspect of economic imperialism.

Another variety of imperialism is *cultural imperialism.* It is still more subtle, sometimes less effective, and certainly less geographic than the other forms, so we do not dwell on it but only mention a few examples. Cultural imperialism aims not at controlling territory or economies but minds. Typically, it does not conquer on its own but prepares a target for later penetration by economic, political, and military weapons. Nazi "fifth columns" in Europe and South America between the wars practiced it and achieved notable successes. The Third (Communist) International was even more

widely successful with its brand of ideological imperialism. Before the Bolshevik Revolution, the Russian tsars exploited the Russian Orthodox Church as an instrument of their foreign policy and gained preponderance in the Balkans in this way. France deliberately used the more attractive elements of French culture as a device for maintaining control in its colonies, but also conducted cultural imperialism in Eastern Europe, the Eastern Mediterranean, and elsewhere.

The result of all this imperialist activity was a kind of hierarchy of colonies: political-military, economic, and cultural, but with much overlapping. British money and French culture competed in the Middle East, for example, and in some cases the local people were quite willing to accept both. The Ottoman Empire disintegrated in part because of the infiltration of European political ideas, especially nationalism and self-determination. American economic power penetrated the British sphere of the Caribbean so thoroughly that Britain began losing its colonies there economically before it began giving them political independence. Soviet and Japanese ideological warfare were instrumental in stimulating independence movements in many European colonies in Asia. The picture of modern imperialism, then, is far more intricate than many ideologists would have us believe.

## Colonial Policies

With few exceptions (notably the United States, Germany, and Japan), the empires were created over several centuries through different circumstances and with no prior plan for their administration. Thus there was generally some adaptation to local conditions, experimentation, and changes that reflected political, social, and economic changes at home. During the nineteenth century there was great and continual rivalry among the major European States (and later others as well) for bases, trading posts, colonies, and spheres of interest around the world. As a result of these rivalries and wars, some territories changed hands many times, and boundaries and alliances (including alliances with indigenous rulers) shifted kaleidoscopically. With all these changes came varying methods of rule, administrators with different personalities, and varying degrees of profitability of the colonies. Economic conditions changed greatly during the colonial period. The Industrial Revolution, with its growing capitalist class; its ever-increasing demands for commodities (foods, fuels, and raw materials); its need for external markets to achieve economies of scale; its generation of surplus population in Europe; its provision of superior transport, weapons, and communications to the imperialist governments contributed mightily to the middle and later stages of nineteenth-century colonialism.

The settlement of Europeans from the metropole in the colonies varied considerably. Generally speaking, the colonies with climates similar to those of the home countries and the colonies with the sparsest aboriginal population tended to attract the most European settlers. This pattern was modified by several factors, including imperial policies, the duration of European rule, and the availability of more attractive settlement areas. After their rule was firmly established, the Europeans tended to suppress slavery and local wars and to introduce modern health and welfare systems. This resulted in a "population explosion" that wrought all kinds of changes in the indigenous societies. Most colonizers introduced into their colonies notions of nationalism and self-determination, European style, though seldom with any premeditation or deliberation. European colonial rulers, especially Spain and Portugal, took various measures to encourage or require colonies to trade exclusively with the "mother country" and to suppress local capital formation and investment in industrialization. The responses of the local peoples to all these variations and changes were nearly as varied and changeable. Variation and change were the two constants of colonial rule.

*Portugal's* vast worldwide empire was greatly reduced in 1822 when a liberal king, Dom João VI, returned home from a long exile in Brazil and left his even more liberal

and popular son, Dom Pedro I, to rule there with permission to separate from Portugal if the people wished. They did, and Dom Pedro became emperor of an independent Brazil. That was only a momentary lapse, however, in Portugal's traditional policy of holding onto its colonies at virtually any cost. For nearly a century and a half thereafter, through war and depression and revolution and burgeoning nationalism, Portugal administered its empire with no concessions whatever. Its rule was autocratic, based on mercantilist economic policies, the political policy of transforming the colonies into "overseas provinces" of Portugal, and the social policy of permitting (though not encouraging) a small native elite to become *assimilados*, Portuguese citizens represented in the Portuguese government. It was steadfast, even obdurate, in resisting pressures for change, supremely confident that its method of carrying out its "civilizing mission" was approved (if not ordained) by heaven itself.

Not even India's forcible takeover of Portugal's colonies there (Goa, Damão, Diu, Dadra, and Nagar Haveli) in 1961 shook its resolve to maintain its empire forever.* Beginning in 1961 Portugal was continually fighting nationalist rebels in Angola, Mozambique, and Portuguese Guinea. Finally, a military coup in Portugal carried out in 1974 by professionals dissatisfied with these colonial wars and convinced of their futility set off a chain of events that led to the evaporation of the 400-year-old African empire in the incredibly brief period of 14 months, from September 1974 to November 1975.

Portuguese colonial policies were most clearly displayed in Africa. Politically, each "province" was administered through the governor general in the capital city. For purposes of administrative control, each territory was divided into subprovinces, where decisions made in Lisbon and transmitted through the governor general were implemented. Economically, however, there was a system of producing regions that affected

---

*Indeed, Portugal did not formally recognize Indian incorporation of its territories until late 1974 and did not formally end its sovereignty there until 1 January 1975!

**A vestige of empire.** This corner of a Portuguese cemetery in Maputo (formerly Lourenço Marques), Mozambique contains the remains of residents who fought in World War I. Most died in the 1950s. It is no longer well maintained. (Martin Glassner)

the population much more directly than the internal political subdivisions. These economic units were in many ways the *de facto* political subdivisions of the provinces, where the Europeans' control over the daily lives of the indigenous population was most strongly felt. This system of the *circunscrição* was modified in the face of international objections, but its impact continues to be felt.

*Belgium* saw its colonial task as a paternal one. Apart from the mandate and trusteeship over Ruanda-Urundi, the Belgians possessed only one dependency: the Congo. Once the personal possession of King Leopold II, the Congo Administration was taken over by the Belgian State as recently as 1907.

The shape and the multicore aspects of Zaïre were discussed in Part Two. These realities faced Belgium as they do the Zaïre government today. Kinshasa (formerly Leopoldville) lies on the periphery of the vast Zaïre (Congo) Basin, and its communications with outlying parts of the country have never been good. The Belgians therefore established six major provinces, each with a capital and each extending over a

section of the highland rim of the Congo as well as over a part of the interior basin. The capital cities became the headquarters of lieutenant governors, whose tasks included the implementation of Brussels policy as conveyed by the governor general from his Leopoldville office. Each province was in turn divided into a number of districts, where lesser officials were in charge.

A glance at even the most general geographic patterns in the Congo reveals the shortcomings, from the Zaïrian viewpoint, of this arrangement. Each province lay astride a major transport route to Leopoldville, but provincial boundaries cut through tribal and linguistic units. Some (such as the Kivu-Katanga boundary) were geometric. The system obviously was created for administrative convenience rather than the integration of the whole country. This was reflected by the events immediately following independence, when the cities of Stanleyville, Luluabourg, and Elizabethville—all provincial capitals—became centers of control for secessionist leaders who took over from the Europeans and ignored Leopoldville's dictates. Thus Belgium unwittingly prepared the country for near-feudal conditions by imposing the administrative framework of the six provinces.

Decisions regarding the Congo were made only in Brussels. In the Congo, neither the European representatives of the Belgian government, the settlers, nor any other European group possessed any political rights or any voice in the fate of the territory. Neither, until a few months before independence, did the indigenous inhabitants. Furthermore, the paternalist approach to colonial administration had its effects in economic, educational, and social spheres: much progress was made at elementary levels but very little at higher ones. While the Portuguese opened a narrow avenue for Africans to attain *assimilado* status, theoretically implying full Portuguese citizenship, the Belgian *evolués* could not aspire to Belgian nationality and equality. Even the "evolved" indigenous population was viewed as needing paternal restrictions.

*France* once adhered to the concept of a *France d'Outre Mer*, an Overseas France, but that idea was largely abandoned with the demise of the Fourth Republic. If the best one-word definition for Belgian colonial policy is paternalism, the most appropriate term for French colonialism is assimilation. Like the Portuguese, but with more to offer and more success, the French brought French culture to their overseas domain, where it remains very much in evidence. Unlike the Belgians, the French desired the quick development of an educated, acculturated elite, which would have French interests at heart and French culture to boast. In Southeast Asia as well as Africa, many modern leaders of the independent "French" emergent States are products of this elite and the system that helped erect it.

French colonial subjects were to be assimilated into the greater French Empire, and they could obtain representation in Paris. A number of overseas territories of France had such representation, and French dependencies were placed in a hierarchy, with their position in terms of closeness to France determined by their historical associations with the motherland.

The politicoterritorial aspects of French colonialism are perhaps most clearly revealed in Africa, where the bulk of French territory was located. Algeria always occupied a special position because of its large settled French population and its proximity and effective communications with European France. The remainder of French Africa was divided into two major groupings, French West Africa and French Equatorial Africa. Within them a number of separate administrative entities were established, often defined in West Africa by geometric boundaries, each focusing on a single core area and capital. French Equatorial Africa consisted of Moyen Congo, Gabon, Ubangi-Shari, and Chad, with their respective capitals of Brazzaville, Libreville, Bangui, and Fort Lamy. The administrative headquarters of French Equatorial Africa, however, were in Brazzaville, so that this city attained a greater importance than the others and grew to greater size. The four States of former French Equatorial Africa (now Congo, Gabon, the Central African Republic, and Chad) are in the same size category as the provinces of the former Belgian Congo, but the French always saw them as individual and separate entities, despite their association for administrative convenience in a "federation." Major colonial policies were transmitted through Brazzaville, but decisions regarding each individual territory were implemented directly through the local capitals.

French West Africa also had its administrative headquarters, Dakar, today the capital of the Republic of Senegal. Again, each individual territory had its own focus, and some core areas are of considerable significance today. Southeastern Côte d'Ivoire, with its impressive capital Abidjan, is a core area of the first order in West Africa. Dakar remains French-influenced Africa's most important port, but Conakry and Porto Novo-Cotonou serve Guinea and Benin (formerly Dahomey), respectively.

*The United Kingdom* was confronted with a wider variety of indigenous cultures and peoples than any other colonial power. British colonial policies were often adapted to the requirements of individual dependencies, but British policy, unlike that of Belgium and France, had a basic premise: indirect rule. British colonial policy was always the least centralized among the European colonial powers.

The principle of indirect rule was intended to prevent the destruction of indigenous culture and organization. In such fields as tribal authority, law, and education, the British often recognized local customs and permitted their perpetuation, initially outlawing only those practices that constituted, in British eyes, serious transgressions. Then the people were slowly introduced to the changes British rule inevitably brought. Prior to World War II the emphasis was especially on slowness; haste in bringing about reforms was seen as evil and dangerous. Despite this attitude, however, the British from very early on professed that their ultimate goal was the independence and self-determination of the peoples under their colonial flag. After World War II the pace of change became faster than anyone had expected. The British response was generally to help dependencies destined for independence develop parliamentary systems of government. Less attention was paid to the principle of indirect rule and more to the development of elected local government.

The variety of British approaches to colonial rule in their numerous colonies also was related to the size and character of the European "settler" population in these dependencies. Indirect rule was usually practiced in territories with a small immigrant European population, little land alienation, and a history of peaceful penetration rather than violence and conquest. In territories where the settler population was large and powerful, the desires of London were often overridden by those of the local whites. Examples include the former colonies of Kenya and Southern Rhodesia. But elsewhere, as in Sudan, Ghana, Sierra Leone, and Uganda, the white colonists presented far fewer obstacles in the path toward independence.

Any territory invaded, conquered, and settled by a white immigrant population became a *colony*, implying a considerable amount of self-determination for the settler population. However, those territories whose indigenous leaders had requested and been granted Crown "protection" became *protectorates*. Although the principle of indirect rule in the white-controlled colonies was mainly replaced by local European control, it was adhered to quite strictly in any territory that had been granted the status of protectorate. Sometimes different parts of the same political entity came under different kinds of administrative control: southern Nigeria, which had much more effective contact with Britain than the north, became a colony, whereas northern Nigeria remained under indirect rule. Britain also administered a number of territories in cooperation with other colonial and noncolonial powers. This joint administration produced the *condominium*, discussed in Chapter 11.

British colonial rule was a conspicuous failure in Ireland, Southern Rhodesia, and the Middle East, and less than completely successful in South Africa and Burma. Nevertheless, considering the immensity, duration, and diversity of their empire, the British did astonishingly well. Their proverbial sense of fair play combined with their realism and genius for political improvisation helped them administer and withdraw from their empire with aplomb. It is noteworthy that of all the territories of the modern British Empire, only Burma and those in the Middle East chose not to join The Commonwealth on attaining independence, and only three Commonwealth members have formally withdrawn (Ireland, Pakistan,* and South Africa*, because it was about to be expelled for its intransigence on the apartheid and South-West Africa issues). Perhaps this is the surest evidence of a point made earlier—that colonialism was not all bad.

If the British Empire was acquired "in a fit of absentmindedness," as one observer once remarked, then the empire of the *United States* was acquired enthusiastically but administered absentmindedly. In a real sense the United States acquired two separate em-

---

*Pakistan rejoined the Commonwealth in 1989, and South Africa was welcomed back in 1994.

pires consecutively: one continental and the other (with one exception) insular. The continental empire was assembled between 1803 and 1867 by peaceful purchase, by negotiation and settlement of territorial disputes with other countries, by voluntary annexation, and by conquest and annexation.* For the most part it was administered in an orderly fashion according to the Constitution, the Northwest Ordinance, and other authority designed to bring the individual territories into the union as states equal to the original ones. The native peoples were subjugated, herded onto reservations, and largely forgotten. In 1912 the last remaining territories in the conterminous United States, Arizona and New Mexico, became states. Alaska and Hawaii became states only in 1959. Except for the fact that in each territory the dominant population group was white American, at least from the time of its formal organization, this first American empire resembled the other continental empires of its time, but now forms a true nation-state.

The second American empire still exists and is quite different from the first. Perhaps its most distinctive feature is that despite its obvious similarities to all other overseas colonial empires, it is never officially (and rarely unofficially) called an "empire." Nor are the individual units ever referred to as colonies. Instead, the United States uses such euphemistic terms as "outlying territories," "insular possessions," "island responsibilities," "dependencies," "territorial areas under U.S. administration," and "territories under U.S. purview," but *never* "colonies" or "empire."

The reason for all this circumlocution is plain enough: The United States is ambivalent at best and perhaps even embarrassed at having a colonial empire, since it had itself fought to break free from a colonial empire and ever since has been critical of empires. This attitude has produced a desire to

ignore the colonies and forget the empire. Few American citizens have been or are now aware of the empire; there has never been the kind of emotional attachment to it that characterized the citizens of the European metropoles. Americans have never gone out to the colonies to settle in any real numbers (except recently to Hawaii); their peoples are of little interest, their status is debated (if at all) largely in Washington, and their existence is of little interest except as sources of sugar, tropical fruits, and exotic music. For this reason, perhaps more than any other, the second American colonial empire (again with the exception of Hawaii) has been administered almost absentmindedly.

Notwithstanding some very real contributions to its colonies in the fields of health, welfare, and education, the United States has a rather mixed record as a colonial administrator. Cuba (which the United States had undertaken not to occupy) was granted independence in 1902—but under a form of American protectorate that lasted until 1933. The Philippines, which had been subjugated only in 1905 after a six-year war against nationalists led by Emilio Aguinaldo, became a commonwealth (self-governing colony) in 1935 and was granted independence in 1946. Puerto Rico was not granted responsible government until 1952, when it, too, became a commonwealth. Until that time Puerto Rico and all the other American colonies, except for Cuba and the Philippines, had been treated with, in Daniel Moynihan's phrase (used in another context), "benign neglect."

Various other forms of colonial status gradually evolved, ranging downward from "incorporated territory," in which the U.S. Constitution was applicable (as in Hawaii after 1900) through "organized unincorporated territories" (Guam and the U.S. Virgin Islands) and "unorganized unincorporated territories" (American Samoa) to simply "island possessions" or some other term (anything but colony!). Most territories were originally under military government (army, navy, or air force), and some still are. Others are or have been under the jurisdiction of the Department of the Interior, as the conti-

---

*The states formed out of lands ceded to the United States by existing states, and the District of Columbia, all colonial in a sense, must be considered in separate categories.

nental territories had been. Within the departments, responsibility for administering the territories has been shifted from one office to another, and there has been reorganization after reorganization. The United States has never had a colonial office or its equivalent, a reflection of the haphazard way the territories were acquired, the indecisive way they have been administered, and the indifference of the American people to them once they were acquired.

*Spain* and *The Netherlands*, like Portugal, exploited their colonies as much as possible, giving very little in return. After they lost the largest and richest portions of their empires, however, the two countries took very different paths. Spain continued to invest almost nothing in its remaining colonies, moderated its rigid rule somewhat, and awaited further developments. Little by little, under mounting pressures for decolonization, Spain relinquished control in one territory after another, with little planning or preparation. Its once vast worldwide empire has now shrunk to five tiny *plazas de soberanía*, or presidencies, on the north coast of Morocco and in the Mediterranean nearby, and these it has long claimed as integral parts of Spain. Spain was an authoritarian and unimaginative ruler, but not always as cruel as its enemies alleged.

The Netherlands had exploited its East Indies in much the same fashion as the Belgians had exploited the Congo. But after the colony won independence as Indonesia, the Netherlands created a partnership with its remaining colonies in 1954, called the Tripartite Kingdom of The Netherlands. It was much like a federation, and Surinam (formerly Dutch Guiana) and the Netherlands Antilles did participate in the affairs of the kingdom, especially its foreign affairs. Surinam, however, moved toward independence and was granted it on request in 1975, leaving the Netherlands Antilles very much a junior partner. Aruba was separated from this territory in 1986 by request and is now at the same level as the remaining Netherlands Antilles.

*Germany* and *Italy* became unified nation-states at about the same time and al-most immediately plunged into the race for colonies, the nineteenth-century status symbol of the Great Power. Germany got a head start, picking up a number of Spanish islands in the Pacific, northeast New Guinea and other Pacific islands, and several territories in west, southwest, and east Africa, all through daring, skillful, or lucky competition with the existing Great Powers. Italy, the last European country to embark on an imperialist adventure, is alleged to have dreamed of recreating the Roman Empire (or at least a reasonable imitation of it). But it was too late to get the most desirable colonies in the first instance and too weak to fight any of the colonial powers for what they already had, so it had to be satisfied with the leftover areas of Africa that no one else wanted. Even so, Italy had a difficult time subduing Libya and was routed by the Ethiopians at the Battle of Adowa in 1896. It finally conquered Ethiopia in 1935–36 with the aid of aircraft, heavy weapons, and mustard gas, thus completing its African empire. Italy in 1912 seized the Dodecanese Islands, in 1939 annexed Albania, in which it had exercised great influence since 1926, unsuccessfully invaded Greece (in 1940), and occupied part of Yugoslavia during World War II.

Neither Germany nor Italy held its empire very long. Germany lost its colonies during and after World War I; Italy during and after World War II. Both ruled their colonies autocratically, with little thought of preparing the inhabitants for self-government. Both were exploitative and indifferent to local cultures. More Italians than Germans migrated to the colonies as settlers, but in neither case were the numbers large. Both countries were interested in the colonies primarily for whatever prestige they could bring, but Germany did have geopolitical motives, and Italy had economic and demographic motives as well.

*Japan's* expansion from its home islands began with the acquisition of the Kuril Islands from Russia in 1875 and its conquest of Korea, southern Manchuria, the Pescadores, and Taiwan in a war with China in 1894–95. Japan then defeated Russia in 1904–05 and received certain rights in Manchuria and

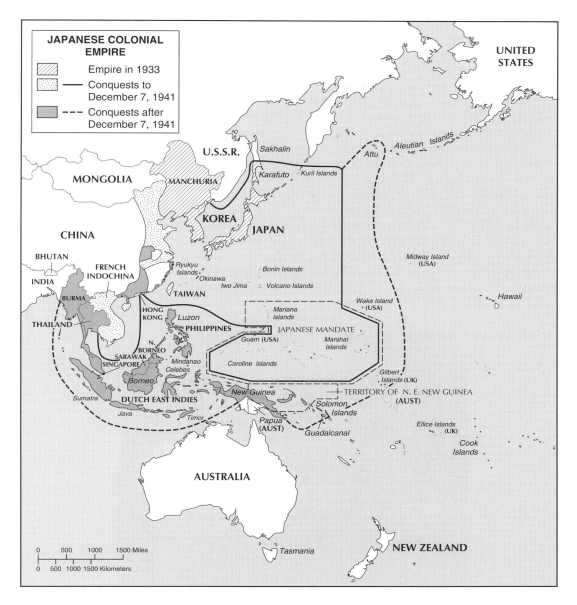

**JAPANESE COLONIAL EMPIRE**

| | |
|---|---|
| (hatched) | Empire in 1933 |
| (dotted) — | Conquests to December 7, 1941 |
| (shaded) - - - | Conquests after December 7, 1941 |

southern Sakhalin. It received a mandate over the former German Pacific islands north of the equator in 1919, occupied Manchuria in 1931, and continued on to occupy much of China and a very considerable portion of Southeast Asia and the Western Pacific.

Japan's motives were almost singlemindedly economic—to gain unrestricted access to the resources of the conquered areas to feed its growing population and new industries. It set up a puppet state in Manchuria (calling it Manchukuo), and elsewhere used some local administrators to help it rule, but on the whole the empire was centrally—and harshly—administered from Tokyo. Some public works were undertaken, but everything was geared toward integrating the colonies into Japan's economy, so the indigenous peoples benefited little from Japanese rule. Japan did try to win friends among the local peoples, partly by capitalizing on their anti-European feelings, but made scarcely any headway. Few (except some diehard Japanese traditionalists) were sorry to see Japan lose its empire entirely during and after World War II.

*Australia* and *New Zealand* became colonial rulers even before they themselves became fully independent. In neither case was this the result of any long-range plan. Instead, territories (mostly islands in the Pacific and Indian oceans) were transferred to them bit by bit over more than half a century by the United Kingdom and the League of Nations. The largest of all these territories was the eastern half of New Guinea, including the Admiralty, Bismarck, northern Solomon, and other islands to the east. This was organized originally as two colonies, German in the north and British in the south. British New Guinea was placed under Australian authority in 1902, and its name was changed to Territory of Papua in 1905. Australia received a League mandate for Northeast New Guinea in 1920. This was converted later into a UN trusteeship, and the two territories were combined in 1945–46 as an administrative union. Papua New Guinea became independent in 1975. The most important of New Zealand's territories was Western Samoa, originally a German colony that New Zealand administered under a mandate and then a trusteeship. It became independent in 1962. Both countries remain responsible for a number of other islands, mostly small in size and population.

All things considered, these two countries have probably been the best of all colonial administrators (along with Norway and Denmark). Notwithstanding some lapses, mis-

**Macau, a fragment of empire.** The photograph shows auspicious phrases and blessings being put up on a wall on Street of Hope. Macau consists of a peninsula, three islets, and two larger islands totaling 16 sq km (6 sq mi) at the mouth of the Si Jiang. It was established in 1557—the first European enclave in China—and became renowned as a center of trade, opium traffic, fishing, manufacturing, and gambling. Portugal was finally granted perpetual occupation and government of Macau by China in 1887. Macau was granted broad autonomy in 1976, and is scheduled to be returned to China in 1999, shortly after nearby Hong Kong is returned to China in 1997. The population of this busy but forlorn remnant of an enormous empire is about 370,000. (Photo courtesy of the Macau Tourist Information Bureau)

takes, and exceptions, they have shown great respect for indigenous traditions and institutions while introducing some of the better features of "Western civilization." They have guided their territories toward self-government and independence without forcing the pace or even determining the islanders' futures for them. There has never been a formal independence movement in any of the territories nor heavy outside pressure for decolonization. The islanders enjoy free access to the metropolitan countries, even as permanent migrants, and most of them are citizens of the administering countries. Niue and Cook Islands (nominally under New Zealand administration) are independent in all but name, while still enjoying special arrangements with New Zealand. Governing small, remote, relatively homogeneous, and not particularly valuable islands is certainly not comparable to governing a Congo, an India, an Algeria, or even a Puerto Rico, but the attitudes and policies of Australia and New Zealand have been commendable nonetheless.

This review of colonial policies has been necessarily brief. We have scarcely mentioned the Ottoman Empire, that fascinating enigma that bridged the gap between ancient oriental and modern European empires. Nor have we mentioned the colonial areas and peoples within independent countries (aside from China and the former USSR), or the role of private companies in creating and administering colonial empires, or many other features of colonialism essential to a complete understanding of the subject. Nevertheless, we have sketched the major outlines of the system, emphasizing variety and change, good and bad, and the dangers of discussing it in slogans and glib generalizations. The same principles are equally valid for our discussion of the decline and eventual demise of the system, to which we now turn our attention.

# 18

# *The Dismantling of Empires*

The imposing colonial structure we have just described disintegrated in far less time than it took to erect it—decades rather than centuries. A trickle of new States became a flood in the 1960s, 44 in just that decade, including 18 in 1960 alone. Yet nothing in human affairs ever happens suddenly; just as imperialism and colonialism had deep roots, so does decolonization.

The process of decolonization began long before the period of modern colonization had ended—before it had fairly begun, in fact. Indeed, some of the same forces that led to the creation of modern empires also led to their demise. Here we can only list the more important forces that became prominent beginning in the mid-nineteenth century and contributed to this process.

## *The Erosion Begins*

Certainly one of the most subtle, yet most critical, factors was the steady evolution of democracy in Western Europe, particularly in the United Kingdom and to a lesser extent in France. Not only were there significant developments in the political system, but there was also a series of social and economic reforms, so that social, political, and economic democracy tended to develop together, though at different rates. The new doctrines in the Western world of progress and evolutionary improvement, one of many offshoots of Social Darwinism, pervaded

thinking in many areas of life. The great flowering of the Industrial Revolution generated remarkable developments in science, medicine, and technology that not only spread around the world, but also led to developments in related areas, among them education, transportation, communications, trade unions, Marxism, and a population growth that, combined with land enclosure movements and other agrarian reforms, sent millions of Europeans out to the colonies.

The spectacular success of Japan in industrializing, in defeating a Great Power of Europe (Russia) in a European-style war, and in starting its own colonial empire demonstrated that independence, industrialization, wealth, and power were not reserved for Europeans. Certainly, the French Revolution was an essential element in the growth of both nationalism and democracy, and it did have a profound influence on the Spanish colonials in the New World, yet it was not until much later that its ideals were transmitted to the French colonies of the modern imperialist age. Of more immediate interest in the colonies of Africa and Asia were the revolutions in Mexico, Russia, and Turkey, in each of which a despotic regime in a poor country was overthrown by (it seemed to them) the downtrodden masses. And the very nationalism that in its most extreme form became imperialism spread in more moderate forms to the victims of imperialism.

The course of human history is punctuated with momentous events that mark true divides between eras. There have not been many of them and they are not always easy to identify, but certainly the First World War is one. It marked the end of the way of life that had evolved in much of the world through the nineteenth century. The frightful carnage of the war was unprecedented; not only were people and cities and forests brought down, but empires as well—Germany, Austria-Hungary, Russia, the Ottoman Empire, gone. Their colonies and minority peoples were transferred to presumably more enlightened guardians or given independence altogether. During the Great War, the horizons of the colonials were expanded as they were called to the colors of the "mother country" and fought in Europe, the Middle East, and elsewhere. European soldiers were sent to the colonies in greater numbers, and battles were fought there with modern weapons and new ideas. Trading patterns were disrupted, and some colonies began to understand how dependent they were on Europe. And the ferment of the war stimulated new interest in self-determination.

The Great Depression and World War II reinforced all these trends and actually intensified them. Dependence, freedom, vulnerability of Europeans, exploitation, nationalism, progress, equality, all were themes brought home to the peoples of the colonies by two world wars and a worldwide depression. In their wake came improved education and communications, bringing to the remotest nations the revolutionary message that they did not have to live in bondage and misery, that there were alternatives, that they could be free, could control their own destinies, could share in the fruits of the new world of abundance created by industrialization. Economists call this "the international demonstration effect;" more commonly it is called "the revolution of rising expectations."

Two world wars and a worldwide depression also left most of the colonial powers, winners and losers alike, dreadfully weakened and no longer capable of ruling huge empires. The weakness was not only military but economic and, in a sense, spiritual. It became clear to all but the old-school imperialists that the world had changed, that colonialism, if not imperialism itself, was now outmoded and had to be replaced by a new version of international and interhuman relations.

This is not to imply that all imperialists gathered under the tree of enlightenment and suddenly became humane and democratic. Far from it. Portugal and Belgium, for example, blithely ignored the new world and made no moves toward giving their colonies even limited self-government until it was too late. Immediately after World War II, France and the Netherlands fought bitter wars against colonials who did not want the prewar masters to return to their colonies just liberated from the Japanese. And the Soviet Union not only annexed whole States and parts of others, but actually colonized Eastern Europe and parts of Asia for a time. Nevertheless, from World War I on, nineteenth-century colonialism was doomed.

To summarize some of the points made in the last chapter: Beginning with the British North America Act of 1867, the United Kingdom adopted the practice of granting gradually increasing degrees of self-government to its colonies. The UK also made extensive use of the protectorate system, in which it recognized the inherent sovereignty of native rulers but performed certain governmental functions for them. These practices were later adopted by other powers, even by France, which established protectorates over Tunisia in 1881 and Morocco in 1912.* There were also many devices used by the imperialists to control foreign territories without incorporating them formally into their colonial systems. They included spheres of influence formally recognized by the other powers; concessions (essentially enclaves) in port cities and coastal areas, chiefly in China; extraterritoriality, usually combined with some other technique, as in Morocco and China; and "advising" local rulers, as the British did in the Trucial States.

---

*Spain also established protectorates over northern and southern zones of Morocco in 1912, but it took more than 20 years to subdue nationalist guerrillas in the northern Rif Mountains and secure the entire territory.

Some of these devices were simply tactics in the game of power politics, of course, and were used where the indigenous cultures were too strong to be subjected to outright colonization; nevertheless, they did represent looser colonial relationships that were relatively easy to modify or abandon as local nationalisms or other pressures began to be felt. The most important and radical departure from colonialism, however, was a product of World War I: the League of Nations mandate.

## The Mandate System of the League of Nations

After World War I the victorious Allies were tempted to continue the pattern of most previous wars by gaining territory at the expense of the losers. They did agree that Austria–Hungary should be split into independent States and that Germany and Turkey should be stripped of their empires. They stopped short, however, of simply dividing up the German and Ottoman colonies among themselves. Influenced by all the new forces described above, reminded by the Arab subjects of the Sultan about Allied promises of freedom in exchange for aid in the war against Turkey, and badgered by Woodrow Wilson, who in a very real sense was then the spokesman for much of mankind, they agreed on a compromise. They established a system of international supervision of the German and Ottoman colonies with themselves as the administering powers, but not as sovereigns. The Allies would assign the colonies, determine their boundaries, and define the terms of the mandates under which they were to be governed. The whole system was to be supervised by the League of Nations, specifically its Permanent Mandates Commission.

Article 22 of the League Covenant spelled out the purpose of the mandate system and, in general terms, its procedures. Its tone was moralistic, even paternalistic, declaring that since the peoples of the former colonies in Africa, the Middle East, and the Pacific were "not yet able to stand by themselves under the strenuous conditions of the modern world, there should be applied the principle that the well-being and development of such peoples form a sacred trust of civilization," and that they should therefore be placed under the "tutelage" of "advanced nations," taking into account the different conditions in the various colonies.

Three essential features of this mandate system are important for our purposes. First, the mandated territories were divided into three classes based on their degree of development. Class A mandates were those former Turkish provinces whose independence would be provisionally recognized until they could sustain independence on their own. Iraq and Palestine were assigned to Britain and Syria to France. Britain partitioned Palestine, creating a territory on the eastern side of the Jordan River for Emir Abdullah as a reward for his help in World War I; it was called Transjordan and received independence in 1946. France partitioned Syria, creating Lebanon, essentially as a State in which Christians could be a majority. Syria and Lebanon became independent in stages between 1941 and 1946. Iraq became independent in 1932.

Most of the German colonies in Africa fell into Class B, in which the mandatory was responsible for the welfare of the people and for their administration. Togoland and the Cameroons (Kamerun) were both divided between Britain and France, Tanganyika was assigned to Britain, and Ruanda-Urundi went to Belgium. Class C mandates were those that could "best be administered under the laws of the mandatory as integral portions of its territories, subject to the safeguards in the interests of the indigenous population" that were specified in the mandates. In this category were South-West Africa; Western Samoa; Northeast New Guinea; the Marshall, Caroline, and Mariana islands (except for Guam); and Nauru Island. The mandatory powers were, respectively, the Union of South Africa, New Zealand, Australia, Japan, and Australia, which administered Nauru and its rich phosphate deposits on behalf of Great Britain, New Zealand, and itself.

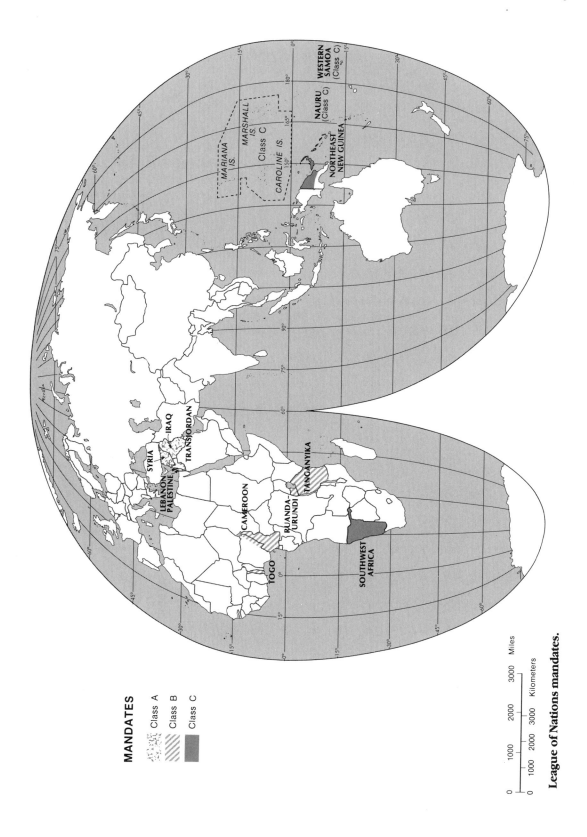

MANDATES

Class A

Class B

Class C

MARIANA IS.

MARSHALL IS.

Class C

CAROLINE IS.

NAURU
(Class C)

WESTERN
SAMOA
(Class C)

NORTHEAST
NEW GUINEA

IRAQ

SYRIA

TRANSJORDAN

LEBANON

PALESTINE

CAMEROON

RUANDA-
URUNDI

TANGANYIKA

TOGO

SOUTHWEST
AFRICA

Miles

0    1000    2000    3000

0    1000    2000    3000    Kilometers

**League of Nations mandates.**

The second important feature of the mandate system was that it actually laid out the responsibilities of the mandatory powers. These fell into three categories: (1) guarantees of freedom of conscience and religion; (2) prohibition of abuses, such as traffic in arms and liquor and slavery and slave trading; and (3) prohibition of fortifications, naval and military bases, and the military training of the natives, except for police and home defense duties. These responsibilities to the local people and to the League, and the contemplation in the Covenant of eventual self-government and independence, distinguished a mandate from a protectorate.

The third major feature of the system also was a departure from all previous colonial policies and techniques: The League had an unqualified right to supervise the mandates to be certain that their terms were being carried out faithfully. The principal means for doing this was the annual report required from each mandatory for each mandated territory. The Permanent Mandates Commission could also receive written petitions and memorials from or concerning the mandated territories.

This whole system unquestionably marked an advance in international relations and international organization, no matter how obvious its weaknesses (such as lack of enforcement machinery) or how severe its critics. It also marked the end of nineteenth-century-style imperialism/colonialism. It was no longer fashionable. The mandate system worked as well as could reasonably be expected, considering that the world was plunged into a severe depression shortly after it began functioning, to be followed immediately by a cruel war.

During the 1920s and 1930s all the forces described were still working. The free trade union movement was burgeoning; unions in many colonies became the nuclei of political parties and independence movements as well as training grounds for their leaders. Nationalism was growing in many colonies, inspired anew by the principle of self-determination that dominated Wilson's Fourteen Points. After the success of the Bolshevik Revolution in Russia, Marxist ideology became more attractive and inspired some colonials around the world to become revolutionaries rather than work for freedom through peaceful means. Mahatma Gandhi and his followers in India provided a different sort of model: nonviolent resistance to colonial rule and patient negotiation for concessions leading to freedom. Education, travel, communications, and the example of the mandates all played their parts. It was World War II, however, that effectively destroyed colonialism.

All the forces that had led to the undermining of the colonial system before and during World War II continued working after the war, and many of them seem to have merged into what we might call a new kind of nationalism, one based not on unity of a culture requiring expression in a territory of its own, but founded largely on anti-colonialism. Among the many peoples thrown together within a colony, the joint goal of independence may have been pursued for different purposes, and in many cases this glue dissolved rapidly after the initial goal was attained.

## *The Role of the United Nations*

The Charter of the United Nations commits that body, in the preamble and more specifically in Article 1, to "equal rights and self-determination of peoples." Articles 73 and 74 deal in more detail with the matter under the heading "Declaration Regarding Non-self-governing Territories." They lay out the responsibilities of colonial powers to the subject peoples, including assisting them to develop self-government, and to the international community, specifically the United Nations. The Charter, moreover, in Articles 75–91 provides for an international trusteeship system to replace the League of Nations mandate system.

The trusteeship system has the same general objectives as the mandate system, except that one of its objectives is *specifically* to promote the trust territories' "progressive development toward self-government or independence." There are, however, some

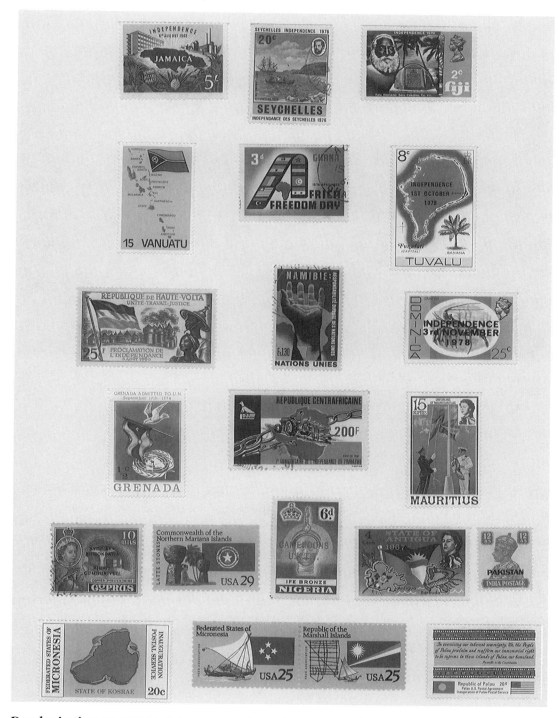

**Decolonization on stamps.** It is customary for each colony as it becomes independent to issue a set of stamps to commemorate the historic event. Frequently the stamps depict the country's history, culture, and modern characteristics. Three of these samples are a little unusual. In the second row is one of three stamps issued by Ghana in 1960 to honor Africa Freedom Day; below that is a United Nations stamp calling attention to Namibia; below that is one issued by the Central African Republic to honor the first anniversary (1981) of the independence of Zimbabwe. The Nigerian stamp in the fifth row is overprinted for use in the Trust Territory of Cameroons. The second stamp in this row and all of those in the bottom row were issued for or by the five new units of the former Trust Territory of the Pacific Islands.

**A trust territory becomes a republic.** The new flag of the Republic of Nauru is raised on 31 January 1968 by the Head of State in front of the Nauruan Administration Offices. Witnessing the ceremony are representatives of Australia, New Zealand, the United Kingdom (the Trustee Powers), and the United Nations. The tiny phosphate-rich island in the South Pacific has only about 6000 people, of whom nearly half are Chinese, Europeans, and other Pacific Islanders. (UN photograph)

very important differences between the two systems. First, the trusteeship system may apply to three categories of territories: "a. territories now held under mandate; b. territories which may be detached from enemy States as a result of the Second World War; and c. territories voluntarily placed under the system by States responsible for their administration." In practice, only the first category (a) has been really important, since only Italian Somaliland was added to the former mandates that had not been terminated, and subparagraph (c) has never been utilized—and is unlikely to be. Second, petitioners from the trust territories are permitted to testify in person before the Trusteeship Council or the General Assembly in addition to submitting written petitions. Third, the Charter provides for periodic visits by UN missions to the trust territories in addition to written annual reports from the administering power. Both devices have been exten-

sively used and have contributed greatly to the success of the trusteeship system.

It has been so successful, in fact, that every trust territory has peacefully become an independent State or part of a neighboring independent State. The last one, the Trust Territory of the Pacific Islands, was designated a "strategic area" under Articles 82–84 and was thus under the jurisdiction of the Security Council rather than the General Assembly. The Trusteeship Council still performed the actual supervision and reported to the Security Council. The administering power, the United States of America, negotiated with the peoples of the Trust Territory a series of agreements during the 1970s and early 1980s on their future status. In 1975, a covenant was concluded to create a Commonwealth of the Northern Mariana Islands in Political Union with the United States and under its sovereignty. The Commonwealth became official in November 1986.

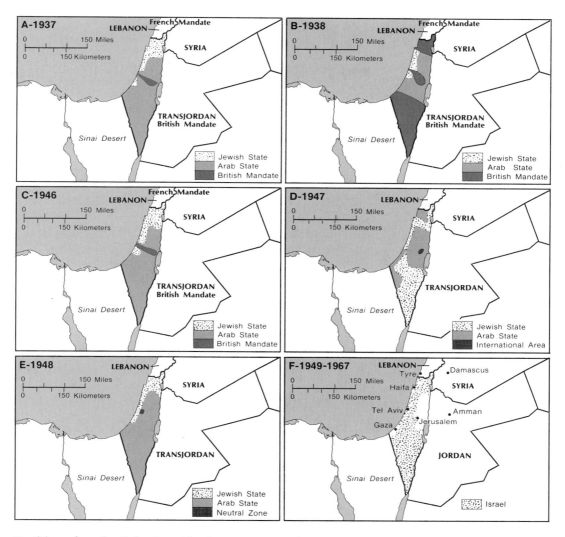

**Partition plans for Palestine.** After Britain partitioned its Palestine mandate to create a throne for Emir Abdullah, called Transjordan, Jewish and Arab nationalism developed in the smaller western portion. Attempting to accommodate them both while maintaining its own position in the Middle East, Britain began proposing a series of partition plans. After World War II, other agencies attempted to mediate with partition plans. None was actually effected.

A. Peel Commission Proposal, 1937.

B. One of two proposals of the Woodhead Commission; the other included a larger Jewish State, 1938.

C. Proposal of the Anglo-American Commission, 1946.

D. Proposal of the United Nations Special Committee on Palestine, 1947. This proposal was voted by the General Assembly and accepted by the Jews of Palestine but rejected by the Arabs.

E. In an attempt to secure agreement of the Arabs to a Jewish State, Count Folke Bernadotte, United Nations mediator, proposed early in 1948 a very much smaller Jewish State. It too was rejected by the Arabs.

F. This is the territory included within the "Green Line," the armistice lines agreed on by Israel and its neighbors after the fighting stopped and they exchanged territory they had occupied. This area, included within *de facto* boundaries, remained intact until 1967 when, during the Six Day War, Israel occupied all of Sinai, Gaza, the "West Bank," and the Golan Heights.

Between 1982 and 1986, "Compacts of Free Association" were negotiated between the United States and the Federated States of Micronesia (Caroline Islands), the Republic of the Marshall Islands, and the Republic of Palau. They provided for considerable self-government but not independence. The first two were approved by the people and the Trusteeship Council. The last, with Palau, was repeatedly rejected by the people because of provisions for the storage of nuclear weapons there and visits by nuclear-powered and/or-armed warships. The Palau constitution was amended, and the compact was approved in a referendum in August 1987. Eventually, both Micronesia and the Marshall Islands became independent, and the trusteeship was terminated in respect of them and the Northern Marianas on 22 December 1990. Palau became independent on 2 October 1994, and the Trusteeship Council was left with no territories to supervise. It is being kept in existence with a skeleton staff in case it is needed again.

There have been two important exceptions to the otherwise excellent performance of the mandate/trusteeship system. Because of conflicts between Arab and Jewish Palestinians in the remaining portion of the Palestine mandate (west of the Jordan River) over the future of the territory, the British were unable or unwilling to replace the mandate with a trusteeship. As the conflict became more intense and the British could offer no satisfactory solution to the problem, they turned it over to the United Nations in 1947 and announced that they would terminate the mandate and leave Palestine in May 1948. In November 1947 the General Assembly voted to partition Palestine again, creating Jewish and Arab States out of it. Jerusalem was to be an international enclave.

The Jews accepted the proposal, although the area proposed was much less than they had hoped for, and they were unhappy about not having Jerusalem, their ancient capital and focus of their religion, included in their new State. The Arabs of Palestine and the Arab States rejected the proposal completely, objecting to any Jewish State in the region whatever its boundaries. Preparations nevertheless went ahead to carry out the UN decision, at least in the area destined for Jewish control. Communal fighting broke out, and the British had a difficult time maintaining a semblance of order until they departed on schedule, even as the armies of six Arab States invaded in a vain attempt to extinguish the new State of Israel. Israel became a member of the United Nations in 1949, but has had to battle challenges to its existence ever since.

The other mandate that was neither terminated with the independence of the territory nor replaced by a trusteeship agreement was that of South Africa over South-West Africa. Indeed, South Africa moved vigorously toward outright annexation of the territory, in violation of the terms of the mandate. The attempt failed, largely because of determined UN opposition and little support from the Western Powers. Nevertheless, South Africa did extend its apartheid system to the territory and governed it as a *de facto* fifth province, with representation (white only) in the South African parliament. South Africa argued that since the League of Nations had expired, so had the mandate, and that it was therefore relieved of any responsibility to the international community for the territory. The International Court of Justice ruled otherwise in an advisory opinion in 1950. South Africa refused to accept the Court's opinion and continued to oppose any form of UN supervision over the territory's affairs.

The United Nations voted in 1966 to terminate the mandate, in 1967 to establish the UN Council for South West Africa to administer the territory until independence, and in 1968 to change the name of the territory to Namibia (from the Namib Desert). It continued, through the General Assembly, Security Council, International Court of Justice, and Secretary-General, to try to persuade South Africa to cooperate—all without success. The Court in 1971 rendered another advisory opinion that the South African presence in Namibia was illegal. In 1978 the Security Council adopted a proposal for free elections in Namibia under UN supervision and

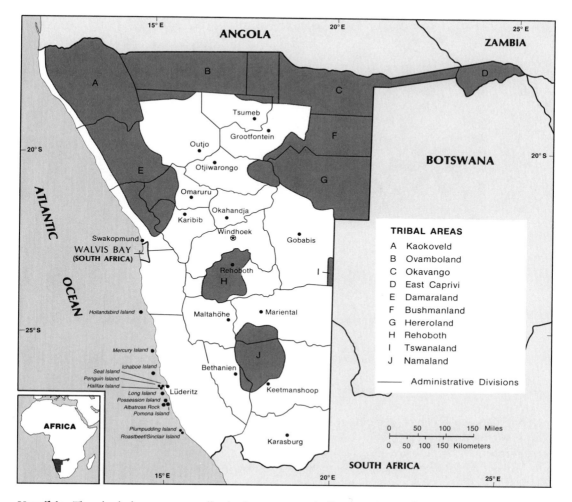

**Namibia.** The shaded areas, generally the least economically attractive in the territory, were set aside by South Africa as "homelands" for the native tribes. The Rehoboth Basters are people of mixed descent who claim territory and special rights to protect their interests. The Caprivi Strip, in the far northeast, is evidence of Germany's drive toward its East African territory in the late nineteenth century. Walvis Bay, which has a splendid natural harbor, was administered as an integral part of South Africa from 1922 to 1990. It was the site of a major South African naval base until 1994. The South Africa-owned offshore islands, with their territorial seas and exclusive economic zones, broke up Namibia's maritime zones until they were transferred to Namibia in 1994.

control. The elections were finally held in 1989, and Namibia became independent in March of 1990.

A small parcel of territory that became a major issue was Walvis Bay. It was a British enclave in the Germany territory, important because it has the best natural harbor in the region. In 1922 Britain transferred it to South Africa, which built a major naval base and other military facilities there and took steps

to incorporate it completely into its territory. Namibia claimed it, however, as well as a string of islands close offshore that broke up its maritime zones. After independence, Namibia and South Africa administered the enclave and the islands jointly, in a kind of condominium arrangement, pending successful completion of negotiations for a permanent settlement. On 1 March 1994 all these territories were transferred to Namibia,

**A vital step in Namibia's long road to independence.** United Nations-supervised elections for a constituent assembly were held in November 1989, and the former mandate became independent in March 1990. This photograph shows voters casting their ballots at the polling station at Ariamsvlei in southern Namibia, watched by officials of UNTAG, the United Nations Transition Assistance Group. (UN Photo 157121/Milton Grant)

and the Namibia-South Africa boundary in the mouth of the Orange River was settled along a median line.

These two exceptions to the success of the mandate-trusteeship system resulted from particular local conditions not found in other territories. As we have already pointed out, however, they were not unique in the colonial world. The Cyprus problem, for example, bears some resemblance to the Palestine problem and still defies solution. Decolonization has certainly not been easy. Considering the number and complexity of the potential problems, however, and the magnitude and historic significance of the dismantling of empires, it is remarkable that it was not much more painful, prolonged, and violent than it has been. Much of the credit for the relative orderliness of the process must go to the United Nations, even without the trusteeship system.

Its first experience with colonies was the matter of the disposition of the former Italian colonies. The UN decided that Libya should become independent almost immediately, in December 1951; that Italian Somaliland should become independent after a ten-year period of trusteeship with Italy as the administering power; and that Eritrea

should be autonomous and federated with Ethiopia.*

Meanwhile, in 1946 the General Assembly had listed those territories that came under the purview of Articles 73 and 74 and had established machinery to carry out their provisions. The major provision was that the administering power was obligated to transmit regularly to the Secretary-General information on the territories. This procedure has been followed rather consistently, and the debates on the reports have been the principal instrument for urging the powers to grant greater self-government to their dependent territories. In 1953 the first territory was removed from the list when the United States informed the UN that Puerto Rico had become a commonwealth (*Estado Libre Asociado*) associated with the United States. Greenland was next in 1954, then Surinam and the Netherlands Antilles, Alaska and Hawaii, and the parade of territories that became fully independent, shown in Table 18-1.

During the General Assembly session in 1960, however, this rather deliberate process

---

*Italy's European conquests had already been returned to their prior sovereigns, Yugoslavia, Albania, and Greece.

**Table 18-1**  *Chronology of Decolonization: Countries Receiving Independence Since 1943*

| Year | Date | No. | Country | Year | Date | No. | Country |
|------|------|-----|---------|------|------|-----|---------|
| 1943 | 22 Nov. | 1 | Lebanon | 1963 | 12 Dec. | 52 | Kenya |
| 1944 | 1 Jan. | 2 | Syria | 1964 | 6 July | 53 | Malaŵi |
| | 17 June | 3 | Iceland | | 21 Sept. | 54 | Malta |
| 1946 | 22 Mar. | 4 | Transjordan (Jordan) | | 24 Oct. | 55 | Zambia |
| | | | | 1965 | 18 Feb. | 56 | The Gambia |
| | 4 July | 5 | Philippines | | 26 July | 57 | Maldives |
| 1947 | 14 Aug. | 6 | Pakistan | | 9 Aug. | 58 | Singapore |
| | 15 Aug. | 7 | India | 1966 | 26 May | 59 | Guyana |
| 1948 | 4 Jan. | 8 | Burma (Myanmar) | | 30 Sept. | 60 | Botswana |
| | | | | | 4 Oct. | 61 | Lesotho |
| | 4 Feb. | 9 | Ceylon (Sri Lanka) | | 30 Nov. | 62 | Barbados |
| | 15 May | 10 | Israel | 1967 | 30 Nov. | 63 | Yemen, South |
| | 15 Aug. | 11 | Korea, South | 1968 | 31 Jan. | 64 | Nauru |
| | 9 Sept. | 12 | Korea, North | | 12 Mar. | 65 | Mauritius |
| 1949 | 8 Mar. | 13 | Vietnam | | 6 Sept. | 66 | Swaziland |
| | 19 July | 14 | Laos | | 12 Oct. | 67 | Equatorial Guinea |
| | 8 Nov. | 15 | Cambodia | 1970 | 4 June | 68 | Tonga |
| | 28 Dec. | 16 | Indonesia | | 10 Oct. | 69 | Fiji |
| 1951 | 24 Dec. | 17 | Libya | 1971 | 14 Aug. | 70 | Bahrain |
| 1956 | 1 Jan. | 18 | Sudan | | 3 Sept. | 71 | Qatar |
| | 2 Mar. | 19 | Morocco | | 2 Dec. | 72 | United Arab Emirates |
| | 20 Mar. | 20 | Tunisia | | | | |
| 1957 | 6 Mar. | 21 | Ghana | 1972 | 4 Apr. | 73 | Bangladesh |
| | 31 Aug. | 22 | Malaysia | 1973 | 10 July | 74 | The Bahamas |
| 1958 | 2 Oct. | 23 | Guinea | 1974 | 7 Feb. | 75 | Grenada |
| 1960 | 1 Jan. | 24 | Cameroon | | 10 Sept. | 76 | Guinea-Bissau |
| | 27 Apr. | 25 | Togo | 1975 | 25 June | 77 | Mozambique |
| | 27 June | 26 | Madagascar | | 5 July | 78 | Cape Verde |
| | 30 June | 27 | Congo (Zaïre) | | 12 July | 79 | São Tomé & Príncipe |
| | 1 July | 28 | Somalia | | | | |
| | 1 Aug. | 29 | Dahomey (Benin) | | 16 Sept | 80 | Papua New Guinea |
| | 3 Aug. | 30 | Niger | | 11 Nov. | 81 | Angola |
| | 5 Aug. | 31 | Upper Volta (Burkina Faso) | | 25 Nov. | 82 | Suriname |
| | | | | | 31 Dec. | 83 | Comoros |
| | 7 Aug. | 32 | Ivory Coast (Côte d'Ivoire) | 1976 | 28 June | 84 | Seychelles |
| | | | | 1977 | 27 June | 85 | Djibouti |
| | 11 Aug. | 33 | Chad | 1978 | 7 July | 86 | Solomon Islands |
| | 13 Aug. | 34 | Central African Rep. | | 1 Oct. | 87 | Tuvalu |
| | | | | | 3 Nov. | 88 | Dominica |
| | 15 Aug. | 35 | Congo-Brazzaville | 1979 | 22 Feb. | 89 | Saint Lucia |
| | 16 Aug. | 36 | Cyprus | | 12 July | 90 | Kiribati |
| | 17 Aug. | 37 | Gabon | | 27 Oct. | 91 | Saint Vincent & the Grenadines |
| | 20 Aug. | 38 | Senegal | | | | |
| | 22 Sept. | 39 | Mali | 1980 | 18 Apr. | 92 | Zimbabwe |
| | 1 Oct. | 40 | Nigeria | | 30 July | 93 | Vanuatu |
| | 28 Nov. | 41 | Mauritania | 1981 | 21 Sept. | 94 | Belize |
| 1961 | 27 Apr. | 42 | Sierra Leone | | 1 Nov. | 95 | Antigua & Barbuda |
| | 19 June | 43 | Kuwait | | | | |
| | 9 Dec. | 44 | Tanganyika (Tanzania) | 1983 | 19 Sept. | 96 | Saint Kitts & Nevis |
| 1962 | 1 Jan. | 45 | Western Samoa | 1984 | 1 Jan. | 97 | Brunei Darusalaam |
| | 1 July | 46 | Burundi | | | | |
| | 1 July | 47 | Rwanda | 1990 | 21 Mar. | 98 | Namibia |
| | 5 July | 48 | Algeria | * | * | * | * |
| | 6 Aug. | 49 | Jamaica | 1993 | 1 Jan. | 117 | Slovakia |
| | 31 Aug. | 50 | Trinidad & Tobago | | 24 May | 118 | Eritrea |
| | 9 Oct. | 51 | Uganda | 1994 | 15 Dec. | 119 | Palau |

*During 1991 all of the union republics of the former Union of Soviet Socialist Republics became independent in stages and were subsequently admitted to the United Nations. The same is true of Slovenia, Croatia, Bosnia and Herzegovina, and Macedonia, all former republics of Yugoslavia; and of the Republic of the Marshall Islands and the Federated States of Micronesia, formerly units of the Trust Territory of the Pacific Islands.

was drastically altered. Early in the session, 17 new members, mostly Francophone, were admitted, creating a new majority of formerly dependent territories. In December the Assembly adopted the Declaration on the Granting of Independence to Colonial Countries and Peoples. It proclaimed "the necessity of bringing to a speedy and unconditional end colonialism in all its forms and manifestations." It also linked very firmly self-determination with independence of dependent peoples and added that "the integrity of their national territory shall be respected."

In 1961 the Assembly created a special committee (known as the Committee of 24) to oversee implementation of this declaration. This committee has been extremely active ever since in accelerating decolonization. Among its activities, it has sent missions to visit a number of territories to investigate conditions or to supervise elections. It even sent a mission to the "liberated" area of Portuguese Guinea in 1972 without the permission of the Portuguese authorities, a historic precedent. It has held many of its meetings away from New York, mostly in Africa, as another way of expediting and publicizing its work. In 1964 it began giving attention to the activities of foreign economic interests in the colonial territories, and in 1967 to military activities and arrangements in the territories. It works with governments, other UN organs, nongovernmental organizations, and "national liberation movements."

It is questionable whether all this activity has actually contributed substantially to decolonization, but it certainly has generated interest in the subject, not only in the colonies but in the world at large. Other UN activities, however, have been of a more practical nature. Examples are the United Nations Temporary Executive Authority that governed West New Guinea between the Dutch departure and the Indonesian arrival, the plebiscite conducted in Bahrain (claimed by both Iran and Saudi Arabia) after the British announced their intention to withdraw, and the prevention of South African annexation of South-West Africa. But it is doubtful if the insistence on completely obliterating colonialism "in all its forms and manifestations" is really the wisest course.

As long as only relatively large territories with some reasonable chance of sustaining themselves as independent States were involved, there were few serious problems. Even such small new countries as Western Samoa (1962), Rwanda and Burundi (1962), and Malta (1964) did not cause much alarm. The first three had been trust territories and were (presumably) groomed for independence, and Malta was European and had a fairly sturdy economy. The admission of the Maldives in 1965, however, initiated a debate about the future of small territories that is discussed in Chapters 7 and 19. Despite reservations, the phrase "self-determination *and* independence" has almost replaced both components used separately. Pressure has continued to give independence to the most minute inhabited island such as many of those listed in Table 18-2. Once a country has become independent, however, "self-determination" is no longer applicable to peoples within it. This raises many questions, some of which we consider in the next chapter.

## Demise of the Soviet Union

In Chapter 12 we discuss the creation and civil divisions of the former Union of Soviet Socialist Republics and note that the union republics and many of the minor civil divisions were based on dominant ethnic groups in each area. At the end of 1991 the whole system collapsed, and each of the union republics became a separate, sovereign State. How can we explain this seemingly sudden disintegration of the world's last large colonial empire? It did not follow the pattern described in this chapter, nor any of the currently fashionable theories about the birth, growth, and decline of States. There is no example in the historical record of such a large and powerful country simply vanishing solely through the peaceful actions of civilian politicians. Nowhere in the literature of geopolitics or power analysis

*Table 18-2   Remaining Dependent Territories*

| Name | Dependency of | Population, 1991 |
|---|---|---|
| American Samoa | United States | 43,000 |
| Anguilla | United Kingdom | 7,000 |
| Aruba | Netherlands | 64,000 |
| Ashmore & Cartier Islands | Australia | — |
| Baker, Howland, & Jarvis Islands | United States | — |
| Bermuda | United Kingdom | 58,000 |
| Bouvet Island | Norway | — |
| British Indian Ocean Territory | United Kingdom | only military |
| British Virgin Islands | United Kingdom | 12,000 |
| Cayman Islands | United Kingdom | 27,000 |
| Channel Islands | United Kingdom | 140,000 |
| Christmas Island | Australia | 2,000 |
| Cocos (Keeling) Island | Australia | 684 |
| Cook Islands | New Zealand | 18,000 |
| Coral Sea Islands | Australia | 3 |
| Faeroe Islands | Denmark | 47,000 |
| Falkland Islands & Dependencies | United Kingdom | 2,000 |
| French Polynesia | France | 195,000 |
| French Southern and Antarctic Lands | France | 150 |
| Gibraltar | United Kingdom | 30,000 |
| Guam | United States | 145,000 |
| Heard & McDonald Islands | Australia | |
| Hong Kong | United Kingdom | 5,856,000 |
| Isle of Man | United Kingdom | 64,000 |
| Jan Mayen | Norway | — |
| Johnston Atoll | United States | 1,000 |
| Kalaallit Nunaat (Greenland) | Denmark | 57,000 |
| Kingman Reef | United States | — |
| Macau | Portugal | 1446,000 |
| Mayotte | France | 75,000 |
| Midway Islands | United States | 453 |
| Montserrat | United Kingdom | 12,000 |
| Netherlands Antilles | Netherlands | 184,000 |
| New Caledonia | France | 172,000 |
| Niue | New Zealand | 2,000 |
| Norfolk Island | Australia | 3,000 |
| Northern Mariana Islands | United States | 3,000 |
| Pitcairn Islands | United Kingdom | 56 |
| Puerto Rico | United States | 3,295,000 |
| St. Helena, Ascension, Tristan da Cunha | United Kingdom | 7,000 |
| St. Pierre and Miquelon | France | 6,000 |
| South Georgia & South Sandwich Islands | United Kingdom | — |
| Svalbard | Norway | 4,000 |
| Swan Islands | United States | — |
| Tokelau | New Zealand | 2,000 |
| Turks & Caicos Islands | United Kingdom | 12,000 |
| U.S. Virgin Islands | United States | 99,000 |
| Wake Island | United States | 195 |
| Wallis & Futuna Islands | France | 17,000 |

N.B. Not included are the French overseas departments, Antarctic claims of various countries, Western Sahara, East Timor, and a number of small islands.

can we find any guidelines to help us understand why it happened.*

Perhaps a clue can be found in a (possibly apocryphal) story that circulated shortly after World War II. It seems that an American asked a Chinese scholar to explain why Japan, hitherto undefeated and still powerful, collapsed so quickly and completely in August 1945, while China, which had many times been invaded and occupied by foreigners and was relatively weak, had survived the Japanese invasion and occupation with its culture largely intact. The scholar replied with an analogy. Japan, he said, was like a block of granite and China was like a pile of sand. If granite is struck a hard blow with a sledgehammer, it shatters. But if a pile of sand is struck an equally hard blow, it simply absorbs the shock and envelops the hammer.

As we have pointed out, the Soviet Union was a jerry-built affair, cobbled together over two decades by persuasion and coercion and held together by the Red Army, the Communist Party, and the settlement of Great Russians in nearly all of the other nationality areas (often receiving the best jobs and other rewards). When the Red Army was humiliated in Afghanistan, the Communist Party exposed as corrupt and inefficient, and the Great Russian minorities blamed for many local problems, the bonds were considerably loosened. The system could no longer withstand the repeated blows of a collapsing economy, political disarray, and foreign influences. First the former Soviet satellite States of Central and Eastern Europe broke away, then the Soviet State itself broke down. It shattered like a block of granite. One may be forgiven for wondering whether the geopoliticians and the power analysts and the political theorists had not overrated the "power" of the USSR just a tad.

What now? It seems at this point (early 1995) that the Balkanization of the Soviet Union has not ended. (A touch of irony, since the Soviets had brought most of the Balkans under their control.) It is clear that the "Commonwealth of Independent States," composed of most of the former union republics (all but the Baltic States), is merely an interim device for temporarily holding off serious strife among the members. It must either evolve into a true, voluntary federation or confederation, create the kind of economic union represented by the European Union (or at least a strong common market), or abandon—at least for a while—any pretense of unity.

Since the history, nature, and decolonization of the Soviet Union were so different from those of the overseas empires discussed in this chapter, its successor States cannot be expected to have a similar future. Very little of what we say in the next chapter applies to the Baltic States—Estonia, Latvia, and Lithuania. They had been part of the Soviet political and economic system for only half a century, with earlier periods of independence and of close relations with the rest of Europe behind them. It may not be long before they are integrated into the new Europe.

Moldova and Azerbaijan could, in some fashion, join Romania and Iran, respectively, when those countries themselves become more attractive partners or hosts. Armenia, Georgia, Ukraine, and Belarus could probably survive as Europe-oriented independent States. Kazakhstan and the Central Asian States will probably face many of the problems discussed in the next chapter, and their future is murky. All 15 of the new States, however, will have to bring some order out of their economic shambles. They have no precedents, no guidelines for doing this, not even the experiences of their former comrades immediately to the west.

Finally, a word about Yugoslavia. Although tribal warfare broke out in both countries and was the proximate cause of both breakups, the fragmentation here has been a bit different from that in the USSR, and the outcomes may be different as well. Slovenia and Croatia resemble the Baltic States in some important respects and could have similar futures. Similarly, Serbia and Bosnia-Herzegovina resemble the Transcau-

*Yet as early as 1979 this author wrote, "One is tempted to comment on the disintegration of the Soviet Empire by exfoliation. . ." See Chapter 22 for the context.

casian countries of Georgia, Armenia, and Azerbaijan, while Montenegro and Macedonia, Kosovo and Vojvodina may be likened to the former Soviet Central Asian republics.

But the analogies can be carried too far. Yugoslavia and the former USSR have had very different histories; the relationships among the civil divisions and between the civil divisions and the central government were very different; they have had different economic systems, different political systems, foreign relations, and cultures. The smaller republics of Yugoslavia were not really colonies of Serbia, as most of the civil divisions of the USSR, including most of those in the Russian federation, were colonies of Russia. And the breakup of Yugoslavia was accompanied by fierce fighting that amounted to civil war. So the breakup of Yugoslavia is not really a case of decolonization. It is more like the many failures of federation that we described in Chapter 10. Still, some of the new States that emerge from the wreckage of Yugoslavia—and even a rump country that may retain the name—are likely to have some of the same difficulties that confront the former colonies in the aftermath of colonialism.

# 19

# *The Aftermath of Colonialism*

Between 1943 and 1990, 98 former dependencies joined the ranks of independent actors on the world stage (99 if we include Zanzibar, which was independent only briefly). The "strenuous conditions of the modern world," as the League of Nations Convenant put it, are even more strenuous today than they were in 1919, and are likely to become more so as still more colonies become independent States in the near future. Few generalizations, however, can be made about the new countries. Not all of them are poor, for example. Kuwait, Israel, Malta, and Iceland at least approach Western European living standards; Nauru, Fiji, Malaysia, Singapore, and the Bahamas are relatively prosperous, more so than many older countries. Many "new" countries are really quite old and only reemerging from a period of colonial domination. India, Lebanon, Syria, Myanmar, Sri Lanka, Israel, Korea, Vietnam, Cambodia, Morocco, Ghana, Samoa, Swaziland, Oman, Tonga, and Fiji are not creations of the colonial period, but have histories measured in centuries and even millennia. Others also have deep roots and rich cultures even if they have not existed as independent political entities before. Nevertheless, we must generalize here and confine ourselves to only a few essentials of this adventure in human history, this miraculous birth and nurturing of new political units containing nearly half the world's population.

## *Nation Building and State Building*

A country just emerging from colonial status has a great deal to do. If it has had a reasonable gestation period with gradually increasing degrees of self-government, if its birth is peaceful and orderly, and if it receives adequate postpartum assistance from its former colonial ruler, other countries, and international agencies, then the transition from colony to State is generally smooth and successful. The State-building process, however, can be difficult, even traumatic, under less than ideal conditions. Just the simple mechanics of running a country, especially a poor one, can be most bewildering to people who lack administrative and political experience.

The most extreme case of this kind, of course, was the former Belgian Congo. Other countries have had similar difficulties attending their birth, although not nearly as severe. Indonesia, Algeria, Guinea, Bangladesh, and others have had problems building a State where none had existed before. But even more difficult, if less dramatic, is the process of nation building; that is, trying to weld a nation out of disparate tribal and ethnic groups thrown together by colonial rulers who had only their own interests in mind. Some new countries seem to have accomplished this task with considerable success in a relatively short time. Others still appear to be having great difficulties in creating nation-states.

Lebanon, after a long period of peace and prosperity, dissolved into civil war. So did Sudan, Myanmar, Angola, Cambodia and Mozambique, and there is no assurance that other new States will be able to avoid the same fate. (Not even older countries are immune from communal tensions: Canada faces the very real threat that Québec will secede; Spain is still fighting Basque separatists; Northern Ireland endured a quarter century of tribal warfare.) Some of the new countries are relatively homogeneous and have had few if any ethnic problems since independence. The rest are still assiduously engaged in nation building. We discuss this in greater detail in Chapter 35.

The nation-building process means generating and nurturing nationalism to replace regionalism or tribalism or localism, creating a sense of identification with and loyalty to this new thing called "the State." Often people who had been told all their lives that "the government" was bad and had to be opposed were suddenly being told that "the government" was good and had to be supported. Many new countries have been fortunate in having talented and charismatic leaders to guide them through the transition period and become symbols of the new State around which most factions could rally. When the hero of independence dies or is deposed, however, especially early in the independence period, the effect can be catastrophic unless a new leader emerges very quickly to replace the old one. Lacking wise and trained leadership, and having no democratic tradition and little experience in self-government, many new countries have soon come under military rule. In this, as in many other aspects of independent statehood, the newer countries of Africa and Asia have emulated those in Latin America. Nowhere, however, have military rulers as a class proven any better than their civilian counterparts.

**Tangible iconography.** Though not really a new State, Bhutan long existed under first Tibetan and then British rule or protection. After attaining formal independence in 1949, Bhutan continued under Indian "guidance" at least until joining the United Nations in 1971. Part of the process of achieving more than symbolic independence for this land-locked country was the construction of an international airport and the acquisition of an air fleet, which in late 1989 consisted of one BAs 146, shown here at Paro, which replaced the original fleet of two Dorniers. It must still overfly Indian territory en route to and from Dhaka or Kathmandu, its only foreign destinations outside India. (Martin Glassner)

Fostering nationalism in the new countries involves but goes beyond much of what Jean Gottmann called *iconography*. Depending on local conditions, the following activities have been typical of new governments in recent decades: resurrecting or fabricating a glorious precolonial history; initiating a new educational system in which new national values are stressed; building the army into a highly visible national symbol, which often works in building roads, disaster relief, public health, or clearing land for pioneer settlement; instituting a national information service to carry the government's message through every possible medium to the people of the country and abroad; developing national sectoral organizations (peasants, chiefs, teachers, manual workers); and the iconography, or symbolism, of national distinctiveness itself—a flag, an anthem, heroic slogans, a steel mill, a national stadium, a national airline, a national costume, replacement of European place and personal names with local ones, a palace of culture or convention hall in the capital, postage stamps carrying nationalistic messages, and similar devices—all designed to replace anticolonialism as centripetal forces to neutralize or overcome the centrifugal forces long held in check by the colonial rulers.

The very emphasis on fostering national unity, however, often leads to neglect of the rights and interests of minorities, and even to their suppression; to the neglect of substance in favor of form or symbol; to investment in extravagant public works, elaborate ceremonies, and conspicuous consumption by the leadership at the expense of essential infrastructure and productive enterprises; even to xenophobia directed against both foreign nationals within the country and other countries themselves. Carried beyond reasonable limits, nationalism can be destructive and retard the development of a new country.

## Economic Development

In most cases, as we have seen, the colonial countries invested in their colonies whatever they felt was necessary to control and administer the territory and to extract its wealth, whether or not that wealth was shared with the territory. After independence, the old relationships between ruler and ruled may be altered or completely severed or scarcely changed at all, except symbolically, but almost invariably there is a new emphasis on development of the country for its own benefit. Some countries found themselves at independence far from rich but still with sound economic structures, money in the treasuries, and good credit ratings. Others, while not as well off as the first group at independence, nevertheless had resources that were very much in demand in the industrialized countries and could look forward with some confidence to a considerable improvement in their living standard.

At the opposite end of the scale are countries so poverty-stricken that the United Nations has created the category of "least developed among the developing countries."* Indeed, some observers feel that some simply can never be self-supporting and will either remain permanent charity cases or go out of business entirely. In between the extremes, the great majority of the former dependencies are struggling against many handicaps, buffeted by political rivalries and conflicts, caught in situations over which they have little or no control, trying to become a little less poor.

Many of the new countries at independence found themselves with economies geared almost entirely toward the former metropolitan countries. Thus the extractive industries—mines, plantations, sawmills, oil wells—tended to be well developed at the expense of manufacturing and services. Transportation systems were designed to connect the capital with important administrative centers, and seaports with important economic centers. Export crops were em-

*The UN, after much experimentation with statistics, no longer uses specific indicators of "development" for the purpose of identifying these least developed countries (LDCs), but now simply defines them as "those low-income countries that are suffering from long-term handicaps to growth, in particular, low levels of human resource development and/or severe structural weaknesses." In 1993 there were 47 countries identified as LDCs.

phasized at the expense of subsistence crops, and food often had to be imported by countries that could, if permitted, feed themselves. Land tenure was often a serious problem, with the best agricultural land held in plantations and *latifundia*. Soil exhaustion and erosion were severe in some places. Trade agreements had to be negotiated that would help the new countries break out of the "colonial economy" pattern of exporting commodities and importing manufactured goods. Vital imports had to be bought in soft currency markets, by barter, or with borrowed money. Populations were growing rapidly, compounding all their economic, social, and political problems. The catalogue of economic problems is very long, but we are concerned here primarily with what is being done to relieve them.

Nearly every new country undertook some degree of national economic planning, even those that rejected a Marxist approach to economic development, and some adopted one or another variety of socialism. Many that did choose these paths have by now liberalized their economies or abandoned socialism altogether. Some have undertaken land reform programs, attempted to settle nomads, invested in massive infrastructure projects, tried to diversify the economy, begun industrializing (in part for political or nationalistic reasons, even if the industries are uneconomic and agriculture is neglected), and joined with other countries in a variety of ways to achieve together what they could not achieve separately.

If the economic problems of new countries are greater than ever before in history, so is the assistance available to them from outside. In most cases the former metropoles have aid programs for their former colonies, with the French program being particularly large. Second, other rich countries have developed aid programs, generally for selected poor countries. Third, there are now a large number of international and regional agencies devoted to assisting developing countries (not just new ones). Finally, developing countries have begun assisting one another, also on both a bilateral and multilateral basis.

Too large a proportion (how much one can only guess) of this aid is wasted on inappropriate projects or equipment or training; too much is squandered on military equipment, supplies, and training; too much is lost through mismanagement and corruption; too much is misapplied because of competition among donors; and even good programs go bad with changes of government or policy or personnel.* The "oil crisis" of 1973–74 and the subsequent quadrupling of petroleum prices dealt a cruel blow to the poor countries of the world, most of which suffered far more than the industrialized rich countries. Some were so badly hurt that the United Nations established a special fund to aid those countries "most seriously affected" by this situation.

Another problem that has become very serious is that of debt servicing. A number of poor countries have such heavy burdens of repayment of loans, both principal and interest, that they are actually sending more money to rich countries than they are receiving from them. Many other economic problems face the new countries, of course, but looming over all of them, both in magnitude and importance, is the incredibly rapid growth of population in most developing countries. In many cases countries that get ample assistance and follow all the rules do indeed increase production, improve their infrastructure, and modernize their economies in general—only to find all of their gains canceled out by an ever-growing population.

But we should not despair. The problems of poor countries are immense but not overwhelming. They are all surviving somehow. Many are making headway and a few have done quite well, moving into the middle class, so to speak. Family planning (or population control) has caught on and in some places has been remarkably successful, contributing materially to higher living standards. Development plans are being rethought and reorganized, and are showing better results. Aid programs are being better

---

*Military expenditures are covered in some detail in Chapters 23 and 30.

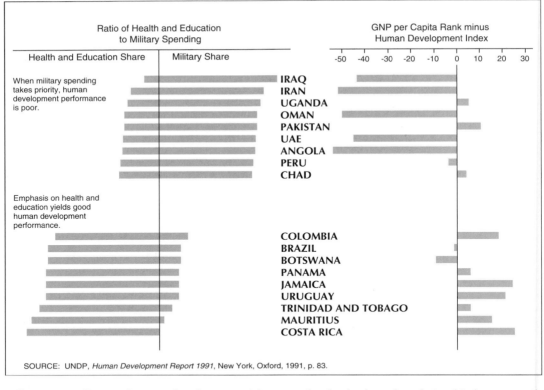

**Military spending vs. human development.** These graphs clearly show the relationship between military expenditures and social and economic progress in developing countries. Since the relationship between the graphs on the left and those on the right is not exact, other factors are important in determining the rate of progress, but diverting funds into armaments is a major one.

coordinated. Poor countries have at last understood that environmental degradation is a problem for them and not just for the industrialized countries and are beginning to adopt environmental protection measures. Gradually, the developing countries have learned that more is to be gained by cooperation with rich countries than by confrontation with them. Whether the rich countries have come to understand that cooperation with the poor countries is better than domination of them, however, is another matter.

## Foreign Relations

A new country has not only domestic problems to wrestle with, but foreign ones as well. They include establishing formal relations with other countries, joining intergov-

ernmental organizations, reviewing treaties to which it has been made a party, dealing with boundary problems, participating in international conferences, negotiating aid agreements and a multitude of technical matters, defining and articulating positions on international issues, even finding and training personnel to do all these things. These are formidable tasks, to say nothing of the problem of paying for them all.

Several small new countries have simply elected not to do all this; they chose instead to keep their foreign relations and economic activities to a minimum. This generally means membership in the United Nations (but little activity therein) and a few other organizations essential to the country, diplomatic relations only with the former colonial ruler and perhaps one other great power and one or two neighbors, a few routine treaty relationships, and very little else. A few have no more than nominal diplomatic relations with other countries. Indeed, Western Samoa postponed applying for UN membership for 14 years after independence, and Nauru and Tuvalu (formerly Ellice Islands) have not yet applied.

Meanwhile, within the United Nations, the difficulties of small island States are receiving considerable attention. Both UNCTAD and ESCAP have special units devoted to relieving their isolation and economic distress. The United Nations Conference on Environment and Development (UNCED) in 1992 recommended convening a special conference on the subject, as a result of which the Global Conference on the Sustainable Development of Small Island Developing States was held in Bridgetown, Barbados, in 1994. In addition, the South Pacific Forum has a Committee on Small Island States, and in January 1992 the first in a series of "Economic Summits" of South Pacific Smaller Island States was held in Rarotonga, Cook Islands. All of this—and other—activity is probably inadequate to cope with the manifold problems generated by small size and isolation, especially in the Pacific. Nevertheless, most new countries, however small, are involved in world affairs. The alternatives to "sovereignty" that we suggested in Chapter 11 are not likely to be adopted very soon.

The first conference of emerging countries was the Afro-Asian Solidarity Conference held in Bandung, Indonesia, in April 1955, whose political aspects are discussed in Chapter 22. Attending it were the People's Republic of China and 28 independent or soon-to-be-independent developing countries from Gold Coast (now Ghana) and Turkey to the Philippines and both Vietnams. It marked the beginning of cooperation among them in economic matters, human rights and self-determination, problems of dependent peoples, and promotion of world peace and cooperation. But cooperation among both new and old poor countries, although much discussed and publicized, was limited primarily to confrontation with the rich countries on decolonization, trade, aid, human rights, and disarmament.

Most of the new countries chose to remain "neutralist" or "nonaligned" in the Cold War between the United States and the Soviet Union. (This is the origin of the term "Third World," which has since taken on so many other meanings.) Others, however, regardless of their public statements, were at birth or soon after clearly aligned with one side or the other. "The spirit of Bandung" faded as the world changed rapidly in many ways during the 1960s.

One problem faced by nearly all new States is their boundaries. Even the island States have problems with maritime boundaries, but our concern here is with land boundaries. As we have seen, many of these boundaries were drawn by the colonial rulers with little regard for the needs or interests of the indigenous peoples. Throughout Africa and Asia today there are scores of boundary and territorial disputes, and even Latin America still has its share. The problems are most numerous and acute, however, in Africa. Considering the number and nature of these problems, it is remarkable that there have not been more border wars in Africa. In part, this has been due to the weakness of most of the contestants, to a lingering spirit of pan-Africanism, and to the need for unity in the face of greater perils, both internal and external, all of which have encouraged neighbors to negotiate their differences quietly and settle them by compromise. Many boundary segments have thus been delimited and at least partially demarcated.

Perhaps a more important influence, however, has been the policy adopted by the Organization of African Unity (enshrined in Article III, paragraph 3 of its 1963 charter and introduced in our Chapter 8) of "respect for the sovereignty and territorial integrity of each state and for its inalienable right to independent existence." The African States, many founded by revolutionary movements and many with radical domestic and foreign policies, have thus pledged themselves to preserve the status quo imposed by the colonial powers. This policy was based on the Roman law principle of *uti possidetis juris*, the same policy adopted by the Spanish American colonies during their wars for independence a century and a half earlier.

Despite the dilemmas posed by several border disputes and attempted secessions, the principle has been upheld: no fragmentation of existing countries; no forcible boundary changes; in a territorial dispute between an African State and a European State, the African is invariably right; all colonial territories must become independent intact. All of this was designed to prevent what the leaders forty years ago commonly referred to as "the Balkanization of Africa."

In applying the principle, the African States were unanimous in opposing the "homelands" policy of South Africa in its own territory and in Namibia, and they refused to recognize Transkei, Bophuthatswana, Venda, and Ciskei as independent. They have also refused to recognize the validity of the 1974 referenda on the Indian Ocean island of Mayotte, in which the mostly Christian inhabitants voted overwhelmingly to remain a French territory, and insist that the island is an integral part of independent and largely Muslim Comoros, with which it was administered by the French until the end of 1975 when the Comoro Islands became independent. Self-determination, it would seem, has its limits in Africa—and elsewhere.

## A New International Economic Order?

As the former colonies began lives separate from their metropolitan countries, they began to realize that political independence, however real or complete, did not ensure economic independence. Furthermore, with independence has come interdependence, skewed, of course, in favor of the rich countries, but still real. Finally, the new States learned very quickly that political independence had not brought them prosperity, and that some of them were actually getting poorer. By the early 1970s a number of trends and developments conjoined to produce an environment in which the poor countries (mostly ex-colonies) felt both the need and the collective strength to demand

a change in the fundamental economic system that had led to the situation depicted in the world map and the cartogram in this chapter.

On both the map and the cartogram, ignoring the oil-rich States of the Middle East, we can readily see three broad bands of countries grouped according to their level of living. In the North, stretching across North America, Europe, and the former Soviet Union, are countries that range from well-to-do to fabulously rich. In the far South are a few more relatively rich countries closely associated with those in the North. In between, in the lower middle and low latitudes, weighted somewhat toward the Southern Hemisphere, is a far larger group of countries that range from lower-middle class to desperately poor. This is the basis for the "North-South" dichotomy in the world, one that is far more deeply rooted, far more enduring, and far more dangerous than the "East-West" confrontation ever was. The poor countries, the "South," resolved to close the gap by working toward a fundamental change in the world economic system.

Many observers consider that the first salvo in this offensive was fired by President Luís Echeverría of Mexico in an address to the General Assembly (GA) of the United Nations in the fall of 1973 in which he urged the creation of a new international economic order. In the following year this proposal was formalized in the GA with the adoption of two documents: the Declaration on the Establishment of a New International Economic Order (NIEO) and the Charter of Economic Rights and Duties of States. The only negative votes on the charter resolution (which carries somewhat greater weight than the declaration) were cast by the United States and five Western European States.

For the next decade the developing countries vigorously campaigned on many fronts for radical changes in the world's economic system. We discuss some of these activities in subsequent chapters. Here, by way of introducing the topic, we simply list some elements of a projected NIEO and then elabo-

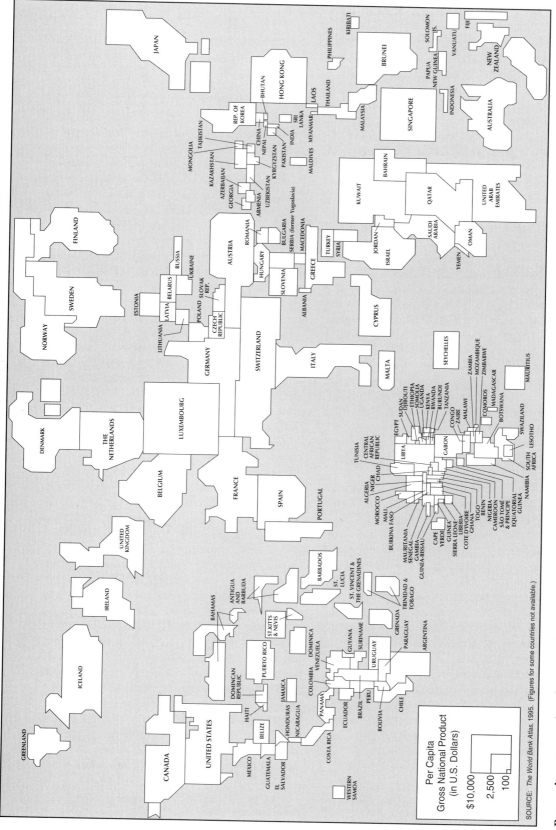

**Per capita gross national product by country.** This cartogram shows each country in a size proportionate to its per capita GNP. The GNP is a very popular measure of prosperity or of economic development, but when used this way it has serious defects. The most important shortcoming is its failure to indicate the distribution of the product within the society. Nevertheless, this egregious disparity of income, whether within or among States, is a major cause of political instability.

SOURCE: *The World Bank Atlas*, 1995. (Figures for some countries not available.)

Per Capita
Gross National Product
(in U.S. Dollars)

$10,000
2,500
100

rate a bit on a few factors not considered later. Among the elements of an NIEO are (in no particular order): strengthening the role of the United Nations in economic and social development, the OPEC oil embargo and price hikes, reforming international trade, expanding the activities of regional organizations in development, rectification of the practices and freight rates of liner shipping conferences, shifting control of marine resources, and controlling the activities of transnational corporations.

Some advances have been made in all of these areas and others, but the gap between rich and poor has continued to widen. The worldwide inflation of the 1970s was followed by a worldwide expansion in the 1980s, then by a worldwide recession in the late 1980s that continued into the 1990s. These phases emphasize again the interdependency of the emerging global economy. The drive for an NIEO surged in the 1970s, faltered in the 1980s, and was essentially dead by 1984.* There are some very important reasons for this, but we can present only two of them here.

### The Continuum of Economic Development

The Human Development Index, fashioned by the United Nations Development Programme (UNDP), shows graphically how the world is *not* divided simply into rich and poor. Instead, all the countries in the world lie along a continuum of economic and social development, along which there are no really large breaks. Moreover, this picture is only a snapshot of a constantly changing situation. Some colonies that became independent with sound economies and money in their treasuries—such as Senegal, Ghana, and Uganda—have been reduced to poverty, while others—notably the "four tigers" of Asia (South Korea, Taiwan, Hong Kong, and Singapore)—have risen into the middle class. Many countries with low GNPs nevertheless have a relatively good quality of life,

or level of human development, and vice versa.

### Little Transfer of Wealth in the Form of Aid

The rich countries cannot be relied upon to provide large sums of money for economic and social development. In 1989, for example, the 18 richest countries, the members of the Organization for Economic Cooperation and Development (OECD), contributed $46.5 billion in all kinds of official development assistance (ODA), including loans, grants, technical assistance, training and commodities. This represents only 0.33 percent of the combined GNP of the OECD countries. This percentage has remained quite consistent, varying only between 0.32 and 0.35 during the decade 1982–92. Individual countries' contributions ranged from Sweden and the Netherlands, at 0.98 and 0.94 percent of their GNP, respectively, to Ireland and the United States, at 0.17 and 0.15 percent, respectively. In 1992 Norway and Sweden were the leaders, contributing 1.16 and 1.03 percent of their GNP, respectively, with the United States and Ireland again last at 0.20 and 0.16, respectively. For the United States, this represents a decline from 0.24 percent of GNP in 1985, when it was also the *least* generous of the OECD countries. On a per capita basis, the United States is niggardly also. In 1987, for example, the United States pledged only 50 cents per capita; Norway pledged $13.50! In 1992 the United States transferred in ODA only 45 *cents* per person, while Luxembourg was transferring 92 *dollars*! Furthermore, more than one-third of total U.S. "foreign aid" at present goes to just two countries: Israel and Egypt. In 1991–92 60 percent of U.S. "aid" went to this region.* Clearly, U.S. aid is more political than humanitarian. A number of factors contribute to these disparities, but American generosity is not among them. These figures could usefully be employed in debates over "foreign aid," most of which is actually spent in the donor countries.

---

*In the mid-1990s some attempts have been made to revive the NIEO movement, but it is much too soon to see any real trends in it.

*All figures are from OECD, *Development Co-operation; 1993 Report.*

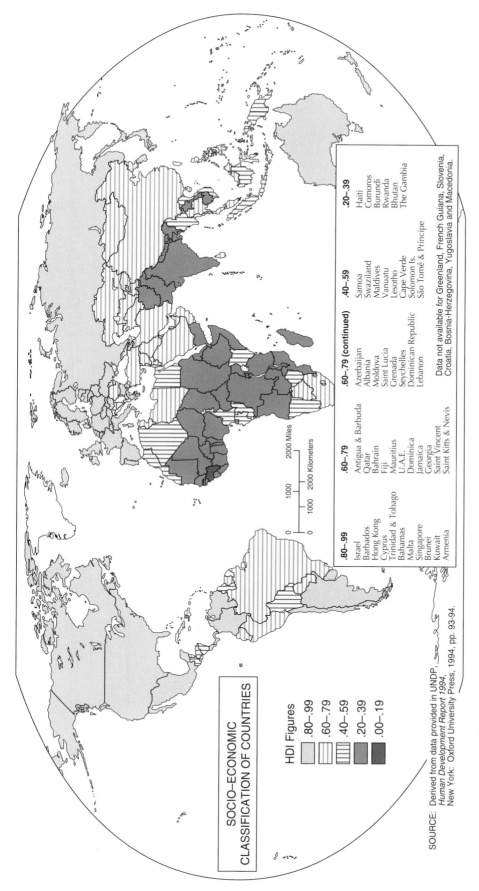

**The Human Development Index.** This index, produced by the United Nations in 1990, is a further refinement of the Physical Quality of Life Index developed by the Washington-based Overseas Development Council and used in *SPG III* and *SPG IV.* It is a composite of life expectancy at birth, educational attainment (adult literacy and mean years of schooling), and adjusted real gross domestic product. In this index, Canada ranks number one, the United States eighth, and Russia thirty-fourth. Guinea still ranks last at .191. Subtracting the HDI rank from the GNP rank, as in the graph on page 280, reveals some very interesting and significant facts hidden by the individual figures. This new tool deserves careful study and extensive use, especially in understanding the distribution of wealth *within* countries.

SOCIO-ECONOMIC CLASSIFICATION OF COUNTRIES

HDI Figures
.80–.99
.60–.79
.40–.59
.20–.39
.00–.19

SOURCE: Derived from data provided in UNDP, *Human Development Report 1994.* New York: Oxford University Press, 1994, pp. 93-94.

.80–.99
Israel
Barbados
Hong Kong
Cyprus
Trinidad & Tobago
Bahamas
Malta
Singapore
Brunei
Kuwait
Armenia

.60–.79
Antigua & Barbuda
Qatar
Bahrain
Fiji
Mauritius
U.A.E.
Dominica
Jamaica
Georgia
Saint Vincent
Saint Kitts & Nevis

.60–.79 (continued)
Azerbaijan
Albania
Moldova
Saint Lucia
Grenada
Seychelles
Dominican Republic
Lebanon

.40–.59
Samoa
Swaziland
Maldives
Vanuatu
Lesotho
Cape Verde
Solomon Is.
São Tomé & Príncipe

.20–.39
Haiti
Comoros
Burundi
Rwanda
Bhutan
The Gambia

Data not available for Greenland, French Guiana, Slovenia, Croatia, Bosnia-Herzegovina, Yugoslavia and Macedonia.

0  1000  2000 Miles
0  1000  2000 Kilometers

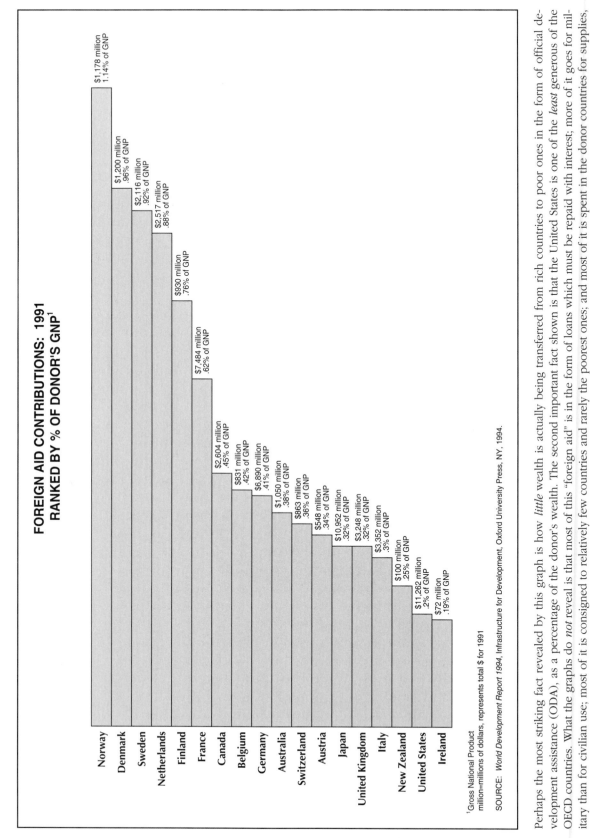

## FOREIGN AID CONTRIBUTIONS: 1991
## RANKED BY % OF DONOR'S GNP[1]

Norway — $1,178 million / 1.14% of GNP

Denmark — $1,200 million / .96% of GNP

Sweden — $2,116 million / .92% of GNP

Netherlands — $2,517 million / .88% of GNP

Finland — $930 million / .76% of GNP

France — $7,484 million / .62% of GNP

Canada — $2,604 million / .45% of GNP

Belgium — $831 million / .42% of GNP

Germany — $6,890 million / .41% of GNP

Australia — $1,050 million / .38% of GNP

Switzerland — $863 million / .36% of GNP

Austria — $548 million / .34% of GNP

Japan — $10,952 million / .32% of GNP

United Kingdom — $3,248 million / .32% of GNP

Italy — $3,352 million / .3% of GNP

New Zealand — $100 million / .25% of GNP

United States — $11,262 million / .2% of GNP

Ireland — $72 million / .19% of GNP

[1]Gross National Product
million=millions of dollars, represents total $ for 1991

SOURCE: *World Development Report 1994*, Infrastructure for Development, Oxford University Press, NY, 1994.

Perhaps the most striking fact revealed by this graph is how *little* wealth is actually being transferred from rich countries to poor ones in the form of official development assistance (ODA), as a percentage of the donor's wealth. The second important fact shown is that the United States is one of the *least* generous of the OECD countries. What the graphs do *not* reveal is that most of this "foreign aid" is in the form of loans which must be repaid with interest; more of it goes for military than for civilian use; most of it is consigned to relatively few countries and rarely the poorest ones; and most of it is spent in the donor countries for supplies, salaries, scholarships, etc.

Although the campaign for an NIEO is at best dormant, its effects are noticeable in many aspects of international relations, and we will see some of them later in this book. Whether the campaign revives remains to be seen, but it seems unlikely for a number of reasons. For one thing, the benefits derived from this drive have certainly not justified the efforts expended on it, and the blame cannot all be assigned to obstructionism by the rich countries. Second, the end of the Cold War undoubtedly means even lower levels of official development assistance, since the political incentive for it has largely evaporated. Third, even when the world economy is healthier, the benefits are unlikely to flow to the poorest countries, but the stoutest advocates of an NIEO are likely to benefit enough to cool their ardor for change. Fourth, other problems have come to the fore since 1973: huge debt burdens, environmental degradation, and increasing protectionism, for example. Fifth, the support of the defunct Warsaw Pact countries has been lost. And so on. Nevertheless, it seems likely that even if the NIEO idea fades into oblivion, there will be a renewed effort to narrow the gap between rich and poor, and the United Nations will probably be the primary vehicle for this effort.

**Technical assistance from the EU.** One of the many sources of economic development assistance for developing countries is the European Union. In the headquarters of the Integrated Rural Development Project in Golfito, Costa Rica are the flags of Costa Rica and the EU, and the project's logo. (Martin Glassner)

## United Nations Assistance

The former colonies are receiving development assistance of many kinds from a variety of international governmental and nongovernmental agencies. Prominent among them are the OECD, the European Union (EU), and OPEC (The Organization of the Petroleum Exporting Countries). Most important, however, is the work of the United Nations, its affiliates, and its specialized agencies.

Although the General Assembly provides policy guidance and overall supervision of development assistance, most of this particular work comes within the purview of the Economic and Social Council and several Secretariat departments. Apart from lingering problems of nuclear weapons and decolonization, the emphasis of the United Nations has, in fact, shifted since the influx of new members in the early 1960s from political and security matters to economic and social matters. This work has become so important and so complex, and is handled by so many UN organs that it is currently being reorganized so as to place it within a more rational and efficient structure. Here we can describe only briefly some of the development work of the United Nations.

The Economic and Social Council (ECOSOC) has established five regional economic commissions that have very broad functions within their regions. They are the United Nations Economic Commissions for Europe (ECE), Latin America and the Caribbean (CEPAL), and Africa (ECA), and the United Nations Economic and Social Commissions for Western Asia (ESCWA) and for Asia and the Pacific (ESCAP). They do not provide money, but instead concentrate on research, publication, training, advice, coordination, stimulation of economic cooperation and integration within the region, and similar activities. The United Nations Development Programme (UNDP) operates in five areas: surveying and assessing development assets or resources of individual countries, stimulating capital investment, training in a wide range of vocational and professional skills, adapting and utilizing modern technology in development projects, and economic and social planning. Very little of its assistance involves financial aid to countries; most is in the form of experts, fellowships, specialized equipment, and technical services. UNDP also coordinates the work of all UN agencies in the individual developing countries.

For its 1992–96 funding cycle, UNDP is concentrating on building national capacity in six specific areas:

1. Eradication of poverty through grassroots participation in development
2. Environmental protection
3. Management development
4. Technical cooperation among developing countries
5. Technology transfer
6. Promotion of women in development

It is no coincidence that, except for environmental protection, all of these areas were neglected during the colonial period, sometimes as a matter of policy.

More specialized assistance is rendered in their areas of competence by the United Nations Conference on Trade and Development (UNCTAD), the United Nations Industrial Development Organization, the United Nations Environment Programme, the World Food Council, and the International Trade Centre. In addition, nearly all the specialized and affiliated organizations provide assistance. In a few cases, such as the International Labour Organisation (ILO) and the Food and Agriculture Organization (FAO), the bulk of the work is devoted to helping developing countries. None of these agencies, except the World Bank Group and the International Monetary Fund (IMF), actually lends or gives money to the poor countries, except small amounts for specific projects from time to time. Another UN agency worthy of mention is UNITAR, the United Nations Institute for Training and Research, a small organ of the General Assembly devoted to enhancing the effectiveness of the organization itself. Its research projects,

however, have included some of special interest to developing countries and to geographers, and its training is devoted almost exclusively to the developing countries. The training includes seminars, fellowships, courses, and exercises for diplomats in training. It is this kind of quiet, undramatic, and relatively inexpensive work being carried on by UN agencies that is so essential to the newer and less experienced countries of the world.

The IMF does provide funds under very strict conditions for the purpose of stabilizing currencies. (The funds must be repaid and are not development funds per se; in fact, they are provided to rich countries experiencing temporary foreign exchange problems as well as to poor countries.) The World Bank Group helps developing countries almost exclusively. It has three main components. The principal one is the International Bank for Reconstruction and Development (IBRD or World Bank). It does preinvestment surveys and comprehensive studies of a country's development needs, helps to formulate overall development plans, and lends money on strict, almost commercial, terms for both specific projects and broader programs. The International Development Association lends smaller amounts of money on concessionary terms (long-term, low-interest, long grace period), generally to the poorest countries and for projects least likely to generate actual cash returns. The International Finance Corporation (IFC) makes loans to private businesses in developing countries that are likely to contribute significantly to their countries' economic development.[*]

To put things in perspective: Of the $33.5 billion in development aid to poor countries (new and old) in 1984, only $15.1 billion came from all multilateral institutions, including, but not limited to, those mentioned

---

[*]Two newer and smaller components of the World Bank Group are the Multilateral Investment Guarantee Agency and the International Centre for Settlement of Investment Disputes, both of which are technical and of little interest to political geographers.

in this section.* The comparable figures for 1992 were $61.3 billion and $17.6 billion.† The real value of UN and most other multilateral aid is not its size, which is obviously relatively small, but its high quality and largely nonpolitical nature. Many developing countries, if they have a choice, prefer multilateral aid for these reasons. The UN also utilizes personnel from developing countries, who may be more acceptable to the receiving countries than personnel from rich countries. And, as mentioned before, there is growing interest in technical cooperation among developing countries, and the United Nations and other multilateral agencies are best equipped to stimulate and assist this encouraging trend.

## Neocolonialism and Neoimperialism

Few sharp divisions can be drawn between States in the world. There are nearly always some that fall in between or overlap two categories. The division between rich and poor, as we have seen, is blurred by those not so poor and those not so rich. Similarly with imperialism and its victims, with colonialism and decolonization. Some countries are only nominally independent, and some former colonies have themselves become imperialists.

The term *neocolonialism* was probably used first by Kwame Nkrumah, president of Ghana, the first black African State to receive independence after being a European colony. When Malaysia became independent five months after Ghana in 1957, Nkrumah ridiculed the move, pointing out that in practice nothing had changed except the flag and some other superficialities. Malaysia was still a British colony because the country had a capitalist economy controlled by the British, British troops were still stationed

in Singapore (then a part of Malaysia), and the country was still very much oriented toward Britain. Thus the arrangement was nothing but a new variety of colonialism. Later on, he and others developed this theme further. Nkrumah later elaborated it in a book whose title emulated Lenin's.* A few excerpts will convey the essentials of the concept.

> The essence of neo-colonialism is that the State which is subject to it is, in theory, independent and has all the outward trappings of international sovereignty. In reality its economic system and thus its political policy is directed from outside. . . . The result of neo-colonialism is that foreign capital is used for the exploitation rather than for the development of the less developed parts of the world. Investment under neo-colonialism increases rather than decreases the gap between the rich and the poor countries of the world. . . . Neo-colonialism is also the worst form of imperialism. For those who practice it, it means power without responsibility and for those who suffer from it, it means exploitation without redress. . . . Neocolonialism is based upon the principle of breaking up former large united colonial territories into a number of small nonviable States which are incapable of independent development and must rely upon the former imperial power for defense and even internal security. Their economic and financial systems are linked, as in colonial days, with those of the former colonial ruler. . . . "Aid," therefore, to a neocolonial State is merely a revolving credit, paid by the neo-colonial master, passing through the neo-colonial State and returning to the neo-colonial master in the form of increased profits. Secondly, it is in the field of "aid" that the rivalry of individual developed States first manifests itself. So long as neo-colonialism persists so long will spheres of interest persist, and this makes multilateral aid—which is in fact the only effective form of aid—impossible.†

---

*1985/86 United Nations *Statistical Yearbook*, p. 338. According to the same source, total development aid from all sources in 1984 amounted to only $9.40 per inhabitant in all developing countries.

†OECD, *Geographical Distribution of Financial Flows to Developing Countries, 1989/1992.*

*Kwame Nkrumah, *Neo-colonialism; The Last Stage of Imperialism*, London, Thomas Nelson, 1965.

†Ibid., Introduction.

### Case Study: Namibia

As an illustration of the kinds of problems faced by newly independent States resulting largely from the colonial experience, it may be helpful to examine briefly the case of Namibia. This is an excerpt from a United Nations press release dated 20 June 1990, which summarizes a Namibian government policy statement on its development program, which was prepared with the help of UNDP.

In its policy statement, the Government defines the Namibian economic system as a mixed economy. Its stated development objectives are: to ensure that every citizen has access to public facilities and services; to raise and maintain the level of nutrition and public health, and the standard of living of the Namibian people; to ensure equal opportunities for women; and to protect and maintain the country's ecosystems and living natural resources. It identifies four priority sectors in which to pursue these objectives: agriculture and rural development; education and training; health care (including provision of potable water); and housing.

The World Bank team which prepared the economic overview text found sharp contrasts in the Namibian economic picture: on the one hand, the country has a relatively high per capita income ($1,200) and well-developed physical infrastructure; however, a tiny minority in Namibia enjoys incomes and health and education services at levels comparable to those of a Western European country, while the vast majority endure living conditions that are barely above subsistence and suffers from highly inadequate public services.

The heritage of the *apartheid* system is seen not only in the income disparities prevalent in Namibia, but also in an expensive, fragmented and unbalanced public administration. Under South African rule, separate services were provided for each of 11 ethnic groups through separate administrations; in 1988, government expenditure amounted to some 56 percent of Namibia's gross national product (GNP). The Government is abolishing this cumbersome and discriminatory system, but has promised to retain the civil servants it employed.

Skewed distribution and unequal development of land, as well as an emphasis on livestock rather than crop production, are identified as key features of the Namibian agricultural profile. In the next three years, the main thrust of government policy in this area will be to increase and diversify crop production in order to improve the country's food self-sufficiency ratio. The Government is also seeking assistance in raising communal farmers' productivity and expanding commercial farming.

Having increased the limit of its territorial waters from six to 200 miles immediately after independence, Namibia is limiting fishing to allow renewal of depleted fish stocks. Once the resource has had time to renew itself, which experts estimate could take from five to 10 years, Namibia will seek foreign investment in local fisheries and fish-processing. Development of processing industries is also envisaged with regard to livestock and local raw materials.

The policy statement explains that 30 to 40 percent of school-age children in Namibia do not attend classes, and 60 percent of teachers are unqualified, with a further 30 percent under-qualified. It attributes this situation to the fragmentation of teaching, curriculum development, teacher training, school administration and student services among 11 ethnic authorities, and draws attention to the fact that, in per capita terms, the resources allocated to the 10 authorities for blacks represented only one tenth of those devoted to white education, stating: ". . . over 99 percent of the untrained teachers and over 80 percent of those who are under-qualified are in the 10 black educational authorities."

Also cited are public health, water, and sanitation problems; many Namibians are seen to suffer from environmentally related preventable diseases. An acute housing shortage is said to be compounded by the arrival of many returnees from exile and by demilitarization.

There is a great deal more to the thesis of neocolonialism, and some of it is exaggerated. Nevertheless, it is undeniably true that many former colonies are still essentially satellites of the great powers and other rich countries. This factor must be taken into consideration in any analysis of current world relations, including geopolitics and power inventories, which we discuss in Part Five.

Perhaps the most obvious and important manifestation of neocolonialism is the current pattern of international trade. We discuss this in more detail later, but a brief summary now may be useful. The great bulk of international trade today (both in volume and in value) is carried on among the industrial (rich) countries. They benefit both from buying and selling and from providing shipping, insurance, banking, and other services necessary to the conduct of international trade. Next in importance is trade between the rich countries and the poor countries. This is still primarily a "colonial economy"; that is, the rich countries sell manufactured goods and semifinished goods to the poor countries (mostly ex-colonies) in exchange for commodities (foods, fuels, and raw materials). Since the rich countries control much of the commodity production in the poor countries, they tend to hold down their prices, despite characteristic short-term price fluctuations. And since they also control their own factories, they tend to keep the prices of manufactured goods high. Thus, the poor countries are caught in a squeeze from which it is very difficult to escape. Both UNCTAD and OPEC represent attempts to escape from this squeeze. Regional economic integration and domestic "bootstrap" operations are others. Nationalization of foreign-owned enterprises is another. Whether any of these efforts will be successful in the long run remains to be seen, but clearly they will be much more successful—and perhaps unnecessary—if the rich countries cooperate with the poor ones.

The new countries, however, are not only victims of imperialism, colonialism, and neocolonialism; some have indulged in these practices themselves. Soon after coming to power in Egypt in 1953, for example, Colonel Gamal Abdul Nasser began preaching pan-Arabism, the unity of all the Arabs from the Atlantic to the Persian Gulf. It seems, though, that he envisioned himself at the head of this vast Arab commonwealth, for one by one he tried to gain control over nearly all his neighbors. Before the British left the Anglo-Egyptian Sudan, he worked diligently to get the Sudanese to agree to Egyptian rule. While Libya was still poverty-stricken, he put enormous pressure on old King Idris to join Egypt—and was on the verge of success when oil was found in Libya. Libya suddenly had its own source of revenue, and Egypt was no longer attractive. Nasser also had an ongoing battle with King Hussein of Jordan (to say nothing of his continuing war against Israel), engineered a federation with Syria that only lasted three years (1958–61), and fought a five-year war in Yemen (1962–67) with up to 70,000 troops and the most modern weapons available to him. The fact that he was not particularly successful does not make Nasser any less imperialistic.

India likewise has been behaving much like an imperialist power from its birth. In 1948, when the Muslim Nizam of Hyderabad acceded to Pakistan, the Indian army simply invaded his large state in southern India and it was absorbed into India. Similarly, when the Muslim people of Kashmir clearly wanted to join Pakistan, India attempted to annex it and fought a war with Pakistan over it, occupying the most valuable two-thirds of it. France yielded its five colonies in the peninsula to India peacefully between 1952 and 1954, but Portugal stubbornly held on to its colonies there. India lost patience, and in 1961 simply invaded and annexed them. The Indian army went to work again in 1971 when it invaded East Bengal to aid separatists there who were fighting the Pakistani army. The Indians withdrew, and Bangladesh emerged from what had been East Pakistan. Even without fighting, India inherited the British role as protector of Sikkim and Bhutan and became the dominant influence in Nepal. In 1971 India graciously permitted Bhutan to become nomi-

**A case of neoimperialism.** President Sukarno of Indonesia tried for years to wrest from the British their territories in Borneo (Kalimintan). In August 1963 a United Nations mission visited Jesselton, British North Borneo, to ascertain the feelings of the people about their future. Their feelings are shown quite clearly in this photo. A month later, their territory joined Singapore, Sarawak, and independent Malaya to form the Federation of Malaysia, but Indonesia continued its attacks on the Borneo territories until 1966. British North Borneo is now Sabah and Jesselton is now Kota Kinabalu. (UN photograph)

nally independent and even join the United Nations, but it is still the dominant outside influence in the country. In 1974 Sikkim was completely absorbed and became a state of India. Since then, India has intervened militarily in Sri Lanka and the Maldives.

Indonesia under President Sukarno was also imperialistic, waging a violent propaganda war against the Netherlands for western New Guinea and even dropping paratroops into the territory. He finally won the territory but lost his prolonged "confrontation" with Britain and then Malaysia to secure control of Sarawak, Brunei, and Sabah (formerly British North Borneo). Then, in December 1975, post-Sukarno Indonesia invaded East Timor as the Portuguese withdrew, and, after heavy fighting against Timorese forces, annexed the territory in July 1976.

Morocco and Mauritania ignored the principles of both self-determination and *uti possidetis juris* when they partitioned Western Sahara between them as the Spanish with-

drew in 1975. (There is a touch of irony in this partnership, since Morocco had a longstanding claim to all of Mauritania as well as the Spanish territory and even delayed Mauritania's admission to the United Nations for a year because of it.) Somalia's irredentist wars against Ethiopia and Kenya, whatever their justification on ethnic grounds, were as imperialistic as Germany's claims to Czechoslovakia's Sudetenland in the 1930s. Vietnam's conquest of Cambodia in 1978–79 and Libya's occupation of a huge strip of northern Chad between 1978 and 1987 were also imperialistic, although Vietnam's action at least had the virtue of deposing the despotic and murderous Khmer Rouge regime.

Iraq's invasion and annexation of Kuwait in August 1990 triggered the remarkable and successful operation in January–February 1991 that evicted the Iraqis from Kuwait. Nevertheless, the Iraqi aggression was only the latest, but almost certainly not the last, case of what we call "neoimperialism," a new phase or cycle of imperialism being

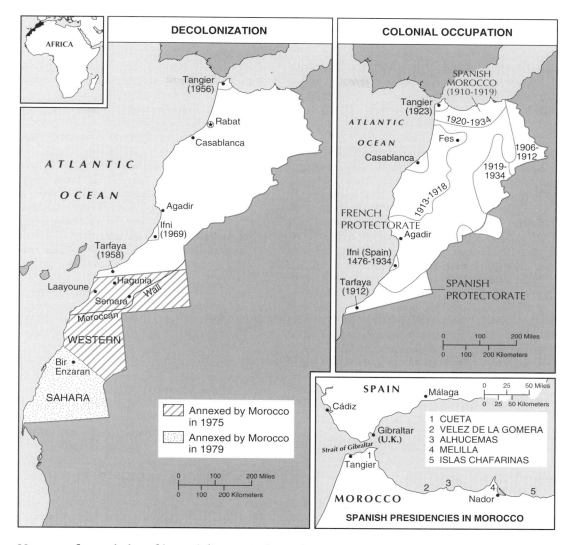

**DECOLONIZATION**

**COLONIAL OCCUPATION**

AFRICA

Tangier (1956)

⊛ Rabat

Casablanca

ATLANTIC

OCEAN

Agadir

Ifni (1969)

Tarfaya (1958)

Laayoune

Haguñia

Semara

Wall

Moroccan

WESTERN

Bir Enzaran

SAHARA

[legend]
▨ Annexed by Morocco in 1975
▨ Annexed by Morocco in 1979

0    100    200 Miles
0    100    200 Kilometers

SPANISH MOROCCO (1910-1919)

Tangier (1923)    1920-1934

ATLANTIC OCEAN    Fes

1906-1912

Casablanca    1919-1934

1913-1918

FRENCH PROTECTORATE    Agadir

Ifni (Spain) 1476-1934

Tarfaya (1912)    SPANISH PROTECTORATE

0    100    200 Miles
0    100    200 Kilometers

**SPAIN**    Málaga

0    25    50 Miles
0    25    50 Kilometers

Cádiz

Gibraltar (U.K.)

Strait of Gibraltar

1
Tangier

2    3

MOROCCO    Nador    4    5

1 CUETA
2 VELEZ DE LA GOMERA
3 ALHUCEMAS
4 MELILLA
5 ISLAS CHAFARINAS

**SPANISH PRESIDENCIES IN MOROCCO**

**Morocco, from victim of imperialism to independent State to neoimperialist.** This country was gradually conquered after Spain and France agreed to establish protectorates over it and Tangier was made an international city. Decolonization came much more quickly (1956–58) and Morocco went on to occupy all of the former Spanish Sahara. Spain still retains five *plazas de soberania* or presidios, essentially exclaves of Spain. The three smaller ones are islands. Note the wall the Moroccans have built in the south to keep out the POLISARIO forces. By 1995 Morocco had still not permitted a UN-supervised plebiscite to ascertain the preference of the Sahrawi people for legal incorporation into Morocco or for independence.

practiced by States that are not even capitalist or rich. Some theories about imperialism will have to be revised.

The new States of the world are interesting enough individually; collectively they are fascinating. It is instructive also to apply to each of them the theories we have developed through the past century in political geography. No discussion of international relations will be of any value from this point on unless it takes into account the effects of the quadrupling of the international community in half a century and the role of the new States individually and collectively.

# BIBLIOGRAPHY FOR PART FOUR

## Books and Monographs

**A**

Abraham, Kinfe, *The Missing Millions; Why and How Africa Is Underdeveloped.* Trenton, NJ: African World, 1995.

Adloff, Richard and Virginia Thompson, *The Western Saharans; Background to Conflict.* Lanham, MD: Barnes & Noble, 1980.

Agarwala, P. N., *The New International Economic Order: An Overview.* London: Pergamon, 1983.

Agbor-Tabi, Peter, *U.S. Bilateral Assistance in Africa: The Case of Cameroon.* Lanham, MD: Univ. Press of America, 1984.

Akiner, Shirin (ed.), *Political and Economic Trends in Central Asia.* London: Tauris, 1992.

Alapuro, Risto and others (eds.), *Small States in Comparative Perspective.* Oslo: Norwegian Univ. Press, 1985.

Aldrich, Robert, *The French Presence in the South Pacific, 1842–1940.* Honolulu: Univ. of Hawaii Press, 1990.

———, *France and the South Pacific Since 1940.* Honolulu: Univ. of Hawaii Press, 1993.

Aldrich, Robert and John Connell, *France's Overseas Frontier.* Cambridge Univ. Press, 1992.

Alexander, Yonah and Robert A. Friedlander (eds.), *Self-Determination: National, Regional, and Global Dimensions.* Boulder, CO: Westview, 1980.

Al-Farsy, Fouad, *Saudi Arabia: A Case Study in Development.* London: Routledge, 1986.

Allan, J. A. (ed.), *Libya Since Independence.* New York: St. Martin's, 1982.

Ansprenger, Franz, *The Dissolution of the Colonial Empire.* London: Routledge, 1989.

Appelbaum, Richard P. and Jeffrey Henderson (eds.), *States and Development in the Pacific Rim.* Newbury Park, CA: Sage, 1992.

Apter, David E., *Rethinking Development; Modernization, Dependency, and Postmodern Politics.* Newbury Park, CA: Sage, 1987.

Arase, David, *Buying Power: The Political Economy of Japanese Foreign Aid.* Boulder, CO: Lynne Reinner, 1995.

Ariff, Mohamed (ed.), *The Pacific Economy; Growth and External Stability.* Sydney: Allen & Unwin, 1991.

Arnold, Guy, *Wars in the Third World.* Mansell, 1991.

Augelli, John P., *The Panama Canal Area in Transition.* AUFS Reports, North America Series, Part I, 1981, No. 3; Part II, 1981, No. 4.

**B**

Babarinde, Olufemi A., *The Lomé Conventions and Development; An Empirical Assessment.* Brookfield, VT: Avebury, 1994.

Balfour-Paul, Glen, *The End of Empire in the Middle East; Britain's Relinquishment of Power in the Last Three Arab Dependencies.* Cambridge Univ. Press, 1990.

Ballard, J. A. (ed.), *Policy-Making in a New State; Papua New Guinea 1972–77.* St. Lucia: Univ. of Queensland Press, 1981.

———, *Uncertain Dimensions: Western Overseas Empires in the Twentieth Century.* Minneapolis: Univ. of Minnesota Press, 1985.

Barone, Charles A., *Marxist Thought on Imperialism: A Critical Survey.* Armonk, NY: Sharpe, 1985.

Bedjaoui, Mohammed, *Towards a New International Economic Order.* New York: Holmes & Meier, 1979.

Bender, Gerald J., *Angola under the Portuguese: The Myth and the Reality.* Berkeley: Univ. of California Press, 1978.

Berat, Lynn, *Walvis Bay: Decolonisation and International Law.* New Haven, CT: Yale Univ. Press, 1990.

Berger, Peter L. and Hsin-Huang M. Hsiao (eds.), *In Search of an East Asian Development Model.* New Brunswick, NJ: Transaction, 1987.

Berger, Susan A., *Political and Agrarian Development in Guatemala.* Boulder, CO: Westview, 1992.

Berman, Bruce, *Control and Crisis in Colonial Kenya; The Dialectic of Domination.* Athens: Ohio Univ. Press, 1991.

Beyan, Amos J., *The American Colonization Society and the Creation of the Liberian State: A Historical Perspective, 1822–1900.* Lanham, MD: Univ. Press of America, 1991.

Bhacker, M. Reda, *Trade and Empire in Muscat and Zanzibar.* London: Routledge, 1993.

Bhagwati, Jagdish N., *The New International Economic Order; The North-South Debate.* Cambridge, MA: MIT Press, 1977.

Bhutto, Zulfikar Ali, *The Myth of Independence.* Oxford Univ. Press, 1969.

Biger, Gideon, *An Empire in the Holy Land; Historical Geography of the British Administration in Palestine 1917–1929.* Jerusalem: Magnes Press, Hebrew Univ. and New York: St. Martin's, 1994.

Bischoff, Paul-Henri, *Swaziland's International*

*Relations and Foreign Policy: A Study of a Small State in International Relations.* Peter Lang, 1990.

Black, Jan Knippers, *The Dominican Republic; Politics and Development in an Unsovereign State.* London: Allen & Unwin, 1986.

———, *Development in Theory and Practice; Bridging the Gap.* Boulder, CO: Westview, 1991.

Blaut, James M., *The Colonizer's Model of the World; Geographical Diffusionism and Euro-centric History.* New York: Guilford, 1993.

Bloomfield, Richard J. (ed.), *Puerto Rico; The Search for a National Policy.* London: Pinter, 1985.

Boahen, A. Adu, *African Perspectives on Colonialism.* Baltimore: Johns Hopkins Univ. Press, 1990.

Brandt, Willy, *North-South: A Programme for Survival.* Cambridge, MA: MIT Press, 1980.

Bremmer, Ian and Ray Taras (eds.), *Nations and Politics in the Soviet Successor States.* Cambridge Univ. Press, 1993.

Brookfield, Harold C., *Colonialism, Development and Independence: The Case of the Melanesian Islands in the South Pacific.* Toronto: Macmillan, 1972.

Budiardjo, Carmel and Liem Soli Liong, *The War Against East Timor.* London: Zed Books, 1984.

Burki, Shahid J., *Pakistan: The Continuing Search for Nationhood.* 2nd ed. Boulder, CO: Westview, 1991

Burnell, Peter, *Economic Nationalism in the Third World.* Boulder, CO: Westview, 1986.

Burnett, Stanton H., *Investing in Security: Economic Aid for Non-Economic Purposes.* Washington, DC: CSIS, 1992.

Burrows, Susan, *Self-Determination: An Interdisciplinary Annotated Bibliography.* New York: Garland, 1986.

**C**

Cabral, Amilcar. *Revolution in Guinea.* New York: Monthly Review Press, 1970.

Carrington, C. E., *An Exposition of Empire.* Cambridge Univ. Press, 1947.

———, *Gibraltar.* London: Royal Institute of International Affairs, 1956.

Carter, Gwendolen M. and Patrick O'Meara (eds.), *African Independence; The First Twenty-Five Years.* Bloomington: Indiana Univ. Press, 1986.

Chan, Kenneth, *Cocos (Keeling) Islands: The Political Evolution of a Small Island Territory in the Indian Ocean.* Honolulu: East-West Center, 1987.

Chapman, Terry M., *The Decolonisation of Niue.* Wellington, N.Z.: Victoria Univ. Press, 1976.

Chikela, Charles O., *Britain, France, and the New African States: A Study of Post Independence Relationships, 1960–1985.* Queenston, Ontario: Mellen, 1990.

Chowdhuri, R. N., *International Mandates and Trusteeship Systems: A Comparative Study.* The Hague: Nijhoff, 1955.

Christopher, A. J., *Colonial Africa.* Totowa, NJ: Barnes & Noble, 1984.

———, *The British Empire at Its Zenith.* London: Croom Helm, 1988.

Church, R.J. Harrison, *Modern Colonization.* London: Hutchinson's Univ. Library, 1951.

Clark, G., *The Balance Sheet of Imperialism.* New York: Columbia Univ. Press, 1936.

Clarke, Colin G. and Anthony J. Payne (eds.), *Politics, Security and Development in Small States.* London: Allen & Unwin, 1987.

Clemens, Walter C., Jr., *Baltic Independence and Russian Empire.* New York: St. Martin's, 1991.

Clyde, P. H., *Japan's Pacific Mandate.* New York: Macmillan, 1935.

Cobban, Alfred, *National Self-Determination.* Oxford Univ. Press, 1944.

Cockram, Gail-Maryse, *South West African Mandate.* Cape Town: Juta, 1976.

Cohen, Michael J., *Palestine and the Great Powers, 1945–1948.* Princeton, NJ: Princeton Univ. Press, 1982.

———, *Palestine to Israel; From Mandate to Independence.* London: Frank Cass, 1988.

Cohen, Robin (ed.), *African Islands and Enclaves.* Beverly Hills, CA: Sage, 1983.

Cooper, Allan D., *The Occupation of Namibia: Afrikanerdom's Attack on the British Empire.* Lanham, MD: Univ. Press of America, 1991.

Copland, Ian, *The Burden of Empire, Perspectives on Imperialism and Colonialism.* Oxford Univ. Press, 1991.

Coulter, John Wesley, *The Pacific Dependencies of the United States.* New York: Macmillan, 1957.

———, *The Drama of Fiji: A Contemporary History.* Rutland, VT: Tuttle, 1967.

Critchlow, James, *Nationalism in Uzbekistan; A Soviet Republic's Road to Sovereignty.* Boulder, CO: Westview, 1991.

Cromer, Earl, *Ancient and Modern Imperialism.* London: John Murray, 1910.

**D**

Dahlberg, Betty S., *The Dutch Caribbean; Prospects for Democracy.* Reading, UK: Gordon & Breach, 1990.

Dalton, Dennis and A. Jeyaratnam Wilson (eds.), *The States of South Asia: Problems of National Integration.* Honolulu: Univ. of Hawaii Press, 1989.

Daly, M. W., *Imperial Sudan; The Anglo-Egyptian Condominium, 1934–56.* Cambridge Univ. Press, 1991.

Damis, John, *Conflict in Northwest Africa; The Western Sahara Dispute.* Stanford, CA: Hoover Institution Press, 1983.

Darby, Phillip, *Three Faces of Imperialism.* New Haven, CT: Yale Univ. Press, 1987.

Davidson, James Wightman. *Samoa mo Samoa: The Emergence of the Independent State of Western Samoa.* Oxford Univ. Press, 1967.

Day, John, *International Nationalism; the Extraterritorial Relations of Southern Rhodesian African Nationalists.* London: Routledge, 1967.

Delius, Peter, *The Land Belongs to Us: The Pedi Polity, the Boers and the British in the Nineteenth Century Transvaal.* Berkeley: Univ. of California Press, 1984.

Dietz, James L., *Economic History of Puerto Rico.* Princeton, NJ: Princeton Univ. Press, 1987.

Dixon, Chris and Michael Hefferman (eds.), *Colonialism and Development in the Contemporary World.* Mansell, 1991.

Dore, Isaak I., *The International Mandate System and Namibia.* Boulder, CO: Westview, 1985.

Dorr, Steven R. and others, *Global Transformation and the Third World.* Boulder, CO: Lynne Rienner, 1992.

Dorrance, John C., *The United States and the Pacific Islands.* New York: Praeger, 1992.

Drakakis-Smith, David and Chris Dixon (eds.), *Economic and Social Development in Pacific Asia.* London: Routledge, 1993.

Drake, Christine, *National Integration in Indonesia: Patterns and Policies.* Honolulu: Univ. of Hawaii Press, 1989.

Dreyer, Ronald, *Namibia and Southern Africa: Regional Dynamics of Decolonization, 1945–1990.* Irvington, NY: Columbia Univ. Press, 1994.

Duffy, James, *Portuguese Africa.* Cambridge, MA: Harvard Univ. Press, 1959.

Dugard, J. (ed.), *The South West Africa/Namibia Dispute: Documents and Scholarly Writings on the Controversy Between South Africa and the United Nations.* Berkeley: Univ. of California Press, 1973.

Duignan, Peter and Lewis Henry Gann (eds.), *Colonialism in Africa 1870–1960.* (Multi-volume) Cambridge Univ. Press, 1975.

Dunn, Rose E., *Resistance in the Desert: Moroccan Responses to French Imperialism.* Madison: Univ. of Wisconsin Press, 1977.

**E**

Elazar, Daniel J., *Two Peoples.... One Land.* Lanham, MD: Univ. Press of America, 1991.

Elsenhans, Hartmut, *Development and Underdevelopment; the History, Economics and Politics of North-South Relations.* New Delhi: Sage, 1991.

Ensign, Margee, *Doing Good or Doing Well? Japan's Foreign Aid Program.* Irvington, NY: Columbia Univ. Press, 1992.

Erlich, Haggai, *Ethiopia and the Challenge of Independence.* London: Pinter, 1986.

**F**

Fairbairn, Teo I.J. and others, *The Pacific Islands; Politics, Economics, and International Relations.* Honolulu: East-West Center, 1991.

Falk, Pamela S. (ed.), *The Political Status of Puerto Rico.* Lexington, MD: Lexington Books, 1986.

Fawcett, Charles B., A *Political Geography of the British Empire.* Boston: Ginn, 1933.

Feinberg, Richard E. and others, *From Confrontation to Cooperation?; U.S. and Soviet Aid to Developing Countries.* New Brunswick, NJ: Transaction, 1991.

Ferdinand, Peter (ed.), *The New States of Central Asia and Their Neighbors.* New York: Council on Foreign Relations, 1995.

Ferguson, Ed and Abdul Sheriff (eds.), *Zanzibar under Colonial Rule.* Athens: Ohio Univ. Press, 1991.

Ferguson, James, *Far from Paradise; An Introduction to Caribbean Development.* London: Latin America Bureau, 1990.

Fetter, Bruce, *Colonial Rule and Regional Imbalance in Central Africa.* Boulder, CO: Westview, 1983.

Fierman, William (ed.), *Soviet Central Asia; The Failed Transformation.* Boulder, CO: Westview, 1991.

Fitzsimmons, M. A., *Empire by Treaty; Britain and the Middle East in the Twentieth Century.* Notre Dame, IN: Univ. of Notre Dame Press, 1964.

Flapan, Simha, *The Birth of Israel; Myths and Realities.* New York: Pantheon, 1987.

Forbes, Dean, *The Geography of Underdevelopment.* London: Croom Helm, 1984.

Forrest, Tom G., *Politics and Economic Development in Nigeria.* Boulder, CO: Westview, 1993.

Forsythe, David P., *United Nations Peace Making; The Conciliation Commission for Palestine.* Baltimore: Johns Hopkins Univ. Press, 1972.

Franda, Marcus, *The Seychelles.* Aldershot, UK.: Gower, 1982.

Frimpong-Ansah, Jonathan H. and others (eds.), *Trade and Development in Sub-Saharan Africa*. Manchester, UK: Manchester Univ. Press, 1991.

Furnivall, J. S., *Netherlands India: A Study in Plural Economy*. New York: Macmillan, 1944.

————, *Colonial Policy and Practice: A Comparative Study of Burma and Netherlands India*. New York: New York Univ. Press, 1956.

**G**

Gale, Roger W., The *Americanization of Micronesia*. Lanham, MD: Univ. Press of America, 1979.

Gann, Lewis Henry and Peter Duignan, *White Settlers in Tropical Africa*. Harmondsworth, UK: Penguin, 1962.

García-Amador, F. V., The *Emerging International Law of Development. A New Dimension of International Economic Law*. Rome: Oceana, 1990.

García Passalacqua, Juan M., *Puerto Rico: Equality and Freedom at Issue*. New York: Praeger, 1984.

Gerner, Deborah J., *One Land, Two Peoples; The Conflict over Palestine*. 2nd ed. Boulder, CO: Westview, 1994.

Geyer, Dietrich, *Russian Imperialism*. New Haven, CT: Yale Univ. Press, 1987.

Geyer, Dietrich and Peter Duignan, *Burden of Empire; An Appraisal of Western Colonialism in Africa South of the Sahara*. New York: Praeger, 1967.

Gifford, Prosser and William Roger Louis (eds.), *France and Britain in Africa: Imperial Rivalry and Colonial Rule*. New Haven, CT: Yale Univ. Press, 1971.

Gill, T.D., *South West Africa and the Sacred Trust 1919–1972*. The Hague: TMC Asser Instituut, 1984.

Gilmore, William C., *Newfoundland & Dominion Status: The External Affairs, Competence and International Law Status of Newfoundland, 1855–1949*. Toronto: Carswell, 1988.

Godlewska, Anne and Neil Smith (eds.), *Geography and Empire*. Oxford Univ. Press, Blackwell, 1994.

Goldman, Robert B., *From Independence to Statehood*. New York: St. Martin's, 1984.

Good, Robert C., *U.D.I.; The International Politics of the Rhodesian Rebellion*. London: Faber & Faber, 1973.

Graham, Norman A., *Seeking Security and Development: The Impact of Military Spending and Arms Transfers*. Boulder, CO: Lynne Rienner, 1993.

Green, Reginald Herbold, *From Sudwestafrika to Namibia; The Political Economy of Transition*. Uppsala: Scandinavian Institute of African Studies, 1981.

**H**

Hammer, Richard M. and others (eds.), *North-South Dialogue: A New International Economic Order*. Institute of Public International Law and International Relations of Thessaloniki, 1982.

Hansen, Holger B. and Michael Twaddle, *Changing Uganda; The Dilemmas of Structural Adjustment and Revolutionary Change*. Athens; Ohio Univ. Press, 1991.

Harrell-Bond, Barbara, *The Struggle for the Western Sahara*. AUFS Reports Africa Series 1981/Nos. 37, 38, 39.

Harrigan, Norwell, *The Inter-Virgin Islands Conference: A Study of a Microstate International Organization*. Gainesville: Univ. Presses of Florida, 1980.

Harris, Nigel, *The End of the Third World; Newly Industrializing Countries and the Decline of an Ideology*. London: Tauris, 1986.

Harrison, Lawrence E., *Underdevelopment Is a State of Mind: The Latin American Case*. Lanham, MD: Univ. Press of America, 1985.

Hart, Jeffrey A., *The New International Economic Order*. New York: St. Martin's, 1983.

Hawley, Donald, *The Trucial States*. New York: Humanities, 1971.

Hedlund, Stefan and Kristian Gerner, *The Baltic States and the End of the Soviet Empire*. London: Routledge, 1993.

Henderson, William Otto, *The German Colonial Empire 1884–1919*. London: Frank Cass, 1993.

Henningham, Stephen, *France and the South Pacific; A Contemporary History*. Honolulu: Univ. of Hawaii Press, 1991.

Hezel, Francis X., *Strangers in Their Own Land; A Century of Colonial Rule in the Caroline and Marshall Islands*. Honolulu: Univ. of Hawaii Press, 1995.

Hinden, Rita, *The United Nations and the Colonies*. London: UN Association, 1950.

Ho, H.C.Y. and L.C. Chou (eds.), *The Economic System of Hong Kong*. Hong Kong: Asian Research Service, 1992.

Holland, R. F., *European Decolonization 1918–1981*. New York: St. Martin's, 1985.

Hook, Steven W., *National Interest and Foreign Aid*. Boulder, CO: Lynne Rienner, 1995.

Howard, Michael C. (ed.), *Ethnicity and Nation-Building in the Pacific*. Tokyo: United Nations Univ. Press, 1990.

————, *Mining, Politics, and Development in the South Pacific*. Boulder, CO: Westview, 1991.

Hudson, W.J. (ed.), *Australia and the Colonial Question at the United Nations*. Honolulu: Univ. Press of Hawaii, 1970.

———, *Australia's New Guinea Question*. Melbourne: Nelson, 1975.

Hunter, Shireen T., *The Transcaucusus in Transition: Nation-Building and Conflict*. Washington, DC: CSIS, 1994.

**I**

Iglesias, Enrique V., *Reflections on Economic Development; Toward a New Latin American Consensus*. Baltimore: Johns Hopkins Univ. Press, 1993.

Imbeau, Louis M., *Donor Aid—The Determinants of Development Allocations for Third World Countries; A Comparative Analysis*. Peter Lang, 1989.

Ingham, Kenneth, *Politics in Modern Africa: The Uneven Tribal Dimension*. London: Routledge, 1990.

Isaacman, Allen and Barbara Isaacman, *Mozambique: From Colonialism to Revolution, 1900–1982*. Boulder, CO: Westview, 1983.

Isaacs, Arnold H., *Dependence Relations Between Botswana, Lesotho, Swaziland and the Republic of South Africa; A Literature Study Based on Johan Galtung's Theory of Imperialism*. Leiden: African Studies Center, 1982.

**J**

John, Sir Rupert, *Pioneers in Nation-Building in a Caribbean Mini-State*. New York: UNITAR, 1979.

Johnson, L. W., *Colonial Sunset: Australia and Papua New Guinea 1970–74*. St. Lucia: Univ. of Queensland Press, 1984.

Johnson, Roberta Ann, *Puerto Rico: Commonwealth or Colony?* New York: Praeger, 1980.

Jomo, K. S., *Industrializing Malaysia*. London: Routledge, 1993.

Jones, F. C., *Japan's New Order in East Asia: Its Rise and Fall, 1937–1945*. Oxford Univ. Press, 1954.

**K**

Kaarsholm, Preben, *Cultural Struggle and Development in Southern Africa*. London: Heinemann, 1991.

Kahler, Miles, *Decolonization in Britain and France*. Princeton, NJ: Princeton Univ. Press, 1984.

Kanduza, Ackson M., *The Political Economy of Underdevelopment in Northern Rhodesia, 1918–1960*. Halifax, NS.: Dalhousie African Studies Series 5, 1986.

Kann, R. A., *The Hapsburg Empire: A Study in Integration and Disintegration*. New York: Praeger, 1957.

Karklins, Rasma, *Ethnopolitics and Transition to Democracy; The Collapse of the USSR and Latvia*. Baltimore: Johns Hopkins Univ. Press, 1994.

Karp, Regina Cowen (ed.), *Central and Eastern Europe; The Challenge of Transition*. Oxford Univ. Press, 1994.

Kawai, T., *The Goal of Japanese Expansion*. Tokyo: Hokuseido Press, 1938.

Kay, Cristóbal, *Latin American Theories of Development and Underdevelopment*. London: Routledge, 1989.

Kelly, Ian, *Hong Kong*. Honolulu: Univ. of Hawaii Press, 1987.

Keltie, John Scott, *The Partition of Africa*. London, 1895.

Kennedy, Charles H. and Neil Nevitte (eds.), *Ethnic Preferences and Public Policy in Developing States*. Boulder, CO: Lynne Rienner, 1986.

Kennedy, Paul M., *The Samoan Tangle*. New York: Harper & Row, 1974.

Khoury, Philip S., *Syria and the French Mandate*. London: Tauris, 1987.

Kofele-Kale, Ndiva (ed.), *An African Experiment in Nation Building: The Bilingual Cameroon Republic Since Reunification*. Boulder, CO: Westview, 1980.

Kolarz, Walter, *Russia and Her Colonies*. London: George Philip, 1952.

Konczacki, Zbigniew A. and others (eds.), *Studies in the Economic History of Southern Africa; Vol. I: The Front-Line States. Vol. II: South Africa, Lesotho and Swaziland*. London: Frank Cass, 1990 and 1991.

Kostiner, Joseph, *The Struggle for South Yemen*. New York: St. Martin's, 1984.

**L**

Lagerberg, Kees, *West Irian and Jakarta Imperialism*. New York: St. Martin's, 1980.

Lapidus, Gail and others, *From Union to Commonwealth*. Cambridge Univ. Press, 1992.

Larrain, Jorge, *Theories of Development: Capitalism, Colonialism and Dependency*. Oxford: Polity Press, 1989.

Lee, John Michael, *Colonial Development and Good Government; A Study of the Ideas Expressed by the British Official Classes in Planning Decolonization, 1939–1964*. Oxford Univ. Press, 1967.

Legum, Colin (ed.), *Zambia; Independence and Beyond. The Speeches of Kenneth Kaunda*. London: Nelson, 1966.

Leibowitz, Arnold H., *Colonial Emancipation in the Pacific and the Caribbean: A Legal and Political Analysis*. New York: Praeger, 1976.

————, *Defining Status; A Comprehensive Analysis of United States Territorial Relations*. Dordrecht: Nijhoff, 1989.

Lewis, Vaughan A. (ed.), *Size, Self-Determination and International Relations; The Caribbean*. Mona, Jamaica: Univ. of the West Indies, Institute of Social and Economic Research, 1976.

Leys, Colin and John S. Saul, *Namibia's Liberation Struggle; The Two-Edged Sword*. Athens: Ohio Univ. Press, 1994.

Lijphart, Arend, *The Trauma of Decolonization: The Dutch and West New Guinea*. New Haven, CT: Yale Univ. Press, 1966.

Lockhart, Douglas G. and others (eds.), *The Development Process in Small Island States*. London: Routledge, 1993.

Loehr, William and John P. Powelson, *Threat to Development; Pitfalls of the NIEO*. Boulder, CO: Westview, 1983.

López, Alfredo, *Doña Licha's Island; Modern Colonialism in Puerto Rico*. Boston: South End Press, 1987.

Louis, William Roger and Robert W. Stookey (eds.), *The End of the Palestine Mandate*. Austin: Univ. of Texas Press, 1986.

Low, D. A., *Eclipse of Empire*. Cambridge Univ. Press, 1991.

Lugard, Lord, *The Dual Mandate in British Tropical Africa*. Edinburgh: Blackwood, 1922.

Lutchman, Harold Alexander, *From Colonialism to Cooperative Republic; Aspects of Political Development in Guyana*. Río Piedras: Univ. of Puerto Rico, 1974.

**M**

Mair, Lucy Philip, *Australia in New Guinea*. 2nd ed. Melbourne: Melbourne Univ. Press, 1970.

Mandaza, Ibbo (ed.), *Zimbabwe: The Political Economy of Transition, 1980–1986*. Atlantic Highlands, NJ: Humanities, 1986.

Mandle, Jay R., *Patterns of Caribbean Development: An Interpretive Essay on Economic Change*. Reading, UK: Gordon & Breach, 1982.

Manning, Patrick, *Francophone Sub-Saharan Arica, 1880–1985*. Cambridge Univ. Press, 1988.

Marks, Siegfried (ed.), *Political Constraints on Brazil's Economic Development*. New Brunswick, NJ: Transaction, 1993.

May, R. J. and Matthew Spriggs (eds.), *The Bougainville Crisis*. Bathurst, NSW, Australia: Crawford House, 1990.

Mazrui, Ali and Michael Tidy, *Nationalism and New States in Africa from about 1935 to Present*. Portsmouth, NH: Heinemann, 1984.

McCormick, Sharon H., *The Angolan Economy*. Boulder, CO: Westview, 1993.

McGregor, Andrew and Mark Sturton, *Fiji: Economic Adjustment, 1987–1991*. Honolulu: Univ. of Hawaii Press, 1992.

————, *Vanuatu: Toward Economic Growth*. Honolulu: Univ. of Hawaii Press, 1992.

McGurn, William, *Perfidious Albion; The Abandonment of Hong Kong 1997*. Lanham, MD: Univ. Press of America, 1992.

McHenry, Donald F., *Micronesia: Trust Betrayed; Altruism vs. Self-Interest in United States Foreign Policy*. Washington, DC: Carnegie Endowment, 1975.

McTague, John J., Jr., *British Policy in Palestine, 1917–1922*. Lanham, MD: Univ. Press of America, 1983.

Mehretu, Assefa, *Regional Disparity in Sub-Saharan Africa: Structural Readjustment of Uneven Development*. Boulder, CO: Westview, 1989.

Melendez, Edwin and Edgardo Melendez, *Colonial Dilemma*. Monroe, ME: South End Press, 1992.

Melkote, Srinivas R., *Communication for Development in the Third World; Theory and Practice*. Newbury Park, CA: Sage, 1991.

Mercer, John, *Spanish Sahara*. London: Allen & Unwin, 1976.

Mikell, Gwendolyn, *Cocoa and Chaos in Ghana*. Washington, DC: Howard Univ. Press, 1994.

Miller, Robert (ed.), *Aid As Peacemaker; Canadian Development Assistance and Third World Conflict*. Ottawa: Carleton Univ. Press, 1992.

Minkel, Clarence W. and Ralph H. Alderman, *A Bibliography of British Honduras 1900–1970*. East Lansing: Michigan State Univ., Latin American Studies Center, 1970.

Misra, B. B., *District Administration and Rural Development in India*. Oxford Univ. Press, 1983.

Mondlane, Eduardo, *The Struggle for Mozambique 1969*. Harmondsworth, UK: Penguin, 1969.

Moorson, Richard, *Walvis Bay, Namibia's Port*. London: International Defence and Aid Fund for Southern Africa, 1984.

Morales Carrion, Arturo, *Puerto Rico and the United States*. San Juan: Editorial Académica, 1990.

Morgan, D. J., *The Official History of Colonial Development*. London: Macmillan, 1980.

Mungazi, Dickson, *Colonial Policy and Conflict in Zimbabwe; A Study of Cultures in Collision, 1890–1979*. London: Taylor and Francis, 1991.

Muñoz, Heraldo (ed.), *From Dependency to Development*. Boulder, CO: Westview, 1981.

Myers, Ramon H. and Mark R. Peattie, *The Japanese Colonial Empire, 1895–1945*. Princeton, NJ: Princeton Univ. Press, 1987.

**N**

Nasser, Gamal Abdul, *Egypt's Liberation: The Philosophy of the Revolution*. Washington, DC: Public Affairs Press, 1955.

Nawaz, Tawfique (comp.), *The New International Economic Order: A Bibliography*. London: Pinter, 1980.

Nelson, Samuel H., *Colonialism in the Congo Basin; 1880–1940*. Athens: Ohio Univ. Press, 1994.

Neuberger, Benyamin, *National Self-Determination in Post Colonial Africa*. London: Pinter, 1986.

Niebuhr, Reinhold, *The Structure of Nations and Empires*. Oxford Univ. Press, 1985.

Nkala, Jericho, *The United Nations, International Law, and the Rhodesian Independence Crisis*. Oxford Univ. Press, 1985.

Nogueira, F., *The United Nations and Portugal; A Study of Anti-Colonialism*. London: Sidgwick & Jackson, 1963.

Noland, Marcus, *Pacific Basin Developing Countries: Prospects for the Future*. Washington, DC: Institute for International Economics, 1991.

Norwine, Jim and Alfonso González, *The Third World; States of Mind and Being*. Boston: Unwin Hyman, 1988.

Nyerere, Julius K., *Freedom and Unity; A Selection from Writings and Speeches, 1952–1965*. Oxford Univ. Press, 1967.

**O**

Offiong, Daniel, *Imperialism and Dependency; Obstacles to African Development*. Washington, DC: Howard Univ. Press, 1994.

Ofuatey-Kodjoe, W., *The Principle of Self-Determination in International Law*. New York: Nellen, 1977.

Orr, Robert M., *The Emergence of Japan's Foreign Aid Power*. Irvington, NY: Columbia Univ. Press, 1992.

Osborne, Robin, *Indonesia's Secret War; The Guerilla Struggle in Irian Jaya*. London: Allen & Unwin, 1985.

**P**

Padmore, George, *Africa: Britain's Third Empire*. London: Dobson, 1949.

———, *Pan-Africanism or Communism? The Coming Struggle for Africa*. New York: Roy, 1956.

Papp, Daniel, *Soviet Perceptions of the Developing World in the 1980s; The Ideological Basis*. Lexington, MA: Lexington Books, 1985.

Peattie, Mark R., *Na'nyō: The Rise and Fall of the Japanese in Micronesia 1885–1945*. Honolulu: Univ. of Hawaii Press, 1987.

Perusse, Roland I., *The United States and Puerto Rico: The Struggle for Equality*. Melbourne, FL: Krieger, 1990.

Peters, Joel, *Israel and Africa*. London: Tauris, 1992.

Peterson, John E., *Oman in the Twentieth Century*. Totowa, NJ: Rowman & Littlefield, 1978.

Phadnis, Urmila, *Ethnicity and Nation-Building in South Asia*. Newbury Park, CA: Sage, 1990.

Phillips, Anne, *The Enigma of Colonialism; British Policy in West Africa*. Bloomington: Indiana Univ. Press, 1989.

Pieragostini, Karl, *Britain, Aden and South Arabia*. New York: St. Martin's, 1991.

Platt, Raye R. and others, *The European Possessions in the Caribbean Area*. New York: American Geographical Society, 1941.

Pomerance, Michla, *Self-Determination in Law and Practice: The New Doctrine in the United Nations*. The Hague: Nijhoff, 1982.

Prochaska, David, *Making Algeria French; Colonialism in Bône, 1870–1920*. Cambridge Univ. Press, 1990.

Pye, Lucian W., *Politics, Personality and Nation Building: Burma's Search for Identity*. New Haven, CT: Yale Univ. Press, 1962.

**R**

Rao, R. P. *Portuguese Rule in Goa, 1510–1961*. London: Asia Publishing House, 1963.

Reitsma, Hendrik-Jan A. and others, The *Third World in Perspective*. Totowa, NJ: Rowman & Littlefield, 1985.

Reno, Philip, *The Ordeal of British Guiana*. New York: Monthly Review Press, 1964.

Rézette, Robert, *The Western Sahara and the Frontiers of Morocco*. Paris: Nouvelles Editions Latines, 1975.

———, *The Spanish Enclaves in Morocco*. Paris: Nouvelles Editions Latines, 1976.

Rivlin, Benjamin, *The United Nations and the Italian Colonies*. New York: Carnegie Endowment, 1950.

Rix, Alan, *Japan's Foreign Aid Challenge*. London: Routledge, 1993.

Roberts, Andrew (ed.), *The Colonial Moment in Africa; Essays on the Movement of Minds and Materials, 1900–1940*. Cambridge Univ. Press, 1990.

Roberts, George O., *The Anguish of Third World*

*Independence: The Sierra Leone Experience.* Lanham, MD: Univ. Press of America, 1982.

Roberts, W. Adolphe. *The French in the West Indies.* New York: Cooper Square Publishers, 1971.

Robinson, A.N.R., *The Mechanics of Independence: Patterns of Political and Economic Transformation in Trinidad and Tobago.* Cambridge, MA: MIT, 1971.

Rodman, Kenneth A., *Sanctity versus Sovereignty; U.S. Policy Toward the Nationalization of Natural Resource Investments in the Third World.* New York: Columbia Univ. Press, 1988.

Rodney, Walter, *How Europe Underdeveloped Africa.* Washington, DC: Howard Univ. Press, 1964.

Rogers, Robert F., *Destiny's Landfall; A History of Guam.* Honolulu: Univ. of Hawaii Press, 1995.

Ross, Angus (ed.), *New Zealand's Record in the Pacific Islands in the Twentieth Century.* Auckland, NZ: Longman, 1969.

Rostow, Walt Whitman, *Rich Countries and Poor Countries.* Boulder, CO: Westview, 1987.

Rotberg, Robert I., *The Founder; Cecil Rhodes and the Pursuit of Power.* Oxford Univ. Press, 1988.

Rudin, H. R., *Germans in the Cameroons.* London, 1938.

Rywkin, Michael, *Moscow's Lost Empire.* Armonk, NY: Sharpe, 1993.

**S**

Saivetz, Carol R., *Soviet-Third World Relations.* Boulder, CO: Westview, 1985.

Sandhu, Kernial S. and Paul Wheatley (eds.), *The Management of Success: The Moulding of Modern Singapore.* Boulder, CO: Westview, 1990.

Saul, John S., *Recolonization and Resistance in Southern Africa in the 1990s.* Trenton, NJ: African World, 1993.

Seligson, Mitchell A. (ed.), *The Gap Between Rich and Poor: Contending Perspectives on the Political Economy of Development.* Boulder, CO: Westview, 1984.

Serapiao, Luis B. and Mohamed A. El-Khawas, *Mozambique in the Twentieth Century: From Colonialism to Independence.* Lanham, MD: Univ. Press of America, 1979.

Shaw, Timothy and Larry A. Swatuk (eds.), *Prospects for Peace and Development in Southern Africa in the 1990s: Canadian and Comparative Perspectives.* Lanham, MD: Univ. Press of America, 1991.

Shinar, Dov, *Palestinian Nation Building and the Role of Communications in the West Bank.* London: Pinter, 1986.

Shorrock, William I., *French Imperialism in the Middle East: The Failure of Policy in Syria and Lebanon.* Madison: Madison: Univ. of Wisconsin Press, 1976.

Sicherman, Harvey, *Palestinian Autonomy, Self-Government, and Peace.* Boulder, CO: Westview, 1993.

Singer, Hans W. and Sumit Roy, *Economic Progress and Prospects in the Third World: Lessons of Development Experience Since 1945.* Aldershot, UK: Edward Elgar, 1993.

Smith, W.D., *The Ideological Origins of Nazi Imperialism.* Oxford Univ. Press, 1986.

Smock, David R. (ed.), *Making War and Waging Peace; Foreign Intervention in Africa.* Washington, DC: U.S. Institute of Peace Press, 1993.

Somerville, Keith, *Foreign Military Intervention in Africa.* New York: St. Martin's, 1990.

South Commission, *The Challenge to the South, The Report of the South Commission.* Oxford Univ. Press, 1990.

Starr, S. Frederick, *The International Politics of Eurasia Vol. 1: The Influence of History on the Foreign Policies of the New States of the Former Soviet Union.* Armonk, NY: Sharpe, 1994.

Starushenko, G. B., *The Principle of National Self-Determination in Soviet Foreign Policy.* Moscow: Foreign Languages Publishing House, 1963.

Stoecker, Helmuth (ed.), *German Imperialism in Africa.* Atlantic Highlands, NJ: Humanities, 1986.

Strachey, John, *The End of Empire.* London: Gollancz, 1959.

Sureda, A. Rigo, *The Evolution of the Right of Self-Determination; A Study of United Nations Practice.* Leiden: Sijthoff, 1973.

Symonds, Richard, *The Making of Pakistan.* London: Faber & Faber, 1950.

**T**

Taagepera, Rein, *Estonia: Return to Independence.* Boulder, CO: Westview, 1992.

Tamarkin, M., *The Making of Zimbabwe; Decolonization in Regional and International Politics.* London: Frank Cass, 1990.

Taylor, John G., *Indonesia's Forgotten War; The Hidden History of East Timor.* London: Zed Books, 1991.

Tessler, Mark, *A History of the Israeli-Palestinian Conflict.* Bloomington: Indiana Univ. Press, 1994.

Thirlwall, A. P., *The Performance and Prospects of the Pacific Island Economies in the World Economy.* Honolulu: Univ. of Hawaii Press, 1992.

Thomas, Clive, *The Poor and the Powerless.* London: Latin America Bureau, 1988.

Thomas, Wolfgang H., *Economic Development in*

*Namibia.* New Brunswick, NJ: Transaction, 1985.

Thompson, Virginia and Richard Adloff, *French West Africa.* Stanford, CA: Stanford Univ. Press, 1958.

———, *The Emerging States of French Equatorial Africa.* Stanford, CA: Stanford Univ. Press, 1960.

———, *Djibouti and the Horn of Africa.* Stanford, CA: Stanford Univ. Press, 1968.

———, *The French Pacific Islands: French Polynesia and New Caledonia.* Berkeley: Univ. of California Press, 1971.

Tinbergen, Jan (ed.), *RIO: Reshaping the International Order.* New York: Dutton, 1976.

Tisch, Sarah J. and Michael B. Wallace, *Dilemmas of Development Assistance; The What, Why, and Who of Foreign Aid.* Boulder, CO: Westview, 1994.

Towsend, M. E., *The Rise and Fall of Germany's Colonial Empire, 1884–1918.* New York: Macmillan, 1930.

Trapans, Jan A. (ed.), *Toward Independence; The Baltic Popular Movements.* Boulder, CO: Westview, 1990.

Tulchin, Joseph S. with Andrew I. Rudman (eds.), *Economic Development and Environmental Protection in Latin America.* Boulder, CO: Lynne Rienner, 1991.

Twaddle, Michael (ed.), *Imperialism and the State in the Third World.* London: Tauris, 1991.

**U**

Urquhart, Brian, *Decolonization and World Peace.* Austin: Univ. of Texas Press, 1991.

**V**

Van Cleve, Ruth G., *The Office of Territorial Affairs.* New York: Praeger, 1974.

Vardys, V. Stanley, *Lithuania; A Rebel Nation.* Boulder, CO: Westview, 1994.

Villari, L., *The Expansion of Italy.* London: Faber, 1930.

Vohra, Dewan C., *India's Aid Diplomacy in the Third World.* New Delhi: Vikas, 1980.

Vucinich, Wayne S., *The Ottoman Empire: Its Record and Legacy.* Melbourne, FL: Krieger, 1979.

**W**

Wainhouse, David Walter, *Remnants of Empires: The United Nations and the End of Colonialism.* New York: Council on Foreign Relations, 1964.

Weeramantry, Christopher, *Nauru, Environmental Damage under International Trusteeship.* Oxford Univ. Press, 1992.

Weinberg, A. K., *Manifest Destiny: A Study of Nationalist Expansion in American History.* Chicago: Quadrangle Books, 1963.

Weinstein, Brian, *Gabon: Nation-Building on the Ogooue.* Cambridge, MA: MIT Press, 1966.

Weinstein, Warren and Thomas H. Henriksen (eds.), *Soviet and Chinese Aid to African Nations.* New York: Praeger, 1980.

Wenner, Manfred W., *Modern Yemen 1981–1966.* Baltimore: Johns Hopkins Univ. Press, 1967.

Wheare, Kenneth C., *The Statute of Westminister and Dominion Status.* Oxford Univ. Press, 1953.

Wiens, Herold J., *China's March Towards the Tropics.* Hamden, CT: Shoe String Press, 1954.

Williams, Patrick (ed.), *Colonial Discourse/Post-Colonial Theory: A Reader.* Irvington, NY: Columbia Univ. Press, 1994.

Woldring, Klass, *Beyond Political Independence; Zambia's Development Predicament in the 1980s.* Amsterdam: Mouton, 1984.

*World Development Report 1994; Infrastructure for Development.* Oxford for the World Bank, 1994.

Woronoff, John, *Asia's "Miracle" Economies.* Armonk, NY: Sharpe, 1986.

Wright, Carol, *Mauritius.* Newton Abbot, UK: David & Charles, 1974.

**Y**

Yeagher, Rodger, *Tanzania: An African Experiment.* Boulder, CO: Westview, 1983.

Young, C., *Politics in the Congo: Decolonization and Independence.* Princeton, NJ: Princeton Univ. Press, 1965.

**Z**

Zahlan, Rosemarie Said, *The Origins of the United Arab Emirates.* London: Macmillan, 1978.

———, *The Creation of Qatar.* Totowa, NJ: Rowman & Littlefield; 1979.

———, *The Making of the Modern Gulf States; Kuwait, Bahrain, Qatar, United Arab Emirates, Oman.* New York: Routledge, 1989.

Zimmerman, Robert F., *Dollars, Diplomacy, and Dependency: Dilemmas of U.S. Economic Aid.* Boulder, CO: Lynne Rienner, 1993.

## *Periodicals*

**A**

Albinski, H. S., "Australia and the Dutch New Guinea Dispute," *International Journal*, 16 (Autumn 1961), 358–382.

Amin, Samir, "On Jim Blaut's 'Fourteen Ninety-Two'," *Political Geography*, 11, 4 (July 1992), 394–395.

Armstrong, Arthur John and Howard Loomis Hills, "The Negotiations for the Future Political Status of Micronesia," *American Journal of International Law*, 78, 2 (April 1984), 484–497.

Ayoob, Mohammed, "The Third World in the System of States: Acute Schizophrenia or Growing Pains," *International Studies Quarterly.*, 33, 1 (March 1989), 67–80.

Azikiwe, Nnamdi, "Essentials for Nigerian Survival," *Foreign Affairs*, 43, 3 (April 1965), 447–461.

**B**

Ballendorf, Dirk Anthony, "The New Freely-Associated States of Micronesia: Their Natural and Social Environmental Challenges." *GeoJournal*, 16, 2 (March 1988), 137–142.

Barber, Hollis W., "Decolonization: The Committee of Twenty-four," *World Affairs*, 138, 2 (Fall 1976).

Barsh, Russel Lawrence, "A Special Session of the UN General Assembly Rethinks the Economic Rights and Duties of States," *American Journal of International Law*, 85, 1 (Jan. 1991), 192–199.

Bassin, M., "Imperialism and the Nation State in Friedrich Ratzel's Political Geography," *Progress in Human Geography*, 11 (1987), 473–495.

Bergsman, Peter, "The Marianas, the United States, and the United Nations: The Uncertain Status of the New American Commonwealth," *California Western International Law Journal*, 6 (Spring 1976), 382–411.

Biddle, William Jesse and John D. Stephens, "Dependent Development and Foreign Policy: The Case of Jamaica," *International Studies Quarterly*, 33, 4 (Dec. 1989), 411–434.

Binaisa, Godfrey L., "Organization of African Unity and Decolonization: Present and Future Trends," *Annals, Am. Acad. Polit. and Soc. Sci.*, 432 (July 1977), 52–69.

Blakeslee, G. H., "The Japanese Monroe Doctrine," *Foreign Affairs*, 11, 4 (1933), 671–681.

Blaut, James M., "Fourteen Ninety-two," *Political Geography*, 11, 4 (July 1992), 355–385.

————, "Response to Comments by Frank, Amin, Dodgshon and Palan," *Political Geography*, 11, 4 (July 1992), 407–412.

Bogan, E. F., "Government of the Trust Territory of the Pacific Islands," *Annals, Am. Acad. of Polit. and Soc. Sci.* (Jan. 1950), 164–174.

Bourde, A., "Comoro Islands: Problems of a Microcosm," *Journal of Modern African Studies*, 3 (May 1965), 91–102.

Boustead, J.E.H., "Abu Dhabi 1761–1963," *Royal Central Asian Journal*, 50 (July 1963), 273–277.

Boyce, Peter J. and Richard A. Herr, "Microstate Diplomacy in the South Pacific," *Australian Outlook*, 28 (April 1974), 24–35.

Bradshaw, York W. and Ana-Maria Wahl, "Foreign Debt Expansion, The International Monetary Fund, and Regional Variation in Third World Poverty," *International Studies Quarterly*, 35, 3 (Sept. 1991), 251-272.

Braganza, Berta M., "On Neo-Colonialism," *Afro-Asian and World Affairs*, 4 (Summer 1967), 87–111.

Branch, James A., "The Constitution of the Northern Mariana Islands: Does a Different Cultural Setting Justify Different Constitutional Standards?" *Denver Journal of International Law and Policy*, 9, 1 (Winter 1980), 35–67.

Broderick, Margaret, "Associated Statehood—a New Form of Decolonisation," *International and Comparative Law Quarterly*, 17 (April 1968), 368–403.

Brunhes, Jean and C. Vallaux, "German Colonization in Eastern Europe," *Geographical Review*, 6 (1918), 465–480.

Burns, R. D., "Inspection of the Mandates, 1919–1941," *Pacific Historical Review*, 37 (Nov. 1968), 445–462.

**C**

Calio, Sonia Alves, "The Brazilian Economic Crisis and Its Impact on the Lives of Women," *Political Geography Quarterly*, 9, 4 (Oct. 1990), 415–424.

Carrington, C. E., "Decolonization: The Last Stages," *International Affairs*, (1962), 29–40.

Chase-Dunn, Christopher, "Resistance to Imperialism: Semi-Peripheral Actors," *Review*, 13 (1990), 1–32.

Christopher, A. J., "'Divide and Rule,' The Impress of British Separation Policies," *Area*, 20, 3 (Sept. 1988), 233–240.

Clark, D. E., "Manifest Destiny and the Pacific," *Pacific Historical Review*, 1 (March 1932), 1–17.

Clark, Don P., "Trade vs. Aid: Distributions of Third World Development Assistance," *Economic Development and Cultural Change*, 39 (July 1991), 829–837.

Clark, Roger S., "The 'Decolonization' of East

Timor and the United Nations Norms on Self-Determination," *Yale Journal of World Public Order*, 7, 1 (Fall 1980), 2–44.

————, "Self-Determination and Free Association: Should the United Nations Terminate the Pacific Islands Trust?" *Harvard International Law Journal*, 21, 1 (Winter 1980), 1–86.

Clarke, Colin G., "Political Fragmentation in the Caribbean: The Case of Anguilla," *Canadian Geographer*, 15, 1 (1971), 13–29.

Coker, Christopher, "Decolonization in the Seventies; Rhodesia and the Dialectic of National Liberation," *Round Table*, No. 274 (April 1979), 122–136.

Collins, B.A.N., "Independence for Guyana," *World Today*, 22 (June 1966), 260–268.

Comyns-Carr, R., "Gibraltar Dispute as Seen from Spain," *World Today*, 21 (Sept. 1965), 364–367.

Coulter, John Wesley, "The United States Trust Territory of the Pacific Islands," *Journal of Geography*, 47 (Oct. 1948), 253–267.

Crawford, John F., "South West Africa: Mandate Termination in Historical Perspective," *Columbia Journal of Transnational Law*, 6 (Spring 1967), 91–137.

Crawford, Robert and Perumala Dayanidhi, "East Timor: A Study in De-Colonization," *India Quarterly*, 33, 4 (Oct. 1977), 419–441.

Crawshaw, Nancy, "The Republic of Cyprus, from Zurich Agreement to Independence," *World Today*, 16, 10 (Dec. 1960), 526–540.

Crocombe, Ron G., "Development and Regression in New Zealand's Island Territories," *Pacific Viewpoint*, 3, 2 (Sept. 1962), 17-32.

Crush, Jonathan, "Colonial Coercion and the Swazi Tax Revolt of 1903–1907," *Political Geography Quarterly*, 4, 3 (July 1985), 179–190.

Cumming, Duncan Cameron, "The Disposal of Eritrea," *Middle East Journal*, 7, 1 (Winter 1953), 18–32.

**D**

Das, Taraknath, "The State of Hyderabad During and After British Rule in India," *American Journal of International Law*, 43, 1 (Jan. 1949), 57–72.

Dean, Vera K., "Geographical Aspects of the Newfoundland Referendum," *Annals, AAG*, 39 (1949), 70.

Demas, William G., "The Caribbean and the New International Economic Order," *Journal of Interamerican Studies*, 20, 3 (August 1978), 229–263.

Dillard, Hardy C., "Status of South-West Africa (Namibia)—Separate Opinion," *International Lawyer*, 6 (April 1972), 409–427.

Dodgshon, Robert A., "The Role of Europe in the Early Modern World-System: Parasitic or Generative?", *Political Geography*, 11, 4 (July 1992), 396–400.

Doumenge, J. P., "Demographic, Economic, Socio-cultural and Political Facts Nowadays in the French Pacific Territories." *GeoJournal*, 16, 2 (March 1988), 143–156.

Dunlop, John Stewart, "The Influence of David Livingstone on Subsequent Political Developments in Africa," *Scottish Geographical Magazine*, 75, 3 (Dec. 1959), 144–152.

**E**

El-Ayouty, Yassin, "The United Nations and Decolonisation, 1960–70," *Journal of Modern African Studies*, 8, 3 (Oct. 1970), 462–468.

El Mallakh, Ragei, "The Challenge of Affluence: Abu Dhabi," *Middle East Journal*, 24, 2 (Spring 1970), 135–146.

Emerson, Rupert, "The United Nations and Colonialism," *International Relations*, 3 (Nov. 1970), 766–781.

**F**

Fallon, Joseph E., "Federal Policy and U.S. Territories; The Political Restructuring of the United States of America," *Pacific Affairs*, 64 (Spring 1991), 23–41.

Fawcett, J.E.S., "Gibraltar; the Legal Issues," *International Affairs*, 43 (April 1967), 236–251.

Fisher, Charles A., "The Malaysia Federation, Indonesia and the Philippines: A Study in Political Geography," *Geographical Journal*, 129, 3 (Sept. 1963), 311–328.

————, "The Vietnamese Problem in its Geographical Context," *Geographical Journal*, 131, 4 (Dec. 1965), 502–515.

Franck, Thomas M., "The Stealing of the Sahara," *American Journal of International Law*, 70, 4 (Oct. 1976), 694–721.

Franck, Thomas M. and Paul Hoffman, "The Right of Self-Determination in Very Small Places," *New York Univ. Journal of International Law and Politics*, 8, 3 (Winter 1976), 331–386.

Frank, Andre Gunder, "The Development of Underdevelopment," *Monthly Review*, 41 (1989), 37–51.

————, "Fourteen Ninety-Two Once Again," *Political Geography*, 11, 4 (July 1992), 386–393.

Freeman, Linda, "The Contradictions of Independence: Namibia in Transition," *International Journal*, 46, 4 (Autumn 1991), 687–718.

Furnivall, J. S., "Twilight in Burma: Independence and After," *Pacific Affairs*, 22, 2 (June 1949), 155–172.

**G**

Gann, Lewis Henry, "Portugal, Africa and the Future," *Journal of Modern African Studies*, 13 (March 1975), 1–18.

Gilmore, William C., "Legal Perspectives on Associated Statehood in the Eastern Caribbean," *Virginia Journal of International Law*, 19, 3 (1979), 489–555.

Ginsburg, Norton, "From Colonialism to National Development," *Annals, AAG*, 63, 1 (March 1973), 1–21.

Gordon, Edward, "Resolution of the Bahrain Dispute," *American Journal of International Law*, 65, 3 (July 1971), 560–568.

Gotlieb, Yosef, "Retrieving Life-place from Colonized Space: Transcending the Encumbrances of the Post-Colonial State," *Political Geography*, 11, 5 (Sept. 1992), 461–474.

Gottmann, Jean, "The Political Partitioning of Our World: An Attempt at Analysis," *World Politics*, 4, 4 (July 1952), 512–519.

Gould, Peter R., "Tanzania 1920–1973: The Spatial Impress of the Modernization Process," *World Politics*, 22, 2 (Jan. 1970), 149–170.

Grahl-Madsen, Atle, "Decolonization; the Modern Version of a Just War," in *German Yearbook of International Law*, 22 (1979), 255–273.

**H**

Haas, Anthony, "Independence Movements in the South Pacific," *Pacific Viewpoint*, 11–12 (1970–1971), 97–120.

Hance, William A., "Economic Potentialities of the Central African Federation," *Political Science Quarterly*, 69, 1 (March 1954), 29–44.

Hargreaves, John, "Pan-Africanism After Rhodesia," *World Today*, 22 (Feb. 1966), 57–63.

Henderson, W. O., "Economic Aspects of German Imperial Colonization," *Scottish Geographical Magazine*, 54 (1938), 150–161.

Henry, Michael, "The Anglo-Spanish Dispute over Gibraltar," *Contemporary Review*, 212 (April 1968), 169–173.

Herr, R. A., "A Minor Ornament: The Diplomatic Decisions of Western Samoa at Independence," *Australian Outlook*, 29 (Dec. 1975), 300–314.

Hickey, John, "Keep the Falklands British? The Principle of Self-Determination of Dependent Territories," *Inter-American Economic Affairs*, 31 (Summer 1977), 77–88.

Higgins, Rosalyn, "International Court and South West Africa: The Implications of the Judgment," *International Affairs*, 42, 4 (Oct. 1966), 573–599.

———, "International Law, Rhodesia and the U.N.," *World Today*, 23 (March 1967), 94–106.

Higgott, Richard, "Structural Dependence and Decolonisation in a West African Landlocked State: Niger," *Review of African Political Economy*, 17 (Jan. 1980), 43–58.

"The High Commission Territories; a Remnant of British Africa," *Round Table*, 213 (Dec. 1963), 26–40.

Holdar, Sven, "The Study of Foreign Aid: Unbroken Ground in Geography." *Progress in Human Geography*, 17, 4 (1993), 453–470.

Honey, Rex D. and S. Abu Kharmeh, "Organizing Space for Development and Planning: The Case of Jordan," *Political Geography Quarterly*, 7, 3 (July 1988), 271–287.

Houphouet-Boigny, Felix, "Black Africa and the French Union," *Foreign Affairs*, 35, 4 (July 1957), 593–599.

House, John, "Unfinished Business in the South Atlantic," *Political Geography Quarterly*, 2, 3 (July 1983), 233–246.

Howe, Marvine, "The Birth of the Moroccan Nation," *Middle East Journal*, 10, 1 (1956), 1–16.

Huff, David L. and James M. Lutz, "The Contagion of Political Unrest in Independent Black Africa," *Economic Geography*, 50, 4 (Oct. 1974), 352–367.

**I**

"Independent Qatar," *Middle East Economic Digest*, 15, 10 (Dec. 1971), 1429–1436.

**J**

Johnson, Nuala C., "Nation-Building, Language and Education: The Geography of Teacher Recruitment in Ireland, 1925–55," *Political Geography*, 11, 2 (March 1992), 170–189.

Jones, N.S. Carey, "The Decolonization of the White Highlands of Kenya," *Geographical Journal*, 131, 2 (June 1965), 186–201.

**K**

Kamanu, Onyeonoro S., "Succession and the Right of Self-Determination; an O.A.U. Dilemma," *Journal of Modern African Studies*, 12, 3 (Sept. 1974), 355–376.

Kamiar, Mohammad, "Changes in Spatial and Temporal Patterns of Development in Iran," *Political Geography Quarterly*, 7, 4 (Oct. 1988), 323–338.

Kearns, Kevin C., "Prospects of Sovereignty and Economic Viability for British Honduras," *Professional Geographer*, 21, 2 (Feb. 1969), 97–103.

Khapoya, Vincent B., "Determinants of African Support for African Liberation Movements; a Comparative Analysis," *Journal of Modern African Studies*, 3 (Winter 1976), 469–489.

Kimble, George H.T., "The Federation of Rhodesia and Nyasaland," *Focus*, 6, 7 (March 1965).

Kirgis, Frederic L. Jr., "The Degrees of Self-Determination in the United Nations Era," *American Journal of International Law*, 88, 2 (April 1994), 304–310.

Knight, David B., "Territory and People or People and Territory?: Thoughts on Post-Colonial Self-Determination," *International Political Science Review*, 6 (1985), 248–272.

Kumar, Satish, "The United Nations Committee of 24; Its Origins and Work," *Africa Quarterly*, 5 (Oct.–Dec. 1965), 174–187.

**L**

Laing, Edward A., "Independence and Islands: The Decolonization of the British Caribbean," *New York Univ. Journal of International Law and Politics*, 12, 2 (Fall 1979), 281–312.

Lane, J. E. and S. Ersson, "Unpacking the Political Development Concept," *Political Geography Quarterly*, 8, 2 (April 1989), 123–144.

Laschinger, Michael, "Roads to Independence; the Case of Swaziland," *World Today*, 21 (Nov. 1965), 486–494.

Lateef, Abdul, "Bahrain; Emerging Gulf State," *Pakistan Horizon*, 26, 1 (1973), 10–15.

Latham-Koenig, A. L., "Ruanda-Urundi on the Threshhold of Independence," *World Today*, 18 (July 1962), 288–295.

Leitner, H. and E. S. Sheppard, "Indonesia: Internal Conditions, the Global Economy and Regional Development." *Journal of Geography*, 86 (1987), 282–291.

Lewis, Gordon K., "British Colonialism in the West Indies: The Political Legacy," *Caribbean Studies*, 7 (April 1967), 3–22.

Lewis, I. M., "Pan-Africanism and Pan-Somalism," *Journal of Modern African Studies*, 1 (1963), 147–162.

Lissitzyn, Oliver J., "International Law and the Advisory Opinion on Namibia," *Columbia Journal of Transnational Law*, 11 (Winter 1972), 50–73.

Lowenthal, David and Colin G. Clarke, "Island Orphans: Barbuda and the Rest, *Journal of Commonwealth and Comparative Politics*, 18, 3 (Dec. 1980), 293–307.

**M**

MacDonald, Barrie, "Secession in Defense of Identity: The Making of Tuvalu," *Pacific Viewpoint*, 16 (1975), 26–44.

Machel, Samora, "The People's Republic of Mozambique; the Struggle Continues," *Review of African Political Economy*, 4 (Nov. 1975), 14–25.

MacLaughlin, Jim, "The Political Geography of 'Nation-Building' and Nationalism in Social Sciences: Structural vs. Dialectical Accounts," *Political Geography Quarterly*, 5, 4 (Oct. 1986), 299–329.

Manning, Edward W., "Sustainable Development, The Challenge" (Presidential Address), *Canadian Geographer*, 34 (Winter 1990), 290–302.

Marston, Geoffrey, "Termination of Trusteeship," *International and Comparative Law Quarterly*, 18, 1 (Jan. 1969), 1–40.

Mazrui, Ali A., "Consent, Colonialism, and Sovereignty," *Political Studies*, 11 (Feb. 1963), 36–55.

———, "Africa's Experience in Nation-Building; Is It Relevant to Papua and New Guinea?" *East Africa Journal*, 7 (Nov. 1970), 15–23.

McDougal, Myres S. and W. Michael Reisman, "Rhodesia and the United Nations; the Lawfulness of International Concern," *American Journal of International Law*, 62, 1 (Jan. 1968), 1–19.

McGee, T. G., "Aspects of the Political Geography of Southeast Asia," *Pacific Viewpoint*, 1, 1 (March 1960), 39–58.

McHenry, D. E., "The United Nations Trusteeship System," *World Affairs Interpreter*, 19 (July 1948), 149–158.

Meinig, Donald W., "The American Colonial Era: A Geographic Commentary," *Proceedings, Royal Geographical Society of Australia, South Australia Branch*, 59 (1957–1958), 1–22.

Melamid, Alexander, "Political Geography of Trucial 'Oman and Qatar,'" *Geographical Review*, 43, 2 (April 1953), 194–206.

Mezerik, A. G. (ed.), "Colonialism and the United Nations," *International Review Service*, 10, 83 (1964), 1–105.

Mihaly, Eugene B., "Tremors in the Western Pacific; Micronesian Freedom and U.S. Security," *Foreign Affairs*, 52, 4 (July 1974), 839–849.

Miller, Jake C., "The Virgin Islands and the United States; Definition of a Relationship," *World Affairs* (Spring 1974), 297–305.

Mingst, Karen A., "The Ivory Coast at the Semi-Periphery of the World-Economy," *International Studies Quarterly*, 32, 3 (Sept. 1988), 259–274.

Mink, Patsy T., "Micronesia; Our Bungled Trust," *Texas International Law Forum*, 6 (Winter 1971), 181.

Mittelman, James H., "Collective Decolonisation and the U.N. Committee of 24," *Journal of Modern African Studies*, 14 (March 1976), 41–64.

Monroe, Elizabeth, "Kuwait and Aden: A Contrast in British Policies," *Middle East Journal*, 18, 1 (Winter 1964), 63–74.

Mushkat, Marion, "The Process of African Decolonization," *Indian Journal of International Law*, 6 (Oct. 1967), 483–508.

**N**

Nijman, Jan, "The VOC and the Expansion of the World-System 1602–1799," *Political Geography*, 13, 3 (May 1994), 211–227.

"Niue Approaches Self-Government," *New Zealand Foreign Affairs Review*, 23 (March 1973), 3–11.

Norman, H. E., "The Genyosha: A Study in the Origins of Japanese Imperialism," *Pacific Affairs*, 17, 3 (1944), 261–284.

Nyerere, Julius K., "Rhodesia in the Context of Southern Africa," *Foreign Affairs*, 44, 3 (April 1966), 373–386.

**O**

Owen, R. P., "The Rebellion in Dhofar—a Threat to Western Interests in the Gulf," *World Today*, 29, 6 (June 1973), 266–272.

Owiti Ayaga, Odeyo, "United Nations and African Decolonization; Ineffectiveness of the Security Council," *Indian Journal of International Law*, 12 (Oct. 1972), 581–605.

**P**

Pal Singh, Surendra, "India and the Liberation of Southern Africa," *Africa Quarterly*, 10 (April–June 1970), 4–8.

Papaioannou, Ezekias, "Neo-Colonialism and Developing Countries," *World Marxist Review*, 14 (May 1971), 86–96.

Pélissier, René, "Spain's Discreet Decolonization," *Foreign Affairs*, 43, 3 (April 1965), 519–527.

Phillips, Dion E., "A New 'Special Relationship' for the US Virgin Islands?," *Transafrica Forum*, 5, 2 (Winter 1988), 39–54.

Pick, H., "Independent Mauritania," *World Today*, 17 (April 1961), 149–158.

Pillai, R. V. and Mahendra Kumar, "The Political and Legal Status of Kuwait," *International and Comparative Law Quarterly*, 11, 1 (Jan. 1962), 108–130.

Plishke, Elmer, "Self-Determination: Reflections of a Legacy," *World Affairs*, 140, 1 (Summer 1977), 41–57.

Ponoma, Jean Baptiste, "The People of Reunion Want Freedom," *World Marxist Review*, 21 (July 1978), 80–83.

Potholm, Christian P., "The Protectorates, the O.A.U. and South Africa," *International Journal*, 22, 1 (Winter 1966–67), 68–72.

Purcell, D. C., Jr., "Economics of Exploitation: The Japanese in the Mariana, Caroline and Marshall Islands, 1915–1940," *Journal of Pacific History*, 11, 3–4 (1976), 189–211.

**R**

Ramazani, Rouhollah K., "The Settlement of the Bahrain Dispute," *Indian Journal of International Law*, 12 (1972), 1–14.

Redclift, Michael, "The Multiple Dimensions of Sustainable Development," *Geography*, 76 (Jan. 1991), 36–42.

Reitsma, Hendrik-Jan A., "Development Geography, Dependency Relations, and the Capitalist Scapegoat," *Professional Geographer*, 34, 2 (May 1982), 125–130.

——, "Geography and Dependency: A Rejoinder," *Professional Geographer*, 34, 3 (August 1982), 337–342.

Richardson, Henry J., III, "Constitutive Questions in the Negotiations for Namibian Independence," *American Journal of International Law*, 78, 1 (Jan. 1984), 76–120.

Riedel, Eibe H., "Confrontation in Western Sahara in the Light of the Advisory Opinion of the International Court of Justice of 16 Oct. 1975; a Critical Appraisal," in *German Yearbook of International Law*, 19 (1976), 405–422.

Robinson, G.W.S., "Ceuta and Melilla: Spain's Plazas de Soberanía," *Geography*, 43 (1958), 266–269.

Rothschild, Donald S., "The Politics of African Separatism," *Journal of International Affairs*, 1 (1961), 18–28.

——, "Rhodesian Rebellion and African Response," *Africa Quarterly*, 6 (Oct.–Dec. 1966), 184–196.

Roucek, J. S., "Geopolitics of Portuguese Colonialism," *Contemporary Review*, 205 (June 1964), 284–296.

Rudebeck, Lars, "Political Mobilisation for Development in Guinea-Bissau," *Journal of Modern African Studies*, 10 (May 1972), 1–18.

**S**

Sady, E. J., "Department of the Interior and Pacific Island Administration," *Public Administration Review*, 10, 1 (1950), 13–19.

Sayre, Francis, B., "Legal Problems Arising from the United Nations Trusteeship System," *American Journal of International Law*, 42 (April 1948), 263–298.

Scarritt, James R., "Review Essay: Explaining the State and State-Society Relations in Africa," *Political Geography Quarterly*, 6, 4 (Oct. 1987), 369–376.

Schurz, Carl, "Manifest Destiny," *Harpers Monthly*, 87 (Oct. 1893), 737–746.

Schweinfurth, Ulrich, "The Future of the Small Islands in the Pacific," *Aussenpolitik*, 28 (1977), 343–356.

Shaudys, Vincent K., "The External Political Geography of Dutch New Guinea," *Oriental Geography*, 5, 2 (1961), 145–160.

Shaw, Malcolm, "The Western Sahara Case," in *British Yearbook of International Law*, 49 (1978), 119–154.

Shaw, Timothy M., "Beyond Any New World Order: The South in the 21st Century." *Third World Quarterly*, 15, 1 (1994), 139–146.

Shilling, Nancy A., "Problems of Political Development in a Ministate: The French Territory of the Afars and the Issas," *Journal of Developing Areas*, 7 (July 1973), 613–634.

Sidaway, James Derrick, "Mozambique: Destabilization, State, Society and Space," *Political Geography*, 11, 3 (May 1992), 239258.

Sidell, Scott, "The United States and Genocide in East Timor," *Journal of Contemporary Asia*, 11, 1 (1981), 44–61.

Simon, D., "Colonial Cities, Post-Colonial Africa and the World Economy: A Reinterpretation," *International Journal of Urban and Regional Research*, 13 (1989), 68–91.

Singh, L. P., "The Commonwealth and the United Nations Trusteeship of Non-Self-Governing Peoples," *International Studies*, 5 (Jan. 1964), 296–303.

Skurnik, W.A.E., "France and Fragmentation in West Africa: 1945–1960," *Journal of African History*, 8, 2 (1967), 317–333.

Slonim, S., "The Origins of the South West Africa Dispute; the Versailles Peace Conference and the Creation of the Mandates System," in *Canadian Yearbook of International Law*, 6 (1968), 115–143.

Smith, Neil, "Theories of Underdevelopment: A Response to Reitsma," *Professional Geographer*, 34, 3 (April 1982), 332–337.

Smith, Roy H. and Michael C. Pugh, "Micronesian Trust Territories; Imperialism Continues?" *Pacific Review*, 4, 1 (1991), 36–44.

Smogorzewski, K. M., "The Russification of the Baltic States," *World Affairs*, 4, 4 (Oct. 1950), 468–481.

Solomon, Neal S., "The Guam Constitutional Convention of 1977," *Virginia Journal of International Law*, 19, 4 (Summer 1979), 725–806.

Somerville, J.J.B., "The Central African Federation," *International Affairs*, 39, 3 (July 1963), 386–402.

Spackman, Ann, "Constitutional Development in Trinidad and Tobago," *Social and Economic Studies*, 14 (Dec. 1965), 283–320.

Stephen, Michael, "Natural Justice at the United Nations: The Rhodesia Case," *American Journal of International Law*, 67, 3 (July 1973), 479–490.

Stewart, Douglas B., "Economic Growth and the Defense Burden in Africa and Latin America," *Economic Development and Cultural Change*, 40 (Oct. 1991), 189–207.

Stokes, E., "Great Britain and Africa: The Myth of Imperialism," *History Today*, 10 (August 1960), 554–563.

Strang, David, "From Dependency to Sovereignty: An Event History Analysis of Decolonization 1870-1987," *American Sociological Review*, 55 (Dec. 1990), 846–860.

———, "Global Patterns of Decolonization, 1500–1987," *International Studies Quarterly*, 35, 4 (Dec. 1991), 429–454.

*Sub-Saharan Africa; Dilemmas in Political and Economic Development*. Special issue of *Journal of International Affairs*, 46, 1 (Summer 1992), whole volume.

Suret-Canale, J., "The French Community at the Hour of African Independence," *International Affairs*, 7 (Jan. 1961), 32–37; 7 (Feb. 1961), 23–30.

**T**

Tangri, R. K., "Rise of Nationalism in Colonial Africa: The Case of Colonial Malawi," *Comparative Studies in Society and History*, 10 (Jan. 1968), 142–161.

Tanner, R.E.S., "The Belgian and British Administrations in Ruanda-Urundi and Tanganyika," *Journal of Local Administration Overseas* (1965), 202–211.

Tata, Robert J., "Poor and Small Too: Caribbean Mini-States," *Focus*, 29, 2 (Nov.–Dec. 1978), 1–11.

*Third World Quarterly; Journal of Emerging Area*. Published in England by Carfax since 1979.

Tomasek, Robert D., "British Guiana: A Case Study of British Colonial Policy," *Political Science Quarterly*, 74, 3 (Sept. 1959), 393–411.

Tregonning, K. G., "The Partition of Brunei," *Journal of Tropical Geography*, 11 (April 1958), 84–89.

Twitchett, Kenneth J., "Colonialism: An Attempt at Understanding Imperial, Colonial and Neo-Colonial Relationships," *Political Studies*, 13 (Oct. 1965), 300–323.

**U**

Umozurike, U. O., "International Law and Self-Determination in Namibia," *Journal of Modern African Studies*, 8 (Dec. 1970), 585–603.

———, "International Law and Colonialism in Africa: A Critique," *Zambia Law Journal*, 3/4, 1/2 (1971/1972), 95–124.

"The United Nations, Self-Determination and the Namibia Opinions," *Yale Law Journal*, 82 (Jan. 1973), 533–558.

**V**

Vance, Robert T., Jr., "Recognition as an Affirmative Step in the Decolonization Process: The Case of Western Sahara," *Yale Journal of World Public Order*, 7, 1 (Fall 1980), 45–87.

Van Wyk, J. T., "The United Nations, South West Africa and the Law," *Comparative and International Law Journal (Southern Africa)*, 2 (1969), 48–72.

Velazquez, C. M., "Some Legal Aspects of the Colonial Problem in Latin America," *Annals, Am. Acad. Polit. and Soc. Sci.*, 360 (July 1965), 110–119.

Venkatavaradan, T., "The Question of Southern Rhodesia," *Indian Yearbook of International Affairs*, 13 (1964), 112–150.

Viviani, Nancy, "Australians and the Timor Issue," *Australian Outlook*, 30 (August 1976), 197–226.

Vogel, Jerome, "Culture, Politics and National Identity in Côte d'lvoire," *Social Research*, 58 (Summer 1991), 439–456.

Von Vorys, Karl, "New Nations: The Problem of Political Development," *Annals, Am. Acad. Polit. and Soc. Sci.*, 358 (1965), 1–179.

**W**

Watt, D. C., "The Decision to Withdraw from the Gulf," *Political Quarterly*, 39 (July 1968), 310–321.

Weiner, Daniel, "Socialist Transition in the Capitalist Periphery: A Case Study of Agriculture in Zimbabwe," *Political Geography Quarterly*, 10, 1 (Jan. 1991), 54–75.

Welensky, Roy, "The United Nations and Colonialism in Africa," *Annals, Am. Acad. of Polit. and Soc. Sci.*, 354 (July 1964), 145–152.

Wellings, Paul, "The 'Relative Autonomy' of the Basotho State: Internal and External Determinants of Lesotho's Political Control," *Political Geography Quarterly*, 4, 3 (July 1985), 191–218.

Whitaker, Paul M., "The Revolutions of 'Portuguese' Africa," *Journal of Modern African Studies*, 8 (April 1970), 15–35.

Whittlesey, Derwent, "Southern Rhodesia—an African Compage," *Annals, AAG*, 46, 1 (March 1956), 1–97.

**Y**

Young, Richard, "The State of Syria: Old or New?" *American Journal of International Law*, 56, 2 (April 1962), 482–488.

**Z**

Zoli, C., "The Organization of Italy's East African Empire," *Foreign Affairs*, 16, 1 (Oct. 1937), 80–90.

# PART FIVE

## *Geopolitics*

Carrier task force, controlling sea lanes and projecting power

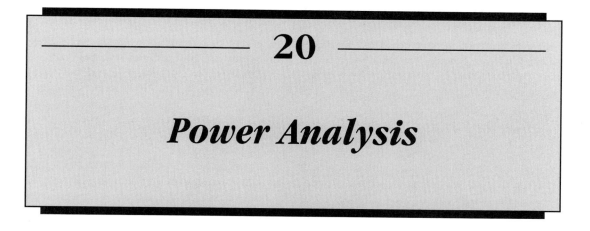

# 20

# *Power Analysis*

Most of the problems of indigenous peoples and of imperialism, discussed in the last four chapters, were generated by policies of governments influenced in part by geopolitical notions. Imperialism, and even geostrategic concepts, as we have seen, have not been limited to large, rich, and powerful countries. Now we examine these concepts in greater detail, beginning with the popular, but little understood, notion of "power."

Three generations of political geographers, political scientists, military intelligence specialists, foreign ministry officials, and others have devoted countless hours and immeasurable energy to the definition, quantification, and analysis of the "power" of many, if not all, States of the world. Not only has it been a popular exercise, even for graduate students, but it has been accepted as a required one in the field of political geography. Indeed, some definitions of "State" even include "power" as one of the requirements. There might have been some value in the ranking of the dozen or so most important States in the world prior to about 1950, and it is still undoubtedly essential for military and civilian leaders of a State to have a reasonably accurate picture of the strengths and weaknesses of potential enemies and allies, but the exercise known as the power analysis or power inventory, as generally practiced, seems of little value beyond the classroom. So many things have changed since World War II, and the pace of change is accelerating so rapidly, that we find it difficult to keep up with the currently significant elements of power in the first place and then to quantify them, assign them realistic values, apply them to individual countries, and rank the countries.

It is quite obvious to anyone who reads even a hometown newspaper that there were until recently two superpowers in the world, with the United States significantly more powerful than the USSR. Below them are perhaps six or eight countries that could bring some power to bear outside their own immediate regions on a sustained basis. Below this group are the remaining 180 or so countries whose ranking would be anyone's guess. No computer could possibly produce a ranking acceptable to the majority of observers as both accurate and useful. Even if it were possible to do so, the ranking would quickly go out of date and have to be recalculated.

For these reasons our discussion of the topic is confined to a survey of the indicators that should be included in any power inventory and to some comments on the difficulty of analyzing them in any meaningful way.* The following topics are by no means exhaustive; they are suggestive only. The organization of the factors is partly arbitrary,

*Readers who are interested in quantification of these indicators are referred to two basic tools: The United Nations *Statistical Yearbook* and the *World Handbook of Political and Social Indicators*, 3rd ed., 2 vols., Charles Lewis Taylor and David A. Judice, New Haven and London, Yale University Press, 1983.

and they are not necessarily listed in order of importance. Finally, we must stress that many factors overlap and most are interrelated, if not interdependent.

## *Territory*

We have already discussed the importance of size and shape of a State and concluded that there is no ideal size or shape. Nevertheless, in measuring the power of an individual State, these factors can be very significant. All other things being equal, reasonably large size affords space for maneuver, defense in depth, variety of resources, and other advantages; a small ratio of perimeter to area makes internal communications and external defense easier than a fragmented or prorupt shape. But a compact shape can leave valuable resources nearby but still outside a State's boundaries, and a large terri-

tory may be vulnerable to attack from different directions and difficult to administer.

"Strategic" or "favorable" location is often cited as an asset to a State, and is sometimes cited by determinists as explaining a particular country's success. We discuss this in more detail later in these chapters on geopolitics, but for now we can point out that since technology, economies, politics, and strategies are constantly changing, so are the values of particular locations. It does make a difference to a State whether it has one seacoast or two or none. Buffer States can benefit from the intermediary role in trade, transportation, and communications, but are also subject to buffeting from both sides. States located at the western edge of the Pacific Ocean, the eastern edge, or the middle clearly derive both advantages and handicaps from these locations. Who ever heard of Diego Garcia 30 years ago? Wasn't Korea

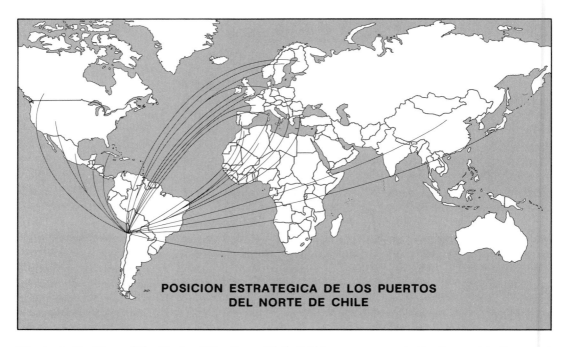

**POSICION ESTRATEGICA DE LOS PUERTOS DEL NORTE DE CHILE**

**"Strategic Position of the Ports of Northern Chile."** This map appears in the Forword to the report *Antofagasta and the North of Chile, Bridge of Exit and Entry of the Central West South American Hinterland, in the Sphere of Action of the G.E.I.C.O.S.* (Interregional Business Group of the South American Central West, a private organization of Chilean, Argentine, Paraguayan, and Bolivian businesspeople trying to develop a new transit corridor through Antofagasta.) It is an example of both the kind of boosterism and of propaganda maps discussed in Chapter 3, and the folly of assigning "strategic" value to any particular place without very sound reasons.

"The Hermit Kingdom" in the past century? Wasn't Valparaíso a major port with a strategic location before the Panama Canal was opened? When the Suez Canal was closed from 1967 to 1974, eastern and southern Africa assumed a new importance. But who now knows or cares much about Château Thierry, Tarawa, Kasserine Pass, Galipoli, Pork Chop Hill, or the A Shau Valley—battlefields where hundreds of thousands died not long ago because of their "strategic importance"? Who can predict which State will occupy a "strategic position" tomorrow?

The "natural resources" of a State are also considered valuable and quantifiable, so they tend to rank high in everyone's power inventory. But what is a "natural resource"? Forty years ago titanium was used primarily as an ingredient in some paints; today mine waste piles are being searched for this metal, which is essential for the skins of high-performance supersonic aircraft. Wars were fought over natural nitrates only a century ago, yet few people today even know what they are. Natural resources would have to include soils, water, forests, and many other elements of the physical environment that are difficult to evaluate. Where in the State's territory are the resources located—out near a hostile border, deep in the inaccessible interior, in the frigid and empty northland? Which resources are currently being utilized, which are known but inaccessible, which suspected but not proven, which too costly to process? What about resources that are needed but totally absent in the national territory? Can they be obtained from friends? At what cost? Even in times of stress? What is the value of resources in the colonies or of investments in the resources of other States? How do we factor in the roles of substitutes, synthetics, conservation, reuse? How do we evaluate resources in a country that squanders them on cosmetics and pet food while another uses them for tools and weapons?

## Population

In any power analysis it is quite reasonable to assume that a country with a large population has a distinct advantage over a country with a small population—all other things being equal. But the fact that all other things are not equal makes it difficult to evaluate population as a power factor. Although sheer numbers are important to some extent in every country, their importance may be enhanced or diminished by such factors as their distribution over the national territory and demographic trends. Qualitative factors may be even more important than quantitative ones. The age/sex ratio, for example, affects the pool of potential military personnel and the proportions of producers and consumers in the society. Immigration and emigration, education levels, health factors, the distribution of wealth, and degree of poverty—all must be considered and evaluated.

Cultural and "racial" homogeneity or heterogeneity may strengthen or weaken a society depending on the peoples involved, their attitudes toward one another, the role of government in fostering integration or segregation, whether the various groups complement or compete with one another. Social organization can be very important, including such things as the degree of socioeconomic stratification and social mobility in the society, the land tenure systems and distribution of good agricultural land, and the status and role of elites. Religious and cultural values are intangibles that can bear heavily on a State's ability to mobilize its strength to pursue some policy. The mobilization might be ideologically motivated: patriotism, communism, democracy, or some other cause might be considered worth working or dying for—or might not.

## Government

The political system of a country invariably figures in its power, but invariably it is also difficult to quantify. Does the government effectively control the whole country; is the ecumene coextensive with the national territory? Is the government popular, likely to be supported and rallied around in times of stress? Or is there widespread opposition either to the government in power or to the whole political system? Are the officials trained and experienced, efficient and honest, or is the government weakened by corruption, nepotism, bloated payrolls, and inefficiency? Is the government capable of providing wise, effective, and inspiring leadership?

### Economy

It is self-evident that wealth in our world is a form of power or can be translated readily into power, so wealthy countries, like popu-

# TRADITIONAL SOCIO-POLITICAL STRUCTURE IN LATIN AMERICA

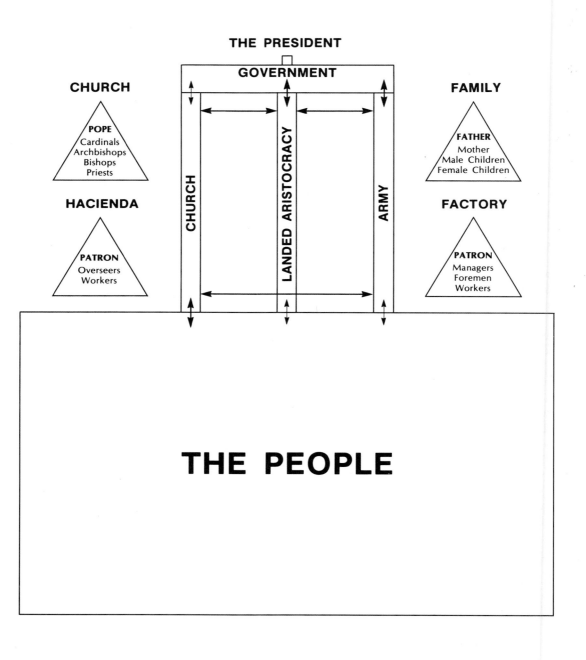

lous countries or those richly endowed by nature, have an automatic advantage over those less fortunate. But wealth, even material wealth, consists of more than just "natural resources" or even GNP (gross national product). If a State is to be powerful, it must have productive capacity, not only to meet its needs, but also to supply surpluses to sell abroad in peacetime and to meet expanded needs in wartime. Science and technology must be well developed, and research must receive adequate investment and other support. Industry must have adequate supplies of efficient, skilled, and dedicated managers and workers. Capital must move readily from where it is to where it is needed, and agriculture must get a fair share. Most societies rest on an agricultural base (no matter how marginal), and in times of stress a country may be cut off from agricultural imports completely. Foreign trade is both a strength and a weakness, and many countries have adopted autarkic policies in order to capitalize on its strength and avoid its weakness. Autarky has no place in the twentieth century, however; all countries are and must be interdependent if they are to prosper.

Banking and insurance systems are vital in any modern economy, yet they are seldom included in power inventories. How can we measure their quality or value in a power system? The question of quality runs through any useful economic analysis, for the quality of goods produced and services performed may be more important than the quantity.

*Nat'l economic self-sufficiency independence of imports*

## Circulation

Transportation and communications—the movement of goods, people, and ideas—are fundamental to any modern society and to any modern State. Without a good circulation system to tie together all sections of the country, all elements of the population, all units of government, all sectors of the economy, and all of them with one another, a State would simply disintegrate. It would project not power but weakness. It must have not only an adequate physical infrastructure, but also efficient management and workers, adequate capitalization, and proper maintenance. This includes all forms of transportation plus radio, telephone, telegraph, and postal systems; a free and vigorous press; and other means of communication. These can be misused, of course, to persuade as well as inform, for mind control as well as mind enrichment. People must be wary of this, especially in times of stress, when they tend to relax their defenses against internal threats in order to concentrate on external ones.

## Military Strength

All these factors are elements of, or can be translated into, military strength. Every conventional power inventory counts every piece of military hardware and every person in military uniform it can identify in a State and assigns values to the totals. But is this enough? Can we say that a country with a

This is a schematic representation of the traditional organization of society in Latin America. In some countries of the region it still exists in very much this form; in others it has disappeared entirely; in most it exists but is much modified. The Church, the landed aristocracy, and the army form an interlocking triumvirate controlling the entire structure. They exploit and repress the masses of the people while supporting and providing leadership for the government. At the apex of this pyramidal structure is the president (or a junta), a symbol of authority whether exercising real power or not. The size of the arrows represents the relative strength of the reciprocal links among the various sectors of society. The relatively rigid pyramidal structure is the product of similar structures in all other important aspects of life, symbolized by the triangles in the upper corners. Even the major pre-Columbian civilizations of the region, of which many of the people are heirs, were hierarchical. Such a structure is inherently conservative, resistant to change. It has hindered economic, social, and political progress in the region, and where it has not been displaced or greatly modified, it must be considered a negative factor in any power analysis.

big, well-equipped army is more powerful than one with a small, poorly equipped one? Of course not. In order to evaluate the effectiveness of a military force, we must reckon with many intangibles. These would include the education and skill of the troops; the quality of their leadership; their motivation, morale, loyalty; their combat experience; the logistical system; and even whether they are up-to-date in military doctrine, strategy, and tactics. (It is no good to be, as was said of the French army, always superbly ready to fight the last war.) The country must have an effective intelligence service and an efficient system for mobilizing reserves.

A country's military strength can be enhanced by having distant bases in strategic places—unless the cost of supplying and protecting them nullifies their value. Its strength can also be enhanced through collective security, having allies pledged to aid one another ("all for one and one for all"), if they all have something tangible to contribute and will remain steadfast in the face of adversity or temptation. How can we measure the potential reliability of an ally under a variety of real or hypothetical circumstances?*

## Foreign Relations

Today no State can remain isolated and be either strong or prosperous. States that have chosen the route of virtual isolation, such as Albania and Myanmar most recently, must rely on their own human and natural resources. This may be good for the soul, but not for the mind or the belly. Today every State, every people, is somehow dependent

on every other one. Even those that have chosen not to join military alliances must still be involved with others. Broadly speaking, it seems that those countries that participate most actively in world affairs—political, economic, and cultural—have the best opportunities to maximize their assets and overcome their handicaps and thus increase their national "power." International trade, cultural links (even with the former "mother country"), participation in transfers of development funds, participation in the work of the United Nations and other world and regional organizations all elevate a country, make it a more valued member of the international community, and perhaps render it less likely to be attacked.

## Analysis

What conclusion can we draw? Only that "power" is exceedingly difficult to define and to measure. After considering the superpowers and perhaps three or four other countries, it becomes almost impossible to evaluate power. Some recent examples may illustrate the point. How do we rank Israel, for example? A country of only 5 million people in a land the size of Vermont with few natural resources and surrounded by hostile forces with overwhelming strength, it has nevertheless won five wars in less than 30 years. Its contributions to the world in science, agriculture, technology, medicine, scholarship, music, and other fields would be impressive in a country many times its size. Its air and shipping lines, banks, development aid programs, educational and cultural exchanges, importers and exporters operate around the world, and it has even been able to project its military power as far away as Uganda, Iraq, and Tunisia.

And how do we rank Cuba, still poor but with a sizable army, parts of which have operated successfully across Africa and in the Arabian peninsula? Or Vietnam, poorer still but strong enough to conquer Cambodia and humiliate China? Or India, which, though still poor, has exploded a nuclear device and has kept its army very busy indeed

---

*These points are clearly illustrated by a classic anecdote reported by U.S. Army Colonel Harry G. Summers: "When the Nixon Administration took over in 1969 all the data on North Vietnam and the United States was fed into a Pentagon computer—population, gross national product, manufacturing capability, number of tanks, ships and aircraft, size of the armed forces, and the like. The computer was then asked, 'When will we win?' It took only a moment to give the answer: '*You won in 1964.*'"

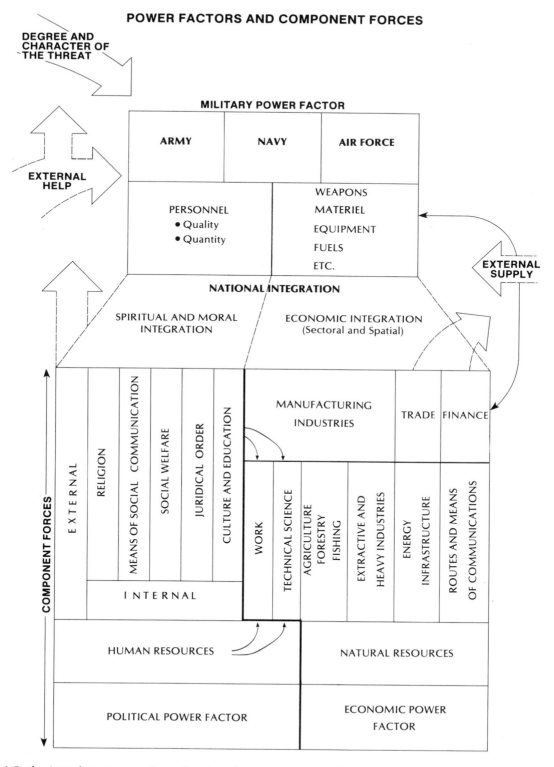

**A Latin American conception of national power analysis.** This diagram, by the retired Argentine general Juan E. Guglialmelli, represents one way of organizing one's thinking about national power. Note that these factors are not weighted in any way. (Translated by Martin Glassner.)

since 1947? Iraq before the Gulf War of early 1991 was considered by some experts to have the world's fourth largest army, equipped with modern and even advanced weaponry, well trained and highly motivated. Yet it was forced to withdraw from Kuwait and then shattered in only five days by an expeditionary force consisting of only fractions of the military forces of its coalition adversaries.

Pakistan was perhaps weakened by its civil war and the breakaway of East Bengal in 1971 to become the new State of Bangladesh, but by how much? Are Pakistan and Bangladesh together equal to the Pakistan of 1970? Was Nigeria weakened or strengthened by the unsuccessful secession effort of Biafra? Iran spent billions of dollars on the most modern conventional weapons available, perhaps in an attempt to recreate the empire of Darius, but none of that helped the Shah to retain power; indeed, it may have cost him his throne, and his successors, using much of that weaponry, were unable to defeat the invaders from neighboring Iraq.

Then there are the States with negligible military power but considerable economic power. Japan, of course, is the prime example of an economic superpower. But it has serious weaknesses in its social and industrial structures and is utterly dependent on the importation of commodities and the export of manufactured goods for its economic health—and perhaps its survival. Is Japan a giant with feet of clay? Even Switzerland is something of an economic power, especially if we factor in its banking system. Is the united Germany of today more or less powerful than the former East and West Germany together? And is Germany more or less powerful—now or potentially—as a result of its participation in the European Union and other supranational organizations than it would be standing alone? The Organization of the Petroleum Exporting Countries has proven the efficacy of concerted action by States that individually have little power. In 1973 to 1974, this cartel, including such countries as Gabon, Ecuador, and the United Arab Emirates, was able to shake the economies and threaten the safety of the United States and Western Europe by withholding exports of oil for a time and quadrupling its price when exports resumed. How then do we measure the "power" of Gabon, Ecuador, and the United Arab Emirates?

Finally, who in 1980 or even 1988 predicted the sudden, rapid, and total disintegration of the Soviet Union, at the time the second most powerful country in the world by any reckoning? Where is all that power now? Where would Russia now rank in a power inventory? And are the former Soviet satellites in Eastern and Central Europe more or less powerful—individually or collectively—now than they were in the heyday of COMECON and the Warsaw Pact?

Perhaps it would be best to cease measuring and ranking "power" and try to measure influence, the ability of a State to influence the behavior of other States. This might be less precise and less geographic, but it would be more rewarding. Nevertheless, the exercise would still have to be repeated periodically to take into account new and altered circumstances. Influence, like power, is constantly changing and quite unpredictable.

### *Ozymandias*

I met a traveler from an antique land
Who said: Two vast and trunkless legs of stone
Stand in the desert. Near them, on the sand,
Half sunk, a shattered visage lies, whose frown,
And wrinkled lip, and sneer of cold command,
Tell that its sculptor well those passions read
Which yet survive, stamped on these lifeless things,
The hand that mocked them and the heart that fed;
And on the pedestal these words appear:
"My name is Ozymandias, king of kings:
Look on my works, ye Mighty, and despair!"
Nothing beside remains. Round the decay
Of that colossal wreck, boundless and bare
The lone and level sands stretch far away.

Percy Bysshe Shelley

# 21

# *Historical Concepts in Geopolitics*

In Part One we discussed the nature of political geography and some of its aspects. Unfortunately, many people, including some geographers, confuse political geography with geopolitics. Geopolitics, however, is only one of the subjects studied by political geographers. It is concerned basically with the study of States in the context of global spatial phenomena, in an attempt to understand both the bases of State power and the nature of States' interactions with one another. Before the Second World War generally and in some countries since then, geopolitics was considered to be the application of geographic information and geographic perspectives to the development of a State's foreign policies. It was called, with some justification, "applied political geography." This concept, however, was distorted during the interwar period by German geographers who twisted some of the basic ideas of geopolitics into the pseudoscience of "*Geopolitik*," the chauvinist, aggressive, and antidemocratic version of geopolitics that led to so much physical and intellectual destruction before and during World War II.

Geopolitics evolved toward the end of the nineteenth century as new developments in science and technology led people to take a broader view of the world than they had previously. The consolidation of the modern State system with the unification of Germany and Italy, the apogee of European imperialist expansion, the appearance of Japan and the United States as new imperialist powers on the fringes of Europe's sphere of interest, rapid population growth and pressures on resources, and differential development all took place in this period and contributed to the new perspectives of scholars and policymakers. Out of this ferment of new thinking (at least new in modern times) came two streams of thought that were geopolitical in nature. One of them emerged from the Social Darwinism fashionable in the period; this was the *organic State theory*. The other was based more on geographic facts and the policies that should be influenced by them; this is often called *geostrategy*. Neither term is entirely satisfactory, as we shall see, but they can help us separate in our minds two very different sets of ideas that were developing simultaneously.

## Organic State Theory

*Friedrich Ratzel* (1844–1904) was a distinguished and prolific scholar, a geographer trained originally in biology, chemistry, and other sciences. He was greatly influenced both by Darwin's discoveries and by Social Darwinism. In his writings, particularly his classic *Politische Geographie* (1896), he used similes and metaphors from biology in his analysis of political science and geography, comparing the State with an organism. We can summarize Ratzel's theory of an organic State in this manner:

The State is land, with man on the land, linked by the State idea and conforming to natural laws, with development tied to the natural environment. Therefore, for example, States, like plants and people, do not do well in desert or polar regions. States need food in the form of *Lebensraum* (living space) and resources, and they constantly compete for them. States, like organisms, must grow or die. They live through stages of youth, maturity, and old age, with possible rejuvenation. The vitality of a State can generally be gauged by its size at a given time. In 1896 Ratzel produced what he called the seven laws of State growth. They are as follows:

1. The space of States grows with the expansion of the population having the same culture.
2. Territorial growth follows other aspects of development.
3. A State grows by absorbing smaller units.
4. The frontier is the peripheral organ of the State that reflects the strength and growth of the State; hence, it is not permanent.
5. States in the course of their growth seek to absorb politically valuable territory.
6. The impetus for growth comes to a primitive State from a more highly developed civilization.
7. The trend toward territorial growth is contagious and increases in the process of transmission.

There is a great deal more to the theory, but even from this sample we can detect that Ratzel had a very deterministic view of the world. Nevertheless, he was a careful scientist, emphasized that his description was only based on an analogy to an organism, and did consider the interrelationships between people and their environment in both directions. He took the position of the detached observer, making no policy recommendations. His American disciple, Ellen Churchill Semple, wrote in the same spirit in her work. Not all of Ratzel's followers were

so careful, however, and some ignored his cautions.

Like Ratzel, *Rudolf Kjellén* (1864–1922) was a university professor, Ratzel at Leipzig, Kjellén at Uppsala. Kjellén, however, was a political scientist and a member of the Swedish parliament. He was a Gemanophile, impressed with the new work in natural science and especially imbued with Ratzel's work in political geography. Unlike Ratzel, however, Kjellén was not a careful scientist. Instead he took Ratzel's analogies literally and insisted flatly that the State *is* an organism. He even titled his most important book *Staten som Lifsform* (*The State as an Organism*, 1916). Here he presented his theory that the State was composed of five organs:

1. *Kratopolitik*—government structure
2. *Demopolitik*—population structure
3. *Sociopolitik*—social structure
4. *Oekopolitik*—economic structure
5. *Geopolitik*—physical structure

Kjellén introduced aspects of the quality of the population, the nation whose aggregate constitutes the body of the State. In addition to moral capacities, there is the will, the cumulative psychological force of the State. The great power of the State is a dynamic, psychological concept. Kjellén saw the State in a condition of constant competition with other States; larger ones would extend their power over smaller ones, and ultimately the world would have only a few very large and extremely powerful States. He envisioned in Europe a superstate controlled by Germany.

This book, in which Kjellén originated the terms *Geopolitik* and *Autarky*, was translated into German in 1917, when the war was already going badly for Germany. At the end of the war, it was seized upon by some German political scientists, geographers, and nationalists who used it to lend the authority of the new evolutionary natural science to the older German political philosophy. It became a tool for rebuilding Germany into a world power and was subsequently used in the same way by Italy and Japan.

## *Geostrategy*

Meanwhile, other scholars were focusing not on the State but on the world and trying to find patterns in State development and behavior. They took a global view of geopolitical affairs and actually recommended policies or strategies to be followed by their governments. The first was *Alfred Thayer Mahan* (1840–1914). Mahan was a naval historian who eventually reached the rank of admiral in the U.S. Navy. He was a prolific writer, producing some 20 books altogether, among them *The Life of Nelson* (1897). His most influential books, however, were *The Influence of Sea Power upon History, 1660–1783* (1890) and *The Influence of Sea Power upon History: the French Revolution and Empire, 1793–1812* (1892). In his books he argued that control of the sea lanes to protect commerce and wage economic warfare was very important to a State. He therefore advocated a big navy. But there were six fundamental factors that affected the development and maintenance of sea power:

1. **Geographical position** (location). Whether a State possesses coasts on a sea or ocean (or perhaps more than one), whether these waters are interconnected; whether it also has vulnerable, exposed land boundaries; whether it can maintain overseas strategic bases and command important trade routes.

2. **Physical "conformation" of the State** (the nature of its coasts). Whether the coastline of a State possesses natural harbors, estuaries, inlets, and outlets; an absence of harbors will prevent a people from having its own sea trade, shipping industry, or navy; the importance of navigable rivers to internal trade but their danger as avenues of penetration by enemies.

3. **Extent of territory** (length of the coastline). The ease with which a coast can be defended.

4. **Population numbers.** A State with a large population will be more capable of building and maintaining a merchant

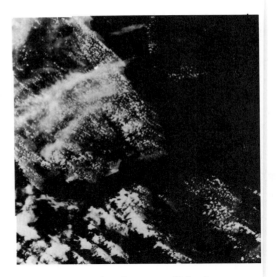

**Singapore, a classic geopolitical outpost.** Founded in 1819 as a trading post on an island off the tip of the Malay Peninsula, Singapore evolved during the heyday of the British Empire as a major naval base and ship-servicing center near the strategic Straits of Malacca, to the left in the photo, along the lines advocated by Admiral Mahan. The island became independent in 1965 and has since become not only the world's largest seaport, but also a major industrial and banking center. (Photo courtesy of NASA)

marine and navy than a State with a small population.

5. **National character.** Aptitude for commercial pursuits; sea power is "really based upon a peaceful and extensive commerce."

6. **Governmental character.** Whether government policy is taking advantage of the opportunities afforded by the environment and population to promote sea power.

Mahan, too, refers (in items 5 and 6) to the question of the national "will," in terms reminiscent of Kjellén's. Indeed, the later German geopoliticians sometimes wrote much as Mahan did. But Mahan was a military man and tended to think in military terms. Moreover, he was writing at the apex of European imperialism, at a time when his own country was beginning to emulate the European imperialists, and was also influenced by Social

Darwinism. He generally took the view that a State could survive only by being fit, and defined fitness chiefly in terms of military strength. But this depended, in turn, on the people's moral and martial fiber.

Mahan was also a practical man. He made specific recommendations for U.S. foreign policy based on his study of history, his military experience, and his geostrategic concepts. He advocated that the United States occupy the Hawaiian Islands, take control of the Caribbean, and build a canal to link the Atlantic and Pacific oceans. President Theodore Roosevelt's administration used several of Mahan's suggestions as the basis of its foreign policy. He was even more influential in Germany, Britain, and Japan.

Of greater interest to the political geographer are the glimpses of Mahan's world view from his later book, *The Problem of Asia*. In this work Mahan recognizes a core area in Asia and Russia's domination of it; he anticipates a struggle between Russian land power and British sea power. Not surprisingly, he presumed that British sea power would be able to contain Russian expansionism. He also predicted that the containment of Russia and the control of China would become the joint concern of the United States, Great Britain, Germany, and Japan. (It is well to remember that he was writing *before* the turn of the present century.) Thus he proposed that Russia be provided warm-water ports in China by guaranteeing it the use of those exits.

While Mahan emphasized sea power, *Sir Halford John Mackinder* (1861–1947), who was much younger than Mahan, felt that the great age of naval warfare was over; that changing technology, especially the railroad, had altered the relationship between sea power and land power. Still, his approach to global strategy was similar to that of Mahan, but with a different emphasis and different forecasts. As professor of geography at Oxford and director of the London School of Economics, Mackinder helped raise geography to a high level in England. He was also a member of parliament from 1910 to 1922 and chairman of the Imperial Shipping Committee from 1920 to 1945.

In 1904 Mackinder read a paper at the Royal Geographical Society in London entitled "The Geographical Pivot of History." This was a true milestone in the geopolitical debate of that period; in fact, the contents of that article (afterward published in the *Geographical Journal*) are still a subject of discussion and evaluation today, nearly a century later.

It is easy to regard Mackinder's paper as sophisticated speculation and to suggest that it has little value as a contribution to political geography. But it is worth remembering the main elements of the world situation when he produced his remarkable piece. Russia was losing a disastrous war with Japan, Germany was still a youthful, organizing State. Yet Mackinder in 1904 expressed the view that there was a Eurasian core area that, protected by inaccessibility from naval power, could shelter a land power that might come to dominate the world from its continental fortress. This Eurasian core area Mackinder called the Pivot Area. Later, he broadened this strategic concept into that of the Heartland. The Heartland's rivers drain into the Arctic, distances to warm-water oceans are huge, and only the Baltic and Black seas could form avenues for sea power penetration, but these are easily defended.

Mackinder reasoned that this Eurasian territory would become the source of a great power that would dominate the Far East, southern Asia, and Europe—most of what he called the "World-Island," which he conceptualized as consisting of Eurasia and Africa. He presumed that the area contained a substantial resource base capable of sustaining a power of world significance. The key, he argued, lay in Eastern Europe, the "open door" to the pivotal Heartland. Thus he formulated his famous hypothesis:

> Who rules East Europe commands the Heartland
> Who rules the Heartland commands the World-Island
> Who rules the World-Island commands the World*

*Democratic Ideals and Reality*, 1919.

**Sir Halford J. Mackinder, 1861–1947.** (Courtesy of Brian Blouet)

In 1924 Mackinder propounded a little-known counterhypothesis: The potentialities of the Heartland could be balanced in the future by Western Europe and North America, which (as Mackinder read the lessons of World War I) "constitute for many purposes a single community of nations." He called the North Atlantic "the Midland Ocean," in the midst of the area from the Volga to the Rockies, which he called "the main geographical habitat of Western civilization."*

In 1943, in the midst of World War II, Mackinder blended all these ideas and modified them in an article titled "The Round World and the Winning of the Peace."† He moved the Heartland east of the Yenesei River and renamed it Lenaland. It would oppose the Midland Basin (the North Atlantic and bordering lands separated by Central Europe and surrounded by deserts and the Arctic). He felt that the Heartland contained

soils and minerals equal to those of North America, but that the two regions would combine against Germany.

Mackinder made a significant contribution to our perspectives of the world, and in a broad sense his assumptions about the Heartland were later substantiated. However, there were three major weaknesses in his work. First, he did not give enough weight to the growing power of North America; second, he failed to explain the seeming contradiction between his thesis of the power of the possessor of the Heartland and the relative weakness of Russia/USSR until World War II; and third, he failed to take into account the growing and very obvious importance of air power and other technological developments. Like Mahan, he oversimplified history and leaned too far in the direction of determinism.

Mackinder had many critics. Prominent among them was *Nicholas John Spykman* (1893–1943). Born in Amsterdam, he studied at Berkeley, became a professor of international relations at Yale in 1920, and became a U.S. citizen in 1928. In his work he emphasized the power relations among States and the impact of geography on politics, but he rejected the German school of *Geopolitik.* In two books, *America's Strategy in World Politics* (1942) and *The Geography of the Peace* (1944), Spykman pointed out two of the basic weaknesses of Mackinder's theories. First, he felt Mackinder overemphasized the power potential of the Heartland; its importance was in fact reduced by the major problem of internal transportation and by access through the barriers that surrounded it. Second, history involving the Heartland was never a matter of simple sea power–land power opposition. Instead, Spykman felt, the real power potential of Eurasia lay in what Mackinder had called the "Inner or Marginal Crescent," and what Spykman called the Rimland. This area is vulnerable to both land and sea power and hence must operate in both modes. Historically, alliances have always been made among Rimland powers or between Heartland and Rimland powers. Spykman, therefore, proposed his own dictum:

---

*The Nations of the Modern World, vol. 2, London, George Philip & Son, 1924, p. 251.

†Foreign Affairs, 21 (July 1943), pp. 595–605.

Who controls the Rimland rules Eurasia;

Who rules Eurasia controls the destinies of the world.*

Spykman advocated that the Allies base their postwar policy on preventing any consolidation of the Rimland. Although there is no evidence that George Kennan (who proposed the "containment" policy of the Cold War era) ever read Spykman, this policy became fundamental in the anticommunist position of the Western powers because of the change in thinking represented by Spykman. Still, however, it was basically a nineteenth-century view of the world.

### Geopolitik

While the two streams of thought in geopolitics were developing, a new school of geopoliticians was forming in Germany, chief of whom was *Karl Haushofer* (1869–1946). A career officer in the Bavarian army, Haushofer served in Japan (1908–10) and rose to the rank of major general in the army general staff, serving throughout World War I. Before the war he took a Ph.D. in geography at the University of Munich. He was embittered by Germany's defeat in the war, blaming it in part on the incompetence of its generals, but also on a too-early start of the war and the lack of links between the State's leaders and the land. He was convinced that Germany should have won the war and wanted somehow to avenge the defeat. He had been impressed by what he had seen of the power and expansionist ambitions of Japan, and also by the power of Britain. He therefore devoted himself to the study of geopolitics.

Haushofer began lecturing at the University of Munich in 1919 and became a professor of geopolitics after the Nazis came to power in 1933. He gathered around him a group of disciples, including journalists who helped spread his ideas. In 1924 he founded the *Zeitschrift für Geopolitik*, a monthly journal in which he and his colleagues propagated their new "science," whose name he had drawn from Kjellén. They also produced a flood of books and other publications, including maps, which they used as a major weapon. They remained active throughout World War II, working to justify Germany's drive for conquest by "science." There was, in truth, little that was scientific in *Geopolitik*.

Haushofer and his group blended together the organic State theory of Ratzel, its refinements and elaborations by Kjellén, and the geostrategic principles of Mahan and Mackinder, added a heavy dose of German chauvinism, willful ambiguity, and mysticism, and created a case for a German policy of expanionism. They used not only maps but also slogans and pictographs to influence people. The notion of *Lebensraum* was drawn from Ratzel but was distorted to justify as "natural" Germany's growth at the expense of its less virile neighbors, Poland and Czechoslovakia.

Kjellén's advocacy of autarky was revived and elaborated, with emphasis on the political character of the German quest for economic self-sufficiency, again at the expense of other countries in southeastern Europe, the Middle East, and elsewhere. Economic policy was, in fact, used as a political weapon by Nazi Germany. Influenced by the Heartland concept, the German geopoliticians advocated an alliance with the Soviet Union and apparently were dismayed when Hitler invaded that vast country. Not only did they develop new versions of geostrategy, all for the benefit of Germany, but they also attempted to create geo-medicine, geo-psychology, geo-economics, and geo-jurisprudence, not as true sciences or legitimate branches of geography, but as weapons for German conquest. There was much, much more, of course, but this sample gives something of the flavor of *Geopolitik*.

How influential was *Geopolitik*? There is still no consensus on this question, but the evidence seems to indicate that its influence was great in some areas, nil in others. Within Germany it pervaded the educational system during the Nazi period, and it helped condition the intelligentsia to a policy of aggression. Its emphasis on planning, especially

*The Geography of the Peace*, New York, Harcourt, Brace, 1944, p. 43.

with regard to natural resources, did influence in part the direction of Nazi expansion. There is no evidence, however, that Hitler, his top advisers, or his generals were notably influenced by *Geopolitik*, even though Rudolf Hess was a disciple of Haushofer,

and *Mein Kampf* contains many examples of Hitler's own geopolitical ideas.

Outside Germany, *Geopolitik* found fertile ground only in Japan, a country with which Haushofer had special ties and about which he wrote six books. Elsewhere it was

greeted with opposition or indifference. A few of its ideas have survived and have been widely adopted: the extensive use of maps to convey ideas, for example, and the need for governments to have a store of accurate and up-to-date information about the earth; concepts of propaganda and psychological warfare, of total war, and of the importance of air power. But *Geopolitik* as such died with the collapse of Nazi Germany in 1945—and none need mourn its passing.

## Resurgence of Geostrategy

Advocacy of air transport and air warfare is not new. It began with the Wright brothers and continued through Lindbergh and Billy Mitchell, Glenn Curtis, Eddie Rickenbacker, and many others around the world. One strategist, however, was the first to advocate forcefully a geopolitical view of the world based on air power. He was *Major Alexander P. de Seversky* (1894–1974). Born in Russia, he served in the Russian navy during World War I and became a naval ace. In 1918 he was sent on a naval mission to the United States, to which he offered his services after Russia dropped out of the war. He served in various capacities, became an American citizen in 1927, invented the world's first fully automatic bombsight, founded an aircraft company that became Republic Aviation Corporation, and continued to advance aviation through his aircraft design, innovative combat strategy and tactics, speed records, production of aircraft, and work in civilian aviation. He was both a practical engineer and a businessman, as well as an imaginative thinker.

Among his many writings, two books had enormous influence on government officials as well as the general public. The first was *Victory Through Air Power* (1942). In it he reviewed the course of the war to that point, declared "the twilight of sea power," deplored the insufficient attention to air warfare being paid by the Allies (especially the United States), and advocated a totally new strategy and organization for victory through air power. The book reads well even today.

After World War II had ended and the Cold War had begun, de Seversky published another innovative book, *Air Power: Key to Survival* (1950). Here he restated even more forcefully his view that land and sea power had been subordinated to air power. He urged the development of massive air superiority for the United States. He advocated defense of the Western Hemisphere, avoidance of small wars as a useless sapping of American strength, and abandonment of overseas bases as costly luxuries. For the first time in such geopolitical writings he used a map drawn on an azimuthal equidistant projection centered on the North Pole to show clearly how close the United States and the Soviet Union really were. It also showed the vast areas of air dominance by both countries and how these areas over-

**A German propaganda map in the Haushofer tradition.** This map, by Friedrich Lange, appeared in *Volksdeutsche Kartenskizzen*, published in Berlin in 1936. "This drawing was done in anger," reads the caption, and goes on to denounce German students who only visit places in "the beleaguered East" rather than settling there "because struggle, real patriotic struggle of the people" is to be found there, "where to be called a German demands courage, has disadvantages, and requires pledging of one's whole personality, the risking of one's economic position, and even the sacrifice of personal freedom. This kind of struggle is going on today on a broad front behind the borders, where German men and women cling to the soil which is their fate in a spirit of steadfastness, spite, and family sense. It takes place all over the East, to the north and south of Czechoslovakia, which brags about being the Slavic fist in Middle Europe. But, to be sure, only he can keep a border who has the right border spirit and is willing to put himself along the border." Such maps helped prepare the way for the German *Drang nach Osten*, (March to the East), which triggered World War II. (Map courtesy of E. J. Huddy, British Library.)

lapped over the North Polar region, in what he called the "Area of Decision."

As a result of this "new" view of the world, the United States and Canada erected at great expense three lines of radar stations and air bases stretching across Alaska and Canada for the defense of North America against attack from the USSR by the shortest routes—over the North Pole. It is easy to criticize this view in an era of intercontinental ballistic missiles and space travel, but at the time it performed a most useful function by tearing us away from our Mercator view of the world, by developing an interim defense system, and by emphasizing defense instead of expansion as the prime goal of geostrategy. The "fortress America" concept was never accepted by policymakers, however, and the United States has become deeply involved in political affairs around the world.

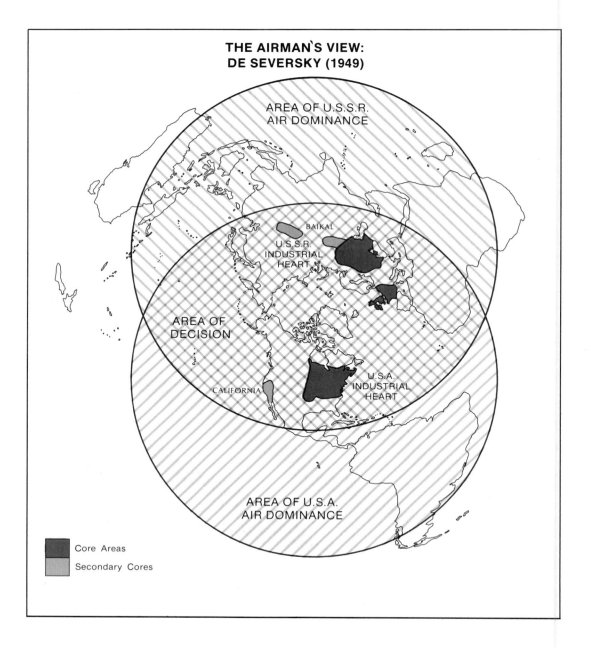

THE AIRMAN'S VIEW:
DE SEVERSKY (1949)

AREA OF U.S.S.R.
AIR DOMINANCE

BAIKAL

U.S.S.R.
INDUSTRIAL
HEART

AREA OF
DECISION

CALIFORNIA

U.S.A.
INDUSTRIAL
HEART

AREA OF U.S.A.
AIR DOMINANCE

Core Areas

Secondary Cores

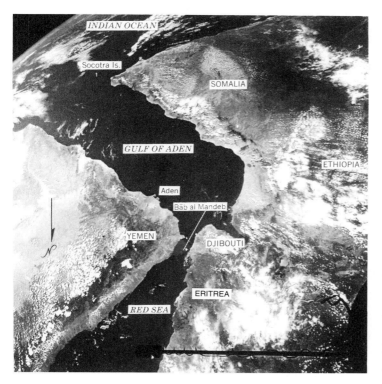

**The Strait of Bab el Mandeb.** One of the most strategic waterways in the world, this strait connects the Red Sea (and beyond it the Suez Canal and the Mediterranean) with the Gulf of Aden and the Indian Ocean. At the top of the photo (southeast of the strait) is the Horn of Africa; surrounding the strait are Aden and Djibouti, formerly major British and French naval bases, and Yemen and Eritrea. (Photo courtesy of NASA)

Various geographers, particularly in the United States, have developed variations on the concepts represented by Mahan, Mackinder, Spykman, and de Seversky, but they have not been very innovative.

One innovator is the American geographer *Saul B. Cohen*. In his *Geography and Politics in a World Divided* he considers the entire world as being divided into "geostrategic regions." Like de Seversky, he takes into account the Americas as well as new technology, and aims at global equilibrium. "The major premise of the work is that the dynamic balance that characterizes relations among States and larger regions is inherent in the ecology of the global political system. This world is organized politically in rational, not random fashion."* He rejects the notion, popular in the immediate postwar period, that spheres of influence are obsolete, even reprehensible. He insists, in fact, that "spheres of influence are essential to the preservation of national and regional expression."* His geostrategic regions are essentially the spheres of influence of the United States, Maritime Europe, the Soviet Union, and China.

He still preserves the concept of maritime and continental powers, but he expands it to include in the "Trade Dependent Maritime World" all of the Americas, Western Europe, Africa except for the northeastern corner, and offshore Asia and Oceania; and includes in the "Eurasian Continental Power" all of the Soviet Union, Eastern Europe, and Eastern and Inner Asia. He still does not take into account, however, the fact that the Soviet Union had become a leading maritime power, with naval, merchant, fishing, and oceanographic fleets ranking among the world's largest.

He classifies South Asia as an Independent Area, and he identifies the Middle East and Southeast Asia as *shatter belts*. The term

---

*Preface to the second edition, New York, Oxford, 1973, p. vi.

*Ibid., p. viii.

*shatter belt* or *shatter zone* has customarily been applied to Central and Eastern Europe, a region of chronic instability in which States appear, disappear, and reappear with frequently changing names and boundaries. Cohen has omitted reference to this original shatter belt in deference to its partition in the 1940s between Soviet and American power systems. His new ones are quite justified in view of their chronic instability, but, as we pointed out in the third edition of *Systematic Political Geography* and repeated in the fourth edition, not the omission of Central and Eastern Europe. The recent stability of this region was deceiving, since ancient rivalries and animosities were only temporarily suppressed by communist togetherness. Even in the former French Indochina, where all three States (Laos, Cambodia, and Vietnam) were under communist rule, all three until very recently have been fighting rebels or one another. Shatter belts do not disappear very readily.

Cohen has a great deal more to say, including recommendations for American policymakers, but a major thesis bears some special mention. He advocates maintaining the unity of Europe and the Maghreb by "subtle economic and political persuasion" rather than by force. This emphasis on cooperation, economic power, persuasion, and propaganda is a realistic summation of actual trends in the postwar period, trends that go on, as it were, under an umbrella (protective or threatening) of nuclear missiles and space satellites.

In his 1991 presidential address to the Association of American Geographers, Cohen updated, modified, and expanded upon these basic themes. Among other changes, he introduced "The Quarter-Sphere of Marginality," the continents centering around the South Atlantic and their bordering oceans, a region "outside the modern economic system [and] marginal in a strategic sense." Another innovation is his "Gateway Regions" and "Gateway States," areas such as Eastern Europe and the Caribbean, that "can help stabilize the system because of their raison d'être as links in an increasingly

interdependent world." There is much more, and this paper deserves careful study.*

A very different view of the world was propounded by *Lin Piao*, late Defense Minister of China. In 1965 he offered a theory of world revolution that viewed the world as similar to a city and the surrounding countryside. The rich, industrialized, largely Western countries represent the city, and the poor, agricultural countries, largely colonies of the Western countries, represent the countryside. The poorer areas will gradually be converted to communism, and, using tactics similar to those described by Cohen but supplemented by guerrilla warfare, will confront and eventually overwhelm the cities.

These are the tactics the Chinese communists used so successfully against the Nationalists and the Japanese during the 1930s and 1940s, and are described by McColl in his studies of "insurgent states." It is also reflected in the later "core-periphery" concept and its variants discussed in this book. Applying them on a world scale, however, seems a bit unrealistic, if only because of the diversity within the developing world and the industrial world's great strength. Insurgencies, in fact, have not all been successful, and even China has gone over to fighting conventional wars when it seems appropriate. Lin 's basic thesis, however, survives in a new incarnation as the core-periphery theory discussed in Chapter 1.

Although all these geostrategic views have serious flaws, they do have the virtue of analyzing the world as a whole rather than as scores of discrete political units. In view of the increasing interdependence of the world, we need a good deal more of this holistic thinking, directed not toward formulation of strategies for confrontation, but toward strategies of cooperation; not a geostrategy of war, but a geostrategy of peace.

*"Global Geopolitical Change in the Post-Cold War Era," *Annals of the Association of American Geographers, 81,* 4 (December 1991), pp. 551–580.

# THE DEVELOPMENT OF GEOPOLITICS

## THE ORGANIC STATE

**FRIEDRICH RATZEL**
1844–1904

"The Laws of the Spatial Growth of States" (1895)
*Political Geography (1896)*

**RUDOLF KJELLEN**
1864–1922

*The State as an Organism (1916)*

## GEOSTRATEGY

Admiral **ALFRED THAYER MAHAN**
1840–1914

*The Influence of Sea Power Upon
History 1660–1783 (1890)*

Sir **HALFORD J. MACKINDER**
1861–1947

"The Geographical Pivot of History" (1904)
*Democratic Ideals and Reality (1919)*

General **KARL HAUSHOFER**
1869–1946

*Journal of Geopolitics (1924–1968)*

**NICHOLAS JOHN SPYKMAN**
1893–1943

*America's Strategy in World Politics (1942)
The Geography of the Peace (1944)*

**ALEXANDER P. DE SEVERSKY**
1894–1974

*Victory Through Air Power (1942)
Air Power: Key to Survival (1950)*

**SAUL B. COHEN**
1928–

*Geography and Politics in a World
Divided (1963)*

**The two streams of thought** that comprise the theoretical foundations of geopolitics are symbolized in this diagram by the leading figures at each stage of development of the field. The organic theory of the State and geostrategy were blended by Haushofer, who added a heavy dose of German chauvinism and liberal portions of racism and militarism to create what he called *Geopolitik*. After the demise of the Third Reich in 1945, *Geopolitik* was no longer seriously advocated publicly. Although Haushofer's *Zeitschrift* resumed publication for several years after the war, it was quite different from the pre–1945 version. With *Geopolitik* died the organic State theory, but geostrategy survived and flourishes today.

# 22

# *Contemporary Geopolitics*

In retrospect, it appears that our devotion of two chapters to geopolitics in the third edition of *Systematic Political Geography (SPG III)*, written in 1979, may have presaged a revival of academic interest in the subject. It has again become acceptable—even respectable—to write on geopolitical theories without being branded a chauvinist or a warmonger. In the United States and Britain especially, geopolitics is again attracting attention and some creativity is being shown. Some of the credit must be given to Henry Kissinger, National Security Adviser and Secretary of State under Presidents Richard Nixon and Gerald Ford, as well as to Charles de Gaulle, Willy Brandt and other pragmatic statesmen, but geopolitics was reviving independently in Italy, West Germany, and France at the same time. New texts specifically on geopolitics appeared in 1985 (G. Parker) and 1986 (O'Sullivan). The new academic work in geopolitics (as distinguished from the Kissinger-type governmental theses) is distinctly less bellicose and imperialistic than the older work was. We may be witnessing (and perhaps participating in) the development of what we urged in *SPG III* and have repeated since: a new kind of geopolitics, a geopolitics of peace. We examine this notion in the next chapter, but first we should point out that while geopolitics was languishing in most of the world, it was flourishing in Latin America.

## *Latin American Geopolitics*

All over Latin America, but especially in the Cono Sur, the Southern Cone of South America, geopolitics is not only a major professional field, but is of great popular interest as well. While in North America and Western Europe books on the subject were produced chiefly by military academies and organizations and geopolitical articles appeared almost exclusively in military journals, in Brazil, Chile, Argentina, Uruguay, Bolivia, Peru, and to a lesser extent Ecuador there has been a steady and voluminous outpouring of writings on the subject by diplomats, professors, and government officials as well as by military men. Daily newspapers contain long, weighty geopolitical analyses; geopolitical books and journals are readily available at public libraries, newsstands, and bookstores. Radio and television programs carry geopolitical themes into the homes even of illiterates.

Most of the themes focus on South America and nearby areas, of course. Among the most prominent have been the South Atlantic as a zone of potential conflict between the United States and the Soviet Union, Brazilian expansionism (real and imagined), the strategic value of the straits at the southern tip of South America, the triangular relationships of Brazil–Peru–Chile or Brazil–Argentina–Chile or Peru–Chile–Argentina, the integration of the "American Antarctic" into

334

South America, and the equivalence of sea and land in calculating national territory and power. The writings are often amply illustrated with maps and diagrams replete with triangles, arrows, arcs, stars, and other figures that illustrate areas of action, directions of movement, and relationships of all kinds.

It is difficult to say how much policy is influenced by this kind of geopolitical thinking, or whether, in fact, geopolitics is simply used to justify policy. In South America there is a constant rotation of people (almost entirely men) among academia, the military, the diplomatic service, and the government, and many, many people have served in all four sectors. Even after retirement some continue to write and to be read and quoted. And not all of them are parochial. Some write on geopolitics in other parts of the world, and the German geopoliticians of the Hitler area are still quoted approvingly. In Chapter 33 we discuss in more detail South American geostrategic views of Antarctica and the Southern Ocean.

There seems also to be a genesis of interest in academic geopolitics elsewhere in the developing world. In April 1990, for example, a weeklong international seminar was held at Panjab University in Chandigarh, India, on Afro-Asian geopolitics. Many of the participants were European, but a significant number came from Africa, Asia, and the Pacific Basin. Even the papers presented by Europeans dealt with the geopolitics of the regions in question. Here at last is a healthy development: the spread of geopolitical studies into regions that were victims of early European geopolitics. We may hope that this trend continues and intensifies. Meanwhile, however, most of the new ideas in the subfield are emerging from Europe and North America.

## Some Contemporary Geopolitical Approaches

In the next chapter we focus on what has emerged in the past decade as the dominant theme in the new geopolitics: the geography of peace and war. Here we examine briefly some contemporary geopolitical notions and in subsequent sections present some of our own.

Geoffrey Parker, of the University of Birmingham, England, is probably the foremost student of geopolitics today. In a 1991 article he summarizes both its history and current developments.[*] He points out that throughout most of the first three quarters of this century the geopolitical concepts that we discussed in the previous chapter, despite many differences, were generally characterized by "an acceptance of the proposition that confrontation was endemic in the system." Within this general perspective, notions of hegemony and dominance have tended to prevail and manifest themselves in many ways. Since the mid-1970s, however, a more humanistic approach has emerged concurrent with both the demise of the colonial empires and the weakening of the power and influence of the two superpowers. Parker raises the difficult question of whether the new approaches are not basically reactions to changes in the world politicogeographical situation.

Peter Taylor, one of the world's leading contemporary political geographers, teaches at Loughborough University of Technology. In the second edition of his textbook he devotes a lengthy chapter to geopolitics.[†] Here he presents detailed discussions of the various theories of cycles in international politics that became popular in the 1980s. They include Modelski's long cycles of global politics, cycles of hegemony, an application of Kondratieff's economic cycle model to British and American roles in the world, and cycles and geopolitical world orders. He then uses the Cold War as a case study of a geopolitical world order, and goes on to raise questions about what will follow the

---

[*]"Continuity and Change in Western Geopolitical Thought During the Twentieth Century," *International Social Science Journal, 43,* 1 (February 1991), pp. 21–33.

[†]*Political Geography: World-Economy, Nation-State and Locality.* London, Longman, 1989, pp. 42–90.

Cold War. Finally, in a section on "geopolitical codes," he uses as examples the U.S. policy of "containment" of the Soviet Union and the "nonalignment" policy of India and other countries.

More recently, "critical geopolitics" has come to the fore. This approach, according to one version, attempts to integrate geopolitics into political discourse aimed at maintaining current power relationships. Gearóid Ó Tuathail describes it as "the social construction of space and time in the practice of statecraft." E. Jeffrey Popke says it "attempts to denaturalize the global order by portraying it as socially and historically constructed." This concept, however formulated or interpreted, is far removed from concepts of sea power versus land power or strategic location or the significance of strategic minerals. Like most contemporary geopolitical thinking, it, too, seems to concentrate almost exclusively on the "politics" and virtually ignores the "geo."

Another problem with many of these theories or approaches is that they tend to ignore reality, especially realities that do not fit the theories. Some of them were rendered obsolete by the end of the Cold War, during which these theories were developed and tended to obscure many people's thinking. The dramatic change in the world political picture that took place during the brief period of 1989–91 (and whose shock waves will reverberate around the world for at least another generation) came as a stunning surprise to most people. Yet, in *SPG III* we observed, "One is tempted to comment on the disintegration of the Soviet Empire by exfoliation, reaching into the Soviet Union itself, with revolts in the Ukraine and elsewhere" (p. 277). This observation was repeated in *SPG IV*, with the addition of "Lithuania, Armenia and elsewhere" (p. 238).

The opposite tactic is equally common. Some theorists seize upon current realities to validate their theories, no matter what grotesque contortions may be necessary to make the facts fit the theories. It is rather like Cinderella's stepsister trying to squeeze her large foot into the tiny glass slipper. Some people, however, are entranced by the

seeming novelty and/or comprehensiveness of the new approaches. Yet the cycle concept, for example, as we point out in Chapter 1, goes back at least to Ibn Khaldun in the fourteenth century.

It is basically for these reasons that Saul Cohen's approach, which builds a theory inductively from observed facts, is likely to prove of greater value in understanding this complex and ever-changing world. We know that States rise and fall in power, however defined, and we have some idea of why. But we still cannot predict what will come next; which State, for example, will supplant the United States (probably in the first third of the twenty-first century) as the dominant world power. It is unlikely to be Japan, the current choice of a plurality of "experts" and ordinary citizens alike.

Let us now consider some geopolitical ideas that have not yet attracted prominent disciples or advocates.

## Some Tentative Geopolitical Concepts

Geographic, political, and historical literature are replete with discussions of *buffer States and buffer zones,* areas of weakness that separate areas of strength, reducing the chances of conflict between them. Classic buffer States include Bolivia, Paraguay, and Uruguay, which almost completely separate Brazil from Chile and Argentina; Poland and the Balkans separating Germany and Russia/USSR; Afghanistan lying between Russia/USSR and British India; the Himalayan kingdoms of Nepal, Sikkim, and Bhutan guarding the major mountain passes between China and India; Mongolia between Russia/USSR and China; and Laos and perhaps Thailand separating British and French power in Asia. These and others have largely, if not completely, lost their buffer functions as a result of rising nationalism and the strengthening of the buffers themselves, changing technology, the role of ideologies other than nationalism, and the general tendency throughout the world to eliminate frontier areas in favor of clearly defined and delimited boundaries.

Even the kind of buffer zone that became prominent after World War II—the ideological buffer—has tended to disintegrate. On the grand scale, the new States that emerged from the dissolving empires combined with a number of older ones, including most of those listed as traditional buffers, to form a neutralist bloc, aligned with neither the Soviet nor the American bloc. This channeled some of the energies of the superpowers into attempts to win friends and influence in the neutralist countries, perhaps contributing to a reduction of tensions between them and avoidance of World War III. Another type of buffer zone developed across southern Africa as black Africans to the north attained independence between 1957 (Ghana) and 1963 (Kenya), while the white minority regimes in Southern Rhodesia and South Africa/South-West Africa retained tight control.

Both buffers have now disappeared. With the achievement of independence by Namibia (ex-South-West Africa) in 1990, largely through the efforts of the United Nations and Soviet-American cooperation, South Africa was forced to choose between withdrawing into the *laager* and accommodating the needs and aspirations of the great majority of its citizens. Fortunately, it chose the latter course. The "nonaligned" bloc has also essentially disappeared. First, many of its members became aligned with one Cold War rival or the other (though rarely admitting it publicly); then new centers of capitalist wealth and power (Western Europe, Japan) developed; and finally the Soviet Union itself weakened and collapsed.

Recently, we have seen the classic type of buffer zone develop on a large scale and similarly wither and vanish. The Soviet Union set about during and after World War II to surround itself with concentric rings of defense against the kinds of invasion it had suffered so often in the past. The innermost ring consisted of territory actually annexed from other countries: Karelia, the Baltic States, eastern Poland, the Carpatho-Ukraine, Bessarabia, Tannu Tuva, southern Sakhalin, and the Kuril Islands. These acquisitions enabled the Soviet Union to regain most of the Tsarist Empire, take most of the

bases Hitler had used for invasion of its territory, create defense in depth for vital and exposed Leningrad, lengthen its coastline, attain common borders with Norway and Hungary, reach the Danube and obtain a seat on the Danube River Commission, and virtually enclose the Sea of Okhotsk, all most useful geopolitical objectives, the value of which it demonstrated repeatedly.

The second ring was composed of States closely aligned with the USSR; some, in fact, were virtually colonies. These included most of the States of Central and Eastern Europe (including East Germany), China, Mongolia, and North Korea. Somewhat less reliable but still useful were States that were either neutral by treaty or neutralist by conviction (or necessity): Finland, Austria, Turkey, Afghanistan. The gaps in this ring were filled by Soviet troops in Iran, Xinjiang, and Manchuria.

Very soon though, beginning with the withdrawal (under UN pressure) of Soviet forces from northern Iran in 1947, this enormous and extensive buffer zone in the "Rimland" began to crumble, and the process is now complete. The Soviets early on withdrew from Xinjiang and Manchuria, from Porkkala in Finland and Saseno in Albania, from Vienna and eastern Austria; they lost control of China, Yugoslavia, Albania, and North Korea. More recently, they have had to relinquish control of all of Central and Eastern Europe, withdraw from Afghanistan, and dissolve COMECON and the Warsaw Pact. As with South Africa, this process of exfoliation has reached into the protected State itself, generating massive internal changes.

A fine exercise in the application of geopolitical theory would be the prediction of where the next buffer States or zones will develop, and something of their nature. Could they be Cohen's Gateway States and Regions? Or the traditional shatter belts?

Two other concepts, now at least temporarily discredited, derive from the Cold War. The dominant and guiding vision of American policymakers long after World War II was that of a *bipolar world*: the Soviet Union and its "satellites" versus the United

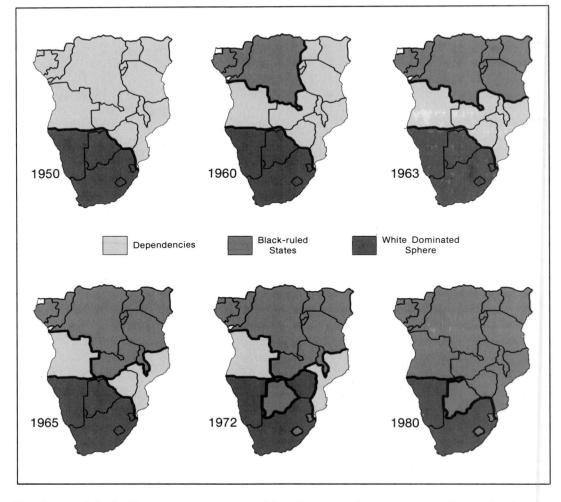

**The decay of the buffer zone in southern Africa.** The zone of European and white settler domination affording the Republic of South Africa and its mandated territory of Namibia (South-West Africa) some degree of protection from native African nationalism to the north steadily shrank during a third of a century as more and more colonies achieved independence, and vanished entirely when Namibia finally became independent in 1990. It is probably no coincidence that democracy in South Africa followed rather quickly and peacefully.

States and its "friends and allies." There was, however, a recognition among more objective observers that not all countries in the world could in any manner be crammed into one or another of these neat categories of "bad guys" and "good guys." Indeed, the Non-Aligned Movement began as early as 1955, at the Bandung Conference, introduced in Chapter 19. Its leaders were Sukarno of Indonesia, Nehru of India, Tito of Yugoslavia, Nasser of Egypt, and (a little

later) Nkrumah of Ghana. In recognition of its split from the Soviet Union, China was (after considerable debate) invited to attend. Their choice of neutralism in the Cold War marked the origin of the term "Third World," a term that later was given so many meanings that it became virtually meaningless.

The reaction of the United States to Bandung was epitomized by Secretary of State John Foster Dulles, who declared flatly, "Neutralism is immoral." With the modest exception of Yugoslavia, it was assumed that any country with a centrally planned economy was a Soviet satellite or, in another then-current phrase, a "captive nation." It was many years before policymakers accepted the Sino-Soviet split as genuine, deep, and abiding and recognized China as a separate force in the world. Meanwhile, led by the Italian Communist Party leader, Palmiro Togliatti, what may have been a reasonably solid bloc of communist parties in and out of power in Europe began fragmenting.

We have now witnessed the conclusion of this process—the end of the Cold War. Yet the concept of nonalignment (or neutralism) may well be revived under new and currently unpredictable circumstances if again two powerful forces dominate at least a large proportion of the earth's surface.

The second concept derives directly from the first; this is the *domino theory*. Like "containment," "massive retaliation," "the balance of terror," and other Cold War doctrines, this one assumed that the Soviets (and all communists and most socialists everywhere) were, and are, unqualifiedly evil, that they were bent on world domination, that they were fiendishly clever, and that any small victory by them would automatically lead to many more. The domino theory originated, apparently, with U.S. Admiral Arthur Radford, who in 1953 urged a carrier-based nuclear strike against the Viet Minh in Vietnam to relieve their pressure on the French at Dien Bien Phu on the theory that a Viet Minh victory there would set off a chain reaction of countries going communist "like a row of falling dominoes." The theory came into prominence during the Vietnam War in the 1960s. The argument went that the United States had to fight and win in Vietnam, for if South Vietnam "went communist," then automatically, like falling dominoes, Cambodia, Laos, Thailand, Burma, and perhaps India would as well. In the other direction, the Philippines, Taiwan, Hawaii, and even mainland United States were similarly threatened. Two decades later, President Ronald Reagan endorsed the theory, asserting that if his attempt to overthrow the

government of Nicaragua was unsuccessful, the Sandinistas would soon overpower El Salvador, then Guatemala and Mexico, and pose a mortal danger to Harlingen, Texas!

Although the theory was based on false assumptions about the nature of local nationalist movements, despite the lack of evidence of any grand Soviet scheme to knock over "a row of dominoes," despite vastly different conditions in countries "threatened" by communism, despite, in some cases, huge problems of distance and logistics from one "domino" to another, despite the obvious fact that many countries actually bordering on the Soviet Union, China, and Cuba did not "fall," despite, that is, the lack of any credible evidence whatever of its validity, it guided much of American foreign policy for a generation.

The theory is not necessarily dead, however. After all, did it not work in reverse as one country after another abandoned communism and socialism? This notion, too, could well be revived in the future.

A third concept that needs geographic analysis has been almost entirely ignored in the geographical literature. This is the notion of *changing orientations*. Three examples of changing orientation illustrate the concept. First is the case of Zambia. Prior to independence, Zambia, as Northern Rhodesia, was a member of the Central African Federation. It remained very much a central African country until after Southern Rhodesia declared its independence in November 1965, whereupon Zambia cut its ties to the south and began strengthening those to the north and east. Zambia thus became, at least temporarily, the southernmost State in East Africa.

A less dramatic but nonetheless important change in orientation has taken place in the Commonwealth Caribbean. Until World War II this region was oriented almost exclusively toward the United Kingdom and to a lesser extent to the rest of the Commonwealth and Western Europe. During the war these ties were severely frayed, and new links were developed with North America. By 1965 it could fairly be said that these new ties were far more important than the traditional ones. Since then there has been evi-

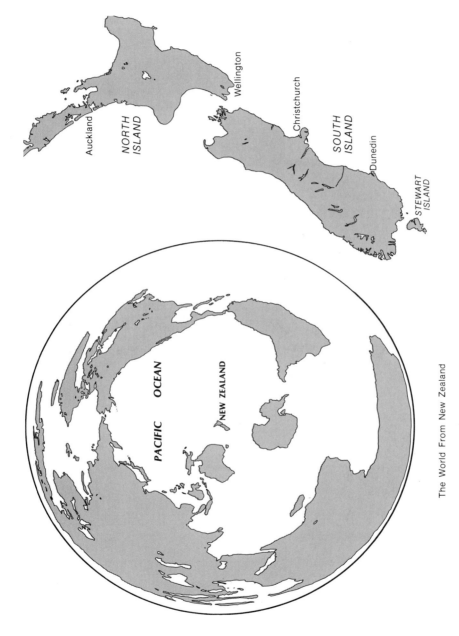

The World From New Zealand

**The centrality of New Zealand.** No longer on the outermost margin of the British Empire, New Zealand, like Australia, has changed its orientation since World War II and now takes a broader view of its place in the world, even developing an independent foreign policy. This new world view is reflected in this official government map.

dence of a second shift in orientation, gradual and still inchoate, but unmistakable: a turning away from North America toward Latin America.

A third example is that of Australia. Until the Second World War, Australia had the same orientations as the Commonwealth Caribbean: toward Britain first, and less importantly toward the rest of the Commonwealth. During the war and for about two decades thereafter, ties with Britain weakened, and Australia began looking northeastward, toward the United States, while still maintaining close relations with New Zealand. More recently, although not abandoning any of these important relationships, Australia has become continually more closely tied to Japan while strengthening relations with nearby Pacific Ocean States. After all, whereas to North Americans Indonesia may be the Far East, to Australians it is the Near North. And now, for the first time, Australia is looking at the Indian Ocean not as a vast void to be traversed en route to or from the British Isles but as an arena for Australian interest and activity.

Now the end of the Cold War has opened up opportunities for a host of reorientations. Perhaps the most important and least publicized case already evident is that of the former predominantly Muslim republics of the USSR. Azerbaijan, Kazakhstan, Kyrgyz Republic, Tajikistan, Turkmenistan, and Uzbek-istan were all, perforce, oriented toward the north and northwest, closely tied to the Soviet economy and polity, with only the weakest of links with neighboring States. Now, while retaining some ties with Moscow through the Commonwealth of Independent States (CIS), they are increasingly turning south and southwest. Turkey, Iraq, Iran, Pakistan, and even Israel are trying to woo them with blandishments of various kinds, and new trade, investment, transportation, cultural and political relationships are being forged.*

In the post–Cold War era we must ask some important geopolitical questions about the orientations of States. Will formerly aligned or nonaligned countries gravitate toward China, Europe, Japan, or still newer power centers? Will territorial expansion be replaced by extraterrestrial expansion? What will be the effects on foreign policies of new technologies, new economic demands, new ideas? What groupings will survive or develop from the welter of regional and subregional groups we see today? Are the various cycle and world systems theories predicting any new orientations, or will they only be summoned to explain them after they have occurred?

---

*See Chapter 29 for some details of the Economic Cooperation Organization in Central and Southwest Asia.

# 23

# The Geography of War and Peace

In the last chapter we discussed the revival of geopolitics as an academic field and some contemporary geopolitical concepts. We have also witnessed in the past decade the growth of what might be considered an even more fundamental study: the relationship between geography and war and peace. Geographers have long contributed to the study of warfare itself, and there is substantial literature in the field. But the study of the geographic factors that can lead to war, or a geographic perspective on the causes of war, is relatively new. Geopolitics, it is often said, is nothing but a justification for imperialism, and imperialism usually leads to war. We discussed imperialism in some detail in Chapter 17, and we saw that there is some truth in the accusation. But it is an oversimplification. Geopolitics does not always lead to imperialism, and imperialism is not the only cause of war. In an introductory text we cannot analyze this observation in great detail. We therefore simply survey the subject, focus on a few topics of special interest to geographers, look at five major world regions to see how some principles apply there, and exhort students to become involved in the search for, and application of, a geopolitics of peace.

## The Arms Race

William Bunge, an American geographer living in Canada (as well as many other scien-

tists), has pointed out in great detail what the effects would be on our species and on our entire planet of the uncontrolled explosion of even one thermonuclear bomb—not to mention a full-scale nuclear war. It should be clear to every thinking person on earth by now that a nuclear war cannot be won and must never be fought. Yet all the major world powers have developed thermonuclear weapons and for decades continued to test new ones and produce them at a furious pace. Even today, a number of smaller countries are working diligently to become nuclear powers. The colossal expenditure of natural resources, productive capacity, time, talent, energy, and capital on preparation for nuclear war is almost unimaginable and—to a rational being—inexcusable. Yet it is only a part of the military expenditures made by countries large and small, rich and poor, old and new all over the world since World War II was supposed to have taught us—again—the folly and futility of war. A few statistics gathered from various sources will help to illustrate the point.

1. During the period 1945–89, the United States alone spent over *ten trillion dollars* ($10,000,000,000,000), in 1988 dollars, on military weapons, equipment, and personnel—more than a third of it in the eight years of the Reagan administration alone.

2. In 1988 alone all the countries of the world together spent over a *trillion dol-*

*lars* ($1,032,400,000,000) for military purposes, and this figure has been falling only slightly recently.

3. In 1986 all the countries of the world together spent an average of about $30,000 to support each soldier, but only about $455 to educate each school-age child.

4. In a single *hour*, the world spent *twice* as much on military materiel as it cost the United Nations to stop a locust plague in Africa in 1986, saving enough grain from the locusts to feed 1.2 million people.

5. The big powers are not entirely to blame for these staggering expenditures, nor even the rich countries together. Military expenditures are increasing fastest in those countries least able to afford them, 800 percent between 1960 and 1986, even adjusting for inflation; between 1978 and 1988, this amounted to $464,000,000 worth of arms imported by poor countries besides their own production!

In Chapter 30 we discuss the arms trade in a different context, but it is clear enough even from these few figures that huge sums are being diverted from productive use and squandered on a vast array of lethal weapons and their support systems. Geographers are not the only ones who can think of better ways to use our finite resources. But the scale of military expenditures is only one aspect of the problem. Another, even more geographic, is their distribution. As indicated in point 5, the poor countries are spending in most years a higher proportion of their GNP on arms than the developed countries; in the decade 1978–88 the rich countries averaged 5.3 percent of GNP, while the comparable figure for the poor countries was 6.0 percent, and much of the money for weapons was borrowed. In 1985 the developing countries accounted for over 17 percent of total world military expenditures; if China is included, the total rises to 21.5 percent. Moreover, it is not the United States, the former Soviet Union, and the other major world

powers alone that have chosen to fight their wars on the soil of developing countries; since 1945 a total of nearly 150 wars have been fought there, most of them between developing countries themselves. No major part of Latin America, Africa, or Asia has been completely free of armed conflict since 1945. Furthermore, in addition to the extraordinarily wide dispersal of sophisticated weapons, the range of ballistic missiles has increased to the point that there is no place at all that they cannot reach. And numerous spying satellites revolving around the earth ensure that very little of consequence on the planet goes undetected. From surveillance or attack, there is no longer any safe place on earth, no place to hide.

### Nuclear-Free Zones and Zones of Peace

The realization of this fact has led many States to propose that certain areas of the earth be declared off limits to nuclear weapons. The first such proposal in modern times came in 1957, when Adam Rapacki, Foreign Minister of Poland, proposed a zone in Central Europe, on both sides of the "Iron Curtain," that would be freed and kept free of all nuclear weapons as a contribution to the reduction of Cold War tensions and of the possibility of the outbreak of a hot war. The United States and its allies rejected the Rapacki Plan, and since then, with one exception, the United States has rejected every proposal for a nuclear-free zone in the inhabited world. Typically, these proposals include prohibition of the manufacture, testing, acquisition, storage, installation, or use of any nuclear weapon of any kind within the region and provide for on-site inspection and verification to ensure that the ban is strictly honored.

The first such nuclear-free zone was established in Latin America by the 1967 Treaty of Tlatelolco. It was not the product of abstract theorizing or a naive desire to change the world, but of a very strong and widespread perception of imminent danger of an introduction of nuclear weapons into the region by an outside power. When Algeria became independent of France in 1962, after a protracted and bitter war, France was forced to remove its nuclear weapon-testing facilities from the Algerian Sahara. It then began building a new facility in French Guiana. France claimed it was to be only a scientific research base from which space vehicles would be launched, but the Latin Americans (and others) had reason to believe otherwise. After Latin America was declared a nuclear-free zone, with inspection and verification to be conducted by the International Atomic Energy Agency, the Kourou station became a joint U.S.-France rocket research base and the French established their new nuclear weapon-testing facilities on Moruroa Atoll in French Polynesia.

Since then other nuclear-free zones have been proposed in various parts of the world: the Balkans, the Adriatic, and the Mediterranean; Africa; Northern Europe; the Middle East; and South Asia among others. None has actually been legally established by treaty, however, and it is known or presumed that nuclear weapons have been produced in, or introduced into, all of these regions. Four areas have, however, been declared nuclear-free zones to date: Antarctica (1959), outer space (1967), the seabed (1970), and the South Pacific (1985–86). The first three are discussed in Part Seven and the last later in this chapter.

Even more sweeping, and more idealistic, are the proposals that have been made over the years for zones of peace. The first of any consequence in modern times was the Declaration of South-East Asia as a Zone of Peace, Freedom, and Neutrality adopted by the Association of Southeast Asian Nations (ASEAN) in November 1971—in the midst of the Vietnam War. It referred specifically to the Treaty of Tlatelolco, but aimed at "neutralization of South-East Asia." The UN General Assembly in November 1986 adopted the Declaration of a Zone of Peace and Co-operation of the South Atlantic. It is much broader and more detailed than the ASEAN declaration, but nowhere mentions "neutralization" of the region. It is still un-

clear what significance either of these declarations will have.*

Meanwhile, the UN General Assembly adopted in December 1971 the Declaration of the Indian Ocean as a Zone of Peace, discussed in more detail below, and King Birendra of Nepal in February 1975 proposed that his country be singled out as a zone of peace. Both of these proposals have been pursued with varying degrees of vigor for two decades, yet neither has actually been implemented. The Nepal proposal is especially interesting since, unlike the "neutral" statuses of Finland, Austria, and Switzerland, it did not derive from wars with neighbors, and, unlike the "neutralist" status of the "nonaligned" countries, it had nothing to do with the rivalry between the United States and the Soviet Union—at least not directly. Instead, it concerned a country that had long been at peace with both of its neighbors, is on good terms with both of them at present, and is not threatened by anyone else.

It is unclear what value all of these proposals for, and declarations of, zones of peace and nuclear-free zones may have in the worldwide search for peace. Perhaps their greatest value lies in the deterrent effect they might have on extraregional powers that might be tempted to intervene in the regions and in the catalytic effect they might have on other countries and regions seeking some permanent relief from the chaos, misery, and bloodshed they have experienced since World War II.

## *Resource Wars?*

For millennia States have fought over territory with the resources and population it contains. Even since the introduction of ideology (going back to the Arab conquests of

*The first meeting of States of the Zone of Peace and Co-operation of the South Atlantic was held at Rio de Janeiro in July 1988. Among other things, the meeting "emphasized the importance of the United Nations Convention on the Law of the Sea as an instrument regulating the uses of the oceans and their resources in a manner consistent with the interests of all nations."

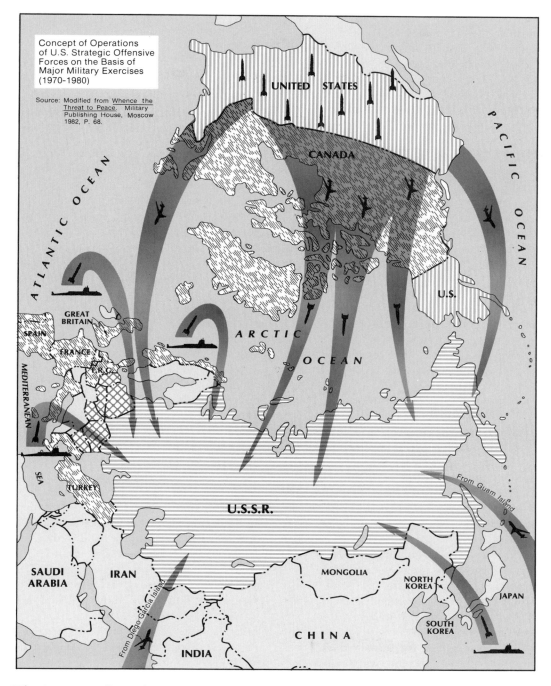

Concept of Operations
of U.S. Strategic Offensive
Forces on the Basis of
Major Military Exercises
(1970-1980)

Source: Modified from Whence the
Threat to Peace, Military
Publishing House, Moscow
1982, P. 68.

**Who is surrounding whom?** In Chapter 3 we pointed out that direction is essentially a matter of perception. We used as an example the perception of the American people that they were surrounded by a chain of their own military bases protecting them from attack by the Soviet Union. Here is a Soviet map showing some of those same bases, supplemented by nuclear-armed submarines, as seen from the Soviet Union. These very different perceptions of the same basic facts helped to create a climate that could have led to war. By presenting spatial facts objectively and analytically, political geographers can help to create a climate that will lead to enduring peace.

the seventh century, if not before, and including the Crusades, Napoleon's campaigns, and the activities of the Comintern*) and of nuclear weapons and "star wars," it is still true that most wars are fought and are likely to continue to be fought to gain and hold territory. While irredentism is still important and there could conceivably be more wars of independence, the danger of wars over resources continues unabated. This was one of the major objectives of the three UN conferences on the Law of the Sea (described in Chapters 31 and 32): to devise a public order of marine space and resources that would reduce, if not eliminate, conflicts over them. To some extent this is also an objective of the "confidence-building measures" currently being pursued within and outside the United Nations and of the work that organization has been doing to foster peaceful decolonization, liberalized world trade, economic integration, and the removal of Antarctica and outer space from territorial competition—all discussed later.

History records numerous conflicts over minerals, including hydrocarbons. In Chapter 17, for example, we pointed out that the quest for assured supplies of commodities, especially of fuels and raw materials, was one of the impulses leading to European conquest and colonization of much of the Southern Hemisphere, while greed for precious metals was probably the most important motivation in the earlier Spanish conquest of most of Latin America. During the world wars of this century, minerals were prime targets of military campaigns and diplomatic intrigues. After the First World War, the U.S. War Department drew up a list of 28 materials, mostly minerals, that had presented supply problems during the war, and the concept of "strategic minerals" became important in the interwar period. Definitions of the term vary considerably and

frequently change, but generally speaking, they include three factors:

1. The mineral is essential for defense.
2. Its supplies are found largely or entirely outside the country.
3. During wartime strict conservation and control of it are necessary.

Our gargantuan consumption of minerals during and since the Second World War enhanced fears that some minerals would be depleted, and the chronic instability of the postwar world enhanced fears of interruption or interdiction of mineral supplies destined for the rich, industrialized countries. The politically motivated, Arab-led embargo of petroleum shipments by members of OPEC to most Western countries and the subsequent quadrupling of the prices of petroleum in 1973–74 nearly resulted in panic in the importing countries (of which, it should be noted again, the most seriously affected were *not* the rich, long-industrialized countries that could well afford the higher prices, but the newly industrializing countries of Latin America, Africa, and Asia that could not). The movement for a New International Economic Order (discussed in Chapter 19), which also began in 1973, included a demand that all "natural resources" in developing countries be under the ownership and control of their own governments, and this heightened the concern of the Western industrialized countries still further.

They took a number of measures to ensure their supply of strategic minerals. They stepped up the search for new sources of supply, not only in the traditional land areas, but in the earth's frontier regions as well: Antarctica, the seabed, and outer space. (See Part Seven for details.) They devised new methods of retaining or even tightening control over supplies and prices; accelerated the search for substitutes for some of the strategic minerals; intensified conservation, recycling, and stockpiling—and lent crucial support to South Africa.

Despite its reprehensible sociopolitical system, its defiance of the UN and most of the world with regard to Namibia, and its

---

*The Comintern (the Third [Communist] International), was an association of communist parties around the world, organized and controlled by the Soviet Union, whose purpose was to protect the Soviet Union against "capitalist encirclement" and, to a lesser extent, to spread communist ideology as widely as possible.

# THE U.S. IS DEPENDENT ON MANY OTHER COUNTRIES FOR MANY OF THE MINERALS IT WANTS AND NEEDS

1991 Net Import Reliance*[1] of Selected Nonfuel Mineral
Materials as a Percent of Apparent Consumption[2]

| Mineral | Percent | Major Sources (1987-90) |
|---|---|---|
| ARSENIC | 100 | France, Chile, Sweden, Mexico |
| BAUXITE and ALUMINA | 100 | Australia, Guinea, Jamaica, Brazil |
| CESIUM (pollucite) | 100 | Canada |
| COLUMBIUM (niobium) | 100 | Brazil, Canada, Fed. Rep. of Germany |
| GRAPHITE | 100 | Mexico, China, Brazil, Madagascar |
| MANGANESE | 100 | Rep. of South Africa, Gabon, France |
| MICA (sheet) | 100 | India, Belgium, Brazil, France |
| RUBIDIUM | 100 | Canada |
| STRONTIUM (celestite) | 100 | Mexico, Fed. Rep. of Gemany, Spain |
| THALLIUM | 100 | Belgium, United Kingdom (U.K.), Fed. Rep. of Germany |
| GEM STONES (natural and synthetic) | 98 | Belgium, Israel, India, U.K. |
| ASBESTOS | 95 | Canada, Rep. of South Africa |
| DIAMOND (industrial stones) | 92 | Ireland, U.K., Rep. of South Africa, Zaire |
| PLATINUM-GROUP METALS | 88 | Rep. of South Africa, U.K., U.S.S.R. |
| TANTALUM | 85 | Fed. Rep. of Germany, Thailand, Brazil |
| COBALT | 82 | Zaire, Zambia, Canada, Norway |
| CHROMIUM | 80 | Rep. of South Africa, Turkey, Zimbabwe, Yugoslavia |
| FLUORSPAR | 79 | Mexico, Rep. of South Africa, China, Canada |
| TUNGSTEN | 75 | China, Bolivia, Fed. Rep. of Germany, Peru |
| NICKEL | 74 | Canada, Norway, Australia, Dominican Republic |
| TIN | 73 | Brazil, China, Indonesia, Malaysia |
| BARITE | 70 | China, India, Mexico, Morocco |
| POTASH | 67 | Canada, Israel, U.S.S.R., Fed. Rep. of Germany |
| ANTIMONY | 57 | China, Mexico, Rep. of South Africa, Guatemala, Bolivia |
| CADMIUM | 54 | Canada, Mexico, Australia, Fed. Rep. of Germany |
| SELENIUM | 52 | Canada, U.K., Japan, Belgium-Luxembourg |
| PEAT | 44 | Canada |
| GYPSUM | 30 | Canada, Mexico, Spain |
| ZINC | 30 | Canada, Mexico, Spain |
| PUMICE and PUMICITE | 22 | Greece, Mexico, Ecuador |
| SILICON | 22 | Brazil, Canada, Venezuela, Norway |
| MAGNESIUM COMPOUNDS | 15 | China, Canada, Greece, Mexico |
| SULFUR | 15 | Mexico, Canada |
| IODINE | 14 | Japan, Chile |
| IRON ORE | 14 | Canada, Brazil, Venezuela, Mauritania |
| NITROGEN | 14 | Canada, U.S.S.R., Trinidad and Tobago, Mexico |
| IRON and STEEL | 12 | European Community (EC), Japan, Canada, Rep. of Korea |
| CEMENT | 11 | Mexico, Canada, Japan, Greece |
| SALT | 11 | Canada, Mexico, Bahamas, Chile |
| VERMICULITE | 10 | Rep. of South Africa, China, Brazil |
| MICA (scrap and flake) | 9 | Canada, India |
| SODIUM SULFATE | 6 | Canada, Mexico |
| PERLITE | 5 | Greece |
| QUARTZ CRYSTAL (industrial) | 5 | Brazil, Namibia |
| LEAD | 4 | Canada, Mexico, Peru, Benelux |

*Estimated.
[1]Net import reliance = imports - exports + adjustments for Government and industry stock changes.
[2]Apparent consumption = U.S. primary + secondary production + net import reliance.

NOTE:
For a number of minerals, net import reliance data are withheld or incomplete. However, commodities for which sufficient data are avaliable to indicate a significant degree of import dependency include: bismuth (Belgium, Mexico, Peru, United Kingdom); ilmenite (Republic of South Africa, Australia, Canada); mercury (China, Spain, Japan, Algeria); pyrophyllite [wonder stone] (Republic of South Africa); rhenium (Chile, Federal Republic of Germany, Sweden, United Kingdom); rutile (Australia, Republic of South Africa, Sierra Leone); tellurium (United Kingdom, Canada, Japan, Philippines); and vanadium (Republic of South Africa, EC, South America, Mexico).

SOURCE:  U.S. Department of the Interior. Bureau of Mines. *Mineral Commodity Summaries. 1992.* P. 3.

vigorous efforts to "destabilize" its neighbors, South Africa claimed—and received—support from the Western democracies on three distinctly geopolitical grounds:

1.  It was both a bulwark against communism, which had taken over much of the continent to the north, and a staunch ally against Soviet expansionism.

2.  It was the guardian of the vital Cape Route, the route around the Cape of Good Hope followed by most supertankers that bring petroleum from the Persian Gulf to Western Europe, and by many other vessels as well.

3.  It was the noncommunist world's major supplier of strategic minerals, producing in 1984, for example, 40 percent of its manganese, 47 percent of its chrome, and about 80 percent of its platinum-group metals.

It was very difficult for the West to abandon South Africa or insist that it abandon mineral-rich Namibia, regardless of apartheid, because of these factors, of which the last was by far the most important.

Does this mean that the need for minerals is the most important element in determining the foreign policies of industrialized countries? Or that they—or anyone else—will again go to war over minerals? It's possible, but not likely. For one thing, there are now so many factors contributing to a breakdown in world order that minerals have simply declined in relative importance. So has the equation between mineral possession and military potential; a review of Chapter 20 on power analysis should make this clear. Third, while instability in the world has grown, so have the mechanisms for preventing and containing armed conflicts of all kinds; despite the scores of conflicts since 1945, none of them has been over minerals, and the world system has not broken down and dissolved into another world war. Finally, the world is now so utterly interdependent, so laced together with innumerable linkages, and so influenced by many countervailing forces that there are few if any insoluble problems and few if any that can be

isolated from all others. This is not to say that no country will ever again go to war in order to ensure its supply of a particular mineral or minerals, only that if a country needs minerals and does not want to go to war, there are many ways of satisfying its needs peacefully. If it wants to go to war, a need for minerals can no longer serve as an excuse. Food or water, perhaps, but not minerals.

## The Middle East

In order to illustrate some of the themes presented in this chapter, and in the book as a whole, we consider very briefly five case studies, regions where issues of war and peace, of interdependence and interrelationships are most apparent. We discuss them in descending order of level of conflict, beginning, of course, with the Middle East, the most tempestuous region of the world since 1945.

To understand the Middle East, one must begin at the beginning, with its location. The terms *Middle East* and *Near East* have been defined and used variously for centuries. Currently, definitions include many combinations of States, ranging from Morocco through Afghanistan and from Turkey through Somalia. All agree, however, that the heart of the Middle East is that area encompassed in the irredentist term *Greater Syria*: Syria, Lebanon, Israel, Jordan, and Iraq, along with Egypt and the Arabian peninsula. The Levant, those countries fronting on the eastern shore of the Mediterranean (Syria, Lebanon, Israel) is the heart of the heart, the great land bridge connecting Asia, Africa, and Europe, the western curve of the ancient Fertile Crescent, highway of armies, merchants, preachers, pilgrims, refugees, students, scholars, and countless others since the beginning of recorded history and probably earlier. Here is an equable climate, water, good soils, some good natural harbors, varied terrain with a fairly high proportion of level land, and even a little oil.

Here, over the centuries, have gathered a bewildering variety of peoples, probably the

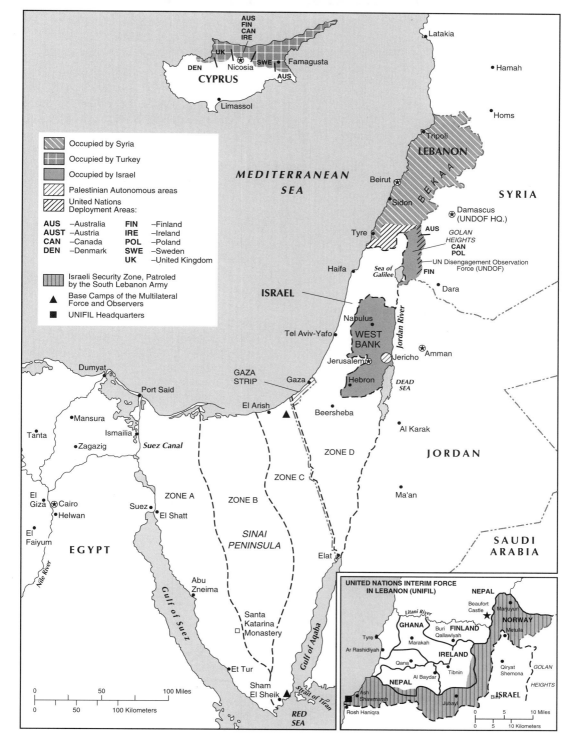

**The Middle East: Occupied areas and peace-keeping forces.** UN troops were withdrawn from Sinai in mid-1979 while Israel was gradually returning the peninsula to Egypt. They were replaced by the Multilateral Force and Observers who continue to monitor the relevant terms of the Israel-Egypt peace treaty. Within Zones A–C, decreasing numbers of Egyptian troops and decreasingly powerful weapons are permitted. In Zone D only lightly armed Israeli infantry are permitted. The MFO consists of one battalion each from Colombia, Fiji, and the United States, and observers in Zone C and an Italian Coastal Patrol Unit based in Sharm el Sheik. Support units come from France, Canada, the Netherlands, New Zealand, Australia, the United States, and Uruguay. There is also a small Civilian Observer Unit of U.S. nationals. The Force's headquarters are in Rome, and it has offices in Cairo and Tel Aviv. UN troops remain in Cyprus, southern Lebanon, and the Golan Heights of Syria, in all cases separating hostile forces.

greatest variety per unit of land of any region of the world. Here there have been wars and lesser conflicts for millennia; here there are wars and lesser conflicts still. The most prominent by far has been the conflict between Israel and its Arab neighbors, not over the territory occupied by Israel, but over the very existence of the State. Yet even the most ardent Arab nationalist would admit that the region knew little peace before the creation of Israel in 1948 or even before the first Zionist settlements in Palestine in the 1890s, and, upon serious consideration, would also admit that should Israel vanish instantly, tonight, there would still be conflicts and even warfare throughout the region for generations to come. The tensions between Christian and Muslim, Arab and Persian, Turk and Arab, Libya and Egypt, Hashemites and other royal houses, Bedouin and farmer, socialist and capitalist, French and British, old and new, oil rich and oil poor, water rich and water poor, Sunni and Shiite, Kurds and their neighbors, monarchist and republican, democrat and dictator, revolutionary and traditionalist, Maronite and Druze, smuggler and policeman, foreigner and native are all independent of Israel.

From 1980 to 1988 Iraq and Iran fought the region's bloodiest war in centuries, a frightful drain of lives and resources that neither party could afford. What were the stakes? Their mutual boundary in the Shatt-al-Arab, control of the head of the Persian Gulf, the Arab residents of western Iran, internal political power in each country, the Shiite majority in Iraq ruled by Sunnis, rival revolutionary ideologies—all these and more. Yet the war had linkages to other conflicts in the region. Both sides have been fighting the Kurds on and off for generations, as have from time to time the Turks. Neither they nor the rulers of the Caucasus nor the Syrians, also hosts to Kurdish settlements for generations, want the Kurds to have their own State or even autonomy within the existing States. Iran was supported by Syria and Libya, which also support terrorist groups operating throughout the region and beyond, but Iran did not sup-

port the expansionist goals of either ally. Libya at various times has invaded Egypt, "federated" with Egypt, occupied northern Chad, and threatened Sudan and Tunisia; Syria has at various times fought with Iraq and Jordan, allied itself with Jordan and Egypt, and occupied most of Lebanon.

We could go on and on with these linkages, exploring how Saudi Arabia and the other conservative monarchies of the Arabian Peninsula are involved, and Yemen, and Sudan, and Pakistan, and Greece—and, of course, Britain, France, the former Soviet Union, the United States, and many other countries. But the point has been made: the Middle East is such a convoluted region with so many complex, interlocking, and ancient rivalries that few theories about peace and war can emerge from it except in tatters. No theorists, for example, predicted the Iraqi invasion of Kuwait in August 1990, the strange coalition of States that formed to re-

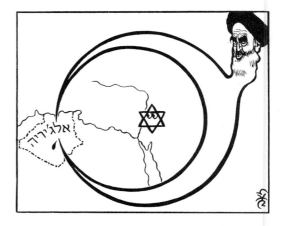

**Another view of North Africa as seen from the Mideast.** This cartoon, like the map in Chapter 3, shows an important and tumultuous part of the world from a perspective unfamiliar to most people in the Western world. Here, Israel watches warily as Iran, in a pincers movement bypassing Israel, tries to take over Algeria, thereby threatening Israel from the west. This cartoon by Ze'ev appeared in the Israeli daily newspaper *Ha'aretz* on 1 July 1992. (Used by permission)

verse that situation, and the Gulf War that followed early in 1991.

The war provided the needed catalyst for beginning the process of settling the Arab-Israel conflict, but, notwithstanding the conclusion of a peace treaty between Jordan and Israel and an agreement between Israel and the Palestine Liberation Organization, both in 1994, it will take heroic efforts by many people over a long period of time to bring peace to this dynamic but troubled region. These efforts include The Middle East/North Africa Economic Summit held in Casablanca late in 1994, which brought together leaders of most States in the region (including Israel) and of intergovernmental organizations and private industry under the patronage of the United States and Russia to discuss matters of common interest and concern. We can hope that it will be followed by many more such efforts to bind together disparate peoples and forces.

## The Caribbean Basin

The Caribbean is a much simpler region to understand, though not without its own complexities. Although for more than three centuries it was the scene of almost continual jockeying and even warfare among Britain, France, Spain, and the Netherlands for control of the region's resources and trade routes, the United States has for more than a century and a half considered it an American lake and resisted any "foreign" intrusion into it. In fact, the parallels between the Caribbean and Eastern Europe since World War II are very close indeed. Just as Yugoslavia broke away from Soviet hegemony in 1948, Cuba broke away from American hegemony in 1959. Thereafter neither superpower hesitated to use its clandestine forces and even its regular troops to maintain its control over its "sphere of influence": the Soviet Union in East Germany, Poland, Hungary and Czechoslovakia; the United States in Guatemala, the Dominican Republic, Grenada, Panama, and Nicaragua—in addition to the bungled invasion of Cuba in 1960.

In Chapter 3 we point out how decision makers are often influenced by their mental maps, their perceptions of the political world. It would be hard to find a better example of this than a statement made by General David C. Jones, then Chairman of the American Joint Chiefs of Staff (the highest-ranking uniformed officer in all the armed services), to the Chicago Bar Association in March 1981. In defending the use of American military advisers to help the conservative government of El Salvador suppress a leftist-led rebellion, he said that the United States must end "erosion in our hemisphere" and "get the Cubans not to play around in our back yard." He went on to invoke the domino theory again, "[If left alone,] El Salvador would probably go under Cuban influence—then you'd see another country and another country." Now, even a child can look at a map of the Caribbean Basin and see that El Salvador, like most other countries, is certainly "in the back yard" of several countries, but the United States is just as certainly not one of them.

The United States does, indeed, have legitimate interests in the region: The Panama Canal, drug trafficking from South America to the United States, petroleum flows from Mexico and Venezuela, heavy private American investment and substantial markets in the region, numerous American citizens residing there, considerable migration to the United States from the region, and various military and civilian installations. It is questionable, however, whether these interests are entirely compatible with the legitimate interests of the peoples of the region or whether they confer on the United States a right to attempt to control all governments and economies around the perimeter of the Caribbean Sea and the Gulf of Mexico.

But the United States is not the only important player on the Caribbean stage; nor is it responsible for all of the region's problems, nor have all of its activities been malevolent, or even consistent; nor is the situation in the region static. As in the Middle East, if the most powerful and controversial country were to be magically removed altogether, the Caribbean would still have very serious problems, though on a smaller scale and at a lower level of intensity. There is in the Caribbean much more of a sense of collegiality, if not of true unity, than in the Middle East, a longer history of serious joint efforts to relieve fundamental economic and social problems, and, of course, far more democratic governments. While there have been clashes around the basin, notably between Haiti and the Dominican Republic and between El Salvador and Honduras, they have been contained in duration, scope, and intensity and have never been a threat to the region.

The various subregional intergovernmental organizations, such as the Central American Common Market, the Caribbean Community (CARICOM), and the Organization of Eastern Caribbean States, while not precisely shining successes, are nevertheless indications of a genuine desire to work together as much as possible. The introduction by the United States of its Caribbean Basin Initiative in early 1982, though clearly a reaction to the accession to power of the Sandinista government in Nicaragua and not entirely altruistic, did, like its more ambitious predecessor, the Alliance for Progress, demonstrate some genuine American interest in assisting the countries of the region to develop economically and socially. To date it has not been notably successful and may, like the Alliance, vanish soon with scarcely a trace left behind, but it is arguably a very great improvement over (though not yet a substitute for) military intervention. All things considered, there is a better chance for a durable peace reasonably soon in the Caribbean than in the Middle East.

## *The Pacific Basin/Pacific Rim*

For many years we were accustomed to thinking of the Pacific Basin as an entity, largely because its size, nearly symmetrical shape, and sprinkling of Polynesian islands seemed to endow it with a kind of unity, if not uniformity, much like the Caribbean Basin. Yet, on closer inspection, we noted far more differences than similarities from one part of the ocean to another, and little unity of any kind. Now, for some inexplicable reason, "Pacific Rim" seems to be the more favored term, perhaps because of the rising importance of the newly industrialized countries (South Korea, Taiwan, Hong Kong, Singapore) on the western edge of the basin or perhaps because of the strong and tightening links among most of the States around the periphery of the basin.

Japan has been the dynamo driving this growing unity of the region, forging economic and cultural ties with many of the countries around the rim, largely because of its constant need for more sources of commodities and more markets for its manufactured goods and services. The United States and Japan remain tightly bound together, notwithstanding some serious disagreements between them. Australia and the United States remain good friends and allies within the framework of the ANZUS treaty, although New Zealand is no longer participating in what was the Australia–New Zealand–United States alliance. Australia and Japan have formed an especially close relationship, based largely on symbiotic economies. Transpacific trade has become so important and is growing so rapidly that it may soon rival transatlantic trade in volume and importance. ASEAN and the South Pacific Forum have at last achieved significance and are bringing tangible benefits to their peoples, while tying them ever more closely together. The U.S.-Canadian relationship continues to be strong, despite familial squabbles from time to time.

The spectacular growth of California combined with recent heavy immigration into the state from most countries in and around the basin have reinforced California's traditional westward orientation, and the state, in turn, seems to be pulling the whole country along in a shift from an Atlantic to a Pacific emphasis. The same is true to a lesser extent of Washington, Oregon, and British Columbia. Australia and New Zealand have forged "closer economic relations," virtually a common market, and retain strong ties with their former colonies and mandates in the South Pacific.

Mexico, and especially Chile, have become active in Pacific affairs, developing new relationships across the ocean. The Permanent Commission of the South Pacific ties together Colombia, Ecuador, Peru, and Chile in a cooperative marine research and conservation effort. In November 1989, the Asia-Pacific Economic Cooperation (APEC) was organized at Australian initiative to enable 11 Pacific Rim countries to work together for an open multilateral trading system and for economic development generally. These and many other developments are tending to validate the vision of Professor Kiyashi Kojima (Hitotsubashi University of Japan) and Hiroshi Kurimoto (UN Economic Commission for Asia and the Far East), who in 1965 first proposed the creation of a Pacific Free Trade Area.

By far the most exciting—and perhaps farthest reaching—project in the entire Pacific Rim is the *Tumen River Area Development Programme.* Focused on the delta of the Tumen River, which for much of its length forms the boundary between China and North Korea and downstream between Russia and North Korea, this project, if it actually reaches fruition, will be one of the world's largest regional development efforts. China, Mongolia, Russia, North Korea, South

Korea, and Japan are collaborating in the project, which is administered by the United Nations Development Programme.

Dating from October 1991, it calls for massive construction projects in a roughly triangular area of some 26,000 square kilometers (10,000 square miles) bounded by Yanji, China; Vladivostok, Russia; and Chongjin, North Korea, within which is to be a free trade area of some 2600 square kilometers (1000 square miles) called the Tumen River Economic Zone. Included in the plan are construction of 11 new ports; exploitation of the region's coal, mineral, and water resources; new transportation facilities; and a new city of 500,000 people. Estimated to cost $30 billion over 20 years, it will for the first time provide China and land-locked Mongolia access to the Sea of Japan. There are problems, of course: lack of infrastructure to start with, potentially serious environmental impacts, and the financial cost. The fact, however, that these six countries—historic enemies—are working together to achieve a common goal is simply astounding, comparable to Britain, France, and Germany collaborating after centuries of conflict among them. Probably only the United Nations could have brought them together.

Yet the forces operating in the region are

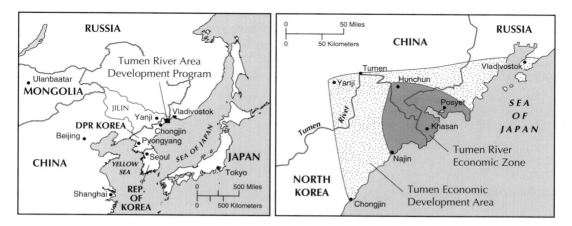

**The Tumen River Area Development Programme.** This area was a political backwater between the end of the Korean War in 1953 and the end of the Cold War in 1991. Since then both its strategic value and its economic potential have been enhanced and have drawn the attention of international agencies and outside powers. If the programme is carried out, even in large part, it could tie the beneficiary countries together so tightly that Northeast Asia may become accepted as another major world subregion.

not all centripetal. The former Soviet Union not only maintains a formidable Pacific fleet, with a major base at Cam Ranh Bay in Vietnam, but has also negotiated aid and fishing agreements with several South Pacific island States. China and Taiwan are still competing for support around the world. Libya has begun to catalyze and support revolutionary movements in a largely peaceful South Pacific. Melanesian nationalists have become more united, vocal, and aggressive in asserting their rights against France, Indonesia, and Australia. The Philippines is still unstable.

There are still enormous disparities in economic well-being and political development within the region, and incredible cultural diversity. Transportation and communications are still poor in the South Pacific. The United States refuses to sign the South Pacific Nuclear-Free Zone Agreement, a strong force for unity in the region. Political tensions on the Asian mainland detract from unity efforts. Russia, the United States, Canada, Japan, and China, the most powerful countries around the Pacific Rim, all have vital interests outside the region. And the Latin American countries of the Pacific Rim (except for Chile) have yet to develop the same commitment to closer relations as their counterparts to the north and west.

Nevertheless, there seems to be evolving some form of closer association in a region that, though heavily armed and beset by rivalries and competition, does have commonalities and an incipient community of interest. During the past 200 years we have seen the center of gravity in the world shift from the Mediterranean to the North Atlantic. It may now be shifting toward the Pacific. We can hope that this shift will be more peaceful than the last.

## The Indian Ocean

Although the Indian Ocean has been crisscrossed by vessels of every description for millennia, it has remained the least known of the three largest oceans. Because no great powers have developed along its shores and no major industrial countries have emerged

on its periphery, it has never become a major battleground except, at times, at its outer margins. Two world wars and the Cold War left it tranquil and of little interest to the world at large. With growing interest in the sea after World War II and growing dependence of the industrialized countries on Persian Gulf petroleum, interest in the Indian Ocean began to awaken. This interest was reinforced by the beginning of the decolonization process on three sides of the ocean. Still, largely under the informal but effective supervision of the Royal Navy and other British forces, it remained outside the sphere of big-power rivalries.

A landmark in its history was the International Indian Ocean Expedition of 1964, sponsored by the International Oceanographic Commission (an agency of UNESCO), the first large-scale, concerted effort to obtain basic scientific information about the ocean. Another was the British decision in 1968 to withdraw from nearly all its remaining important positions "east of Suez." Since then Mauritius, Bahrain, Qatar, the United Arab Emirates, Mozambique, the Comoros, the Seychelles, Eritrea, Djibouti, Bangladesh, and land-locked Swaziland have joined the ranks of independent States in and around the ocean.

In 1965 the British created the British Indian Ocean Territory out of a number of uninhabited and sparsely inhabited islands in midocean, of which they still retain the Chagos Archipelago. Australia and France also retain some small islands in the eastern, southern and western sectors. Otherwise, only small or mid-size countries occupy the land in and around the Indian Ocean.

The withdrawal of the British created a classic power vacuum in a huge and increasingly strategic region. The United States and the Soviet Union began sending naval units into the ocean on show-the-flag missions. India and Iran began building navies that could operate outside their coastal waters. Australia turned hesitantly from the Pacific and began debating building a navy base at Cockburn Sound in Western Australia. Intensive jockeying began for control of the Bab-el-Mandeb, the southern entrance

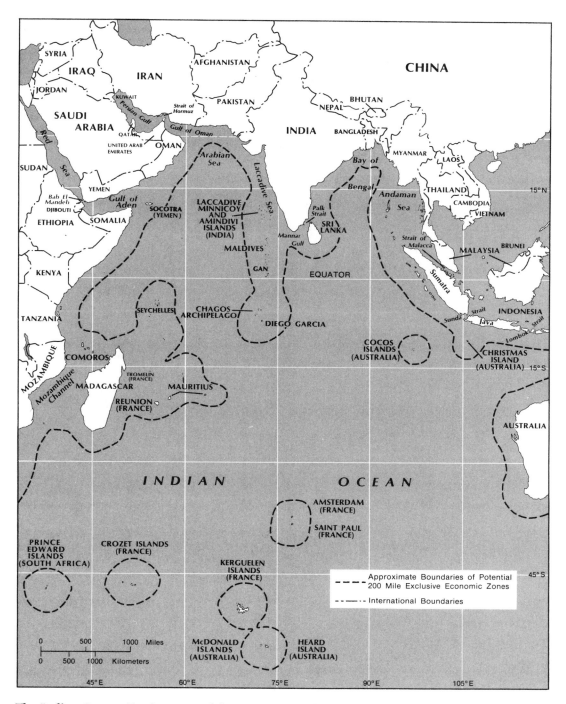

**The Indian Ocean: Testing ground for a new geopolitics of peace.** Note the large proportion of the ocean that will come under national jurisdiction if 200-mile exclusive economic zones are claimed by all of the States in and around the basin. (See Chapter 32.) Note also the formerly remote, but now "strategic," location of Diego Garcia island and the critically important straits of Bab el Mandeb, Hormuz, and Malacca around the northern arc of the ocean. This northern arc is a part of Spykman's Rimland.

to the Red Sea, with Israel more than just an interested bystander. Japan grew nervous about its vital oil-supply route from the Persian Gulf. China began making friends in East Africa. The Soviet Union for a time had the use of shore facilities in Somalia for its navy and offered to lease a former Royal Air Force base in the Maldives. Iran helped Oman suppress a rebellion in Dhofar province. The closure of the Suez Canal from 1967 to 1974 diverted considerable maritime traffic along the east coast of Africa. The United States and Britain set up joint communications and refueling stations on Aldabra Island in the Seychelles and Diego Garcia in the Chagos Archipelago, almost precisely in midocean. Iran occupied the disputed Tunb and Abu Musa Islands near the Strait of Hormuz in 1971. The United States in 1976 began upgrading its austere facility on Diego Garcia into a full-fledged carrier task force support base.

While all this maneuvering was going on, nearly every country bordering the Indian Ocean experienced internal unrest, up to and including actual warfare, and a number engaged in localized international wars. Clearly, the Indian Ocean had become "destabilized." The situation has been watched carefully at the United Nations and around the world. There is ample evidence that no one wants the Indian Ocean to become a new arena for big-power rivalry, but the middle-rank powers of the littoral—India, Iran, South Africa, Australia—all have their own interests to protect. Indonesia, Thailand, Pakistan, Saudi Arabia, and Kenya may also, in time, begin to project themselves into the Indian Ocean, complicating what is at present a relatively simple situation.

To prevent these rivalries from developing into conflicts and to keep the region peaceful, three approaches are being pursued simultaneously. First, bilateral negotiations between the United States and the Soviet Union/Russia have been conducted irregularly since 1977. Second, there is an incipient sense of regional community developing, exemplified by the Indian Ocean Marine Affairs Co-operation, which held its first conference in Colombo, Sri Lanka, in July 1985,

and by the South Asian Association for Regional Cooperation, which held its first formal meeting, also in Colombo, in August 1983. Third, negotiations have taken place within the United Nations, which in 1971 passed a Declaration of the Indian Ocean as a Zone of Peace. The UN continues trying to implement this declaration through a variety of devices. One of them was a July 1979 "meeting of littoral and hinterland States" that was to lead toward a general conference on the Indian Ocean. But for a number of reasons the conference continues to recede into the future and the emphasis now is on confidence-building measures and economic and environmental cooperation.

Will these moves work? Will the reluctance of the major powers to make significant military commitments in the Indian Ocean and the opposition of the littoral States to such commitments sustain the present "balance of no power" situation? Can the Indian Ocean remain a power vacuum in a turbulent world? Is a new kind of geostrategy at work here, a geostrategy of peace? It's much too early to tell, but certainly what we are witnessing is an advance over the ethnocentric strategies discussed in the two preceding chapters, strategies based on power. And it is better than the evolving and growing—and related—conflicts between rich and poor and between energy-rich and energy-poor. Perhaps there is a sign of hope in the Indian Ocean that sanity can prevail even where vital interests of many countries, large and small, are at stake.

### The Arctic

This is the area de Seversky dubbed "the Area of Decision," which became so strategically important during the Second World War and the Cold War. We discussed it in Chapter 21, but now we see it from a different perspective.

Like the Indian Ocean, the Arctic Ocean has never been an arena for great power clashes, but it has not been the scene of any intense national rivalries either. Its location between the world's two most powerful countries has not changed, but their attitude toward one another has. Since the collapse of the Soviet Union in 1991, earlier trends toward cooperation across and around the Arctic Ocean have accelerated, and it is now reasonable to consider it an emerging world region.

In Chapter 16 we discussed the Inuit Circumpolar Conference, and in Chapter 29 we discuss the Nordic Council, two early manifestations of regional cooperation. Another is the Northern Regions Conference, initiated in the early 1970s by the government of Hokkaido, Japan. At that time, the focus was on the human environment, and the conference was attended by officials from the United States, Canada, Finland, Sweden, and Norway. Since then, sessions have been held in Alberta and in Alaska, each time with a different theme and greater participation. It is not a formal organization, but rather a forum for discussion of a wide range of topics of common interest. Since it now includes Russia, China, and four other participants in addition to the original seven, it could serve as the nucleus of a new regional organization. There is also a Circumpolar Universities Association, whose fourth Circumpolar Universities Cooperation Conference was held in Prince George, British Columbia in February 1995. Canada and the Nordic countries have appointed "ambassadors to the Arctic," officials responsible for circumpolar Arctic issues.

The region is also being laced together by a considerable body of agreements among the circumpolar governments on a wide range of topics. Among them, two deserve special mention. In June 1991 the eight Arctic States (the five Nordic States plus Canada, the United States, and the former Soviet Union) adopted the Declaration on the Protection of the Arctic Environment and the Arctic Environmental Strategy. These two documents formalize the already impressive cooperation in Arctic environmental matters and provide a point of departure for further cooperation.

Then, in January 1993 representatives of the Nordic States, Russia, and the European Community (now the European Union), meeting in Kirkenes, Norway, signed a declaration establishing the Barents Euro-Arctic Council (BEAC). Other European countries sent observers to the meeting, as did Canada, the United States, and Japan. Initiated by Norway, the BEAC will feature a Regional Council of local government representatives from northern Norway, Sweden, Finland and northwest Russia, as well as a Saami (Lapp) representative, which will supervise its activities. These activities will initially concentrate on environmental problems, smuggling and other criminal problems, and—perhaps most important and certainly most geopolitical—developing Russia's old Northern Sea Route into a new highway of commerce between Northeast Asia and Europe, with all the ancillary economic development of the land area in the High Arctic. This alone is another truly remarkable change from the Cold War period, when the Northern Sea Route was shrouded in secrecy and was the scene of occasional confrontations between Soviet and U.S. naval vessels.

It is too early to tell whether all this cooperative activity in a former zone of confrontation will lead to a formal regional organization or to a new Arctic consciousness among the ordinary people of the region, but already it is serving as an example of what can be done constructively in a still-strategic region with genuine security concerns. It could be one of the key elements in the evolution of a new geopolitics of peace.

## War and Peace and Geography

Having introduced the subject of war and peace, considered the possibility of resource wars, and glanced at five important geopolitical regions, we can now review some of the major themes in the contemporary literature on the subject. Few of these writings, it should be noted, are by geographers, a deficiency that should be rectified in the near future. We *broadly* follow here the rough outline of an analysis by John O'Loughlin and Herman van der Wusten, American and Dutch geographers, respectively, in the December 1986 issue of *Progress in Human Geography*.

1. **Geography of war.** Broadly speaking, there have been three major approaches to the geography of war. Traditionally, international relations theory has been characterized by emphasis on power relationships among States. Foreign policies of individual States, sometimes analyzed in their historical, cultural, geographic, and ideological contexts, determine the outcomes of given situations and the long-term trends in world affairs. Typical of this *traditionalist* view is the vision of the world as split between East and West (socialist and capitalist economic systems) and between North and South (rich and industrialized, poor and agricultural), yielding a quadripartite division of the world, with competition among the four areas the dominant theme. We have already seen, however, that while such competition surely does exist, it is seldom clear-cut and seldom decisive.

   *Behavioralism* has dominated international relations research for the past 30 years or so, with its quantitative techniques and emphasis on the relationship between a State's domestic qualities and its foreign policy. Here geographic factors, such as location, population, and alliances, join with economic attributes, political orientation, and arms races as predictors of conflict. Deriving from this approach are some of the newer geopolitical theories, notably that of Saul Cohen, described in Chapter 21. This approach is also incomplete because it does not adequately consider the historical evolution of States and tends to concentrate too much on particular features of the political world while neglecting others.

   More recently, *structuralist* perspectives have attracted considerable attention. Otherwise known as *world-systems* theory, these perspectives derive largely from the work of Immanuel Wallerstein and George Modelski. They take a global and historical view of the world rather than one centered on States and the contemporary scene in order to explain how the world works today. Although they differ in some respects, they both emphasize the links between the dynamic global economic and political systems, the cycles of international hegemony, and the occurrences of conflict. Wallerstein's view tends to be rather Marxian, and includes such features as the core-periphery model of the world and both the class struggle between proletariat and bourgeoisie and the political struggle between different bourgeois. Modelski's contribution has been the long-cycle model of successive long periods of world dominance by single States.

2. **Geography of peace.** For lack of a better practical definition, peace can best be defined as the absence of armed conflict. This does not preclude competition, even rivalry, but it does preclude violence. The pattern of peace studies so far has generally tended to follow the same pattern as war studies. The relatively few scholars active in the field concentrate on analyzing policies that result in the avoidance of violent conflict and hence the maintenance of peace. Generally speaking, both for States and groups of States, there are two ways to avoid war: dissociation and association.

   *Dissociation* means essentially separation of States from other States that might engage in violence. Geographical isolation helps, of course, but (as we have seen and shall see again later) is no guarantee of peace and in any case does not result from government policy. Neutrality, nonalignment, deterrence, partition, peace-keeping forces, and political/economic isolation are policies designed to prevent conflict by reducing contact. Because of the many features of the contemporary world already discussed in this book and to be discussed later (e.g., long-range missiles, economic interdependence, global environmental problems), these policies are seldom entirely successful.

   *Association* is based on interaction,

sharing, or cooperation. This means essentially the kinds of relationships covered in Part Six: international law, economic integration, regional and worldwide functional cooperation, efforts toward intercultural understanding, and so on. But proximity alone, even common membership in organizations, does not ensure peace. There must be serious and sustained efforts to resolve differences peacefully, or proximity may simply make war more tempting and more likely.

In summary, then, we can say that geographers can and should contribute to the achievement and maintenance of world peace in both their personal and professional activities. One way to do this is through the various peace academies being opened around the world. Another is through the University for Peace, created by the United Nations in 1981 but operating in Costa Rica outside the UN framework, to promote peace through research and education. Another is to get involved in some of the activities described in the next few chapters.

# BIBLIOGRAPHY FOR PART FIVE

## *Books and Monographs*

**A**

Abdulghani, Jasim, *Iraq and Iran; the Years of Crisis.* Baltimore: Johns Hopkins Univ. Press, 1985.

Ahmad, Zakaria H. and others, *Pacific Asia in the 1990's.* New York: Routledge, 1992.

Alaolmolki, Nozar, *Struggle for Dominance in the Persian Gulf—Past and Present and Future Prospects.* Peter Lang, 1991.

Alford, Jonathan (ed.), *Sea Power and Influence: Old Issues and New Challenges.* Montclair, NJ: Allanheld, Osmun, 1980.

Aliboni, Roberto, *The Red Sea Region; Local Actors and the Superpowers.* Syracuse, NY: Syracuse Univ. Press, 1985.

Allen, Philip M., *Security and Nationalism in the Indian Ocean.* Boulder, CO: Westview, 1987.

Amirahmadi, Hooshang and Nader Entessar (eds.), *Iran and the Arab World.* New York: St. Martin's, 1992.

Anderson, Ewan W., *Strategic Minerals: The Geopolitical Problems for the United States.* New York: Praeger, 1988.

Anderson, Thomas P., *Geopolitics of the Caribbean: Mini-states in a Wider World.* New York: Praeger, 1984.

Archer, Clive and David Scrivener (eds.), *Northern Waters: Security and Resource Issues.* Barnes & Noble, 1986.

Aronson, Geoffrey, *Israel, Palestinians and the Intifada; Creating Facts on the West Bank.* New York: Kegan Paul, 1990.

**B**

Bagley, Bruce M. (ed.), *Contadora and the Diplomacy of Peace in Central America,* 2 vols. Boulder, CO: Westview, 1987.

Bailey, Sydney D., *The Making of Resolution 242.* Dordrecht: Nijhoff, 1985.

Bakhash, Shaul B., *The Politics of Oil and Revolution in Iran.* Washington, DC: Brookings, 1982.

Ball, Desmond and Cathy Downes (eds.), *Security and Defence; Pacific and Global Perspectives.* Allen & Unwin, 1991.

Bar-Siman-Tov, Yaacov, *Israel, the Superpowers and the War in the Middle East.* New York: Praeger, 1987.

——, *Israel and the Peace Process 1977–1982; in Search of Legitimacy for Peace.* Albany, NY: SUNY Press, 1994.

Bateman, Michael and Raymond Riley (eds.), *The Geography of Defence.* Barnes & Noble, 1987.

Bello, Walden, *People and Power in the Pacific: The Struggle for the Post–Cold War Order.* San Francisco: Food First, 1992.

Ben-Dor, Gabriel and David B. Dewitt (eds.), *Conflict Management in the Middle East.* Lexington, MA: Lexington Books, 1987.

Ben-Rafael, Eliezer, *Israel-Palestine.* Westport, CT: Greenwood, 1987.

Benvenisti, Meron and Shlomo Khayat, *The West Bank and Gaza Atlas.* Boulder, CO: Westview, 1990.

Bergesen, Helge Ole and others, *Soviet Oil and Security in the Barents Sea.* New York: St. Martin's, 1987.

Betts, Richard K., *Nuclear Blackmail and Nuclear Balance.* Washington, DC: Brookings, 1987.

Blight, James G. and Thomas G. Weiss, *The Suffering Grass: Superpowers and Regional Conflict in Southern Africa and the Caribbean.* Boulder, CO: Lynne Rienner, 1992.

Blomley, Nicholas K., *Law, Space, and the Geographies of Power.* New York: Guilford, 1994.

Blouet, Brian, *Halford Mackinder: A Biography.* College Station: Texas A&M Press, 1987.

Boulding, Kenneth E., *Conflict and Defense: A General Theory.* Lanham, MD: Univ. Press of America, 1988.

——, "Toward a Theory of Peace," in R. Fisher (ed.), *International Conflict and Behavioral Science.* New York: Basic Books, 1964, 70–87.

——, *Stable Peace.* Austin: Univ. of Texas Press, 1978.

Braveboy-Wagner, Jacqueline, The *Caribbean in World Affairs: The Foreign Policies of the English-Speaking States.* Boulder, CO: Westview, 1987.

Braveboy-Wagner, Jacqueline A. and others, *The Caribbean in the Pacific Century: Prospects for Caribbean-Pacific Cooperation.* Boulder, CO: Lynne Rienner, 1992.

Brenchley, Frank, *Britain and the Middle East.* New York: St. Martin's, 1991.

Bresnan, John, *From Dominoes to Dynamos; The Transformation of Southeast Asia.* New York: Council on Foreign Relations, 1994.

Brigham, Lawson W. (ed.), *The Soviet Maritime Arctic.* London: Pinter, 1991.

Broek, Jan O.M., "The German School of Geopolitics," in Russell H. Fitzgibbon (ed.), *Global Politics.* Berkeley: Univ. of California Press, 1944, 167–177.

Brown, James and William P. Snyder (eds.), The *Regionalization of Warfare.* New Brunswick, NJ: Transaction, 1984.

Brown, Neville, *The Strategic Revolution; Thoughts for the Twenty-First Century*. London: Brassey's, 1992.

Brundtland, Arne and John Skogan, *Soviet Sea Power in Northern Waters*. New York: St. Martin's, 1990.

Bruton, Henry J., *The Promise of Peace: Economic Cooperation Between Egypt and Israel*. Washington, DC: Brookings, 1981.

Bryan, Anthony T. (ed.), *The Caribbean; New Dynamics in Trade and Political Economy*. New Brunswick, NJ: Transaction, 1994.

Bryan, Anthony T. and Andrés Serbin (eds.), *Distant Cousins: The Caribbean—Latin American Relationship*. New Brunswick, NJ: Transaction, 1994.

Bunge, William, *The Nuclear War Atlas*. New York: Blackwell, 1987.

Butts, Kent Hughes, and Paul R. Thomas, *The Geopolitics of Southern Africa; South Africa As a Regional Superpower*. Boulder, CO: Westview, 1986.

## C

Cable, James, *Gunboat Diplomacy, 1919–1979: Political Applications of Limited Naval Force*. New York: St. Martin's, 1986.

Caflisch, Lucius and F. Tanner (eds.), *The Polar Regions and Their Strategic Significance*. Geneva: PSIS, 1989.

Caldwell, John, *A Military Assessment of Chinese Military Power Projection Capabilities, 1993–2000*. Washington, DC: CSIS, 1994.

Caldwell, Nathaniel French, Jr., *Arctic Leverage; Canadian Sovereignty and Security*. Westport, CT: Greenwood, 1990.

*The Casablanca Report; The Middle East/North Africa Economic Summit, Casablanca, Morocco, Oct. 30–Nov. 1, 1994*. New York: Council on Foreign Relations, 1995.

Castle, Emery and others, *U.S. Interests and Global Natural Resources; Energy, Minerals, Food*. Baltimore: Johns Hopkins Univ. Press, 1983.

Chandra, Nehru Satish and others (eds.), *The Indian Ocean and Its Islands: Strategic, Scientific, and Historical Perspectives*. Newbury Park, CA: Sage, 1993.

Chase-Dunn, Christopher, *Global Formation: Structures of the World-Economy*. Oxford: Blackwell, 1989.

Chay, J. and T. E. Ross (eds.), *Buffer States in World Politics*. Boulder, CO: Westview, 1986.

Chayes, Abram, *The Cuban Missile Crisis*. Oxford Univ. Press, 1974.

Child, Jack, *Geopolitics and Conflict in South America: Quarrels Among Neighbors*. New York: Praeger, 1985.

———, *Conflict in Latin America*. New York: St. Martin's, 1986.

———, *The Central American Peace Process, 1983–1991: Sheathing Swords, Building Confidence*. Boulder, CO: Lynne Rienner, 1992.

Cohen, Michael J., *The Origins and Evolution of the Arab-Zionist Conflict*. Berkeley: Univ. of California Press, 1987.

Cohen, Saul B., *The Geopolitics of Israel's Border Question*. Boulder, CO: Westview, 1987.

Coll, Alberto R. and Anthony C. Arend (eds.), *The Falkland War: Lessons for Strategy, Diplomacy and International Law*. London: Allen & Unwin, 1985.

Collins, Joseph J., *The Soviet Invasion of Afghanistan*. Lexington, MA: Lexington Books, 1985.

Cordesman, Anthony H., *The Gulf and the West; Strategic Relations and Military Realities*. Boulder, CO: Westview, 1988.

Crawford, *Raw Materials and Pacific Economic Integration*. Vancouver: UBC Press, 1986.

Curtis, Michael (ed.), *The Middle East*. New Brunswick, NJ: Transaction, 1986.

## D

Dalby, Simon, *Creating the Second Cold War; The Discourse of Politics*. London: Pinter, 1990.

Day, Arthur R., *East Bank/West Bank*, New York: Council on Foreign Relations, 1986.

Deeb, Mary-Jane, *Libya's Foreign Policy in North Africa*. Boulder, CO: Westview, 1987.

Deger, Saadet and Somrath Sen, *Military Expenditure; The Political Economy of International Security*. Oxford Univ. Press, 1991.

Demangeon, A., *America and the Race for World Dominion*. New York: Doubleday, Page, 1921.

Demko, George J. and William B. Wood (eds.), *Reordering the World; Geopolitical Perspectives on the 21st Century*. Boulder, CO: Westview, 1994.

De Seversky, Alexander P., *Victory Through Air Power*. New York: Simon & Schuster, 1942.

———, *Air Power: Key to Survival*. New York: Simon & Schuster, 1950.

———, *America: Too Young to Die*. New York: McGraw-Hill, 1961.

Deudney, David, *Whole Earth Security: A Geopolitics of Peace*. Washington, DC: Worldwatch Institute, 1983.

DeWitt, David and others (eds.), *Building a New Global Order; Emerging Trends in International Security*. Oxford Univ. Press, 1994.

Dickinson, Robert E., The *German Lebensraum.* London: Routledge, 1943.

Dirlik, Arif (ed.), *What Is in a Rim?; Critical Perspectives on the Pacific Region Idea.* Boulder, CO: Westview, 1993.

Dorpalen, A., *The World of General Haushofer.* New York: Holt, Rinehart, 1942.

**E**

Elazar, Daniel J., *The Camp David Framework for Peace: A Shift Toward Shared Rule.* Lanham, MD: Univ. Press of America, 1979.

Ellison, Herbert J. (ed.), *Japan and the Pacific Quadrille.* Boulder, CO: Westview, 1987.

Epstein, Joseph M. *Strategy Force Planning: The Case of the Persian Gulf.* Washington, DC: Brookings, 1987.

Erisman, H. Michael, *Pursuing Postdependency Politics: South-South Relations in the Caribbean.* Boulder, CO: Lynne Rienner, 1992.

**F**

Fairgrieve, James, *Geography and World Power.* 8th ed. New York: Dutton, 1941.

Farid, Abdel Majid (ed.), *The Red Sea.* New York: St. Martin's, 1984.

Faringdon, Hugh, *Strategic Geography: NATO, the Warsaw Pact and the Superpowers.* 2nd ed. London: Routledge, 1989.

Fauriol, Georges A., *Foreign Policy Behavior of Caribbean States: Guyana, Haiti and Jamaica.* Lanham, MD: Univ. Press of America, 1984.

———, *U.S.–Caribbean Relations into the Twenty-First Century.* Washington, DC: CSIS, 1995.

Fernandez, Damian J., *Cuba's Foreign Policy in the Middle East.* Boulder, CO: Westview, 1987.

Fifield, Russell H. and G. Etzel Pearcy, *Geopolitics in Principle and Practice.* Boston: Ginn, 1944.

Frieden, Jeffry A. and David A. Lake (eds.), *International Political Economy; Perspectives on Global Power and Wealth.* 2nd ed. New York: St. Martin's, 1991.

Fry, Greg (ed.), *Australia's Regional Security.* Allen & Unwin, 1991.

Fuller, Graham E., *The "Center of the Universe;" The Geopolitics of Iran.* Boulder, CO: Westview, 1991.

Fuller, Graham E. and others, *Turkey's New Geopolitics; From the Balkans to Western China.* Boulder, CO: Westview, 1993.

Furst, Andreas, *Europe at Sea; Maritime Policies of Great Britain, France and Germany.* Oxford Univ. Press, 1995.

**G**

Gaddis, J. L., *Strategies of Containment: A Critical Approach to Postwar American National Security Policy.* Oxford Univ. Press, 1982.

Galtung, Johan (ed.), *Essays in Peace Research.* Copenhagen: Chr. Ejlers, Vol. I, 1975; Vol. II, 1976.

———, *The True Worlds.* New York: Free Press, 1979.

———, *Peace and Development in the Pacific Hemisphere.* Honolulu: Univ. of Hawaii Press, 1989.

Gamba-Stonehouse, Virginia, *Strategy in the Southern Oceans; A South American View.* New York: St. Martin's, 1989.

Garfinkle, Adam, *Israel and Jordan in the Shadow of War.* New York: St. Martin's, 1991.

Gause, F. Gregory, III, *Oil Monarchies; Domestic and Security Challenges in the Arab Gulf States.* New York: Council on Foreign Relations, 1994.

George, Alexander L. (ed.), *Avoiding War; Problems of Crisis Management.* Boulder, CO: Westview, 1991.

Gerges, Fawaz, *The Superpowers and the Arab Regional System 1955–1967.* Boulder, CO: Westview, 1994.

Ghee, Lim T. and Mark J. Valencia (eds.), *Conflict over Natural Resources in South-East Asia and the Pacific.* Oxford Univ. Press, 1991.

Girot, Pascal and Eleanore Kofman (eds. & trans.), *International Geopolitical Analysis; A Selection from Herodote.* New York: Routledge, 1987.

Glassner, Martin Ira (ed.), *Global Resources: Challenges of Interdependence.* New York: Praeger, 1983.

———, "Bolivia's Orientation: Toward the Atlantic or The Pacific?", Chapter 10 in Phillip Kelly and Jack Child, eds., *Geopolitics of the Southern Cone and Antarctica,* Boulder, CO: Lynne Rienner, 1988.

Goldberg, David H. and Paul Marantz, The *Decline of the Soviet Union and the Transformation of the Middle East.* Boulder, CO: Westview, 1994.

Goldstein, J., *Long Cycles.* New Haven, CT: Yale Univ. Press, 1988.

Goodby, James E. (ed.), *Regional Security after the Cold War.* Oxford Univ. Press, 1994.

Gordon, Lincoln, *Eroding Empire: Western Relations with Eastern Europe.* Washington, DC: Brookings, 1987.

Gottmann, Jean (ed.), *Centre and Periphery.* Beverly Hills, CA: Sage, 1980.

Grant, Shelagh D., *Sovereignty or Security?: Government Policy in the Canadian North, 1936–1950.* Vancouver: UBC Press, 1988.

Graubard, Stephen R., *Eastern Europe . . . Central Europe . . . Europe.* Boulder, CO: Westview, 1991.

Gray, Colin S., *The Geopolitics of the Nuclear Era: Heartland, Rimlands and the Technological Revolution.* New York: Crane, Russak, 1977.

———, *Maritime Strategy, Geopolitics, and the*

*Defense of the West.* New York: National Strategy Information Center, 1986.

———, *The Geopolitics of Super Power.* Lexington: Univ. Press of Kentucky, 1987.

Graz, Liesl, *The Turbulent Gulf.* London: Tauris, 1992.

Green, James R. and Brent Scowcroft (eds.), *Western Interests and U.S. Policy Options in the Caribbean Basin.* Boston: Oelgeschlager, Gunn & Hain, 1984.

Griffith, Ivelaw L. (ed.), *Strategy and Security in the Caribbean.* New York: Praeger, 1991.

Grinter, Lawrence E. and Young W. Kihi (eds.), *Security, Strategy, and Policy Responses in the Pacific Rim.* Boulder, CO: Lynne Rienner, 1989.

Grugel, Jean, *Politics and Development in the Caribbean Basin; Central America and the Caribbean in the New World Order.* Bloomington: Indiana Univ. Press, 1995.

Grundy, Kenneth W., *South Africa; Domestic Crisis and Global Challenge.* Boulder, CO: Westview, 1991.

Gunnarsson, Gunnar and Alan K. Henrikson (eds.), *The Strategic High North in a Changing World.* Lanham, MD: Univ. Press of America, 1991.

Gyorgy, Andrew, *Geopolitics: The New German Science.* Berkeley: Univ. of California Press, 1944.

**H**

Haglund, David G. (ed.), *The New Geopolitics of Minerals: Canada and International Resource Trade.* Vancouver: UBC Press, 1989.

Harden, Sheila, *Small Is Dangerous: Micro States in a Macro World.* London: Pinter, 1985.

Harkavy, Robert E. and Stephanie G. Neuman (eds.), *The Lessons of Recent Wars in the Third World.* Lexington, MA: Lexington Books, 1986.

Harris, Lillian Craig, *China Considers the Middle East.* London: Tauris, 1992.

Hauner, Milan, *The Soviet War in Afghanistan: Patterns of Russian Imperialism.* Lanham, MD: Univ. Press of America, 1991.

Heine, Jorge and Leslie F. Manigat, *The Caribbean and World Politics: Cross Currents and Cleavages.* New York: Holmes & Meier, 1988.

Higgott, Richard and others, *Pacific Economic Relations in the 1990s: Conflict or Cooperation?* Boulder, CO: Lynne Rienner, 1993.

Hillel, Daniel, *The Rivers of Eden; The Struggle for Water and the Quest for Peace in the Middle East.* Oxford Univ. Press, 1994.

Horensma, Pier, *The Soviet Arctic.* New York: Routledge, 1991.

**I**

Ince, Basil A. and others (eds.), *Issues in Caribbean International Relations.* Lanham, MD: Univ. Press of America, 1983.

Isard, Walter, *Arms Races, Arms Control, and Conflict Analysis: Contributions from Peace Science and Peace Economics.* Cambridge Univ. Press, 1988.

Ismael, Tareq Y. and Jacqueline S. Ismael, *The Gulf War and the New World Order: International Relations of the Middle East.* Gainesville: Univ. Press of Florida, 1994.

**J**

Jacobsen, Carl G., *Strategic Power—USA/USSR.* New York: St. Martin's, 1990.

Jawatkar, K. S., *Diego Garcia in International Diplomacy.* Bombay: Popular, 1983.

Jervell, Sverre and Kare Kyblom, *The Military Buildup in the High North: American and Nordic Perspectives.* Lanham, MD: Univ. Press of America, 1986.

Jervis, Robert and Jack Snyder (eds.), *Dominoes and Bandwagons; Strategic Beliefs and Great Power Competition in the Eurasian Rimland.* Oxford Univ. Press, 1991.

Jockel, Joseph T., *Security to the North.* East Lansing: Michigan State Univ. Press, 1991.

Johnston, Ronald J. and Peter J. Taylor (eds.), *A World in Crisis? Geographical Perspectives.* Oxford: Blackwell, 1989.

Jones, Stephen B., *Australia and New Zealand and the Security of the Pacific.* New Haven, CT: Yale Institute of International Studies, 1944.

———, *The Arctic: Problems and Possibilities.* New Haven, CT: Yale Institute of International Studies, Memorandum No. 29, 1948.

———, *Theoretical Studies of National Power.* New Haven, CT: Yale Univ. Press, 1955.

**K**

Keal, Paul and Andrew Mack (eds.), *Security and Arms Control in the North Pacific.* Allen & Unwin, 1990.

Kearsley, Harold J., *Maritime Power in the Twenty-First Century.* U.K.: Dartmouth, 1992.

Kelly, Philip and Jack Child (eds.), *Geopolitics of the Southern Cone and Antarctica.* Boulder, CO: Lynne Rienner, 1988.

Kemp, Geoffrey, *The Control of the Middle East Arms Race.* Washington, DC: Brookings, 1992.

————, *Strategic Geography and the New Middle East.* Washington, DC: Carnegie Endowment, 1992.

Kennedy, Paul M., *The Rise and Fall of the Great Powers.* New York: Random House, 1987.

Khouri, Fred J., *The Arab-Israeli Dilemma.* 3rd ed. Syracuse, NY: Syracuse Univ. Press, 1985.

Kidron, Michael, *The War Atlas: Armed Conflict-Armed Peace.* New York: Simon & Schuster, 1983.

Kipp, Jacob (ed.), *Central European Security Concerns, Bridge, Buffer or Barrier?* London: Frank Cass, 1993.

Kipper, Judith and Harold H. Saunders (eds.), *The Middle East in Global Perspective.* Boulder, CO: Westview, 1991.

Kissinger, Henry A., *Nuclear Weapons and Foreign Policy.* New York: Harper, 1957.

Kliot, Nurit and Stanley Waterman, *The Political Geography of Conflict and Peace.* London: Belhaven, 1991.

Knox, Paul L. and John A. Agnew, *The Geography of the World-Economy.* New York: Routledge, 1989.

Korn, David A., *Stalemate; Israel, Egypt, and Great Power Diplomacy in the Middle East, 1967–1970.* Boulder, CO: Westview, 1992.

Kwan, C. H., *Economic Interdependence in the Asia-Pacific Region.* New York: Routledge, 1994.


**L**

Langdon, Frank C. and Douglas A. Ross (eds.), *Superpower Maritime Strategy in the Pacific.* New York: Routledge, 1990.

Larson, David L. (ed.), *The "Cuban Crisis" of 1962: Selected Documents, Chronology and Bibliography.* Lanham, MD: Univ. Press of America, 1986.

Laursen, Finn, *Small Powers at Sea: Scandinavia and the New International Marine Order.* Kluwer, 1993.

Lax, Marc D., *Selected Strategic Minerals: The Impending Crisis.* Lanham, MD: Univ. Press of America, 1991.

Levie, Howard S., *The Code of International Armed Conflict,* 2 vols. New York: Oceana, 1986.

Lim Joo Jock, *Geo-strategy and the South China Sea Basin.* Singapore: Singapore Univ. Press, 1979.

————, *Territorial Power Domains, Southeast Asia, and China.* Singapore: Institute of Southeast Asian Studies, 1984.

Lowe, J. Y., *Geopolitics and War; Mackinder's Philosophy of Power.* Lanham, MD: Univ. Press of America, 1981.

Lukacs, Yehuda and Abdalla M. Battah (eds.), *The Arab-Israel Conflict—Two Decades of Change.* Boulder, CO: Westview, 1988.

Luttwak, Edward N. and Robert G. Weinland, *Sea Power in the Mediterranean: Political Utility and Military Constraints.* Lanham, MD: Univ. Press of America, 1979.


**M**

Mackinder, Halford T., *Britain and the British Seas.* Oxford: Clarendon; New York: Appleton, 1902.

————, *Democratic Ideals and Reality; A Study in the Politics of Reconstruction.* London: Constable; New York: Holt, 1919.

Mahan, Alfred Thayer, *The Influence of Sea Power upon History, 1660–1783.* Boston: Little, Brown, 1890.

————, *The Influence of Sea Power upon History; The French Revolution and Empire, 1793–1812.* Boston: Little, Brown, 1892.

————, *The Interest of America in Sea Power, Present and Future.* Boston: Little, Brown, 1898.

———— *The Problem of Asia and Its Effect upon International Politics.* Boston: Little, Brown, 1900.

Maingot, Anthony P., *The United States and the Caribbean.* Boulder, CO: Westview, 1994.

Malik, Hafeez, *Central Asia; Its Strategic Importance and Future Prospects.* New York: St. Martin's, 1993.

Mandel, Robert, *Conflict over the World's Resources: Backgrounds, Trends, Case Studies, and Considerations for the Future.* New York: Greenwood, 1988.

Mandelbaum, Michael (ed.), *The Strategic Quadrangle; Russia, China, Japan and the United States in East Asia.* New York: Council on Foreign Relations, 1995.

Manz, Beatrice F. (ed.), *Central Asia in Historical Perspective.* Boulder, CO: Westview, 1994.

Mattern, Johannes, *Geopolitik—Doctrine of National Self-Sufficiency and Empire.* Baltimore: Johns Hopkins Univ Press., 1942.

McCord, William, *The Dawn of the Pacific Century; Implications for Three Worlds of Development.* New Brunswick, NJ: Transaction, 1991.

McDowall, David, *Palestine and Israel.* London: Tauris, 1989.

McGwire, Michael, *Military Objectives in Soviet Foreign Policy.* Washington, DC: Brookings, 1987.

McInnes, Colin and Mark Rolls (eds.), *Security in Asia and the Pacific Rim.* London: Frank Cass, 1994.

McLuhan, Marshall and Quentin Fiore, *War and Peace in the Global Village*. New York: Mc-Graw-Hill, 1968.

McMillen, Donald Hugh (ed.), *Asian Perspectives on International Security*. New York: St. Martin's, 1984.

*The Middle East*. 8th ed. Washington, DC: Congressional Quarterly, 1994.

Misra, R. N., *Indian Ocean and India's Security*. Delhi: Mittal, 1986.

Mitchell, William, *Winged Defense: The Development and Possibilities of Modern Air Power—Economic and Military*. New York: Putnam, 1925.

Modelski, George (ed.), *Exploring Long Cycles*. Boulder, CO: Lynne Rienner, 1987.

———, *Long Cycles in World Politics*. New York: Macmillan, 1987.

Modelski, George and W. R. Thompson, *Seapower in Global Politics, 1494–1993*. Seattle: Univ. of Washington Press, 1988.

Morris, Benny, *1948 and After, Israel and the Palestinians*. Oxford Univ. Press, 1991.

———, *Israel's Border Wars, 1949–1956 Arab Infiltration, Israeli Retaliation, and the Countdown to the Suez War*. Oxford Univ. Press, 1994.

Morris, Michael A. and Victor Millán (eds.), *Controlling Latin American Conflicts: Ten Approaches*. Boulder, CO: Westview, 1983.

Mouzelis, N. P., *Politics in the Semi-Periphery*. London: Macmillan, 1986.

Myers, David J., *Regional Hegemons; Threat Perception and Strategic Response*. Boulder, CO: Westview, 1991.


**N**

Nathan, K. S. and M. Pathmanathan (eds.), *Trilaterlism in Asia*. Honolulu: Univ. of Hawaii Press, 1986.

Nelson, Daniel N. and John R. Lampe (eds.), *East European Security Reconsidered*. Baltimore: Johns Hopkins Univ. Press, 1993.

Nemetz, Peter H. (ed.), *The Pacific Rim; Investment, Development and Trade*. Vancouver: UBC Press, 1987.

Nugent, Jeffrey and others, *Bahrain and the Gulf; Past Perspectives and Alternate Futures*. New York: St. Martin's, 1985.


**O**

Oberg, Jean (ed.), *Nordic Security in the 1990s*. London: Pinter, 1992.

Ogunbadejo, Oye, *The International Politics of Africa's Strategic Minerals*. London: Pinter, 1985.

O'Loughlin, John (ed.), *Dictionary of Geopolitics*. Westport, CT: Greenwood, 1994.

Olson, William J. (ed.), *U.S. Strategic Interests in the Gulf Region*. Boulder, CO: Westview, 1987.

Osherenko, Gail and Oran R. Young, *The Age of the Arctic: Hot Conflicts and Cold Realities*. Cambridge Univ. Press, 1989.

O'Sullivan, Patrick, *Geopolitics*. London: Croom Helm, 1986.

O'Sullivan, Patrick and Jesse W. Miller, *The Geography of Warfare*. London: Croom Helm, 1983.


**P**

Paddison, Ronan, *The Political Geography of Power*. New York: St. Martin's, 1983.

Paone, Rocco Michael, *Strategic Nonfuel Minerals and Western Security*. Lanham, MD: Univ. Press of America, 1992.

Parker, Geoffrey, *Western Geopolitical Thought in the Twentieth Century*. New York: St. Martin's, 1985.

———, *The Geopolitics of Domination*. London: Routledge, 1988.

Parker, Richard, *North Africa: Regional Tensions and Strategic Concerns*. New York: Praeger, 1984.

Parker, W. H., *Mackinder: Geography as an Aid to Statecraft*. Oxford University Press, 1982.

Payne, Anthony and others, *Grenada*. New York: St. Martin's, 1986.

Payne, Anthony and Paul Sutton (eds.), *Modern Caribbean Politics*. Baltimore: Johns Hopkins Univ. Press, 1993.

Pearce, Jenny, *Under the Eagle; U.S. Intervention in Central America and the Caribbean*. Boston: South End Press, 1982.

Peltier, Louis C. and G. Etzel Pearcy, *Military Geography*. Princeton, NJ: Van Nostrand, 1966.

Pepper, David and Alan Jenkins (eds.), *The Geography of Peace and War*. New York: Blackwell, 1986.

Peretz, Don, *Intifada: The Palestinian Uprising*. Boulder, CO: Westview, 1990.

———, *Palestinians, Refugees, and the Middle East Peace Process*. Arlington, VA: U.S. Institute of Peace, 1993.

Pick, Otto (ed.), *Ending the Cold War in Europe*. New York: St. Martin's, 1991.

Pierre, Andrew J., *Third World Instability: Central America as a European-American Issue*. New York: Council on Foreign Relations, 1984.

Platt, Alan (ed.), *Arms Control and Confidence Building in the Middle East*. Arlington, VA: U.S. Institute of Peace, 1992.

Portugali, Juval, *Implicate Relations: Society and Space in the Israeli-Palestinian Conflict.* Kluwer, 1992.

Preeg, Ernest H., *Cuba and the New Caribbean Economic Order.* Boulder, CO: Westview, 1993.

————, *Cuba and the Caribbean.* Washington, DC: CSIS, 1994.

Prestowitz, Clyde V., Jr. and others (eds.), *Power-economics.* Lanham, MD: Univ. Press of America, 1991.

**Q**

Quandt, William B., *Camp David: Peacemaking and Politics.* Washington, DC: Brookings, 1986.

Quddus, Syed Abdul, *Afghanistan and Pakistan: A Geopolitical Study.* Lahore, Pakistan: Feroz, 1982.

**R**

Rabinovich, Itamar, *The Road Not Taken, Early Arab-Israeli Negotiations.* Oxford Univ. Press, 1991.

Radvanyi, Janos (ed.), *The Pacific in the 1990s: Economic and Strategic Change.* Lanham, MD: Univ. Press of America, 1990.

Ravenhill, John (ed.), *No Longer an American Lake?* Allen & Unwin, 1990.

Richardson, Bonham C., *The Caribbean in the Wider World, 1492-1992.* Cambridge Univ. Press, 1992.

Rikhye, Indar Jit, *Afghanistan, Iran and Iraq; External Involvement and Multilateral Options.* Oxford Univ. Press, 1989.

Rizvi, Hasan-Askari, *Pakistan and the Geostrategic Environment.* New York: St. Martin's, 1992.

Rotfeld, Adam Daniel (ed.), *Global Security and the Rule of Law.* Oxford Univ. Press, 1994.

Rusinow, Dennison I., "Danubia From the Danube." *UFSI Reports*, Part I 1987/No. 19 Europe [DIR-1-'87], Part II 1987/No. 20 Europe [DIR-2-'87].

Rustow, Dankwart A., *Turkey: America's Forgotten Ally.* New York: Council on Foreign Relations, 1987.

**S**

Saivetz, Carol R., *The Soviet Union and the Gulf in the 1980's.* Boulder, CO: Westview, 1987.

Sanders, David, *Losing an Empire, Finding a Role; British Foreign Policy Since 1945.* New York: Macmillan, 1990.

Sanders, Thomas G., *Brazilian Geopolitics: Securing the South and North.* UFSI Reports 1987/No. 23, Latin America [TGS-10-'87].

Satloff, Robert B. (ed.), *The Politics of Change in the Middle East.* Boulder, CO: Westview, 1993.

Schoenhals, Kai P. and Richard A. Melanson, *Revolution and Intervention in Grenada.* Boulder, CO: Westview, 1985.

Schulz, Donald E. and Douglas H. Graham, *Revolution and Counterrevolution in Central America and the Caribbean.* Boulder, CO: Westview, 1984.

Segal, Gerald, *Rethinking the Pacific.* Oxford Univ. Press, 1991.

Seger, Robert and Doris D. MaGuire (eds.), *Letters and Papers of Alfred Thayer Mahan.* Annapolis, MD: Naval Institute Press, 1975.

Serbin, Andres, *Caribbean Geopolitics: Toward Security Through Peace?* Boulder, CO: Lynne Rienner, 1990.

Shalev, Aryeh, *The Intifada; Causes and Effects.* Boulder, CO: Westview, 1991.

Shikaki, Khalil, *Intifada and the Transformation of Palestinian Politics.* UFSI Reports, Africa/Middle East 1989-90/No. 18.

Sivard, Ruth Leger, *World Military and Social Expenditures.* Washington, DC: World Priorities, annual.

Sloan, Geoffrey R., *Geopolitics in United States Strategic Policy, 1890-1987.* Brighton, UK: Wheatsheaf Books, 1988.

Slowe, Peter M., *Geography and Political Powers; The Geography of Nations and States.* London: Routledge, 1990.

So, Alvin Y., *Social Change and Development; Modernization, Dependency, and World-System Theories.* Newbury Park, CA: Sage, 1990.

Spiegel, Steven (ed.), *Conflict Management in the Middle East.* Boulder, CO: Westview, 1992.

Sprout, Harold H., *Towards a Politics of the Planet Earth.* New York: Van Nostrand, Reinhold, 1973.

Spykman, Nicholas John, *America's Strategy in World Politics.* 1942; reprinted: Hamden, CT: Shoe String Press, 1970.

Stephan, Paul B., III and Boris M. Klimenko, *International Law and International Security: Military and Political Dimensions: A U.S.-Soviet Dialogue.* Armonk, NY: Sharpe, 1992.

Stern, Eliahu and Yehuda Hayuth, "Developmental Effects of Geopolitically Located Ports," in Brian Stewart Hoyle and David Hilling (eds.), *Seaport Systems and Spatial Change.* New York: Wiley, 1984, 239–255.

Stockholm International Peace Research Institute. *SIPRI Yearbook, World Armaments and Disarmament.* Oxford Univ. Press.

Stokke, Olav Schram and Ola Tunander (eds.),

*The Barents Region; Cooperation in Arctic Europe.* Newbury Park, CA: Sage, 1994.

Suter, Keith, *An International Law of Guerrilla Warfare; The Global Politics of Law-Making.* New York: St. Martin's, 1984.

Sutton, Paul and Anthony Payne (eds.), *Size and Survival, the Politics of Security in the Caribbean and the Pacific.* London: Frank Cass, 1993.

**T**

Tardanico, Richard, *Crises in the Caribbean Basin.* Newbury Park, CA: Sage, 1987.

Taylor, Alan R., *The Superpowers and the Middle East.* Syracuse, NY: Syracuse Univ. Press, 1993.

Taylor, Griffith, *Canada's Role in Geopolitics.* Toronto: Ryerson, 1942.

———, *Our Evolving Civilization: An Introduction to Geopacifics. Geographical Aspects of the Path Toward World Peace.* Oxford Univ. Press, 1946.

———, "Geopolitics and Geopacifics," in Griffith Taylor (ed.), *Geography in the Twentieth Century.* New York: Philosophical Library, 1951, 587–608.

Taylor, Peter J., *Britain and the Cold War: 1945 As Geopolitical Transition.* London: Pinter, 1990.

Tillema, Herbert K., *International Armed Conflict Since 1945; A Bibliographic Handbook of Wars and Military Interventions.* Boulder, CO: Westview, 1991.

Tunander, Ola, *Cold Water Politics: The Maritime Strategy and Geopolitics of the Northern Front.* London: Sage, 1989.

**V**

Valenta, Jiri and Herbert Ellison, *Grenada and Soviet/Cuban Policy.* Boulder, CO: Westview, 1986.

Väyrynen, Raimo and others (eds.), *The Quest for Peace: Transcending Collective Violence and War Among Societies, Cultures and States.* London: Sage, 1987.

**W**

Wallerstein, Immanuel, *Modern World System.* 2 vols., New York: Academic Press, 1974 and 1980.

———, *The Politics of the World Economy.* Cambridge Univ. Press, 1984.

———, *Geopolitics and Geoculture; Essays on the Changing World-System.* Cambridge Univ. Press, 1991.

Weigert, Hans W., *Generals and Geographers: The Twilight of Geopolitics.* Oxford Univ. Press, 1942.

Weiss, Thomas G. (ed.), *Collective Security in a Changing World.* Boulder, CO: Lynne Rienner, 1993.

Wellmann, Christian, *The Baltic Sea Region: Conflict or Cooperation.* Boulder, CO: Westview, 1994.

West, Philip and Frans A.M. Alting von Geusau, *The Pacific Rim and the Western World.* Boulder, CO: Westview, 1987.

Westermeyer, William E. and Kurt M. Shusterich (eds.), *United States Arctic Interests: The 1980's and 1990's.* New York: Springer, 1984.

Westing, Arthur H., *Warfare in a Fragile World: Military Impact on the Human Environment.* London: Taylor & Francis, 1980.

———, *Global Resources and International Conflict.* Oxford Univ. Press, 1986.

Whittlesey, Derwent, *German Strategy of World Conquest.* New York: Holt, Rinehart, 1942.

Wiens, Herold J., *Pacific Island Bastions of the United States.* Princeton, NJ: Van Nostrand, 1962.

Williams, Colin H. (ed.), *The Political Geography of the New World Order.* London: Belhaven, 1993.

Wriggins, W. Howard and others, *Dynamics of Regional Politics; Four Systems on the Indian Ocean Rim.* Irvington, NY: Columbia Univ. Press, 1994.

**Y**

Yasmeen, Samina (ed.), *Political and Strategic Changes in the Indian Ocean Region: Implications for Australia.* Nedlands, Western Australia: Indian Ocean Centre for Peace Studies, 1993.

Yishai, Yael, *Land or Peace?; Wither Israel?* Stanford, CA: Hoover Institution Press, 1987.

Young, Oran R., *Arctic Politics; Conflict and Cooperation in the Circumpolar North.* Univ. Press of New England, 1992.

**Z**

Zoppo, C. E. and C. Zorgbibe (eds.), *On Geopolitics; Classical and Nuclear.* Nijhoff, 1985.

## Periodicals

**A**

Agnew, John, "The U.S. Position in the World Geopolitical Order After the Cold War," *Professional Geographer*, 44, 1 (Feb. 1992), 7–10.

Amin, S., "Democracy & National Strategy in the Periphery," *Third World Quarterly*, (1987), 1129–1156.

Ashley, R. K., "The Geopolitics of Geopolitical Space; Toward a Critical Theory of International Politics," *Alternatives*, 12 (1987), 403–434.

*Arms Control.* Published thrice a year by Frank Cass & Co., London.

**B**

Beck, Nathaniel, "The Illusion of Cycles in International Relations," *International Studies Quarterly*, 35, 4 (Dec. 1991), 455–476.

Besch, Edwin W., "How the Technology Explosion Is Changing World Power Relationships," *Defense and Foreign Affairs*, 19 (March 1991), 8–11.

Block, F., "Capitalism Versus Socialism in World Systems Theory," *Review*, 13 (1990), 265–286.

Blouet, Brian W., "The Maritime Origins of Mackinder's Heartland Thesis," *Great Plains–Rocky Mountain Geographical Journal*, (1973), 6–11.

———, "Halford Mackinder's Heartland Thesis, *Great Plains–Rocky Mountain Geographical Journal*, 5 (1976), 2–6.

———, "The Political Career of Sir Halford Mackinder," *Political Geography Quarterly*, 6, 4 (Oct. 1987), 355–368.

Boswell, Terry and Mike Sweat, "Hegemony, Long Waves, and Major Wars: A Time Series Analysis of Systemic Dynamics, 1496–1967," *International Studies Quarterly*, 35, 2 (June 1991), 123–150.

Bowman, Isaiah, "The Military Geography of Atacama," *Educational Bi-monthly*, 6 (1911), 1–21.

———, "Political Geography of Power," *Geographical Review*, 32 (1942), 349–352.

———, "The Strategy of Territorial Decisions," *Foreign Affairs*, 24, 2 (Jan. 1946), 177–194.

Bradshaw, Michael J. and Nicholas J. Lynn, "After the Soviet Union: The Post–Soviet States in the World System." *Professional Geographer*, 46, 4 (Nov. 1994), 439–449.

Brunn, Stanley D., "A World of Peace and Military Landscapes," *Journal of Geography*, 86 (1987), 253–262.

Brush, John E., "Peace Research and Geography," *Professional Geographer*, 16 (1964), 49.

Bull, Hedley, "Kissinger: The Primacy of Geopolitics," *International Affairs*, 56, 3 (1980), 484–487.

*Bulletin of the Atomic Scientists.* Chicago: Educational Foundation for Nuclear Scientists. Published 10 times per year.

*Bulletin of Peace Proposals*, published quarterly by Sage. Bundy, William P., "Elements of Power," *Foreign Affairs*, 56, 1 (Oct. 1977), 1–26.

**C**

Chai, Caroline and Winberg Chai (eds.), *Symposium: In Search of Peace in the Middle East. Asian Affairs* (New York) 18 (Spring 1991), whole issue.

Chase-Dunn, Christopher, "World-state Formation: Historical Processes and Emergent Necessity," *Political Geography Quarterly*, 9, 2 (April 1990), 108–130.

Child, Jack, "Geopolitical Thinking in Latin America," *Latin American Research Review*, 14, 2 (1979), 89–111.

———, "Latin American 'Lebensraum': The Geopolitics of the Ibero-American Antarctica," *Applied Geography*, 10 (1990), 287–306.

Cohen, R. and P.A. Wilson, "Superpowers in Decline: Economic Performance and National Security," *Comparative Strategy*, 7 (1988), 99–132.

Cohen, Saul B., "A New Map of Global Geopolitical Equilibrium: A Developmental Approach," *Political Geography Quarterly*, 1, 3 (July 1982), 223–241.

———, "The World Geopolitical System in Retrospect and Prospect," *Journal of Geography*, 89, 1 (1990), 2–12.

———, "Policy Prescriptions for the Post Cold War World," *Professional Geographer*, 44, 1 (Feb. 1992), 13–16.

Cole, J. P., "The World of Jan Kowalewski: Pawns in Other Peoples' Games," *Scottish Geographical Magazine*, 106 (1990), 66–74.

Crone, Gerald R., "A German View of Geopolitics," *Geographical Journal*, 111, 1–3 (1948), 104–108.

*Current Research on Peace and Violence*, Tampere, Finland-Tampere Peace Research Institute. Published quarterly.

**D**

Dalby, Simon, "American Security Discourse: The Persistence of Geopolitics," *Political Geography Quarterly*, 9, 2 (April 1990), 171–188.

———, "Critical Geopolitics: Discourse, Difference and Dissent," *Environment and Planning D*, 9, (1991), 261–283.

———, "The 'Kiwi Disease': Geopolitical Discourse in Aotearoa/New Zealand and the South Pacific," *Political Geography*, 12, 5 (Sept. 1993), 437–456.

Dayan, Moshe, "Israel's Border and Security Problems," *Foreign Affairs*, 33, 2 (Jan. 1955), 250–267.

Debre, Michel, "France's Global Strategy," *Foreign Affairs*, 49, 3 (April 1971), 395–406.

Delcoigne, G., "An Overview of Nuclear Weapon-Free Zones," *International Atomic Energy Bulletin*, 24 (1982), 50–55.

De Seversky, Alexander P., "The Twilight of Seapower," *American Mercury*, 52 (June 1941), 647–658.

———, "Air Power Ends Isolation," *Atlantic Monthly*, 168 (Oct. 1941), 407–416.

Dodds, Klaus-John, "Geopolitics, Experts and the Making of Foreign Policy." *Area*, 25 (1993), 70–74.

———, "Geopolitics, Cartography and the State in South America," *Political Geography*, 12, 4 (July 1993), 361–381.

———, "Geopolitics in the Foreign Office: British Representations of Argentina 1945–1961." *Transactions*, Institute of British Geographers, NS 19 (1994), 273–290.

———, "Geopolitics and Foreign Policy: Recent Developments in Anglo-American Political Geography and International Relations." *Progress in Human Geography*, 18, 2 (1994), 186–208.

Drysdale, Alistair, "Political Conflict and Jordanian Access to the Sea," *Geographical Review*, 77 (1987), 86–102.

**E**

Earney, Fillmore C.F., "The Geopolitics of Minerals," *Focus*, 31, 5 (May–June 1981), 1–16.

Ewell, J., "The Development of Venezuelan Geopolitical Analysis Since World War II," *Journal of Interamerican Studies and World Affairs*, 24 (1982), 295–320.

**F**

Fahlbush, Michael and others, "Conservatism, Ideology and Geography in Germany 1920–1950," *Political Geography Quarterly*, 8, 4 (Oct. 1989), 353–368.

Falah, Ghazi, "Israeli 'Judaization' Policy in Galilee and Its Impact on Local Arab Urbanization," *Political Geography Quarterly*, 8, 3 (July 1989), 229–254.

———, *Some Geographical Aspects of the Israeli-Palestinian Conflict*. Special issue of *Geographical Journal* 21, 4 (August 1990), whole volume.

Fisher, Charles A., "The Expansion of Japan: A Study in Oriental Geopolitics," *Geographical Journal*, 115, 1–3 (Jan. 1950), 1–19; 115, 4–6 (April 1950), 179–193.

———, "Containing China? I The Antecedents of Containment," *Geographical Journal*, 136, 4 (Dec. 1970), 534–556. II "Concepts and Applications of Containment," 137, 3 (Sept. 1971), 281–310.

Furniss, Edgar S., Jr., "The Contribution of Nicholas John Spykman to the Study of International Politics," *World Politics*, 4, 3 (April 1952), 382–401.

**G**

Galtung, Johan, "The Geopolitics of War: Total War and Geostrategy," *Journal of Politics*, 5, 4 (1943), 347–362.

Gilbert, E. W. and W. H. Parker, "Mackinder's 'Democratic Ideals and Reality' After 50 Years," *Geographical Journal*, 135, 2 (June 1969), 228–231.

Glassner, Martin Ira, "Management of Marine Resources as a Binding Force in the Eastern Caribbean," *Ocean and Shoreline Management*, 20, 1 (Feb. 1993), 63–88.

Goldfrank, W., "Current Issues in World-System Theory," *Review*, 13 (1990), 251–254.

Goldstein, Joshua S., "The Possibility of Cycles In International Relations," *International Studies Quarterly*, 35, 4 (Dec. 1991), 477–480.

Gordon, Bernard K., "The Asian-Pacific Rim; Success at a Price," *Foreign Affairs*, 70, 1 (1991), 142–159.

Gorman, Stephen M., "Geopolitics and Peruvian Foreign Policy," *Inter-American Economic Affairs*, 36, 2 (Autumn 1982), 65–88.

Gyorgy, Andrew, "The Application of German Geopolitics: Geo-Science," *American Political Science Review*, 37 (1943), 677–686.

**H**

Hadar, Leon T., "The United States, Europe and the Middle East," *World Policy Journal*, 8 (Summer 1991), 421–449.

Hall, A. R., "Mackinder and the Course of Events," *Annals, AAG*, 45, 2 (June 1955), 109–126.

Harvey, David, "The World-Systems Theory Trap," *Studies in Comparative International Development*, 22, 1 (1987), 42–47.

Hayes, John D., "Peripheral Strategy—Mahan's Doctrine Today," *Proceedings, United States Naval Institute* (Nov. 1953).

Hensel, Paul R. and Paul F. Diehl, "Testing Empirical Propositions About Shatterbelts, 1945–76," *Political Geography*, 13, 1 (Jan. 1994), 33–51.

Herrick, Francis H., "World Island and Heartland—The Strategical Theories of Sir Halford John Mackinder," *South Atlantic Quarterly*, 43 (June 1944), 248–255.

Hoffman, George W., "The Shatter-Belt in Relation to the East-West Conflict," *Journal of Geography*, 51 (1952), 265–275.

———, "Nineteenth-Century Roots of American World Power Relations," *Political Geography Quarterly*, 1, 3 (June 1982), 279–292.

Houbert, Jean, "The Indian Ocean Creole Islands: Geo-politics and Decolonisation." *Journal of Modern African Studies*, 30, 2 (1992), 465–484.

**I**

*The Indian Ocean Review.* Published quarterly since 1988 by the Centre for Indian Ocean Regional Studies, Curtin Univ. of Technology, Perth, Western Australia.

*The International Dynamics of the Commonwealth Caribbean.* Special issue of *Journal of Interamerican Studies and World Affairs*, 31, 3 (Fall 1989), whole volume.

**J**

Jackson, W.A. Douglas, "Mackinder and the Communist Orbit," *Canadian Geographer*, 6, 1 (Spring 1962), 12–21.

*Japan, The United States, and Pacific Ocean Resources.* Special issue of *Ecology Law Quarterly*, 16, 1 (1989), whole volume.

Jones, Stephen B., "The Power Inventory and National Strategy," *World Politics*, 6, 4 (July 1954), 421–452.

———, "Views of the Political World," *Geographical Review*, 45, 3 (July 1955), 309–326.

———, "Global Strategic Views," *Geographical Review*, 45, 4 (Oct. 1955), 492–508.

*Journal of Conflict Resolution.* Published quarterly by Sage.

*Journal of Peace Research.* Sage for the International Peace Research Institute, Oslo. Published quarterly since 1964.

**K**

Karsh, E., "Geopolitical Determinism: The Origins of the Iran–Iraq War," *Middle East Journal*, 44 (1990), 256–268.

Kearns, G., "History, Geography and World-Systems Theory," *Journal of Historical* 14 (1988), 281–292.

Kelly, Philip L., "Escalation of Regional Conflict: Testing the Shatterbelt Concept," *Political Geography Quarterly*, 5, 2 (April 1986), 161–180.

Kennan, George (writing under pseudonym "X"), "The Sources of Soviet Conduct," *Foreign Affairs*, 25 (1947), 566–582.

Kennedy, Edward M., "The Persian Gulf: Arms Race or Arms Control?" *Foreign Affairs*, 54 (Oct. 1975), 14–35.

Kipnis, Baruch A., "Geopolitical Ideologies and Regional Strategies in Israel," *Tidjschrift voor Economische en Sociale Geografie*, 78 (1987), 125–138.

Kish, George, "Political Geography into Geopolitics: Recent Trends in Germany," *Geographical Review*, 32, 4 (Oct. 1942), 632–645.

Klink, Frank F., "Rationalizing Core-Periphery Relations: The Analytical Foundations of Structural Inequality in World Politics," *International Studies Quarterly*, 34, 2 (June 1990), 183–210.

Kliot, Nurit, "Israel vs. Palestine: Competition for Resources," *Focus* (Fall 1988), 30–33.

Knight, David B., "Changing Orientations: Elements of New Zealand's Political Geography," *Geographical Bulletin*, 1 (1970) 21–30.

Kost, Klaus, "The Conception of Politics in Political Geography and Geopolitics in Germany Until 1945," *Political Geography Quarterly*, 8, 4 (Oct. 1989), 369–386.

Krause, Keith, "Military Statecraft: Power and Influence in Soviet and American Arms Transfer Relationship," *International Studies Quarterly*, 35, 3 (Sept. 1991), 313–336.

Kristof, Ladis K.D., "The Geopolitical Image of the Fatherland: The Case of Russia," *Western Political Quarterly*, 20, 4 (1967), 941–954.

Krushchev, Nikita S., "On Peaceful Coexistence," *Foreign Affairs*, 38, 1 (Oct. 1959), 1–18.

Kugler, J. and A.F.K. Organski, "The End of Hegemony," *International Interactions*, 15 (1989), 113–128.

**M**

Mackinder, Halford J., "The Physical Basis of Political Geography," *Scottish Geographical Magazine*, 6, 2 (Feb. 1890), 78–84.

———, "The Geographical Pivot of History," *Geographical Journal*, 23, 4 (April 1904), 421–444.

Mahan, Alfred Thayer, "The United States Looking Outward," *Atlantic Monthly*, 66 (1890), 816–824.

———, "Hawaii and Our Future Sea Power," *Forum*, 15 (March 1893), 1–11.

———, "The Future in Relation to American Naval Power," *Harper's Magazine*, 91 (1895), 767–775.

———, "The Problem of Asia," *Harper's Magazine*, 100 (1900), 536–547.

———, "The Panama Canal and Sea Power in the Pacific," *Century Magazine*, 82 (1911), 240–248.

———, "Strategic Features of the Caribbean Sea and the Gulf of Mexico," *Harper's New Monthly Magazine*, 95 (1911), 680–691.

———, "Importance of Command of the Sea," *Scientific American*, 105 (Dec. 9, 1911), 512.

———, "The Panama Canal and the Distribution of the Fleet," *North American Review*, 200 (1941), 406-417.

———, "The Round World and the Winning of the Peace," *Foreign Affairs*, 21, 4 (July 1943), 595–605.

Maingot, Anthony P., *Trends in U.S.–Caribbean Relations, Annals, Am. Acad. of Polit. and Soc. Sci.*, 533 (May 1994), whole issue.

McColl, Robert W., "The Insurgent State: Territorial Bases of Revolution," *Annals, AAG*, 59, 4 (Dec. 1969), 613–631.

———, "Geographical Themes in Contemporary Asian Revolutions," *Geographical Review*, 65, 3 (July 1975), 301–310.

Mearsheimer, J., "Back to the Future: Instability in Europe after the Cold War," *International Security*, 15 (1990), 5–56.

Meinig, Donald W., "Heartland and Rimland in Eurasian History," *Western Political Quarterly*, 9, 3 (Sept. 1956), 553–569.

Midlarsky, Manus I., "Boundary Permeability as a Condition of Political Violence," *Jerusalem Journal of International Relations*, 1, 2 (Winter 1975), 53–69.

*Military Balance*. London: International Institute for Strategic Studies. Published annually.

Modelski, George, "The Premise of Geocentric Politics," *World Politics*, 22, 4 (July 1970), 617–635.

———, "The Long Cycle of Global Politics and the Nation-State," *Comparative Studies in Society and History*, 20 (1978), 214–235.

Moll, Kenneth L., "A.T. Mahan, American Historian," *Military Affairs*, 27 (1963), 131–140.

Momsen, Janet Henshall, "Caribbean Conflict: Cold War in the Sun," *Political Geography Quarterly*, 3, 2 (April 1984), 145–151.

———, "Canada–Caribbean Relations: Wherein the Special Relationship?", *Political Geography*, 11, 5 (Sept. 1992), 501–513.

Moore, John Norton, "Grenada and the International Double Standard," *American Journal of International Law*, 78, 1 (Jan. 1984), 145–168.

———, "The Secret War in Central America and the Future of World Order," *American Journal of International Law*, 80, 1 (Jan. 1986), 43–127.

Morris, Michael A., "Maritime Geopolitics in Latin America," *Political Geography Quarterly*, 5, 1 (Jan. 1986), 43–55.

Morrison, Joel A., "Russia and Warm Water: A Fallacious Generalization and Its Consequences," *Proceedings, United States Naval Institute*, 78, 11 (1952), 1169–1179.

**N**

Newman, David, "Civilian and Military Presence as Strategies of Territorial Control: The Arab—Israel Conflict," *Political Geography Quarterly*, 8, 3 (July 1989), 215–228.

Newman, David and Juval Portugali, "Israeli-Palestinian Relations as Reflected in the Scientific Literature," *Progress in Human Geography*, 11 (1987).

Nijman, Jan, "The Limits of Superpower: The United States and the Soviet Union Since World War II," *Annals, AAG*, 82, 4 (Dec. 1992), 681–695.

Noble, Allen G. and Elisha Efrat, "Geography of the Intifada," *The Geographical Review*, 80, 3 (July 1990), 288–307.

*The Nordic Region. Annals, Am. Acad. of Polit. and Soc. Sci.* 512 (Nov. 1990), whole volume.

**O**

O'Keefe, Phil and others, "A Sad Note for Grenada," *Political Geography Quarterly*, 3, 2 (April 1984), 152–159.

O'Loughlin, John, "Superpower Competition and the Militarization of the Third World," *Journal of Geography*, 86 (1987), 269–275.

O'Loughlin, John and Herman van der Wusten, "Geography, War and Peace: Notes for a Contribution to a Revived Political Geography," *Progress in Human Geography*, 10, 4 (Dec. 1986), 484–510.

Openshaw, S. and P. Steadman, "On the Geography of a Worst Case Nuclear Attack on the Population of Britain," *Political Geography Quarterly*, 1, 3 (July 1982), 263–278.

Ossenbrügge, Jurgen, "Territorial Ideologies in West Germany 1945–1985: Between Geopolitics and Regionalist Attitudes," *Political Geography Quarterly*, 8, 4 (Oct. 1989), 387–400.

O'Sullivan, Patrick, "Antidomino," *Political Geography Quarterly*, 1, 1 (Jan. 1982), 57–64.

———, "A Geographical Analysis of Guerrilla Warfare," *Political Geography Quarterly*, 2, 2 (April 1983), 139–150.

Ó'Tuathail, Geróid, "The Language and Nature of the 'New Geopolitics'—the Case of U.S.–Salvador Relations," *Political Geography Quarterly*, 5, 1 (Jan. 1986), 73–85.

———, "The Critical Reading/Writing of Geopolitics: Re-reading/Writing Wittfogel, Bowman and Lacoste," *Progress in Human Geography*, 18, 3 (1994), 313–332.

ÓTuathail, Geróid and John Agnew, "Geopolitics and Discourse: Practical Geopolitical Reasoning in American Foreign Policy," *Political Geography*, 11, 2 (March 1992), 190-204.

**P**

Paasi, Anssi, "The Rise and Fall of Finnish Geopolitics," *Political Geography Quarterly*, 9, 1 (Jan. 1990), 53–66.

Parker, Geoffrey, "Continuity and Change in Western Geopolitical Thought During the Twentieth Century," *International Social Science Journal*, 127 (Feb. 1991), 21–33.

*Peace and Disarmament; Academic Studies.* Moscow: Scientific Research Council on Peace and Disarmament. Published annually.

*Peace Research.* Oakville, Ontario: Canadian Peace Research Institute. Published quarterly.

Pepper, David M., "Geographical Dimensions of NATO's Evolving Military Strategy," *Progress in Human Geography*, 12 (1988), 157–178.

Pepper, David and Alan Jenkins, "A Call to Arms: Geography and Peace Studies," *Area*, 14 (1983), 202–208.

Popke, E. Jeffrey, "Recasting Geopolitics: The Discursive Scripting of the International Monetary Fund," *Political Geography*, 13, 3 (May 1994), 255–269.

Portugali, Juval, "Jewish Settlement in the Occupied Territories: Israel's Settlement Structure and the Palestinians," *Political Geography Quarterly*, 10, 1 (Jan. 1991), 26–53.

Posony, Stefan T. and Leslie Rosenzweig, "The Geography of the Air," *Annals, Am. Acad. of Polit. and Soc. Sci.*, 299 (May 1955), 1–11.

Puri, Madan Mohan, "Geopolitics in the Indian Ocean," *International Studies*, 23, 2 (April 1986), 155-168.

**R**

*Resources Policy; The International Journal of Minerals Policy and Economics*, Butterworth-Heinemann. Published quarterly since 1974.

Roberts, Adam, "Prolonged Military Occupation: The Israeli-Occupied Territories Since 1967," *American Journal of International Law*, 84, 1 (Jan. 1990), 1–43.

Roucek, Joseph S., "The Geopolitics of Danubia," *American Journal of Economics and Sociology*, 5 (1946), 211-230.

———, "Geopolitics of Poland," *American Journal of Economics and Sociology*, 7, 4 (1948), 421–427.

———, "The Geopolitics of the Baltic States," *American Journal of Economics and Sociology*, 8, 2 (Jan. 1949), 171–175.

———, "The Geopolitical Implications of the Macedonian Problem," *World Affairs Interpreter*, 21, 1 (1950), 95–107.

—, "Geopolitical Trends in Central-Eastern Europe," *Annals, Am. Acad. of Polit. and Soc. Sci.*, 271 (1950), 11–19.

———, "Moscow's European Satellites," *Annals, Am. Acad. of Polit. and Soc. Sci.*, 271 (Sept. 1950), 1–253.

———, "The Geopolitics of the U.S.S.R.," *American Journal of Economics and Sociology*, 10, 1 (Oct. 1950), 17–26; 10, 2 (Jan. 1951), 153–159.

———, "The Geopolitics of the Aleutians," *Journal of Geography*, 40, 1 (Jan. 1951), 24–29.

———, "The Geopolitics of Greenland," *Journal of Geography*, 50, 6 (Sept. 1951), 239–246.

———, "Some Geographic Factors in the Strategy of the Kuriles," *Journal of Geography*, 51 (1952), 297–301.

———, "The Geopolitics of the Adriatic," *American Journal of Economics and Sociology*, 11, 2 (Jan. 1952), 171–178.

———, "The Geopolitics of Tibet," *Social Education*, 15, 4 (May 1952), 217–218, 299.

———, "The Geopolitics of Albania," *World Affairs Interpreter*, 23, 1 (1952), 83–96.

———, "Geopolitics and Air Power," *Air Univ. Quarterly*, 6 (1952), 52–73.

———, "Notes on the Geopolitics of Singapore," *Journal of Geography*, 7 (Feb. 1953), 78–81.

———, "The Geopolitics of the Mediterranean," *American Journal of Economics and Sociology*, 12, 4 (July 1953), 346–354; 13, 1 (Oct. 1953), 71–86.

———, "The Geopolitics of Pakistan," *Social Studies*, 44, 7 (Nov. 1953), 254–258.

———, "The Geopolitics of the Philippine Islands," *World Affairs Interpreter*, 25, 1 (1954), 79–90.

———, "The Geopolitics of Thailand," *Social Studies*, 45, 2 (Feb. 1954), 57–63.

———, "The Geopolitics of Spain," *Social Studies*, 46, 3 (March 1955), 89–93.

———, "The Geopolitics of the United States," *American Journal of Economics and Sociology*, 14, 2 (Jan. 1955), 185–192; 14, 3 (April 1955), 287–303.

———, "The Geopolitics of Yugoslavia," *Social Studies*, 47, 1 (Jan. 1956), 26–29.

———, "The Geopolitics of Afghanistan," *Social Studies*, 48, 4 (1957), 127–129.

———, "The Development of Political Geography and Geopolitics in the United States," *Australian Journal of Politics and History*, 3, 2 (1958), 204–217.

———, "Tibet and Its Geopolitical Aspects," *United Asia*, 12, 4 (1960), 377–380.

———, "Cuba in Its Geopolitical Setting," *Contemporary Review*, 198 (1960), 663–667.

———, "Vietnam in Geopolitics," Contemporary Review, 200 (1961), 420–423.

———, "The Geopolitics of the Congo," *United Asia*, 14, 1 (Jan. 1962), 81–85.

———, "Thailand's Geopolitics," *United Asia*, 14 (1962), 381–387.

———, "Albania in Geopolitics," *Contemporary Review*, 201 (1962), 17–20.

———, "Yemen in Geopolitics," *Contemporary Review*, 202 (1962), 310–317.

———, "Nigerian Geo-Politics," *United Asia*, 14, 3 (March 1962), 182–185.

———, "The Geopolitics of Mongolia," *United Asia*, 15, 3 (1963), 240–244.

———, "The Geopolitics of Asia," *United Asia*, 15, 4 (1963), 290–296.

———, "The Geopolitics of Japan," *United Asia*, 15, 4 (1963), 297–301.

———, "The Geopolitics of Korea," *United Asia*, 15, 4 (1963), 302–304.

———, "The Geopolitics of Formosa (Taiwan)," *United Asia*, 15, 4 (1963), 305–308.

———, "Venezuela in Geopolitics," *Contemporary Review*, 203, Pt. I, (Feb. 1963), 84–87; 203, Pt. II, (March 1963), 126–132.

———, "Haiti in Geopolitics," *Contemporary Review*, 204 (July 1963), 21–29.

———, "Iraq in Geopolitics," *Contemporary Review*, 204 (Sept. 1963), 116–123.

———, "Peru in Geopolitics," *Contemporary Review*, 204 (1963), 310–315; 205 (1964), 24–31.

———, "Changing Geopolitical Patterns Along the Persian Gulf," *Il Politico* (Univ. of Pavia), 29, 2 (1964), 440–474.

———, "Ecuador in Geopolitics," *Contemporary Review*, 205 (Feb. 1964), 74–82.

———, "The Geopolitics of South-East Asia," *France-Asia*, 19, 182 (1964), 1077–1094.

———, "Cambodia in Global Geopolitics," *International Review of History and Political Science*, 1, 1 (1964), 13-32.

———, "Portugal in Geopolitics," *Contemporary Review*, 205 (1964), 476–488.

———, "The Pacific in Geopolitics," *Contemporary Review*, 206, 1189 (Feb. 1965), 63–76.

———, "Chile in Geopolitics," *Contemporary Review,* 206 (1965), 127–141.

———, "The Dominican Republic in Geopolitics," *Contemporary Review*, 206, (1965), 289–294; 207 (1965), 12–21.

———, "Geopolitical Trends Behind the Iron Curtain," *Central Europe Journal*, 14, 9 (1966), 267–279.

———, "Geopolitics of Zambia," *New Africa*, 9 (July 1967), 9–12.

———, "Africa in Geopolitics," *International Review of History and Political Science*, 4, 2 (1967), 76–96.

———, "The Geopolitics of Hong Kong," *Revue du Sud-Est Asiatique* (Brussels), 2 (1967), 155–189.

———, "Burma in Geopolitics," *Revue du Sud-Est Asiatique et I'Extr[ax]eme Orient*, 1 (1968), 47–82.

———, "South Africa in Geopolitics," *Africa Quarterly*, 8, 2 (1968), 166–176.

———, "Okinawa in Geopolitics," *Study of Current English*, 24, 11 (1969), 8–18, 24; 12 (1969), 15–21.

———, "Cambodia in Geopolitics," *Revue du Sud-Est Asiatique*, (1970), 197–223.

———, "South-East Asia in Global Geopolitics," *Il Politico* (Univ. of Pavia), 35, 3 (1970), 472–495.

———, "Modern Implications of the Geopolitical Heartland Theory for Eurasia and the Pacific Ocean," *International Behavioral Scientist*, 21 (1970), 19–46.

———, "The Pacific in World Geopolitics," *International Review of History and Political Science*, 8, 3 (1971), 1–30.

———, "The Indian Ocean in Global Geopolitics," *International Review of History and Political Science*, 8, 4 (1971), 57–77.

———, "The Geopolitics of the Adriatic Sea," *Il Politico* (Univ. of Pavia), 36, 3 (1971), 569–594.

———, "Geopolitics of Indonesia," *International Behavioral Scientist*, 2 (1972), 49–72.

Rowley, Gwyn, "The West Bank: Native Water-Resource Systems and Competition," *Political Geography Quarterly*, 9, 1 (Jan. 1990), 39–52.

*RUSI and Brassey's Defence Yearbook.* London: Royal United Services Institute. Published annually.


**S**

Schwartz, Benjamin I., "The Maoist Image of World Order," *International Affairs*, 21, 1 (1967), 92–102.

Semmel, Bernard, "Sir Halford Mackinder: Theorist of Imperialism," *Canadian Journal of Economic and Political Sciences*, 24, 4 (Nov. 1958), 554–561.

Sidaway, James, "Geopolitics, Geography and Terrorism in the Middle East." *Environment and Planning D*, forthcoming.

Slessor, Sir John, "Air Power and World Strategy," *Foreign Affairs*, 33, 1 (Oct. 1954), 43–53.

*Small Wars and Insurgencies.* Published thrice a year since 1990 by Frank Cass in London.

Soustelle, Jacques, "France and Europe: A Gaullist View," *Foreign Affairs*, 30 (1952), 545–553.

Spencer, D. S., "A Short History of Geopolitics," *Journal of Geography*, 87 (1988), 42–48.

Spykman, Nicholas John, "Geography and Foreign Policy," *American Political Science Review*, 32, 1 (Feb. 1938), 28–50; 32, 2 (April 1938), 213–236.

Spykman, Nicholas John and Abbie A. Rollins, "Geographic Objectives in Foreign Policy," *American Political Science Review*, 33, 3 (June 1939), 391–410; 33, 4 (August 1939), 591–614.

———, "Frontiers, Security, and International Relations," *Geographical Review*, 32, 3 (July 1942), 436–447.

Starr, Harvey and Randolph N. Siverson, "Alliances and Geopolitics," *Political Geography Quarterly*, 9, 3 (July 1990), 232–248.

Stefansson, Vihjalmur, "The Arctic," *Air Affairs*, 3 (1950), 391–402.

Stone, Adolf, "Geopolitics as Haushofer Taught It," *Journal of Geography*, 52, 4 (April 1953), 167–171.

*Strategic Analysis*. New Delhi: Institute for Defence Studies and Analysis. Published monthly.

*Strategic Studies*. Islamabad: Institute of Strategic Studies.

Swearingen, Will D., "Geopolitical Origins of the Iran–Iraq War," *Geographical Review*, 78 (1988), 405–416.

**T**

Takeuchi, K., "Geopolitics and Geography in Japan Reexamined," *Hitotubashi Journal of Social Studies*, 12 (1980), 14–24.

Tambs, Lewis A., "Latin American Geopolitics. A Basic Bibliography," *Revista Geográfica*, 73 (1970), 71–106.

Taylor, Frank F., "Peacekeeping in Paradise: The Arming of the Eastern Caribbean," *Transafrica Forum* (Rutgers Univ.), 3, 3 (Spring 1986), 49–54.

Taylor, Peter J., "World-Systems Analysis and Regional Geography," *The Professional Geographer*, 40, 3 (August 1988), 259–265.

Teggart, Frederick J., "Geography as an Aid to Statecraft; an Appreciation of Mackinder's 'Democratic Ideals and Reality,'" *Geographical Review*, 8, 4–5 (Oct.–Nov. 1919), 227–242.

Treverton, Gregory F., "Elements of a New European Security Order," *Journal of International Affairs*, 45 (Summer 1991), 91–112.

**U**

Unstead, J. F., "H. J. Mackinder and the New Geography," *Geographical Journal*, 113, 1 (Jan. 1949), 47–57.

**V**

Vagts, Detlev F., "International Law Under Time Pressure: Grading the Grenada Take-Home Examination," *American Journal of International Law*, 78, 1 (Jan. 1984), 169–172.

van der Wusten, Herman, "A New World Order (No Less)," *Professional Geographer*, 44, 1 (Feb. 1992), 19–22.

van der Wusten, Herman and John O'Loughlin, "Claiming New Territory for a Stable Peace: How Geography Can Contribute," *Professional Geographer*, 38, 1 (Feb. 1986), 18–28.

Vartanov, Raphael V. and Alexei Yu. Roginko, "New Dimensions of Soviet Arctic Policy; Views from the Soviet Union," *Annals, Am. Acad. of Polit. and Soc. Sci.*, 512 (Nov. 1990), 69–78.

Vitkovskiy, V., "Political Geography and Geopolitics: A Recurrence of American Geopolitics," *Soviet Geography*, 22 (1981), 586–593.

**W**

Wallerstein, Immanuel, "World-Systems Analysis: The Second Phase," *Review*, 13 (1990), 287–293.

Weigert, Hans Werner, "German Geopolitics; a Workshop for Army Rule," *Harpers*, 183 (Nov. 1941), 586–597.

———, "Haushofer and the Pacific," *Foreign Affairs*, 20, 4 (July 1942), 732–742.

———, "Iceland, Greenland and the United States," *Foreign Affairs*, 23, 1 (Oct. 1944), 112–122.

———, "The Northward Course," *Virginia Quarterly Review*, 20, 1 (1944), 114–122.

———, "Mackinder's Heartland," *American Scholar*, 15, 1 (Winter 1945–46), 43–54.

———, "U.S. Strategic Bases and Collective Security," *Foreign Affairs*, 25, 2 (Jan. 1947), 250–262.

Whebell, C.F.J., "Mackinder's Heartland Theory in Practice Today," *Geographical Magazine*, 42 (1970), 630–636.

Wiskemann, Elizabeth, "The 'Drang nach Osten' Continues," *Foreign Affairs*, 17, 4 (July 1939), 764–773.

**Z**

Zabur, J., "Geopolitics: The Struggle for Space and Power," *Revista Brasileira de Geografia*, 4, 4 (Dec. 1942), 849–852.

# PART SIX

## *Contemporary International Relations*

United Nations Headquarters in New York

# 24

# *International Law*

There are a number of different approaches to international relations. The systems approach is one, in which international society is considered as a system with subsystems and fixed rules binding them together and making them behave in particular ways. One problem with this approach is that international relations are neither systematic nor predictable. Another view is that international relations are based solely on power: each State acts in accordance with its own perceptions of self-interest, mainly the acquisition and retention of power, and the real decision makers are the most powerful States. But we have seen how difficult it is to define and measure power and how the components of power are continually changing. We prefer to consider international relations as relations among members of a community.

Like most communities, it is organized, but the organization is not systematic. Some members have a large measure of independence and others are still dependent, but all are interdependent. Some members are more powerful than others, however power is defined, but the balance of power is continually shifting, and even small and weak States often play important roles in international affairs. The community has grown gradually with no plan and no steady progression of development. The whole is bound together by an intricate network of bilateral and multilateral relationships of all kinds: religious, political, economic, histori-

cal, cultural, legal, ideological, and all of these are overlapping and interrelated. In Parts Four and Five we discussed several of these relationships under the general headings of geopolitics and of imperialism, colonialism, and decolonization; we now examine more of these relationships that are geographic in nature.

## *What Is International Law?*

Every society or community must be governed by a set of general principles and specific rules if it is to function at all. The international community is no exception. It is governed by a complex network of principles, treaties, judicial decisions, customs, practices, and writings of experts that are binding on States in their mutual relations. This is what we call international law, and it has been evolving for centuries. While international law is not created by a particular legislature and enforced by an executive with police powers as domestic law is, it is binding nevertheless. And it is usually effective. Despite the scoffing of cynics, the plain fact is that States do recognize their interdependence, do understand the need for world order, and do operate according to international law most of the time. Naturally, as in every society, there are lawbreakers. But in the international community no State can long remain a chronic lawbreaker. Even if formal sanctions are not applied against it, a

State that refuses to accept and abide by the rules simply isolates itself from the rest of the community and suffers from the lack of normal intercourse every State needs. Law is the only alternative to anarchy; law is demanded by the community of interests among States.

Having said this, we must recognize that "international law" was originally created by a relative handful of States, primarily in Western Europe. Although it does contain some components of other traditions, it is basically derived from Roman law, the Anglo-Saxon common law, and Christian theology. When the dismantling of empires began in earnest after World War II, many new States realized that they were expected to accept and adhere to a body of law in whose creation they had played no role. In some cases they had strong indigenous legal traditions that they valued highly and that they wanted to project into the international sphere. In others, Western law represented imperialism (indeed, had often been used to justify and buttress imperialism) and must *ipso facto* be rejected. Despite these feelings, however, on the whole the new States respect, utilize, and help develop international law, while objecting to certain aspects of it that they deem contrary to their interests. These interests are often different from, sometimes contrary to, the community of interests recognized by the older Western States. Conflicts are thus inevitable, but they are being resolved both by traditional methods and through the political process, principally in the United Nations. It is in the United Nations, in fact, where international law is currently being most vigorously developed.

## The United Nations and International Law

Article 13 of the UN Charter charges the General Assembly with responsibility for, among other things, "encouraging the progressive development of international law and its codification." Overall responsibility for legal matters is assigned to the Sixth Committee of the General Assembly, but the actual work is being done chiefly by the International Law Commission (ILC), established in 1947 by the General Assembly. The 34 members of the ILC are distinguished authorities drawn from all major legal traditions who do not represent governments but function in their personal capacities. The new States not only participate actively in the work of the Sixth Committee and the ILC, but look to them to broaden international law so that it becomes truly universal. Among the many topics dealt with in great detail by the ILC in its first four decades, two are of special interest to geographers: international rivers and the Law of the Sea. We discuss international rivers later in this chapter and the Law of the Sea in Chapters 31 and 32.

Another UN organ helping to develop international law is the International Court of Justice, successor to the Permanent Court of International Justice of the League of Nations. The ICJ, or World Court, is composed of 15 judges elected by the General Assembly and the Security Council. Like the members of the ILC, they represent "the main forms of civilization and the principal legal systems of the world." They act in their individual capacities and not as representatives of their governments. The Court sits in The Hague and considers cases brought to it by States either for judgment or for advisory opinions.

Other UN bodies participate, some of them rather marginally, in international lawmaking, including the General Assembly itself and the specialized agencies. More important, however, are the international lawmaking conferences. Most of them have dealt with strictly legal matters, such as diplomatic and consular immunities, State succession, and treaties, but at least four fall largely within the field of political geography. These are the three UN conferences on the Law of the Sea and the UN Conference on Transit Trade of Land-locked Countries. Unlike conferences of experts gathered to discuss current problems and perhaps pass resolutions, these conferences are designed to produce *conventions* (multilateral treaties) which, if duly ratified, are binding on at least the signatories and become part of interna-

**The World Court in session.** The judges hear an oral pleading in a case in the Peace Palace in The Hague. (United Nations photograph number 186867)

---

### *The World Court and Political Geography*

Among the cases considered by the World Court during its first half century, the following are of special interest in political geography: the Corfu Channel (*United Kingdom* v. *Albania*, 1948–49), Norwegian Fisheries (*UK* v. *Norway*, 1951), Miniquiers and Ecrehos (*France* v. *UK*, 1953), Antarctica (*UK* v. *Argentina and Chile*, 1956), Right of Passage over Indian Territory (*Portugal* v. *India*, 1957–60), Sovereignty over Certain Frontier Land (*Belgium* v. *Netherlands*, 1959), Arbitral Award Made by the King of Spain (*Honduras* v. *Nicaragua*, 1960), Temple of Preah Vihear (*Cambodia* v. *Thailand*, 1962), South-West Africa (*Ethiopia* v. *South Africa; Liberia* v. *South Africa*, 1966), Northern Cameroons (*Cameroon* v. *UK*, 1963), North Sea Continental Shelf (*Federal Republic of Germany* v. *Denmark; FRG* v. *Netherlands*, 1969), Fisheries Jurisdiction (*UK* v. *Iceland; FRG* v. *Iceland*, 1974), Western Sahara, 1975, Aegean Sea Continental Shelf (*Greece* v. *Turkey*, 1978), Continental Shelf (*Libya* v. *Tunisia*, 1982), Gulf of Maine (*United States and Canada*, 1984), Continental Shelf (*Libya* v. *Malta*, 1985), Frontier Dispute (*Burkina Faso/Mali*, 1986), Land, Island, and Maritime Frontier Dispute (*El Salvador/Honduras*, 1987), Maritime Delimitation in the Area Between Greenland and Jan Mayen (*Denmark* v. *Norway*, 1988), Certain Phosphate Lands in Nauru (*Nauru* v. *Australia*, 1989), Arbitral Award of 31 July 1989 (*Guinea-Bissau* v. *Senegal*, 1989), Territorial Dispute (*Libya/Chad*, 1990), East Timor (*Portugal* v. *Australia*, 1991), Maritime Boundary (*Guinea-Bissau* v. *Senegal*, 1991), Passage Through the Great Belt (*Finland* v. *Denmark*, 1991), Maritime Delimitation and Territorial Questions Between Qatar and Bahrain, 1991, Gabčíkovo-Nagymaros Project (*Hungary* v. *Slovakia*, 1994), and Land and Maritime Boundary between Cameroon and Nigeria (1994). Of these cases, all but the Corfu Channel, Right of Passage, Northern Cameroons, South-West Africa, Phosphate Lands, Great Belt and Danube River cases were boundary or territorial disputes, and of these seven, all but the Corfu Channel, Phosphate Lands, Great Belt and Danube River involved colonies or mandates. Thirteen cases involved the Law of the Sea. We may expect the Court to be asked to deal even more in the future with cases in these categories. With eleven cases on its docket in 1994, the Court has the heaviest caseload in its history. Also, for the first time cases are before it from every continent except Antarctica. All of this represents a vote of confidence in the Court, in international law, and in the peaceful settlement of disputes, even by former colonies.

tional law. It is possible that we will see fewer lawmaking conferences in the future, with more of this work being done by the Sixth Committee. But we are unlikely to see any reduction in the overall contribution of the United Nations to the codification and progressive development of international law. On the contrary, as more States join the international community, as all States become ever more tightly bound together by mutual dependence, and as the need for universal law becomes both more accepted and more urgent, we can expect a great increase in UN activity in this field.

## Conflicts and Conflict Resolution

There are and probably always will be conflicts and disputes among States. They have many origins and take many shapes. Some are bilateral, and some involve groups of States on both sides. Some are ancient, some relatively new. Some are fairly simple, others intricately complex. All, however, can be resolved without resort to war if the parties are willing. Conflict resolution (or, as it is called in international law, *pacific settlement of disputes*) has long been a popular subject for political scientists. Geographers have tended to concentrate on the issues instead of on the process of settlement. If geographers are to understand a conflict and analyze it usefully, however, they must be familiar with the procedures as well as the issues.

By far the most important method of settling a dispute is *bilateral negotiations*. The parties use diplomatic channels to begin with. Sometimes the normal diplomatic efforts will be supplemented or supplanted by negotiations between cabinet ministers and even heads of government. On occasion imaginative devices are used: a private citizen acting as go-between, a meeting in a train parked on a bridge straddling an international river, an official going secretly in disguise to talk with a hostile chief of State. Generally, though, the negotiations themselves are quite mundane, even if they are accompanied by flamboyant publicity and ostentatious saber-rattling. Most disputes are

settled in this way; only if the dispute is particularly intractable or the parties are unwilling to talk with one another is a third party called in to help.

There are five standard types of third-party participation: *good offices, conciliation, mediation, arbitration,* and *judicial proceedings*. The disputing parties may choose any one of them, use a combination or variation of them, or progress from one to another. In any case, they determine the third party and establish the ground rules, either in general or in detail.

The simplest form of third-party participation, the one in which the third party is least directly involved in the dispute, is called *good offices*. The third party expedites bilateral negotiations by performing such services for the disputants as providing a neutral site for the negotiations; supplying interpreters, office space, secretarial services, and the like; transmitting messages between the parties; doing basic research and providing factual information to the parties; even providing entertainment and sightseeing so as to create and maintain a relaxed and friendly atmosphere for the negotiations. This work is done every day around the world, generally quietly with little or no publicity. Many countries, especially Switzerland, provide good offices, but by far the most frequent and useful provider of the service is the United Nations. This is, in fact, one of its least publicized but most important services to world public order. Even while countries are shooting at each other, they can negotiate quietly at UN headquarters or elsewhere under UN auspices.

A third party can intervene rather modestly in the negotiations by offering *conciliation*. A conciliator will consider the positions of both sides and offer a compromise solution to the problem. He or she does not participate in the negotiations, undertake detailed studies, or pass judgment. The conciliator's function is to facilitate the resolution of a dispute by offering a face-saving solution to the parties. Very close to conciliation, but more formal and active, is *mediation*. A mediator studies the case in more detail, participates actively in the negotiations, and of-

---

### Modes of Redress Short of War

There are some less pacific methods of dealing with international grievances or disputes. In international law these are known as "modes of redress short of war." The mildest response is called *retorsion*, which is applied when the act or acts complained of are unfriendly but not illegal. Such acts may include discriminatory tariffs, immigration restrictions, port rules, currency control, and the like. Typically, the retort is reciprocity; that is, the complainant responds in kind. A more vigorous response to an alleged unfriendly or illegal act is *retaliation*. A State has a large arsenal of retaliatory weapons available to it, and it may exercise them as it deems appropriate within the limits of its own ability and commitments. It may take such diplomatic action as recalling its ambassador, closing its embassy, or breaking diplomatic relations. Economic action might involve raising tariffs or imposing quotas against the goods of the other party or selective or general embargoes or boycotts. And an aggrieved State might take military action: a brief raid into the other party's territory by land, sea, and/or air to destroy selected military targets or take military prisoners, or perhaps capture of one of the other party's vessels at sea. Finally, it may resort to *reprisal*, a form of retaliation far in excess of the acts complained of. Reprisal borders on aggression and may sometimes be considered illegal, though the distinction is unclear. As with the modes of pacific settlement of disputes, the less damaging the mode of redress short of war, the more frequently it is used. Reprisal is a last resort short of war.

---

fers a formal proposal for solution of the problem. Usually, only the more difficult problems require third-party intervention in the first place and, therefore, the tendency is for the parties that have agreed on such third-party intervention to prefer mediation to conciliation. The number of situations requiring conciliation or mediation is much smaller than those requiring only good offices, but they tend to be more politicized and involve higher stakes for the parties—and sometimes for the world.

If a dispute is more legal than political in nature (and it is frequently difficult to disentangle the two) and if it has been protracted and particularly trying for both parties, they may resort to *arbitration*. This is a more formal, time-consuming, and expensive undertaking and, consequently, is less frequently utilized than any method discussed so far. But many arbitrations over the years have had far-reaching influence, not only on the parties, but also on the evolution of international law, even though technically they are not precedents. The parties to the dispute agree in advance whether the results of the arbitration are to be advisory only or actually binding on them. Usually they agree on binding arbitration. Then they choose an umpire or arbitrator (who may be a sovereign, a distinguished judge, or a tribunal or panel). If there is more than one arbitrator, the parties typically choose one or two each and they choose another to serve as president or chairman.* The parties then submit to the arbitrator(s) a *compromis*, a formal statement of the principles and rules of law that the parties agree are applicable in this particular case. The arbitrator then takes testimony, studies it, and renders a decision. If there is a question of damages or compensation involved, the arbitrator may also issue an award.

Unlike good offices, conciliation, and mediation, arbitration aims at justice, not compromise. It still represents, however, the parties' attempt to arrive at an amicable resolution of a problem under their own rules by arbitrators of their own choice. *Judicial proceedings*, on the other hand, are formal adversary proceedings before a permanent court following established rules. They are typically the last resort, used after all other

---

*Arbitrators are sometimes chosen from among the judges on call from the Permanent Court of Arbitration, founded at The Hague in 1899. In this case each party selects two from the list, and they select a fifth to be the umpire.

methods of pacific settlement of a dispute have been rejected or have failed. The proceedings may take place before a national court, a regional court (such as the Court of Justice of the European Union), or the International Court of Justice. Naturally, being sovereign States, one or both parties may ignore or reject either arbitral or judicial decisions, but generally they are respected.

## International Rivers and Lakes

Some aspects of international law are of special interest to geographers because of their spatial dimension. We have already discussed a number of them and cover others in subsequent chapters. Here we examine a subject that overlaps many others: international rivers and lakes. Generally, the same principles and rules apply to lakes as to rivers; we mention lakes only when there is a special reason to do so.

*International rivers* are rivers shared by two or more countries. The international boundary may follow the river or cut across it. If the river serves as a boundary, the actual boundary may be on the left bank (looking downstream), the right bank, or somewhere in between. If it flows across two or more States, there are upstream and downstream riparians that may have different interests in the river. (Sometimes part of a river serves as a boundary and another part traverses one or more States.)

*Internationalized rivers* are those that by treaty or other formal arrangement have been opened to navigation by vessels of States in addition to those of the riparians, even if they lie entirely within the territory of a single country. Among them are the Scheldt, Rhine, Niger, Danube, Congo, Zambezi, Amazon, Plata, and St. Lawrence. The ones most heavily used for international navigation have international commissions to supervise and regulate their use. Navigation has traditionally been the most important use made of international rivers, but other uses have become so important in the last half century that the International Law Commission undertook a thorough study of

"the law of the non-navigational uses of international watercourses" and in 1994 submitted draft treaty articles on the subject to the General Assembly.

From earliest times people have used rivers to transport themselves and their goods. Water transport is particularly useful and economical for heavy and bulky goods. Rivers large and small were the world's first highways. With increasing population and industrialization, with cheaper and more efficient land and air transport, the proportion of goods carried on internal waterways (canals as well as rivers) has decreased. The *volume* of waterborne commerce has nevertheless increased dramatically and is more important than ever.

Over the centuries both natural and political barriers to river navigation have gradually been removed, beginning in Europe with its high density of population, industry, and waterways. The waterways of North America are also intensively utilized and have been reshaped and managed, generally more elaborately and on a larger scale than in Europe. Elsewhere, rivers tend to be used more for local navigation by small vessels, often very small ones, and the engineering works are not as elaborate.

The natural obstacles to international river traffic, however, have been much easier to remove than the artificial ones. States have frequently forbidden nonnationals to navigate on national and international rivers within their territories. Where navigation by foreigners was permitted, it was often subjected to tolls, taxes, and restrictions. As trade grew in importance, the commercial advantages of free navigation gradually assumed greater importance than absolute territorial sovereignty. As early as the eleventh and twelfth centuries, territories in Europe, particularly in Italy, began the internationalization of rivers. By the late eighteenth century, however, tolls and restrictions were still common. The principle of free transit was reestablished by the armies of revolutionary France, which opened the River Scheldt, closed by the Dutch since 1648, to the traffic of all riparian States. In 1795 Spain and the United States negotiated the Treaty of

San Lorenzo el Real that granted Americans the right to navigate the Mississippi and other Spanish rivers to the sea. This gave substance to President Jefferson's claim that "the ocean is free to all men, and their rivers to all their inhabitants." Gradually, the concept of freedom of the seas was applied to rivers, considered by some to be long, narrow arms of the sea.

At the end of the Napoleonic Wars, the Congress of Vienna codified the principles of freedom of navigation on international waterways that had been established in recent years and applied them specifically to the Scheldt, the Meuse, and the Rhine. Through the nineteenth century the principles of free transit on rivers gradually became established, and by the end of the century transit duties in Europe had virtually disappeared. In Latin America the newly independent States opened their territories to the ships and traders of all States. The Congo Act of 1885 opened the Congo River and all its tributaries and the Niger River to the trade of all States. By World War I, most navigable waterways, natural and artificial, in Europe, East Asia, Africa, and the Americas had been internationalized. Today there are few political problems in river navigation; where they do exist, they tend to be part of a much larger pattern of hostility. Some restrictions on foreign navigation on both national and international rivers are still practiced by such States as China and Russia, even where the rivers have been internationalized by treaty, but these restrictions are often informal and difficult to document. On the whole, however, river navigation is freer now than it has been for centuries.

Nonnavigational uses of rivers are many and varied and are becoming more so all the time. The earliest civilizations—Egypt, Mesopotamia, the lower Indus, Southeast Asia, east China—all developed in river valleys using the river water for irrigation. Irrigation has spread around the world from these culture hearths, even to regions that are relatively humid. In some places, such as the southeastern United States, this irrigation merely supplements rainfall and is used largely for luxury crops, but in other areas, such as Egypt, the river water is absolutely essential for the maintenance of even bare subsistence for the people.

All around the world, even in recent years, there have been disputes over the uses of international rivers. One, which brought two large neighbors to the very brink of war, was over the Indus River. During their rule in India, the British developed a sophisticated multipurpose system for the upper Indus Valley, utilizing the main stream, six major tributaries, and a number of smaller tributaries. When India was partitioned in 1947, the new international boundary cut right through this integrated system, wreaking havoc on irrigation, flood control, hydroelectricity, and other services in the basin. In April 1948, India cut off the flow of water from the eastern tributaries of the Indus into the canals leading into Pakistan. Service was restored a month later, but the dispute over ownership of the Indus waters dragged on until both parties finally agreed in 1960 to a World Bank plan for dividing them between the two countries. Since then, a permanent Indus commission has supervised implementation of the Indus Waters Treaty, but the enormous cost and disruption to India and Pakistan of creating and operating two parallel river-development schemes could have been avoided if the neighbors had formed a joint commission immediately in 1947 to operate and improve the system they inherited from the British.

Similar but less virulent disputes over the waters of the Nile, Colorado, Columbia, Ganges, Euphrates, and other rivers have been settled only after arduous bilateral negotiations and in some cases third-party participation. As the world's population grows larger and richer, it places greater and greater demands on the Earth's finite supply of fresh water. Inevitably, as the larger streams are utilized, smaller and smaller streams are likely to be the focus of competing demands for their water.

The Jordan waters dispute between Israel and Jordan, Lebanon, and Syria is now more than 40 years old. Both Israel and international experts have repeatedly urged cooperative development of the whole Jordan

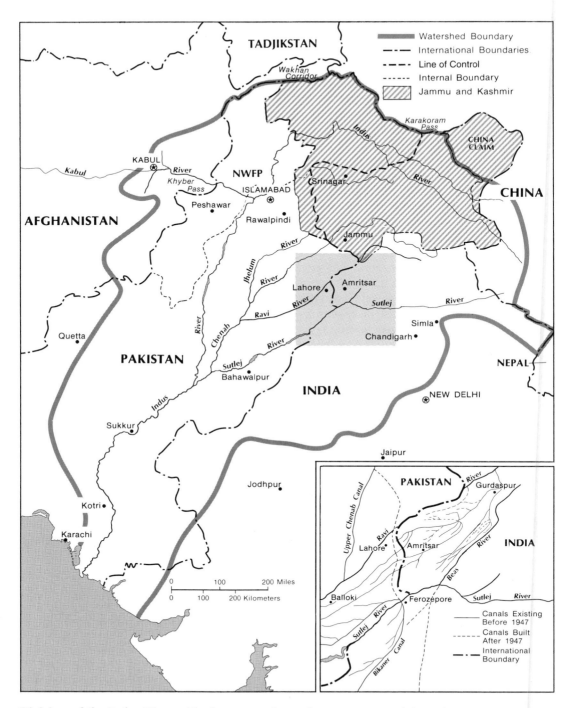

**Division of the Indus Rivers.** The large map shows the main stream of the Indus River and its principal tributaries. Note the importance of Jammu and Kashmir in this respect as well as the area's strategic location between China and India and Pakistan. Note also the relationship of the Khyber Pass and Pakistan's North-West Frontier Province to this area and the location of Pakistan's capital of Islamabad. Amritsar is a center of the Sikh religion, and Chandigarh is capital of the Indian states of Punjab and Haryana. The inset map of the irrigation canal system is generalized but shows clearly the effect of the 1947 partition on what had been an integrated system.

system for the benefit of all the countries in the basin, but the Arab countries have steadfastly refused to cooperate with Israel for political reasons. As a result, Israel and Jordan have developed parallel water-diversion systems for irrigation and other uses. In their 1994 peace treaty, however, the two neighbors agreed on allocation of the water of the Jordan and Yarmouk rivers and to cooperate in the management of the water resources of the region.

An even smaller and much more obscure stream, the Río Lauca, was the cause of a major crisis between Chile and Bolivia in the early 1960s. The Lauca rises in Chile high in the Andes and flows for about 225 kilometers into Lake Coipasa in Bolivia. In 1939, Chile announced a plan to divert some of the river's water down into the valley of the Río Azapa to provide irrigation and domestic water for the city of Arica. Bolivia protested then and on a number of occasions thereafter as plans were developed and then executed. In 1962 the diversion ac-

tually began. Bolivia broke off diplomatic relations with Chile and requested an urgent special meeting of the Council of the Organization of American States (OAS), charging Chile with "geographic aggression!" Bolivia was dissatisfied with the OAS action and quit the organization for a time. The diversion continues, but the basic issue has not been resolved.

More recently, Syria and Iraq, which agree on very little else, have vigorously protested Turkey's plans to dam both the Tigris and Euphrates rivers. As recently as 1993, the prime minister of Turkey threatened to cut the flow of Euphrates water to Syria unless Syria ceased its support of Kurdish rebels in Turkey. It may be a very long time before the three countries agree on a water-sharing formula. Since there is not yet any generally accepted international law on the subject of nonnavigational uses of international rivers (other than a broad principle that the upstream riparian should not use a river in such a way as to harm a downstream ripar-

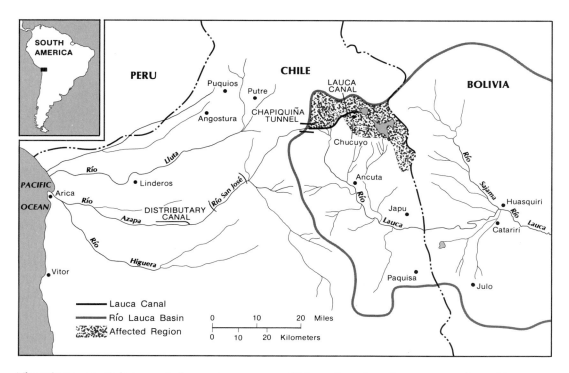

**The Río Lauca.** Bolivia created a controversy over Chilean diversion of some water from this stream, but soon linked the diversion with its desire for a *salida al mar*, or outlet to the sea. Bolivia never demonstrated any damage done to its territory or people by the diversion.

ian), conflicts involving them are bound to continue.

Other nonnavigational uses of rivers include fishing, flood control, hydroelectric power generation, recreation, waste disposal, industrial processes, timber floating, domestic consumption, drainage, construction, and aquaculture. Various traditions around the world cover some of these uses, but the traditions often conflict with one another. So do many of the uses of rivers. The problem is serious enough in the United States, where water law in the Southwest is very different from water law in the Northwest and Northeast, but it is far more severe at the international level. In Europe most major problems have been worked out over generations, but in Latin America generations of work by experts have failed to produce a generally accepted code for regulating the use of international river waters. In Africa and Asia there has not even been a serious attempt at such standardization. This is why the International Law Commission has developed a draft convention on the subject, and why its work is so important.

A final point concerns the concept of the *river basin*, regardless of political boundaries, as an ecological unit that should be developed according to a comprehensive, integrated, multipurpose plan. All the present and potential uses of the main stream, its tributaries, and its drainage basin should be accounted for as thoroughly as possible so as to minimize conflicts over them. Such conflicts include competing uses, waste of water, soil erosion, water and soil pollution, salinization and waterlogging of overirrigated land, destruction of wildlife habitats, overfishing, unnecessarily destructive floods, and other ills caused by unwise use of this scarce natural resource. This concept is really quite old, but its first modern expression was the Tennessee Valley Authority, still the model for similar schemes the world over.* The Papaloapan in Mexico, the Cauca in Colombia, the Damodar in India, the Helmand in Afghanistan, and the Volta in Ghana are just a few examples. Developing

*See Chapter 14 for details of TVA.

*international* river basins in this manner, however, is far more difficult because of the political problems entailed.

Two ambitious schemes for multipurpose cooperative development of *international* river basins are those for the Plata and the lower Mekong. The first involves Brazil, Paraguay, Bolivia, Argentina, and Uruguay, which share the basin of the Paraguay–Paraná–Plata river system. In 1941 the five countries met in Montevideo and agreed on a resolution expressing their aspirations for regional development. Nothing more happened until the 1950s, when the countries individually and in pairs began serious development work there. Finally, all five countries met in Buenos Aires in 1967 and established an intergovernmental coordinating committee to begin planning a variety of cooperative projects. Since then many bilateral projects have been completed and more are under way, all within the general framework of the Plata Basin program. It has the active support of the OAS and other regional and international agencies, but its course has not been smooth. There have been numerous squabbles among the participants, particularly Brazil, Paraguay, and Argentina, over everything from overall priorities to the location of highway bridges to the type of electric current to be generated by a hydroelectric project. Nevertheless, it goes on.

Even larger is the project for developing the lower basin of the Mekong, the largest unbridged (until April 1994) and undammed river in the world, shared by Thailand, Laos, Vietnam, and Cambodia. It was initiated by the UN Economic Commission for Asia and the Far East (ECAFE, now ESCAP) and the Colombo Plan in the mid-1950s. As the years went by and plans developed, more sponsors joined in: 25 countries plus several international agencies and private foundations. Detailed plans and cost estimates were ready by 1960, and field surveys began soon after. Before work had gone very far, the United States got heavily involved in what had been small-scale fighting in South Vietnam. As the scale of the war increased, and it spread through much of Laos and Cambodia, the work slowed down and nearly

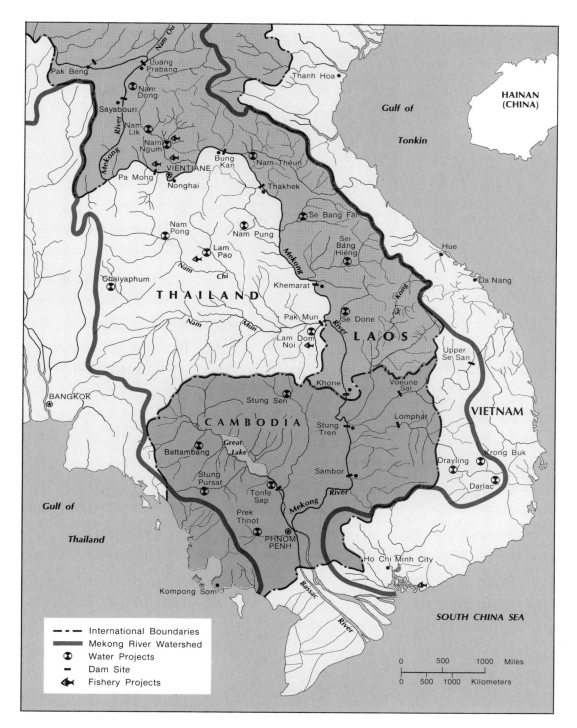

**The Lower Mekong Basin scheme.**

**The Mekong Scheme in Laos.** At the top is a Canadian survey team measuring topographical levels of the Nam Ngum River, a tributary of the Mekong. The project is now well advanced. (UN) Below is the original headquarters of the Lao National Mekong Committee in Vientiane. (Martin Glassner)

ceased. The Mekong Committee,* composed of the four riparian States, does the planning. In 1977, Laos, Thailand, and Vietnam, because of the absence of Cambodia, agreed to form an interim committee, and the pace of the work picked up. Cambodia rejoined the committee in 1991 and has been actively participating in its work.

*Officially, the Committee for Coordination of Investigations of the Lower Mekong Basin. Its headquarters is in Bangkok.

Development work continues in the fields of hydrology, meteorology, mapping by remote sensing, economic and social studies, social planning and resettlement, environmental studies and planning, irrigation, drainage, flood and salinity intrusion control, hydroelectricity, navigation, agriculture and aquaculture, geological surveys, water-borne diseases, control of erosion and sedimentation, and agro-industrial development. In January 1986 the committee ran a UNDP workshop for Nile Basin countries in Bangkok to illustrate the advantages of the

Mekong approach to the Nile riparians. In November 1994 the four lower basin States accepted a new draft agreement on cooperation for the sustainable development of the region. It includes establishment of a Mekong River Commission and a Basin Development Plan, the whole project to be conducted according to international law and open to the participation of China and Myanmar. This new arrangement, which will continue to be assisted by the United Nations Development Programme and Asian Development Bank, promises to resume and expand the cooperation that was so productive and inspiring in the mid-1960s.

Other multinational basin-development organizations include those for the Niger and Senegal rivers and Lake Chad in Africa and the Amazon in South America. We can be reasonably certain that more of these schemes will evolve, for the benefits of such cooperation are so great and so obvious that not even the most ardent nationalists can deny them.

# 25

# International Trade

International trade, in a sense, has existed for several thousand years, since the first organized human communities began trading surpluses with other communities. We have both written and archaeological evidence of elaborate trading systems functioning in all of the early culture hearths and of goods being traded among them, often over impressive distances. Geographers, economists, anthropologists, and historians have studied in detail the trading systems of West Africa, South America, the South Pacific, and other regions before their exposure either to Europeans or to modern technology. Modern international trade, however, really began with the Industrial Revolution and the decline of mercantilism.

## The Evolution of International Trade

*Mercantilism* was the dominant economic theory in Europe during the late Middle Ages and into the eighteenth century. It was based on the notion that wealth consisted only of gold and silver. Naturally, a country could get rich only by accumulating large stores of gold and silver bullion. There were three ways of doing this: first, by stealing it or by conquering countries that had large stores of it; second, by finding and exploiting new sources of the precious metals; and third, by exchanging goods and services for gold and silver. The first method was em-

ployed from time to time but proved impractical as a general policy; the second led to the great Age of Discovery and the creation of the first large European colonial empires; the third produced stagnation and decay, even for some countries with large empires. Mercantilism was an essentially restrictive economic policy that made a few individuals and governments rich but did nothing to improve the lot of the many millions of ordinary people in the world. Gold and silver produce nothing; only investment produces anything, and it is production of goods and services that improves people's lives, not hoarding of gold and silver. The exchange of surplus production increases the value of the initial investment and multiplies people's real incomes.

The first major attack on mercantilism was launched by the English philosopher David Hume in 1752 when he demonstrated the fallacy of the "favorable balance of trade" concept. His friend Adam Smith followed up in 1776 with an elaborate treatise called *Wealth of Nations* that expounded the concept of international division of labor, or specialization and international trade. This was a direct attack on autarky and a plea for free trade, which would bring some benefit to everyone rather than great benefit to a few. Smith was followed by David Ricardo, John Stuart Mill, and other classical English economists who developed further the theory of *comparative* *advantage*.* This is the theoretical underpinning of free trade. The indisputable advantages of specialization and trade led to the free trade movement of the nineteenth century and the burgeoning world trade that developed from it. Today most people in the world can afford *some* imported product, and *some* of what most people produce is exported. A far cry indeed from the situation 300 years ago when most people subsisted on what they could produce themselves and only a few rich and powerful people could benefit from international trade.

Not all governments, however, adopted the *laissez-faire* policies so important for free trade. Some, in fact, persisted with mercantilist policies for generations. Antonio Salazar, dictator of Portugal and erstwhile economics professor, based his *Estado Novo* on mercantilist principles. It brought stability but not prosperity to Portugal and its empire during his tenure as Minister of Finance (1926–40) and Prime Minister (1932–68). Few countries ever adopted a completely *laissez-faire* attitude toward international trade; in most, the liberalization of trade was limited by protectionist policies or other considerations. Still, comparative advantage, economies of scale, and differential demand for goods in countries with different cultures were powerful forces generating the expansion of trade. Gradually, the "colonial economy" (the exchange of the manufactured goods of Western Europe for the foods, fuels, and raw materials of Africa, Asia, and the Americas) was surpassed in importance by trade among the industrialized countries.

As the industrializing countries became richer as a result of their control over manufacturing, commodities, and trade, they

---

*This theory is best illustrated by the hypothetical example of the best lawyer in town who is also the best typist in town. He and everyone else will benefit most if he specializes in the law and hires a typist. Even though he has an *absolute* advantage over the typist in typing, the typist has a *comparative* advantage in that specialty. So even if a country can produce *everything* more efficiently than every other country, it can maximize its assets and grow more rapidly if it specializes in those things in which it has a comparative advantage and imports other things from somewhat less efficient producers.

began demanding and producing ever-more sophisticated and expensive products. They found that the only feasible sources of the goods they wanted were other industrialized countries that, through no coincidence at all, were also the only countries rich enough to buy the new manufactured goods they were producing. The bulk of world trade then began to flow among the rich countries themselves. This pattern was reinforced by the restrictions placed by the colonial States on manufacturing and trade by their colonies, by protective tariffs imposed by some industrializing States, by subsidization of local agriculture, by development of substitutes for some commodities, and by other practices and developments. Today we find this pattern still dominant, even though many poor countries really are "developing" and becoming more important traders in their own right rather than serving simply as appendages of a colonial ruler. Table 25-1 shows this pattern clearly. Over 73 percent of all goods traded among States and territories originated in the developed market economy (noncommunist) States in 1992, with only 3.4 percent coming from the more developed socialist countries, including the former Soviet Union. The poorer countries of the world contributed only 22 percent.

The table shows a number of other interesting things, including who trades with whom. The developed market economy countries, for example, did 77.7 percent of their trading among themselves, and bought only 19.1 percent of their total imports from the developing countries, including the Asian socialist ones, of which China is by far the largest. Even the poor countries bought almost two-thirds of their imported goods from the rich countries, and the major petroleum exporters (in whose cases the terms "poor" and "developing" need new definitions) bought 72.6 percent of their imports of all kinds in the rich industrialized countries—which, of course, are the best customers for their principal export. Note also the dramatic fall in the total trade of the former socialist European countries, and the rapid growth of trade between the developing countries and the Asian socialist countries, chiefly China.

***Table 25-1***   *Origin and Destination of World Trade (%) 1976, 1982, 1989, and 1992*

| | | | Destination | | | | |
| | | | | Socialist Countries | | Developing Countries | |
| Origin | Year | World | Developed Market Economy Countries | Eastern Europe | Asia | Total | Major Petroleum Exporters |
|---|---|---|---|---|---|---|---|
| Developed Market | 1976 | 65.1 | 68.6 | 34.7 | 56.4 | 67.9 | 84.0 |
| Economy Countries | 1982 | 63.6 | 68.2 | 29.0 | 57.8 | 63.1 | 77.4 |
| | 1989 | 70.1 | 77.6 | 22.9 | 44.5 | 62.1 | 69.0 |
| | 1992 | 73.6 | 77.7 | 48.1 | 39.5 | 62.5 | 72.6 |
| Socialist Countries | 1976 | 8.8 | 3.7 | 53.3 | 32.1 | 5.2 | 4.1 |
| Eastern Europe | 1982 | 9.1 | 4.0 | 59.5 | 21.6 | 4.7 | 4.9 |
| | 1989 | 7.6 | 2.3 | 62.9 | 16.3 | 5.8 | 6.7 |
| | 1992 | 3.4 | 2.2 | 31.5 | 8.4 | 2.3 | 1.9 |
| Asia | 1976 | 0.8 | 0.4 | 1.4 | — | 1.6 | 0.9 |
| | 1982 | 1.2 | 0.8 | 1.5 | — | 2.3 | 0.9 |
| | 1989 | 1.8 | 0.9 | 3.1 | 0.9 | 4.5 | 1.1 |
| | 1992 | 2.2 | 1.0 | 4.5 | 0.7 | 5.6 | 1.4 |
| Developing Countries | 1976 | 25.3 | 27.3 | 10.5 | 11.5 | 25.3 | 11.1 |
| Total | 1982 | 26.1 | 27.0 | 10.0 | 20.1 | 29.8 | 16.7 |
| | 1989 | 20.4 | 19.0 | 11.0 | 38.1 | 26.6 | 25.1 |
| | 1992 | 22.2 | 19.1 | 15.9 | 51.2 | 28.6 | 24.1 |
| Major Petroleum | 1976 | 14.0 | 16.0 | 2.3 | 0.0 | 12.8 | 1.0 |
| Exporters | 1982 | 12.2 | 13.7 | 1.9 | 0.5 | 12.8 | 2.1 |
| | 1989 | 4.4 | 4.1 | 1.2 | 0.9 | 6.4 | 4.4 |
| | 1992 | 4.8 | 4.5 | 2.3 | 1.9 | 5.6 | 3.8 |

*Source:*   United Nations Conference on Trade Development, *Handbook of International Trade and Development Statistics,* 1978, 1984, 1991, 1993.

The world trade graphs illustrate in more detail the dominance of trade by the rich countries and the relatively small proportion of trade composed of commodities.

## GATT, UNCTAD, and WTO

The system of world trade that had matured in the late nineteenth century was disturbed by World War I, battered by the Great Depression, and destroyed by World War II. All along, however, it had been weakened by autarkic policies adopted by various countries to restrict trade for a number of reasons. There were, of course, surviving mercantilists, but more important were desires to protect infant industries, build up uneconomic industries for strategic reasons, con-

serve scarce foreign currencies, give preferences to political allies or former colonies, develop monopolies and cartels to increase profits by manipulating prices, maintain strong merchant fleets, and achieve other limited objectives, all of which effectively brought to an end the limited experiment with free trade promoted by Britain. High protective tariffs, including especially the U.S. Smoot-Hawley Tariff of 1930, were only the most publicized instrument of this economic nationalism that escalated a bank failure in Austria and a stock market crash in the United States into a great worldwide depression, illustrating the interdependence of much of the world, even then.

Toward the end of World War II, Western businessmen, statesmen, and economists felt the urgent need to rebuild the world econ-

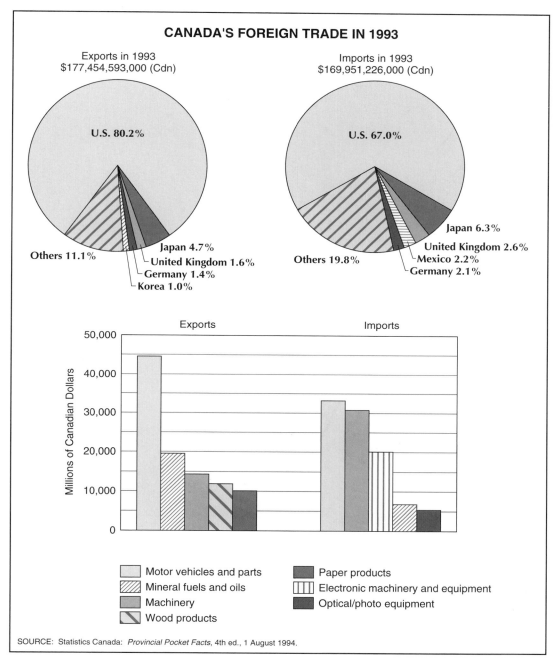

**CANADA'S FOREIGN TRADE IN 1993**

Exports in 1993
$177,454,593,000 (Cdn)

U.S. 80.2%

Others 11.1%
Japan 4.7%
United Kingdom 1.6%
Germany 1.4%
Korea 1.0%

Imports in 1993
$169,951,226,000 (Cdn)

U.S. 67.0%

Japan 6.3%
United Kingdom 2.6%
Others 19.8%
Mexico 2.2%
Germany 2.1%

Exports        Imports

Millions of Canadian Dollars

Motor vehicles and parts       Paper products
Mineral fuels and oils         Electronic machinery and equipment
Machinery                      Optical/photo equipment
Wood products

SOURCE: Statistics Canada: *Provincial Pocket Facts*, 4th ed., 1 August 1994.

**Canada's foreign trade in 1985.** Note the overwhelming dependence of Canada on trade with the United States, and the similarities of the major components of the largest two-way flow of trade in the world. These factors have encouraged negotiation of a full free trade area between the two countries, a form of preferential trading discussed at length in the next chapter. They also illustrate the dominance in world trade of rich countries' trade with one another.

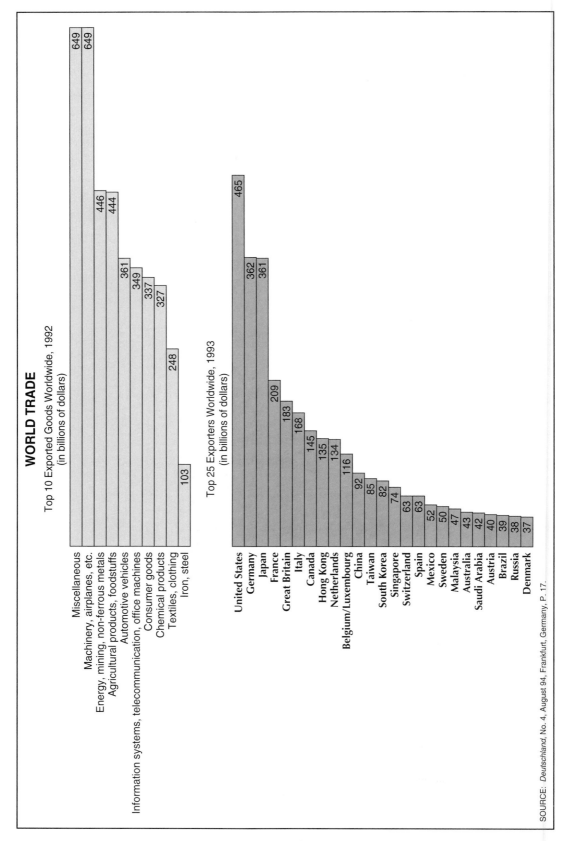

**WORLD TRADE**

Top 10 Exported Goods Worldwide, 1992
(in billions of dollars)

|  |  |
|---|---|
| Miscellaneous | 649 |
| Machinery, airplanes, etc. | 649 |
| Energy, mining, non-ferrous metals | 446 |
| Agricultural products, foodstuffs | 444 |
| Automotive vehicles | 361 |
| Information systems, telecommunication, office machines | 349 |
| Consumer goods | 337 |
| Chemical products | 327 |
| Textiles, clothing | 248 |
| Iron, steel | 103 |

Top 25 Exporters Worldwide, 1993
(in billions of dollars)

|  |  |
|---|---|
| United States | 465 |
| Germany | 362 |
| Japan | 361 |
| France | 209 |
| Great Britain | 183 |
| Italy | 168 |
| Canada | 145 |
| Hong Kong | 135 |
| Netherlands | 134 |
| Belgium/Luxembourg | 116 |
| China | 92 |
| Taiwan | 85 |
| South Korea | 82 |
| Singapore | 74 |
| Switzerland | 63 |
| Spain | 63 |
| Mexico | 52 |
| Sweden | 50 |
| Malaysia | 47 |
| Australia | 43 |
| Saudi Arabia | 42 |
| Austria | 40 |
| Brazil | 39 |
| Russia | 38 |
| Denmark | 37 |

SOURCE: *Deutschland*, No. 4, August 94, Frankfurt, Germany, P. 17.

omy, including the trading system. In July 1944 they gathered at Bretton Woods, New Hampshire for this purpose. There they designed the International Bank for Reconstruction and Development (IBRD or World Bank) and the International Monetary Fund (IMF or Fund). As successful as these agencies have been, however, they could not alone promote a return to a system of reasonably free trade and payments in the postwar world. In 1947 two separate but interrelated sets of negotiations began: one to create an International Trade Organization, which would enable private enterprise to play an active role in trade liberalization; the other to develop a tariff agreement among 23 of the most important trading countries. The Havana Charter establishing the ITO was signed in 1948, but only one country (Australia) ever ratified it and it never went into effect. *The General Agreement on Tariffs and Trade (GATT)* had meanwhile been signed in Geneva in 1947; it was ratified and did go into effect. It was enormously successful in its mission of encouraging and guiding the expansion of world trade.

GATT was essentially two things: a complex network of over 200 interrelated bilateral trade agreements, and a series of rules, all designed to lower tariffs and eliminate nontariff barriers to trade. Its "temporary" secretariat in Geneva took on the characteristics of a permanent organization; it developed loose links with the United Nations; its semiannual meetings served as forums within which new proposals for trade liberalization could be explored; and it incorporated many of the provisions of the Havana Charter into its own rules and functions. The original 23 adherents to GATT were joined by well over 100 more, and still others participated under special arrangements. Even a number of former socialist countries joined, another sure sign of its indispensable role in facilitating the trade of countries just entering the world trading system.* The large for-

*In view of our discussion of the status of Taiwan in Chapter 11, it is interesting to note that in January 1990 the "Republic of China" applied for membership in GATT under the name of "Customs Territory of Taiwan, Penghu, Kinmen and Matsu."

mal tariff-cutting conferences of GATT, held approximately every five years, reduced or eliminated tariffs on many thousands of items. The rules against discriminatory trade practices, dumping (selling export goods below cost), and other nontariff barriers to trade are usually followed—although Japan, the United States, and other countries do violate them from time to time—and the rules of GATT governed some 85 percent of international trade.

During the nineteenth century a new technique was developed to supplement the traditional principle of reciprocity. This was the "most-favored-nation" clause incorporated into a bilateral trade agreement. This MFN clause provided that each partner would grant to the other whatever tariff and other trade concessions had been granted or would be granted in the future to any third country. This practice helped somewhat in reducing tariffs but did not become generally accepted until GATT made it the foundation of its whole system. The MFN clause and the GATT tariff-cutting rounds replaced the U.S. Reciprocal Trade Agreements and were the principal instruments in a general but still quite inadequate worldwide reduction of tariffs. MFN status must be negotiated and is frequently (as in the United States at present) linked to political concessions.

GATT provided for two major exceptions to its rules about tariff reduction and nondiscrimination. It permitted developing countries to use tariffs to protect infant industries for limited periods, and it permitted groups of countries to grant concessions to their members not granted to outsiders, provided the result was an expansion rather than a contraction or simply a diversion of trade. Both devices are widely and for the most part successfully used.

There is no doubt that GATT contributed hugely to the incredible expansion of world trade during its existence. But, as we see in Table 25-1 and the World Trade graphs, the benefits of this trade expansion have not been evenly distributed. They have accrued largely to the industrial countries, those that were already rich. The "colonial economy" still hampers to some extent the development of the poor countries of the world. There has long been a general tendency for prices of manufactured goods to rise more rapidly than the prices of commodities in world trade, thus worsening the terms of trade of those countries that are primarily producers of commodities and consumers of manufactured goods; that is, the poor countries, generally former colonies.*

All these trends became evident to the newly emerging countries in the late 1950s and 1960s. They saw that the rich were getting richer, called GATT "a rich man's club," and refused, many of them, to become parties to it. They felt that GATT's mandate was too restrictive to help them in development and demanded a new organization that could. As their numbers increased, so did their voting strength in the United Nations. By 1964 there were 75 of them, and they were able to pressure the rich countries (including the USSR) to agree—reluctantly—to a special UN conference on the subject. We discuss this conference and its results shortly, but first we should consider some of the major trade issues of the 1980s and 1990s with which the world must deal.

The latest round of trade liberalization negotiations under GATT began with a preliminary meeting of ministers and other officials from more than 70 GATT members in Punta del Este, Uruguay in September 1986. This "Uruguay Round" lasted 7 1/2 years because it had to tackle and resolve some exceedingly difficult problems. This was because of two basic factors: first, the easy things had been done already, and the difficult residual conflicts had to be faced; and second, the world trading pattern had changed considerably since the Tokyo Round (1973–79), and the negotiators faced new and more challenging problems.

The huge growth in international trade is related to four closely interrelated trends, all of which have political causes and/or effects. First, the *mobility of capital* of the cur-

rent period of history is quite unprecedented and could not have been foreseen even half a century ago. But like trade itself, most of this international investment (some two-thirds) has been concentrated in the developed countries. Second, international firms (transnational corporations) have increasingly engaged in *intraindustry trade*; that is, trade in products of the same industrial sector, so that countries have been bound more closely together. Third, there has been a growing *tendency toward oligopoly* in international trade, thus reducing price competition and increasing the importance of technological innovation and other factors in worldwide competition. Fourth, some transnational firms have created a new kind of international division of labor within their own organizations, so that much of the growth of trade is accounted for by *trade among components of the same company.* All four of these trends are linked by the transnational corporation, to which we return later in this chapter.

Other issues with which the Uruguay Round had to grapple were:

1. Nearly all of GATT's trade expansion work had been in manufactured goods, but the problems of *agricultural production* and distribution (discussed in more detail in Chapter 37) had become too great to be evaded any longer. The negotiators had to wrestle with the costly, inefficient, and trade-distorting agricultural policies of Western Europe and the United States especially, in particular their more than $50 billion of subsidies to farmers to encourage overproduction and then the dumping of surplus agricultural products on saturated markets in which those who are really hungry are too poor to buy these surplus foods and fibers.

2. The only important manufactured goods not covered by the rules of GATT were textiles and clothing. Under the Multifibre Agreement (MFA), some 50 percent of trade in these products has been regulated through bilateral quotas, an important technique for protecting the tex-

---

*Since 1973, of course, petroleum has been a major exception to this general tendency, but few other commodities producers have been able to exercise the same politicoeconomic power as OPEC, the Organization of the Petroleum Exporting Countries.

tile and garment industries of the rich countries from "market disruption" by low-cost suppliers, such as the poor countries and those of Central and Eastern Europe. These latter countries wanted the trade liberalized by phasing out the MFA and bringing the trade under GATT.

3. Trade in *services*, such as banking, tourism, insurance, shipping, computer software, communications, and so on, had grown spectacularly, especially among the rich countries, even as much manufacturing had passed from them to some newly industrializing countries (such as Brazil, South Korea, Taiwan, Singapore, and Mexico). The rich countries wanted services brought under the GATT system, but many developing countries did not.

4. Largely on the insistence of the United States, the Uruguay Round included drafting of rules to protect *intellectual property* (e.g., patents, trademarks, copyrights) from piracy and counterfeiting.

5. Again largely because of an American initiative, the negotiators considered distortions of international trade caused by certain *investment policies*, such as local content and trade-balancing requirements.

All of these issues were further complicated by such relatively new—and probably temporary but nonetheless vexing—situations as OPEC-dictated crude petroleum prices, which are far lower than they were in the late 1970s and early 1980s, and the colossal and unprecedented deficit in the U.S. balance of payments, which has led to protectionist pressures not seen in over half a century. In addition, there is still a wide range of more or less traditional distortions of international trade that we list at the end of this chapter.

But some constructive and hopeful developments have been taking place. For example, the powerful entry of China into the world trading system; the growing strength of some economic integration arrangements (discussed in detail in Chapter 26); the grow-

ing, but still small, trade among developing countries (the so-called South–South trade); and efforts by bodies other than GATT to reduce barriers to trade and to use trade as a stimulus to development. Among these bodies are the OECD (covered in Chapter 27) and UNCTAD, to which we now turn. After we review some of the basic features of UNCTAD, we will see how some of its concerns were addressed in the Uruguay Round, from which emerged the World Trade Organization.

*The United Nations Conference on Trade and Development* was held in Geneva from 23 March to 16 June 1964. The rich countries refused to make any significant concessions, but the Conference achieved enough for it to be established in December 1964 as a permanent organ of the General Assembly. Since then UNCTAD has met in various venues every three to four years. Between sessions of the Conference, its work is carried on by its permanent body, the Trade and Development Board, whose headquarters is also in Geneva. Nearly every country in the world belongs to it (including some that do not belong to the United Nations, such as the Holy See [Vatican City] and Switzerland). The core of its membership, however, is the *Group of 77*, or the developing countries, named for the original 75 plus two later arrivals that were instrumental in convening UNCTAD in 1964. (Today the Group of 77 has approximately 130 members, and it functions as a caucus or negotiating group within the United Nations.)

UNCTAD's *raison d'être* has been to initiate changes in international relations leading to a new international economic order based on new concepts, among which is the use of trade policy as an instrument of economic development. This is not very different from the policies of the individual European countries that in the past used trade as instruments both of their economic development and of their imperialist policies, but today conditions are very different from those of a century and more ago. Since UNCTAD's mission is so much greater and more complex than GATT's, its secretariat is also much larger and its activities more diverse.

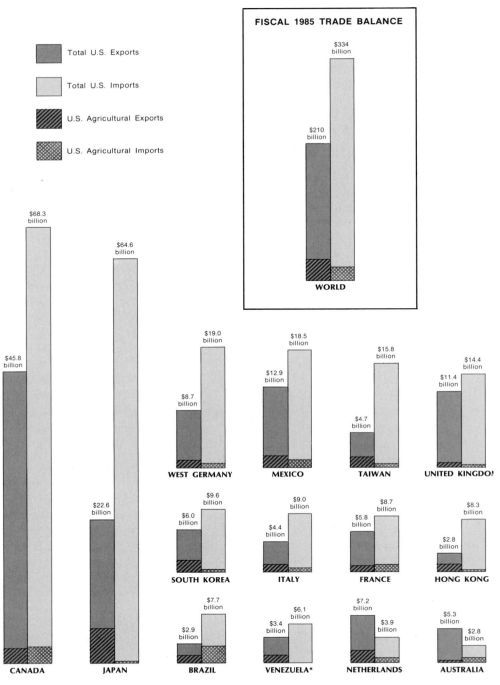

*U.S. agricultural imports from Venezuela were valued at less than $50 million.

Source: Department of Agriculture.
Farmline, September 1986. P. 15.

**Aspects of U.S. merchandise trade deficits, 1985.** These graphs indicate clearly both the voracious appetite of the United States for material goods and the broad spread of its trade deficits. Not only is Japan *not* the only country that sells more to the United States than it buys from it, but Japan *buys* more from the United States than any other country except Canada. Furthermore, the figures are for trade in goods only, but the U.S. profits greatly from trade in services and intellectual property, which explains its positions on these issues in the Uruguay Round. Although specific figures have changed since these graphs were prepared, the pattern has not. Furthermore, the U.S. trade deficit could be significantly reduced if its citizens simply abandoned such absurdities as importing water from France.

**Palais des Nations, Geneva.** This was the headquarters of the League of Nations and currently serves as the European office of the United Nations. UNCTAD headquarters is located in the high-rise annex at the top of this United Nations photograph.

We cannot detail UNCTAD's activities here, but it might be useful simply to summarize some of them in outline form.

I.  Restructuring of international trade to reduce dependency of some countries on others.
  A.  Increasing the value of exports of the developing countries.
    1.  Sponsorship of commodity agreements (between the major producers and major consumers of a commodity to trade fixed quantities of particular types and qualities at fixed prices over a specified number of years).
    2.  For manufactures, the Generalized System of Preferences (GSP) under which the developed countries, through bilateral and multilateral agreements, grant certain import preferences to the manufac-

tured and semimanufactured goods of developing countries.*
  B.  Regulation of trade through principles and policies on trade and related problems of economic development.
    1.  The Charter of Economic Rights and Duties of States (adopted by the General Assembly in 1974); the Declaration and Programme of Action on the Establishment of a New International Economic Order (adopted by the General Assembly in 1974).
    2.  Sectoral regulation, as in the field of shipping, through aiding poor countries to develop merchant marines and helping obtain special freight rates for poor countries.

*The United States in 1976 became the *nineteenth* developed country to implement a GSP.

**II.** Restructuring of international cooperation to strengthen it and reorient it for development.

    **A.** Expansion of international relations by promoting trade among States with varying economic and social systems and levels of development.

    **B.** Strengthening of international relations by devising and implementing policies and methods that are better adapted to the real needs of poor countries.

        **1.** Both to encourage all forms of technical assistance and to provide technical assistance in certain sectors of trade and development (UNCTAD's only actual operational function).

        **2.** More appropriate solutions for certain countries, such as promoting special measures in favor of land-locked States, consideration of the special needs of developing island countries, and promotion of special measures to help the least-developed countries, discussed in Chapter 19.

In its work, UNCTAD cooperates in many ways with other organs of the United Nations and its specialized agencies as well as with other international governmental and nongovernmental organizations. Especially noteworthy here was its cooperation with GATT that partly, at least, replaced their original rivalry. Not only did they coordinate their work, particularly in multilateral trade relations, but they also jointly operated the UNCTAD/GATT International Trade Centre in Geneva. This center provides to developing countries a trade promotion advisory service, a training service for export promotion specialists, and a market development advisory service.

**UN technical assistance.** One of UNCTAD's many activities is technical assistance in the improvement of transportation in Africa and Asia, particularly for the land-locked countries. This is the headquarters in Blantyre, Malaŵi of one such project, which covers the 10 countries in the Southern African Development Community, including Namibia, which joined SADC and the project shortly after this photograph was taken. (Martin Glassner)

It cannot be said that UNCTAD has been a rousing success. The GSP, for example, though having some symbolic value, has not led to the rich countries' purchase of a significantly larger proportion of their imports from poor countries. In fact, as Table 25-1 shows clearly, this flow actually diminished significantly between 1976 and 1989. The commodity agreements have had mixed results. The International Tin Agreement collapsed in 1985; those for coffee and cocoa are barely alive; those for wheat, olive oil and table olives, and sugar call only for co-operation rather than for fixing quotas and prices; and those for tungsten, iron ore, copper, and bauxite provide only for studies of the market situation and related matters. At present, only the agreements on tropical timber and natural rubber resemble the previous price-and-quota-setting agreements.

UNCTAD's efforts to develop a code of conduct for transnational corporations (covered later in this chapter) were unsuccessful, as were most of its other efforts to bring about a New International Economic Order. From its inception through the 1980s, it served chiefly as a useful forum for North–South economic negotiations, but it also provided important technical assistance and other services in the areas of shipping, ports, insurance, aid to land-locked countries, trade facilitation, and others.

At the eighth session of the Conference (UNCTAD VIII), held in Cartagena de Indias, Colombia in February 1992, the first since the disintegration of the Soviet Union and the end of the Cold War, the major issue was the role of UNCTAD itself. It was virtually free of the polemics and confrontations between North and South, East and West, that had characterized earlier sessions. Taking into account the many changes that had taken place in the world since UNCTAD VII in Geneva in 1987, the Conference agreed to revitalize the organization by reducing its range of activities and reorganizing its structure. Its focus was shifted from international negotiation to development of pragmatic policies and measures within individual countries by providing more technical assistance to help member countries adapt to the

new agreements being formulated in the Uruguay Round of trade negotiations under GATT. In general, UNCTAD was streamlined, largely depoliticized, and better prepared to work in partnership with the new World Trade Organization.

The Trade Negotiations Committee of the Uruguay Round of GATT concluded its work in December 1993, after more than seven years of strenuous and at times acrimonious and highly political negotiations. Its product, a 26,000-page package of agreements, was formally adopted in Marrakesh, Morocco, on 15 April 1994. Among the most important of the agreements are:

1. The General Agreement on Tariffs and Trade 1994 (GATT 1994), which includes the 1947 GATT and related documents; the Marrakesh Protocol, which, with its five appendixes, spells out the concessions made on trade in goods; and 19 understandings and agreements.

2. General Agreement on Trade in Services.

3. Agreement on Trade-related Aspects of Intellectual Property Rights.

4. Understanding on Rules and Procedures Governing the Settlement of Disputes.

5. Trade Policy Review Mechanism.

6. Four Plurilateral Trade Agreements.

In addition, the package contains a long list of Ministerial Declarations and Decisions related to a variety of administrative, procedural, and substantive matters. The entire package is formally known as the *Agreement Establishing the World Trade Organization.*

When the Uruguay Round began in 1986, there was no thought of replacing GATT with another organization. In 1990, however, Canada and the European Communities proposed one as a mechanism for implementing the ultimate agreements more effectively. The World Trade Organization (WTO) came into existence formally on 1 January 1995. It will be years, however, before the transition from GATT to WTO is complete, before all of the agreements and undertakings are fully implemented, before the relationships of WTO with the United Nations and other intergovernmental organizations are worked out, and before negotiations resume on the unfinished business of the Uruguay Round and on new questions raised by the 1994 agreement.

These include a uniform code for international investments, immigration policies (especially for labor migration, covered in our Chapter 37), internationally recognized labor standards, remaining subsidies in the entertainment industry, better health and safety standards, wider opening of markets in the industrialized countries for important exports of the poor countries—in such areas as agriculture, textiles, and labor-intensive services—and some genuine environmental concerns.

It is easy to be cynical about the results of the Uruguay Round and about the new WTO, and to dwell on their deficiencies and shortcomings, as many undoubtedly will. Seen in historical context, however, the achievements have been monumental. To summarize some of them briefly:

1. Trade in goods is to be liberalized very considerably. Tariffs on industrial products are to be reduced, including those on such sensitive goods as textiles, clothing, transportation equipment, leather footwear, and travel products. Tariffs on agricultural products are to be reduced, and, much more important, export subsidies for agriculture are to be reduced even more.

2. More important than tariff and subsidy reduction are the agreements to reduce still further many nontariff barriers to trade favored by the rich countries and to ban entirely such relatively new restrictive trade practices as "voluntary" export restraints, orderly marketing arrangements, and similar measures—all designed to protect industries in rich countries.

3. The MFA (Multifibre Agreement) is to be phased out over 10 years, during which the products subject to MFA quotas will be gradually integrated into GATT, re-sulting in strengthening the more efficient producers of clothing and textiles and reducing costs to consumers.

4. Protection is provided for the first time for intellectual property rights, which should boost research and development activities and investments.

5. Agreement on trade-related investment measures prohibits certain arrangements that restrict trade, such as local content and trade-balancing requirements.

6. The new agreement clarifies and strengthens trade rules in a number of areas, including antidumping measures and industrial subsidies.

7. The developing countries, though gaining less than they had hoped for and even losing in some cases as preferential access to markets is gradually eroded by the new rules, on the whole are likely to enjoy a long-term net gain greater than any they had experienced under GATT 1947. During the Uruguay Round, UNCTAD provided both direct and indirect assistance to developing country negotiators, financed by the United Nations Development Programme. This assistance is likely to continue as UNCTAD works out and works within cooperation arrangements with the World Trade Organization.

8. In the long run, nearly everyone will benefit from a huge increase in international trade. The WTO is substantially different from the loose organization of GATT. If allowed to carry out its assigned tasks, the WTO, in partnership with UNCTAD, can truly nurture something revolutionary, a historic new international economic order far beyond anything envisioned in the 1970s.

## Economic Sanctions

In the last chapter we mentioned economic sanctions as one of the types of sanctions and other modes of redress short of war available to States with grievances against other States. They deserve more attention, however, and since most economic sanctions involve international trade in some way, this seems like an appropriate place to discuss them.

The most important economic sanctions against an alleged offender are the *embargo* on exports to that country and the *boycott* of imports from the country. Both embargoes and boycotts may be partial or total, may be applied by one or more States or by an intergovernmental organization against one or more States, may or may not involve third States (through secondary and tertiary boycotts), and are enforced and evaded with varying degrees of effectiveness. The same is true of other kinds of sanctions besides embargoes and boycotts. They include such measures as freezing the assets of the object State that are located in other countries, imposing a ban on granting of credit and other financial transactions, prohibiting the planes and ships of the object State to carry goods or passengers of the boycotting State or to call at its seaports and airports, and banning advertising, insurance, patents, licenses, technical or managerial assistance, and other activities that could aid the economy of the object of the sanctions.

The United States has imposed embargoes since colonial times, notably in 1807 against Britain, in 1940 against Japan, since the early 1950s against China, Cuba, and other communist-ruled countries and in the 1980s against Panama and Nicaragua. The League of Nations imposed economic sanctions against Italy, the United Nations against Rhodesia, South Africa, and Iraq, and the Organization of American States and the Organization of African Unity against countries in their regions. In fact, economic sanctions seem to have become more common in recent years and are especially favored by the United States and by the smaller States of Africa—though, of course, against very different targets. Surveying the history of such sanctions, one can only reach the conclusion that their results have been mixed.

By far the most comprehensive, sophisticated, polemical, and protracted sanctions in modern times have been those imposed by the Arab countries against Israel. Their object is not simply to redress a grievance or to punish a transgression or even to coerce some desired behavior; their object is to destroy Israel if possible while strengthening themselves. Although the secrecy surrounding both the application and evasion of sanctions precludes a definitive evaluation, it is clear that Israel has not been destroyed, nor have the Arab countries achieved noteworthy net economic or political gains as a result of the sanctions.

The same is true of the United Nations-sponsored sanctions against Rhodesia. Through the entire period from Southern Rhodesia's unilateral declaration of independence in November 1965 and the Security Council's imposition of sanctions in April 1966 to the final achievement of negotiated, internationally recognized independence under majority rule in April 1980, the sanctions never were implemented effectively. Portugal refused to cooperate until 1975, and France and other countries (including the United States) were less than enthusiastic about enforcing the sanctions. Even Britain's famed Beira Patrol, while stopping and searching 52 tankers in the Indian Ocean between 1966 and the day of Mozambique's independence in 1975, never found a single tanker bound for Beira with crude oil destined for Rhodesia. Yet Rhodesia managed to sell tobacco and chrome, buy petroleum and all manner of other goods, build important domestic industries, maintain disguised diplomatic missions abroad, and otherwise thwart most of the sanctions.

More recently, United Nations-sponsored sanctions against Iraq in the early 1990s similarly failed to achieve their stated or unstated objectives: President Hussein did not willingly withdraw from Kuwait, abide by the truce agreement that ended the Gulf War, or cooperate with UN teams inspecting and destroying his nuclear weapons facilities. Although the sanctions undoubtedly caused some suffering among the Iraqi people, they were quite openly circumvented by some neighboring countries, and there is no evidence yet that they forced Hussein to change any of his policies or tactics.

**Billboard on the Malecon of Havana.** Economic sanctions have had mixed results around the world and have rarely if ever achieved their desired effects. In some cases they only strengthen the resolve of the object of sanctions, as in the case of Cuba. This billboard on the Malecon (shoreline boulevard) of Havana in 1991 shows an armed Cuban on the island shouting at a threatening Uncle Sam in Florida: "Messers. Imperialists, We Have Absolutely No Fear of You!" This defiance has already lasted for several years after the withdrawal of Soviet support. (Martin Glassner)

Since then, the Security Council has imposed at least arms embargoes and in some cases other economic sanctions against the remnant of Yugoslavia (Serbia and Montenegro), Libya, Somalia and Liberia, all in 1992, and Angola in 1993. All but the Libya case involved civil strife, a relatively new stimulus for international intervention, in which economic sanctions might be useful.

Selected sanctions imposed in very special circumstances and effectively enforced can influence the behavior of the object State to some degree and can enhance the prestige of the party imposing the sanctions. But there is little evidence that sanctions alone can intimidate any but the smallest and weakest States (against whom sanctions are rarely applied anyway). Nevertheless, sanctions do have sufficient value that they will continue to be invoked for as far into the future as we can see. Most important, by offering an aggrieved country an alternative to war, sanctions may and apparently do prevent war—or at least hasten the conclusion of a war. This alone makes sanctions worthwhile, even if they are futile in other contexts.

## Transnational Corporations*

Transnational firms have received considerable attention—mostly unfavorable—from the press and public for their roles in overthrowing and installing governments in Latin America. The United Fruit Company in Central America in the interwar period and the International Telephone and Telegraph Company in Chile in the early 1970s are only the most notorious examples. During the first two decades after World War II, transnational corporations (TNCs), some with a century of experience behind them, proliferated and expanded rapidly. Coinciding with decolonization, this movement conjured up images of a new form of colonialism.

Indeed, the size, skills, and mobility of those new empires were seen by many crit-

*The term *multinational corporation* has been abandoned by the United Nations in favor of "transnational" for a number of reasons. Outside the United Nations the trend seems also to be in favor of *transnational*, though the two terms are sometimes used interchangeably. There is still no precise, generally accepted definition of either term.

ics as constituting a threat to the sovereignty of small countries, especially those recently emerging from colonial status without the skill, experience, and power to deal with the corporate giants on anything like equal terms. Many of the larger firms, in fact, are richer, more powerful and even larger by several measures than many States, as illustrated by Table 25-2. Demand developed for some devices to control them and to prepare new (or even older) developing countries to stand up to them.

As a result of these new pressures, and of several well-publicized excesses of TNCs, the international community began to take them seriously. Studies were undertaken, principles developed, and codes of conduct proposed. In 1974 ECOSOC, the United Nations Economic and Social Council, created a Commission on Transnational Corporations, which, with its secretariat, the Centre on Transnational Corporations, labored to negotiate a comprehensive code of conduct for TNCs. The commission was transferred to UNCTAD in 1994 and renamed the Commission on International Investment and Transnational Corporations, and work on a code of conduct was suspended. As a symbol of changed attitudes toward TNCs in the international community, the commission now concentrates on facilitating cooperation between TNCs and host developing coun-

tries, on assisting in bringing the benefits of TNC innovations and technical prowess to the poorer countries, and on integrating its work with other UNCTAD activities.

Earlier, in 1976, the OECD adopted a Declaration on International Investment and Multinational Enterprises, an essential part of which is the OECD Guidelines for Multinational Enterprises. These guidelines set voluntary standards of behavior, provide guidance to such enterprises, and help to ensure that their operations are in harmony with the policies of the countries in which they operate. They were reviewed and slightly modified in 1979, 1982, 1984, and 1991, all through the latter part of the East–West and North–South confrontations, and are still useful in a new era in which developing countries are no longer denouncing and excluding TNCs but competing for their investment capital and skills.

While still formidable, the political power of TNCs is declining and was probably never as great as legend would have it. Under pressures from intergovernmental organizations, national governments, and the general public (aroused more by the Nestlé baby formula scandal than by any government overthrow), TNCs have begun behaving more responsibly. They are also better understood now than they were in the 1960s. We know now, for example, that most

**Table 25-2**  *Sales of Transnational Corporations and GDP\* of Countries (in billions of US$)*

| Corporation/Country | Sales/GDP | Corporation/Country | Sales/GDP* |
|---|---|---|---|
| Belgium | $148.7 | Norway | $90.3 |
| Austria | 126.9 | *Exxon* | 87.3 |
| *Mitsui & Co.* | 117.0 | *Royal Dutch Shell* | 78.4 |
| *General Motors* | 110.0 | *Nissho Iwai* | 72.9 |
| *C. Itoh* | 108.5 | *IBM* | 59.7 |
| Denmark | 107.6 | *Mobil* | 54.4 |
| Finland | 104.8 | Greece | 52.8 |
| *Sumitomo* | 103.6 | *Toyota Motor Co.* | 50.4 |
| *Marubeni* | 96.1 | *Sears, Roebuck* | 50.3 |
| *Mitsubishi* | 93.3 | *Hitachi* | 44.7 |
| *Ford Motor Co.* | 92.5 | New Zealand | 42.0 |
| | | Portugal | 41.8 |

\*Gross domestic product is the total of goods and services produced domestically in a year. (GNP minus foreign trade in goods and services equals GDP.)

*Sources:* "The Global 1000," *Business Week*, July 17, 1989, and OECD, *Main Economic Indicators*, October 1989.

**An overseas operation of a transnational corporation.** This synthetic detergent plant in Ain-Sebaa, north of Casablanca, Morocco produced 75 to 80 percent of the detergents used in Morocco, an indication of the kind of influence wielded by foreign-owned corporations in many developing countries. (Martin Glassner)

transborder capital flows are among countries that are already rich, not from rich to poor countries; that most foreign facilities of TNCs headquartered in rich countries are located in other rich countries; and that, while the United States is the most important investor in other countries, it is also the most important host to foreign direct investment (FDI) and locus of foreign branches and subsidiaries of TNCs.

The global economy is now so complex, so tightly interconnected and steadily growing more so, that import and export figures become more difficult to interpret, corporate responsibility becomes more diffuse, and heroes and villains become more difficult to identify. An American automobile assembled in the United States bearing an American nameplate typically contains components made in two dozen or more countries. American firms operating in Japan not only sell their products to Japanese consumers, but also provide goods and services to Japanese industry and even export to the United States. Despite the strenuous efforts of the U.S. government to destroy the government of Nicaragua, some 40 U.S.-based

transnationals were still operating profitably in Nicaragua in late 1986. Japanese firms have established factories south of the U.S. border to take advantage of Mexico's *maquiladora* program. They import components from Japan and other countries, assemble them or finish them in Mexico, and ship the finished products to the United States and elsewhere. Examples abound.

Another striking development that has helped to defuse criticism of transnationals is the growth of TNCs based in developing countries. By 1987 there were about 1000 of them, operating in scores of countries, both rich and poor. Based in such countries as South Korea, Hong Kong, Taiwan, Singapore, Argentina, Brazil, Venezuela, Mexico, Kuwait, Saudi Arabia, and India, they engage in the same activities as their counterparts from richer countries, and from similar motivations. While still minuscule compared with the OECD transnationals, these "Third World" TNCs already dominate FDI in Asian developing countries and the textile industry, and in other regions and sectors their influence is growing rapidly. The political implications of this phenomenon, to say nothing of the economic ones, are considerable.

Even the States with centrally planned economies joined this activity that they once reviled. By 1989 State enterprises of countries belonging to COMECON, the Soviet-led Council for Mutual Economic Assistance, had established over 800 companies in developing countries and in all but one of the OECD countries. In keeping with their ideological posture, they claimed that they were not motivated by a desire for profits and should therefore be exempt from any international regulation. In fact, however, they were profit-oriented and earned large quantities of hard currencies desperately needed at home. Again, the political implications of this phenomenon were very great indeed.

More significant observations about FDI:

1. The indigenization policies of many developing countries in the 1970s, a type of economic nationalism (autarky) and one aspect of the push for an NIEO, was a failure. The techniques used, such as requiring most of the staffs of TNC operations in a country to be locals, requiring joint ventures with local entrepreneurs, requiring local raw materials and components to be used in manufacturing, restricting repatriation of profits to the TNC's home base, and so on had the effect not of controlling them but of driving them away altogether.

2. Partly as a result of autarkic policies, FDI in developing countries fell during the 1980s. Investment capital was diverted to the newly industrializing countries (NICs) and to the OECD countries. Currently, in the mid-1990s, capital is moving to those developing countries, such as Indonesia, Malaysia, and Thailand, that are hospitable to TNCs.

3. Transnational corporations, even when they operate in developing countries, have—however grudgingly or superficially—been demonstrating concern for such matters as environmental protection, nondiscrimination in employment, product safety, training and promotion of local people, voluntary joint ventures

with local capital, and respect for the land and customs of indigenous peoples. They are now better citizens in their host countries than they were a generation ago.

4. Since the 1960s developing countries as a group (with major differences among them) have reoriented their economies toward manufacturing, and exports of manufactured goods have grown considerably. TNCs have played an important role in this process, not only by exporting manufactures at a higher rate than their host countries, but also by indirect means, such as contracting for local goods and services, spurring local manufacturers to become more efficient, and so on.

5. "Foreign direct investment has been increasing far more rapidly in the 1980s than both world trade and world output, and promises to continue to do so in the future. This suggests that [FDI] is increasingly becoming an engine of growth in the world economy. Furthermore, the composition of foreign direct investment and the major actors associated with it have seen a significant change over the last decade. Behind these changes are the strategies of [TNCs] which . . . are furthering the emergence of regional clusters of countries around the three poles of the Triad."*

6. "National boundaries have become increasingly irrelevant in the definition of market and production spaces, while regions other than countries are emerging as the key economic policy arenas. The dynamic interplay between the globalization of the world economy and economic integration is likely to have profound effects on the activities of TNCs. The role of developing countries, in

*World Investment Report 1991; The Triad in Foreign Direct Investment. United Nations publication E.91.II.A, preface by Peter Hansen, Executive Director of the United Nations Centre on Transnational Corporations. The Triad refers to Japan, the European Union, and the United States.

turn, is bound to change, both as host and as home countries for FDI."*

The serious, detailed, academic study of TNCs has only begun. Like the subject of sanctions, this political/economic/geographic topic requires study and analysis by political geographers. Neither subject has yet been adequately addressed by them, and there is much scope here for energetic, creative work. It can yield very great rewards.

## Remaining Distortions in International Trade

While there has been, very broadly speaking, a general movement toward freer trade since World War II, many restrictive practices are still utilized. Some of them have economic motivations, but many are essentially political. Their net effect, whatever their objectives, is to reduce the total volume of trade artificially and to divert it out of its natural channels into artificial ones. The following list of restrictive trade practices is by no means exhaustive. Many others are extremely imaginative, technical, or obscure, but no less harmful.

By far the most common is the *protective tariff.* By discouraging competition (in some cases eliminating it altogether through excessively high tariffs), the government of the importing State also protects the local producer from the normal penalties a free market levies against an inefficient, high-cost, obsolete, careless, or dishonest producer. The consumer then suffers from both high prices and poor quality when denied access to foreign goods in a free competitive system. A protective tariff can be justified, but only for infant industries during the period of their infancy. A case can be made for protection of high-cost, inefficient industries (including primary ones) that bring important social or security benefits to a country if the added cost is considered tolerable by the so-

*Regional Economic Integration and Transnational Corporations in the 1990s: Europe 1992, North America, and Developing Countries.* United Nations publication E.90.II.A.14, preface by Peter Hansen.

ciety. A modest revenue tariff can also be important to a small, poor country for which it may be a principal source of income. But high tariffs imposed to "save local jobs" or "combat low-wage imports" usually result in retaliation. Economists call this a "beggar-thy-neighbor" policy, and have demonstrated conclusively that in a trade war resulting from such "protection" everyone is left exposed and no one benefits except the inefficient producers and the intermediaries.

Even if tariffs are kept low—through GATT or other arrangements—there are numerous nontariff barriers to trade and other practices designed to restrict competition so as to give an unfair advantage to one exporter over another. They include (in no particular order) such things as import and export *quotas,* or quantitative limits on particular items being traded; *preemptive buying,* or buying quantities of something just to keep another party from obtaining it; *licensing,* or requiring importers to obtain (or buy) licenses to import military, luxury, or other special goods; *exchange controls,* designed to keep to a minimum the outflow of scarce foreign exchange, especially "hard" currencies; unusual or unnecessarily high *standards* for protection of the importing country's health, safety, and morals; cumbersome import *procedures* and excessive *documentation;* complex and expensive *packaging requirements;* confusing or unreasonable *classification* of imported goods and customs *nomenclature;* extraordinary *export incentives; dumping; taxes* imposed on imports and exports in lieu of tariffs; *State trading,* practiced generally by socialist countries and to a limited extent by many other countries; *control on exports* of arms, nuclear materials, goods in short supply, and strategic technology and products; "*voluntary*" *restraints on exports* that are in fact coerced by importers to reduce competition; and many more. Most of these nontariff barriers to trade, as mentioned earlier, are greatly restricted or banned altogether by GATT 1994 and associated agreements, but it will be years before the WTO can expunge them.

There are also many illegal, unethical, or just unsavory practices that raise questions about trade statistics and trade agreements. For one thing, there is a great and growing flow from rich countries to poor countries of outmoded, defective, unsafe, and toxic products banned in the home country. This includes tobacco products, pharmaceuticals and agricultural chemicals that can be—and sometimes are—lethal. Then there is a counterflow from poor countries to rich ones of counterfeit, substandard, and even dangerous goods, including imitation name-brand goods of all kinds and defective parts for automobiles and aircraft. There is also a huge and immensely profitable trade in endangered species of plants and animals and their products, and of stolen art and antiquities. Restrictions on all of these kinds of trade are not only desirable in their own right, but could help to stimulate legitimate trade.

Finally, *smuggling* of all kinds is a major problem around the world. In Chapter 30 we discuss in some detail the worldwide clandestine trade in arms and drugs; what we speak of here is what is generally referred to officially as "unrecorded" or "informal" trade. This is used to bypass customs duties or outright bans on imports of certain products. Along borders everywhere in the world there are exchanges of goods and services by local residents back and forth across these borders. This is generally not a problem; often, in fact, it performs a valuable service where governments are unable to provide adequately for their citizens in remote border areas. It also stimulates the local economy, cements ties between relatives and friends, and creates international good will. What is of concern, however, is the large-scale transfer of contraband for commercial purposes. This can seriously damage government economic policies—or, if the government is a party to the smuggling, can generate huge profits for corrupt officials. It is estimated, for example, that one-quarter to one-third of Paraguay's actual foreign trade is unrecorded, consisting of goods smuggled in both directions across the borders of Brazil and Argentina. Regardless of the reasons for smuggling, it does represent a serious distortion of international trade.

**Smuggler's goods.** This jewelry was being smuggled from Hong Kong into Canada concealed in a tight corset called a bodypacker. It was confiscated by alert customs officers in Vancouver. (Photo courtesy of Canadian Department of National Revenue, Customs and Excise)

GATT, UNCTAD, the UN Commission on International Trade Law, and other agencies have been trying to combat these practices, so far with only limited success. Other practices that affect trade are not necessarily harmful; in fact, if conceived, designed, and implemented wisely, they can actually help to free and expand trade. They can be grouped under the headings of preferences, discussed below, and economic integration, to which we devote the next chapter.*

## *Preferential Trading*

Among the economic weapons in the armories of many States is that of giving preferred treatment to particular trading partners. Preferences became important during the nineteenth century but have achieved even more importance in the past half century. Some preference systems have deliber-

---

*Another important and growing trade practice that has both political causes and implications is countertrade. This is a practice whereby a supplier commits contractually to reciprocate and undertake certain specified commercial initiatives that compensate and benefit the buyer; that is, goods are paid for with other goods or services or some other thing of value rather than cash. These arrangements include barter, buyback, counterpurchase, and offset, and have become common in trade involving developing, planned, and transitional economies.

ately been *ad hoc,* others have been abandoned as conditions have changed, and still others survive and thrive. Among the minor or obsolete ones of recent times have been the special trading arrangements between the United States and the Philippines, part of the "independence package" that led other ex-colonies to question openly whether the Philippines was really sovereign; the sugar quotas granted by the United States to selected exporting countries and reallotted periodically; the various currency blocs that were so prominent in the first decade and a half after World War II (the sterling, ruble, franc, and dollar blocs); and the greatly skewed trading partnerships between South Africa and the land-locked countries of southern Africa and between India and the Himalayan kingdoms. By far the most important, however, have been the Commonwealth Preference System and COMECON.

The *Commonwealth Preference System* (originally Imperial Preference) was one of the most important features of the Commonwealth from the Imperial Conference in Ottawa in 1932 until the entry of Britain into the European Communities in 1973. Under this system, Britain granted free entry to the goods of Commonwealth countries and imposed tariffs on competing goods of nonmembers, thus abandoning its traditional policy of free trade. On their part, the other Commonwealth countries agreed to assess lower tariffs on many goods they imported from one another than on similar goods imported from nonmembers. There were exceptions to this system, breaches of it, and changes in it over the years, but it survived even the creation of GATT, which made, as we have seen, provision for such preferential arrangements. The dependence of many Commonwealth countries on the protected British market, especially for their agricultural exports, was one of the major factors delaying Britain's entry into the European Economic Community. Not until they had begun developing other markets and Britain was able to negotiate a gradual phasing out of its Commonwealth preference obligations did the merger take place. There is still, however, a strong tendency for Commonwealth countries to trade with one another, and they account for one-quarter to one-third of total world trade.

Except for China, all the countries with centrally planned economies ("socialist" countries) were members or observers of *COMECON,* the Council for Mutual Economic Assistance. These countries encountered difficulties in trading with countries with market economies because their pricing systems had to be arbitrary; their currencies were not convertible into "hard" currencies (such as U.S. dollars, sterling, yen, Swiss francs, or German marks); and they had both political and economic reasons for turning inward and trading very little with outsiders. Until recently they had very little to sell to hard currency countries and so could not earn the foreign exchange necessary to buy goods outside their bloc. Therefore, whatever trading they did was largely among themselves by barter. There was also an ideological imperative for this arrangement and a fear (probably legitimate) of being penetrated by capitalism and imperialism if they traded with capitalist countries.

Nevertheless, COMECON never developed a formal, compulsory trading system or any mechanism for enforcing one, and trade among the European members (including the USSR) dropped steadily, from well over 80 percent in the early 1950s to about 66 percent in 1961 to 58 in 1973 and just over half in 1984. The comparable figures for the Asian socialist countries (including China) were always much lower. Intrabloc trade, however, was still greater than the trade among members of most organizations that were actually attempting economic integration, until COMECON was dissolved in 1991. We discuss the organization at greater length in Chapter 27.

As mentioned earlier, preferential trading schemes that still survive, including most-favored-nation (MFN) arrangements, which are often used to achieve political objectives, are to be phased out under the 1994 agreement that created the World Trade Organization. Regional integration schemes, however, will still be allowed, and to them we devote the next chapter.

# *Economic Integration*

The basic arguments in favor of economic integration are quite simple. Two or more countries can, by combining their economies, reduce their costs of production by producing for a larger market, thus achieving economies of scale; provide a greater variety of goods and more opportunities for both managers and workers in a larger economy; speak with a louder voice in international economic (and perhaps political) affairs; and achieve a variety of ancillary benefits. These benefits have been realized in varying degrees around the world. Any disadvantages to economic integration seem to be felt primarily by individuals or small sectors of society rather than by the society as a whole. Achieving economic integration, however, whatever its benefits, is exceedingly difficult. There are many obstacles, but the most important is the reluctance of States to surrender any of their sovereignty, any control over economic decision making, any political options. Consequently, economic integration did not make very impressive progress until quite recently.

There are four levels, or degrees, of economic integration. *A free trade area* is the simplest. Here, two or more countries agree to eliminate tariffs and other barriers to trade between or among them so that goods flow freely across their mutual boundaries. Each member, however, retains its own tariffs and other trade restrictions in respect of nonmembers (third parties). The free trade area may be limited to certain classes of goods, such as agricultural products, or even to individual products, such as the old U.S.-Canada free trade area in automobiles and automobile parts. A *customs union* is a free trade area plus a common external tariff. That is, the members agree not only to trade freely among themselves but also to form one unit for trading with nonmembers. The customs duties received at the ports of entry of all members are commonly pooled and either used for common purposes or apportioned among the members or both.

A *common market* is a customs union plus the free movement within the group of capital and labor as well as goods. This means that an investor may invest capital in any of the member countries without discrimination, a firm may seek loans or investments or workers in any of the member countries, and a worker may seek work and be employed in any of the member countries as long as he or she is a citizen of one of them. Complete economic integration is achieved through an *economic union* in which the members have not only a common market, but also common economic and monetary policies, a common currency, common banking and insurance systems, uniform taxes and corporation laws, and so on. Economic union without political union is probably impossible, except for very small countries, such as Luxembourg or Liechtenstein, since common political decisions must be made before these steps can be taken and this is most difficult with multiple constitu-

**The Liechtenstein-Austria boundary—visual evidence of an economic union.** Since Liechtenstein and Switzerland share a single economy, they maintain joint customs posts on the border with Austria, as here in Schaanwald, Liechtenstein. On the other hand, on the Swiss-Liechtenstein border there are no customs or immigration posts at all. (Martin Glassner)

tions, parliaments, cabinets, central banks, and pressure groups.

Typically, countries agree to form free trade areas, customs unions, or common markets gradually, in stages, usually according to a fixed schedule. Thus, there are degrees of these forms of integration. A common market may even be working toward an economic union and have economic institutions beyond the free circulation of goods, labor, and capital. Groups may also fail to reach their goals and yet retain their original names. Thus the name of a group may be misleading; it does not necessarily indicate its degree of integration or its success in reaching its goals. Countries that wish to integrate can get considerable help from the UN regional commissions.

## United Nations Regional Commissions

As we point out in Chapter 19, one of the most vital functions of the United Nations is assisting member countries, their dependencies, and even nonmembers with their economic growth and development. Many of the UN organs and specialized agencies are devoted to this effort, as is much of the UN budget. Of all the UN organs working in the economic area, the regional commissions, part of the Economic and Social Council, are among the most important and least known. All of them are concerned to some extent with social as well as economic matters, and two of them even have "social" in their names. They all have broadly similar organizational systems and activities, but they dif-

**Africa Hall, Addis Ababa, Ethiopia.** This is the headquarters of the United Nations Economic Commission for Africa. Nearby, out of sight at upper left, is the kind of squalid slum that the UN economic commissions are working to eliminate. (United Nations photograph)

fer in many ways according to the needs and resources of the regions they serve. All of them, in varying degrees, are concerned with economic integration (the one for Latin America being most involved in this activity and the one for Western Asia least of all). This is why they are discussed in some detail in this chapter.

The activities of these UN regional commissions are so diverse and widespread and so intimately integrated into the daily lives of people nearly everywhere in the world that they are at least indirectly associated with nearly every subject covered in this book. Much of the basic theoretical work on economic integration, for example, was done by economists of the commission for Latin America and the Caribbean (still commonly known by its original Spanish initials, CEPAL). The commissions do research and publish important studies, statistics, and analyses of regional and national economic problems; provide guidance and technical assistance in the formulation and execution of development and integration programs; sponsor seminars, workshops, and training programs; sponsor meetings, conferences, and groups of experts to grapple with economic problems; and otherwise facilitate the economic growth and health of their members. Typically, many of these activities are carried out through subsidiary bodies and in subregional offices within the region. The commissions also cooperate with other intergovernmental organizations in furthering their work. While some have been more effective than others, all deserve more attention and credit than they have received.

In order that we may understand better the processes, problems, and possibilities of economic integration, we now examine a few examples from several regions.

---

## United Nations Regional Commissions

### UN Economic Commission for Europe (ECE)

Founded: 1947
Headquarters: Geneva

34 members, including Belarus, Ukraine, Albania, Cyprus, Switzerland, Malta, United Kingdom, and Canada.

### UN Economic and Social Commission for Asia and the Pacific (ESCAP)

Founded: 1947
Headquarters: Bangkok

38 members, including Nauru, Tuvalu, South Korea, France, Netherlands, United Kingdom, and United States; 9 associate members, all dependent territories of members.

### UN Economic Commission for Latin America and the Caribbean (ECLAC; CEPAL)

Founded: 1948
Headquarters: Santiago

40 members, including Canada, France, Netherlands, Portugal, Spain, United Kingdom, United States, and the Bahamas, 4 associate members, all dependent territories of members.

### UN Economic Commission for Africa (ECA)

Founded: 1958
Headquarters: Addis Ababa

53 members, all independent African States.

### UN Economic and Social Commission for Western Asia (ESCWA)

Founded: 1974
Headquarters: Baghdad

14 original members, all Arab countries of Southwest Asia, plus Egypt and the Palestine Liberation Organization, both admitted in 1977.

## *Europe*

The best known economic group in the world has been the EEC, the European Economic Community (popularly but quite inaccurately known as the European Common Market or simply as "the common market," as if there were no others), now only a part of the *European Communities.* The European Communities (informally called the Community and abbreviated EC) is far more than a common market. The 1957 Treaty of Rome that created it clearly envisages and lays the groundwork for *political* as well as economic union of Western Europe, and the members have made very impressive progress in that direction. The institutions of the EC resemble a government and perform many of the functions of a government. Most of its activities are concentrated in its Brussels headquarters—giving Brussels a big advantage in the competition for the honor of being the future "capital of Europe"—but important activities are also located in the other competitors, with the Court of Justice in Luxembourg and the Parliament in Strasbourg. More than 150 countries have accredited diplomatic representatives to the EC, and the EC maintains some 90 offices abroad. It functions as a unit in international trade negotiations and has signed the United Nations Convention on the Law of the Sea. It formulates common policies on many international issues, though it still has difficulties agreeing on some domestic policies. It is by far the most powerful trading unit in the world and is approximately equal to the United States in total economic strength.

All this power and unity did not develop overnight, however. The nucleus of the EC is the Economic Union of Belgium and Luxembourg (UEBL), which dates from 1922. In 1944 Belgium, Luxembourg, and the Netherlands signed a Customs Union Treaty in London that became fully effective on 1 January 1948. Then in 1960 the Treaty of Benelux Economic Union came into force. Now *Benelux* functions as a unit within the EC, having its own institutions (including a Court of Justice, founded in 1974), and may be moving toward becoming a single State.

After World War II the Marshall Plan provided a catalyst for putting into practice the new ideal of European unity, with Benelux not only as a nucleus, but also as a model. The Organization for European Economic Cooperation (OEEC) was formed in 1948 by 18 European countries to administer Marshall Plan funds from the United States in an orderly and rational manner. This gave them the experience and the confidence to attempt economic integration. The first fruit of this new unity drive was the European Coal and Steel Community (ECSC), a product of "the Schuman Plan" (named for French Foreign Minister Robert Schuman), founded in 1951 by France, West Germany, the Netherlands, Italy, Luxembourg, and Belgium. This essentially created a common market for these countries' coal and iron ore resources and production and their iron and steel industries. Its High Authority was located in Luxembourg.

The European unity movement gathered strength and in 1957 both the European Atomic Energy Community (Euratom) and the European Economic Community were created, with the same membership as the ECSC. The three communities worked successfully side by side. The EEC by 1 July 1968 (18 months ahead of schedule), had eliminated tariffs on intracommunity trade and had established a common external tariff. On 1 July 1967, the three communities merged their institutions and now have one commission, council, parliament, and court. The United Kingdom, Ireland, and Denmark formally joined the Communities on 1 January 1973. The Norwegian government had also wanted to join at that time, but its people rejected membership in a 1972 referendum, largely because they wanted to protect their fishing industry from the competition of other Community fishermen.* The Europe of Six became the Europe of Nine. In May

*Objection to the Common Fisheries Policy was also the reason that the Faeoe Islands refused to join when Denmark did in 1973 and why Greenland, which had joined, decided in 1982 to withdraw from the EC effective from February 1985. This is another clear indication of the high degree of autonomy exercised by these"dependencies" of Denmark.

1979, Greece—long an associate member—joined the EC as a full member; Spain and Portugal joined in 1986.

The pace of the integration movement has now accelerated considerably. In 1985 the Commission issued a white paper spelling out its program for completing the internal market by 1 January 1993. This was followed in 1986 by the Single European Act, which supplemented the existing treaties, committed the members to implement the white paper, and created a single economic and social area in Europe. In May 1990, on the initiative of France, the EC agreed to create the European Bank for Reconstruction and Development to help transform the centrally planned economies of Europe and the USSR into market-oriented economies. It became operational in April 1991.

The next stage in the movement for complete European unity began on 7 February 1992 when, after long and difficult negotiations, the Treaty on European Union was signed in Maastricht, The Netherlands. This historic treaty builds on the existing ones, amending and adding to those creating the ECSC, Euratom, and the EEC. The last of these has officially been renamed the European Community and assumes responsibility for many additional functions in such areas as the environment, research and technology, health, visa policy, transportation, communications, energy, education, and training.* Economic and monetary union, with a common European currency, is to be achieved by 1999 at the latest.† All of this is referred to now as "Pillar One." It is to be supplemented by Pillar Two, a common foreign and security policy, with the Western European Union, an existing military alliance, to back it up; and Pillar Three, juris-

diction over justice and home affairs, notably in the areas of immigration, fraud, drugs, and terrorism. Finally, there are special rules on social policy, from which several members have exempted themselves. Also in the Maastricht package are provisions for strengthening the European Parliament and for a common European Union citizenship.

Clearly, the creation of the *European Union* by the Maastricht Treaty marks a major step forward. It is a somewhat hesitant step, however. The Norwegian people in 1994 again rejected entry into the EC/EU in a referendum. Britain has not only exempted itself from the EU's social policy but is also opposed to a common currency and other binding mechanisms. The expansion of the group to 15 on 1 January 1995 by the admission of Sweden, Finland, and Austria, while certainly welcome, raises questions about the admission of still more members. Turkey has been knocking on the door for many years; Malta and Cyprus are also eager to join; several of the Central and Eastern European States have expressed a desire to join. Inevitably, in any integration process, there is a potential conflict between expansion of membership and deepening of integration; each can make the other more difficult and even weaken the whole structure. Furthermore, even if all goes according to plan, the EU will still not be a complete economic union by 1999. A common taxation system, for example, has not yet been agreed upon. Political union is even farther away.

But look at what has been accomplished! Germany and France, those age-old enemies, are not only working harmoniously together, but are even planning a joint army unit. The regional policy has contributed greatly toward reducing the economic disparities among the various regions, bringing special aid to the backward areas of southern Italy, western and southwestern France, northern Netherlands, eastern Germany, half of Ireland, and parts of England, Scotland, and Wales. Common policies in many areas have been put into effect, and, perhaps most important of all, the EC has avoided the pitfall of becoming an exclusivist, inward-

---

*The proper, legal usages of the terms EC and EU are still a bit muddled and still need to be clarified. Much depends on the legal competence of each body to perform various functions. In this book, for simplicity, we use EU except when EC is clearly the appropriate term.

†The European Currency Unit (ecu) is already in use as a unit of account and some goods and services are priced in ecus. The eventual common currency will probably be called the ecu.

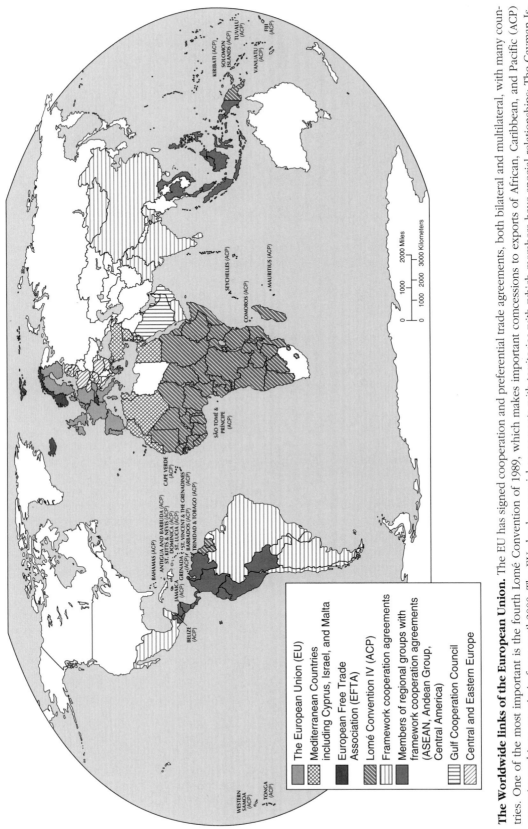

**The Worldwide links of the European Union.** The EU has signed cooperation and preferential trade agreements, both bilateral and multilateral, with many countries. One of the most important is the fourth Lomé Convention of 1989, which makes important concessions to exports of African, Caribbean, and Pacific (ACP) countries and is to remain in force until 2000. The EU also has special agreements with territories with which members have special relationships: The Cayman Islands, Greenland, British Virgin Islands, Netherlands Antilles; French Polynesia, Anguilla, Mayotte, the British Indian Ocean Territory, and New Caledonia.

416

oriented "rich man's club" by signing many special agreements with other countries. The EU has even ventured into common intervention in the tragic imbroglio in Bosnia and Herzegovina, with armed forces from several members participating in UNPROFOR, the United Nations Protection Force, and even administering the hotly contested and sharply divided city of Mostar in the name of the European Union. This could be a precedent for further constructive action by the EU outside its territory. It has come very far indeed since 1951.

*The European Free Trade Association* (EFTA) was created through the 1960 Stockholm Convention by seven Western European countries that believed in the creation of a single European market but that for various reasons were unable to join the EEC. The strongest member was the United Kingdom, which, with its Commonwealth commitments, the use of its currency (the pound sterling) as a worldwide reserve currency, and its reluctance to accept the political provisions of the Treaty of Rome, still needed the advantages of a wider trading group to help it to recover the losses sustained by participation in World War II and the loss of most of its empire. The other original members were Austria, Norway, Portugal, Denmark, Sweden, and Switzerland, with Finland as an associate member. EFTA's goals were more modest than those of the EEC, and it reached them more rapidly. It achieved free trade in industrial goods by the end of 1966, three years ahead of schedule. It was not designed as a competitor to the EEC; rather it stems from the same dreams and aspirations, and from the same experience within the OEEC. From the beginning it aimed at strengthening its members until they reached a point at which they could join the EEC individually or collectively.

This policy has been quite effective; Britain, Sweden, Finland, Austria, Denmark, and Portugal have joined the EC/EU, and EFTA and the EC have long had a close treaty relationship.

In the early 1960s, it was common to refer to Europe as being "at sixes and sevens," a reference both to the EEC and EFTA and to the uncertainties then prevailing about the future of Europe in the face of a perceived Soviet threat, the dominance of the United States in the world economy, economic distress in many parts of Europe, declining coal production and increases in energy needs, changes being wrought by decolonization, and political turmoil in more than one country. Since then, however, the EC and EFTA have survived and surmounted crisis after crisis. Today, notwithstanding their weaknesses and problems, these organizations are strong and confident and so obviously beneficial—not only to their members, but also to the world at large—that there is no really strong and organized opposition to either. In October 1991, in fact, the EC and EFTA agreed to create a European Economic Area (EEA), effective 1 January 1993, the day the EC single market became effective. This is more than a free trade area of 19 countries, yet less than a common market. Still, it represents a remarkable step toward European unity. Unfortunately, the successes of integration in Europe have not been replicated elsewhere in the world, despite many attempts to do so.

## Latin America

Outside Europe, Latin America has made the most progress toward economic integration. Common features in the histories, cultures, and outlooks of most of its countries have led them to many attempts at cooperation and even federation during their century and a half of independence. But isolation, competition, internal politics, and nationalism have conspired to frustrate most of those attempts. Since World War II the emphasis has been on economic development and on the cooperation and integration necessary to achieve it. The UN Economic Commission for Latin America and the Caribbean has been by far the most dynamic of the economic commissions and has been led by some of the world's most talented and articulate economists.

Another regionwide organization devoted, among other things, to promoting economic integration is the *Latin American Economic System* (SELA). One of its immediate forerunners was the Special Committee on Latin American Coordination, established in 1963 by the Organization of American States. In 1974 Mexico and Venezuela promoted a more highly structured economic system. An agreement to create such a system was signed in Panama City in October 1975 by 25 Latin American and Caribbean countries. Its headquarters is in Caracas. It is responsible for promoting cooperation to achieve the integral, self-sustaining, and independent development of the area. The organization aims to create Latin American multinational enterprises; encourage the conversion of raw materials in the region, industrial complementation, intraregional trade, and the export of manufactures; promote the development of transport, communications, and tourism; support the integration processes in the region; and coordinate positions and strategies on economic and social matters, among other activities. At its annual meeting in 1991, the Council of SELA placed the greatest emphasis on technical cooperation among the members. SELA has not yet achieved any notable integration among its members, but it has encouraged the subregional efforts that have been made.

Of the organizations aimed specifically at economic integration and thus bearing directly on international trade, the most ambitious so far has been the *Central American Common Market* (CACM), with headquarters in Guatemala City. Between 1951 and 1957, the five Central American States—Guatemala, Honduras, El Salvador, Nicaragua, and Costa Rica—signed six bilateral free trade agreements among themselves. By stages during the next few years these countries agreed on closer and closer trade relationships until four of them signed the General Treaty on Central American Integration in December 1960, to which Costa Rica acceded in 1962. By 1969, about 95 percent of trade among the members had been liberalized, and the common external tariff had been applied to 97.5 percent of tariff items. They also established 14 autonomous economic cooperation bodies, including the Central American

Bank for Economic Integration.

The CACM, however, has not yet adopted a common agricultural policy, its scheme for establishing subregional industries rather than competing industries in each country has not made much progress, and several of its other projects are foundering. The reasons for its failure to make significant progress after an inspiring beginning are many, among them the war of July 1969 between El Salvador and Honduras and the civil war in Nicaragua in the 1980s. Other reasons include the gravitation of new industries toward the more developed parts of the area, the lack of coordination of development policies, and balance-of-payments problems. All can be attributed to a lack of commitment by the members to subdue their own petty nationalisms and cooperate with one another for their mutual benefit. Perhaps they are simply too poor to bear the financial as well as the political and psychological costs of such integration. European integration, after all, began in earnest only after centuries of rivalry, hostility, and war, and is stimulated and sustained both by real internationalist sentiments and by a good deal of cash, both of which are absent in Central America—and in most of the rest of the world.

During the latter half of 1991 the five countries began serious efforts to revive the CACM with a view toward negotiating *en bloc* with the United States for a free trade agreement. In 1993 a common external tariff was adopted, and in 1994 all members except Costa Rica agreed to work toward full economic integration—and by the turn of the century, full political integration. Despite this apparent revival of a moribund group, skepticism remains warranted.

The *Latin American Free Trade Association* (LAFTA) was the largest of the subregional organizations aimed at economic integration in Latin America, and the least successful. Just as in Central America, the South American countries during the early postwar years built up a network of bilateral agreements that enabled them to expand trade with one another, but they proved inadequate to achieve any lasting benefits, and trade among them actually fell appreciably between 1955 and 1961.

With the aid of CEPAL, Mexico and six South American States formulated and signed in 1960 the Treaty of Montevideo that established LAFTA. The main objective of the treaty was the establishment of a free trade area among the members by 1973, not a very difficult one to achieve, it would seem. Yet it never was reached. Tariff-cutting negotiations among the members were on an item-by-item basis and painfully slow. Most of the tariff cutting was in manufactured consumer goods and capital goods, products that had never been traded extensively among them. Protection of local industries took precedence over cooperation or the expansion of intraregional trade, partly because of the very real and very considerable differences in the members' levels of economic development. By 1980, intra-LAFTA trade still amounted to only 14 percent of members' total trade. The organization was finally recognized as a failure, and it was allowed to expire in 1980.

During the 1960s the six Pacific-oriented members of LAFTA grew impatient with the glacial pace of integration and formed (within LAFTA and under its rules) the *Andean Group*. After a promising beginning, it too faltered. Chile withdrew after a military coup overthrew the democratic government in 1973. After that, the Andean Group continued with its constructive work. In a renewed burst of enthusiasm, the members agreed to form an Andean Free Trade Area by 1992 and then an Andean Common Market by 1994. Trade among them, however, remains rather modest.

For similar reasons and under similar circumstances, the States of the Cono Sur (the Southern Cone of South America), again without Chile, signed in March 1991 the Treaty of Asunción—with the customary hoopla and fanfare. The new grouping is called the *Southern Common Market*, or *Mercosur*. Its ambitious goals included not only a full common market by 1 January 1995, but also coordination of macroeconomic policies. By that date, however, only a partial customs union had been achieved

and a full customs union postponed until 2002. Given the history of integration efforts in Latin America, one is entitled to regard this latest one with just a bit of skepticism. There is one important difference between this one and its predecessors, however: On 19 June 1991 the four members signed an agreement with the United States (the "Rose Garden Agreement") that establishes a framework for negotiating a free trade area among them. This was the first step in implementing President Bush's 1990 Enterprise for the Americas initiative, which would create a nearly hemispheric free trade zone. More on this shortly.

LAFTA was immediately succeeded by the *Latin American Integration Association (ALADI)*, created by the former members of LAFTA (Mexico and all of the States of South America except Guyana and Suriname) through a new 1980 Treaty of Montevideo, which is much less ambitious than the 1960 treaty. It lays down no strict rules, no timetables, and no restrictions on members' trade arrangements with third countries. It permits the conclusion of trade agreements with nonmember developing countries.

It got off to a slow start. Its first five years were dedicated almost completely to renegotiating the remnants of LAFTA. This process has so far yielded numerous bilateral trade agreements, but only a single multilateral one, the Regional Tariff Preference. Even this is not very daring, since it anticipates only very modest tariff reductions for trade among member countries. Far from an increase in intraregional trade, the period 1982–86 actually saw a decrease of 40 percent in an already low figure. In the early 1990s there was a renewed burst of integration efforts, bilateral and multilateral. Tariffs and nontarifff barriers fell nearly everywhere in Latin America in response to stimuli from GATT, the EU, and the United States, and as a reaction to the restrictive trade practices that contributed so much to the economic decline of the region in the 1980s, Latin America's "lost decade." Despite all this activity in removing barriers to trade, reducing government controls, relaxing restrictions on international movement of persons and cap-

***Table 26-1***    *Trade Flows Between Different Countries and Regions of the Americas*

| | | | | Exports to (in percent) | | | | |
|---|---|---|---|---|---|---|---|---|
| Country/Bloc | Total Exports (in billions of US$) | U.S. | Mexico | Central America and Caribbean | Andean Pact | Mercosur | Chile | Rest of World |
| U.S. | $447.40 | — | 9.07% | 1.87% | 2.45% | 2.15% | 0.55% | 83.91% |
| Mexico | 42.70 | 76.40 | — | 1.31 | 1.03 | 1.18 | 0.47 | 19.61 |
| Central America and Caribbean | 10.79 | 47.83 | 1.55 | 10.11 | 0.46 | 0.58 | 0.10 | 39.37 |
| Andean Pact | 30.36 | 42.33 | 1.27 | 3.68 | 6.20 | 3.58 | 1.53 | 41.40 |
| Mercosur | 50.79 | 16.92 | 2.78 | 0.90 | 4.33 | 13.80 | 3.28 | 57.98 |
| Chile | 9.96 | 16.57 | 0.93 | 0.54 | 5.40 | 9.95 | — | 66.60 |

*Source:* IMF Direction of Trade Statistics.

ital, and in general in moving toward freer economies, trade among countries in the region remains minuscule (as shown in Table 26-1) and the region's economy is not recovering as rapidly as expected. Nevertheless, optimism is currently the dominant mood, as typified by the conclusion reached by a 1994 CEPAL study that a hemisphere free trade area is feasible—with no date specified. It would be wise to remember that, as pointed out earlier, Latin America is not Europe and what works in Europe may not necessarily work in Latin America—or elsewhere in the world.

## North America

As we saw in the last chapter, the single largest trading relationship in the world is that between the United States and Canada. About a quarter of all U.S. exports go to Canada, and nearly 80 percent of Canadian exports go to the United States. Each is the other's best customer and most important supplier. In addition, the largest bilateral flow of foreign direct investment in the world is between these two countries. The two countries are similar in size (though widely disparate in population), history, culture, political system, and level of living. They share the proverbial "world's longest undefended boundary." Large numbers of Canadians and Americans live in one another's country more or less permanently.

They share the defense of North America, international parks, and a general outlook on life. Yet it was not until 1965 that they took the first timid step toward economic integration. In that year they initiated free trade across the border in automobiles and automobile parts.

Two decades later they took the next, bolder, step toward integration when they began negotiating a complete free trade area. By the end of 1987 the negotiations were successfully concluded, and on 2 January 1988, the American president and the Canadian prime minister signed the Canada-United States Free Trade Agreement. The pact was greeted with little opposition and some enthusiasm in the U.S. Senate, general indifference and almost total ignorance among the American people, and raging controversy in Canada. It became, in fact, the dominant issue in the Canadian elections in November 1988. Prime Minister Brian Mulroney and his Progressive Conservative party won the election (though with a reduced majority in Parliament), and the agreement was duly ratified in both countries.

The agreement provides for elimination of all tariffs on bilateral trade over a ten-year period, including those on agricultural products. Import licenses and quotas will be liberalized or eliminated. The Canadian embargo on used car imports will end. The United States will be given nondiscriminatory access to Canadian energy supplies, and Canada will have access to Alaska's North

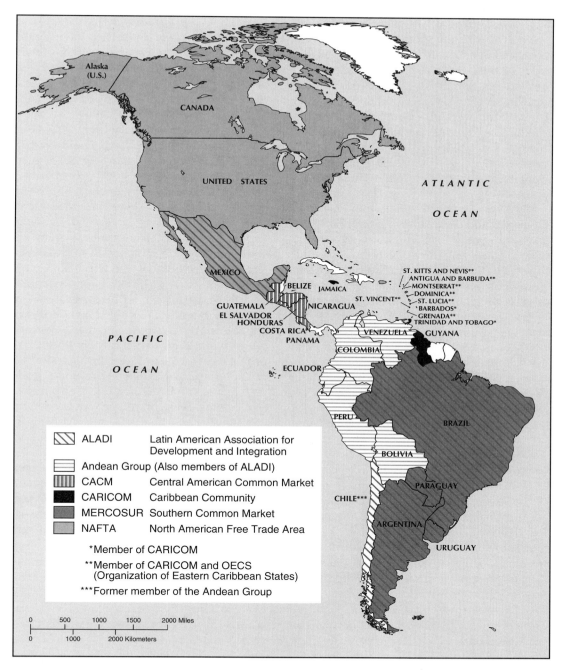

| | ALADI | Latin American Association for Development and Integration |
|---|---|---|
| | | Andean Group (Also members of ALADI) |
| | CACM | Central American Common Market |
| | CARICOM | Caribbean Community |
| | MERCOSUR | Southern Common Market |
| | NAFTA | North American Free Trade Area |

*Member of CARICOM

**Member of CARICOM and OECS
(Organization of Eastern Caribbean States)

***Former member of the Andean Group

**Organizations for economic integration in the Americas.** In addition, there are varying degrees of economic integration among the following countries: Mexico, Venezuela, and Colombia; Chile and Mexico, Bolivia and Venezuela; Venezuela and six Central American States.

Slope oil. Citizens of each country providing services in the other will receive national treatment, and transborder movement for work in this area will be facilitated. Investments will generally be given national treatment in both countries. Certain cultural areas will be exempted from free trade, but others will be liberalized. Both countries will retain current laws concerning dumping and subsidies, but a new dispute settlement mechanism will deal with claims of unfair trade practices.

Why was Canadian opposition so strong? Generally because of the centuries-long love-hate feelings of Canadians toward the United States and their fear of being swallowed up by their giant neighbor—economically, culturally, and eventually politically. Most manufacturers and other large businesses favored the agreement, but some smaller businesses and the labor unions opposed it, fearing loss of business and jobs. There was some fear that the generous Canadian social welfare system would be reduced to U.S. levels. Generally, industrialized Ontario opposed the pact because it would eliminate the tariffs protecting Ontario industries, but the other provinces generally favored it because it would enable them to import less costly goods from the United States. In the end Canada overcame the fears and followed the recommendation of a 1985 royal commission that it take "a leap of faith."

Already the free trade area has had the effects of shutting down inefficient factories in Canada, opening up new opportunities for Canadian resource-based and specialized industries, and attracting more Canadian companies into the United States, particularly near the border. Now, however, the free trade area has been expanded by the admission of Mexico.

Mexico has changed very considerably since the mid-1980s. The country's economy has experienced radical reforms, foreign trade has been liberalized since the country joined GATT in 1986, and the manufacturing sector has greatly expanded. Under the *maquiladora* program (discussed in Chapters 25 and 39), Mexican manufacturing and

assembly in the border zone have expanded phenomenally, and industries have developed further south as well, including high-technology ones in the Guadalajara area. Mexican-U.S. trade has also grown spectacularly. Free trade talks began between the two countries in November 1990.

There was opposition on both sides. Mexican manufacturers feared being overwhelmed by larger and more efficient American counterparts; there is still very deep and emotional opposition to opening the Mexican oil industry to American investment; freer agricultural trade will endanger the *ejido*, a communal land tenure system enshrined in the constitution; some workers' rights will have to be modified, since currently they tend to hinder production and complicate calculation of business costs.

In the United States there was concern about low Mexican wages putting Americans out of work; the very serious environmental problems in Mexico, especially air and water pollution and hazardous waste; the already inadequate infrastructure on both sides of the border; lack of reciprocal access for cross-border trucking; and many constraints on agricultural trade.

Nevertheless, the North American Free Trade Agreement (NAFTA) was negotiated by representatives of the United States and Mexico beginning in November 1990, with Canadian delegates joining in June 1991. The massive pact was completed in late 1992. It was generally welcomed in both Canada and Mexico, but, in contrast with the situation in 1987, when only Canada was involved, it was massively controversial in the United States. Eventually, however, it was ratified by all three countries and took effect on 1 January 1994.

Briefly, the NAFTA provides for elimination of all tariffs among the partners within 15 years, but producers of some sensitive agricultural products in Mexico and the United States will be able to use the full 15-year adjustment period. The domestic content requirement for automobiles will be raised from 50 to 62.5 percent to discourage nonmembers from shipping parts to Mexico for assembly into cars destined for the

United States and Canada. For the first time in such arrangements, intellectual property is included; Mexico has been opened to U.S. and Canadian banks, brokerage firms, and insurance companies; and U.S. and Canadian firms will be allowed to compete in the Mexican telecommunications sector. Trinational commissions will oversee administration of environmental and labor laws. Despite Mexico's economic and social problems, which could lead to a destablization of its political system, in the long run NAFTA is likely to bring great benefits to most residents of all three countries.

Meanwhile, in June 1991 the Rose Garden Agreement was signed by the United States and the Mercosur countries, laying the groundwork for a trade agreement between them. Soon, however, this initiative was submerged in a much grander vision—a Free

**Economic integration on stamps.** Regional integration is honored on stamps of many countries. These represent the Andean Group, Central American Common Market, Caribbean Free Trade Association (defunct), Belgium-Netherlands-Luxembourg Customs Union, European Free Trade Association, European Communities, Airbus Industrie, Caribbean Common Market, University of East Africa and the Economic Community of West African States. (Martin Glassner)

Trade Area of the Americas that would extend from Point Barrow to Cape Horn and incorporate all existing integration agreements. This vision was given some substance in December 1994 at the "Summit of the Americas" in Miami, at which representatives of all 34 independent States of the Western Hemisphere (except Cuba, of course) discussed and signed a 34-page Declaration of Principles and Plan of Action. Going far beyond free trade, this remarkable statement of objectives, if ever translated into real actions and institutions, could transform the hemisphere and its role in the world in important ways. Considering the fate of countless grand visions in the history of the Americas, not the least of which was the similar one of Simón Bolívar, we would be wise to restrain our enthusiasm and await developments. But this one could possibly be realized some day. After all, most human progress begins with grand visions.

## *Africa*

Progress in developing trading groups and integrating economies in Africa has been even more difficult and less impressive than in Latin America. All the factors we mentioned earlier in our discussions of imperialism, decolonization, nation building, and the colonial economy have conspired against economic integration in Africa, despite the romantic ideology of pan-Africanism that was so popular in the late 1950s and early 1960s.

There has been no lack of effort toward integration in Africa, however. For a time, just a decade, in fact, the *East African Community* was considered a model to be followed by groups of countries elsewhere in Africa. Its origin was a customs union between Kenya and Uganda in 1917. After Britain obtained a mandate over Tanganyika, that territory was gradually assimilated into the union, becoming a full member in 1927. Throughout the colonial period Britain encouraged a kind of "functional approach" to economic integration in East Africa, gradually developing a large number of important common services throughout

the area. In 1947 the British established in Nairobi the East Africa High Commission (EAHC) to coordinate approaches in the territories to common problems. It functioned quite well, given the difficulties it faced.

With the approach of independence and the persistence of some problems, however, it was considered wise to replace the EAHC with an East African Common Services Organization (EACSO), which came into existence formally two days after Tanganyika became independent in 1961. EACSO, despite shortcomings and inequities, survived rising African and local nationalisms to provide very good services to the three countries, even after Kenya and Uganda became independent. It was replaced in 1967 by the East African Community, with headquarters in Arusha, Tanzania, and offices of the various common services dispersed from Nairobi around all three countries.

Despite many assets and high hopes, however, the Community started going downhill soon after its creation. The many unresolved issues and problems of both the common services and the budding common market simply were not solved. It gradually disintegrated and finally collapsed in 1977.

In Chapter 28 we present a chart showing the names and membership of many African organizations with a variety of economic and other objectives, including, in a few cases, economic integration. Of these, perhaps the *Economic Community of West African States* (ECOWAS) has the best chance for long-term success. It was founded in 1975 by 15 countries meeting in Lagos, and it began functioning there in 1977, the year of the death of the East African Community. Nearly two decades later it was still vigorous, though still very far from achieving true integration. If the provisions of its treaty are faithfully carried out, West Africa will one day be a far better place in which to live, for it is sweeping and visionary, though tempered by realism.

Another organization aiming at economic integration is the *PTA/COMESA*. One outcome of the efforts of the UN Economic Commission for Africa to force the pace of cooperation among African States in trade,

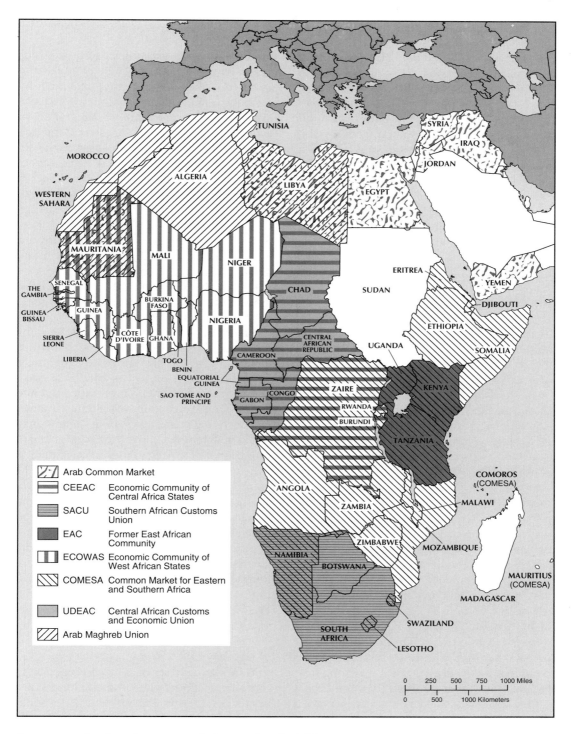

**Some organizations for economic integration in Africa.**

transport, and industry was the *Preferential Trade Area for Eastern and Southern African States* (PTA). Founded in June 1982 in Lusaka by 12 States, it has grown in both activity and membership (20 in 1995) as it moves gradually toward a common market and, eventually, economic union. In addition to trade liberalization and facilitation, it has also helped improve the physical infrastructure of the region and develop agriculture and industry. For a variety of reasons, it has not been successful in developing a significant internal market, and only a tiny fraction of members' trade is with one another. It has also had limited success in its other endeavors.

Nevertheless, the leaders of the member States decided in January 1992 to transform PTA into COMESA, the *Common Market for Eastern and Southern Africa*, and the treaty establishing COMESA was signed in November 1993. At the same time, it began working harder for a merger with the Southern African Development Coordination Conference (SADCC), which in August 1992 became the Southern African Development Community, with new objectives including free trade among its members.* Despite a growing convergence of objectives and considerable overlap of membership—and some outside pressure—SADC has resisted a merger. The two organizations established a Joint Committee of Ministers to recommend ways to harmonize and rationalize their activities, but by 1995 this had still not been achieved. In fact, as a practical matter, the SADC members of PTA had effectively withdrawn by not ratifying the COMESA treaty. All of this complex maneuvering clearly illustrates that conditions in Africa are very different from those in Europe and that their paths toward unity and prosperity must also be different. It is likely to be a very long time before the dream expressed in the Abuja Treaty for establishing an African Common Market is realized.

*SADCC/SADC is discussed in more detail in Chapter 28.

## Arab States, Asia, and the Pacific

In the vast area extending from Morocco across North Africa through all Asia and across the Pacific, there was not until very recently a single organization formed primarily to achieve economic integration or even sharply increased trade among the members. The nearest approach was the so-called *Arab Common Market*. The Agreement on Arab Economic Unity was signed in June 1957 but did not enter into force until June 1964. By May 1976, 12 States had become parties to it, but only seven of them have since taken even initial steps toward forming a free trade area. These seven are Libya, Mauritania, Yemen, Egypt, Iraq, Jordan, and Syria. Despite their abolition of most tariffs and quotas on their mutual trade in 1971 (Jordan excepted because of its special need for protection of industries), many obstacles limit the effectiveness of this move. As a result, bilateral trade agreements have tended to be more flexible and effective than the multilateral tariff reduction system among Arab countries. Some progress has been made among them in mining, livestock development, investment, and financial cooperation, but trade among them is still minuscule.

In February 1989 in Marrakesh, Morocco, the heads of Mauritania, Morocco, Algeria, Tunisia, and Libya proclaimed the *Arab Maghreb Union*. It aims to create a common market; joint ventures in manufacturing, transport, and services; joint development of resources; and other cooperative activities. It is too early to judge its significance, but given the history of attempts at Arab unity and the many thorny issues still to be resolved by the members, one would be wise to be skeptical.

The one really serious integration effort in this area is taking place in the South Pacific. Australia and New Zealand had had a series of preferential bilateral trade agreements over a long period of time. They culminated in the 1966 New Zealand Australia Free Trade Agreement (NAFTA). By the late 1970s more than 80 percent of the trade across the Tasman Sea had been freed from tariffs and quantitative restrictions. It did not meet the growing needs of the two countries, however, and in 1980 they agreed on "closer economic relationships." The result was the *Australia New Zealand Closer Economic Relations Trade Agreement* (commonly called the CER Agreement), which took effect on 1 January 1983. By 1990 the two countries had removed all remaining restrictions on trade between the two countries, five years earlier than planned. Since then they have liberalized trade in services, begun reducing shipping costs, and taken other measures to coordinate their economies better. They are very cautious, however, and each country still wants certain protections, subsidies, and restrictions that it considers necessary for its own economy. Nevertheless, by 1994 all restrictions to trade between them had been abolished, and trade had more than doubled since 1983.

Economic cooperation has made some progress in Asia and the Pacific, but economic integration has made none at all, except for that between Australia and New Zealand. Only in December 1991 and January 1992, respectively, did SAARC (the South Asian Association for Regional Cooperation) and ASEAN (the Association of Southeast Asian Nations), both discussed in Chapter 28, make the first timid moves toward real intraregional trade liberalization. The United Nations Economic and Social Commission for Asia and the Pacific (ESCAP) may be able to stimulate more effective integration, but considering the region's history of rivalries and even hostilities among the countries and their traditional lack of any real cooperation among them—or even perception of many common interests—progress in this direction will be very difficult.

## Economic Integration for the Twenty-first Century

At the beginning of this chapter, we pointed out the benefits to be gained from economic integration, but most of the chapter has been devoted to a description of false starts, glacial progress, and even outright failures.

There are, in fact, legitimate and even serious objections, obstacles, and costs involved in such integration. Aside from mindless chauvinism, these would include fears of loss of control over the undesirable aspects of international trade, such as trade in endangered species and products made from them, art and archaeological treasures, counterfeit currency, munitions and terrorist weapons, illegal drugs and psychotropic substances, defective or dangerous products, and so on. Integration may hamper or even preclude the use of some economic sanctions as a device for persuading States to conform to acceptable norms of behavior, especially in such areas as human rights, labor standards, environmental protection, and cooperation in the use of international rivers.

In allowing free market forces to control the flow of goods, people, and investment funds, even with what may be considered appropriate safeguards, integration may lead to severe distress in countries with a comparative advantage in few products or products of little value internationally. It could lead to runaway monopolies, price-fixing, "robber barons," and other evils of the early stages of capitalism, which remain threats emanating from some transnational corporations. Certainly, integration leads to realignment of industries, adjustments in agriculture, people being forced to abandon traditional occupations and learn new ones, and other kinds of social and economic disruption.

Yet, as we have seen, since 1989 there has been a great upsurge in economic integration worldwide. This is one more symptom, added to those discussed in the Prologue, of our entry into a new century which, among other things, is characterized by a global economy. This is a radically new situation, never before experienced even in the days of the great European empires, when the world economy was controlled by only a few countries. Autarky has been obsolete for generations. Trade is vital to every society. Economic integration at some level, with appropriate safeguards and regulation, is essential to economic health. Ultimately, after a certain level of economic and social development has been reached, failure to integrate will result in stagnation, even backsliding into poverty, perhaps even total disintegration.

The whole field of economic integration among developing countries is a fruitful area for investigation by political geographers. They can explain the links among politics, economics, and the physical environment that tend to inhibit integration and even cooperation, and recommend solutions. It would be a most worthwhile contribution.

# 27
# *Global Intergovernmental Organizations*

We have stressed in this book that today the dominant form of political organization is the State and its ideal aspect, the nation-state. We have also discussed other forms of political organization—tribe, city-state, feudal realm. In each case the form did not completely satisfy the needs and aspirations of the participants, and they joined together in a larger association, retaining their identities but cooperating and even surrendering some of their sovereignty for the common good. The Iroquois Confederacy, the Hanseatic League, and the Holy Roman Empire of the German Nation are only the most prominent examples of organizations designed to attain goals unattainable by any unit individually. Farther back in history are more examples: the Greek city-states are the best known. None encompassed more than a small portion of the earth's land or population, and none was able to withstand severe pressure for very long. They show us nonetheless that human beings cannot best be served exclusively by parochial political organizations, no matter how grand or powerful. A century and a half ago, in 1842, Alfred Tennyson included in his long poem *Locksley Hall* a remarkable prediction of the twentieth century that expresses the aspirations and the dreams of many centuries:

For I dipt into the future, far as human eye could see,
Saw the Vision of the world, and all the
wonder that would be;
Saw the heavens fill with commerce, argosies
of magic sails,

Pilots of the purple twilight, dropping down
with costly bales;
Heard the heavens fill with shouting, and
there rain'd a ghastly dew
From the nations' airy navies grappling in the
central blue;
Far along the world-wide whisper of the
south-wind rushing warm,
With the standards of the peoples plunging
thro' the thunderstorm;
Till the war-drum throbb'd no longer, and the
battle flags were furl'd
In the Parliament of man, the Federation of
the world.

The post–World War II era has been one of unprecedented activity in the field of supranationalism. Furthermore, in their determination to associate, States are doing what they had not been willing to do on such a scale previously: They are giving up some of their sovereignty in return for security, economic advantage, cultural strength, or whatever benefit they perceive to accrue from their involvement.

## *The League of Nations*

The League of Nations was primarily designed to curb aggressive war and was created to act as an instrument of collective security through joint repudiation of any aggressor. The horrors of World War I contributed to a desire for such an agency, and the United States was among those countries that worked hardest for its establishment. It did not spring up full-blown after the war, however. For generations philosophers, poets, and moralists had been proposing such an organization, and during the war even practical politicians and statesmen prepared carefully reasoned proposals for one. During the nineteenth century a number of peace conferences took place, the Hague Court of Arbitration was founded, and international bureaus, such as the Universal Postal Union and the International Institute of Agriculture, were established to deal with particular work in which international cooperation was essential. During the war the Allies cooperated on a scale hitherto unknown among "sovereign States." They also laid much of the groundwork for the League, so that when the Paris Peace Conference convened in 1919 the League Covenant was drafted in a few days under the leadership of President Wilson.

But the League received its first setback at the very beginning. In the United States, the proposal to join the organization was defeated, mainly through the hostility of Republican senators to the efforts of President Wilson, who had strongly championed U.S. participation. The members had hoped that the threat of total economic boycott would stifle any would-be aggressor; the absence of the United States from the organization eliminated that possibility and dealt a severe blow to the confidence of the membership. In all, 63 States were members of the League of Nations, though, in fact, the total membership at any single time did not reach 63. But despite these troubles, the League of Nations represented a tremendous leap forward in international relations, for it embodied a principle that until that time had been a matter only for theoretical discussion among political scientists and historians.

The League of Nations was officially terminated in 1946 after the United Nations had begun to function. Its contribution to the latter organization in terms of principles and practices is such that the *United Nations* in many ways is a renewal of the same aspirations expressed by the League.

In evaluating the work of the League of Nations, we can hardly do better than to quote Lord Robert Cecil, one of its founders and staunchest supporters and representative of Britain at the final session of the League Assembly on 9 April 1946:

> It is common nowadays to speak of the failure of the League. Is it true that all our efforts for those twenty years have been thrown away? . . . The work of the League is purely and unmistakably printed on the social, economic and humanitarian life of the world. But above all that, a great advance was made in the international organization of peace. . . . It was not, indeed, a full-fledged federation of the world—far from it—but it was more than the pious aspiration for peace embodied in those partial alliances which had closed many great struggles. . . . We saw a new world centre, imperfect materially, but enshrining great hopes, an Assembly representing some fifty peace-loving nations, a Council, an international civil service, a World Court of International Justice, so often before planned but never created, an International Labour Office to promote better conditions for the workers. And very soon there followed that great apparatus of committees and conferences striving for an improved civilisation, better international co-operation, a larger redress of grievances and the protection of the helpless and oppressed.
>
> Truly this was a splendid programme, the very conception of which was worth all the efforts which it cost . . . but, as we know, it failed in the essential condition of its existence—namely, the preservation of peace—and so, rightly or wrongly, it has been decided to bury it and start afresh. . . .
>
> The League is dead: Long live the United Nations!

### The United Nations

The United Nations was founded in 1945 under conditions similar to those attending the birth of the League of Nations. It had the benefit of the League's experience, its successes and failures, and many of its staff and delegates. Its roots are deeper and its foundations firmer than those of its predecessor. Its membership has finally approached universality, and no member has yet resigned or

# THE UNITED NATIONS SYSTEM

**GENERAL ASSEMBLY**

**TRUSTEESHIP COUNCIL**

**SECURITY COUNCIL**

**SECRETARIAT**

**INTERNATIONAL COURT OF JUSTICE**

**ECONOMIC AND SOCIAL COUNCIL**

---

● Main and other sessional committees
● Standing committees and ad hoc bodies
● Other subsidiary organs and related bodies

▲ UNRWA United Nations Relief and Works Agency for Palestine Refugees in the Near East

■ IAEA International Atomic Energy Agency
■ WTO World Trade Organization

▲ INSTRAW International Research and Training Institute for the Advancement of Women
▲ UNCHS United Nations Centre for Human Settlements (Habitat)
▲ UNCTAD United Nations Conference on Trade and Development
▲ UNDCP United Nations International Drug Control Programme
▲ UNDP United Nations Development Programme
▲ UNEP United Nations Environment Programme
▲ UNFPA United Nations Population Fund
▲ UNHCR Office of the United Nations High Commissioner for Refugees
▲ UNICEF United Nations Children's Fund
▲ UNIFEM United Nations Development Fund for Women
▲ UNITAR United Nations Institute for Training and Research
▲ UNU United Nations University
▲ WFC World Food Council

---

## Security Council

Peace-keeping operations

▲ UNTSO United Nations Truce Supervision Organization, June 1948 to date
▲ UNMOGIP United Nations Military Observer Group in India and Pakistan, January 1949 to date
▲ UNFICYP United Nations Peace-keeping Force in Cyprus, March 1964 to date
▲ UNDOF United Nations Disengagement Observer Force, June 1974 to date
▲ UNIFIL United Nations Interim Force in Lebanon, March 1978 to date
▲ UNIKOM United Nations Iraq-Kuwait Observation Mission, April 1991 to date
▲ UNAVEM II United Nations Angola Verification Mission II, June 1991 to date
▲ ONUSAL United Nations Observer Mission in El Salvador, July 1991 to date
▲ MINURSO United Nations Mission for the Referendum in Western Sahara, September 1991 to date
▲ UNPROFOR United Nations Protection Force, March 1992 to date
▲ UNOMIG United Nations Observer Mission in Georgia, August 1993 to date
▲ UNMIH United Nations Mission in Haiti, September 1993 to date
▲ UNOMIL United Nations Observer Mission in Liberia, September 1993 to date
▲ UNAMIR United Nations Assistance Mission for Rwanda, October 1993 to date
▲ UNMOT United Nations Mission of Observers in Tajikistan, December 1994 to date

● Military Staff Committee
● Standing committees and ad hoc bodies

---

## Economic and Social Council

▲ WFP World Food Programme
▲ ITC International Trade Centre UNCTAD/GATT

● Functional Commissions
Commission for Social Development
Commission on Crime Prevention and Criminal Justice
Commission on Human Rights
Commission on Narcotic Drugs
Commission on Science and Technology for Development
Commission on Sustainable Development
Commission on the Status of Women
Population Commission
Statistical Commission

● Regional Commissions
Economic Commission for Africa (ECA)
Economic Commission for Europe (ECE)
Economic Commission for Latin America and the Caribbean (ECLAC)
Economic and Social Commission for Asia and the Pacific (ESCAP)
Economic and Social Commission for Western Asia (ESCWA)

● Sessional and Standing Committees
● Expert, Ad Hoc and Related Bodies

---

■ ILO International Labour Organisation
■ FAO Food and Agriculture Organization of the United Nations
■ UNESCO United Nations Educational, Scientific and Cultural Organization
■ WHO World Health Organization
World Bank Group
■ IBRD International Bank for Reconstruction and Development (World Bank)
■ IDA International Development Association
■ IFC International Finance Corporation
■ MIGA Multilateral Investment Guarantee Agency
■ IMF International Monetary Fund
■ ICAO International Civil Aviation Organization
■ UPU Universal Postal Union
■ ITU International Telecommunication Union
■ WMO World Meteorological Organization
■ IMO International Maritime Organization
■ WIPO World Intellectual Property Organization
■ IFAD International Fund for Agricultural Development
■ UNIDO United Nations Industrial Development Organization

---

▲ United Nations programmes and organs (representative list only)
■ Specialized agencies and other autonomous organizations within the system
● Other commissions, committees and ad hoc and related bodies

been expelled.* Its activities are more varied than those of the League, its structure more complex, its budget and secretariat far larger, its role as a public forum more effectively utilized, and its political and security functions more vigorously carried out. There are many reasons for its relative success and durability, despite many problems and weaknesses, but the most important are undoubtedly the active participation of the United States and the importance of UN multilateral diplomacy and assistance to the developing countries.

We have already discussed some UN organs and activities—its role in the decolonization process and in assisting the new States; the International Court of Justice and the International Law Commission; GATT and UNCTAD; the regional economic commissions—and we discuss others later. Many others have a spatial dimension and are thus of particular interest to geographers. They have varying memberships, fields of activity, and degrees of effectiveness, in all of which political factors are important. Some of these aspects are included in the diagram of the UN system.

One major activity of the United Nations (as was true of the League of Nations) is to sponsor large international conferences devoted to specific topics. We have already mentioned the lawmaking conference; the other type is a conference of government officials and experts (who are sometimes the same people) who discuss a problem, study reports on it, and ultimately make recommendations to States and to the United Nations on courses of action that should or could be taken to deal with the problem.

There have been scores of such conferences, and there will be many more. Among those dealing with geographic topics have been those on food, population, the human environment, human settlements (Habitat), technical cooperation among developing countries, desertification, water, science and technology for development, energy, outer space, and pollution of the Mediterranean, as well as many sponsored by the intergovernmental agencies and other UN affiliates.

The United Nations has received much criticism, almost since its founding, much of it justified. There have been countless proposals for reform and restructuring. It has neither solved the world's great problems nor brought an end to war, for the UN only represents the collective will of its members. With nationalism still the strongest political force in the world, it is unreasonable to expect that States would grant the United Nations truly supranational authority. It is very far indeed form being a world government. The significance of the United Nations, in the present context, lies in the fact that it reflects, however imperfectly, the continued desire on the part of States for mutual action, open communications, and maintenance of international peace and security. States that have faced the censure of the United Nations nevertheless continue to participate and contribute. This surrender of even the smallest measure of sovereignty is a completely new phenomenon on the politicogeographical scene, and a long stride along the road toward "the Parliament of man, the Federation of the world."

## Peace-keeping Operations

One of the most important, most politicogeographical and, until 1988, one of the least publicized and appreciated activities of the United Nations is its peace-keeping operations. Belated and well-deserved honors came to the peace-keepers when in 1988 the Nobel Prize for Peace was awarded to them. They also deserve much more attention from political geographers and the general public than they have received so far.

---

*President Sukarno of Indonesia, reacting to the UN's failure to support his claims on Malaysia's Borneo territories, informed that body on 20 January 1965 that Indonesia was withdrawing "at this stage and under the present circumstances." After his downfall, the new government on 19 September 1966 announced its decision "to resume full cooperation with the United Nations and . . . participation in its activities." In addition, South Africa was excluded from participation in the General Assembly from 1974 to 1993 and from several other UN bodies and specialized agencies, but remained a member.

## *Nongovernmental Organizations*

More than 700 nongovernmental organizations (NGOs) that are in consultative status with the Economic and Social Council play a very important and growing role in the work of the United Nations. They provide a vehicle for citizens, including students, to make their opinions known directly to the United Nations, to learn about obscure details of UN work in areas of special interest to them, and to help in drafting proposals for UN action. A representative sample of those concerned with geographic matters would include:

International Federation of Agricultural Producers
International Planned Parenthood Federation
Associated Country Women of the World
Carnegie Endowment for International Peace
International Association of Ports and Harbours
International Catholic Migration Commission
International Commission on Irrigation and
   Drainage
International Council of Environmental Law
International Union for Conservation of Nature
   and Natural Resources
American Association for the Advancement
   of Science
International Association on Water Pollution
   Research
International Association of Physical
   Oceanography

United Towns Organization
Inter-American Planning Society
International Association for Water Law
International Civil Airport Association
International Law Association
International Road Transport Union
International Union of Railways
Society for International Development
International Union of Local Authorities
World Population Society
Friends of the Earth
Sierra Club
International Chamber of Commerce
International Geographical Union
International Society for
   Photogrammetry

The term "peace-keeping" has no precise, generally accepted definition that can be used to delineate categorically those activities by outside parties designed to keep the peace or to bring about or maintain a cessation of hostilities between two or more factions within a country or between two or more countries. For our purposes, however, we limit the term to the insertion of civilian and/or military personnel by a third party consisting of either a recognized intergovernmental organization or a coalition of countries with the consent and cooperation of the parties to the conflict. Thus we rule out such recent single-country interventions as those of India in Sri Lanka and the Maldives, Tanzania in Uganda, Vietnam in Cambodia, Turkey in Cyprus, the Soviet Union in Czechoslovakia and Afghanistan, and the United States in Grenada and Panama, even if, as in the Grenada and Czechoslovakia

cases, the intervening forces are accompanied by or followed by token units of other countries.

There is considerable variety among peace-keeping operations, however defined. The League of Nations, for example, sent commissions of inquiry to the Aaland Islands, Albania, Corfu, Memel, Mosul, Greece-Bulgaria, and Manchuria; a military commission of control to the Lithuania-Poland border; and an administrative commission to Leticia (Colombia); sponsored plebiscites to resolve a number of boundary and territorial disputes; and instituted sanctions against Italy for its conquest of Ethiopia. Only the first group of actions can be considered "peace-keeping" as the term is used now, and their results were decidedly mixed.

Since the Second World War there have been a number of such operations mounted

**United Nations peace-keeping forces in action.** The picture shows members of the Fijian contingent of the United Nations Interim Force in Lebanon inspecting a motorbike at a checkpoint south of Tyre. (Photo by UN/Inge Lippmann)

by established intergovernmental organizations and by ad hoc coalitions of States. These are summarized in tabular form in Chapter 28. By far the largest number of such operations, however, and probably on the whole the most important and most successful, have been those conducted by the United Nations. These are summarized in Table 27-1. We cannot examine the matter of peace-keeping in detail here for lack of space, except to note that it is being expanded to include *peace-making*—at great risk and with uncertain results—and humanitarian intervention in cases of civil strife or massive human rights violations. But all of these operations, however valuable, cost money and as of 31 January 1995, $2.2 billion of the $3.6 billion owed the UN by its members were for peace-keeping assessments. The United States is still the world's largest debtor to the UN, owing some 40% of the total debt.

The bibliography at the end of Part Six includes a number of good studies by political scientists and others that provide useful background for the kind of politicogeographical studies so badly needed now. Good analyses of the subject by political geographers could assist in making future peace-keeping operations of all kinds more efficient and more successful.

## The Commonwealth

We have already mentioned the *Commonwealth* (originally the British Commonwealth of Nations) in connection with decolonization and international trade, but it is such an extraordinary organization that it deserves some additional attention. It is not a federation, not an empire, not an alliance, not a trade group, not, in fact, like any other organization in the world. It is a free association of 50 countries on every continent, as diverse a group of States as can be imagined.* All they have in common is their recognition of the British sovereign as Head of the Commonwealth, their former status as British territories, use of the English language and legal system, and agreement that Commonwealth membership brings them important benefits. These benefits no longer include the Commonwealth preference system, as we have seen, or free migration into Britain (virtually halted since 1963), or the traditional protection afforded by the Royal Navy. Even the Commonwealth Fund for Technical Cooperation is not large by world standards. Although it is generally considered to have come into existence at the Im-

*South Africa, with its new democratic system and majority rule, was readmitted on 1 June 1994.

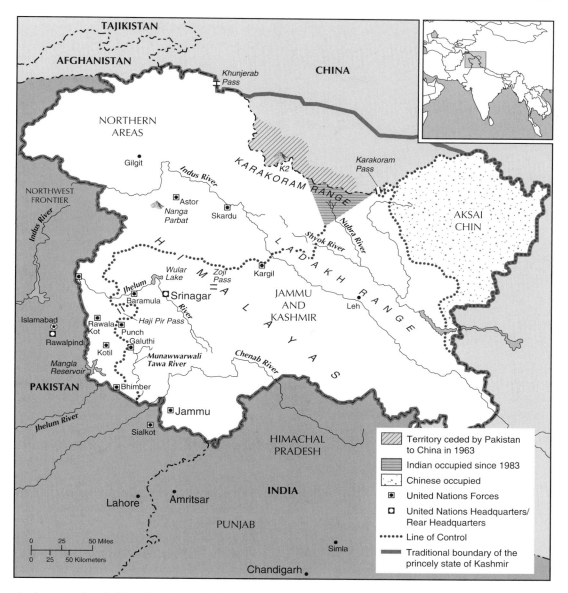

**Kashmir, a divided land.** India and Pakistan have fought several bitter wars over this formerly princely state and its future is still in doubt, although its *de facto* partition along a *Line of Control* may be permanent. United Nations peace-keeping forces have been on duty here for more than 40 years. The area north and west of the Line of Control is occupied by Pakistan; the rest, by far the most valuable portion, by India.

perial Conference of 1926, it is actually a product of gradual evolution; it has never been formally organized, has no constitution or charter, and until 1965 never had a head-quarters or secretariat. What then is it? How is it organized? What does it do?

There is no longer a hierarchy of mem-

bership in the Commonwealth. All members are equal. When a British territory becomes independent, it may apply for full Commonwealth membership and, if it does, it is (almost automatically) admitted as an equal, no matter how small or weak or poor it may be. The members exchange opinions in a

*Table 27-1*   *Chronology of United Nations Peace-keeping Operations*

| Identifier | Formal Name | Dates | Area of Operation (if not specified in name) | Headquarters |
|---|---|---|---|---|
| UNSCOB | UN Special Committee on the Balkans | Oct. 1947–Dec. 1951 | Chiefly Greece | Salonika |
| UNTSO | UN Truce Supervision Organization in Palestine | June 1948–Present | Israeli Border Areas | Jerusalem |
| UNMOGIP | UN Military Observer Group in India and Pakistan | Jan. 1949–Present | Jammu and Kashmir | Rawalpindi (winter) Srinagar (summer) |
| POC | Peace Observation Commission | Jan. 1952–May 1954 | Greece | New York |
| UNEF | UN Emergency Force | Nov. 1956–June 1967 | Sinai & Gaza Strip | Gaza |
| UNOGIL | UN Observer Group in Lebanon | June–Dec. 1958 | Lebanon-Syria Border | Beirut |
| UNSF | UN Security Force in West New Guinea (Now Irian Jaya)  (Operated in connection with UNTEA, the UN Temporary Executive Authority) | Oct. 1962–Apr. 1963 | | Hollandia (now Jayapura) |
| UNYOM | UN Yemen Observation Mission | July 1963–Sep. 1964 | | San'a |
| UNFICYP | UN Peace-Keeping Force in Cyprus | Mar. 1964–Present | | Nicosia |
| DOMREP | Mission of the Representative of the Secretary-General in the Dominican Republic | May 1965–Oct. 1966 | | Santo Domingo |
| UNIPOM | UN India-Pakistan Observation Mission | Sep. 1965–Mar. 1966 | India-Pakistan Border | Lahore, Amritsar |
| UNEF II | Second UN Emergency Force | Oct. 1973–July 1979 | Sinai | Ismailia |
| UNDOF | UN Disengagement Observer Force | June 1974–Present | Golan Heights | Damascus |
| UNIFIL | UN Interim Force in Lebanon | Mar. 1978–Present | Southern Lebanon | Naqoura |
| UNGOMAP | UN Good Offices Mission Afghanistan & Pakistan | Apr. 1988–Mar. 1990 | | Kabul, Islamabad |
| UNIIMOG | UN Iran-Iraq Military Observer Group | Aug. 1988–Feb. 1991 | | Baghdad, Tehran |
| UNAVEM | UN Angola Verification Mission | Jan. 1989–June 1991 | | Luanda |
| UNTAG | UN Transition Assistance Group | Apr. 1989–Mar. 1990 | Namibia | Windhoek |
| ONUVEN | UN Observer Group to Verify the Electoral Process in Nicaragua | Aug. 1989–Feb. 1990 | | Managua |
| CIAV | UN International Support and Verification Mission (in cooperation with OAS) | Sep. 1989–Oct. 1990 | Nicaragua | Managua |

| | | | |
|---|---|---|---|
| ONUCA | UN Observer Group in Central America | Nov. 1989–Jan. 1992 | Tegucigalpa |
| ONUVEH | UN Observer Group to Verify the Elections in Haiti | June 1990–Jan. 1991 | Port-au-Prince |
| UNIKOM | UN Iraq-Kuwait Observation Mission | Apr. 1991–Present | Umm Qasr, Iraq |
| UNAVEM II | Second UN Angola Verification Mission | June 1991–Present | Luanda |
| ONUSAL | UN Observer Mission in El Salvador | July 1991–Present | San Salvador |
| MINURSO | UN Mission for the Referendum in Western Sahara | Sep. 1991–Present | Laayoune |
| UNAMIC | UN Advance Mission in Cambodia | Nov. 1991–Mar. 1992 | Phnom Penh |
| UNPROFOR | UN Protection Force in Yugoslavia | Feb. 1992–Present | Belgrade, Sarajevo, Zagreb; Serbia, Bosnia, Croatia |
| UNTAC | UN Transitional Authority in Cambodia | Mar. 1992–Sep. 1993 | Phnom Penh |
| UNOSOM I | First UN Operation in Somalia | Apr. 1992–Apr. 1993 | Mogadishu |
| ONUMOZ | UN Operation in Mozambique | Dec. 1992–Dec. 1994 | Maputo |
| UNOSOM II | Second UN Operation in Somalia | May 1993–Mar. 1995 | Mogadishu |
| UNOMUR | UN Observer Mission Uganda-Rwanda | June 1993–Sep. 1994 | Kabale, Uganda |
| UNOMIG | UN Observer Mission in Georgia | Aug. 1993–Present | Sukhumi |
| UNOMIL | UN Observer Mission in Liberia | Sep. 1993–Present | Monrovia |
| UNMIH | UN Mission in Haiti | Sep. 1993–Present | Port-au-Prince |
| UNAMIR | UN Assistance Mission for Rwanda | Oct. 1993–Present | Kigali |
| UNASOG | UN Aouzou Strip Observer Group | May 1994–June 1994 | Aouzou administrative post |
| UNMOT | UN Mission of Observers in Tajikistan | Dec. 1994–Present | Dushanbe |

*Notes:* 1. Definitions of "peace-keeping operations" vary. They may or may not include small civilian missions or liaison groups. This list must therefore not be considered authoritative, but only indicative. It is nevertheless the most complete such list compiled to date.

2. Large-scale military operations in Korea (1950s) and the Congo (now Zaire) (1960s) are not included.

3. Some of the operations utilize small field offices away from the headquarters. These are not listed here.

4. As of January 1995, 35 States have committed themselves to make standby arrangements to provide peace-keeping forces for the UN. Canada, Ireland, Australia, and the Scandinavian countries have long maintained standby forces.

5. As of early 1995 1,262 people from many countries had died while on active duty with UN peace-keeping operations, and many more had been wounded.

6. The Nobel Prize for Peace was awarded to United Nations peace-keeping forces in 1988 in recognition of their contributions to the maintenance of world peace.

7. Additional information may be obtained from the latest edition of *The Blue Helmets: A Review of United Nations Peace-keeping*, published in New York by the United Nations from time to time.

friendly, informal, and intimate atmosphere; they do not agree on positions, make collective decisions, or take collective action. Even the secretariat, located in London, has no career service, is forbidden to perform executive functions, and is formally limited to servicing governmental and other meetings, preparing papers on matters of common interest, giving advice, and assisting in areas of functional cooperation. It has, however, by common consent grown to perform executive functions and generate stronger influence, still informal, within the group. The first secretary general was from Canada, the second from Guyana, and the post has continued to be held by citizens of various member countries.

The principal activity of the Commonwealth is the biennial Heads of Government Meeting, with senior government officials meeting in the alternate years. There are also regular meetings of ministers responsible for education, finance, health, law, food production and rural development, and youth matters. Commonwealth high commissioners (ambassadors) in London form a useful group for special purposes, such as overseeing the sanctions against Rhodesia and supervising the budget of the secretariat. The Commonwealth frequently sends missions to observe elections and help resolve national disputes in member countries, and is even beginning to take positions on major world issues.

There are scores of other significant activities and institutions in agriculture, forestry, commerce, communications, education, human rights, law, science and technology, and other practical affairs, but it is primarily a framework for consultation, mutual education, and mutual support. Formal activities and organization are relatively new and still not essential to its success. Its durability is astonishing in an era of such rapid and profound change. It has survived depression and world war, the dismantling of the British Empire, squabbles and even wars between members (India–Pakistan, Uganda–Tanzania), Britain's entry into the EC, sharp disagreements about Rhodesia and South Africa, and many other potentially shattering experiences. It is not difficult to imagine the Commonwealth continuing indefinitely,

ever-changing, ever-adapting, serving as a halfway house between the loneliness of isolation and the often bewildering complexity of the United Nations. It is truly a singular institution.

### The French Community

The *French Community* has a very different history. It was born with the Fifth Republic, itself largely the product of the series of disasters that befell France in its efforts to retain a hold over its far-flung colonial empire. Prior to the French Community, the French empire consisted of a very complex organizational structure known as the French Union, in which overseas but self-governing territories were represented to some degree in the Paris government. However, so many changes took place in the empire during and after World War II that almost constant revision was necessary. In 1958, with the crisis in Algeria dragging on for several years and "the wind of change" having an impact in other parts of Africa, the Community was erected and French overseas possessions were given a choice: to become independent outside any French Community, to attain independence within the Community framework, or to continue as dependencies as before. When the vote was taken, all African territories except French Guinea and French Somaliland chose autonomy and membership in the Community. French Guinea chose complete independence outside the French sphere; French Somaliland chose to remain a French colony.

Under the Fifth Republic, independent States that were members of the Community did not have the representation in the French Parliament enjoyed by self-governing French overseas territories. In turn, the French president did not have the powers in the member States once enjoyed by the Parliament. Nevertheless, France continued to have considerable influence in its former empire.

But the relationship between France and the countries with which it constituted the Community, as described in the 1958 Constitution, has been replaced by a series of contractual agreements that describe the area of cooperation between each country

---

### Some International Nongovernment Organizations

*The Inter-Parliamentary Union*, founded in 1889 to promote personal contacts among the members of the world's parliaments, currently has 72 member groups, and its secretariat is in Geneva. *The International Council of Scientific Unions* was founded in 1931 to coordinate international cooperation in theoretical and applied sciences. The academies or research councils of 66 countries belong as well as 17 international unions. Its headquarters is in Paris. It has committees on Antarctic, oceanic, space, and water research. *The International Air Transport Association (IATA)* has headquarters in Montreal and Geneva. At present more than 90 international airlines are full members. IATA was founded in 1948 to promote civil aviation and enable airlines to collaborate on technical and marketing problems. Until 1978 it also set most international airline fares.

---

and France. The institutions of the Community have fallen into disuse, although the Community has never been formally dissolved, and it is impossible to determine when the various bodies ceased functioning.

This was no great loss, for the French Community was never more than a poor imitation of the Commonwealth and was primarily an instrument for retaining French dominance over its former colonies. It did not have the long historical evolution or the flexibility of self-government and mutual assistance of the Commonwealth. Perhaps France never really had its heart in it. As a highly centralized State, France probably could not really accommodate itself to the decentralization attendant upon the loss of its empire. It seems unlikely that the Commonwealth will ever be successfully imitated.

### Organization for Economic Cooperation and Development

In the wake of the devastation of World War II, the United States offered Europe generous economic assistance in rebuilding. Formally known as the European Recovery Program, it was popularly called the Marshall Plan after U.S. Secretary of State George C. Marshall, who proposed it in 1947. The plan was worked out cooperatively with the 18 European countries that accepted the U.S. invitation to join. (All the communist countries rejected the invitation, Poland and Czechoslovakia with regret.) They formed the Organization for European Economic Cooperation (OEEC) to administer the program at their end.

As we noted in Chapter 26, the OEEC provided Europeans with the first opportunity in history to work cooperatively on a major and continuing peaceful project affecting the vital interests of all of them. It continued functioning well even after the EEC and EFTA divided its members into the Six, the Seven, and the five nonparticipants in either grouping. It laid the groundwork for the UN Economic Commission for Europe, and it carried out its principal mission with efficiency and élan by the end of the 1950s. It was so successful that everyone concerned felt it would be a pity to scrap it, to break up a winning team. And so in 1961 it was converted into the *Organization for Economic Cooperation and Development (OECD)*.

The fundamental differences between the two organizations are apparent from the change in name. First, it was no longer strictly a European organization; the United States and Canada were charter members; Japan joined in 1964; Finland, which had been an associate, became a full member; and Australia and New Zealand joined as full members. Mexico became the twenty-fifth member in March 1994 and Korea will likely join in 1995.* Second, its new emphases were on development instead of on recovery, including aid of various kinds to the

---

*Poland, Hungary, and the Czech Republic are expected to join by 2000. The Commission of the European Union also participates in the work of the OECD.

poor countries of the world, and on expansion of world trade. With headquarters in Paris and an elaborate structure, its activities are many and varied. Most are highly technical, but some are of interest to geographers. They include committees on tourism, maritime transport, trade, international investment and transnational enterprises, research, various industries, fisheries, agriculture, and nuclear energy. Perhaps most interesting, and certainly one of the most potent of the OECD organs, is the Development Assistance Committee (DAC), its principal vehicle for channeling, monitoring, and evaluating aid from the OECD to developing countries. The economic forecasts of the OECD may be open to challenge, its technical accomplishments may be uneven, its relationships with the EU, EFTA, and ECE obscure, and the motives and effectiveness of its development assistance questionable, but the OECD remains a powerful and influential force in world affairs.

Unlike GATT, the OECD had no poor members until Mexico's admission; it has been very much a "rich man's club." In its wealth and unity lie its power. In the Conferences on International Economic Cooperation during the mid-1970s (the so-called North–South dialogues), the OECD countries were on one side and the Group of 77 on the other. The conferences were inconclusive. The parties were no more able to reach understandings about such fundamental issues as industrialization of poor countries, trade concessions by the rich to the poor, the flow of aid from the North to the South, the distribution and effective use of all forms of energy, and overall economic relations between North and South than they have been in the United Nations, UNCTAD, GATT, or elsewhere. Nevertheless, the OECD continues to render valuable service to the world, rich and poor alike.

## Other International Organizations

Among the numerous other international governmental and nongovernmental organizations, a few that are both geographic and

influential should be mentioned briefly. The *Organization of the Petroleum Exporting Countries* is one. It was organized in 1960 by Iran, Iraq, Kuwait, Saudi Arabia, and Venezuela to unify and coordinate members' petroleum policies and to safeguard their interests. The founders have been joined by Algeria, Ecuador, Gabon, Indonesia, Libya, Nigeria, Qatar, and the United Arab Emirates. Its secretariat is located in Vienna. OPEC established a special fund in 1976 to provide financial assistance to other developing countries on easy terms as a contribution toward the redistribution of wealth from the rich countries to the poor, but most aid has gone to Jordan, Syria, and the Palestine Liberation Organization.

OPEC's quadrupling of oil prices in 1973 and the embargo of 1973–74 (the "oil weapon") were initially aimed at Israel and its friends; the continuation of OPEC's supply and pricing policies became part of the drive for a New International Economic Order.* More recently, however, for several reasons including conservation in the consuming countries, OPEC has lost power and influence and has been unable to maintain petroleum prices at more than half their peak levels.

The *Group of 77*, consisting basically of the developing countries of the world, now numbers approximately 130 members and takes an active role in shaping common positions on many current issues. It is a negotiating group within the United Nations and other global forums, but has no tangible projects or activities. The same is true of the *Nonaligned Movement* (NAM), which grew out of the Bandung Declaration of 1955, was formalized at the Belgrade Conference of 1961, and currently has 110 members, all of them also in the Group of 77. The criteria for membership in both are rather loose; the 77 contains a number of countries that are hardly poor, and the NAM contains a num-

*The Organization of Arab Petroleum Exporting Countries (OAPEC) was founded in 1968, currently has 10 members, and has headquarters in Kuwait. Although officially it sponsors a number of joint economic undertakings, its activities are primarily political.

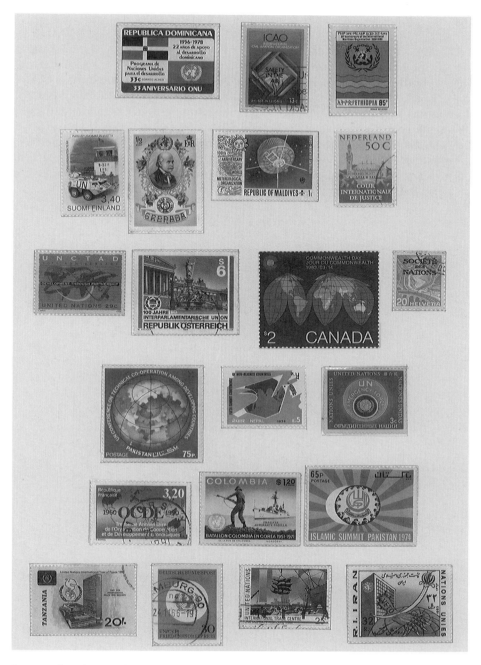

**International organization on stamps.** This is a popular theme on stamps, especially the United Nations and its specialized agencies. Only a small sample is shown here. These stamps honor the United Nations Development Programme, International Civil Aviation Organization, International Maritime Organization, World Health Organization, World Meteorological Organization, International Court of Justice, United Nations Conference on Trade and Development, Interparliamentary Union (an INGO), The Commonwealth, League of Nations (defunct), United Nations Conference on Technical Cooperation Among Developing Countries, Non-Aligned Movement, United Nations Emergency Force in Sinai, Organization for Economic Cooperation and Development, Colombian Battalion that fought with United Nations Forces in Korea, Organization of the Islamic Conference, United Nations, United Nations Children's Fund, and the UNCTAD/GATT International Trade Centre. The last one, denouncing the veto in the UN Security Council, was issued by Iran in 1983. (Martin Glassner)

ber of countries that were clearly aligned with one or another of the superpowers. Nevertheless, both groups have been powerful voices speaking for at least the leadership of countries that are small, poor, and weak. Since the end of the Cold War, however, the NAM has lost its *raison d'être* and is likely soon to fade into oblivion.

The *Organization of the Islamic Conference* was organized in 1969 and now has 46 members in Africa, Asia, and around the Indian Ocean. It has a secretariat in Jeddah, Saudi Arabia, but no formal structure. Its aims are very broadly to promote Islamic solidarity among members and cooperation in many fields. Despite annual meetings of ministers, however, and numerous organs, institutions, affiliates, resolutions, and pronouncements, it has few concrete achievements to its credit.

# 28

# *Regional and Subregional Organizations*

Developing side by side with the international or worldwide organizations just described have been a great many groups of States within particular regions. Regardless of how we classify them functionally, we must recognize that they all perform multiple functions, thus overlapping all categories, and all are fundamentally political, even if their stated purpose is economic or cultural or social. All are groups of political bodies (States) organized as a result of political decisions to achieve political goals.

There are many reasons for the existence of regional organizations. Some observers feel that true world organization is impossible to achieve all at once and that therefore regional organizations can serve as "building blocks," gradually forging greater links among themselves until the ultimate goal is reached. Others feel that regional organization *is* the ultimate goal, that only groups based on a clear community of interests with a common heritage and a common approach to solving problems are necessary or possible. Another view is that regional organizations are a compromise, a "halfway house" between the State and a world government, and are, therefore, to be accepted as experiments or trial runs preparing us for the real thing. Regardless of the theoretical bases for them, however, they have existed for a long time, continue to proliferate, and disappear when no longer needed or desired. In the survey that follows, we include only regional organizations of some size and

durability and those whose functions are broadly geographic. There are hundreds more that are smaller, more technical, or less substantial. There are also scores of organizations of sub-State units: of individual government ministries, subdivisions of States, technocrats and nongovernmental professionals, and other citizens, all organized on a regional or subregional basis.

## *Organizations of General Competence*

The largest and generally most active of the regional organizations, apart from those devoted to economic integration, are those commonly called "political" or "political–cultural," but which in reality have economic and social functions as well. Foremost among them in nearly every respect is the *Organization of American States* (OAS), whose headquarters is in Washington.

During the more than a century and a half since most of the Spanish-American colonies became independent, a unique network of relationships has developed among them, many of which have traditionally included the United States and many that have recently been joined by the independent countries of the Commonwealth Caribbean. This network is commonly called the Inter-American System. There are skeptics who doubt the efficacy of the system. Many observers recognize the need for changes in it, and some question whether the needed changes are possible or even worthwhile. Nevertheless, nowhere else in the world has there ever been such a large and diverse group of independent States that have voluntarily associated themselves in so many ways for so long.

There were a number of unsuccessful attempts to form an organization of the newly independent States of the Western Hemisphere during the nineteenth century, but not until 1890 did the effort bear fruit. At an inter-American meeting in Washington that year, the Pan American Union was born. During the next half century it developed a number of functional organs, served as a framework for organizing hemispheric defense, aided in the settlement of disputes among its members, and otherwise proved useful to both the United States and the smaller and poorer Latin American States. It was reorganized at Bogotá in 1948 and transformed into the secretariat of the new Organization of American States. The principal functions of the OAS originally were collective defense and pacific settlement of disputes within the hemisphere. Its principal cornerstones are the Inter-American Treaty of Reciprocal Assistance (the Rio Treaty of 1947), the Charter of the OAS (1948), and the American Treaty on Pacific Settlement (1948). The last of these binds its signatories to utilize all the methods of pacific settlement that we discussed in detail in Chapter 24. The Rio Treaty organized a military alliance that we discuss briefly later in this chapter. Here we review the structure and work of the OAS stemming from its charter, as amended by the Protocol of Buenos Aires in 1967 and the Protocol of Cartagena de Indias in 1985.

The OAS today is organized much like a small United Nations. Its supreme organ is the General Assembly, which meets annually to decide general policies and actions of the organization and to coordinate the work of the other organs. The Meeting of Consultation of Ministers of Foreign Affairs is utilized frequently on an ad hoc basis "to consider problems of an urgent nature" (OAS Charter) or to serve as an "Organ of Consultation" in case of a threat to international peace and security (Rio Treaty). Directly responsible to the General Assembly are the Permanent Council, which deals with any matter referred to it by the General Assembly or the Meeting of Consultation and performs certain mandated functions; the InterAmerican Economic and Social Council (CIES);* the Inter-American Council for Education, Science and Culture (which is similar to UNESCO); the Inter-American Commission on Human Rights; the Inter-American Juridical Committee; the General Secretariat (formerly the Pan American Union); the Inter-American Court of Human Rights; and the specialized conferences, held as needed to deal with special aspects of inter-American cooperation. Many conferences have dealt with geographic subjects, including agriculture, ports and harbors, highways, natural resources, and Indian affairs.

The OAS, like the UN, has specialized organizations and other entities affiliated with it. Among them are the Inter-American Indian Institute (Mexico City), the Inter-American Institute for Cooperation on Agriculture (San José), the Pan American Health Organization (Washington; formerly the Pan American Sanitary Bureau and now also functioning as the regional arm of WHO), and the Pan American Institute of Geography and History (Mexico City). The PAIGH has committees on geography, history, geophysics, and cartography. Its geography publications are of high quality and should be more widely known.

Today all 35 independent States of the Western Hemisphere are members of the OAS, and the EU and 29 individual countries have the status of Permanent Observer. (Cuba was suspended from active participation in 1962 but remains a member.) Its activity in many practical matters, such as development of the Plata Basin, settlement of disputes, human rights, regional and subregional population and development projects, and peacekeeping has been considerable, but it is far from being central to the everyday concerns of its members. What

---

*The Alliance for Progress was operated through CIES until the Alliance died in 1974.

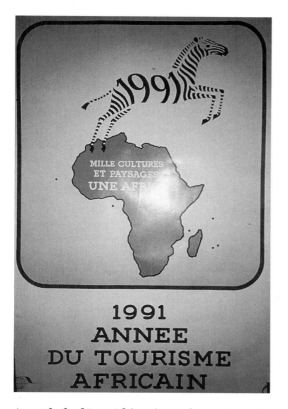

**A symbol of Pan-Africanism.** This poster celebrating Africa Tourism Year is on a wall of a very
small restaurant in Masaka, Uganda. Most African
countries also cooperate in continental sports
competitions, music festivals and other manifestations of unity, but economic and political unity
are still very far in the future. (Martin Glassner)

matters most, however, is probably the perception—irrespective of the reality—of
"hemispheric unity," of a "special relationship," among the diverse States and peoples
of the hemisphere, a true Inter-American
System, epitomized by the OAS. Despite
many problems and uncertainties, it is likely
to continue indefinitely and be of value to its
members and to the world community.

A comparable, but much younger and less
developed, organization has been formed in
Africa: the *Organization of African Unity*
(OAU). Founded in 1963 by 30 States, all but
a very few newly independent, it gradually
replaced a long list of regional and subregional groups, many of them ephemeral,

which had formed during the first decade of
decolonization. Its spiritual father was
Kwame Nkrumah, President of Ghana and
vigorous advocate of pan-Africanism. Its objectives were furtherance of African unity
and international cooperation in Africa; coordination of political, economic, cultural,
health, scientific, and defense policies; and
the eradication of colonialism in Africa. Its
early years were dominated by the anticolonialism struggle, and there is no doubt that
it did hasten decolonization.

Another important achievement has been
its stabilizing influence in the continent and
its assistance in the settlement or containment of many disputes among its members.
As we have seen, its insistence on the principle of *uti possidetis juris* has prevented
many violent conflicts over the frequently arbitrary boundaries inherited from the colonial powers. Where fighting has broken out
between neighbors (Algeria–Morocco, Somalia–Ethiopia, Ghana–Burkina Faso), OAU
action has been successful; it has been unsuccessful in cases of civil war (Congo, Nigeria, Sudan) and extracontinental intervention
(Shaba, Ogaden, Angola). A third major
achievement of the OAU has been to formulate and represent to the world African positions on important world issues.

The headquarters of the OAU is in Addis
Ababa, which is also the headquarters of the
UN Economic Commission for Africa, with
which it sometimes works very closely. Its
principal organ is the Assembly of Heads of
State and Government, held annually and
preceded by semiannual ministerial meetings. There are a permanent secretariat and
four specialized commissions responsible to
the Assembly directly or indirectly through
the Council of Ministers. They deal respectively with mediation, conciliation, and arbitration; defense; economics and transport;
and education, science, and culture. Its numerous other organs are responsible to  its
secretary general; they include functional
agencies and a number of regional offices,
some of which have rather specialized functions.

The OAU has survived many crises in its
nearly three decades of existence, and that is

**Table 28-1** *Non-UN Inter-Governmental Peace-keeping Operations*

| Dates | Location | Sponsoring Organization or Participating States | Title | Comments |
|---|---|---|---|---|
| 1948–59 | Central America | Organization of American States | | Six small missions, some ineffective |
| May 1953–July 1954 | Burma | United States, Thailand, Burma, Nationalist China | Joint Military Committee | Supervised evacuation of Nationalist Chinese troops |
| 1954–65 | Indochina | Canada, India, Poland | International Commission for Supervision & Control | Helpful for the first few years |
| Sep. 1961–Feb. 1963 | Kuwait | League of Arab States | Arab League Security Force | Collective self-defense |
| Oct.–Nov. 1962 | Cuba | Organization of American States | Combined Quarantine Force (naval forces) | Helpful in assisting USA in Cuban Missile Crisis |
| 1963 | Cyprus | Greece, Turkey, United Kingdom | | Inadequate |
| May 1965–Aug. 1966 | Dominican Republic | Organization of American States | Inter-American Peace Force | Token forces to assist U.S. occupation |
| July 1969–Dec. 1971 | El Salvador, Honduras | Organization of American States | Committee of Seven | Intervention in the "Soccer War" |
| Jan. 1973–Apr. 1975 | Vietnam | Canada, Hungary, Indonesia, Poland | International Commission for Control & Supervision | Ineffective |
| July 1976 | El Salvador, Honduras | Organization of American States | Committee of Seven | |
| Aug. 1976 | El Salvador, Honduras | Organization of Eastern Caribbean States | | |
| 1978–1979 | Shaba Province, Zaïre | Mostly Moroccan troops | Inter-African Intervention Force | Successfully restored order after uprising |
| Dec. 1979–Mar. 1980 | Rhodesia/Zimbabwe | Commonwealth | Commonwealth Observer Group | Successfully oversaw transition to independence |
| 1980 | Uganda | Commonwealth | | Observed Elections |

| Date | Location | Organization | Force | Result |
|---|---|---|---|---|
| Nov. 1981–June 1982 | Chad | Organization of African Unity | Neutral OAU Force in Chad | Failure |
| 1982 | Lebanon | League of Arab States | Arab Deterrent Force | Token units, soon withdrawn, leaving Syrian troops in control |
| Apr. 1982–Present | Sinai | Colombia, Fiji and United States; Canada, France, Italy, Australia, Netherlands, New Zealand, United Kingdom, Uruguay | Multinational Force and Observers | Very successful |
| Aug.–Sep. 1982 | Beirut | France, Italy, United States | Multinational Force | Supervised evacuation of PLO |
| Sep. 1982–Mar. 1984 | Beirut | France, Italy, United Kingdom, United States | Multinational Force II | Failure |
| 1989–Present | Nicaragua | Organization of American States (in cooperation with the UN) | International Support and Verification Mission (CIAV) | |
| Aug. 1990–Present | Monrovia | Economic Community of West African States | ECOWAS Monitoring Group (ECOMOG) | Successful intervention in Liberian civil war |
| 1991 | Yugoslavia | European Communities | | Civilian and military observers; ineffective |
| Late 1993–Present | Tajik-Afghan border | Commonwealth of Independent States | CIS Collective Peace-keeping Forces | Russian troops with a few Uzbeks. Cooperates with the UN mission and a Russian-Kazakh-Kyrgiz force. |
| Apr.–Aug. 1994 | West Bank | Israel and the PLO agreed to let the UN Security Council invite Norway, Denmark and Italy. | Temporary International Presence in the City of Hebron (TIPH) | Provided security in city after a crazed Israeli massacred Arabs in a mosque. |

*Note:* This list is not as complete as the one for the United Nations because of the difficulty of assembling scattered bits of information, in addition to the basic problem of definition. It should be considered only indicative and not definitive. Nevertheless, it is the first and the most compete and accurate such compilation ever published. It should be noted that both the Commonwealth and OAS frequently observe or monitor elections and referenda at the request of host governments, and other regional and subregional organizations do so less frequently.

probably its most significant achievement. Beyond survival and the modest successes achieved in decolonization and peace-keeping, the OAU has so far accomplished little. There are a number of reasons for this, of which four are probably most important. First, the diversity of its membership is much greater than that of the OAS, and there is no history of cultural affinity or political cooperation to underpin current efforts. Second, all nonpolitical activities had to be subordinated to the struggles against colonialism and apartheid. Third, in the OAU Charter the members zealously guarded their newly won sovereignty by granting the organization few real powers. And fourth, the continent simply lacks the funds, technical skills, and experience to deal effectively with the other problems on a continental basis.

The latest—but assuredly not the last—crisis to plague the OAU is the question of the former Spanish Sahara (Western Sahara). After a lengthy and divisive debate, the Sahrawi Arab Democratic Republic was admitted to the OAU in 1982 as the fifty-first member, even though it is only an insurgent state. In protest, Morocco withdrew from the OAU in November 1984. Nevertheless, the OAU will very likely limp along, perhaps even gain some strength, especially since peace and stability have come at last to Ethiopia and Eritrea, South Africa, and Mozambique, and ultimately turn its attention more to economic development, health, science, education, and culture.

A regional group that straddles two continents but consists of a single dominant ethnic group is the *League of Arab States.* Among the 22 current members (including the Palestine Liberation Organization), some are more Arab than others, some have important non-Muslim minorities, and in some the Arabic language is used regularly only by a small proportion of the population. Nevertheless, the members are bound together by perceived common "race," religion, and language. But there are also many things that continue to divide the Arab peoples. Political systems differ greatly, and some Arab governments don't like the governments that rule in others. Perhaps the

greatest unifying element has been the general hostility to the State of Israel and "the forces of Zionism."

During World War I some Arab leaders hoped for the establishment of a great Arab State, but they saw their hopes destroyed by the many diverse forces that still prevail there: tribalism, poverty, illiteracy, vested interests. With the breakup of the Ottoman Empire and the elimination of Turkish control in the Middle East, Britain and France were put in charge of mandates over various parts of the Arab world. Arab nationalism arose after World War I just as African nationalism did after World War II. The French withdrew from Syria, the British from Iraq. Opposition to a British-sponsored Jewish State grew, and Arab unity began to develop.

The Arab League was founded in 1945 by Egypt, Iraq, Syria, Lebanon, Transjordan (now Jordan), Saudi Arabia, and Yemen. It has a council, a number of special committees, and a permanent secretariat, all located in Cairo until 1979 when League headquarters was transferred to Tunis as a reaction to the Israel–Egypt peace treaty. Also affiliated more or less loosely with it are the Arab Educational, Cultural and Scientific Organization; the Civil Aviation Council of Arab States; the Arab Centre for Studies of Arid Zones and Dry Lands; the Council of Arab Economic Unity, which supervises the Arab Common Market; and perhaps a dozen other groups. Prior to the Yom Kippur War of 1973 (Egypt and Syria against Israel), however, the League did little practical work in any of these fields. There was some mediation of disputes among members, and small peace-keeping forces were occasionally employed (as in Kuwait in 1961), but its energies were primarily devoted to political, economic, and propaganda warfare against Israel.

Recently, however, it has become more constructive, serving as a conduit for petrodollars from its more prosperous members to the poorer ones, sending a peace-keeping force into Lebanon (consisting almost entirely of Syrian troops), sponsoring scholarship and training programs for Palestinian Arab refugees, and engaging in some social and cultural activities.

Very different from all of these organizations is the *Council of Europe*. A product of the postwar popular movement for a united Europe, which spawned many international governmental and nongovernmental organizations in the region, it is primarily a consultative political body. It was founded in 1949 by the Benelux countries, Denmark, France, Ireland, Italy, Norway, Sweden, and the United Kingdom. Since then 23 other countries have joined, including Malta, Cyprus and Turkey, and several of the formerly socialist States of Central and Eastern Europe, making it the largest organization of its kind in Europe. Its organs are a Council of Ministers with powers of decision and of recommendation to governments, and a Parliamentary Assembly, composed of members of the parliaments of the member States. Both are served by a secretariat in Strasbourg. Its work in the human rights field is outstanding, and its European Court of Human Rights can serve as a model for similar courts elsewhere.

Its overriding function is to further European unity. In pursuance of this aim, it has concluded about 80 treaties covering particular aspects of European cooperation. It has no real power or authority, yet it has been influential in many fields and serves to some extent as the "conscience" of Europe. Its value is indicated by the fact that it has not only survived 40 years of rivalries, crises, and changes, but it is still attracting new members.

Much smaller but with deeper roots and more day-to-day activity is the *Nordic Council*. The Nordic countries have a long history of cooperation and coordination, although there has not in recent times been any strong move toward economic or political integration of Iceland, Norway, Denmark, Sweden, and Finland. So many joint institutions exist in the region that their exact number is unknown; certainly there are at least 22 permanent committees of national civil servants and 83 other permanent governmental organs besides the council itself. This is probably the most informally integrated region in the world today.

The Nordic Council was established in 1952. It has an elaborate structure consisting of a Council of Ministers, five standing committees, a secretariat (in Oslo), a cultural secretariat (in Copenhagen), and subsidiary organs. The five members' parliaments, secretariats, cabinets, and agencies all participate in its decision making and work. The council is supervised and directed by its Presidium (located in Stockholm). This largely decentralized and intricate network seems to enhance cooperation and coordination in particular fields of interest, including foreign policy, but the various links themselves are not coordinated and the members disagree on many matters. They are still very much individual "sovereign" States. Moreover, now that Sweden and Finland have joined Denmark in the EU, the future of the Nordic Council is cloudy.

The disintegration of the Soviet Union and the organizations it had created to bind it and the other socialist States together—COMECON and the Warsaw Pact (both discussed later in this chapter)—cast adrift both the former Soviet republics and the associated European States. All of them almost immediately began seeking other associations and alliances, with a distinct preference for existing European groupings. When they were not instantly welcomed into those groups, they began, individually and in clusters, to join other existing groups and to form new ones. Some had specific economic objectives; others were concerned with pragmatic cooperation in a variety of fields. Some had, and have, an air of permanence about them; others are clearly transitional, way-stations providing some shelter and succor pending admission into more substantial regional organizations. Here we can list only a few of them to illustrate the point—again: Few States today wish or are able to "go it alone."

The *Alps-Adria Working Community**\** (1979) was originally composed only of Slovenia, Croatia, and the border provinces of Italy and Austria. This early attempt to link both sides of the "Iron Curtain" is

---

*Officially, Work Community of Provinces, Regions and Republics of the Eastern Alps, commonly known as Alps-Adria.

unique because it has as members only first-order civil divisions of States, all equal. Now including Bavaria and two additional Hungarian provinces, it fosters cooperation in a wide range of activities.

The *Central European Initiative* (1989), originally a "Quadragonale" composed of Italy, Austria, Hungary, and Czechoslovakia, now has nine members: Italy, Austria, Hungary, Poland, Croatia, Slovenia, Bosnia and Herzogovina, and the Czech and Slovak republics, with Belarus, Bulgaria, Romania, and Ukraine waiting eagerly to join. It fosters cooperation in infrastructure projects, environmental protection, protection of ethnic minorities, agriculture, energy, tourism, migration, and other areas.

The *Black Sea Economic Cooperation* (1992) consists of Azerbaijan, Moldova, Russia, Romania, Turkey, Armenia, Bulgaria, Georgia, and Ukraine. These countries are also cooperating in a variety of areas, but with the expressed goal of working toward a (very distant) common market.

The *Council of the Baltic Sea States* (1992) consists of Denmark, Estonia, Finland, Germany, Latvia, Lithuania, Norway, Poland, Russia, and Sweden. Its stated purpose is "to serve as a forum for guidance and overall coordination among the participating states," without a formal institutionalized framework.

All of these groups (and some others) are functioning with the encouragement of the European Union, the European Free Trade Association, and the United Nations Economic Commission for Europe. We may expect the picture to be very different a decade from now. Meanwhile, there is no doubt that these groups have helped the Central and Eastern European countries survive a very difficult transition to democracy, free market economies, and voluntary associations with other States.

As the USSR was gasping on its deathbed in December 1991, 11 of the 15 union republics signed documents creating the *Commonwealth of Independent States*, a loose association with ill-defined objectives. In general, they were all afraid that political separation would mean economic collapse if the benefits of having been part of a single economic unit for two generations were suddenly terminated. The three Baltic States—Estonia, Latvia, and Lithuania—had been part of the USSR for a much shorter period, had experience as independent States, and were closely linked to Europe, so they felt no need or desire to remain within Russia's embrace, no matter how gentle. Georgia, beset by civil war and other problems, remained aloof until essentially blackmailed into joining in 1993. Meanwhile, the CIS

**A symbol of Pan-Europeanism.** This unusual billboard, consisting of the flags of European countries from Ireland to Russia, carries the slogan "Europe-Common House." It is located in a traffic circle in Chisinau, Moldova (formerly Kishinev, Moldavian SSR) and symbolizes Mikhail Gorbachev's proposal while he was still President of the USSR. The proposal has far more supporters in the East than in the West. (Martin Glassner)

members were gradually working out trading and other economic relations among themselves.

Dissatisfied with the slow progress in economic cooperation, however, the five Central Asian States announced the creation of a Central Asian Commonwealth in January 1993. The Slavic CIS members reacted in July by calling for the creation of their own economic union. The CIS remains in being, with some objectives and functions other than strictly economic ones, but its future is murky. Clearly, the present situation is unstable and unsatisfactory. It remains to be seen whether the centrifugal or centripetal forces operating here are the more powerful.

The *Organization of Central American States* (ODECA) was founded in 1951 as the latest of several attempts to reestablish the unity of Central America that existed in colonial times. Under its new charter of 1965, ODECA's highest authority is the Conference of Foreign Ministers, except when the Heads of Government meet. Under them are executive, legislative, and economic councils and the Central American Court of Justice. Its secretariat is in San Salvador. ODECA was deeply involved in the dispute between Honduras and El Salvador; it placed a cease-fire supervision unit on the border of the two countries in August 1976 in its first major peace-keeping operation. It remains to be seen whether ODECA will be more successful than its nineteenth-century predecessors in breaking down the parochialism of its members and forging unity among Guatemala, Honduras, Nicaragua, El Salvador, and Costa Rica.

The *Caribbean Community* (CARICOM) is an outgrowth of CARIFTA, the Caribbean Free Trade Association, which itself emerged in 1968 out of the wreckage of the Federation of the West Indies. CARIFTA's goals were quite modest, centering on the expansion of trade among the members under harmonious and equitable terms, and the balanced and progressive economic development of the economies of the area. It had a council and a secretariat, located in Georgetown, Guyana. Perhaps because of its modest goals, progress in meeting them

was good and its leaders were encouraged to take a further step toward economic integration. The decision to establish a common market and a wider community in the Caribbean was made in 1972, and they came into being with the Treaty of Chaguaramas of 1973.

Currently, CARICOM has 13 members, all of the former members of CARIFTA (which it replaced), and the Caribbean Common Market has 9 members. It has much broader objectives than CARIFTA, both economic and political, and more mechanisms to help the members reach them. They include a common external tariff, harmonization of tax systems, rationalization of agriculture, harmonization of monetary policies, and joint action in relation to industrial development programs, transport, and tourism. The members also attempt to coordinate their political policies in various international fora and conferences.

The structure of CARICOM is relatively simple. Below the three principal organs are the institutions responsible for various functional areas, such as health, agriculture, and foreign affairs. It has associated institutions, much like the specialized agencies of the United Nations. These include two regional universities, the East Caribbean Common Market, the Regional Shipping Council, and five others. Finally, there are the Community Secretariat (located in Georgetown) and many integration institutions that predate CARICOM. There are two recognized groups of members: the more developed countries (MDCs) are Jamaica, Trinidad and Tobago, Barbados, and Guyana; the rest are the less-developed countries (LDCs). Despite all efforts to equalize benefits received from integration by these two groups, it seems that most still gravitate toward the MDCs. In addition, the same centrifugal forces that destroyed the West Indies Federation are still at work, and the centripetal forces of shared interests and viewpoints and a common Caribbean identity are still not very strong. While there are some hopeful signs that CARICOM will fulfill its promise and unify the Commonwealth Caribbean (and perhaps other Caribbean countries as well),* the obstacles to unity are great. As the late Errol Barrow, then Prime Minister of Barbados, observed in 1964, "We live together very well, but we don't like to live together together."

In 1981 six members of the Arab League formed the Cooperation Council for the Arab States of the Gulf, more commonly known as the *Gulf Cooperation Council* (GCC). The six—Bahrain, Kuwait, Oman, Qatar, Saudi Arabia, and the United Arab Emirates—are all oil rich (except for Oman) and ruled by conservative monarchs. While officially dedicated to a broad program of cooperation in trade, industry, agriculture, transport, and a variety of social and cultural affairs, there is little doubt that the GCC's primary interest is in common defense against external threats, most immediately from the fanatically Shiite revolutionary regime of Iran. Many of its activities seem designed to strengthen the partners against a very powerful Iranian challenge, and it was simply unable to act constructively after one of its members, Kuwait, was annexed by Iraq in August 1990. During the Gulf War that followed, Saudi Arabia was a major participant, Bahrain was an important staging base, and the smaller members made token contributions to the campaign to liberate Kuwait, but the GCC did nothing at all. Even after the war, the GCC was unable to move significantly on the matter of joint security. At the annual "summit" meeting in December 1991, the members simply agreed to "pursue" the matter. It seems that they prefer to rely on third parties rather than one another for their security. Having failed its first real test, the GCC does not have a very promising future.

In February 1989, the *Arab Co-operation Council* (ACC) was founded by Iraq, Egypt, Yemen, and Jordan with the aim of achieving coordination in agriculture, economics, finance, trade, and industry. It is too early to tell whether it will ever achieve anything, but it may very well be one more casualty of the Gulf War and inter-Arab rivalries generally.

There is no organization in Asia comparable to the OAS, the OAU, or the Arab League. The one nearest to it is the *Association of Southeast Asian Nations* (ASEAN). Founded in 1967 by Indonesia, Malaysia, the Philippines, Thailand, and Singapore, it was joined by newly independent Brunei in 1984. ASEAN replaced the Association of Southeast Asia (Malaysia, the Philippines, and Thailand) and the proposed Maphilindo (Malaysia, Philippines, Indonesia) and has displaced the Asia and Pacific Council (ASPAC), which, with nine members, was the largest Asian political organization. As in Africa, many others have been proposed or formed but few have survived for any length of time. Founded originally as a device for economic cooperation, ASEAN accomplished very little and remained low-key until after the communist takeover of Indochina in 1975. Since then its political, social, and cultural activities have become more important, and the organization as a whole has "come alive." It still has no charter, however, and a central secretariat was finally established in Jakarta only in 1976. It has a number of committees composed of various officials of ASEAN members, most of which have subsidiary bodies.

Its major organ has been its annual conference of foreign ministers, and its heads of government did not meet until 1976. Its committees consider, among other things, matters relating to communications, science and technology, transport, food production, and commerce and industry. There is as yet no dispute-settlement machinery, nor is there, as some members want, a common policy for having the region designated a "zone of peace" or for standing staunchly against communism.

Its chief activities remain economic, but although there have been many agreements in principle and statements of intent, there have been few accomplishments in economic matters. Originally drawn together by common fears and weaknesses, ASEAN was long prevented by mutual distrust and narrow nationalism from achieving its full potential. The distrust seems to be fading after two decades of working together, and

---

*One hopeful sign is the reduction of the common external tariff in 1995 from 45% to 35% and its scheduled reduction to 20% in 1998.

ASEAN has begun taking positions and actions in regard to such matters as energy and Indochinese refugees.

More important, it has recently been moving forward on several other fronts. After the failure of the ASEAN Preferential Trading Arrangements (1976) to expand trade among the members, they agreed in 1992 to work toward a free trade area, to be effective by 2007. Also in 1992 they signed a Treaty of Amity and Cooperation in Southeast Asia, signalling an intensified concern with regional security. Then in 1993 they established the ASEAN Regional Forum, which held its first meeting in Bangkok in 1994. In addition to the six members, the Forum includes Australia, New Zealand, Canada, the European Union, South Korea, Japan, China, the United States, and Russia, with Papua New Guinea, Laos, and Vietnam as observers. How the regional security situation has changed in less than 20 years! Not only is Vietnam, once the major nemesis of ASEAN, now an observer, but it is scheduled to become a full member in 1995. ASEAN must now be ranked as one of the world's most successful subregional organizations.

The newest major subregional organization of general competence is the *South Asian Association for Regional Cooperation* (SAARC). It was formally launched in August 1983 on the separate initiatives of Bangladesh and Nepal, and its heads of government first met in Dhaka in December 1985. Its history suggests that the leaders of its seven members—Bangladesh, Bhutan, India, Maldives, Nepal, Pakistan, and Sri Lanka—have approached the novel adventure with varying degrees of optimism and caution. These countries, after all, like those farther east, have no history of regional cooperation, except through the Colombo Plan and ESCAP, both of which are very much larger organizations in which the differences among them can be more readily camouflaged or ignored.

SAARC is organized with a Council of Ministers as its highest policy-making body, a Standing Committee, a Programming Committee, a secretariat, and 12 technical committees responsible for such matters as

meteorology, sports, science, transport, agriculture, health, rural development, telecommunications, culture, arts, and postal services. Its headquarters was inaugurated in Kathmandu, Nepal, in January 1987, and a Bangladeshi diplomat took over as Secretary-General for a two-year term. Since then the organization has been encouragingly active. It has adopted a Regional Convention on the Suppression of Terrorism and one on narcotic drugs and psychotropic substances; established a food security reserve; sponsored a number of specialized conferences; initiated major studies on regional problems, including drugs; and is exploring other possibilities for regional cooperation. Economic integration and coordination of foreign policies, however, were not even on the agenda until 1991, and not until 1993 was a rather timid SAARC Preferential Trading Arrangement adopted.

Considering the deep and long-standing frictions among some of the members and the difficulty of some of their bilateral problems, it is remarkable that the organization has come into being at all. It is unusual among such subregional groupings in that it is so completely dominated by one country, India—one, moreover, that has not been particularly enthusiastic about it. The degree of its ultimate success may very well be proportional to the degree of participation and support by India.

In the South Pacific there are two major regional organizations—the *South Pacific Commission* (SPC) and the *South Pacific Forum*—and several smaller ones. The Commission is the older and larger. It originated in 1947 when Australia, France, New Zealand, the Netherlands, the United Kingdom, and the United States signed the Canberra Agreement. The Netherlands withdrew in 1962 after giving up its half of New Guinea. Since then the organization has evolved dramatically. It now has 27 members of equal status, 22 island States and territories, and 5 metropolitan countries. It is now an international technical assistance agency with an advisory and consultative role, providing information, training, assistance, and advice in many social, economic,

**Southern African Development
Community
Gaborone, Botswana**

**South Asian Association for Regional
Cooperation
Kathmandu, Nepal**

**Common Market for Eastern and Southern
Africa, Lusaka, Zambia**

**Council of the Entente, Abidjan, Côte d'Ivoire**

and cultural fields. It does not, however, have any political functions or coloration, it does not operate large aid programs or common services, and it has only a modest secretariat staff in Noumea, New Caledonia, and small operations in Sydney, Australia, and Suva, Fiji.

The Commission is guided by the annual South Pacific Conference, its decision-making body, in which every member has one vote. In November 1986 it concluded the Convention for the Protection of the Natural Resources and Environment of the South Pacific Region, a potentially important step to-

**Organization of Eastern Caribbean States
Castries, St. Lucia**

**Council for Mutual Economic Assistance
Moscow, Russian Federation**

**Headquarters buildings of six subregional organizations.** (All photographs by Martin Glassner).

ward preserving a fragile environment in danger of destruction. All in all, the Commission seems to have profited from the mistakes of the Caribbean Commission (later called the Caribbean Organization and now defunct), and it has tried to avoid being viewed as an instrument of neocolonialism. It is likely to continue providing valuable services and stimulating a regional consciousness among the peoples of the region.

The South Pacific Forum is very different. It is the annual meeting of heads of government of the independent and self-governing States of the South Pacific, currently totaling 15. It has no formal charter or rules of procedure, and functions by consensus. It provides an opportunity for informal discussions on a wide range of issues and problems. It has tended to emphasize political and security matters since its founding in

1971, including such thorny issues as independence for New Caledonia, French nuclear testing in French Polynesia, and apartheid in South Africa. It endorsed the establishment of the South Pacific Nuclear-Free Zone and a South Pacific Regional Environment Programme, and it deals in various ways with fisheries, petroleum, shipping, telecommunications, and trade problems.

The Forum Secretariat is located in Suva, Fiji. Its aim is to facilitate cooperation and consultation among the members on trade, economic development, tourism, and related matters. In January 1987 the South Pacific Regional Trade and Economic Cooperation Agreement (SPARTECA) entered into force. Under its terms Australia and New Zealand grant many concessions to imports from the other members in order to help redress a chronic trade imbalance. The Secretariat is

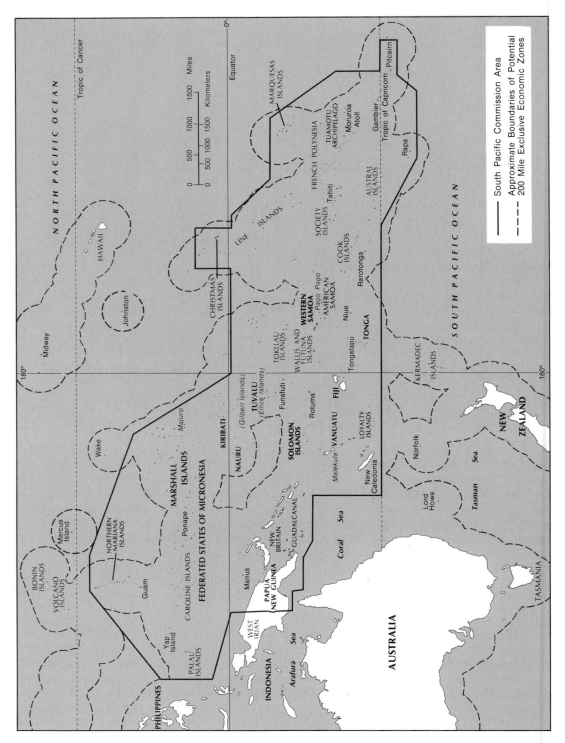

**The jurisdiction of the South Pacific Commission.** The outer limit is not a boundary; it simply indicates the area within which the islands participate in activities of the Commission.

also heavily involved in regional transport services, energy, disaster relief, telecommunications, fisheries, and other matters. It is gradually increasing both the scope and importance of its activities. Other important regional organizations include the Forum Fisheries Agency and the South Pacific Applied Geoscience Commission. Generally, it is fair to say that the South Pacific organizations have been quite successful and have served their people well.

Africa has not yet developed major organizations of general competence other than the OAU. Instead there is a large number of subregional organizations, many with overlapping memberships and functions. In time these will probably evolve into fewer and stronger bodies; meanwhile they are difficult to follow. The chart of selected African subregional organizations shows the names and memberships of some of these groups.

## *Economic Organizations*

We have devoted a whole chapter to organizations designed to foster economic integration in several regions and have discussed several devoted to development of international rivers and lakes, but there are many other organizations with different economic goals. Among them are a large number of regional and subregional development banks and institutions for financial cooperation. Instead of trying to cover all of them, we discuss briefly only three of the more prominent ones and mention two new ones in an attempt to convey their variety and significance.

We discussed *COMECON* in Chapter 25, in connection with international trade. Here we look at it somewhat more comprehensively. The *Council for Mutual Economic Assistance* (or CMEA or CEMA) was founded in 1949 largely as a response to the Marshall Plan and the OEEC. Its headquarters was in Moscow. Original members were the Soviet Union and the "socialist" countries of Eastern Europe, with Mongolia and North Korea as observers. Albania was expelled (*de facto*) in 1961, Mongolia was admitted in

1962, Yugoslavia became an associate in 1964, Cuba and Vietnam were admitted in 1972 and 1978, respectively.* It was not designed as a device for economic integration; in fact, it had no clear, detailed objective or plan and went through a number of stages, each characterized by a different emphasis and organization.

At first Stalin tried to use it for classic colonial purposes: to exploit the resources of each country for the benefit of the Soviet Union, sending them Soviet products in return at very high prices, all on a barter basis. This gave way to an attempt to make each of the Eastern European countries a miniature replica of the USSR, with its own basic and subsidiary industries, agricultural system, and so on. The organization did not really become active until 1956, when it shifted its emphasis to coordination of the individual economic plans of its members. This evolved in the early 1960s into a plan for "the international socialist division of labor," with each country assigned to produce certain specialties for the whole group. In 1964 Romania rebelled against this plan, declaring that it "did not want to become the vegetable garden of Eastern Europe." This led to other rebellions and a failure of the specialization scheme in particular and multilateral approaches in general. Emphasis shifted to bilateral trading agreements with some efforts at coordination, but with several members undertaking their own internal reforms, this became increasingly difficult.

After the Warsaw Pact invasion of Czechoslovakia in August 1968 and Soviet installation of a more compliant government there, COMECON members reacted more strongly than ever against any supranationalism (except for the Soviet Union), but there was still no agreement on what COMECON should be. As it bogged down in drift, delay, and debate, members turned more and more to trade with the West. In the summer of 1971 the council agreed on a program of "integration," which really meant closer cooperation and coordination of economic activities

*Observers in 1987 were Angola, China, Ethiopia, North Korea, and Laos.

## Selected African Subregional Organizations

| Country | General Competence | | | | | | Predominantly | | | | Economic | | | | | | Development Oriented | | | | | | | Other | |
|---|---|---|---|---|---|---|---|---|---|---|---|---|---|---|---|---|---|---|---|---|---|---|---|---|---|
| | African & Mauritian Common Organization* | African & Malagache Union | Mano River Union | Council of the Entente | Union of Central African States | Organization of the Senegal River States | Economic Community of West African States | Common Market for Eastern and Southern Africa | Customs Union of West African States | Economic Community of the Great Lakes | Central African Customs & Economic Union | Equatorial Customs Union | Customs Union of Southern Africa | Economic Community of the States of Central Africa | West African Economic Union | African Petroleum Producers Association† | Southern African Development Community | Organization for the Development of the Senegal River | Authority for Integrated Development of Liptako-Gourma | Organization for the Development of the Gambia River | Niger Basin Authority | Lake Chad Basin Commission | Kagera Basin Organization | CILSS (Club du Sahel) | Intergovernmental Authority on Drought and Development†† |
| Angola | | | | | | | | X | | | | | | | | X | X | | | | | | | | |
| Benin | X | X | | X | | | X | | X | | | | | | X | X | | | | | X | | | | |
| Botswana | | | | | | | | | | | | | X | | | | X | | | | | | | | |
| Burkina Faso | X | X | | X | | | X | | X | | | | | | X | | | | X | | X | | | X | |
| Burundi | | | | | | | | X | | X | | | | X | | | | | | | | | X | | |
| Cameroon | | X | | | | | | | | | X | X | | X | | X | | | | | X | X | | | |
| Cape Verde | | | | | | | X | | | | | | | | | | | | | | | | | X | |
| Central African Republic | X | X | | | | | | | | | X | X | | X | | | | | | | | X | | | |
| Chad | X | X | | | X | | | | | | X | X | | X | | | | | | | X | X | | X | |
| Comoros | | | | | | | | X | | | | | | | | | | | | | | | | | |
| Congo | X | X | | | | | | | | | X | X | | X | | | | | | | | | | | |
| Côte d'Ivoire | X | X | | X | | | X | | X | | | | | | X | X | | | | | X | | | | |
| Djibouti | | | | | | | | X | | | | | | | | | | | | | | | | | X |
| Equatorial Guinea | | | | | | | | | | | X | | | X | | | | | | | | | | | |
| Eritrea | | | | | | | | X | | | | | | | | | | | | | | | | | X |
| Ethiopia | | | | | | | | X | | | | | | | | | | | | | | | | | X |
| Gabon | X | X | | | | | | | | | X | X | | X | | X | | | | | | | | | |
| The Gambia | | | | | | | X | | | | | | | | | | | | | X | | | | X | |
| Ghana | | | | | | | X | | | | | | | | | | | | | | | | | | |

|  |
|---|
| Guinea |
| Guinea-Bissau |
| Kenya |
| Lesotho |
| Liberia |
| Madagascar |
| Malaŵi |
| Mali |
| Mauritania |
| Mauritius |
| Mozambique |
| Namibia |
| Niger |
| Nigeria |
| Rwanda |
| São Tomé & Príncipe |
| Senegal |
| Sierra Leone |
| Somalia |
| South Africa |
| Swaziland |
| Tanzania |
| Togo |
| Uganda |
| Zaïre |
| Zambia |
| Zimbabwe |

N.B. The classification of these organizations is somewhat arbitrary; in reality, many of them have multiple functions, many of which are not actually performed. Some of the organizations, in fact, are moribund, whereas there are many smaller organizations that are active. In addition, there are many subregional banks and other financial institutions. Clearly, there is a need for joint action among African States.

*Dissolved in 1985.

†Also includes Algeria, Egypt, and Libya.

††Also includes Sudan.

rather than a merger of economies as already described.

COMECON functioned thereafter primarily as an instrument to encourage cooperation and coordination among its members, but each member developed its own economy independently. As their communist governments were replaced by democratic ones, however, emphasis shifted more toward trade with the West, but this trade had to be limited because of a number of very real problems. Finally, COMECON was formally dissolved in 1990.

The *Colombo Plan for Cooperative Economic and Social Development in Asia and the Pacific* was established in 1951 as a result of a meeting of Commonwealth foreign ministers held in Colombo in 1950. The original Asian members were British Borneo and Malaya (now joined into Malaysia), Cambodia, Ceylon (now Sri Lanka), India, Laos, Pakistan, Singapore, and Vietnam (no longer a member). They have since been joined by 13 others.* There are six extraregional members, which are the primary donors of aid: the United Kingdom, the United States, Japan, Canada, Australia, and New Zealand. (Israel was also an active member for a time.) Its purpose is explained in its formal title, but this may be somewhat misleading. The Colombo Plan is not a master plan at all but rather a mechanism for review and coordination of bilateral aid programs and projects worked out between donor and recipient countries. Its highest body is its Consultative Committee, which meets biennially in different countries. The Colombo Plan works closely with ESCAP, with which it has been a partner in the Asian Highway and the Lower Mekong Basin scheme, and with other international and regional organizations.

Britain has been the leader and chief donor of funds, but the other donors have also contributed substantially. A most encouraging feature is that some recipients, notably India, the Philippines, Indonesia, South Korea, Pakistan, Singapore, and Thailand, have become donors as well. Projects include large multipurpose dams and other infrastructure elements, but most aid is in the form of technical assistance and training. Most training takes place outside the region, but local facilities are being developed and increasingly used. In addition to four specific types of training activities, the secretariat has recently added assistance to member countries in the area of natural disasters and environmental degradation.

In addition to the Consultative Committee, the Colombo Plan's organs are the Council, consisting of the diplomatic representatives of members who are resident in Colombo and who advise the Consultative Committee and monitor the implementation of its decisions by two other organs; the Bureau, which serves as the secretariat; and the Drug Advisory Programme, a leader in the regional approach to fighting drug abuse. Finally, there is the Staff College for Technical Education in Manila.

Although the numbers of participants are roughly equal, it would be unreasonable to expect the Colombo Plan to have had the same salutary effect that the OEEC had in Europe because of the very different conditions in the two regions. But the very fact that the food situation has improved somewhat has been a great achievement, and many of the Colombo Plan States have irrigation works, cement factories, and other industrial concerns to show for their participation. Its quiet success is due to its businesslike, nonpolitical functioning, as indicated by the great diversity in its membership and its durability. Like the Commonwealth, it is a splendid example of a diverse group of countries doing quiet, constructive work with very little publicity. It deserves more.

A subregional organization that has shown remarkable imagination, flexibility, achievement, and durability under great stress is the *Southern African Development Community* (SADC). Founded in 1980 as the *Southern African Development Coordination*

*Afghanistan, Bangladesh, Bhutan, Fiji, Indonesia, Iran, South Korea, Maldives, Myanmar, Nepal, Papua New Guinea, the Philippines, and Thailand.

*Conference* (SADCC) as an instrument for reducing the dependency of the independent States of southern Africa on the Republic of South Africa, it has since grown and matured and expanded its vision. It now consists of all nine of the former British and Portuguese territories in the subregion plus Namibia, which joined immediately after attaining independence, and South Africa, which joined in August 1994.

SADC is unusual in another respect. It is organized on a sectoral basis, with each member being responsible for one or more sectors of the economy and society. These include such matters as agriculture, energy, health, science and technology, education, industry and so on. Each has a formal organization located in the country's capital to do the actual work. The initial emphasis was placed on transport and communications, the responsibility of Mozambique. The Southern Africa Transport and Communications Commission, located in Maputo, played a major role in neutralizing South Africa's destabilizing and disruptive practices in these areas, largely by developing alternative routes to and from the sea for its six land-locked members. Nearly all of the problems of land-locked States, reviewed in Chapter 29, have been experienced here, and here some very creative ways of dealing with them are being worked out.

SADCC was not aiming for economic integration, but rather for rational economic and social development throughout its area. However, much of the groundwork for closer association had been laid, and in August 1992 it was transformed into SADC, a much more formal organization with more ambitious objectives. With peace in the region (except in Angola), the participation of South Africa, and a solid foundation of experience, the future for SADC looks bright.

The *Regional Cooperation for Development* (RCD), which consisted of Turkey, Iran, and Pakistan, was established in 1964 and did some good work in such fields as shipping, telecommunications, manufacturing and tourism. But after the Iranian Revolution early in 1979, the Soviet invasion of

Afghanistan near the end of that year, and political instability in Pakistan, the organization became dormant.

In 1985 it was revived by the same three countries. Restructured at that time and again in 1990, it now has a Council of Ministers, a Council of Deputies, a Regional Planning Council, seven Technical Committees, and a secretariat, located in Tehran. Given a new name, the *Economic Cooperation Organization* (ECO), to reflect its new orientation, it is now active in investment and development banking, commercial banking, trade expansion through a free trade area (ultimately), joint industrial ventures, shipping, telecommunications, civil aviation, training, tourism, agriculture, scientific cooperation, a data bank, and a South West Asia Postal Union. Its sudden expansion in 1992 by the admission of Afghanistan, Azerbaijan, and the five former Soviet Central Asian republics has created a major group of 300 million people and very considerable resources; it is likely to become a significant force in a most strategic but unstable region.

Finally, two relatively new organizations should be mentioned. First is the *Arab Maghreb Union* (UMA), introduced in Chapter 26. It is the culmination of three decades of rhetoric about Maghreb unity, and a decade or more of dreaming and planning before that. The members are Morocco, Algeria, and Tunisia (the core of the Maghreb), plus Libya and Mauritania. Like the ECO, it has great ambitions, not only in the scope of its activities, but also in its open-ended membership. It is no accident that when sounded out, its French abbreviation sounds much like the Arabic word *ummah*, meaning the whole of the Muslim nation. It is too early to tell if this will be the first successful economic grouping in the Arab world.

Very different is an even newer organization, the *Asia–Pacific Economic Cooperation* (APEC), first proposed in 1979 by Australia and finally born in November 1991. It has at present 17 members, all located around the Pacific Rim, with the possibility of more later. Although it includes the United States, Japan, China, and a dozen smaller countries,

**Regional and subregional organizations on stamps.** Since these organizations are so important to so many countries, it's not surprising that many of them are honored on stamps. This selection illustrates the Council of Europe, UN Economic Commission for Africa, South Pacific Commission, League of Arab States, Organization of African Unity, Regional Cooperation for Development (defunct), Organization of Senegal River States, African Development Bank, Cooperation Council for the Arab States of the Gulf, Organization of American States, North Atlantic Treaty Organization, Mano River Union, Council of the Entente, Colombo Plan, Association of Southeast Asian Nations, South Asian Association for Regional Cooperation, UN Economic Commission for Latin America and the Caribbean, Southeast Asia Treaty Organization (defunct), Southern African Development Coordination Conference (now SADC), Malaysia-Philippines-Indonesia (defunct), UN Economic Commission for Asia and the Far East (now ESCAP), European Organization for Nuclear Research, and African and Malgache Union. (Martin Glassner)

its goals are still unclear. Apparently, it is designed to support ASEAN but thwart any new economic groupings in the region. Whatever its ultimate direction and fate, it seems to be the first step in development of the Pacific Basin as a new major focus of world-class political and economic activity, as we suggested in Chapter 23.

## *Military Alliances*

It is very likely that the first inter-State organization in history was a military alliance. Since then there have been hundreds of them, perhaps thousands. Our own observations of contemporary alliances reveal again and again the essential truth of the old adage that a State has no permanent friends or permanent enemies, only permanent interests. But even a cursory glance at the current world situation should make one wonder if States even know what their permanent interests are!

The array of alliances developed for the bipolar world of the 1940s and 1950s fell into disarray during the 1960s and 1970s as new centers of power emerged. China, Western Europe, Japan, and the Group of 77 have all become powers in their own right because of their economic, demographic, numerical, intellectual, or moral strength, of necessity diluting the dominance of the United States and the Soviet Union and distracting them from their concentration on each other. More recently, the collapse of communism and its major institutions has made even NATO of the Cold War days obsolete, and the Warsaw Pact was dissolved in 1991. Nevertheless, they deserve brief reviews.

The *North Atlantic Treaty Organization* (NATO) was created in 1949 in Washington by the United States, Canada, and ten Western European countries. It was joined by Greece and Turkey in 1952, the Federal Republic of Germany in 1955, and later Portugal and Spain. Unlike the military alliances created before 1949, it has its own military force composed of national units from all member countries (except Iceland, which has no military forces) under unified command. It is a formidable military power, by far the most powerful in the world. Unlike most other military alliances, it has a *raison d' être* stronger than that of simply uniting for collective security. Most of the members have a real community of interest expressed in numerous institutions, traditions, and organizations, some of which we have already discussed. How long will it last? No one can predict, of course, but probably for some time, although its character will change.

The *Warsaw Treaty Organization* (Warsaw Pact) was created in May 1955 by the Soviet Union and its Eastern European satellites, not as a reaction to NATO, as some people believe, but in response to the rearming of West Germany and its admission into NATO, which they perceived as a real potential threat. It functioned much like NATO and had headquarters in Moscow. Albania, because of its alignment with China after the Sino-Soviet split, was suspended in 1962 and withdrew in 1968. Unlike NATO, the Warsaw Pact engaged in military action—though not of the kind envisaged in 1955—when token forces from East Germany, Poland, Bulgaria, and Hungary (but not Romania) accompanied the Soviet forces that invaded Czechoslovakia in August 1968 and crushed the new liberal communist government of Alexander Dubcek. After that the Warsaw Pact continued many of its activities, but at a reduced level and with little apparent enthusiasm. It was formally terminated in 1991.

The *Western European Union* (WEU; Benelux, France, Britain, Italy, and West Germany) was formed in 1955 as an outgrowth of the 1948 Brussels Treaty of collaboration in economic, social, and cultural matters and for collective self-defense. Its principal purpose was to oversee West Germany's adherence to its agreement to limit production of armaments. Its social and cultural activities were transferred to the Council of Europe in 1960, and its military functions have been incorporated into those of NATO. It still performs some minor eco-

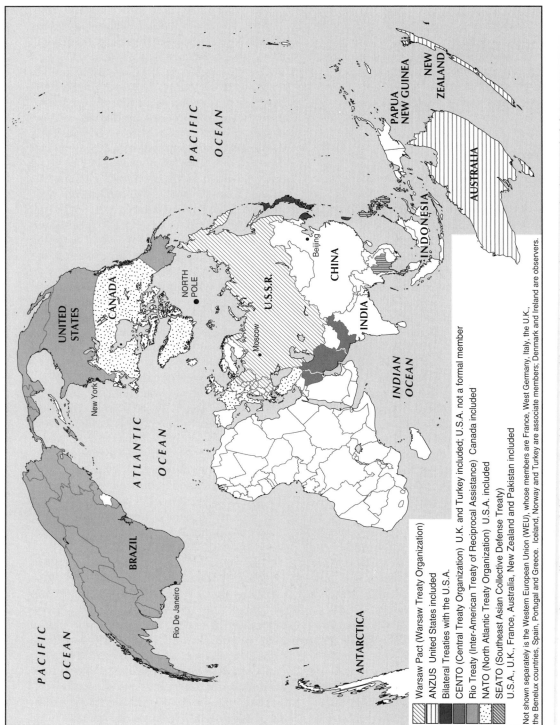

**Military alliances of the post–World War II world.** Note how they almost completely encircled the USSR. SEATO, CENTO, and the Warsaw Pact are defunct now, the U.S. alliance with Taiwan has been terminated, and New Zealand is no longer active in ANZUS.

Warsaw Pact (Warsaw Treaty Organization)

ANZUS  United States included

Bilateral Treaties with the U.S.A.

CENTO (Central Treaty Organization)  U.K. and Turkey included; U.S.A. not a formal member

Rio Treaty (Inter-American Treaty of Reciprocal Assistance)  Canada included

NATO (North Atlantic Treaty Organization)  U.S.A. included

SEATO (Southeast Asian Collective Defense Treaty)
U.S.A., U.K., France, Australia, New Zealand and Pakistan included

Not shown separately is the Western European Union (WEU), whose members are France, West Germany, Italy, the U.K., the Benelux countries, Spain, Portugal and Greece.  Iceland, Norway and Turkey are associate members; Denmark and Ireland are observers.

nomic and political functions. Now that Germany has been reunited and there is no longer a major military threat looming in the East, WEU has been suggested as a replacement for NATO since it is more readily adaptable to expanded economic and political functions than a scaled-down NATO. Since this would also have the effect of excluding the United States and Canada from active participation in these activities, it seems unlikely that this suggestion will be adopted very soon. It will remain in the wings, however.

It is possible that both NATO and the Warsaw Pact will be replaced by the *Organization for Security and Cooperation in Europe,* originally a very loose association of the 35 countries that gathered in Helsinki in July 1975 to try to develop nonmilitary means of maintaining peace and human rights in all of Europe—East, Central, and West. In the spring of 1990 the group met in Bonn and agreed to shift their emphasis to economic activities based on free markets and to strengthening democracy. By January 1992, 13 of the 15 former Soviet union republics had joined, and there was more serious discussion of formalizing and strengthening the group. It has gradually developed an institutional structure and has dispatched numerous observer, fact-finding, or conflict-preventing missions to various trouble spots in the former Soviet Union and Macedonia. Its High Commissioner for National Minorities has begun to achieve respect. In December 1994, its name was changed from "Conference on Security…" to "Organization for…," a sign of its growing confidence and authority. It could ultimately evolve into a truly pan-European organization.

## *Analysis*

What does all this mean? What can we learn from this lengthy survey of dozens of intergovernmental organizations? The principal lesson to be learned is probably that nearly all countries in the world today recognize

that they cannot stand alone, no matter how rich or powerful, or how poor and isolated. As we have seen, many organizations have been unable to withstand various stresses or have been left behind by the march of events and have died, only to be resurrected in a somewhat different form, stronger than before. We may even be observing a progression of the world from interdependency into an era of regionalization.

It is easy to exaggerate the importance of all of these organizations, to focus on form and ignore substance. Merely joining an organization does not automatically render a State peaceful, democratic, extroverted, or wise. Yet international law and custom do place upon States the obligation to honor their undertakings, if freely accepted. There is, in fact, a tendency for rogue or aberrant States to be socialized somewhat by the discipline of peaceful group interaction. All States, moreover, gain much by working together with others to achieve common goals. New States especially find the group helpful in mitigating the shock of birth and the inevitable postpartum stresses.

But there is a potential danger lurking in this growing regionalization. National rivalries could be replaced by regional rivalries, bellicose nations or States could carry regional partners along the path of aggression, and we could have bigger and far more destructive wars than ever. This danger, however, is well understood, and there are mechanisms for avoiding it, many of which are covered in this book. There is one more that should be noted here: Many countries belong to more than one group, and therefore share both responsibilities and rewards with different partners. Canada and the United States, for example, form a kind of global geopolitical hub, joined as they are with States to the west across the Pacific, to the south all the way to Cape Horn, to the north across the frozen Arctic, and to the east across the Atlantic and all of Europe to the old Curzon Line and perhaps beyond.

Larger States and those that are centrally located may have larger interests and farther horizons than smaller and more remote ones, but there have always been important exceptions to this generalization. China, for example, has always tended to be introverted, while New Zealand has been internationalist nearly all through its history. We have observed before that no place on earth is really remote any longer; there is no place to hide. "No man is an island," wrote John Donne in his immortal poem *Devotions* nearly 400 years ago, "entire of itself." Today the same can be said of States. No State is an island, not even an island State.

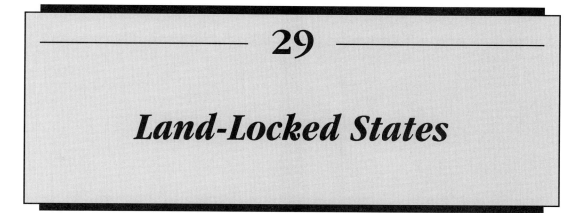

# 29

# *Land-Locked States*

In Part II we examined many characteristics of States, including size, shape, location, and boundaries. Another characteristic of some States is land-lockedness, the complete absence of a seacoast. This characteristic is shared by interior civil divisions of many coastal States, of course, but their problems are largely economic and technical and thus of little interest to *political* geographers. A land-locked *State*, however, faces many problems deriving from the fact that an *international* boundary lies between it and the sea. Land-lockedness, then, is an international issue, and one of the more perplexing ones today. There are two aspects to the problem: access to and from the sea, and use of the sea, especially its resources.

Countries with very short coastlines, such as Zaïre, Jordan, and Iraq, often have problems of access, but at least they do have some sovereign territory along the sea that they can utilize constructively. Truly land-locked States have no choice but to transit the territory of another country en route to and from the sea. If the transit State has adequate transportation and port facilities and if it is friendly and cooperative, this need not be a problem. But frequently, for one reason or another, transit is taxed, impeded, curtailed, or suspended, and the interior State is nearly or completely cut off from the world.

If we count Vatican City as a State, then there are at present 42 land-locked States in the world.* Of the 42 States, 15 are in Eu-

rope and are rather well-to-do by world standards. Of the remaining 27, which are all poor or "developing" countries, 15 are among the 47 "least developed" countries by UN reckoning, the poorest of the poor. It is no accident that a third of the world's poorest States have no coastline. A location in the interior of a continent, especially one as large as Africa or Asia, isolates a country from the main world flows of goods, people, and ideas. Interior countries lack the "window on the world" that a seaport provides. Once isolation was valued, for it provided some measure of security and protection against a wide range of dangers from hurricanes and tsunamis to pirates and foreign invaders. Now, however, isolation is a disadvantage.

## *Characteristics of Land-Locked States*

Those land-locked States closest to the sea are located mostly in Europe, smallest of all the continents and the one with the best circulation system. Only Bolivia, Lesotho, Swazi-

*Tibet and Sikkim, formerly at least quasi-States, have been absorbed by China and India, respectively. The former Indian state of Jammu and Kashmir has been divided between India and Pakistan and has no international personality. If Montenegro separates from Serbia, as Eritrea did from Ethiopia, then Serbia, like Ethiopia, will revert to its former land-locked status.

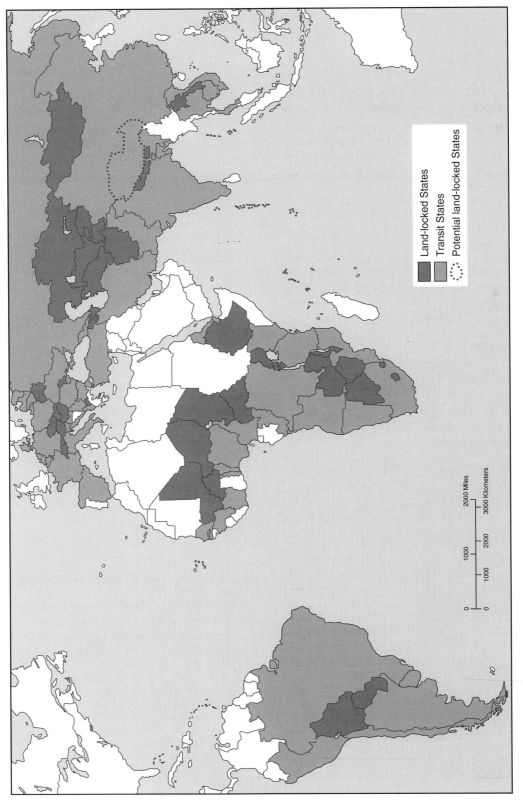

**Land-locked and transit States.**

land and Malawi on the other continents can be considered reasonably close to the sea. But this advantage is largely canceled out for these States by high transportation costs resulting from difficult terrain or sparsely settled country that neither produces nor consumes very much.

Typically, partly because of topographic features, partly because of the poverty of most land-locked *and* transit States, the transport systems used by land-locked States are inadequate in facilities, maintenance, and management. Frequently, there are serious imbalances in the direction or seasonal flow of goods as well. Communications, which are essential to keep transit traffic flowing smoothly, often are inadequate or unreliable, making traffic management very difficult. Delays in transit caused by these circumstances not only increase such costs as insurance, storage, interest on loans, penalties, and similar direct charges, but also increase the risk— and the actuality—of loss, theft, damage, and deterioration of goods. Merchants at both ends of a trading relationship can lose a great deal of money when delays in transit mean goods arrive too late to be useful or saleable. Construction projects can be delayed and machinery and vehicles immobilized, all for want of materials, equipment, or parts that are lost, damaged, or delayed in transit.

Of all the land-locked States outside Europe, all but three have three or more neighbors with which they must remain on good terms if they are to be reasonably certain of free access to and from the sea. The more coastal neighbors a land-locked country has, the wider a choice of routes to the sea it should have. In practice, though, few really have more than one or two practicable routes, and only Bolivia and Zambia (and to a lesser extent Afghanistan and Rwanda) can readily choose between ports on two oceans. Some land-locked countries have land-locked neighbors, which might seem to be something of an asset, but the existence of more than one State between the land-locked State and the sea merely complicates the situation.

Even more important is the distance and nature of the terrain between the *economic* core of a land-locked State and the core areas of its neighbors or between its core and a good seaport. In some cases, notably in South America and West Africa, the core area of a land-locked State is closer to that of another State or to a seaport than to the extremities of the State itself. This is largely due to the European colonial practice of providing transportation and communications between coastal ports and the interior to facilitate the export of cash crops, minerals, ivory, and slaves. It was and still is helpful in terms of access to the sea, but is an overall weakness of too many African land-locked States; this weakness is a distinct handicap in bargaining for the right to *use* these facilities.

**Uganda as a Transit State.** This truck, clearly marked "Transit Goods," is one of many plying the roads of Uganda connecting Rwanda, Burundi, and Shaba and Kivu provinces of Zaïre with Mombasa, Kenya. Like many other land-locked countries, Uganda provides transit routes and facilities for both land-locked and transit States. (Martin Glassner)

All the land-locked States may be considered militarily weak, even compared with middle-rank powers such as Canada or Israel. Few of them are truly modern, united nation-states in complete control of their national territory and guided by the State idea. Fewer still outside of Europe are economically strong. Thus, although some of the land-locked States are useful to the larger coastal States because of their location, minerals, or votes in the United Nations, in terms of the traditional indicators of national power we discussed earlier, they are uniformly weak. Their bargaining power in bilateral negotiations with transit States derives principally from international law and the possibilities of reciprocity and mutual economic benefits. In multilateral negotiations they could have the advantage of being able to present a solid front but, as we demonstrate shortly, this advantage is more hypothetical than real. They still must rely on persuasion rather than power.

Among the developing land-locked States, few are able to ship the bulk of their imports and exports over one route by one mode of transport. Usually there are expensive, time-consuming, and risky breaks-in-bulk and transshipments—often several in one journey. Moreover, with very few exceptions, land-locked States have little or no control over the availability, suitability, or operating efficiency of the transport systems and port facilities outside their borders. Furthermore, even though transit trade can be profitable for the transport systems of transit States, the bargaining power of the land-locked States is considerably reduced in many cases because they compete with their transit States in the production of the same few products, particularly in Africa. As for imports, the profitability of transit trade is often reduced by smuggling. This varies in type, degree, and technique depending on the prevailing economic conditions and policies in both the land-locked and transit States at a given time, but there is no doubt that smuggling can seriously injure economies that are fragile to begin with.

Few of these problems, however, as we pointed out at the beginning of this chapter, are peculiar to land-locked countries alone; many are suffered by the interior districts of developing coastal States as well. What compounds these problems for the land-locked States is that between them and the sea is at

**Transit problems of developing land-locked States.** At the left is one of the better portions of the Tribhuvan Rajmarg, the Indian Army-built highway between Kathmandu and the Indian border that is Nepal's principal route to and from the sea. In the "hills" south of Kathmandu, the longest straight stretch of road is less than two kilometers long. In the photo, the two heavily loaded Nepali trucks cannot pass on the road, and one of them must pull off the road to let the other go by. On the right is one of the three ferries that carried vehicles across the Mekong River between Thanaleng, Laos (background), near Vientiane, and Nongkhai, Thailand. This was a major bottleneck on the principal route for Lao imports and exports until the Friendship Bridge between the two towns, financed by Australia, was opened in April 1994. Tank trucks are the only means for importing petroleum products into Laos. (Martin Glassner)

least one international border, sometimes more. This means that the transit States can, and often do, impose complicated transit and customs formalities, excessive documentation, and even, at times, higher charges for transit traffic than for domestic traffic. Goods in transit, therefore, have to endure cumbersome procedures and high costs not only at the ports, but also at the border, and this increases the time and expense involved in transit.

These and other problems can be reduced, although never eliminated, if the land-locked State either has several alternative routes to the sea or is on excellent terms with its transit State or States. Happily, as we have seen, one or both of these conditions prevails in the case of nearly all land-locked States. There are, however, many examples of transit routes being deliberately blocked for varying periods. Clearly, political as well as economic factors are important in determining the true significance of "land-lockedness" in individual cases.

There is one more characteristic that distinguishes land-locked from coastal States: They are not only dependent on transit States for their access to and from the sea, but also they *know* they are so dependent. There is a distinct psychological factor involved that must always be taken into account when considering land-locked States in terms of history, politics, economics, geography, or international law. The urge to the sea, the feeling of national claustrophobia, has been strongest in the large interior States of Europe and in Bolivia. It was evident in Ethiopia before it acquired Eritrea and still is to a lesser degree in Afghanistan. It is less conspicuous in Africa than elsewhere, but these countries became land-locked *only* upon the attainment of independence. Whether a feeling of geographic strangulation will develop among the interior States of Africa remains to be seen. Certainly Mali, Uganda, Zambia, Malaŵi, Lesotho, Zimbabwe, and Swaziland have felt the effects of having their routes to and from the sea controlled by hostile powers, and Niger and Chad were innocent victims of the Nigerian civil war.

### *Relevant International Law*

Since ancient times land-locked territories have often faced obstructions, restrictions, tolls, or heavy transit fees on goods and persons en route to or from the sea. There is no "right of innocent passage" on land (or in the air) as there is at sea. Gradually, however, insistence on absolute sovereignty began to give way to a recognition of the advantages of a free flow of trade. By the eleventh and twelfth centuries of the Christian era, territories in Europe, particularly in Italy, were giving treaty rights to land-locked territories, and the internationalization of rivers was beginning. Nevertheless, restrictions and heavy tolls were still common by the late eighteenth century.

Gradually, through the nineteenth century, the principle of free transit through straits, on rivers, and overland became firmly established as recognition grew that such transit was necessary for the development of commerce and industry to the benefit of both land-locked and coastal States. Railroads and other new forms of transportation were instrumental in generating greater competition among industries, necessitating a freer flow of goods, and in reducing the advantages of waterways, hence increasing the bargaining power of the land-locked States. By the end of the century transit duties in Europe had virtually disappeared, the Latin American States had reversed Spanish trade policies and opened their territories to unrestricted trade, and most navigable waterways in Europe, Africa, East Asia, and South America had been internationalized. But access to the sea was still not accepted as a specific right of land-locked States.

President Woodrow Wilson of the United States included "a free and secure access to the sea" for Serbia and Poland among his Fourteen Points for a just and permanent peace settlement in Europe after World War I. The League of Nations Covenant for the first time included "freedom of communications and transit" as a worldwide goal and standard. In pursuance of this provision (Article 23 [e]) of the Covenant, a series of conferences during the 1920s produced both multilateral conventions and bilateral treaties aimed at the facilitation of free transit. Of these the most important were the Barcelona Convention of 1921 and the Convention and Statute on the International Regime of Maritime Ports, signed at Geneva on 9 December 1923.

In the period after World War I a relatively new concept—corridors to the sea—enjoyed a measure of acceptance, and a number of States and territories had their boundaries adjusted to allow for very short frontage on the open sea or a navigable river. Finland, Iraq, Transjordan, Palestine, and Colombia all received such corridors, as had several territories in Africa in the late nineteenth century. But the most famous was the Polish Corridor. Corridors present many problems of self-determination, security, and transit across the corridors themselves, and they no longer seem to be a practical means of assuring access to the sea for land-locked countries. Even the transit conferences of the 1920s did not consider corridors in their deliberations. They did, however, cover in great detail navigable waterways and overland transit.*

During the interwar period other important agreements were negotiated concerning railways, power lines, pipelines, ports, and other transport facilities, all in conformity with and tending to reinforce the principles of free transit codified in the Barcelona Convention. Aircraft, however, present special problems, and air transport is a major class of exceptions to these principles.

## Developments Since World War II

Several important historical events and trends conjoined shortly after World War II to produce an atmosphere and approach to-

ward access to the sea quite different from those just described. For one thing, territorial reorganizations in Europe eliminated the Polish and Finnish corridors and enhanced international cooperation in both water and overland transit. An elaborate system of internationalized rivers and canals, free zones and free ports, and special transit arrangements had produced a situation so adequate that there have been few recent difficulties with transit, even between communist and noncommunist countries (except for the special case of West Berlin before the reunification of Germany in 1990). Therefore, Europe has not been important in discussions of access to the sea, except as an example of how land-locked and transit States can develop harmonious and mutually satisfactory arrangements.

At the same time, the world trading system was being reorganized to meet the needs of reconstruction and economic expansion following the war. Both the GATT and the Havana Charter reaffirmed and in some respects strengthened the Barcelona Convention's provisions for freedom of transit. Article V of GATT and Article 33 of the Havana Charter incorporated transit provisions into an elaborate system for expanding world trade rather than the more limited approaches that had prevailed in the past. Attitudes were changing.

We may identify two basic approaches and one lesser approach to the question of access to the sea, all rooted deeply in international law. The first approach is that of freedom of transit, which has already been discussed in some detail. The second view is that access to the sea derives directly from the "freedom of the seas." These two approaches are closely linked; indeed, some writers consider one simply an aspect of the other. A third concept, one that has never been really important and is scarcely mentioned today, is that sea access constitutes an international servitude.

The early postwar period, however, saw the rapid evolution of a relatively new concept, embodied in GATT and the Havana Charter; namely, that access to the sea is essential for the expansion of international

*In 1975 Bolivia and Chile began negotiating a proposed corridor for Bolivia along the Peruvian border to the sea. The negotiations ended inconclusively in 1979 due to opposition within Bolivia and to the refusal of Peru to assent to the proposed transfer of territory, as required by the 1929 Treaty of Santiago.

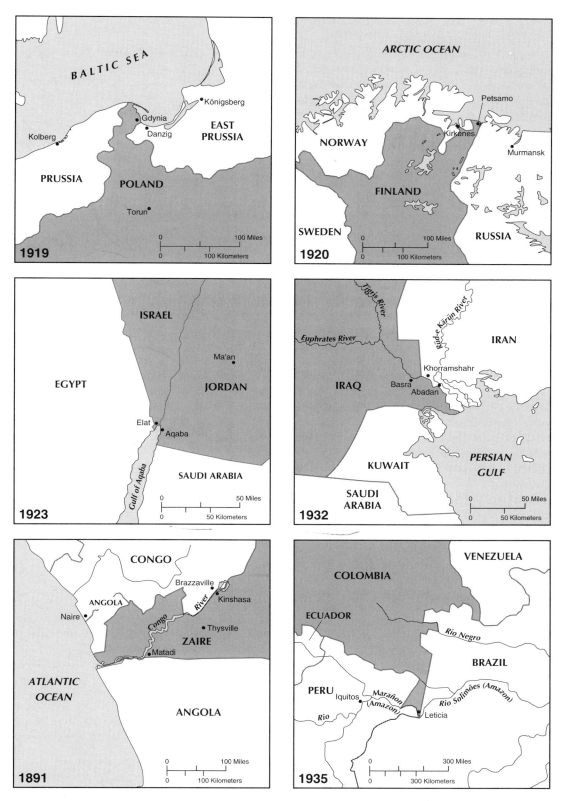

Examples of territorial corridors created to provide access to the sea.

trade and economic development. Indeed, in January 1956, when ECAFE considered the problems of its land-locked members (Afghanistan, Laos, and Nepal), its recommendations referred to the "needs" of these States rather than "rights," a term that pervades the classical literature on the subject. One result of ECAFE's interest was the beginning of a drive within the United Nations to deal with access to the sea in a comprehensive and definitive manner.

This activity has been stimulated chiefly by the achievement of independence by many former European colonies in Asia and Africa, their entry into the United Nations, and their decision to use the United Nations as a prime vehicle for protecting their new "sovereignty" and supporting their economic development. Most definitely, the mid-1950s saw the fading away of traditional concepts of international law regarding land-locked States—which were replete with such words as "rights," "obligations," "sovereignty," and "natural law"—and the emergence of newer, more practical arguments for "a free and secure access to the sea."

Early in 1958 the United Nations Conference on the Law of the Sea assigned the question of access to the sea to its Fifth Committee and ultimately adopted the committee's recommendations as Article 3 of the Convention on the High Seas, one of four conventions produced by the conference.

ECAFE continued to pursue the matter, in Manila in 1963 and Tehran in March 1964. At the latter meeting a resolution was unanimously passed stressing the "critical importance" of access to the sea in economic development and urging that the subject be considered at the forthcoming UN Conference on Trade and Development (UNCTAD). After long and complex discussions of access to the sea, UNCTAD finally adopted eight principles and several recommendations, similar to those adopted in 1958 but with some significant differences. Regional transit agreements were encouraged, and exemption from most-favored-nation clauses was urged. Other provisions covered special facilities and rights of land-locked States.

## The Convention on Transit Trade of Land-Locked States

On the recommendation of UNCTAD, a Committee on Preparation of a Draft Convention Relating to Transit Trade of Land-locked Countries met at UN Headquarters during October and November 1964 and, as had been the case since 1956, the African and Asian land-locked States led the discussions. Its draft convention was presented the following summer to the UN Conference on Transit Trade of Land-locked Countries. This was the first conference convened under UNCTAD, which had become a permanent agency of the United Nations. The conference was attended by delegates of 58 countries and observers of 10. Their deliberations again covered basic principles and technical details, ancient traditions and new circumstances. The issues were so complex, in fact, and the participants split so many ways, that the convention produced by the conference was a compromise of compromises. Nevertheless, it is most significant.

For the first time in history, an international lawmaking conference had dealt exclusively with the question of access to the sea, particularly for the developing land-locked States of Africa, Asia, and South America. This certainly gave a measure of "status" to the problems of those States, but did not solve them. The convention did not change the geographical condition of any land-locked country. Despite its importance as a "datum plane," it was only one more step in the continuing drive of the land-locked States to reduce to a minimum the difficulties inherent in their mediterranean ("middle of the land") location.

## Progress Since 1965

Since 1965 the developing land-locked States have tried to improve their situation in many different ways. We can group these into five major categories.

1. There has been a greater emphasis on *internal development*, with particular stress laid on improved transportation

networks and facilities. As improved transport contributes to the general economic development of a country, some handicaps of isolation and internal weakness are reduced, living standards rise (if the improvements are not canceled out by population growth), and the State becomes a more cohesive unit. At the same time its demand for imports increases and it must export more to pay for them, so its need for improved transit across coastal States is increased, making it both more valuable as a customer on transit routes and more vulnerable to disruptions in its transit.

2. *Bilateral negotiations* for improved transit remains the most important and most common method of obtaining it. International law provides a framework for such negotiations and sets minimum standards, but does not (at least not yet) guarantee a *right* of free transit. Even if it did, the individual land-locked and transit States would still have to work out detailed arrangements based on local conditions. In these negotiations the coastal State invariably has the advantage. Even when reasonably satisfactory agreements are concluded, they can be breached in times of stress. The Afghan–Pakistan border, to cite an unusual case, was closed in 1950 and 1955 for short periods and again from September 1961 to July 1963. Bilateral

arrangements, while necessary, are a slender reed on which to lean.

3. Both transit and land-locked States, as well as the international community in general, have been making concerted efforts in recent years to *improve transport facilities and seaports* for everyone's benefit. Congested seaports have been a worldwide problem, and improved transport is an essential element in any country's economic development. At the same time some land-locked States are vigorously developing arrangements and facilities for *alternative transit routes* through countries other than the traditional transit State. Here the classic case in modern times is Zambia, which, after Southern Rhodesia declared independence in 1965, began seeking a more secure, though longer and more expensive, route to the sea through Tanzania. Afghanistan diverted most of its transit trade from Pakistan to the Soviet Union, Rwanda normally uses routes through Uganda and Kenya rather than Zaïre as formerly, and so on. Although time-consuming and expensive, these methods do give the interior State more flexibility, hence more security and greater bargaining power, than reliance on a single transit State or inadequate facilities. In 1989, however, when India closed all but two border crossings with Nepal and placed many restrictions on Nepali tran-

**Mali transit area in port of Dakar, Senegal.** Many land-locked countries have areas in sea and river ports of transit countries designated for their exclusive use. In some cases, these may consist of single sheds or warehouses, and in some cases may be free zones, outside the customs territory of the transit state. In Dakar, Mali does not have any formal jurisdiction, but has had no recent problems with the administration of its transit area. (Martin Glassner)

sit traffic, causing great distress in Nepal, that country had no choice but to negotiate a settlement of the dispute with India. Transit through China, its only other neighbor, was, and remains, infeasible.

4.  As we noted in Chapter 26, many postwar efforts at *regional economic integration* have involved land-locked countries. If the degree of integration is great enough, border formalities are greatly reduced or eliminated for goods in transit, and the land-locked country may even share in the ownership and control of the port and transport facilities. Uganda had been the best example of this type, since it was a full partner in the East African Common Services Organization and the East African Community. But after the Israeli rescue of hostages from the Entebbe airport in July 1976, the border between Uganda and Kenya was closed for some time. As relations between the two partners continued to deteriorate and the East African Community disintegrated, so did Uganda's transit arrangements. Economic integration, while highly desirable for many reasons, is still no substitute for political integration as a means of solving the political problem of guaranteed access to and from the sea.

5.  The land-locked States have continued their *efforts in the United Nations* and its various organs and specialized agencies both to establish a legal right of transit and to gain special considerations and assistance to ameliorate their special problems. UNCTAD has been especially active in developing "special measures related to the particular needs of the land-locked developing countries." The World Bank, the United Nations Development Programme, ESCAP, CEPAL, the Asian Development Bank, and other agencies have been rendering practical assistance in various ways. Efforts to establish a legal right of transit were concentrated in the Third United Nations Conference on the Law of the Sea (UN-

CLOS III) and in the Seabed Committee that preceded it.

## *The Land-Locked States at UNCLOS III*

During the lives of both the Seabed Committee (1967–73), which acted as a preparatory committee for UNCLOS III, and of the Conference itself (1973–82), the land-locked States worked diligently to achieve two objectives: a guaranteed right of *free transit to and from the sea* and the greatest possible *access to the resources of the sea*. The United Nations Convention on the Law of the Sea contains a section (Part X) titled "Right of Access of Land-locked States To and From the Sea and Freedom of Transit." The right of access is thus conceded, but not a right of transit, without which access is meaningless. The coastal States are still unwilling to surrender even a small part of their sovereignty to assure free transit to their less fortunate neighbors. To emphasize this point within Part X, one article provides that "the terms and modalities for exercising freedom of transit shall be agreed between the land-locked States and the transit States concerned through bilateral, subregional or regional arrangements" and that "Transit States, in the exercise of their full sovereignty over their territory, shall have the right to take all measures necessary to ensure that the rights and facilities . . . for land-locked States shall in no way infringe their legitimate interests."

Other articles, among other things, exempt the special rights and facilities of land-locked States from the application of most-favored-nation clauses; prohibit customs duties, taxes, and most charges from being placed on traffic in transit; provide that free zones or other customs facilities may be established at ports for traffic in transit; require that delays or technical difficulties of traffic in transit should be avoided or eliminated; and that "ships flying the flag of land-locked States shall enjoy treatment equal to that accorded to other foreign ships in maritime ports." This last provision, incidentally, is neither facetious nor hypothetical. Switzerland has long had an efficient and profitable merchant fleet operating chiefly out of Genoa and Basel; Bolivia, Nepal, Paraguay, Czechoslovakia, Hungary, Swaziland, Uganda, and other land-locked States also operate or have considered high-seas fleets.

On the whole, considering the gains and losses, this convention represents a slight net gain for the land-locked States compared with the 1965 convention. It does not, however, ensure the "free, uninterrupted and continuous" movement of traffic in transit referred to in the 1965 convention.

The nine land-locked former Soviet Republics are learning this first-hand as they negotiate transit arrangements with coastal neighbors and in most cases with one another. The fighting between Armenia and Azerbaijan and Kazakh-Russian rivalries have impeded the flow of traffic in transit. Kazakhstan and Azerbaijan need new pipelines to the Black and Mediterranean seas and perhaps to the Persian Gulf in order to market their petroleum in Japan and the West for hard currencies, but this means crossing unfriendly or hostile territory. Individually and collectively, the Central Asian States are also exploring new trade routes to the south, through Afghanistan and Pakistan. Meanwhile, they are working through the CIS to maintain both markets in and transit routes through Russia, and the European Union has been working with them since 1993 to develop the Transport Corridor Europe–Caucasia–Asia (TRACECA) and the Transport Corridor Europe–Asia.

Similarly, the breakup of Yugoslavia created one new land-locked State (Macedonia) and one (Bosnia and Herzogovina) with only a tiny seacoast. Because of hostile neighbors, Macedonia must rely on a tenuous and circuitous route to the sea through Bulgaria, while the civil war in Bosnia is prolonged by the Bosnian government's insistance that any territorial settlement must leave it with adequate territorial access to the sea.

### Access to the Sea's Resources

In December 1970 the United Nations General Assembly declared that the seabed and ocean floor beyond the limits of national jurisdiction and its resources are "the common heritage of mankind, irrespective of the geographical location of States, whether land-locked or coastal." Through the debates in UNCLOS III and the successive draft treaty articles considered, this declaration was not challenged. The land-locked States are to share more or less equitably in both the machinery created to manage the international seabed area and in the benefits to be derived from seabed mining. Nevertheless, because of profound legal and economic uncertainties, it seems most unlikely that any developing land-locked State—or any poor country for that matter—will derive any appreciable benefit from seabed mining until well into the twenty-first century. Of much greater practical importance to them is the sharing of benefits of resource exploitation within "the limits of national jurisdiction."

These limits include the continental shelves to the outermost edge of the continental margin, *and* a 200-nautical-mile exclusive economic zone, measured from the coastline. All the resources in these areas, living and nonliving, belong to the coastal States. Thus, within a quarter century, the land-locked States have, in effect, been driven another 188 or 197 nautical miles *back* from the resources of the sea. They have been denied their traditional free access to all the world's known offshore oil and natural gas, all the presently exploitable minerals, most of the readily accessible manganese nodules, and 85 to 95 percent of the world's current fish catch. In UNCLOS III they succeeded only in gaining a token right to a portion of the surplus fish within the exclusive economic zone; that is, the fish that the coastal State cannot or chooses not to harvest up to the allowable catch based on conservation principles. The land-locked States—some 25 percent of the world's independent States and 10 percent of its population—will derive little or no benefit from the intensive exploitation of our last great frontier on earth.

Although they failed to gain much from UNCLOS III, the land-locked States will undoubtedly continue their efforts to improve their situation by the methods outlined here:

**Land-locked States on stamps.** Bolivia issued the top four stamps in 1979 to commemorate the centennial of the loss of its coastal province to Chile at the beginning of the War of the Pacific. Since this loss, it has waged an unremitting struggle to regain some kind of *salida al mar*, or outlet to the sea. The next stamp honors the acquisition of coastal Eritrea in 1952 by Ethiopia, giving this land-locked country a seacoast for the first time in modern history. It has again become land-locked, however, since Eritrea attained independence, but Eritrea has granted Ethiopia free transit facilities and a free port in Assab. The bottom stamp from Mozambique shows the major transit routes of the land-locked States of southern Africa. (Martin Glassner)

internal development, bilateral negotiations (both for transit and access to the economic zones of coastal States), improved and diversified transit routes and facilities, regional economic integration, and assistance from international agencies. They are also likely to turn more to air transport, joint ventures for marine resource development, and other devices for overleaping the territory that lies between them and the sea and its resources. It is possible also that, as we indicated in our discussion of "strategic location," some of the poorest and most isolated States of today could become the Switzerlands of tomorrow. At this time, however, this prospect seems quite remote.

# 30

# *Outlaws and Merchants of Death*

Since 1979, when the third edition of *Systematic Political Geography* was written, a relatively new element has been added to the array of forces active in international relations discussed in the preceding chapters. Individuals and small groups worldwide have been engaging in a variety of unlawful acts designed to enrich themselves or to achieve some limited political goal, but their actions have international ramifications. Piracy, terrorism, drug trafficking, and arms trafficking are not really new. Each has a long and dishonorable history, and there is an abundance of literature on each. None has been of professional interest to geographers, however, certainly not to political geographers, until now. What are new and what now make these activities of interest to political geographers are four factors:

1. The incredible scale of most of these activities and their reach into every sector of society.

2. The worldwide distribution of all of these activities with their locations and transport routes constantly shifting in response to changing circumstances.

3. The significant and growing links among them, and the formation of alliances that increase the destructiveness of each.

4. The very considerable impact they are having on policies and decisions and activities of governments at all levels.

All four factors cause distortions in international trade and in tourism and other travel. They divert resources of all kinds, in rich and poor countries alike, from economic and social development to security and law enforcement and, of course, to the illegal activities themselves. They influence government actions, corrupt individuals, and erode democratic institutions. For the remainder of this century—and probably well beyond—no discussion of international relations can be complete or realistic without the inclusion of outlaws and merchants of death.

## *Piracy*

Since the late 1970s there has been a sharp increase of incidents labeled as piracy, including air piracy, by governments, commentators, and news media. Few if any of these incidents are piracy in any legal sense, as we will see, but they do resemble the popular image of piracy in some respects and deserve our attention.

During the period 1980–84, more than 400 incidents of "piracy" were officially reported, though this figure does not include attacks on refugees from Indochina (the so-called boat people), and in any case probably represents only a small portion of such incidents. Many go unreported for a variety of reasons: insurance, fear of retribution, possible exposure of other illegal activity, and so on.

Most acts of "piracy" today are thefts of high-value, low-bulk cargo, such as electronic equipment, from ship cargoes; of cash and valuables from merchant ships, their crews, and passengers; and even of ships' equipment. Most take place in port or in territorial waters and are thus indisputably within the jurisdiction of the State in whose waters they occur. Most take place in areas of poverty and high unemployment, where crime is one of the few alternatives to starvation. Much of it is carefully planned and carried out by organized gangs (usually small, but ranging up to 50 or more) operating from shore. Most of the thieves are armed with knives or machetes, rarely guns, and prefer to flee rather than fight if confronted by armed guards. Often the local police cannot or will not deal with the problem, nor can the small, unarmed crews, so some shipowners have taken to hiring private guards from the ports—who often belong to criminal gangs and actually facilitate the thefts.

The character of piracy in individual locations varies in detail from these generalizations. They are most appropriate for West Africa. Nigerian ports were infested with ship thieves until the government began to take decisive action against them in the mid-1980s. They then shifted their activities to other ports farther west. These depredations continue and often involve gratuitous violence that intimidates even local police.

The situation is different in the Strait of Malacca and adjacent Phillip Channel, where ships must slow down and become easy prey for pirates in canoes and small boats. Since more than 200 ships pass through the straits each day, there are plenty of targets, and over 150 incidents were reported from there in 1981–84 alone. The number reached a peak in 1991, when 200 were reported, and has dropped slightly since then. Normally, violence is not used there unless resistance is offered. Singapore controls piracy very well and there are few incidents in its waters, but Indonesia does not, and the nearby Riau Archipelago is a major pirate base. A new technique is for robbers to board a ship at sea and after stealing all they can, they tie up the crew and leave the ship to steam ahead at full power with no one in

control. The dangers of collision and major oil spills are very real.

In the Gulf of Thailand, the South China Sea, and the Sulu Sea, piracy is characterized by great cruelty and repeated violent attacks, especially on helpless refugees in small craft. In many cases Thai fishermen are part-time pirates, but they are also victims of pirates. There is more well-organized, skillful robbery of cargo vessels now; some ships are even hijacked temporarily until their cargoes are unloaded—very professionally. Off the coast of Brazil, there are occasional reports of piracy, but only in the port of Santos is it more than sporadic, and even there it is not a serious problem. Piracy around the Caribbean is still different. There, gangs involved in narcotics trafficking steal yachts and other small boats to use in their smuggling operations, primarily on the coast of Colombia and among the Leeward and Windward Islands.

This brief summary shows a clear pattern: Rarely are ships attacked by "pirates" on the high seas outside the jurisdiction of any State. Now that most coastal States have declared exclusive economic zones of up to 200 nautical miles, crimes committed within these zones cannot be considered piracy since the zones have been removed from the classification of high seas. Even aircraft are rarely hijacked over international waters, and thus the term "air piracy" is probably not applicable in most cases. In fact, the term has not yet been used in criminal proceedings.

The International Civil Aviation Organization (ICAO, Montréal) has sponsored global treaties and provided technical assistance to governments and airlines in their efforts to combat "air piracy." Under the Tokyo Convention of 1963, the State of registry of an aircraft is competent to try such cases, and The Hague Convention of 1970 refers only to "the offense" of unauthorized seizure of an aircraft. The Montréal Convention of 1971 similarly avoids the word "piracy," and there has been no serious attempt since then to define aircraft hijacking legally as piracy.

There is, in fact, no public international law defining piracy, only national laws based on individual policies and situations. The nearest approaches to an international definition were in the 1958 Geneva Convention on the High Seas and in the 1982 United Nations Convention on the Law of the Sea (discussed in more detail in Chapter 32), which simply repeats the 1958 formulation verbatim. This definition is exceedingly difficult to interpret and leaves many gaps unfilled; it is therefore almost useless in combating "acts of violence or detention, or any act of depredation, committed for private ends by the crew or passengers of a private ship or aircraft."

Instead of relying on the Law of the Sea to combat piracy, however defined, States, in addition to using their own law enforcement mechanisms, are cooperating internationally. The International Maritime Bureau (IMB), sponsored by the International Chamber of Commerce in London, works with the International Maritime Organization (formerly the Intergovernmental Maritime Consultative Organization), also in London, to register and investigate both maritime fraud (a serious and growing problem) and piracy.* Various shipowners' associations do similar work. In 1982, the United Nations High Commissioner for Refugees (UNHCR) began an antipiracy campaign, largely by helping Thailand to patrol the Gulf of Thailand. In 1983 UNCTAD also began investigating "maritime fraud, including piracy." In 1985 the UNHCR shifted its emphasis to land-based enforcement of criminal laws, and the United States established a Regional Antipiracy Unit within the Refugee Section of its embassy in Bangkok.

All of this antipiracy activity might have had the desired effect, for the rate of pirate attacks dropped in 1986 and fell to just three in Southeast Asian waters in all of 1989. But in 1990 the number soared to 32. The local Shipowners' Association appealed to ASEAN for help. In 1992 the IMB established a Regional Piracy Centre in Kuala Lumpur to collect and disseminate information on the subject. Since 1993 IMO has been more active in combating both piracy and maritime fraud.

Piracy is likely to continue to interfere with, and increase the cost of, international trade; be used for political purposes at times; and require international cooperation for its suppression. Political geographers could assist in this endeavor with careful studies and analyses of modern piracy.

### Terrorism

Like piracy, terrorism is illegal everywhere but still not adequately defined in international law. Rarely, if ever, is a person charged with "terrorism"; rather, the charge under domestic law is likely to be murder, kidnapping, assault, or some other, common crime or crimes. Nevertheless, terrorism is of interest to us because it is politically motivated and is generally carried out across State boundaries. It differs from traditional guerrilla warfare (a topic studied to a very limited degree by a few political geographers) chiefly in that it is very small-scale, sporadic violence committed against civilians, often randomly, in order to intimidate governments and achieve some political goal, rather than organized attacks by irregular or paramilitary forces against military personnel or facilities in order to achieve a military goal. Terrorists probably contributed to the success of some colonial wars in the 1945–80 period, but nowhere since then have they achieved their own stated objectives.

Terrorists can usefully be divided into two major categories: separatists and ideologists. We have already discussed nationalism and irredentism, and later we discuss religious, linguistic, and ethnic minorities in some detail. Here we only point out that, as we indicated in our discussion of non-state nations in Chapter 11, many aggrieved minority groups around the world have resorted to terrorism in order to call attention to their desire to separate from the State in which they find themselves, or at least to win greater autonomy within the State. The

---

*The IMB adopted its own definition of piracy in 1992: "Piracy is the act of boarding any vessel with the intent to commit theft or other crime and with the capability to use force in furtherance of the act."

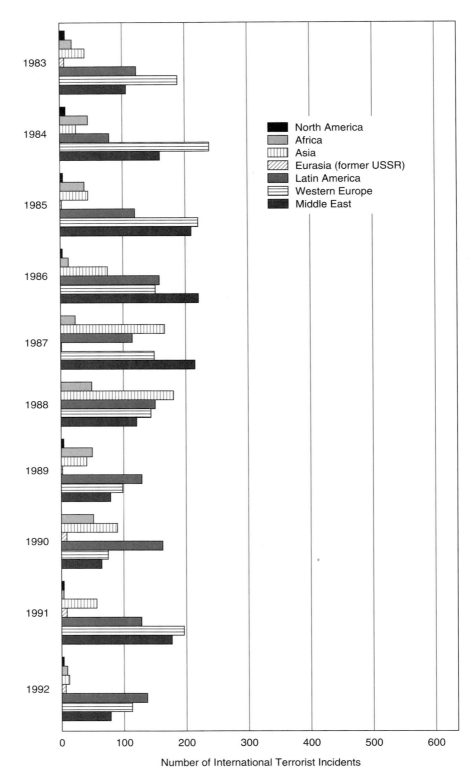

SOURCE: U.S. Department of State. *Patterns of Global Terrorism: 1993.*

**International terrorist attacks, 1983–92.** Such incidents are likely to continue to fluctuate from year to year but remain a problem for some time. International cooperation against terrorism, however, will probably result in its gradual reduction.

South Moluccans, Croatians, Bretons, Corsicans, Irish nationalists, Kurds, Tamils in Sri Lanka, Philippine Muslims (Moros), Karens and Kachins in Myanmar (Burma), Basques in Spain, Sikhs in Punjab, Puerto Rican nationalists, Palestinians, blacks in southern Sudan, Armenians, numerous groups in Lebanon, and many other minority peoples around the world have spawned terrorists.

The other major category, ideologists, is represented by radical leftists in France, West Germany, Italy, Japan, El Salvador, Guatemala, Peru, the Philippines, and elsewhere, and by radical rightists in El Salvador, Colombia, France, Israel, Haiti, and other countries. There have also been rebels against the national governments of Angola, South Africa, Mozambique, Myanmar, India, Nicaragua, and other countries, who combine nationalism, political ideology, and probably other motivations in their struggles. Some groups in all categories have been defeated, some are quiescent, some are active now, and some are nascent. Whatever their motivations, size, location, tactics, or other characteristics, they pose many problems not only for governments, but also for private citizens who become their victims.

Although most of these terrorists carry out their activities against governments (often indirectly), some attacks are directed against private corporations and even charitable organizations. In addition, many of them are sponsored or aided by governments. This State-sponsored terrorism (euphemistically called "low-intensity conflict") is not a new phenomenon either, but it has become dramatically more significant recently. There is reasonably impressive evidence, for example, that Libya has aided terrorists in Chile, Colombia, El Salvador, Guatemala, New Caledonia, Northern Ireland, Pakistan, the Philippines, Portugal, and Thailand as well as various Palestinian groups. Similarly, Syria, Iran, Cuba, and South Yemen have been accused by the United States of supporting terrorism in various countries around the world, mostly in the Middle East. South Africa for many years sponsored terrorist activities in nearby States in an effort to "destabilize" their regimes. The United States has

engaged in similar activities, chiefly in Latin America. Government support of terrorism includes such things as training, weapons, cash, passports, transportation, intelligence information, and even diplomatic cover—including use of the diplomatic pouch, diplomatic passports, and protection in embassies. This support complicates an already difficult problem.

Combating terrorists is not easy. Terrorists can attack anyone or anything at any time. But governments cannot protect everyone and everything all the time. Terrorists can also utilize the latest technology, which often enables them to do more damage at less cost with less risk than earlier methods. Modern transportation and communications not only enhance the mobility and security of terrorists, but also magnify the impact of their actions through extensive media coverage. State terrorism; that is, terrorism carried out directly by States, is even more difficult to deal with, even if State actions can be legally defined as terrorism and even if the evidence of State terrorism is conclusive, neither of which happens very often. In 1983, North Korean agents in Rangoon killed 16 South Korean officials by setting off a bomb near them. But what kind of sanctions could be carried out against North Korea and by whom? Other countries, including the major powers, have also been accused of terrorism, but the actions in question are usually excused as self-defense, retaliation, or aid to a friendly government.

There has been some progress in combating terrorism through international cooperation. Some information on terrorists is exchanged through Interpol; there are occasional binational antiterrorist operations; the ICAO conventions (referred to earlier in connection with piracy) have helped reduce aerial hijackings; and the United Nations produced in 1985 the International Convention Against the Taking of Hostages. But mostly it has been a matter of individual States taking such actions individually as they see fit. The most flamboyant of such actions to date was the bombing by the United States of assorted targets in Libya in 1986, but there is no evidence that it had the desired effect. Israel's invasion of Lebanon in 1982 and the subsequent expulsion of the Palestine Liberation Organization from the country disrupted Palestinian terrorism for a time—but only for a time. It was not until June 1987 that the first high-level antiterrorist meeting of major industrial States (nine of them) took place in Paris to discuss collaboration against terrorism.

We seem to be in a new phase of terrorism now. Although the relevant statistics are even less reliable than most statistics, there does seem to have been a gradual decline of *political* terrorism in one region after another as (one way or another) they are cleared of terrorists of the more familiar types. They are gradually being replaced, however, by terrorists belonging to extremist *religious* factions. Militant defense of doctrinal orthodoxy, attempts to convert unwilling people, and tactics to gain political power and impose their own brand of religion on the body politic are all growing now. Around the world Hindus, Sikhs, Shiite and Sunni Muslims, Jews, and some Christian sects are engaging in it. This kind of terrorism is no less destructive of life, property and spirit than the political type, and may be even more difficult to combat.

## *Illicit Drug Traffic*

This is big business. It is estimated that in 1990 alone, $40.4 *billion* worth of drugs, including marijuana, cocaine, heroin, LSD, amphetamines, and other controlled substances, were sold in the United States, the world's biggest market, but certainly not the only one. Worldwide, the figure can only be guessed at, but it is probably triple this amount.* Besides this huge amount of money diverted from productive and con-

---

*All statistics concerning illicit narcotics, marijuana, and other controlled substances must be regarded with considerable skepticism because of the nature of the industry and the real incentives for all reporters, including governments, to falsify such figures. Those emanating from the United Nations and other intergovernmental agencies may be more reliable, but they are generally based on government figures, which are themselves suspect.

structive activities in nearly every inhabited area of the world, drug production, processing, and distribution have generated inflation in poor countries, corrupted governments, ruined development plans, caused innumerable deaths (not only from the drugs themselves, but also from bullets and bombs), made a mockery of many trade figures, and caused untold suffering and misery.

The principal drug-producing areas of the world are Mexico, some Caribbean islands, the lower Andean slopes and valleys, Turkey, the Iran–Afghanistan–Pakistan "Golden Crescent," and the Myanmar–Thailand–Laos "Golden Triangle." The chief consuming areas are the United States and a number of countries in Western Europe. While this "North–South" trade does help ever so slightly to redress the imbalance of material wealth in the world, its negative effects are far more consequential. The distinction between producer and consumer, for example, is rapidly disintegrating. The United States now produces a very large proportion of the controlled drugs it consumes, while everywhere that there is drug production, processing, or trafficking, local drug abuse begins. Drugs have even infiltrated societies once thought too moral, too tightly controlled, or too remote to succumb to the lure. Today there are few places in the world free of drug abuse.

How can we account for the size and pervasiveness of the drug business? Fundamentally, of course, it's a product of a seemingly insatiable demand in societies that can afford to indulge in a relatively cheap escape from difficult problems, real or imagined. We cannot here go into the sociological causes of drug abuse, but we can look briefly into the supply side of the equation.

In some parts of the world, drugs, like alcoholic beverages, have been used but seldom seriously abused for centuries; they are part of the local culture. They include hallucinogens in northern Mexico and southwestern United States, coca in the Altiplano and central Andes, hashish in the Middle East, and opium in Asia from Turkey to China. In addition, very small amounts of these drugs

**Coca leaves for sale in Bo-livia.** The peoples of the central Andes have long chewed these leaves to assuage their chronic hunger, thirst, cold, and fatigue and to make their lives endurable in this harsh environment. More recently, however, coca has become a cash crop to satisfy the demand in rich countries for cocaine. Still more recently, addiction among the producers has become a serious problem. (UN photograph)

are used widely for legitimate medicinal purposes: morphine (derived from opium), marijuana, and amphetamines, for example. Drug abuse and drug trafficking became a serious international problem in the nineteenth century. Chinese resistance to European importation of opium into China led to the Opium War of 1839–42 between China and Great Britain, and the first international drug-control treaty, the International Opium Convention, was signed in 1912. But only since the 1950s has drug trafficking assumed overwhelming proportions.

Coca, marijuana, and opium poppies are all easy to grow, adaptable to different physical environments, and immensely profitable for poor farmers who are ill-prepared to grow legitimate crops profitably. Coca, for example, the shrub whose leaves produce cocaine, produces five crops a year with very little cultivation and generates up to ten times the profits of other crops. Peasants in many producing countries, especially in South America, have resorted to violence to protect their crops from drug enforcement officers. Drug-plant eradication and crop-substitution programs have frequently foundered and even failed completely because of peasant resistance, little cooperation from corrupted or intimidated local officials, envi-

ronmental concerns, or violent opposition from the traffickers themselves. There is evidence, however, that these programs are beginning to be more successful, notably in Turkey, Pakistan, and Thailand.

But as production is reduced in one place, it often simply moves to another. Marijuana production, for example, began in Thailand about 1982, but the government clamped down on it quickly. It thereupon moved across the Mekong to Laos, where the government not only protects, but actually helps the growers. Marijuana is now the leading export from Laos to the United States. In April 1984 the Colombian minister of justice was murdered by drug chieftains, prompting the government, at last, to mount a reasonably successful campaign against them.

The same is true with the distribution system. As one route is disrupted by authorities, another is developed. For example, the traditional route from Pakistan to the United States and Europe through the Middle East was disrupted by the turmoil in Lebanon, law enforcement efforts in Iran, and perhaps the Gulf War and other factors, so a new route developed across Africa from Nairobi or Addis Ababa through West African ports. And as American surveillance and interdiction become more effective in the Caribbean,

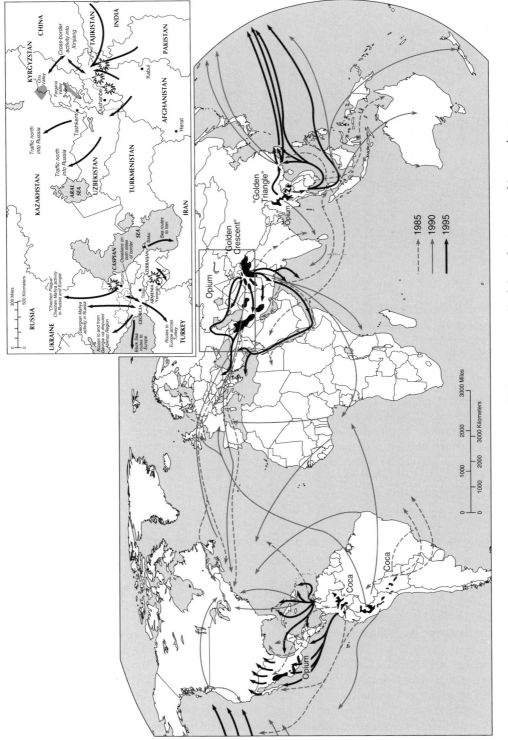

**Principal drug sources, markets, and trafficking routes.** All of these factors are subject to change as supply, demand, and antidrug activities change. There is no sign yet, however, of a drop in the total demand for or supply of drugs.

more drugs are sent from South America overland and by air through Mexico directly to California. A huge new market for drugs has developed in Central and Eastern Europe and the former Soviet Union since 1991, and new transport routes now run through the entire region. Central Asia is now an important producing area as well as a new conduit for drugs produced in "the Golden Crescent."

Even the choice of drugs produced is subject to change in response to market demands and drug control efforts. Heroin, for example, a derivative of opium, is now resurging as the "drug of choice" in the United States, because the "Medellín and Cali Cartels" in Colombia have been hard-hit by government enforcement efforts. In response, opium poppies are now being grown extensively in southwestern Colombia, where coca shrubs once grew. The "drug lords" are happy to accommodate the market, for heroin is about ten times as profitable as cocaine.

We would like to stress a point made previously: The United States is the world's largest market for illicit drugs (though Europe is rapidly catching up, and a number of poor countries have joined the sordid competition), and it has also become a major producer. Most of the marijuana consumed in the country is apparently now produced domestically, mostly in the West, and in Hawaii it is probably the single most impor-

tant crop. It is grown widely around the country, however, and the latest scientific techniques are being used to adapt the plants to all growing conditions. We may wonder whether this is the most constructive way to use our resources.

Slowly, international cooperation is beginning to develop in combating drug production and distribution. The United States has won the reluctant support of some countries, notably Colombia, Bolivia, and Turkey, for bilateral control efforts. Various regional intergovernmental organizations are also beginning to take the problem seriously. In 1986 alone, it was considered by the Council of Arab Ministers of the Interior, ASEAN, the Commission of the EC, the Council of Europe, the OAS, the OECD, the Organization of the Islamic Conference, SAARC, and others, as well as by several regional and interregional conferences convened by the United Nations. In 1987 the UN held in Vienna a major International Conference on Drug Abuse and Illicit Trafficking, which generated intensified activity.

In the following year the UN held a plenipotentiary conference in Vienna that adopted the United Nations Convention Against Illicit Traffic in Narcotic Drugs and Psychotropic Substances, whose 34 articles listed practical measures that national authorities can take against drug trafficking. These measures are quite comprehensive, but are restricted to the flow between sup-

**A drug-free country?** Most of the islands lying between South America, where cocaine and some other drugs are produced, and the major markets in North America are used as transit points for the narcotraffickers. In the process, local people in the islands are becoming involved both as traffickers and as users. This sign outside the police station/post office in the village of St. David, Grenada is evidence that some island governments are taking the matter seriously. (Martin Glassner)

ply and demand, and are still only recommendations to, not obligations of, States.

The United Nations had adopted conventions dealing with the problem in 1961 and 1971, but their application by States was quite uneven; hence the need for a new convention. The UN has also reorganized its own drug-fighting agencies. In 1991 it combined three of them into one, the UN International Drug Control Programme (UNDCP). Its activities are governed by the policy-making UN Commission on Narcotic Drugs. It is too early to evaluate all of these international efforts, but we can identify some very real successes.

First, they have proven quite effective in controlling the diversion of *legal* drugs into illicit channels; there has been minimal diversion for some time. Second, they have supported national and regional drug-control programs in various ways: technical, financial, and moral. Third, they have helped to coordinate these various national programs and those of various intergovernmental agencies. Unfortunately, they get very little publicity. They tend to be obliterated in the media by the more flamboyant and politically attractive activities of individual States. "Drug wars," whether successful or not (and few have been), win more generous appropriations and more votes than quiet diplomacy, dogged police work, and international cooperation.

## Arms Traffic

In Chapter 23 we discussed the general problem of the almost unbelievable buildup of armaments in the world, both nuclear and conventional, and the perils of nuclear war. Here we are concerned with the almost equally difficult-to-believe, and perhaps even more frightening, problem of the dispersion of sophisticated conventional arms to virtually every part of the world. Competition among arms producers is so great and control over arms sales so weak that today nearly anyone can get nearly any kind of weapon he or she wants for nearly any purpose at "affordable" prices. Broadly

speaking, there are two categories of international arms traffic: government-to-government transfers, including sales, loans, and gifts of weapons and government-approved, brokered, and subsidized private sales to other governments; and the private, usually clandestine, generally small-scale, international arms market. We briefly examine each of them.

The proliferation of sellers as well as buyers is a major new factor in the arms trade. China now ranks fourth in arms exports to developing countries, with Egypt, North Korea, Spain, Israel, and Brazil also ranking in the top 15. Developing countries, in fact, are the fastest growing arms exporters in the world, and most of their customers are other developing countries. It is questionable whether in the long run either buyers or sellers profit from such trade.

The figures given so far are all for "major conventional weapons." To them must be added many, many more billions of dollars worth of smaller personal and crew-served weapons (many extremely deadly), military equipment of all kinds, anti-personnel mines, ammunition and explosives, spare parts, repair and modernization services, and so on. Then there is the export of military personnel. Nepal has traditionally earned significant amounts of hard currency by exporting tough "Gurkha" soldiers to serve in the Indian and British armies, but Cuba, Pakistan, North Korea, and other developing countries have sent their own regular military personnel abroad to serve as instructors, guards, advisors, and even (as in the case of Cuba) to engage in combat.

The rate of increase of world military expenditures, especially by developing countries, has slowed since 1982, apparently owing to a number of factors: (1) the drop in oil prices and the worldwide recession of the 1980s have reduced the disposable funds available for arms purchases; (2) many developing countries are already saturated with weapons; and (3) a number of developing countries, as we have seen, have become producers and even exporters of arms. But Pakistan, Afghanistan, Iran, India, and a few other countries are still in-

## TOP SIX SUPPLIERS AND BUYERS
## OF CONVENTIONAL ARMS: 1988-1992

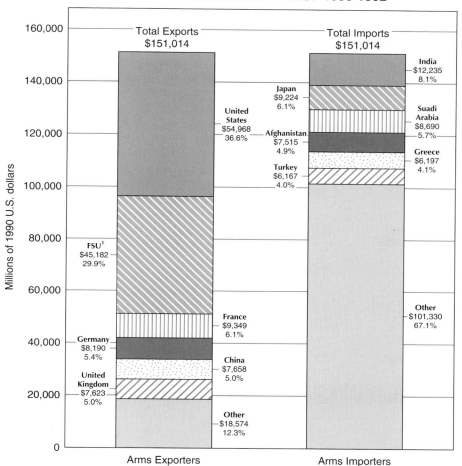

[1]Former Soviet Union

*Sources: United Nations Human Development Report 1994,* and *Statistical Abstract of the United States 1993,* U.S. Department of Commerce, Economics and Statistics Administration, Bureau of the Census

**Suppliers and buyers of arms.** This graph should be read in conjunction with those in Chapter 19 (on "foreign aid") and in Chapter 25 (on international trade). It shows clearly how lucrative is the official arms trade for the sellers, how much "foreign aid" is diverted away from development to the military, and why even some developing countries are becoming significant merchants of death.

creasing their arms purchases. Whenever there is discussion of arms control or disarmament, it must be remembered that notwithstanding all the rhetoric from the "Third World" or "nonaligned States" or "peace-loving" developing countries about nuclear-free zones or nuclear proliferation, there is little support among them for conventional arms control.

Of great and growing importance is the apparent trend of governments, especially in Western Europe, toward relaxing their own controls over the arms trade and gradually reducing their subsidies and supervision of

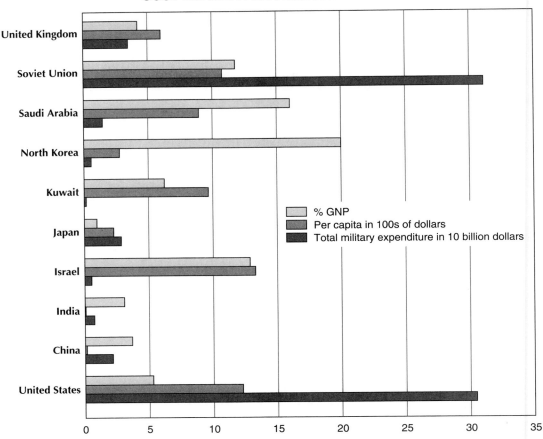

### MILITARY EXPENDITURES OF SELECTED COUNTRIES: 1989
### COST BURDEN PER PERSON AND ON ECONOMIES

*Sources*:    *United Nations Human Development Report 1994*, and *Statistical Abstract of the United States 1993*, U.S. Department of Commerce, Economics and Statistics Administration, Bureau of the Census

**Military expenditures, per capita and as a percentage of GNP.** Many countries, rich, poor and middle-class, impose heavy burdens on their citizens for military expenditures, not always for legitimate self-defense or international peacekeeping.

arms production. Cooperation among Western European governments in arms production and distribution is growing, but so is competition among private suppliers, to say nothing of the new producers entering the market. There is also a trend now toward the transfer of arms-production technology, for both political and economic reasons, and this accelerates the decentralization of weapons production in the world.

All of these trends reinforce the importance of the private trade in conventional arms. Today, with the proper connections and enough cash, any revolutionary group, any separatist group, any local warlord, or any fanatical hate group can buy on the open or black market virtually any conventional weapon short of sophisticated aircraft, naval vessels, armor, and artillery. Weapons are openly advertised in widely circulated magazines, and international arms fairs showcase the latest military equipment available for sale to virtually anyone, with few questions asked. In an already unstable world, such activities can hardly lead to greater stabilization.

Most recently, the end of the Cold War and the breakup of the Soviet Union and Yugoslavia have had profound effects on the international arms trade. First, reduction of military forces in the NATO and ex–Warsaw Pact countries has meant a drop in military production there, but has thrown huge quantities of now-surplus, sometimes sophisticated materiel onto the international market. Second, the successor States and other newly independent countries are shopping for materiel for their own "essential" armed forces. Third, more suppliers are entering the market as the Cold War restraints on transfer of military technology weaken (including those on chemical, biological, and nuclear weapons) and production costs fall. Fourth, drug traffickers and terrorists (of old and new varieties) are seeking more sophisticated weapons. The merchants of death are not likely to go out of business very soon.*

What should or can be done about the situation? The decision makers are all working for individual governments, and few of them seem very concerned about it. Perhaps political geographers can increase their awareness of the problem through careful studies and recommendations.

## Linkages

While each of these types of deadly activity is interesting in itself and in varying degree significant in international relations, what makes them especially susceptible to politicogeographical analysis is the linkages among them. These linkages are becoming so complex and pervasive that, if the trend continues, it may not be long before there is one global network of criminal activity ranging from mugging and petty theft to major revolutionary operations. This is a frightening prospect, and for that reason alone likely to be thwarted by governments acting more and more in concert. Meanwhile, however, the criminals are affecting—even upsetting—many of the politicogeographical patterns discussed in this book.

Since all of these criminal activities involve the use of weapons, we begin our discussion of linkages with the international traffic in arms, both government-to-government transfers and the commercial market. Generally, modern pirates are lightly armed and do not yet appear to be participating significantly in the arms trade, but terrorists and drug traffickers certainly are. Except for small arms and simple explosive devices, most of the weapons they use must be provided directly or indirectly by governments. Some are stolen from military armories and ammunition dumps, and some are surplus sold at auction or otherwise openly available. More and more, however, terrorists of various persuasions and purveyors of drugs of various degrees of lethality are deliberately supplied with new and sometimes fairly sophisticated weapons, ammunition, explosives, communications equipment, and other paraphernalia. Particularly in South America and South Asia, there have been numerous all-out battles between narcotics traffickers and politically motivated terrorists, sometimes allied in the same operation, and soldiers and/or police and other law enforcement authorities. In these battles the bad guys sometimes have more and bigger weapons and the good guys don't always win.

There is even substantial evidence that some governments are engaging in drug traffic themselves. Not just an occasional corrupt official or trafficker infiltrated into the government, but the government as a matter of policy controlling the flow of drugs through its territory and extracting commissions for its services, or the government raising cash by selling drugs, or affording protection to the traffickers in return for cash payments, intelligence about rival gangs, or assistance in other areas. In a number of weak, poor, unstable countries, renegade or

---

*One conventional weapon that has killed many thousands of people is the land mine. Scores of millions of them have been planted in some 69 countries (perhaps 10 million in Cambodia alone) and are still killing hundreds of people annually. The UN Development Programme is operating a major mine survey and clearance project, especially in Cambodia and Mozambique, but it will take decades to remove them all. Meanwhile, more will undoubtedly be planted.

mutinous troops have gone into the drug business, taking their weapons with them and sometimes forming alliances with local revolutionaries. Very soon after being expelled from China to Burma in 1949, for example, several divisions of Chinese Nationalist troops fought local rebels, bandits, and opium traffickers until they gradually took over most of the illegal activities there and settled down as warlords in much of northern Burma. Since then similar patterns have been noted in Uganda, Afghanistan, and elsewhere.

Much more important, the former Soviet Union and its allies and, as noted earlier, Iran, Syria, and Libya, are known to have supplied material and other assistance to various groups seeking to destabilize unfriendly or neutral governments. The United States also engages in these activities, though perhaps on a smaller scale, and on a still smaller scale, so do, or have, Israel, South Africa, Turkey, Iraq, Taiwan, India, and perhaps other countries.

The linkages among insurgents, terrorists, drug producers, drug traffickers, and common criminals are well documented. In Myanmar, Afghanistan, Colombia, Peru, Bolivia, Jamaica, the Philippines, and elsewhere, they exchange drugs for weapons, provide intelligence and protection for one another, help one another "launder" money, and otherwise collaborate, ignoring ideological differences in the joint pursuit of profits. The former Soviet Union has become a major arena for all of these activities, and more. A frightening new trend has developed here: the formation of international linkages of gangs engaged in all of these activities, with formal alliances among them and allocations of territory.

It is doubtful that the average user of cocaine or marijuana or any other controlled substance in the United States or Canada or Western Europe truly understands that a portion of the money he or she pays for a moment's pleasure is used to destroy the lives of other people, not only slowly through the provision of more drugs, but violently through murders of rivals, assassination of government officials, slaughter of uncooperative peasants, and killing of law enforcement officers no older than themselves.

Terrorists have also engaged in actions that could be considered piracy. The Irish Republican Army, the POLISARIO Front of Western Sahara, the Basque ETA, the Moro Liberation Front, and the New People's Army in the Philippines have all attacked ships in port and at sea. In 1985 Arab terrorists seized the Italian cruise ship *Achille Lauro* and held it for weeks, killing an American Jewish passenger and terrorizing the others until they were finally allowed to escape without having achieved any of their objectives, except gaining more publicity.

A far larger and potentially more dangerous kind of linkage was revealed in the early 1990s by the BCCI scandal. The Bank of Credit and Commerce International, an Arab-owned transnational corporation operating worldwide, was exposed as having been deeply involved for many years in financing terrorist groups and arms merchants and in laundering drug money. Its tentacles reached widely, into both legitimate and criminal enterprises, and into many governments. We are unlikely ever to learn the full extent of its activities and its accomplices, but for our purposes we already know enough to determine that the BCCI story would be appropriate for analysis by a political geographer. And it is very likely that there is more than one BCCI operating in the world that has not yet been exposed.

To summarize: International criminal activity, whether labeled as drug running, arms trafficking, terrorism, or piracy, is likely to be closely linked with all of the other categories and aided, actively or passively, by some governments. There are those who try to justify any or all of these activities or excuse them on sociological or ideological grounds, but their impact on civilized society is incalculable and inexcusable. Bringing the question down to an individual level, we can do no better than to paraphrase Golda Meir, late prime minister of Israel, commenting on Arab terrorism: It is possible that somewhere there is a cause worth dying for, but there can surely be no cause worth killing for.

# BIBLIOGRAPHY FOR PART SIX

## *Books and Monographs*

**A**

Abbott, John, *Politics and Poverty; A Critique of the Food and Agriculture Organisation of the United Nations.* London: Routledge, 1992.

Abegaz, Berhanu and others (eds.), *The Challenge of European Integration; Internal and External Problems of Trade and Money.* Boulder, CO: Westview, 1994.

*The Academic Council on the United States System,* Thomas J. Watson, Jr. Institute for International Studies, Brown Univ., Providence, RI 02912, publishes occasional authoritative monographs on the United Nations that can be very useful.

Africa Research Centre for the Black Caucus, *The Sanctions Weapon; A Summary of the Debate over Sanctions Against South Africa.* Cape Town: Buchu Books, 1989.

Agnew, John A. and Paul Knox, *The Geography of the World-Economy.* New York: Edward Arnold, 1989.

Agosin, Manuel R. and Diana Tussie (eds.), *Trade and Growth; New Dilemmas in Trade Policy.* Halifax: Dalhousie Univ. Press, 1993.

Ahiakpor, James C.W., *Multinationals and Economic Development; The Integration of Competing Theories.* London: Routledge, 1991.

Albinski, Henry S., *ANZUS, the United States and Pacific Security.* Lanham, MD: Univ. Press of America, 1987.

Alexander, Yonah, *International Terrorism: Political and Legal Documents.* Norwell, MA: Kluwer, 1992.

Alexander, Yonah and Charles K. Ebinger (eds.), *Political Terrorism and Energy: The Threat and Response.* New York: Praeger, 1982.

Alexander, Yonah and Allan Nanes (eds.), *Legislative Responses to Terrorism.* Dordrecht: Nijhoff, 1986.

Alexander, Yonah and Alan O'Day (eds.), *The Irish Terrorism Experience.* Hanover, NH: Dartmouth, 1991.

Alexander, Yonah and Eugene Sochor, *Aerial Piracy and Aviation Security.* Dordrecht: Nijhoff, 1990.

Alexandrowicz, Charles Henry, *The Law-Making Functions of the Specialised Agencies of the United Nations.* London: Angus & Robertson, 1973.

Ali, Almeen, *Land-locked States and International Law with Special Reference to the Role of Nepal.* New Delhi: South Asian Publishers, 1989.

Allan, Richard, *Terrorism; Pragmatic International Deterrence and Cooperation.* Boulder, CO: Westview, 1991.

al-Sowayegh, Abdulaziz, *Arab Petro-Politics.* New York: St. Martin's, 1984.

Amate, C.O.C., *Inside the OAU; Pan-Africanism in Practice.* New York: St. Martin's, 1986.

Amin, Samir and others (eds.), *SADCC; Prospects for Disengagement and Development in Southern Africa.* Tokyo: United Nations Univ. Press, 1987.

Anand, R. P., *International Law and the Developing Countries.* Norwell, MA: Kluwer, 1987.

Anania, Giovanni and others (eds.), *Agricultural Trade Conflicts and GATT; New Dimensions in U.S.-European Agricultural Trade Relations.* Boulder, CO: Westview, 1994.

Anderson, Irvine H., *Aramco, the United States and Saudi Arabia.* Princeton, NJ: Princeton Univ. Press, 1987.

Anderson, Kym and Richard Blackhurst, *Regional Integration and the Global Trading System.* New York: St. Martin's, 1993.

Andersson, Thomas, *Multinational Investment in Developing Countries.* New York: Routledge, 1991.

————, *Managing Trade Relations in the New World Economy.* New York: Routledge, 1993.

Anthony, Ian, *The Naval Arms Trade.* Oxford Univ. Press, 1990.

————, *Arms Trade Regulations.* Oxford Univ. Press, 1991.

Archer, Clive, *Organizing Europe: The Institutions of Integration.* New York: Edward Arnold, 1994.

Artis, Michael and Norman Lee (eds.), *The Economics of the European Union; Policy and Analysis.* Oxford Univ. Press, 1994.

Asler, Fen and Christopher J. Maule (eds.), *Canada Among Nations 1990–91; After the Cold War.* Ottawa: Carleton Univ. Press, 1991.

Atkins, J. Pope, *Latin America in the International Political System.* 3rd ed. Boulder, CO: Westview, 1994.

Avery, William P. (ed.), *World Agriculture and the GATT.* Boulder, CO: Lynne Rienner, 1993.

**B**

Baek, Kwang-Il and others (eds.), *The Dilemma of Third World Defense Industries: Supplier Control or Recipient Autonomy?* Boulder, CO: Westview, 1989.

Baer, M. Delal and Guy F. Erb (eds.), *Strategic Sectors in Mexican-U.S. Free Trade*. Washington, DC: CSIS, 1991.

Baer, M. Delal and Sidney Weintraub (eds.), *The NAFTA Debate: Grappling with Unconventional Trade Issues*. Boulder, CO: Lynne Rienner, 1994.

Bagley, Bruce M. and William O. Walker (eds.), *Drug Trafficking in the Americas*. New Brunswick, NJ: Transaction, 1994.

Bahbah, Bishara and Linda Butler, *Israel and Latin America*. New York: St. Martin's, 1986.

Bailey, David and others, *Transnationals and Governments; Recent Policies in Japan, France, Germany, the United Staes and Britain*. New York: Routledge, 1994.

Bailey, Martin, *Oilgate; Sanctions Scandal*. London: Hodder & Stoughton, 1979.

Bakan, Abigail and others, *Imperial Power and Regional Trade; the Caribbean Basin Initiative*. Waterloo, Ont.: Wilfrid Laurier Univ. Press, 1992.

Ball, M. Margaret, *The OAS in Transition*. Durham, NC: Duke Univ. Press, 1969.

———, *The Open Commonwealth*. Durham, NC: Duke Univ. Press, 1971.

Barry, Donald (ed.), *Toward a North American Community? Canada, the U.S. and Mexico*. Boulder, CO: Westview, 1995.

Bayard, Thomas O. and Kimberly Ann Elliott, *Reciprocity and Retaliation: An Evaluation of Aggressive Trade Policies*. Washington, DC: Institute for International Economics, 1994.

Beamish, Paul W., *Multinational Joint Ventures in Developing Countries*. New York: Routledge, 1988.

Beckman, Robert C. and others, *Acts of Piracy in the Malacca and Singapore Straits. Maritime Briefing*, 1, 4 (1994). Durham, NC: International Boundaries Research Unit.

Beliaev, Igor and John Marks (eds.), *Common Ground on Terrorism; Soviet-American Cooperation Against the Politics of Terror*. New York: Norton, 1991.

Berber, Friedrich J., *Rivers in International Law*. London: Stevens, 1959.

Bercovitch, Jacob and Jeffrey Z. Rubin, *Mediation in International Relations; Multiple Approaches to Conflict Management*. New York: St. Martin's, 1993.

Bergsten, C. Fred, *Pacific Dynamism and the International Economic System*. Washington, DC: Institute for International Economics, 1993.

Berliner, Diane T. and others, *Trade Protection in the United States: 31 Case Studies*. Washington, DC: Institute for International Economics, 1986.

Bethlehem, D.L. (ed.), *The Kuwait Crisis; Sanctions and Their Economic Consequences*. Cambridge: Grotius, 1991.

Biersteker, Thomas J., *Multinationals, The State and Control of the Nigerian Economy*. Princeton, NJ: Princeton Univ. Press, 1987.

Biles, Robert E. (ed.), *Inter-American Relations: The Latin American Perspective*. Boulder, CO: Lynne Rienner, 1988.

Billet, Bret L., *Investment Behavior of Multinational Corporations in Developing Areas; Comparing the Development Assistance Committee, Japanese, and American Corporations*. New Brunswick, NJ: Transaction, 1991.

Blair, David J., *Trade Negotiations in the OECD*. New York: Routledge, 1993.

Blandford, David and others (eds.), *North-South Grain Markets and Trade Policies*. Boulder, CO: Westview, 1993.

Block, Alan A. and Alfred W. McCoy (eds.), *The War on Drugs*. Boulder, CO: Westview, 1992.

Bloed, Arie, *The External Relations of the Council for Mutual Economic Assistance*. Dordrecht: Nijhoff, 1988.

———, *The Conference on Security and Cooperation in Europe: Analysis and Basic Documents 1972–1993*. Norwell, MA: Kluwer, 1993.

Bokor-Szegö, Hanna, *The Role of the United Nations in International Litigation*. Amsterdam: North-Holland, 1978.

Bonser, Charles F. (ed.), *Toward a North American Common Market; Problems and Prospects for a New Economic Community*. Boulder, CO: Westview, 1991.

Bouzas, Roberto and Nora Lustig, *Trade Liberalization, Economic Integration, and the Enterprise for the Americas Initiative*. New Brunswick, NJ: Transaction, 1993.

Boxill, Ian, *Ideology and Caribbean Integration*. Kingston Jamaica: Univ. of the West Indies, 1993.

Bradley, Bill and others, *The American Challenge in World Trade: U.S. Interests in the GATT Multilateral Trading System*. Washington, DC: CSIS, 1989.

Broches, Charles F. and Michael S. Spranger, *The Politics and Economics of Columbia River Water*. Seattle: Washington Sea Grant Program, May 1985.

Brown, Bartram S., *The United States and the Politicization of the World Bank*. London: Routledge, 1992.

Brown, Michael Barratt, *Fair Trade; Reform and Realities in the International Trading System*. London: Zed Books, 1993.

Brown-John, C. Lloyd (ed.), *Federal-Type Solutions and European Integration.* Lanham, MD: Univ. Press of America, 1994.

Bruhács, J., *The Law of Non-Navigational Uses of International Watercourses.* Norwell, MA: Kluwer, 1993.

Brus, Marcel and others (eds.), *The United Nations Decade of International Law; Reflections on International Dispute Settlement.* Norwell, MA: Kluwer, 1991.

Brzoska, Michael and Thomas Ohlson, *Arms Transfers to the Third World, 1971–85.* Oxford Univ. Press, 1987.

Bueno, Gerardo and others (eds.), *The Dynamics of North American Trade and Investment: Canada, Mexico, and the United States.* Stanford, CA: Stanford Univ. Press, 1991.

Bugajski, Janusz, *Nations in Turmoil: Conflict and Cooperation in Eastern Europe.* 2nd ed. Boulder, CO: Westview, 1995.

Bushnell, P. Timothy and others (eds.), *State Organized Terror.* Boulder, CO: Westview, 1991.

**C**

Cafruny, Alan W. and Glenda G. Rosenthal (eds.), *The State of the European Community, Vol. 2: The Maastricht Debates and Beyond.* Boulder, CO: Lynne Rienner, 1994.

Carlton, David (ed.), *Controlling the International Transfer of Weaponry and Related Technology.* U.K.: Dartmouth, 1995.

Carter, Barry E., *International Economic Sanctions: Improving the Haphazard U.S. Legal Regime.* Cambridge Univ. Press, 1988.

Cassese, A., *Terrorism, Politics and the Law: The Achille Lauro Affair.* Oxford Univ. Press, 1989.

Cavanagh, John and others (eds.), *Trading Freedom: How Free Trade Affects Our Lives, Work, and Environment.* San Francisco: Food First, 1992.

Červenka, Zdenek (ed.), *Land-locked Countries of Africa.* Uppsala: Scandinavian Institute of African Studies, 1973.

Chaliand, Gerard, *Terrorism: From Popular Struggle to Media Spectacle.* Atlantic Highlands, NJ: Humanities, 1987.

Chauhan, Balbir R., *Settlement of International Water Law Disputes in International Drainage Basins.* Berlin: Erich Schmidt, 1981.

Chetley, Andrew, *The Politics of Baby Foods: Successful Challenges to an International Marketing Strategy.* London: Pinter, 1986.

Chiang, Pei-heng, *Non-governmental Organizations at the United Nations.* New York: Praeger, 1981.

Child, Jack (ed.), *Regional Cooperation for Development and the Peaceful Settlement of Disputes in Latin America.* Dordrecht: Nijhoff, 1987.

Clark, Robert P., *The Basque Insurgents: ETA, 1952–1980.* Boulder, CO: Westview, 1984.

Clements, Frank A., *Arab Regional Organizations.* New Brunswick, NJ: Transaction, 1992.

Clutterbuck, Richard, *Terrorism, Drugs and Crime in Europe After 1992.* New York: Routledge, 1990.

Cohn, Theodore H., *The International Politics of Agricultural Trade; Canadian-American Relations in a Global Agricultural Context.* Vancouver: UBC Press, 1990.

Colas, Bernard (ed.), *Global Economic Cooperation; A Guide to Agreements and Organizations.* Tokyo: United Nations Univ. Press, 1994.

Cole, John, *The Geography of the European Community.* New York: Routledge, 1993.

Conybeare, John A.C., *Trade Wars: The Theory and Practice of International Commercial Rivalry.* New York: Columbia Univ. Press, 1987.

Cowhey, Peter F. and Jonathan D. Aronson, *Managing the World Economy; the Consequences of Corporate Alliances.* New York: Council on Foreign Relations, 1993.

Coyle, William T. and others (eds.), *Agriculture and Trade in the Pacific.* Boulder, CO: Westview, 1992.

Crow, Ben, *Sharing the Ganges; The Politics and Technology of River Development.* Beverly Hills, CA: Sage, 1995.

Cuddington, John T. and others, *Tariffs, Quotas and Trade: The Politics of Protectionism.* San Francisco: ICS Press, 1987.

Cuthbertson, Ian M., *Redefining the CSCE.* Boulder, CO: Westview, 1992.

Cutler, A. Claire and Mark W. Zacher (eds.), *Canadian Foreign Policy and International Economic Regimes.* Vancouver: UBC Press, 1992.

**D**

Damrosch, Lori Fisler (ed.), *The International Court of Justice at a Crossroads.* Dobbs Ferry, NY: Transnational, 1987.

———, *Enforcing Restraint; Collective Intervention in Internal Conflicts.* New York: Council on Foreign Relations, 1993.

Davidson, Scott, *Grenada: A Study in Politics and the Limits of International Law.* Avebury, UK: Gower, 1987.

Dawson, Andrew H., *The Geography of European Integration.* London: Belhaven, 1993.

Degenhardt, Henry W. (compiler), *Treaties and Alliances of the World*. Detroit: Gale Research Co., 1987.

Deger, Saadet, *Military Expenditure in Third World Countries*. New York: Routledge, 1986.

de Lattre, Anne and Arthur M. Fell, *The Club du Sahel; An Experiment in International Cooperation*. Paris: OECD, 1984.

Delors, Jacques, *Our Europe*. New York: Routledge, 1992.

Demas, William G., *Seize the Time: Towards OECS Political Union*. St. Michael, Barbados, 1987.

Deng, Francis M. and I. William Zartman (eds.), *Conflict Resolution in Africa*. Washington, DC: Brookings, 1991.

Destler, I. M., *American Trade Politics*. 2nd ed. Washington, DC: Institute for International Economics, 1992.

Detter, Ingrid, *The International Legal Order*. U.K.: Dartmouth, 1994.

Dicke, Detlev Chr. and Ernst-Ulrich Petersmann (eds.), *Foreign Trade in the Present and a New International Economic Order*. Fribourg, Switzerland: Univ. Fribourg Press, 1988.

Dicken, Peter, *Global Shift: The Internationalization of Economic Activity*. New York: Guilford, 1992.

Diebold, William, Jr. (ed.), *Bilateralism, Multilateralism and Canada in U.S. Trade Policy*. Cambridge, MA: Ballinger, 1988.

Diehl, Paul F., *International Peacekeeping*. Baltimore: Johns Hopkins Univ. Press, 1993.

Dinan, Desmond, *Ever Closer Union?: An Introduction to the European Community*. Boulder, CO: Lynne Rienner, 1994.

Dinstein, Yoram (ed.), *International Law at a Time of Perplexity; Essays in Honour of Shabtai Rosenne*. Dordrecht: Nijhoff, 1988.

Dombrowski, Franz Amadeus, *Ethiopia's Access to the Sea*. Leiden and Cologne: Brill, 1985.

Dorn, Nicholas and others, *Traffickers; Drug Markets and Law Enforcement*. New York: Routledge, 1992.

Doxey, Margaret P., *International Sanctions in Contemporary Perspective*. New York: St. Martin's, 1987.

Doyle, Michael W., *UN Peacekeeping in Cambodia: UNTAC's Civilian Mandate*. Boulder, CO: Lynne Rienner, 1995.

Drake, Daniel and Meric S. Gertler (eds.), *The New Era of Global Competition; State Policy and Market Power*. Montreal: McGill-Queen's, 1991.

Drummond, Ian M. and Norman Hillmer, *Negotiating Freer Trade; The UK, USA, Canada and the Trade Agreements of 1938*. Waterloo, Ont.: Wilfrid Laurier Univ. Press, 1992.

Dubner, Barry Hart, *The Law of International Sea Piracy*. Dordrecht: Nijhoff, 1980.

Dupree, Louis, *Pakistan 1964–1966, Part IV: The Regional Cooperation for Development*. AUFS Reports, South Asia Series, Vol. 10, No. 8, 1966.

Durch, William J. (ed.), *The Evolution of U.N. Peacekeeping; Case Studies and Comparative Analysis*. New York: St. Martin's, 1993.

**E**

Ellen, Eric F. (ed.), *Violence at Sea; A Review of Terrorism, Acts of War and Piracy, and Countermeasures to Prevent Terrorism*. Paris: ICC Publishing (International Chamber of Commerce), 1986.

Elliott, Kimberly A. and Gary C. Hufbauer, *The Costs of U.S. Trade Barriers*. Washington, DC: Institute for International Economics, 1992.

Elliott, Kimberly A. and others, *Economic Sanctions Reconsidered*. 2nd ed. Washington, DC: Institute for International Economics, 1990.

Encarnation, Dennis J., *Dislodging Multinationals; India's Strategy in Comparative Perspective*. Ithaca, NY: Cornell Univ. Press, 1989.

Esedebe, P. Olisanwuche, *Pan-Africanism; The Idea and Movement 1776–1991*. 2nd ed. Washington, DC: Howard Univ. Press, 1994.

Esty, Daniel C., *Greening the GATT: Trade, Environment and the Future*. Washington, DC: Institute for International Economics, 1994.

**F**

Fermann, Gunnar, *Bibliography on International Peacekeeping*. Norwell, MA: Kluwer, 1993.

Fifer, J. Valerie, *Bolivia: Land, Location, and Politics Since 1825*. Cambridge Univ. Press, 1972.

Findlay, Trevor, *Cambodia; The Legacy and Lessons of UNTAC*. Oxford Univ. Press, 1994.

Fishel, Wesley R., *The End of Extraterritoriality in China*. Berkeley: Univ. of California Press, 1952.

Fisher, Bart S. and Kathleen M. Harte (eds.), *Barter in the World Economy*. New York: Praeger, 1985.

Fitzmaurice, John, *Damming the Danube; Gabcikovo and Post-Communist Politics in Europe*. Boulder, CO: Westview, 1995.

Fitzmaurice, M. and others (eds.), *The Changing Political Structure of Europe; Aspects of International Law*. Norwell, MA: Kluwer, 1991.

Flynn, Stephen and Gregory M. Grant, *The Transnational Drug Challenge and the New World Order*. Washington, DC: CSIS, 1992.

Focas, Spiridon G., *The Lower Danube River.* East European Monographs. New York: Columbia Univ. Press, 1987.

Fodor, Neil, *The Warsaw Treaty Organization.* New York: St. Martin's, 1990.

Folke, Steen and others, *South-South Trade and Development; Manufacturers in the New International Divisions of Labour.* New York: St. Martin's, 1993.

Fontaine, Rober W., *The Andean Pact: A Political Analysis.* Lanham, MD: Univ. Press of America, 1977.

Forsythe, David P. (ed.), *The United Nations in the World Political Economy.* New York: St. Martin's, 1989.

French, Hilary F., *Costly Tradeoffs; Reconciling Trade and the Environment.* Washington, DC: Worldwatch Institute, 1993.

Friman, H. Richard, *Patchwork Protectionism: Textile Trade Policy in the United States, Japan, and West Germany.* Ithaca, NY: Cornell Univ. Press, 1990.

**G**

Galtung, Johan, *Europe in the Making.* London: Taylor & Francis, 1989.

Gambari, Ibrahim A., *Political and Comparative Dimensions of Regional Integration; The Case of ECOWAS.* Atlantic Highlands, NJ: Humanities, 1991.

George, Alexander (ed.), *Western State Terrorism.* New York: Routledge, 1991.

George, Stephen, *An Awkward Partner; Britain in the European Community.* Oxford Univ. Press, 1994.

Gereffi, G., *The Pharmaceutical Industry and Dependency in the Third World.* Princeton, NJ: Princeton Univ. Press, 1983.

Gibb, Richard and Wieslaw Michalak (eds.), *Continental Trading Blocs: The Growth of Regionalism in the World Economy.* New York: John Wiley, 1994.

Gilroy, Bernard Michael, *Networking in Multinational Enterprises: The Importance of Strategic Alliances.* Columbia: Univ. of South Carolina Press, 1992.

Glassner, Martin Ira, *Access to the Sea for Developing Land-locked States.* Dordrecht: Nijhoff, 1970.

———, "CARICOM and the Future of the Caribbean," in Gary S. Elbow, ed. *International Aspects of Development in Latin America: Geographical Perspectives.* Muncie, IN: CLAG Publications, 1977, 111–117.

———, "The Transit Problems of Land-locked States; the Cases of Bolivia and Paraguay," *Ocean Yearbook 4.* Chicago: Univ. of Chicago Press, 1983, 366–389.

———, *Bibliography on Land-locked States.* 4th ed. Dordrecht: Nijhoff, 1995.

Godana, Bonaya Adhi, *Africa's Shared Water Resources: Legal and Institutional Aspects of the Nile, Niger and Senegal River Systems.* London: Pinter, 1985.

Golubev, G. and A. Biswas, *Large Scale Water Transfers.* London: Taylor & Francis, 1985.

Gorove, Stephen, *Law and Politics of the Danube.* Dordrecht: Nijhoff, 1964.

Govitrikar, Vishwas P. and others (eds.), *Electronic Highways for World Trade: Issues in Telecommunications and Data Services.* Boulder, CO: Westview, 1989.

Graham, Edward M. and Paul R. Krugman, *Foreign Direct Investment in the United States.* Washington, DC: Institute for International Economics, 1994.

Grant, Richard L. and others, *Asia Pacific Economic Cooperation: The Challenge Ahead.* Washington, DC: CSIS, 1990.

*The Great Tin Crash; Bolivia and the World Tin Market.* London: Latin America Bureau, 1987.

Green, L.C., *International Law—A Canadian Perspective.* Toronto: Carswell, 1988.

Grieco, Joseph M., *Between Dependency and Autonomy: India's Experience with the International Computer Industry.* Berkeley: Univ. of California Press, 1984.

———, *Cooperation Among Nations: Europe, America, and Non-tariff Barriers to Trade.* Ithaca, NY: Cornell Univ. Press, 1990.

Grilli, Enzo R., *The European Community and the Developing Countries.* Cambridge Univ. Press, 1992.

Grinspun, Ricardo and Maxwell A. Cameron, *The Political Economy of North American Free Trade.* New York: St. Martin's, 1993.

Grosse, Robert, *Multinationals in Latin America.* London: Routledge, 1989.

Gruhn, Isebill V., *Regionalism Reconsidered: The Economic Commission for Africa.* Boulder, CO: Westview, 1979.

Grunwald, Joseph and Kenneth Flamm, *The Global Factory: Foreign Assembly in International Trade.* Washington, DC: Brookings, 1985.

**H**

Haas, Michael, *The Asian Way to Peace; A Story of Regional Cooperation.* New York: Praeger, 1989.

————, *The Pacific Way; Regional Cooperation in the South Pacific.* New York: Praeger, 1989.

Hallstein, Walter, *United Europe: Challenge and Opportunity.* Cambridge, MA: Harvard Univ. Press, 1962.

Hamilton, Geoffrey (ed.), *Red Multinationals or Red Herrings? The Activities of Enterprises from Socialist Countries in the West.* London: Pinter, 1986.

Han, Henry H., *Problems and Prospects of the Organization of American States.* Peter Lang, 1987.

————, *Terrorism and Political Violence: Limits and Possibilities of Legal Control.* New York: Oceana, 1993.

Hanlon, Joseph, *SADCC in the 1990s; Development on the Front Line.* London: The Economist Intelligence Unit Special Report No. 1158, September 1989.

Hanlon, Joseph and Roger Omond, *The Sanctions Handbook.* New York: Penguin, 1987.

Hansson, Gote (ed.), *Trade, Growth and Development; The Role of Politics and Institutions.* New York: Routledge, 1993.

Haq, Khadija (ed.), *Linking the World; Trade Policies for the Future.* Islamabad: North-South Roundtable, 1988.

Hardt, John P. and Young C. Kim (eds.), *Economic Cooperation in the Asia-Pacific Region.* Boulder, CO: Westview, 1990.

Harris, Richard G., *Trade, Industrial Policy and International Competition.* Toronto: Univ. of Toronto Press, 1987.

Harrod, Jeffrey and Nico Schrijver (eds.), *The U.N. under Attack.* Aldershot, U.K.: Gower, 1988.

Hart, Michael M., *Canadian Economic Development and the International Trading System.* Toronto: Univ. of Toronto Press, 1987.

Hart, Michael M. and others, *Decision at Midnight; Inside the Canada–U.S. Free Trade Negotiations.* Vancouver: UBC Press, 1994.

Hartland-Thunberg, Penelope, *China, Hong Kong, Taiwan and the World Trading System.* Washington, DC: CSIS, 1990.

Hathaway, Dale E. and William E. Miner (eds.), *World Agricultural Trade: Building a Consensus.* Washington, DC: Institute for International Economics, 1988.

Haus, Leah A., *Globalizing the GATT; The Soviet Union's Successor States, Eastern Europe, and the International Trading System.* Washington, DC: Brookings, 1992.

Hayes, J. P., *Economic Effects of Sanctions on Southern Africa.* Aldershot, U.K.: Gower, 1987.

Hayward, David J., *International Trade and Regional Economies: The Impacts of European Integration on the United States.* Boulder, CO: Westview, 1995.

Heinz, John, *U.S. Strategic Trade; An Export Control System for the 1990s.* Boulder, CO: Westview, 1991.

Henig, Ruth B. (ed.), *The League of Nations.* New York: Harper and Row, 1973.

Henkin, Louis and others, *Right v. Might: International Law and the Use of Force.* New York: Council on Foreign Relations, 1989.

Henning, C. Randall and others, *Reviving the European Union.* Washington, DC: Institute for International Economics, 1994.

Historical Dictionaries Series on International Organizations. Volumes issued to date: *The European Community, International Monetary Fund, International Organizations in Sub-Saharan Africa, European Organizations, International Tribunals, The International Food Agencies, Refugee and Disaster Relief Organizations.* Metuchen, NJ: Scarecrow Press, 1993–.

Hoffmann, Stanley and Robert O. Keohane, *The New European Community; Decisionmaking and Institutional Change.* Boulder, CO: Westview, 1991.

Holland, Stuart and George Irvin (eds.), *Central America: The Future of Economic Integration.* Boulder, CO: Westview, 1989.

Honig, Jan W., *NATO; An Institution under Threat?* Boulder, CO: Westview, 1991.

Hudson, Manley O., *The Permanent Court of International Justice.* New York: Macmillan, 1934.

Hufbauer, Gary Clyde and Jeffrey J. Schott, *North American Free Trade: Issues, Recommendations, Results.* Washington, DC: Institute for International Economics, 1993.

————, *Western Hemisphere Economic Integration.* Washington, DC: Institute for International Economics, 1994.

Hufbauer, Gary C. and others, *Economic Sanctions Reconsidered; History and Current Policy.* 2nd ed. Washington, DC: Institute for International Economics, 1990. Two volumes.

Hume, Cameron, *Ending Mozambique's War; The Role of Mediation and Good Offices.* Washington, DC: U.S. Institute of Peace Press, 1994.

Hume, Cameron R., *The United Nations, Iran, and Iraq; How Peacemaking Changed.* Bloomington: Indiana Univ. Press, 1994.

Hunter, Shireen, *Gulf Cooperation Council; Problems and Prospects.* Washington, DC: CSIS, 1984.

———— (ed.), *The PLO after Tripoli.* Lanham, MD: Univ. Press of America, 1984.

Hurwitz, Leon and Christian Lequesne (eds.), *The State of the European Community: Politics, Institutions, and Debates into the 1990s.* Boulder, CO: Lynne Rienner, 1991.

Hyland, Francis P., *Armenian Terrorism; the Past, the Present, the Prospects.* Boulder, CO: Westview, 1991.

**I**

Ives, Jane H. (ed.), *Export of Hazard: Transnational Corporations & Environmental Control Issues.* London: Methuen, 1985.

**J**

James, Harvey S. Jr. and Murray Weidenbaum, *When Businesses Cross International Borders: Strategic Alliances and Their Alternatives.* Washington, DC: CSIS, 1993.

Jamieson, Alison (ed.), *Terrorism and Drug Trafficking in Europe in the 1990s.* U.K.: Dartmouth, 1994.

Janis, Mark W., *International Courts for the Twenty-First Century.* Norwell, MA: Kluwer, 1992.

Jeffery, Charlie and Roland Sturm (eds.), *Federalism, Unification and European Integration.* London: Frank Cass, 1993.

Jenkins, Rhys, *Transnational Corporations and the Latin American Automobile Industry.* Pittsburgh: Univ. of Pittsburgh Press, 1987.

————, *Transnational Corporations and Uneven Development.* London: Methuen, 1987.

Jockel, Joseph T., *Canada and International Peacekeeping.* Washington, DC: CSIS, 1994.

Johnson, D. Gale and others, *Agricultural Policy and U.S.–Taiwan Trade.* Washington, DC: AEI Press, 1993.

Johnson, Phyllis and David Martin, *Apartheid Terrorism; The Destabilization Report.* Bloomington: Indiana Univ. Press, 1989.

Johnston, Mary Troy, *The European Council: Gatekeeper of the European Community.* Boulder, CO: Westview, 1994.

Jones, Patrice F., *The Brazilian Defense Industry: A Case Study of Public-Private Collaboration.* Boulder, CO: Westview, 1991.

Jopp, Mathias and others (eds.), *Integration and Security in Western Europe; Inside the European Pillar.* Boulder, CO: Westview, 1991.

Judge, David (ed.), *A Green Dimension for the European Community; Political Issues and Processes.* London: Frank Cass, 1993.

**K**

Kaempfer, William H. and Anton D. Lowenberg, *International Economic Sanctions.* Boulder, CO: Westview, 1992.

Kaplan, William and Donald McRae (eds.), *Law, Policy and International Justice.* Montreal: McGill-Queens, 1993.

Katz, Mark N. (ed.), *Soviet-American Conflict Resolution in the Third World.* Arlington, VA: U.S. Institute of Peace Press, 1991.

Kennedy, S., *The Pan-Angles: A Consideration of the Federation of the Seven English-Speaking Nations.* London: Longmans, Green, 1914.

Khan, Khushi M. (ed.), *Multinationals of the South.* New York: St. Martin's, 1986.

Khong, Cho Oon, *The Politics of Oil in Indonesia: Foreign Company-Host Government Relations.* Cambridge Univ. Press, 1986.

Kliot, Nurit, *Water Resources and Conflict in the Middle East.* New York: Routledge, 1994.

Köchler, Hans (ed.), *Terrorism and National Liberation.* Peter Lang, 1988.

Korhonen, Pekka, *Japan and the Pacific Free Trade Area.* New York: Routledge, 1994.

Koul, Autar K., *The Legal Framework of UNCTAD in World Trade.* Leiden: Sijthoff, 1977.

Krause, Keith, *Arms and the State: Patterns of Military Production and Trade.* Cambridge Univ. Press, 1992.

Krueger, Anne O. (ed.), *Development with Trade; LDCs and the International Economy.* Woodbridge, VA: Sequoia Institute, 1988.

————, *American Trade Policy, A Tragedy in the Making.* LaVergne, TN: AEI, 1995.

Kwan, C. H., *Economic Interdependence in the Asia-Pacific Region; Towards a Yen Bloc.* New York: Routledge, 1994.

**L**

Laffan, Brigid, *Integration and Co-operation in Europe.* New York: Routledge, 1992.

Lall, Arthur, *The UN and the Middle East Crisis, 1967.* New York: Columbia Univ. Press, 1968.

Lambert, Joseph J., *Terrorism and Hostages in International Law.* Cambridge: Grotius, 1990.

Lammers, J. G., *Pollution of International Watercourses: A Search for Substantive Rules and Principles of Law.* Dordrecht: Nijhoff, 1984.

Langdon, Frank, *The Politics of Canadian-Japanese Economic Relations, 1952–1983.* Vancouver: UBC Press, 1984.

Larby, Patricia M. and Harry Hannam, *The Commonwealth.* New Brunswick, NJ: Transaction, 1993.

Laursen, Finn and Sophie Vanhoonacker, *The Intergovernmental Conference on Political Union; 1990–1991: Rome, Italy, and Maastricht, Netherlands*. Norwell, MA: Kluwer, 1992.

Lauterpacht, Hersh, *Recognition in International Law*. Cambridge, 1947.

———, *The Development of International Law by the International Court*. London: 1958.

Lavigne, Marie, *East-South Relations in the World Economy*. New York: Praeger, 1988.

*The League of Nations in Retrospect: Proceedings of the Symposium*. Berlin: de Gruyter, 1983.

Lee, Lai To, *The Reunification of China; PRC–Taiwan Relations in Flux*. New York: Praeger, 1991.

Lee, Rensselaer W., III, *The White Labyrinth: Cocaine and Political Power*. New Brunswick, NJ: Transaction, 1989.

Le Marquand, David G., *International Rivers: The Politics of Cooperation*. Vancouver: Univ. of British Columbia, Westwater Research Center, 1977.

Lemco, Jonathan, *Canada and the Crisis in Central America*. New York: Praeger, 1991.

Leonardi, Robert (ed.), *The Regions and the European Community*. London: Frank Cass, 1993.

Leonardi, Robert and Raffaella Y. Nanetti (eds.), *The Regions and European Integration; The Case of Emilia-Romagna*. London: Pinter, 1991.

Leventhal, Paul and Yonah Alexander (eds.), *Preventing Nuclear Terrorism*. Lexington, MA: Lexington Books, 1987.

Levitt, Geoffrey N., *Democracies Against Terror: The Western Response to State-Supported Terrorism*. New York: Praeger, 1988.

Linnemann, Hans (ed.), *South-South Trade Preferences*. Beverly Hills: Sage, 1992.

Lintner, Bertil, *Cross-Border Drug Trade in the Golden Triangle (S. E. Asia)*. International Boundaries Research Unit, Univ. of Durham. Territory Briefing No. 1, 1991.

Liu, F. T., *United Nations Peacekeeping and the New Use of Force*. Boulder, CO: Lynne Rienner, 1992.

Lodge, Juliet, *The European Community and New Zealand*. London: Longwood, 1982.

———, *The European Community and the Challenge of the Future*. New York: St. Martin's, 1993.

Looney, Robert E., *The Political Economy of Latin American Defense Expenditures; Case Studies of Venezuela and Argentina*. Lexington, MA: Lexington Books, 1986.

Louscher, David J. and Michael D. Salomone (eds.), *Marketing Security Assistance*. Lexington, MA: Lexington Books, 1987.

Low, Patrick, *Trading Free; The GATT and U.S. Trade Policy*. Washington, DC: Brookings, 1993.

Luard, Evan, *The United Nations*. New York: St. Martin's, 1985.

Luciani, Giacomo, *The Oil Companies and the Arab World*. New York: St. Martin's, 1984.

Lundahl, Mats and Lennart Peterson, *The Dependent Economy; Lesotho and the Southern African Customs Union*. Boulder, CO: Westview, 1991.

Lustig, Nora and others (eds.), *North American Free Trade; Assessing the Impact*. Washington, DC: Brookings, 1992.

**M**

Maachou, Abdelkader, *OAPEC; Organization of Arab Petroleum Exporting Countries*. New York: St. Martin's, 1982.

MacDonald, Robert W., *The League of Arab States*. Princeton, NJ: Princeton Univ. Press, 1965.

MacDonald, Scott B., *Dancing on a Volcano: The Latin American Drug Trade*. New York: Praeger, 1988.

———, *Mountain High, White Avalanche: Cocaine and Power in the Andean States and Panama*. Washington, DC: CSIS, 1989.

MacGaffey, Janet, *The Real Economy of Zaire: The Contribution of Smuggling and Other Unofficial Activities to National Wealth*. Philadelphia: Univ. of Pennsylvania Press, 1991.

Mackinlay, John, *The Peacekeepers: An Assessment of Peacekeeping Operations at the Arab-Israeli Interface*. London: Unwin Hyman, 1989.

Makarcxyk, Jerzy, *Principles of a New International Economic Order: A Study of International Law in the Making*. Dordrecht: Nijhoff, 1988.

Malamud-Goti, Jaime E., *Smoke and Mirrors: The Paradox of the Drug Wars*. Boulder, CO: Westview, 1992.

Maloney, Clarence, *SAARC: The Nations of South Asia Begin to Cooperate*. UFSI Reports Asia 1988-89/No. 9.

Mamuddin, Hasan, *The Charter of the Islamic Conference and Legal Framework of Economic Co-operation Among Its Member States*. Oxford Univ. Press, 1987.

Marchildon, Geoffrey and Duncan McDowall (eds.), *Canadian Multinationals and International Finance*. London: Frank Cass, 1992.

Marlin-Bennet, Renée, *Food Fights; International Regimes and the Politics of Agricultural Trade Disputes*. New York: Gordon and Breach, 1993.

Martin, John M. and Anne T. Romano, *Multinational Crime: Terrorism, Espionage, Drugs and Arms Trafficking.* Beverly Hills: Sage, 1992.

Martin, Linda G. (ed.), *The ASEAN Success Story; Social, Economic, and Political Dimensions.* Honolulu: Univ. of Hawaii Press, 1987.

Martin, Philip, *Migration and Trade: The Case of NAFTA.* Washington, DC: Institute for International Economics, 1993.

Martz, Mary Jeanne Reid, *The Central American Soccer War: Historical Patterns and Internal Dynamics of OAS Settlement Procedures.* Athens: Ohio Univ. Center for International Studies, 1978.

Mason, Mark, *American Multinationals and Japan; The Political Economy of Japanese Capital Controls, 1899–1980.* Cambridge, MA: Harvard Univ. Press, 1992.

Mastanduno, Michael, *Economic Containment: CoCom and the Politics of East-West Trade.* Ithaca, NY: Cornell Univ. Press, 1992.

Mazzeo, Domenico (ed.), *African Regional Organizations.* Cambridge Univ. Press, 1984.

McCoubrey, Hilaire and Nigel D. White, *International Organizations and Civil Wars.* U.K.: Dartmouth, 1995.

McKibbin, Warwick J. and Jeffrey D. Sachs, *Global Linkages; Macroeconomic Interdependence and Cooperation in the World Economy.* Washington, DC: Brookings, 1991.

McKinney, Joseph A. and M. Rebecca Sharpless, *Implications of a North American Free Trade Region.* Ottawa: Carleton Univ. Press, 1992.

McMahon, J. A., *Agricultural Trade, Protectionism and the Problems of Development.* London: Belhaven, 1991.

McMillan, Carl H., *Multinationals from the Second World.* New York: St. Martin's, 1987.

McWhinney, Edward, *Judicial Settlement of International Disputes; Jurisdiction, Justiciability and Judicial Law-Making of the Contemporary International Court.* Norwell, MA: Kluwer, 1991.

Menkveld, Paul A., *Origin and Role of the European Bank for Reconstruction and Development.* Norwell, MA: Kluwer, 1991.

Merari, Ariel and Shlomo Elad, *The International Dimension of Palestinian Terrorism.* Boulder, CA: Westview, 1987.

Merrills, John G., *International Dispute Settlement.* 2nd ed. Cambridge: Grotius, 1991.

Michel, Aloys A., *The Indus Rivers: A Study of the Effects of Integration.* New Haven, CT: Yale Univ. Press, 1967.

Mickolus, Edward F. and Potter A. Flemming (eds.), *Terrorism, 1980–1987: A Selectively Annotated Bibliography.* Westport, CT: Greenwood, 1988.

Mirza, Hafiz, *Multinationals and the Growth of the Singapore Economy.* New York: St. Martin's, 1986.

Molle, Willem, *The Economics of European Integration: Theory, Practice, Policy.* Brookfield, VT: Dartmouth, 1990.

Morales, Edmundo, *Cocaine: White Gold Rush in Peru.* Tucson: Univ. of Arizona Press, 1989.

Morgenstein, Felice, *Legal Problems of International Organizations.* Cambridge: Grotius, 1986.

Moss, Ambler H. (ed.), *NAFTA: Assessments of the North American Free Trade Agreement.* New Brunswick, NJ: Transaction, 1994.

Mueller, Gerhard and Freda Adler, *Outlaws of the Ocean: The Complete Book of Crime on the High Seas.* New York: Hearst Marine Books, 1985.

Muñoz, Heraldo (ed.), *Environment and Diplomacy in the Americas.* Boulder, CO: Lynne Rienner, 1992.

Muñoz, Heraldo and Robin Rosenberg (eds.), *Difficult Liaison: Trade and the Environment in the Americas.* New Brunswick, NJ: Transaction, 1993.

Mytelka, Lynn Krieger, *Regional Development in a Global Economy; The Multinational Corporation, Technology and Andean Integration.* New Haven, CT: Yale Univ. Press, 1979.

**N**

Nakhleh, Emile A., *The Gulf Cooperation Council.* New York: Praeger, 1986.

Naldi, Gino J., *Documents of the Organization of African Unity.* London: Mansell, 1989.

————, *The Organization of African Unity; An Analysis of Its Role.* London: Mansell, 1989.

Nanetti, Raffaella Y., *The Rise of the Periphery: Development Planning in the Regions of the European Community.* London: Frank Cass, 1994.

Nau, Henry R., *Domestic Trade Politics and the Uruguay Round.* New York: Columbia Univ. Press, 1988.

Nelson, Brent F. and Alexander C.G. Stubb, *The European Union: Readings on the Theory and Practice of European Integration.* Boulder, CO: Lynne Rienner, 1994.

Newbigin, Marion I., *Geographical Aspects of the Balkan Problems in Their Relation to the Great European War.* New York: Putnam, 1915.

Nierop, Tom, *Systems and Regions in Global Politics; An Empirical Study of Diplomacy, International Organization and Trade, 1950–1991.* New York: Wiley, 1994.

Nkonoki, Simon R., *Regional Development Planning of the Kagera River Basin in Eastern Africa under the Kagera Basin Organisation (KBO); A Case Study of Hydropower Development Planning and Related Environmental Impacts.* Bergen, Norway: DERAP Publications, November 1983.

Noland, Marcus, *Pacific Dynamism and the International Economic System.* Washington, DC: Institute for International Economics, 1993.

Nollkaemper, André, *The Legal Regime for Transboundary Water Pollution: Between Discretion and Constraint.* Norwell, MA: Kluwer, 1993.


**O**

Odén, Bertil (ed.), *Southern Africa After Apartheid.* Uppsala: Scandinavian Institute of African Studies, 1993.

Ofuatey-Kodjoe, W. (ed.), *Pan-Africanism: New Directions in Strategy.* Univ. Press of America, 1986.

Ojo, Olusola, *Africa and Israel: Relations in Perspective.* Boulder, CO: Westview, 1988.

Okolo, Julius (ed.), *West Africa: Regional Cooperation and Development.* Boulder, CO: Westview, 1988.

Osiander, Andreas, *The States System of Europe, 1640–1990; Peacemaking and the Conditions of International Stability.* Oxford Univ. Press, 1994.


**P**

Page, Sheila and others, *Trading with South Africa; The Policy Options for the EC.* London: Overseas Development Institute, 1992.

Painter, James, *Bolivia and Coca: A Study in Dependency.* Boulder, CO: Lynne Reinner, 1993.

Palankai, Tibor, *The European Community and Central European Integration.* Boulder, CO: Westview, 1991.

Palmer, Ronald D. and Thomas J. Rockford, *Building ASEAN: 20 Years of Southeast Asian Cooperation.* New York: Praeger, 1987.

Papadopoulos, Andrestinos N., *Multilateral Diplomacy Within the Commonwealth: A Decade of Expansion.* Dordrecht: Nijhoff, 1982.

Parker, Geoffrey, *A Political Geography of Community Europe.* London: Butterworths, 1983.

Payne, Anthony J., *The Politics of the Caribbean Community 1961–79: Regional Integration Among New States.* New York: St. Martin's, 1980.

Pearson, Charles S. (ed.), *Multinational Corporations, Environment and the Third World: Business Matters.* Durham, NC: Duke Univ. Press, 1987.

Pelcovits, Nathan A., *The Long Armistice: UN Peacekeeping and the Arab-Israeli Conflict, 1948–1960.* Boulder, CO: Westview, 1993.

Peng, Martin Khor Kok, *The Uruguay Round and Third World Sovereignty.* Penang, Malaysia: Third World Network, 1990.

———, *The Future of North-South Relations: Conflict or Cooperation?* Penang, Malaysia: Third World Network, 1992.

Petersmann, Ernst-Ulrich and Meinhard Hilf (eds.), *The New GATT Round of Multilateral Trade Negotiations: Legal and Economic Problems.* 2nd rev. ed. Norwell, MA: Kluwer, 1991.

Peterson, Dean F. and A. Berry Crawford (eds.), *Values and Choices in the Development of the Colorado River Basin.* Tucson: Univ. of Arizona Press, 1978.

Peterson, Erik R., *The Gulf Cooperation Council: Search for Unity in a Dynamic Region.* New York: Routledge, 1988.

———, *Multinational Corporations in the United States.* Washington, DC: CSIS, 1994.

Philip, George, *The Political Economy of International Oil.* Irvington, NY: Columbia Univ. Press, 1994.

Phillips, Peter W.B., *Wheat, Europe and the GATT.* London: Belhaven, 1990.

Pinder, John, *European Community, The Building of a Union.* Oxford Univ. Press, 1991.

Pogany, Istvan, *The Arab League and Peacekeeping in the Lebanon.* Aldershot, U.K.: Avebury, 1987.

Pomfret, Richard, *Mediterranean Policy of the European Community: A Study of Discrimination in Trade.* New York: St. Martin's, 1986.

Potholm, Christian P. and Richard A. Fredland (eds.), *Integration and Disintegration in East Africa.* Lanham, MD: Univ. Press of America, 1980.

Pradhan, Gajendra Mani, *Transit of Land-locked Countries and Nepal.* Jaipur: Nirala, 1990.

Preeg, Ernest H., *Trade Policy Ahead: Three Tracks and One Question.* Washington, DC: CSIS, 1995.

Pryce, Roy and John Pinder (eds.), *Maastricht and Beyond; Building a European Union.* New York: Routledge, 1994.

Pugh, Michael, The *ANZUS Crisis, Nuclear Visiting and Deterrence*. Cambridge Univ. Press, 1989.

Purcell, Susan Kaufman and Francoise Simon (eds.), *Europe and Latin America in the World Economy*. Boulder, CO: Lynne Rienner, 1994.

**R**

Ra'anan, Uri and others, *Hydra of Carnage; International Linkages of Terrorism*. Lexington, MA: Lexington Books, 1987.

Raghavan, Chakravarthi, *Recolonization: GATT, the Uruguay Round and the Third World*. Penang, Malaysia: Third World Network and Zed Books, 1990.

Raine, Linnea P. and Frank J. Cilluffo (eds.), *Global Organized Crime: The New Empire of Evil*. Washington, DC: CSIS, 1994.

Ramazani, Ruohollah K., *The Gulf Cooperation Council; Record and Analysis*. Charlottesville: Univ. Press of Virginia, 1988.

Ramcharan, B. G. and L. B. Francis (eds.), *Caribbean Perspectives on International Law and Organizations*. Dordrecht: Nijhoff, 1989.

Ramsaran, Ramesh, *The Commonwealth Caribbean in the World Economy*. New York: Macmillan, 1989.

Ray, John and Geoffrey Till, *The East-West Relations in the 1990's*. New York: St. Martin's, 1990.

Rees, Judith and Peter Odell, *The International Oil Industry*. New York: St. Martin's, 1987.

Renner, Michael, *Critical Juncture; The Future of Peacekeeping*. Washington, DC: Worldwatch Institute, 1993.

Renninger, John P., *The Future Role of the United Nations in an Interdependent World*. Dordrecht: Nijhoff, 1989.

Rikhye, Indar Jit, *Strengthening UN Peacekeeping: New Challenges and Proposals*. Washington, DC: USIP Press, 1992.

Roberts, Adam and Benedict Kingsbury, *United Nations, Divided World; The UN's Roles in International Relations*. Oxford Univ. Press, 1994.

Robinson, Peter and others (eds.), *Electronic Highways for World Trade; Issues in Telecommunications and Data Services*. Boulder, CO: Westview, 1989.

Ronzitti, Natalino (ed.), *Maritime Terrorism and International Law*. Dordrecht: Nijhoff, 1990.

Rotfeld, Adam Daniel (ed.), *From Helsinki to Helsinki and Beyond; Analysis and Documents of the Conference on Security and Co-operation in Europe, 1973–93*. Oxford Univ. Press, 1994.

Rubin, Alfred P., *The Law of Piracy*. Newport, RI: Naval War College Press, 1988.

Rubin, Barry (ed.), *The Politics of Counterterrorism: The Ordeal of Democratic States*. Lanham, MD: Univ. Press of America, 1990.

Rubin, Seymour J. and Gary Clyde Hufbauer (eds.), *Emerging Standards of International Trade and Investment: Multinational Codes and Corporate Conduct*. Totowa, NJ: Rowman & Allanheld, 1984.

Rubner, Alex, *The Might of the Multinationals; The Rise and Fall of the Corporate Legend*. Westport, CT: Greenwood, 1990.

Rugman, Alan M., *Multinationals and Canada–United States Free Trade*. Columbia: Univ. of South Carolina Press, 1990.

Runge, C. Ford and others, *Freer Trade, Protected Environment; Balancing Trade Liberalization and Environmental Interests*. New York: Council on Foreign Relations, 1994.

**S**

Saasa, Oliver S., *Joining the Future; Economic Integration and Co-operation in Africa*. Nairobi: ACTS Press, 1991.

Sahnoun, Mohamed, *Somalia; The Missed Opportunities*. Washington, DC: U.S. Institute of Peace Press, 1994.

Saliba, Samir N., *The Jordan River Dispute*. Dordrecht: Nijhoff, 1968.

Salvatore, Dominick, *Protectionism and World Welfare*. Cambridge Univ. Press, 1993.

Sanderson, Steven E., *The Politics of Trade in Latin American Development*. Stanford, CA: Stanford Univ. Press, 1992.

Sandwick, John A., *The Gulf Cooperation Council*. Boulder, CO: Westview, 1987.

Sarna, Aaron J., *Boycott and Blacklist: A History of Arab Economic Warfare Against Israel*. Totowa, NJ: Rowman & Littlefield, 1986.

Sauvant, Karl P., *The Group of 77; Evolution, Structure, Organization*. New York: Oceana, 1981.

———, *The Collected Documents of the Group of 77*. New York: Oceana, 1981–.

———, *International Transactions in Services: The Politics of Transborder Data Flows*. Boulder, CO: Westview, 1986.

Savary, Julian, *French Multinationals*. London: Pinter, 1984.

Sbragia, Alberta M. (ed.), *Euro-Politics; Institutions and Policymaking in the "New' European Community*. Washington, DC: Brookings, 1991.

Schive, Chi (Hsueh, Ch'i), *The Foreign Factor; The Multinational Corporation's Contribution to the Economic Modernization of the Republic of China.* Stanford, CA: Hoover Institution Press, 1990.

Schmid, Alex P. and Ronald D. Crelenstein (eds.), *Western Responses to Terrorism.* London: Frank Cass, 1993.

Schoenberg, Harris O., *A Mandate for Terror; The United Nations and the PLO.* New York: Shapolsky, 1988.

Schott, Jeffrey J. (ed.), *The Uruguay Round: An Assessment.* Washington, DC: Institute for International Economics, 1994.

Schwartz, Herman M., *States Versus Markets; History, Geography, and the Development of the International Political Economy.* New York: St. Martin's, 1994.

Scudder, Thayer, *African River Basin Development.* Boulder, CO: Westview, 1991.

Sesay, Amadu and others, *The OAU After Twenty Years.* Boulder, CO: Westview, 1984.

Shaw, Timothy M. and Julius Emeka Okolo (eds.), *The Political Economy of Foreign Policy in ECOWAS.* New York: St. Martin's, 1994.

Shaw, Timothy M. and Yash Tandon (eds.), *Regional Development at the International Level,* Vol. II: *African and Canadian Perspectives.* Lanham, MD: Univ. Press of America, 1985.

Shepherd, George W., Jr. (ed.), *Effective Sanctions on South Africa; The Cutting Edge of Economic Intervention.* New York: Praeger, 1991.

Sherry, George L., *The United Nations Reborn: Conflict Control in a Post-Cold War World.* New York: Council on Foreign Relations, 1990.

Siekmann, Robert C.R., *Basic Documents on United Nations and Related Peace-Keeping Forces.* Dordrecht: Nijhoff, 1985.

———, *National Contingents in United Nations Peace-Keeping Forces.* Norwell, MA: Kluwer, 1991.

Simai, Mihaly, *The Future of Global Governance; Managing Risk and Change in the International System.* Washington, DC: U.S. Institute of Peace Press, 1994.

Sinclair, Ian, *The International Law Commission.* Cambridge: Grotius, 1987.

Skeet, Ian, *OPEC; 25 Years of Prices and Politics.* Cambridge Univ. Press, 1988.

Skjelsbaek, Kjell and Anthony McDermott (eds.), *The Multinational Force in Beirut, 1982–1984.* Gainsville: Univ. Presses of Florida, 1991.

Slinn, Peter, "The Role of the Commonwealth in the Peaceful Settlement of Disputes," in W.E. Butler (ed.), *The Non-Use of Force in International Law.* Dordrecht: Nijhoff, 1989, 119–135.

Smith, Dale L. and others (eds.), *The 1992 Project and the Future of Integration in Europe.* Armonk, NY: Sharpe, 1992.

Smith, Hugh (ed.), *International Peacekeeping: Building on the Cambodian Experience.* Canberra: Australian Defence Studies Centre, 1990.

Smith, Peter H. (ed.), *Drug Policy in the Americas.* Boulder, CO: Westview, 1992.

———, *The Challenge of Integration; Europe and the Americas.* New Brunswick, NJ: Transaction, 1994.

Sobell, Vladimir, *The CMEA in Crisis; Toward a New European Order?* Westport, CT: Greenwood, 1990.

Sohn, Louis B. (ed.), *International Organization and Integration; Annotated Basic Documents of International Organizations and Arrangements.* Dordrecht: Nijhoff, 1986.

———, *Rights in Conflict; The United Nations and South Africa.* New York: Transnational, 1993.

Spero, Joan Edelman, *The Politics of International Economic Relations.* 4th ed. New York: St. Martin's, 1990.

Sprout, Harold and Margaret Sprout, *Man-Milieu Relationship Hypotheses in the Context of International Politics.* Princeton, NJ: Princeton Univ. Center for International Studies, 1956.

———, *Foundations of International Politics.* Princeton, NJ: Van Nostrand, 1962.

Stairs, Denis and Gilbert R. Winham, *The Politics of Canada's Economic Relationship with the United States.* Toronto: Univ. of Toronto Press, 1987.

Staley, Robert Stephens II, *The Wave of the Future: The United Nations and Naval Peacekeeping.* Boulder, CO: Lynne Rienner, 1992.

Stamp, L. Dudley, *The British Commonwealth.* London: Longmans, 1951.

Steger, Debra P., *A Concise Guide to the Canada-United States Free Trade Agreement.* Toronto: Carswell, 1988.

Stewart, George R. and others (eds.), *International Trade and Intellectual Property; The Search for a Balanced System.* Boulder, CO: Westview, 1994.

Stoetzer, O. C., *The Organization of American States.* New York: Praeger, 1965.

Stone, Frank, *Canada, the GATT and the International Trade System.* Montreal: Institute for Research on Public Policy, 1984.

Strack, Harry R., *Sanctions; The Case of Rhodesia.* Syracuse, NY: Syracuse Univ. Press, 1978.

Strange, S., *States and Markets: An Introduction to International Political Economy.* Oxford: Blackwell, 1988.

Subedi, Surya Prasad, *Land-locked Nepal and International Law.* Kathmandu: Kokila Gautam, 1989.

Swann, Dennis (ed.), *The Single European Market and Beyond; A Study of the Wider Implications of the Single European Act.* New York: Routledge, 1992.

**T**

Tabory, Mala, *The Multinational Force and Observers in the Sinai—Organization, Structure, and Function.* Boulder, CO: Westview, 1986.

Takamiya, Susumu and Keith Thurley (eds.), *Japan's Emerging Multinationals.* Tokyo: Univ. of Tokyo Press, 1985.

Tarazona-Sevillano, Gabriela, *Sendero Luminoso and the Threat of Narcoterrorism.* Westport, CT: Greenwood, 1990.

Taylor, Michael and Nigel Thrift (eds.), *Multinationals and the Restructuring of the World Economy.* London: Croom Helm, 1986.

Taylor, Peter J., *World Government.* Oxford Univ. Press, 1990.

Taylor, Philip, *Nonstate Actors in International Politics; from Transregional to Substate Organizations.* Boulder, CO: Westview, 1984.

Teclaff, Ludwik A., *Water Law in Historical Perspective.* Buffalo, NY: William S. Hein, 1985.

Tennyson, Brian Douglas (ed.), *Canadian-Caribbean Relations; Aspects of a Relationship.* Sydney, N.S.: Center for International Studies, Univ. College of Cape Breton, 1990.

Thakur, Ramesh, *Peacekeeping in Vietnam.* Edmonton: Univ. of Alberta Press, 1984.

———, *International Peacekeeping in Lebanon.* Boulder, CO: Westview, 1987.

Thakur, Ramesh and Carlyle A. Thayer, *Reshaping Regional Relations.* Boulder, CO: Westview, 1993.

Thoumi, Francisco E., *Political Economy and Illegal Drugs in Colombia.* Boulder, CO: Lynne Rienner, 1994.

Toro, Celia, *Mexico's "War" on Drugs: Causes and Consequences.* Boulder, CO: Lynne Rienner, 1995.

Tovias, Alfred, *Foreign Economic Relations of the European Community: The Impact of Spain and Portugal.* Boulder, CO: Lynne Rienner, 1990.

Tow, William T., *Subregional Security Cooperation in the Third World.* Boulder, CO: Lynne Rienner, 1990.

Trebilcock, Michael J. and Robert Howse, *The Regulation of International Trade; Political Economy and Legal Order.* New York: Routledge, 1995.

Turner, Barry and Gunilla Nordquist, *The Other European Community.* New York: St. Martin's, 1982.

Tussie, Diana and David Glover, *The Developing Countries in World Trade: Policies and Bargaining Strategies.* Boulder, CO: Lynne Rienner, 1993.

**U**

Urquidi, Victor L., *Free Trade and Economic Integration in Latin America.* Berkeley: Univ. of California Press, 1964.

**V**

van Brabant, Jozef M., *Economic Integration in Eastern Europe: A Reference Book.* New York: Routledge, 1989.

van der Veer, Peter, *Changing Patterns in the International Arms Trade.* UFSI Reports General 1989–90/No. 4.

Vasciannie, Stephen C., *Land-locked and Geographically Disadvantaged States in the International Law of the Sea.* Oxford Univ. Press, 1990.

Vautier, Kerrin M. and others (eds.), *CER and Business Competition; Australia and New Zealand in a Global Economy.* Auckland: Commerce Clearing House, 1990.

Verleger, Philip K., Jr., *Global Oil Crisis Intervention.* Washington, DC: Institute for International Economics, 1991.

Villagrán de León, Francisco, *The OAS and Democratic Development.* Washington, DC: USIP Press, 1992.

———, *The OAS and Regional Security.* Washington: USIP Press, 1993.

Villar, Roger, *Piracy Today; Robbery and Violence at Sea Since 1980.* London: Conway Maritime Press, 1985.

Vitányi, Béla, *The International Regime of River Navigation.* Alphen aan den Rijn: Sijthoff & Noordhoff, 1979.

**W**

Wallace, Cynthia Day, *Foreign Direct Investment in the 1990s: A New Climate in the Third World.* Dordrecht: Nijhoff, 1990.

Wallace, Helen (ed.), *The Wider Western Europe; Reshaping the EC/EFTA Relationship.* London: Pinter, 1991.

Wallace, William (ed.), *The Dynamics of European Integration.* London: Pinter, 1992.

Waqif, Arif A. (ed.), *Regional Cooperation in Industry and Energy; Prospects for South Asia.* Newbury Park, CA: Sage, 1991.

Wardlaw, Grant, *Political Terrorism; Theory, Tactics and Counter Measures.* 2nd ed. Cambridge Univ. Press, 1989.

Waterbury, John, *Hydropolitics of the Nile Valley*. Syracuse, NY: Syracuse Univ. Press, 1979.

Weintraub, Sidney (ed.), *Integrating the Americas; Shaping Future Trade Policy*. New Brunswick, NJ: Transaction, 1994.

Weiss, Thomas G., *Multilateral Development Diplomacy in UNCTAD*. New York: St. Martin's, 1986.

————, *The United Nations and Civil Wars*. Boulder, CO: Lynne Reinner, 1995.

Wells, Clare, *The UN, UNESCO and the Politics of Knowledge*. New York: St. Martin's, 1987.

Wells, Robert N., Jr. (ed.), *Peace by Pieces; United Nations Agencies and Their Roles*. Metuchen, NJ: Scarecrow, 1991.

Whalley, John, *Canada and the Multilateral Trading System*. Toronto: Univ. of Toronto Press, 1987.

————, *Canada–United States Free Trade*. Toronto: Univ. of Toronto Press, 1987.

————, *Canada's Resource Industries and Water Export Policy*. Toronto: Univ. of Toronto Press, 1987.

———— and others, *Canadian Trade Policies and the World Economy*. Toronto: Univ. of Toronto Press, 1987.

———— assisted by Colleen Hamilton, *The Future of the World Trading System*. Washington, DC: Institute for International Economics, 1991–92.

———— and Peter Uimonen, *Trade and the Environment: Setting the Rules*. Washington, DC: Institute for International Economics, 1993.

Willemin, Georges and Roger Heacock, *The International Committee of the Red Cross*. Dordrecht: Nijhoff, 1984.

Wilson, Patricia A., *Exports and Local Development: Mexico's New Maquiladoras*. Austin: Univ. of Texas Press, 1992.

Wionczek, Miguel S. (ed.), *Economic Cooperation in Latin America, Africa, and Asia*. Cambridge, MA: MIT, 1969.

Wistrich, Ernest, *The United States of Europe*. 2nd ed. New York: Routledge, 1993.

Wolf, Aaron T., *Hydropolitics Along the Jordan River; Scarce Water and Its Impact on the Arab-Israeli Conflict*. Tokyo: United Nations Univ. Press, 1995.

**Y**

Yannopoulos, George N., *Greece and the EEC*. New York: St. Martin's, 1986.

Young, Thomas-Durell, *ANZUS*. Boulder, CO: Westview, 1992.

Yunker, James A., *World Union on the Horizon: The Case for Supranational Federation*. Washington, DC: World Resources Institute, 1993.

**Z**

Zacher, Mark and Jack Finlayson, *Developing Countries and the Commodity Trading Regime*. New York: Columbia Univ. Press, 1988.

Zacklin, Ralph and others (eds.), *The Legal Regime of International Rivers and Lakes*. Dordrecht: Nijhoff, 1981.

Zartman, I. William, *Ripe for Resolution: Conflict and Intervention in Africa*. New York: Council on Foreign Relations, 1985.

## *Periodicals*

**A**

Acharya, Amitav, "The Association of Southeast Asian Nations; 'Security Community' or 'Defence Community'," *Pacific Affairs*, 64 (Summer 1991), 159–178.

Aghrout, Ahmed and Keith Sutton, "Regional Economic Union in the Maghreb," *Journal of Modern African Studies*, 28, 1 (1991), 115–139.

Alagappa, Muthiah, "Regional Arrangements and International Security in Southeast Asia; Going Beyond ZOPFAN," *Contemporary Southeast Asia*, 12 (March 1991), 269–305.

Alexander, Dean C., "Maritime Terrorism and Legal Responses," *Denver Journal of International Law and Policy*, 19 (Spring 1991), 529–567.

Aminoff, Nicholas A., "The United States–Israel Free Trade Area Agreement of 1985," *Journal of World Trade*, 25 (Feb. 1991), 5–42.

Anderson, James, "Problems of Inter-State Economic Integration; Northern Ireland and the Irish Republic in the European Community," *Political Geography*, 13, 1 (Jan. 1994), 53–72.

**B**

Baer, M. Delal, "North American Free Trade," *Foreign Affairs*, 70 (Fall 1991), 132–149.

Bagley, Bruce M. and William O. Walker III (eds.), *Drug Trafficking Research Update*. Special issue of *Journal of Interamerican Studies and World Affairs*, 34, 3 (Fall 1992), whole volume.

Barton, Thomas Frank, "Outlets to the Sea for Land-locked Laos," *Journal of Geography*, 59, 5 (May 1960), 206–220.

Biger, Gidon, "Physical Geography and Law: The Case of International River Boundaries," *Geo-Journal*, 18 (1988), 341–347.

Birtles, Terry G., "A Single Western Europe? Implications of the Changing Division of External-Relations Powers Between the European Community and Member States," *Political Geography Quarterly*, 9, 2 (April 1990), 131–145.

Block, Alan (ed.), *The Politics of Cocaine*. Special issue of *Crime, Law and Social Change*, 16 (July 1991), whole volume.

Boutros-Ghali, Boutros Y., "The Arab League, 1945–1955," *International Conciliation*, 498 (May 1954), 387–448.

———, "The Addis Ababa Charter," *International Conciliation*, 546 (1964).

Bowen, Robert E., "The Land-locked and Geographically Disadvantaged States and the Law of the Sea," *Political Geography Quarterly*, 5, 1 (Jan. 1986), 63–69.

Brada, Josef C., "The Political Economy of Communist Foreign Trade Institutions and Policies," *Journal of Comparative Economics*, 15 (June 1991), 211–238.

Buchan, David, "The Constraints of the European Community," *Political Quarterly*, 62 (April 1991), 186–192.

Bucholtz, Barbara K., "Coase and the Control of Transboundary Pollution; The Sale of Hydroelectricity under the United States–Canada Free Trade Agreement of 1988," *Boston College Environmental Affairs Law Review*, 18 (Winter 1991), 279–317.

**C**

Caflisch, Lucius C., "Land-Locked States and Their Access to and from the Sea," *British Yearbook of International Law*, 49 (1978), 71–100.

Cárdenas, Emilio J., "The Treaty of Asunción: A Southern Cone Common Market (Mercosur) Begins to Take Shape." *World Competition* (Geneva), 15, 4 (June 1992), 65–77.

Castells, Manuel (ed.), *Transnational Corporations, Industrialization and Social Restructuring in the ASEAN Region*. Special issue of *Regional Development Dialogue*, 12 (Spring 1991), 1–199.

Choi, E. Kwan and Harvey E. Lapan, "Optimal Trade Policies for a Developing Country under Uncertainty," *Journal of Development Economics*, 35 (April 1991), 243–260.

Clapham, C., "The Political Economy of Conflict: The Horn of Africa," *Survival*, 32 (1990), 403–420.

Clark, Gordon L., "NAFTA—Clinton's Victory, Organized Labor's Loss," *Political Geography*, 13, 4 (July 1994), 377–384.

*Colloquium on Terrorism as an International Crime. Israel Yearbook on Human Rights*, 19 (1989), whole volume.

Corbridge, Stuart, "Perversity and Ethnoregionalism in Tribal India: The Politics of the Jharkand." *Political Geography Quarterly*, 6, 3 (July 1987), 225–240.

**D**

Da Costa, Peter, "A New Role for ECOWAS," *Africa Report*, 36 (Sept.–Oct. 1991), 37–40.

Dubner, Barry Hart, "Piracy in Contemporary National and International Law," *California Western International Law Journal*, 21, 1 (1990–91), 139–149.

**E**

East, W. Gordon, "The Geography of Land-locked States," *Transactions and Papers, Institute of British Geographers*, 28 (1960), 1–22.

Ehrenfeld, Rachel, "Narco-terrorism and the Cuban Connection," *Strategic Review*, 16, 3 (Summer 1988), 55–64.

Eisner, Robert and Paul J. Pieper, "Real Foreign Investment in Perspective," *Annals, Am. Acad. Polit. and Soc. Sci.*, 516 (July 1991), 22–35.

El-Jafari, Mahmoud, "Non-tariff Trade Barriers; The Case of the West Bank and Gaza Strip," *Journal of World Trade*, 25 (June 1991), 15–32.

Ellen, Eric, "Contemporary Piracy," *California Western International Law Journal*, 21, 1 (1990–91), 123–127.

**F**

Fair, Denis, "West Africa's River Basin Organizations." *Africa Insight*, 21, 4 (1991), 257–262.

Francioni, F., "Maritime Terrorism and International Law: The Rome Convention of 1988," *German Yearbook of International Law*, 31 (1988), 263.

**G**

Gambari, Ibrahim A., "The OAU and Africa's Changed Priorities," *Transafrica Forum*, 6, 2 (Winter 1989), 3–14.

Gardner, Richard N., "The Comeback of Liberal Internationalism," *Washington Quarterly*, 13 (Summer 1990).

Gibb, Richard A., "The Effect on the Countries of SADCC of Economic Sanctions Against the Republic of South Africa," *Transactions, Institute of British Geographers*. New Series 12, 4 (1987), 398–412.

Gill, S., "Two Concepts of International Political Economy," *Review of International Studies*, 16, (1990), 369–381.

Gilmore, William C., "Legal and Institutional Aspects of the Organisation of Eastern Caribbean States," *Review of International Affairs*, 11, 4 (Oct. 1985), 311–328.

Glassner, Martin Ira, "The Río Lauca: Dispute over an International River," *Geographical Review*, 60, 2 (April 1970), 192–207.

———, "International Law Regarding Land-locked States," *Nepal Review*, 11, 8 (June 1970), 388–393.

———, "The Status of Developing Land-locked States Since 1965," *Lawyer of the Americas*, 5, 3 (Oct. 1973), 480–498.

———, "Developing Land-locked States and the Resources of the Seabed," *San Diego Law Review*, 11, 3 (May 1974), 633–655.

———, "Land-locked Nations and Development," *International Development Review*, 19, 2 (Sept. 1977), 19–23.

———, "CARICOM: A Community in Trouble," *Focus*, 29, 2 (Nov. 1978), 12–16.

———, "The Land-locked States at the Third United Nations Conference on the Law of the Sea," *Proceedings, AAG*, Committee on Marine Geography, (1978), 119–122.

———, "Transit Rights for Land-locked States and the Special Case of Nepal," *World Affairs*, 140, 4 (Spring 1978), 304–314.

———, "Land-locked States and the 1982 Law of the Sea Convention," *Marine Policy Reports* (Univ. of Delaware), 9, 1 (Sept. 1986), 8–14.

*Global Governance*, published thrice a year since 1995 by Lynne Rienner.

Gooding, G., "Fighting Terrorism in the 1980's: The Interception of the Achille Lauro Hijackers," *Yale Journal of International Law*, 12 (1987), 158.

Goulding, Marrack, "The Evolution of United Nations Peacekeeping," *International Affairs*, 69 (1993), 451–464.

Grant, Richard, "Against the Grain: Agricultural Trade Policies of the U.S., the European Community and Japan at the GATT," *Political Geography*, 12, 3 (May 1993), 247–262.

———, "The Geography of International Trade," *Progress in Human Geography*, 18, 3 (1994), 298–312.

Griffiths, Ieuan L., "The Quest for Independent Access to the Sea in Southern Africa," *Geographical Journal*, 155, 3 (Nov. 1989), 378–391.

Gupta, Vijay, "Land-locked Uganda—Constraints on Political Economy," *Ind-Africana* (Delhi), 4, 2 (Oct. 1991), 1–14.

**H**

Halbertstam, Malvina, "Terrorism on the High Seas: *The Achille Lauro*, Piracy and the IMO Convention on Maritime Safety," *American Journal of International Law*, 82, 2 (April 1988), 269–310.

———, "Terrorist Acts Against and On Board Ships," *Israel Yearbook on Human Rights*, 19 (1989), 331.

Hall, Kenneth O. and Byron W. Blake, "The Emergence of the African, Caribbean, and Pacific Group of States: An Aspect of African and Caribbean International Cooperation," *African Studies Review*, 22, 2 (Sept. 1979), 11–125.

Halverson, Karen, "Foreign Direct Investment in Indonesia; A Comparison of Industrialized and Developing Country Investors," *Law and Policy in International Business*, 22, 1 (1991), 75–105.

Hansen, Peter and Victoria Aranda, "An Emerging International Framework for Transnational Corporations," *Fordham International Law Journal*, 4, 4 (1990–91).

Harkavy, Robert E. and Stephanie G. Neuman (eds.), *The Arms Trade: Problems and Prospects in the Post–Cold War World. Annals, Am. Acad. of Polit. and Soc. Sci.*, 535 (Sept. 1994), whole issue.

Hay, I. and J. E. Bell, "Small Places and Big States; Changing Global Relations in a New Global Environment," *Tijdschrift voor Economische en Sociale Geografie*, 81 (1990), 322–331.

Heisler, Martin O. (ed.), *The Nordic Region: Changing Perspectives in International Relations. Annals, Am. Acad. of Polit. and Soc. Sci.*, 512 (Nov. 1990), whole issue.

Highet, Keith, "The Peace Palace Heats Up: The World Court in Business Again?," *American Journal of International Law*, 85 (1991), 646–654.

Hussey, Antonia, "Regional Development and Cooperation Through ASEAN," *Geographical Review*, 81 (Jan. 1991), 87–98.

**I**

*International Organization*. Published quarterly by MIT Press Journals, Cambridge, MA.

**J**

James, Alan, "The UN Force in Cyprus," *International Affairs*, 65, 3 (Summer 1989), 487–500.

Jenkins, Rhys, "Transnational Corporations, Competition and Monopoly," *Review of Radical Political Economics*, 21 (Winter 1989), 12–32.

*Journal of Common Market Studies*. Published quarterly by Blackwell, Oxford.

*Journal of Economics and International Relations*. Asian Research Service, Hong Kong. Published quarterly since 1987.

*Journal of European Integration*. Saskatoon: Univ. of Saskatchewan, Department of Political Science. Published semiannually since 1977.

*Journal of World Trade*, published bimonthly by Werner in Geneva.

Joyner, Christopher C., "Suppression of Terrorism on the High Seas: The 1988 IMO Convention on the Safety of Maritime Navigation," *Israel Yearbook on Human Rights*, 19 (1989), 343.

Junz, Helen B., "Integration of Eastern Europe into the World Trading System," *American Economic Review*, 81 (May 1991), 176–180.


**K**

Kafando, Talata, "The Case of Liptako-Gourma," *Ceres*, 37 (Jan. 1974), 45–48.

Kain, Ronald Stuart, "Bolivia's Claustrophobia," *Foreign Affairs*, 16, 4 (July 1938), 704–713.

Kapstein, Ethan B., "The Brazilian Defense Industry and the International System," *Political Science Quarterly*, 105 (Winter 1990-91), 579–596.

Karno, Valerie, "Protection of Endangered Gorillas and Chimpanzees in International Trade: Can CITES Help?" *Hastings International and Comparative Law Review*, 14, 4 (1991).

Kellas, James G., "European Integration and the Regions," *Parliamentary Affairs*, 44 (April 1991), 226–239.

Khadka, Narayan, "The Crisis in Nepal-India Relations," *Journal of South Asian and Middle Eastern Studies*, 15, 1 (Fall 1991), 54–92.

Kirsch, P., "The 1988 ICAO and IMO Conferences: An International Consensus Against Terrorism," *Dalhousie Law Journal*, 12 (1989), 5.

Kleinman, S. B., "Terror at Sea: Vietnamese Victims of Piracy," *American Journal of Psychoanalysis*, 50 (Dec. 1990), 452–462.

Kliot, Nurit, "Geography of Hostages: The Case of Lebanon," *1983 Yearbook of the Israeli Geographical Association* (1984).

Kwarting, Charles, "Difficulties in Regional Economic Integration: The Case of ECOWAS," *Transafrica Forum*, 5, 2 (Winter 1988), 17–26.


**L**

Lanciaux, Bernadette, "Ethnocentrism in U.S./Japanese Trade Policy Negotiations," *Journal of Economic Issues*, 25 (June 1991), 569–580.

Landau, Georges D., "The Treaty for Amazonian Cooperation: A Bold New Instrument for Development," *Georgia Journal of International and Comparative Law*, 10, 3 (1980), 463–489.

Laurent, Pierre-Henri, *The European Community; To Maastricht and Beyond. Annals, Am. Acad. of Polit. and Soc. Sci.*, 531 (Jan. 1994), whole issue.

Lee, Margaret C., "Political and Economic Implications of Sanctions Against South Africa: The Case of Zimbabwe," *Journal of African Studies*, 15, 3–4 (Fall–Winter 1988), 52–60.

——, "SADCC and Post-Apartheid South Africa," *Transafrica Forum*, 6, 3–4 (Spring–Summer 1989), 99–111.

Lewis, Vaughan A., "Evading Smallness: Regional Integration as an Avenue Towards Viability," *International Social Science Journal*, 30, 1 (1978), 57–72.

Licklider, Roy, "The Power of Oil: The Arab Oil Weapon and the Netherlands, the United Kingdom, Canada, Japan, and the United States," *International Studies Quarterly*, 32, 2 (June 1988), 205–226.

Ligthart, Henk and Henk Reitsma, "Portugal's Semi-Peripheral Middleman Role in Its Relations with England, 1640–1760," *Political Geography Quarterly*, 7, 4 (Oct. 1988), 353–362.

*Low Intensity Conflict and Law Enforcement*. Published thrice a year since 1992 by Frank Cass, London.


**M**

Malan, Theo, "The Preferential Trade Area (PTA) of Southern and Eastern African Countries," *Africa Institute Bulletin*, 13 (1982), 104–106.

Maluwa, Tiyanjana, "Legal Aspects of the Niger River under the Niamey Treaties," *Natural Resources Journal*, 28, 4 (Oct. 1988), 671–697.

Mandel, Robert, "Sources of International River Basin Disputes." *Conflict Quarterly*, 12, 4 (Fall 1992), 25–56.

Marks, Susan, "Transit Rights of Lesotho," *Commonwealth Law Bulletin*, 16, 1 (Jan. 1990), 329–348.

Martínez Puñal, Antonio, "The Rights of Landlocked and Geographically Disadvantaged States in Exclusive Economic Zones," *Journal of Maritime Law and Commerce*, 23, 3 (July 1992), 429–459.

Mason, T. David and Dale A. Krane, "The Political Economy of Death Squads; Toward a Theory of the Impact of State-sanctioned Terror," *International Studies Quarterly*, 33, 2 (June 1989), 175–198.

McFadden, Eric J., "The Collapse of Tin: Restructuring a Failed Commodity," *American Journal of International Law*, 80, 4 (Oct. 1986), 811–830.

McKeown, Timothy J., "A Liberal Trade Order? The Long-Run Pattern of Imports to the Advanced Capitalist States," *International Studies Quarterly*, 35, 2 (June 1991), 151–172.

McMichael, Philip, "World Food System Restructuring under a GATT Regime," *Political Geography*, 12, 3 (May 1993), 198–214.

Menefee, Samuel Pyeatt, "Scourges of the Sea: Piracy and Violent Maritime Crime," *Marine Policy Reports*, 1, 1 (1989), 13–36.

Mirvahabi, Farin, "The Rights of the Land-locked and Geographically Disadvantaged States in Exploitation of Marine Fisheries," *Netherlands International Law Review*, 26, 2 (1979), 130–162.

*Multilateralism and Bilateralism in Trade Policy*. Symposium in *The Economic Journal*, 100 (Dec. 1990).

*Multilateralism: Old and New*. Special issue of *International Journal*, 45, 4 (Autumn 1990).

Murphy, Alexander B., "The Emerging Europe of the 1990s," *Geographical Review*, 81 (Jan. 1991), 1–17.

Mwase, Ngila R.L., "Zambia, the TAZARA and the Alternative Outlets to the Sea," *Transport Reviews*, 7, 3 (July 1987), 191–206.

———, "Non-physical Barriers to Traffic Flow and the PTA Programme of Action: Problems and Prospects," *Transport Reviews*, 13, 1 (Jan. 1993), 25–44.

**N**

Nadelmann, Ethan A., "The DEA in Latin America: Dealing with Institutionalized Corruption," *Journal of Interamerican Studies and World Affairs*, 29, 4 (Winter 1987–88), 1–39.

———, "U.S. Drug Policy: A Bad Export," *Foreign Policy*, 70 (Spring 1988), 83–108.

———, "International Drug Trafficking and U.S. Foreign Policy," *The Washington Quarterly* (Fall 1989), 87–104.

Nierop, Tom, "Macro-regions and the Global Institutional Network, 1950–1980," *Political Geography Quarterly*, 8, 1 (Jan. 1989), 43–66.

Noyes, John E., "An Introduction to the International Law of Piracy," *California Western International Law Journal*, 21, 1 (1990–91), 105–121.

**O**

O'Loughlin, John, "Geo-economic Competition in the Pacific Rim: The Political Geography of Japanese and U.S. Exports, 1966–1988," *Transactions, Institute of British Geographers*, 18 (1993), 438–459.

O'Loughlin, John and Herman van der Wusten, "The Political Geography of Panregions," *Geographical Review*, 80 (1990), 1–20.

**P**

Parker, Ron, "The Senegal-Mauritania Conflict of 1989; A Fragile Equilibrium," *Journal of Modern African Studies*, 29 (March 1991), 155–171.

Pazzanita, Anthony G., "Legal Aspects of Membership in the Organization of African Unity: The Case of the Western Sahara," *Case Western Reserve Journal of International Law*, 17, 1 (Winter 1985), 123–158.

Pillai, K. Raman, "Tensions Within Regional Organizations: A Study of SAARC," *Indian Journal of Political Science*, 50, 1 (Jan. 1989), 18–27.

Polland, V. K., "A.S.A. and A.S.E.A.N. 1961–1967: Southeast Asian Regionalism," *Asian Survey*, 10, 3 (March 1970), 244–255.

Pounds, Norman J.G., "A Free and Secure Access to the Sea," *Annals, AAG*, 49, 3 (Sept. 1959), 256–268.

**Q**

Quester, George H., "Finlandization as a Problem or an Opportunity?" *Annals, Am. Acad. of Polit. and Soc. Sci.*, 81 (March 1991), 127–151.

**R**

Reitsma, Henk and Leo De Haan, "Northern Togo and the World Economy," *Political Geography*, 11, 5 (Sept. 1992), 475–484.

Robson, Peter, "The Mano River Union," *Journal of Modern African Studies*, 20, 4 (Dec. 1982), 613–628.

Rothstein, Robert L., "Condemned to Cooperate: U.S. Resource Diplomacy," *SAIS Review*, 5, 1 (Winter 1985), 163–177.

Rubin, Alfred P., "Revising the Law of 'Piracy'," *California Western International Law Journal*, 21, 1 (1990–91), 129–137.

Rubin, Seymour, "Transnational Corporations and International Law: An Uncertain Partnership," *Chinese Yearbook of International Law and Affairs*, 4 (1984), 39–49.

Russett, Bruce and James S. Sutterlin, "The U.N. in a New World Order," *Foreign Affairs*, 70 (Spring 1991), 69–83.

**S**

Sackey, James A., "The Structure and Performance of CARICOM: Lesson for the Development of ECOWAS," *Canadian Journal of African Studies*, 12, 2 (1978), 259–277.

Sanjian, Greogry S., "Great Power Arms Transfers: Modeling the Decision-Making Process of Hegemonic, Industrial, and Restrictive Exporters," *International Studies Quarterly*, 35, 2 (June 1991), 173–194.

Sayigh, Yezid, "The Gulf Crisis; Why the Arab Regional Order Failed," *International Affairs*, 67 (July 1991), 487–507.

Schachter, Oscar, "United Nations Law in the Gulf Conflict," *American Journal of International Law*, 85, 3 (July 1991), 452–473.

Schlager, Erika B., "The Procedural Framework of the CSCE: From the Helsinki Consultations to the Paris Charter, 1972–1990," *Human Rights Law Journal*, 12 (July 1991), 221–237.

Schmidt, Elizabeth, "The Sanctions Weapon: Lessons from Rhodesia," *Transafrica Forum*, 4, 1 (Fall 1986), 3–16.

Schuyler, George W., "Perspectives on Canada and Latin America; Changing Context—Changing Policy?" *Journal of Interamerican Studies and World Affairs*, 33 (Spring 1991), 19–58.

Schwartzberg, Joseph E., "Editorial: Towards a More Representative and Efective Security Council," *Political Geography*, 13, 6 (Nov. 1994), 483–491.

Serbin, Andrés, "The CARICOM States and the Group of Three; A New Partnership Between Latin America and the Non-Hispanic Caribbean?" *Journal of Interamerican Studies and World Affairs*, 33 (Summer 1991), 53–80.

Sesay, Amadu, "The Limits of Peace-keeping by a Regional Organization; The OAU Peace-keeping Force in Chad," *Conflict Quarterly*, 11 (Winter 1991), 7–26.

Silva, Jesus and Richard K. Dunn, "A Free Trade Agreement Between the United States and Mexico: The Right Choice?" *San Diego Law Review*, 27 (1990), 937-992.

Smith, José, "The Beira Corridor Project," *Geography*, 73, 3 (June 1988), 258-261.

Spykman, Nicholas John, "Frontiers, Security and International Organizations," *Geographical Review*, 32, 3 (1942), 436–447.

Storper, M., "The Limits to Globalization: Technology Districts and International Trade," *Economic Geography*, 68 (1992), 60–93.

Subedi, Surya Prasad, "The Marine Fishery Rights of Land-locked States with Particular Reference to the EEZ," *International Journal of Estuarine and Coastal Law*, 2, 4 (1987), 227–239.

Suret-Canale, J., "The French Community at the Hour of African Independence," *International Affairs*, 7 (Jan. 1961), 32–37; 7 (Feb. 1961), 23–30.

Sweeney, Jane Chace, "State-Sponsored Terrorism: Libya's Abuse of Diplomatic Privileges and Immunities," *Dickinson Journal of International Law*, 5, 1 (Fall 1986), 133–165.

*Symposium on North American Free Trade*. World Economy, 14 (March 1991), 53–111.

**T**

Tarlton, C. D., "The Styles of American International Thought: Mahan, Bryan, and Lippman," *World Politics*, 17, 4 (July 1965), 584–614.

*Terrorism*. Published bimonthly by Taylor & Francis.

*Terrorism and Political Violence*. Published quarterly since 1989 by Frank Cass in London.

Thacher, Peter S., "Multilateral Cooperation and Global Change," *Journal of International Affairs*, 44 (Winter 1991), 433–455.

Thrift, Nigel, "Muddling Through: World Orders and Globalization," *Professional Geographer*, 44, 1 (Feb. 1992), 3–7.

Toledano Lardeo, Armando, "The EEA Agreement: An Overall View," *Common Market Law Review*, 29, 6 (Dec. 1992), 1199–1213.

**U**

*The United Nations After the Gulf War*. World Policy Journal, 8 (Summer 1991), 537–574.

**V**

Vail, Jeffrey, "Halting the Elephant Ivory Trade," *Wisconsin International Law Journal*, 9, 1 (Fall 1990).

van der Wusten, Herman and Tom Nierop, "Functions, Roles and Form in International Politics," *Political Geography Quarterly*, 9, 3 (July 1990), 213–231.

Vasciannie, Stephen, "Landlocked and Geographically Disadvantaged States and the Question of the Outer Limit of the Continental Shelf," *British Yearbook of International Law*, 58 (1987), 271–302.

**W**

Ward, Michael, Review Essay: "Recognizing a Hegemon," *Political Geography Quarterly*, 4, 4 (Oct. 1985), 343–346.

Watkins, Kevin, "Agriculture and Food Security in the GATT Uruguay Round," *Review of African Political Economy*, (March 1991), 38–50.

Weintraub, Sidney (ed.), "The New U.S. Economic Initiative Toward Latin America," *Journal of Interamerican Studies and World Affairs*, 33 (Spring 1991), 1–18.

———, *Free Trade in the Western Hemisphere. Annals, Am. Acad. of Polit. and Soc. Sci.*, 526 (March 1993), whole issue.

Weisfelder, Richard F., "Collective Foreign Policy Decision-making Within SADCC; Do Regional Objectives Alter National Policies?" *Africa Today*, 38, 1 (1991), 5–17.

Will, M. Marvin, "A Nation Divided; The Quest for Caribbean Integration," *Latin American Research Review*, 2 (1991), 3–37.

Wise, Mark and Gregory Croxford, "The European Regional Development Fund: Community Ideals and National Realities," *Political Geography Quarterly*, 7, 2 (April 1988), 161–182.

Wright, Esmond, "The 'Greater Syria' Project in Arab Politics," *World Affairs*, 5, 3 (July 1951), 318–329.

**Y**

*Yearbook of International Organizations*. London: Taylor & Francis.

**Z**

Zhukov, S. I., "The Geography of the USSR's Foreign Trade with Bordering Socialist Countries," *Soviet Geography*, 31 (1990), 46–53.

# PART SEVEN

## *Our Last Frontiers*

Movement of hazardous cargoes: A Japanese ship unloading plutonium from France in Tokai

# 31

# *The Law of the Sea*

*Those who go down to the sea in ships,*
*Who do business in great waters;*
*They see the works of the Lord*
*And His wonders in the deep.*
From Psalm 107

So wrote the psalmist more than 2000 years ago when seafaring was already an ancient and respected calling. King Solomon nearly a thousand years earlier maintained a large and profitable merchant navy. The ancient Phoenicians and Polynesians ranged farther over the oceans of the world than we can imagine in what seem to us to be the frailest of sailing craft. The ancient Greeks not only viewed the sea as the binding agent of their fragmented homeland, but concentrated so much on the sea that they became vulnerable to overland attack. The Roman Empire, on the other hand, managed to combine sea power with effective organization on land. The Mediterranean became what the Greeks had never been able to make it, an interior sea, a Roman lake.

The Romans ventured far beyond the Mediterranean, for we have descriptions of Indian Ocean coasts written by Roman sea captains. But in the centuries that followed, the oceans remained an insurmountable barrier to Europe. Chinese vessels reached the eastern coast of Africa while the Dark Ages settled over the European continent. Some may even have reached America. Not until the ascent to power of the Spanish and Portuguese kingdoms on the Iberian Peninsula, almost a dozen centuries after the fall of

Rome, did the oceans begin once again to form a link between Europe and other parts of the world. Since then the link has been permanent, and the oceans have carried European power to all parts of the globe. In search of land and resources, the Spanish and Portuguese vessels returned home with new knowledge of the earth. Soon there were competitors for the domination of the sea routes to spices, gold, and slaves. The Hollanders successfully challenged the Portuguese; the English challenged the Dutch. Once the oceans had contained Europe—now they were the routes to wealth and power.

Meanwhile, ancient rules pertaining to the use of the sea were being compiled, translated, and circulated in Europe, beginning in the seventh century (probably) with the *Lex Rhodia* (Rhodian Sea Law). During the next few centuries other compilations appeared in France, England, Scandinavia, and The Netherlands. The best known and most influential of these early versions of the Law of the Sea was the *Consolato del Mare* (Consulate of the Sea), which was written in Catalan, probably in Barcelona in the late thirteenth or early fourteenth century. Little more than a century later, however, this slow evolutionary process was greatly accelerated, propelled by European exploration, discovery, conquest, annexation, and colonization of much of the world far beyond Europe's shores.

## Seaborne Empires

We may recall from our discussion of colonial empires in Chapter 17 that development of overseas "spheres of influence" began almost immediately. So intense was the competition for control in the Americas that Spanish and Portuguese interests were limited by the Treaty of Tordesillas even before the end of the fifteenth century. The treaty, in fact, constitutes a first attempt to define and delimit a geometric boundary. Known as the "Pope's Line," it divided the yet unexplored world into a Spanish and a Portuguese realm, the boundary lying along the meridian 370 leagues west of the Cape Verde Islands. The Portuguese therefore established themselves in Brazil and along the west coast of Africa and made contact with the Arabs plying the eastern coasts, while the Spanish colonized much of the Americas, the Philippines, and numerous intervening Pacific islands.

The Dutch laid claim to major portions of Southeast Asia and established settlements and fortifications in a multitude of other places, including Cape Town. Danes, Brandenburgers, English, French, Germans, Italians—all entered the scramble for overseas possessions at one time or another.

While the oceans had thus become avenues of conquest and power, they had also become a threat to coastal communities, not only in the colonized world but also in Europe itself. Hitherto, frontiers on land acted as separators, and it was never difficult to interpret the intentions of any army moving into such frontiers, whether such an army was one of benign occupation or of conquest. But a group of vessels approaching offshore might simply be passing toward some distant land or might suddenly attack, having approached unmolested to within a few hundred meters of the coast!

Through all these centuries, while some countries were stoutly defending the freedom of the seas, others were trying to assert various kinds of jurisdiction or even complete sovereignty over portions of the sea. The Greek city-states, the Roman Republic, and Venice and other Italian city-states were the most active in the Mediterranean in laying the foundations of the Law of the Sea, the core of which was the concept of the territorial sea.

## The Territorial Sea

In the thirteenth and fourteenth centuries the Norwegians, Danes, English, and Dutch exercised control over various parts of the North Sea and North Atlantic Ocean. They did so by treaty or ordinance, and usually the areas claimed were not precisely defined. In the fourteenth century activity in the North Sea greatly increased. The Dutch

---

### The Nautical Mile

The nautical mile is widely used in air and sea navigation and the measurement of maritime boundaries. It is almost always used in Law of the Sea matters and is used uniformly in this book when we refer to distances at sea. It is derived from one minute (1/60 of one degree) of arc on the earth's surface. Since the earth is not a true sphere and its arcs are not uniform, a standard nautical mile approximating one minute of arc had to be defined. By international agreement, based on a 1929 proposal by the International Hydrographic Bureau, it was set at 6076.12 feet or about 1.15 statute (English or land) miles or 1.852 kilometers. Wind velocity and speeds at sea or in the air are expressed in knots: A knot is one nautical mile per hour. It is thus incorrect to use the phrase "knots per hour," since a knot is a measure of both distance and time. In Chapters 31 and 32 of this book, *all* references to miles are understood to mean nautical miles.

fishing fleet expanded, and fishing vessels from Flanders, France, and England joined in the search for the best fishing grounds. Sporadic friction occurred, but it was not until the end of the sixteenth century that the question of countries' rights on neighboring waters developed into a full-scale legal battle.

In the 1590s, the Danes, to whom power had gone from Norway, decided to abandon the "closed sea" practice they had inherited from the Norwegians. Instead they announced that a belt of water, eight miles in width, lying around their possession of Iceland, constituted Danish territorial waters. Some of northwestern Europe's best fishing grounds lie off the Norwegian, Icelandic, and Faeroe coasts, and soon the Danish crown defined similar belts around all these possessions. It was not long before they increased these widths; in the mid-seventeenth century they claimed as much as 24 miles. The English, meanwhile, had drawn baselines along the British coast from promontory to promontory, cutting off large portions of the coastal waters. Some of these baselines were more than 50 miles long. Furthermore, the English demanded that the Dutch and others fishing near their coasts obtain permits to do so.

Thus began the legal debate that ultimately produced many of the principles to which States still adhere today. The Dutch, who did not have large coastal bays and indentations, argued that closing portions of the sea constituted violation of "customary" international law. Early in the seventeenth century a Dutch jurist, Hugo Grotius (Huigh Cornets deGroot, 1583–1645), published a treatise, the significant part of which was called *The Free (Open) Sea.* The British legal expert John Selden soon replied under the title *The Closed Sea.*

As the seventeenth century wore on, the argument grew in intensity. Indeed, it was a major contributor to four naval wars fought between the English and the Hollanders, while the legal debate raged on in the intervening periods. The Dutch objected to restrictions of any kind other than a narrow belt of territorial sea; the British stood firm

on the principle of extensive closed seas. But time was on the side of the Hollanders. Like the Dutch, the English came to depend to an ever greater extent on their growing merchant fleet, and they wanted that fleet to face as few obstacles as possible. Thus by the end of the seventeenth century, the English had yielded to some extent to Dutch desires, and the debate now came to focus on the manner by which the extent of any belt of territorial sea should be determined.

Again, two schools of thought existed. The question previously had revolved around "free" or "closed" seas; it now concerned the principle of effective occupation as opposed to that of a continuous (if narrow) belt of territorial sea along all coasts. Cornelius Van Bynkershoek, a Dutch jurist, was probably the most vocal and energetic legal expert ever to become involved in this question. As a result, many innovations are attributed to him that actually seem to have originated with others. Van Bynkershoek's *De Dominio Maris Dissertatio*, first published in 1702 and revised in 1744, summarized the entire question as it stood at that time, established a terminology that is still in use today, and brought the Dutch standpoint forcefully to the attention of the English. The English favored the concept of a continuous belt of equal width rather than Van Bynkershoek's shore-domination principle, and about this time the English idea came to be generally accepted. The Danes, French, and Italians preferred the equal-width zone of territorial water also, and now the problem was one of determining the exact breadth of such a belt.

The Danes in 1745 announced that they would claim a width of 1 marine league, equal to 4 nautical miles. The Dutch agreed to observe a belt of equal width, but they insisted on using the cannon shot as their determinant. Abbé Fernando Galiani, an Italian writer, is credited with the proposal that 3 miles be used as the generally acceptable standard width. Galiani apparently based his suggestion on what he viewed as the maximum range of the cannon of that day, but military experts have determined that the range of eighteenth-century cannons probably did not exceed 1 mile. Whatever the origins of the 3-mile limit, after the United States and Britain adopted it for their maritime boundaries around the turn of the nineteenth century, it rapidly became the most common claim in many parts of the world. The Scandinavians, however, continued to claim 4-mile territorial seas, and Spain, Portugal, and the Ottoman Empire claimed 6 miles.

During much of the nineteenth century and the first half of the twentieth, the situation remained relatively stable by virtue of European colonial control and the application of European claims to overseas possessions. There were disputes over European fishing grounds, and several separate treaties were signed among interested States, but until World War II those States that thrived partly or largely on their overseas trade maintained the 3- or 4-mile limit. In 1951 80 percent of the merchant-shipping tonnage of the world was registered in countries that subscribed to a 3-mile limit, and another 10 percent to States adhering to a 4-mile limit. Today most coastal States are claiming 12 miles for their territorial sea.

### Other Aspects of the Traditional Law of the Sea

Once the international community accepted the concept that every coastal State was entitled to a territorial sea and most coastal States had adopted 3 or 4 nautical miles as the breadth of their territorial seas, the other issues of maritime jurisdiction became either easier to resolve or less important. They included methods of measuring the territorial sea; protection of fishing grounds; jurisdiction over customs, fiscal, immigration, and health matters; neutrality and security jurisdiction outside the territorial sea; and the status of bays and straits. The Law of the Sea, like other branches of international law, continued to evolve at a leisurely pace. All this was changed by World War I.

As we have already seen, World War I destroyed the nineteenth-century world. Out of the ruins came, among other things, the League of Nations. As part of its effort to develop and codify international law, the League sponsored a number of lawmaking conferences. One of the most important was the Conference for the Codification of International Law, held at The Hague in 1930. Among the many matters debated there were several issues involving the Law of the Sea. They included a uniform 3-mile territorial sea; the claims of certain States to broader territorial seas; acceptance of the principle of a contiguous zone within which the coastal State could exercise jurisdiction in customs, health, and security matters; proposals for extensive exclusive fishery zones; and a newer "common patrimony" concept. The Conference failed to resolve these mat-

**Russian trawler in the Bering Sea.** Vessels of this type, sailing chiefly under Russian and Japanese flags, are primarily used to process fish caught by smaller vessels carried on board. Their voyages commonly last more than a year, and their canned or frozen catches may be sold in ports far from home. (Photo courtesy of the U.S. National Marine Fisheries Service)

**Drilling for oil off Australia.** About a third of all of the petroleum being produced in the world today comes from wells drilled into the continental shelf, and this proportion is likely to increase in the future. Australia has become a significant oil producer, mostly from offshore wells. This one is in the Bass Strait off the coast of Victoria state. (Photo courtesy of UN/Australian Information Service)

ters, but it did perform a useful function by identifying and partially defining them, since they were to grow steadily in importance. The major confrontation at The Hague was between distant-water fishing States, such as Japan and Britain, which wanted a narrow territorial sea and no exclusive fishing zones, and those States that wanted to keep foreign fishing fleets as far from their shores as possible. This signaled a major shift in emphasis in the evolving Law of the Sea from commerce and security to the resources of the sea.

Through the 1930s and throughout World War II, the Law of the Sea continued to evolve slowly, despite excessive claims to maritime jurisdiction by a small number of countries, led by the Soviet Union, which in 1927 claimed a 12-mile territorial sea. This deliberate pace was ended abruptly on 28 September 1945. On that day President Harry Truman of the United States issued two proclamations. The first was a moderate one, announcing that the United States would regulate fisheries in those areas of the high seas contiguous to its coasts, but that there would be no restrictions on navigation in those zones. The second proclamation, however, was stunning. It asserted U.S. jurisdiction over the resources of the continental shelf and of the subsoil of the shelf. A subsequent policy statement defined the

outer limit of U.S. jurisdiction as the 200-meter isobath.* Motivated though it was by a perceived need to tap known and suspected offshore oil deposits so as to reduce dependence on imports for a ravenous oil-based economy and by the need to establish federal control over this resource in the face of claims by several coastal states, it nevertheless set off what we call "The Great Sea Rush" of the twentieth century.

Looking back on it now, the terms of "The Truman Proclamation," as it is now called, seem quite moderate, since they included preservation of all traditional high seas rights in the water above the shelf (superjacent water). It did, however, call the world's attention to the fact that out beyond the coastal zone was something of value besides fish and that there was nothing to prevent a coastal State from simply grabbing it.† If the United States could do it, why couldn't others? They could—and did. Only a month later Mexico also claimed both exclusive fishing and continental shelf rights. Ar-

---

*An isobath is a line connecting points of equal water depth, an underwater contour line. A depth of 200 meters is roughly 100 fathoms or about 660 feet.

†By this one presidential act, the United States acquired exclusive rights to the vast mineral and living resources of 2.4 million square kilometers of land under the sea (700,000 square miles).

gentina in 1946 claimed not only its extraordinarily broad continental shelf, but also the superjacent waters. In the following year Peru and Chile went even further and claimed *sovereignty* over a belt of sea extending 200 nautical miles from their coasts. They claimed that since they had little or no continental shelf, they were entitled to the rich fisheries of the Humboldt Current as compensation. Since then claims to maritime jurisdiction have multiplied and escalated so rapidly and so far that the Law of the Sea has been transformed.

### The United Nations Conferences on the Law of the Sea

We may recall that one of the functions of the United Nations is to continue the work of the League in developing and codifying international law. One subject assigned to the International Law Commission was the preparation of draft conventions on several aspects of the Law of the Sea in 1958. Much of the customary Law of the Sea was thus codified into three conventions: one on the Territorial Sea and the Contiguous Zone; one on the High Seas; and one on Fishing and Conservation of the Living Resources of the High Seas. The Convention on the Continental Shelf codified a new doctrine that had been adopted by so many States so rapidly that it has been referred to as an "instant custom."

Why was the 1958 conference so much more successful than The Hague Conference of 1930, even though many of the issues were the same? Much had changed in the interim. The 1930 conference was dominated by lawyers who tended to be doctrinal and legalistic and who felt no sense of urgency. The 1958 conference was attended by 86 States (compared with 47 in 1930), including many new ones in Asia, Africa, and the Middle East, and their delegations included experts in many fields besides law. And the times were different. In 1930, the world was quiet; in 1958, the delegates were faced with the continental shelf doctrine, population pressures, proliferating claims to maritime

resources and territory, the relatively greater strength of the United Nations, rapidly advancing technology, increased expectations of the poor countries of benefits from the sea, and, of course, the Cold War between the United States and the Soviet Union. The Law of the Sea was not, has never been, and never can be divorced from the tangle of political, economic, and social developments in the world at large.

Despite its very considerable success, the 1958 conference left a number of matters unsettled, including two of the most controversial ones: the breadth of the territorial sea and the establishment of exclusive fishing zones. In order to resolve these and other issues, the Second United Nations Conference on the Law of the Sea was held in Geneva in the spring of 1960. It was short (six weeks), sharp, and unsuccessful. A U.S. compromise proposal on the territorial sea—a 6-mile territorial sea plus a 6-mile fishing zone—was defeated by one vote. If only a single country of the 87 present had voted for it instead of against it, much of the subsequent history of the world would have been very different.

The four 1958 conventions and the 1965 UN Convention on Transit Trade of Landlocked States entered into force and became the new datum plane for the Law of the Sea. But the requisite ratifications had scarcely been deposited with the Secretary-General of the United Nations when another climactic event led to its total rethinking and renegotiation.

On 17 August 1967, Arvid Pardo, Permanent Representative (ambassador) of Malta to the United Nations, proposed to the Secretary-General that the agenda of the forthcoming twenty-second session of the General Assembly include an item entitled "Declaration and treaty concerning the reservation exclusively for peaceful purposes of the seabed and of the ocean floor, underlying the seas beyond the limits of present national jurisdiction, and the use of their resources in the interests of mankind." In an accompanying memorandum, he suggested that the seabed and ocean floor be declared a "common heritage of mankind."

He proposed the creation of an international agency that would exercise complete jurisdiction over this area. Arvid Pardo's proposal set off a chain of events that transformed the Law of the Sea more in two decades than in the thousands of years of its evolution before 1967.

In December 1967 the General Assembly established an ad hoc Committee on the Peaceful Uses of the Seabed and Ocean Floor Beyond the Limits of National Jurisdiction, which came to be known as the Seabed Committee. In 1969 the committee was enlarged (ultimately to 91 members) and became essentially a preparatory committee for the Third United Nations Conference on the Law of the Sea (UNCLOS III). The Seabed Committee labored on for four more years. It was unable to produce a draft convention on the Law of the Sea, but it did do a lot of research, its members and the public generally learned a great deal about the sea, and it defined issues and narrowed the range of options for dealing with them. While it was working, interest in the sea was growing both within the United Nations and in the world outside. A revised Law of the Sea was becoming more urgent. Why?

There are a number of reasons. *First*, the 1958 and 1965 conventions left a number of important matters unsettled, had not been

**Polymetallic (manganese) nodules from the seabed.** Nodules are composed of fine-grained oxide material apparently precipitated out of the sea water. They vary widely in their composition as well as in their physical and chemical properties, but all are polymetallic, containing about 26 elements altogether. The most important commercially are manganese, copper, cobalt, and nickel. It is estimated that 22 billion tons of nodules lie on the ocean floor, but the most important concentration apparently is in the northeast Pacific, between Hawaii and Mexico. (Photo by UN/S. Lewin)

ratified or acceded to by enough States to be firmly established as binding on the entire community of States, and were being outdistanced by events. *Second*, decolonization had led to the admission to the United Nations of 41 new countries between 1958 and 1967, and the Group of 77 had organized as a pressure group within the UN. They wanted to participate in both the exploitation of new resources and the progressive development of international law. *Third*, the potential value of polymetallic nodules, discovered by the British oceanographic research vessel HMS *Challenger* in 1873–76, was first understood and publicized in the early 1960s.* *Fourth*, the "population explosion" and the equally spectacular "technology explosion" were putting enormous pressure on finfish, shellfish, marine mammals,

**Arvid Pardo.** At the signing ceremony for the United Nations Convention on the Law of the Sea, Montego Bay, Jamaica, December 1982, he was an observer but had no official role. (Martin Glassner)

*They are not rocks, but round or potato-shaped lumps of precipitates from the seawater that litter the floors of the continental shelves, the ocean basins, and even some freshwater lakes. The potential value of their manganese, copper, cobalt, and nickel alone is estimated in the trillions of dollars.

and seaweed, with some species being harvested to the brink of extinction.

*Fifth*, the superpowers were turning the sea into a colossal military playing field. Their nuclear submarines were ranging everywhere, and even their surface ships were intruding into waters hitherto free of their rivalries, such as the polar seas and the Indian Ocean, and they were developing techniques for placing on the seafloor not only sensing devices, but also nuclear and other weapons of mass destruction. *Sixth*, the ecosystem of the sea was being seriously damaged by modern technology in addition to overfishing, as dramatized by the 1967 *Torrey Canyon* oil-spill disaster, yet there was no international mechanism for adequately handling this problem.

*Seventh*, the International Geophysical Year (1957–58), the International Indian Ocean Expedition (1959–65), and other remarkable ventures in scientific cooperation not only produced a great deal of new information about the sea, but graphically demonstrated how little we know about it even now. They also demonstrated the value of international as well as interdisciplinary cooperation in marine scientific research.

*Eighth*, more countries were extending their jurisdiction farther and farther out to sea, unrestrained by any provisions of international law. Since annexing land of neighbors was no longer feasible, they were turning for expansion to the defenseless and nearly friendless sea. By the end of 1967 about a dozen countries were already claiming some sort of jurisdiction *beyond* 12 nautical miles, and six more joined them by 1970. The trend was clear.

All these factors led the General Assembly to adopt in December 1970 a "Declaration of Principles" that essentially enshrined Arvid Pardo's common heritage proposals. Another resolution authorized UNCLOS III. But already the common heritage concept was being undermined by the very States that were giving it such vociferous verbal support in the Seabed Committee and the General Assembly. Through 1972 and 1973 a series of regional meetings in Latin America and Africa endorsed the demand of Chile, Ecuador, and Peru for some kind of 200-mile economic zone within which a coastal State would have exclusive rights to *all* resources, living and nonliving. By the time UNCLOS III opened with a brief organizational session in New York in December 1973, these States had managed to acquire enough support so that one of the major conference issues had already been settled in principle. Nevertheless, the Conference still had a great deal of work to do.

The Third United Nations Conference on the Law of the Sea was the largest, longest,

**A petroleum tanker disaster at sea.** In December 1976, the *Argo Merchant*, a tanker loaded with petroleum, ran aground off the coast of Massachusetts, then broke up and sank. A succession of such disasters spurred UNCLOS III to devise new rules for the preservation of the marine environment, at least insofar as ship-based pollution is concerned. Nationalism and other political considerations, however, have prevented agreement on truly effective measures to protect the "indivisible ecological whole" of ocean space. (Photo courtesy of NOAA)

**Caracas welcomes UNCLOS III.** "Welcome to Our Land, People of the Sea" reads this colorful billboard designed for the first substantive session of the Third United Conference on the Law of the Sea. About 5000 people participated in this 10-week session in the magnificent new Parque Central conference facilities, and the Venezuelan government did everything possible to assure its success. (Martin Glassner)

and most complex diplomatic conference in history, and without question one of the most important. We should first say a little about the Conference itself, though a brief review can convey only a glimmer of its magnitude and complexity.

First, the number of States participating—135 to 150 per session—was triple the number at The Hague Conference of 1930 and nearly double the number at Geneva in 1958. Many newcomers were not only inexperienced in creating international law, but had no maritime experts or maritime tradition. Second, the objective of the Conference was to produce a single "generally acceptable" convention covering *all* aspects of the Law of the Sea, old and new. The agenda contained 92 "subjects and issues" that somehow had to be incorporated into a treaty, and their complexity, interrelatedness, and importance were almost overwhelming. Third, the organization of the Conference, its rules of procedure, and the dynamics of conference diplomacy were more than a little unusual. No single country or group or interest, in fact, dominated the Conference, and traditional alignments of States were not often evident. Science and technology, when introduced at all, were used as instruments of national policy.

There were 11 sessions of the Conference, from December 1973 to December 1982. The first substantive session (of ten weeks),

which met in Caracas in the summer of 1974, was the longest. The others were held in New York and Geneva, and a formal signing ceremony was held in Montego Bay, Jamaica. The Conference seemed to be dominated by three interlocked themes: a general determination to establish some kind of international regime for the seabed beyond the outermost limits of national jurisdiction; a vigorous nationalism, expressed primarily, but certainly not exclusively, in the drive by coastal States to secure as much of the sea and its resources as they possibly could; and the effort of the Group of 77 to use the emerging new Law of the Sea as an instrument to help them achieve a New International Economic Order.

The Conference participants were organized in a great variety of formal and informal groups, subgroups, working groups, consulting groups, and so on. The most important were the regional groups: the Latin American, Asian, African, and Eastern European groups, and the Western European and Others Group, which included nearly all the States left over. Other groups overlapped these and one another: the Islamic Group, Arab Group, EC, and NATO, for example, and other overlapping groups were issue-oriented, such as the Archipelagic States Group, Coastal States Group, Oceanic Group, the Territorialists (those advocating a 200-mile territorial sea), and others.

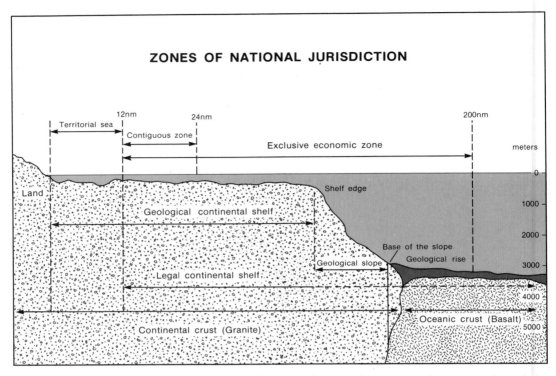

**Zones of national jurisdiction in the sea.** This is a schematic diagram not drawn to scale and designed only to illustrate certain terms and concepts.

***Table 31-1*** *Group of Land-locked and Geographically Disadvantaged States at the Third United Nations Conference on the Law of the Sea, with Dates of Admission to the Original Group*

| | | | |
|---|---|---|---|
| Afghanistan* | Ethiopia 10 April 1975 | Laos* | Sudan |
| Algeria, April 1976 | Finland | Lesotho* | Swaziland* |
| Austria* | The Gambia 10 April | Liechtenstein* | Sweden |
| Bahrain | 1975 | Luxembourg* | Switzerland* |
| Belgium | German Democratic | Malaŵi* | Syria, June 1977 |
| Bhutan* | Republic (now | Mali* | Turkey, April 1976 |
| Bolivia* | merged with | Mongolia* | Uganda* |
| Botswana* | Germany) | Nepal* | United Arab Emirates |
| Bulgaria 7 May 1975 | Germany, Federal | Netherlands | 10 April 1975 |
| Burundi* | Republic of (now | Niger* | Upper Volta (now |
| Byelorussian SSR* | Germany) | Paraguay* | Burkina Faso)* |
| (now Belarus) | Greece, April 1976 | Poland | Zaïre |
| Cameroon, June 1977 | Holy See*† | Qatar, 10 April 1975 | Zambia* |
| Central African | Hungary* | Romania April 1979 | Zimbabwe* (observer, |
| Republic* | Iraq | Rwanda* | June 1977; full |
| Chad* | Jamaica 10 April 1975 | San Marino* | member, March |
| Czechoslovakia* | Jordan 10 April 1975 | Singapore | 1981) |
| | Kuwait | | |

*Land-locked.
†Withdrew summer 1976.

**The final session of UNCLOS III, Montego Bay, Jamaica, 10 December 1982.** At the far right, H. E. Mr. Javier Pérez de Cuellar, Secretary-General of the United Nations, is addressing the Conference. On the dais are the Conference officials; in the center is H. E. Ambassador Tommy T. B. Koh of Singapore, President of the Conference. In the foreground are some of the delegates. (Martin Glassner)

Most bizarre of all to a geographer was the Group of Land-locked and Other Geographically Disadvantaged States, shown in table 31-1. We have already discussed land-locked States, and both their characteristics and problems seem clear enough. But what was (or is) a "geographically disadvantaged State"? No one seems to know. Many definitions have been offered, but none has been generally accepted. It seems to mean a country that stands to gain little or nothing from a 200-mile exclusive economic zone; in practice, however, admission to the group was largely based on political criteria, not geographic ones. The Holy See (Vatican City) withdrew from the group in 1976, purportedly because it was becoming too political. Israel, which qualified on at least two counts and applied for admission, was rejected for political reasons. Nevertheless, the Land-locked and GDS (geographically disadvantaged States) Group was at times a potent lobby for their interests and managed to win some useful concessions in the Conference.

In 1982 UNCLOS III adopted overwhelmingly the United Nations Convention on the Law of the Sea. Only three other States joined the United States in voting against it, all for different reasons. On the very first day that it was open for signature (10 December 1982), 119 States and two other entities

signed it and 38 more signed within the following two years. By the date the Convention entered into force, 16 November 1994, 155 States and five other entities had signed it, and 68 had ratified, acceded or succeeded to it. Although these figures do not include the United States, Germany, or the United Kingdom, a mattter we discuss in the next chapter, they do indicate that the 1982 Convention does represent a consensus on the developing Law of the Sea, and most countries—even nonsignatories—are gradually adjusting their domestic legislation to conform to it.

Much of this developing law derives from the traditional concepts and practices discussed previously. Some of it, in fact, is retained verbatim from the 1958 and 1965 conventions. Although we tend to stress the innovations, the dramatic departures from tradition, we must remember that everything being done now in the Law of the Sea is really a continuation of a very long tradition, two thousand years of legal development and six thousand years of seafaring.

Nevertheless, the Third United Nations Conference on the Law of the Sea will without question be recorded in the history books of the future as a milestone in political affairs. It produced what is essentially a constitution for the sea, nearly three quarters of the earth's surface.

# 32

# *The Political Geography of the Sea*

*I must go down to the seas again, to the lonely seas and the sky,*
*And all I ask is a tall ship and a star to steer her by,*
*And the wheel's kick and the wind's song and the white sails' shaking,*
*And a gray mist on the sea's face, and a gray dawn breaking.*
John Masefield, *Sea Fever*

Now that we understand something of the background of the Law of the Sea, we can examine more closely those elements of it that form the core of the political geography of the sea. In addition, we consider in this chapter some aspects of our topic that are not included in the United Nations Convention on the Law of the Sea. We also include some current examples of contemporary maritime problems that illustrate some of the major themes. We refer to and quote the Convention frequently and note how it is similar to or different from previous provisions of the Law of the Sea.

## The Territorial Sea

As we pointed out in Chapter 31, the concept of a territorial sea has long been incorporated into international law, but there has never been any agreement on its breadth. Generally speaking, the major maritime and naval powers wanted—and claimed—a narrow territorial sea so as to have as few restrictions as possible on their vessels navigating around the world, while those with very small merchant and naval fleets preferred to have a broad buffer zone between potentially hostile or exploitative foreign

vessels and their own coasts. The compromise reached is a uniform 12-mile territorial sea, measured from the applicable baseline. Territorial waters are the sovereign territory of the coastal State just as is the land and, like the land, its sovereignty over the territorial sea extends from the core of the earth to the heavens. There is one major restriction, however, on a coastal State's sovereignty over its territorial waters: it must grant to foreign vessels, including warships, the right of innocent passage. This is a very old restriction, and an essential one if commerce and navigation are to continue linking together the peoples of the world across the great ocean highways. The new definitions of the terms *innocent* and *passage*, though, are more elaborate, and that of *innocent* is much more precise than the traditional one, which was: "Passage is innocent as long as it is not prejudicial to the peace, good order or security of the coastal State." Submarines passing through the territorial sea are still required to navigate on the surface showing their flag. Aircraft still do not have a right of innocent passage over the territorial sea.

Many other provisions are included in this section of the Law of the Sea Convention (LOSC), with several relating to the determination of the baseline from which the breadth of the territorial sea (up to 12 nautical miles) is measured. The normal baseline is "the low-water line along the coast as marked on large-scale charts officially recognized by the coastal State." However, "In localities where the coastline is deeply indented and cut into, or if there is a fringe of islands along the coast in its immediate vicinity," straight baselines may be drawn according to rather detailed instructions that take into account bays, estuaries, ports and roadsteads, and other coastal features. Shoreward of the baseline, whether normal or straight, all waters, salt or fresh, are considered *internal waters* (with one exception), in which there is no right of innocent passage. On the whole, these provisions for the territorial sea appear to be a fair and practical compromise between the interests of international navigation and coastal State protection.

A coastal State retains the right to establish a *contiguous zone* on the seaward side of the territorial sea of not more than 24 miles from its baseline. Within this zone "the coastal State may exercise the control necessary to (a) Prevent infringement of its customs, fiscal, immigration or sanitary regulations within its territory or territorial sea; (b) Punish infringement of the above regulations committed within its territory or territorial sea." The contiguous zone is not part of the State's territory, but is instead a portion of the sea within which the State is permitted to exercise limited jurisdiction.

No significant problems have arisen recently with regard to the principles underlying these three maritime zones. There have been, however, many apparent violations of the rules governing the drawing of straight baselines. First, some countries have claimed straight baselines where they do not appear to be warranted. Second, lines are drawn that violate the requirement that such lines follow the general configuration of the coast. The purpose of these apparent violations of the rules, of course, is twofold. The area of internal waters is increased, and the baseline for measuring other zones is pushed farther to sea, leading to larger areas of national jurisdiction. Some of these excesses have been challenged, some seem harmless and are ignored, and a few appear to be justified because of special local circumstances.

Perhaps the most egregious of these excessive claims is that by Libya to the entire Gulf of Sirte (Sidra, Surt). In 1973 Libya announced that the entire gulf south of 32°30'N was Libyan internal waters. The bases for this claim are fuzzy. The semicircle rule clearly rules out consideration of the area as a bay, and Libya's historical justification for the claim does not meet the requirements for a "historic bay." Only Syria and Burkina Faso have accepted the claim. It has been protested by Malta, the former Soviet Union, six NATO countries from Britain to Turkey, and by the United States (which in 1981 and 1986 took vigorous defensive action against Libyan aircraft and missile bases that were attacking U.S. aircraft over the gulf). It is noteworthy that the claim has not been recognized by any major country, or even by any other Arab country. Such claims are likely to be abandoned, in fact if not formally, in the face of opposition from the international community.

Another heartening trend is evident. Many observers (including the author) had been concerned since the Caracas session of UNCLOS III in 1974 that States that had already claimed broader territorial seas would insist on enforcing their claims. About a score of countries, mostly in Latin America and West Africa, were even claiming *200-mile* territorial seas. At UNCLOS III they tried to have their claims legitimated and came to be known as the Territorialist Group. They failed, of course, and since the UN Convention on the Law of the Sea (LOSC) was opened for signature in 1982, States claiming broader territorial seas, even some Territorialists, have one by one been abandoning these claims and conforming to the new 12-mile limit. Now only 11 States make such a claim, and five claim between 20 and 50 miles. The Philippines is a special case, since various claims over the years run up to 285 miles, and some of the claims are disputed. It seems that peer pressure can be very effective, even at the international level. There is no assurance, however, that this trend will not be reversed in the future.

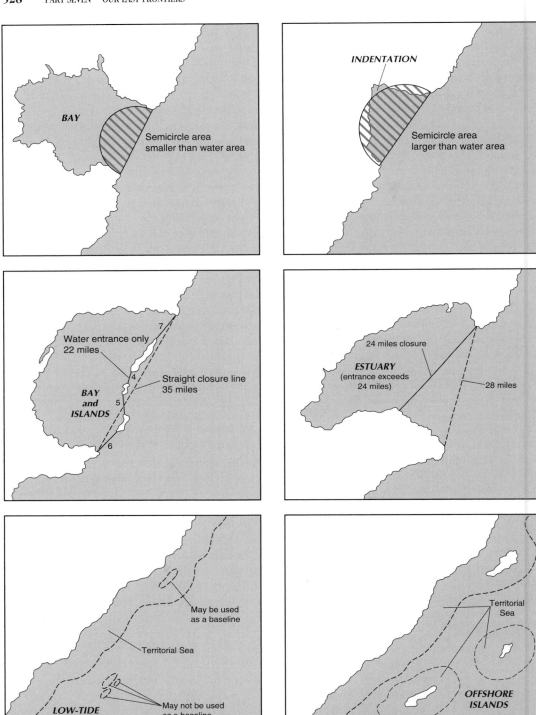

**Baselines of the territorial sea.** These diagrams show how straight baselines are drawn along irregular coastlines according to the rules spelled out in the 1958 Convention on the Territorial Sea and the Contiguous Zone. These rules have been retained in the United Nations Convention on the Law of the Sea.

## Straits Used for International Navigation

One of the major reasons the larger maritime powers opposed expansion of the territorial sea to 12 miles was their concern that more than 100 straits used for international navigation that were between 6 and 24 miles wide would fall under national control and become subject to closure by the straits States. They insisted that high seas rights, especially unrestricted navigation and overflight, should be allowed in these straits, whereas some States bordering the straits wanted the rule of innocent passage to prevail in them. The 1958 Territorial Sea Convention has only one sentence on the subject; the LOSC has 11 articles spelling out an elaborate new regime called *transit passage.* It is a real compromise in which the straits States won control over fishing, pollution, traffic movement, and other potential hazards to them, and the maritime States won overflight of aircraft, including military aircraft, and avoided special restrictions on nuclear propulsion, nuclear cargoes, and submarines. The most important of these straits are shown on the front endpaper map.

There are still debates in the professional literature and in diplomatic and military circles over this compromise, so two observations might be appropriate here. First, the "Israel Clause" in the 1958 Convention was carried over into the LOSC. This clause provided for nonsuspendable innocent passage within straits between "one part of the high seas and another part of the high seas *or the territorial seas of a foreign State*" (emphasis added). This provision could only apply to the Strait of Tiran, between the Gulf of Aqaba and the Red Sea. Now, however, such passage is governed by the 1979 Treaty of Peace between Israel and Egypt, which both parties agree is compatible with the LOSC. The peace treaty provides for "unimpeded and non-suspendable freedom of navigation and overflight" for all States. This is a much more liberal regime than that of transit passage or of innocent passage.

Second, there has been only one noteworthy strait State violation adversely affecting transit passage to date. In September 1988 an Indonesian naval official closed the Lombok and Sunda Straits to foreign navigation and overflight during air and sea military exercises that lasted for a month. The United States, Japan, Australia, and a number of European countries protested in various ways. Apparently, the Indonesian navy had closed the straits without consulting the foreign ministry, which apologized for the incident and promised that it would not hap-

**A vital international strait.** This waterway connects the Kattegat to the left and the Sound to the right. In the foreground is the Danish port of Helsingor, protected by Kronborg Castle (made famous by Shakespeare as Elsinore in his *Hamlet*). In the background is the larger Swedish city of Helsingborg. (Photo courtesy of The Danish Tourist Board)

pen again. Since Indonesia had been one of the most outspoken proponents of continuation of innocent passage in straits, this incident bodes well for the ultimate acceptance of the transit passage regime as customary international law. In fact, the general trend in State practice is toward such acceptance.

## Islands

We all learned as children that an island is a "piece of land surrounded by water," but such a simple definition could hardly suffice in a world in which over half a million such "pieces of land," no matter how small, might each qualify for its own territorial sea, contiguous zone, continental shelf, and exclusive economic zone. Therefore the definition of "island" in the Convention is more restrictive: "An island is a naturally formed area of land, surrounded by water, which is above water at high tide." Then a new paragraph reads, "Rocks which cannot sustain human habitation or economic life of their own shall have no exclusive economic zone or continental shelf." They may still have territorial seas and contiguous zones, however, and the entire section on islands leaves many questions still unanswered.

These questions, deriving from gaps and ambiguities in the definitions, have led many countries to claim flyspecks as islands and others to challenge these claims. The map of disputed islands in the East and South China Seas, found in Chapter 8, locates some of them; there are numerous others.

Perhaps the ultimate in questionable island claims is Okinotorishima (Douglas Reef) in the Pacific south of Japan. This reef now consists of only two rocks, the larger of which is only 4.5 meters (15 feet) long. Japan has claimed it since 1931. If these rocks remain above water, Japan can claim an EEZ around them of 125,000 square nautical miles (over 283,000 square kilometers or 144,000 square miles)! However, the coral has been eroding badly. So Japan in 1988 built reinforcements of concrete and iron connecting the two rocks. Is the result "a naturally formed area of land?"

## Archipelagic States

Although discussed at The Hague Conference in 1930, the question of archipelagos was avoided in Geneva in 1958. But Indonesia and the Philippines (and later Fiji) lobbied so hard for a new regime for them that UNCLOS III devised one that is really quite new in international law.

First, "archipelagic State" is defined as one "constituted wholly by one or more archipelagos and may include other islands," while an archipelago is "a group of islands, including parts of islands, interconnecting waters and other natural features which are so closely interrelated that such islands, waters and other natural features form an intrinsic geographical, economic and political entity, or which historically have been regarded as such." An archipelagic State is entitled to draw straight baselines connecting "the outermost points of the outermost islands and drying reefs of the archipelago provided that within such baselines are included the main islands and an area in which the ratio of the area of the water to the area of land, including atolls, is between one to one and nine to one." Also, generally, "the length of such baselines shall not exceed 100 nautical miles."

These restrictions effectively limit the status of "archipelagic State" to relatively few countries. Indonesia, Fiji, the Philippines, the Bahamas, and 12 others had claimed archipelagic State status by the end of 1994, though not all had specified archipelagic baselines. The waters enclosed by the baselines (except for internal waters) are designated *archipelagic waters* over which the State has sovereignty but within which "ships of all States enjoy the right of innocent passage" subject to provisions for sea lanes within the archipelagic waters and lists of rights and duties of the State and of ships and aircraft passing through. The territorial sea and contiguous zone are to be measured outward from the baseline. Although additional area has been removed from the high seas by these provisions, on the whole they also represent a reasonable compromise.

The chief violations of these provisions are apparently being committed by island States that are exploiting the ambiguities of the definitions of "archipelagic State," including such matters as the term drying reefs and the measurement of baseline segments and land-to-water ratios. However, even this statement is hesitant, for often the statements made by island States about their maritime boundaries do not conform to the LOSC language and are far from models of clarity. In any case, there do not yet appear to have been any notable violations of innocent passage or high seas navigation rights. But this cloudy, though tolerable, situation has the potential for generating serious disputes.

## Enclosed or Semi-enclosed Seas

One of the more enigmatic parts of the Convention deals with the "enclosed or semi-enclosed sea," defined as "a gulf, basin, or sea surrounded by two or more States and connected to the open seas by a narrow outlet or consisting entirely or primarily of the territorial seas and exclusive economic zones of two or more coastal States." States bordering such bodies of water are exhorted to cooperate in a variety of ways in marine affairs. The definition of the terms is certainly open to question, but a more fundamental question is why the terms are used at all, why any distinction is made between these and any other parts of the sea, why this part is in the Convention at all. It is possible that the reason will become clear in time; it is also possible that the delegates at UNCLOS III simply turned an academic exercise into a legal issue so as to leave no possible marine feature uncovered by the Convention they were writing.* Various explanations have been offered in the professional literature, including some by participants, but none is very convincing.

## The Exclusive Economic Zone

Although the concept of an exclusive economic zone (EEZ) has roots deep in history, its modern scope is quite new and revolutionary. Each coastal State now is entitled to a zone, seaward of the territorial sea and adjacent to it, within which it has "sovereign rights for the purpose of exploring and exploiting, conserving and managing the natural resources, whether living or non-living, of the sea-bed and subsoil and the superjacent waters," as well as jurisdiction over a number of other activities in the zone. The breadth of the EEZ is not to exceed 200 nautical miles from the baseline. This part of the Convention has 21 articles, some of them quite long, that describe the numerous detailed rights and few vague responsibilities of the coastal States in their EEZs. They alone have jurisdiction over conservation of fish, marine mammals, and other "living resources" in their zones, including anadromous, catadromous and highly migratory species.* Land-locked States are granted greatly restricted and probably meaningless rights to certain surplus fish within the EEZ.

The exclusive economic zone has been subtracted from the high seas and the "common heritage of mankind," thus shrinking this historic international area by one-third! Within 200 nm are all the world's known offshore oil and natural gas deposits, all the presently exploitable minerals, most of the readily accessible polymetallic nodules, 85 to 95 percent of the world's current fish catch, nearly all marine plants, and nearly all potentially exploitable sources of unconventional energy (waves, tides, thermal layers). More than half the combined EEZ of the

---

*Anadromous fish are those, such as salmon, that are born in freshwater streams and migrate out to sea where they spend the greater part of their lives before returning to their birthplaces to spawn. Catadromous fish are those, such as some eels, that are born at sea and migrate into freshwater lakes and streams to live until they return to the sea to spawn. Highly migratory species are those, such as tunas and sharks, that range far through the sea, sometimes thousands of kilometers, on a fairly regular schedule.

---

*The academic exercise referred to is Lewis M. Alexander, "Regionalism and the Law of the Sea: The Case of Semi-enclosed Seas," *Ocean Development and International Law Journal*, 2, 2 (Summer 1974), 151–186.

world goes to only ten countries, of which six of the first seven are already rich (in order of size of maritime territory gained: the United States, Australia, New Zealand, Canada, Russia, and Japan). All the top five gainers of submarine petroleum are also rich (United States, Russia, Britain, Norway, and Australia). Thus, most poor countries, which began the drive for an EEZ and finally won acceptance for it, stand to gain relatively little from it.

It seems highly unlikely at this point that the creation of the EEZ will contribute materially to a reduction of the great and growing gap between rich and poor in the world, or to the wise preservation of the sea as "an indivisible ecological whole," or to the replacement of petty nationalisms with real and substantial international cooperation.

All is not negative, however. Around the world, coastal States are taking seriously their opportunities and responsibilities to

**The United States exclusive economic zone.** Most of the EEZs do not yet have boundaries agreed on with the respective neighbors of the United States; those in the Pacific Northwest are still being negotiated with Canada and pose particularly difficult problems. (*Source:* Office of the Geographer, U.S. Department of State.)

preserve and protect the marine environment and its living creatures. Some of this work is being done in individual EEZs, some regionally. Some is systematic, some haphazard. In some cases enforcement is strict, in some lax. It is too early to try to assess quantitatively the effectiveness of the EEZ concept worldwide in reducing frictions among States and protecting the marine environment. There does, however, seem at the moment to be a slight net gain for the world community as a result of all this activity. The trends, moreover, seem to be positive. Regional cooperation is growing, and more and more States are harmonizing their national legislation with the relevant provisions of the LOSC, even those that have not yet ratified or even signed it.

By now, most States that are able to claim EEZs have already done so. Some claim only exclusive fishing zones, or zones for pollution control, and there are few examples of gross violation of the basic principles. The major remaining problem is that of delimiting the EEZs between opposite or adjacent States. This was not a very serious matter when national jurisdiction extended only 3, 4, or 6 miles out to sea. Now, with territorial seas of 12 miles and EEZs extending as far as 200 nautical miles, even insignificant differences in the angles of boundaries create large disputed areas far from shore. As the demand for marine resources and other uses of the sea increases and technology improves to permit more varied and intensive use of the sea, quarrels over these marine boundaries and zones are likely to increase. So far, however, roughly half of all potential maritime boundaries have been settled peacefully and there have not yet been any violent clashes over the remainder (except for some disputed islands). We may hope that these trends continue and that national jurisdiction does not "creep" as had been feared.

## *The Continental Shelf*

Prior to 1945, the continental shelf was a geological feature known only to geologists and to some fishermen, engineers, and sea-

men. By 1958 it was not only well known, but so important to so many coastal States that the first United Nations Conference on the Law of the Sea granted the coastal States exclusive rights to the resources of the shelf and its subsoil. The Convention on the Continental Shelf is the shortest of the four Geneva conventions of 1958 and the most tentative. It defines the shelf as "the seabed and subsoil of the submarine areas adjacent to the coast but outside the area of the territorial sea, to a depth of 200 meters or, beyond that limit, to where the depth of the superjacent waters admits of the exploitation of the natural resources of the said area." Both portions of this definition—the obsolete 200-meter isobath and the absurd "exploitability clause"—have been abandoned in the 1982 Convention in favor of a more precise but far more generous (to a coastal State) definition: "the sea-bed and subsoil of the submarine areas that extend beyond its territorial sea throughout the natural prolongation of its land territory to the outer edge of the continental margin, or to a distance of 200 nautical miles." This definition assures that coastal States have exclusive rights to much of what had been considered "the common heritage of mankind."

The remaining nine articles on the continental shelf grant to the coastal States most of the exclusive rights they obtained in the EEZ, but one article requires coastal States to pay a kind of royalty on nonliving resources produced from their shelves beyond 200 miles, with the money to be distributed to the poorest countries. It is unlikely that any such royalties will actually be forthcoming for decades, but the principle is important.

Aside from the question of delimitation of shelf boundaries between opposite and adjacent States, few problems have arisen regarding the continental shelf. We are now in the second generation of experience with the legal doctrine and seem to be comfortable with it. Currently, there are 55 continental shelf boundaries in force, which were negotiated or adjudicated, and another 76 single maritime boundaries or shelf/fishing boundaries in force. But these are the easy ones. As indicated by the current caseload of

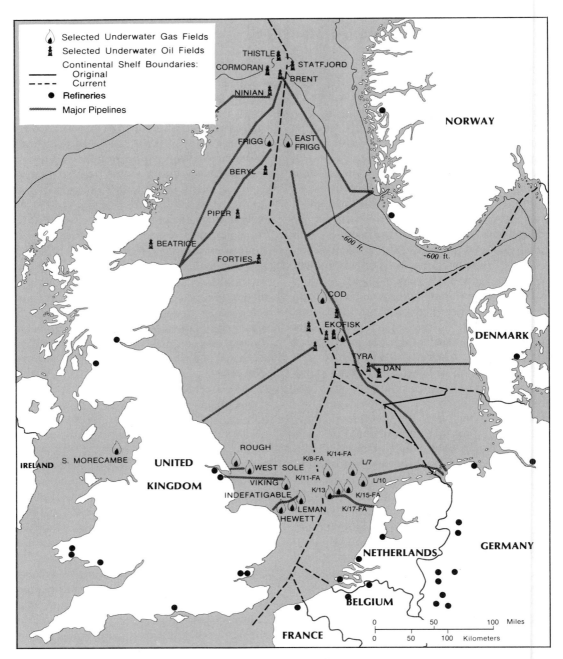

**The North Sea continental shelf.** Note the deep trench off the coast of Norway; this was deliberately ignored by Norway and the United Kingdom when they decided on a median line as their shelf boundary. Because of the configuration of its coast, Germany received a very small portion of the shelf and none of the area most promising for hydrocarbon deposits. It then brought the Netherlands and Denmark before the World Court, which ruled that equity as well as the median line should be taken into consideration. The parties then negotiated new boundaries for the German sector. Only selected oil and gas wells and pipelines are shown; there are many more.

the ICJ in The Hague, discussed in Chapter 24, many of the more difficult ones remain to be resolved.

The United States alone will ultimately have about 28 maritime boundaries with other States and territories, of which only 9 had been resolved by mid-1995. Still to be negotiated are three boundary segments with Canada, three adjacent to American Samoa, three with Kiribati, and others with Cuba, Haiti, Jamaica, The Netherlands (for Saba), the Dominican Republic, Japan, the Federated States of Micronesia, the Marshall Islands, and The Bahamas. A boundary treaty with the former Soviet Union was signed on 1 June 1990, but had not yet been ratified as of July 1995.

## The High Seas

The two-thirds of the sea remaining after internal, territorial, and archipelagic waters and EEZs have been subtracted and assigned to coastal States continue to be high seas and essentially the property of the international community. Many traditional high seas rights and responsibilities are retained, but the old "freedom of the seas" is dead. This is not necessarily bad, for in truth it is obsolete; today it would mean freedom to overfish and freedom to pollute the sea. The new definition of "freedom of the high seas" is both more precise and more narrow than the traditional one expressed in the 1958 High Seas Convention. On the whole, the new provisions preserve the traditional high seas rights that tend to serve community interests, such as navigation, and some have even been enhanced. Although not perfect, from a geographer's standpoint the Convention provisions on the high seas seem satisfactory.

## The International Seabed Area

We may recall Arvid Pardo's proposal of 1967 that the "seabed and ocean floor and subsoil thereof beyond the limits of present national jurisdiction" be declared "the common heritage of mankind." It should be used only for peaceful purposes and its resources exploited for the benefit of the international community, particularly its poorest members. This triggered the massive and thorough review and revision of the Law of the Sea that followed. By far the greatest innovation—and the most complex and controversial issue—is the establishment of a common heritage regime for the International Seabed Area. The main features of this regime are as follows:

1.  The Area and its resources cannot be appropriated by any State.

2.  The Area is to be used exclusively for peaceful purposes.

3.  All rights in the resources of the Area "are vested in mankind as a whole."

4.  International cooperation is encouraged in marine scientific research, protection and conservation of the marine environment, and transfer of technology relating to mineral resource exploration and exploitation.

5.  Developing countries are assisted to participate effectively in these activities.

6.  An International Seabed Authority is created to administer the mineral resources of the Area.

7.  The Authority must undertake scientific research in the Area, establish a system of equitable sharing of benefits from the Area, and participate in deep-sea mining on its own account as well as in association with States and private entities.

Much of the Convention is devoted to detailed provisions concerning the structure and operations of the Authority, which do not concern us here. It seems that the poor countries of the world are really going to get very little out of this arrangement until well into the twenty-first century, but at least it should result in the establishment of an international agency that, however unwieldy, will manage the mineral resources of over half the earth's surface for the ultimate benefit of all the peoples of the earth. This might be a giant step toward the replacement of the currently dominant European State system with a form of political organi-

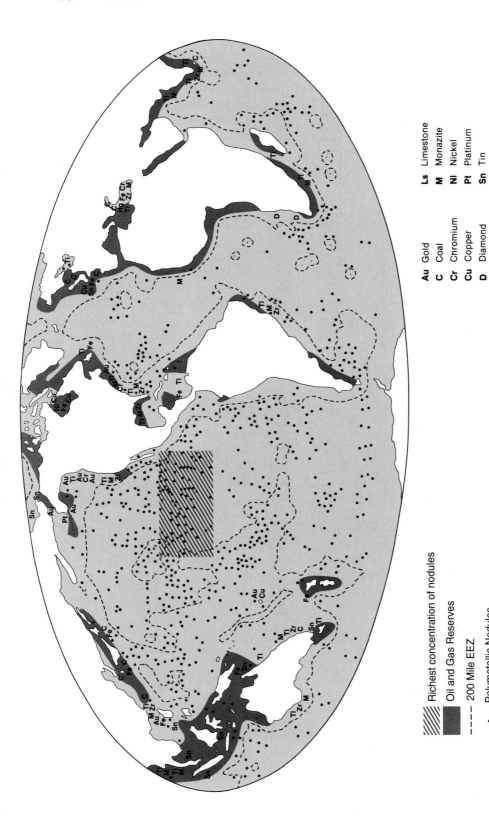

| | | |
|---|---|---|
| | Richest concentration of nodules | |
| | Oil and Gas Reserves | |
| - - - | 200 Mile EEZ | |
| • | Polymetallic Nodules | |

| | | | | |
|---|---|---|---|---|
| **Au** | Gold | **Ls** | Limestone |
| **C** | Coal | **M** | Monazite |
| **Cr** | Chromium | **Ni** | Nickel |
| **Cu** | Copper | **Pt** | Platinum |
| **D** | Diamond | **Sn** | Tin |
| **Fe** | Iron | **Ti** | Titanium |
| **Hg** | Mercury | **Zr** | Zirconium |

**Mineral resources of the sea.** The polymetallic nodules and the hydrocarbon resources alone are worth trillions of dollars. In 1974 Deepsea Ventures filed with the U.S. Department of State mining claims for operations in the area of the richest nodule deposits. The State Department took no action, and to date there has been no commercial mining of nodules anywhere. The United Nations, however, has officially registered as "pioneer investors" four Western consortia of companies with seabed mining claims and others from France, Japan, Poland, the Republic of Korea, China, India and the former Soviet Union, and has resolved all overlapping claims.

zation more suited to our future needs and conditions. Meanwhile, however, it still allows for the expansion of national jurisdiction into the international area and is very far from a common heritage regime that can be effectively implemented. We have more to say about it at the end of this chapter.

---

In 1978 Francis T. Christy, Jr., composed this ditty for the American Society of International Law. Christy was head of Resources for the Future and an early critic of the traditional Law of the Sea.

### *Ode to the Grotian Ocean*

Good gracious, dear Grotius
Your law is atrocious,
Your mare liberum must end.

The cannon shot rule
Is a rule for a fool
In these days of the ICBM.

Van Bynkershoek 's wishes
Aren't good for the fishes
When everyone has his own fleet.

Economists smirk
At McDougal and Burke*
'Cause their freedom's a right to deplete.

The maritime powers
Have long had their hours
In using the ocean for free.

Now the 77
Are in 7th heaven
Repealing the law of the sea.

And Selden is seen
As fully redeemed,
With nodules increasing the stakes.

Dear Grotius, my gracious!
The oceans aren't spacious.
They're nothing but coastal states' lakes.

*Myres S. McDougal and William T. Burke, *The Public Order of the Oceans; A Contemporary International Law of the Sea*. New Haven and London: Yale University Press, 1962.

## *Preservation of the Marine Environment*

Instead of treating ocean space as "an indivisible ecological whole,"* UNCLOS III partitioned it—horizontally, vertically, and functionally—into innumerable international, regional, national, and private jurisdictions with no centralized control over them and no overall plan for the preservation of the integrity of the marine environment. The Conference, partly because of its limited mandate, also dealt not with environmental considerations as a whole, but only with pollution of the sea, and even here only with the 10 to 20 percent of pollution that originates from ships. (See Table 32-1 for details of some oil spills.) Although the developing countries have overcome their original unconcern with environmental matters (and even at times opposition to environmental protection proposals because they feared interference with their economic development in general and industrialization in particular), the Conference was long engaged in the difficult job of allocating responsibility for regulations and enforcement among flag States, port States, and coastal States. The *Torrey Canyon* disaster in 1967 awakened public interest in marine pollution, and the *Amoco Cadiz* disaster in 1978 not far away spurred UNCLOS III, meeting at the time in Geneva, into tightening the relevant provisions of the Convention. The result is quite inadequate to ensure long-term protection for the marine environment, but it is the best arrangement negotiable at present and does represent real, if modest, progress.

## *Marine Scientific Research*

It is hard for the average person to comprehend how highly politicized marine scientific research has become in much of the

---

*Ocean space includes the surface of the sea, the water column, the seabed and its subsoil. Some would include the airspace over the sea as a part of ocean space as well. The quotation is from Elisabeth Mann Borgese, one of the foremost leaders in the development of the new Law of the Sea and public awareness of it.

*Table 32-1    Major Oil Spills, 1967–91*

| No. | Date | Spill | Location | Volume (millions of gallons) |
|---|---|---|---|---|
| 1 | 1979–80 | Ixtoc I/well blowout | Mexico | 139–428* |
| 2 | 1983 | Nowruz oil field/well blowout(s) | Persian Gulf | 80–185 |
| 3 | 1983 | *Castillo de Bellver*/broke, fire | South Africa | 50–80* |
| 4 | 1978 | *Amoco Cadiz*/grounding | France | 67–76 |
| 5 | 1979 | *Aegean Captain/Atlantic Empress* | Off Tobago | 49* |
| 6 | 1980–81 | D-103 Libya, well blowout | Libya | 42 |
| 7 | 1979 | *Atlantic Empress*/fire | Barbados | 41.5* |
| 8 | 1967 | *Torrey Canyon*/grounding | England | 35.7–38.6* |
| 9 | 1980 | *Irene's Serenade*/fire | Greece | 12.3–36.6* |
| 10 | 1972 | *Sea Star*/collision, fire | Gulf of Oman | 35.3* |
| 11 | 1981 | Kuwait National Petroleum tank | Kuwait | 31.2 |
| 12 | 1976 | *Urquiola*/grounding | Spain | 27–30.7* |
| 13 | 1970 | *Othello*/collision | Sweden | 18.4–30.7 |
| 14 | 1977 | *Hawaiian Patriot*/fire | N Pacific | 30.4* |
| 15 | 1979 | *Independenta* | Turkey | 28.9 |
| 16 | 1978 | No. 126 well/pipe | Iran | 28 |
| 17 | 1975 | *Jakob Maersk* | Portugal | 25* |
| 18 | 1985 | BP storage tank | Nigeria | 23.9 |
| 19 | 1985 | *Nova*/collision | Iran | 21.4 |
| 20 | 1978 | BP, Shell Fuel Dept. | Zimbabwe | 20 |
| 21 | 1971 | *Wafra* | South Africa | 19.6* |
| 22 | 1989 | *Kharg 5*/explosion | Morocco | 19 |
| 23 | 1974 | *Metula*/grounding | Chile | 16 |
| 24 | 1983 | *Assimi*/fire | Off Oman | 15.8* |
| 25 | 1970 | *Polycommander* | Spain | 3–15.3 |
| 26 | 1978 | Tohoku storage tanks/earthquake | Japan | 15 |
| 27 | 1978 | *Andros Patria* | Spain | 14.6 |
| 28 | 1983 | *Pericles GC* | Qatar | 14 |
| 29 | 1985 | Ranger, TX/well blowout | Texas | 6.3–13.7 |
| 30 | 1968 | *World Glory*/hull failure | South Africa | 13.5 |
| 31 | 1970 | *Ennerdale*/struck granite | Seychelles | 12.6 |
| 32 | 1974 | Mizushima refinery/tank rupture | Japan | 11.3 |
| 33 | 1973 | *Napier* | SE Pacific | 11* |
| 34 | 1980 | *Juan A, Lavalleja* | Algeria | 11 |
| 35 | 1989 | *Exxon Valdez*/grounding | Alaska | 10.8 |
| 36 | 1978 | Turkish Petroleum Corporation | Turkey | 10.7 |
| 37 | 1979 | *Burmah Agate*/collision, fire | Texas | 1.3–10.7* |
| 38 | 1971 | *Texaco Oklahoma*, 120 mi. offshore | North Carolina | 9.2–10.7 |
| 39 | 1972 | *Trader* | Mediterranean | 10.4 |
| 40 | 1976 | *St. Peter* | SE Pacific | 10.4 |
| 41 | 1977 | *Irene's Challenge* | Pacific | 10.4 |

*Source:* Congress of the United States, Office of Technology Assessment, "Coping with an Oiled Sea: An Analysis of Oil Spill Response Technologies," OTA-BP-0-63, Washington, U.S. Government Printing Office, March 1990.

Tanker spills from the Iran–Iraq War were not generally available.

*Fire burned part of spill.

Observations:

1. Oil spills can occur nearly anywhere: in any country, along any coast, on any important sea lane, at any time, from wells or vessels of any nationality.

2. The two largest spills and many smaller ones have been from wells, not ships.

3. The size of the spill is not necessarily an indicator of damage done. Much oil may burn or evaporate and dissipate in the atmosphere, spreading pollutants very thinly; small spills in ecologically fragile areas can do more damage than large spills in more resilient areas; spills far out at sea are less harmful to the environment than smaller ones close to shore. Two oil spills in 1989 illustrate the point. The *Exxon Valdez* grounding in Prince William Sound near Valdez, Alaska, did very great damage to the wildlife of the region. Far more environmentally damaging was the far smaller and far less publicized grounding of the Argentine tanker *Bahía Paraíso* off the west coast of the Antarctic Peninsula, which spilled only an estimated 175,000 gallons of oil but killed numerous birds and mammals. The oil may take a century to break down completely in this frozen region.

4. All oil spills taken together account for only a fraction of all marine pollution, most of which derives from normal, everyday human activity that does not attract the attention of the news media.

world and in UNCLOS III. We tend to accept it unquestioningly as a positive good; it is part of our culture and we have reaped the benefits of it. Yet few of the poor countries of the world (the vast majority of all the countries) have had any first-hand experience with marine scientific research or any real understanding of it. They often suspect—and quite understandably, considering their colonial past—that a mysterious ship lurking off their coast doing inexplicable things with mysterious instruments is preparing for some kind of assault on either their "sovereignty and territorial integrity" or on their offshore resources. The struggle in UNCLOS III over a regime for marine scientific research was bewildering. What emerged is a modified consent regime; that is, a State wishing to do any kind of scientific research in any of the areas of coastal State jurisdiction (and remember there are six of them now) must obtain the advance permission of the coastal State, though this consent may be implied if the coastal State does not deny it within a specified time period. There are many other provisions relating to cooperation in such research, sharing of results, and so on. Marine scientific research is being shackled with so many restrictions and cumbersome procedures that in the future it may be less rewarding and far more expensive than it was before 1982.

## Other Aspects of the Political Geography of the Sea

UNCLOS III avoided several aspects of the political geography of the sea which eventually will have to be addressed through some other agency. First, the status of the Southern Ocean is not defined at all; in fact, there are no limits specified for the international areas—the seabed or the high seas. They are simply whatever is left over after coastal States have taken whatever portions of the sea they wish within the limits set by the United Nations Convention on the Law of the Sea and other treaties that may be binding on them. However, even within these limits, some states may claim continental shelves as far out as 350 nautical miles or

more from their baselines, and States have considerable freedom in drawing straight baselines. We may expect more problems in the Southern Ocean as States begin to exploit its resources more intensively.

The Conference also avoided the whole question of the military uses of the sea. While the phrase "exclusively for peaceful purposes" appears frequently in the Convention, it is nowhere defined. Aside from assuring warships their traditional immunities and permitting submarines to pass submerged through international straits, there is no reference to arms control or disarmament at sea, or the creation of "zones of peace" or "nuclear-free zones." Although it does not necessarily hinder moves toward disarmament, the Convention does not encourage it either.

Another serious omission, though an understandable one considering the dominance of nationalism in UNCLOS III, is the total absence of any reference to national obligations to preserve the integrity of the marine environment on the continental shelf, in territorial waters, in archipelagic waters, or in internal waters. These are precisely the marine areas most intensively utilized now, and they will be far into the future. In addition, the failure of a State to meet its obligations relating to the conservation and management of the living resources of the EEZ is excluded from the jurisdiction of the Law of the Sea Tribunal provided for in the 1982 Convention, so a State might consider itself free to ignore even the very limited obligations it undertakes in this area.

Two examples of environmental problems that have become prominent since 1982 serve to illustrate the penalties for failure to consider the marine environment holistically: driftnet fishing and "straddling stocks." Large-scale drift nets are rather small-mesh fishing nets made of very fine nylon filament set out by large ships (usually at night) over distances up to some 30 nautical miles and extending, typically, through the upper zone of the sea where most finfish and marine mammals are concentrated. They are not anchored but drift with currents until reeled in with their catch. Since the nets are quite strong and nearly invisible, this catch is gen-

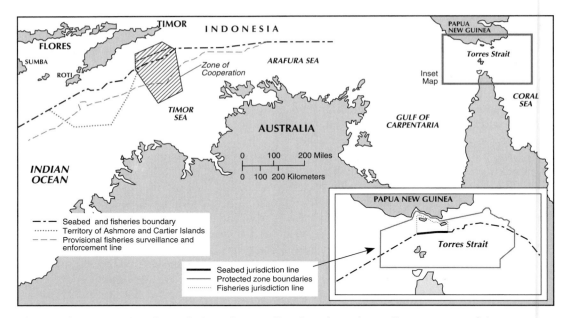

**The northern maritime boundaries of Australia.** These boundaries illustrate some of the numerous very difficult problems generated by the extension of national jurisdiction far out to sea, and how they can be resolved creatively and peacefully. Australia and Indonesia negotiated their maritime boundaries in the Arafura and Timor Seas in 1971 and 1972, respectively, but this left a gap south of the Portuguese colony on the eastern half of Timor Island. Indonesia invaded, occupied, and annexed the colony in late 1975 to mid-1976, but this action has not been accepted by the United Nations. The two countries finally agreed in late 1988 on a provisional settlement of the Timor Gap issue. A Zone of Cooperation was devised in which revenues from exploitation of petroleum and natural gas would be shared in different proportions in the three areas of the zone. The Provisional Fisheries Surveillance and Enforcement Line had been agreed upon in 1981. In the Torres Strait between Australia and Papua New Guinea, the major issue was not oil and gas, but the ancestral fishing rights of the Torres Strait Islanders, who are culturally and ethnically distinct Australian nationals. The Torres Strait Agreement, signed in 1978, provides for an exception to the general pattern of single maritime boundaries in order to protect these rights.

erally both large and indiscriminate, with considerable by-catch, even of birds, in addition to the target species.

The practice has been most common in the Pacific Ocean, in areas where stocks are widely dispersed and where the migratory patterns, distribution, and abundance of targeted species are poorly understood. In addition to an enormous wastage of marine life through the process of discarding less valuable species (on the order of 40 percent of total catch), driftnetting also causes inestimable damage from hundreds of miles of nets accidentally or deliberately cut and left to drift aimlessly for years, becoming "ghost nets." The United Nations has tried to deal with this problem in a series of resolutions, conferences, workshops, and the like, since

1989—so far with uncertain but probably limited results.

"Straddling stocks" are those stocks of fish extending across the outer limits of States' EEZs. Some coastal States, especially those abutting the major fisheries of the Bering Sea, Southeast Pacific, and northwest Atlantic, have been trying to extend their conservation jurisdiction beyond the EEZ in order to regulate or prohibit foreign vessels fishing in the high seas for these stocks. This is clearly an infringement on the right to fish in the high seas and could be a cover for the coastal States' desire to preserve these stocks for exploitation by their own fishing industries, but it may be necessary to preserve stocks that otherwise could be obliterated by harvesting just outside the EEZ.

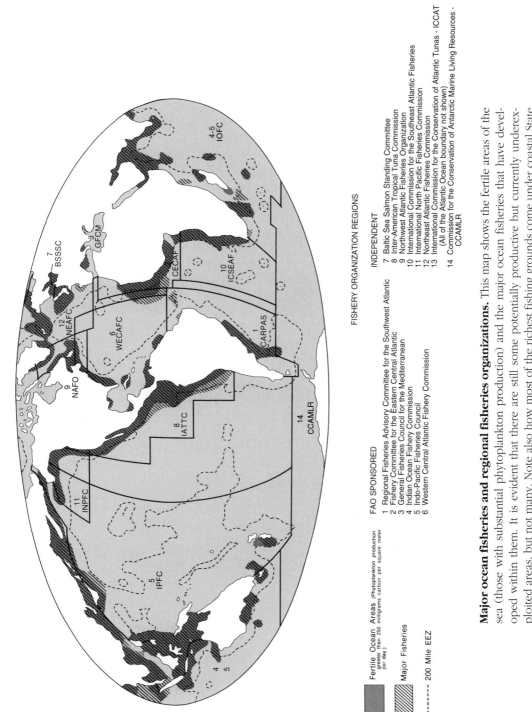

FISHERY ORGANIZATION REGIONS

**FAO SPONSORED**
1 Regional Fisheries Advisory Committee for the Southwest Atlantic
2 Fishery Committee for the Eastern Central Atlantic
3 General Fisheries Council for the Mediterranean
4 Indian Ocean Fishery Commission
5 Indo-Pacific Fisheries Council
6 Western Central Atlantic Fishery Commission

**INDEPENDENT**
7 Baltic Sea Salmon Standing Committee
8 Inter-American Tropical Tuna Commission
9 Northwest Atlantic Fisheries Organization
10 International Commission for the Southeast Atlantic Fisheries
11 International North Pacific Fisheries Commission
12 Northeast Atlantic Fisheries Commission
13 International Commission for the Conservation of Atlantic Tunas - ICCAT
   (All of the Atlantic Ocean boundary not shown)
14 Commission for the Conservation of Antarctic Marine Living Resources -
   CCAMLR

Fertile Ocean Areas (Phytoplankton production greater than 250 milligrams carbon per square meter per day.)

Major Fisheries

------- 200 Mile EEZ

**Major ocean fisheries and regional fisheries organizations.** This map shows the fertile areas of the sea (those with substantial phytoplankton production) and the major ocean fisheries that have developed within them. It is evident that there are still some potentially productive but currently underexploited areas, but not many. Note also how most of the richest fishing grounds come under coastal State jurisdiction with adoption of 200-mile EEZs. Finally, a comparison of this map with the one of marine minerals on page 536 will reinforce the point that nearly all the currently recoverable marine resources are to be found within 200 nm of the coast. This means intensive and often conflicting uses of the EEZs for a long time to come, and the virtual inaccessibility of these resources to land-locked countries.

Both of these problems are very complex, but they illustrate the folly of the negotiators at UNCLOS III who settled on the principle of "maximum sustainable yield" (MSY) to guide conservation of the "living resources" of the sea rather than the more conservative principle of "optimum sustainable yield" (OSY), especially since we still have insufficient information about fisheries biology to know what the actual MSY or OSY might be. The failure of existing conservation efforts to prevent a precipitous decline in some of the world's formerly most productive fisheries in the early 1990s is a direct result of the decisions made by individual governments to pursue at UNCLOS III treaty provisions that would assure themselves of maximum shares of oceanic resources with little regard for long-term community benefit. After three years of difficult negotiations, however, the United Nations Conference on Straddling Fish Stocks and Highly Migratory Fish Stocks produced in August 1995 a convention that balances the interests of all States and should bring some order into high seas fishing, thus filling a serious gap left by UNCLOS III.

Another potential problem might develop as more countries follow the lead of the United States in establishing marine sanctuaries or marine parks. If all vessels are banned from them, except for scientific and tourist vessels, some States might object that freedom of navigation is being unduly restricted. Other jurisdictional and functional problems could arise.

And what about sovereignty over newly formed islands, such as the one that is slowly building up from sediments being dumped into the Bay of Bengal by the Ganges and Brahmaputra rivers? Or others that emerge as a result of tectonic activity, as Surtsey did in 1964?

No mention is made either of the practice of several countries (and others may follow before long) of warning vessels out of enormous tracts of the sea from time to time when they have wished to test long-range missiles or other weapons, a blatant violation of the freedom of the high seas. Nor is

there reference to the establishment of air defense identification zones extending hundreds of kilometers out to sea, which could be construed as a similar violation of the freedom of the skies over the high seas.

The status of ice shelves, such as those adjacent to Antarctica, is not covered either, and the whole question is still unclear.

Another matter of special interest to political geographers is marine regions. In the LOSC there are seven references to regional arrangements or agreements, six to regional organizations, three to regional rules or standards, six to regional programs or centers, and 13 occurrences of "region" or "regional" standing alone. Nowhere, however, is regional activity of any kind required, nor is "region" anywhere defined. Nevertheless, the encouragement provided in the Convention to regional cooperation and the logical imperatives of the marine environment have impelled States to seek regional solutions to many problems. Marine regional organizations are being created similar to the terrestrial organizations described in Chapter 29.

Until recently, the largest classes of both agreements and organizations concerned with marine affairs have been those devoted to the conservation of the "living resources" of the sea, such as those sponsored by the UN Food and Agriculture Organization (FAO) in Rome. The second largest classes concern prevention and control of marine pollution, most of them sponsored either by the UN's International Maritime Organization (IMO) in London or by the UN Environment Programme (UNEP) in Nairobi. In the early 1970s the latter organized a Regional Seas Programme. Now known as the Oceans and Coastal Areas Programme, it has helped coastal States develop action programs for 13 areas of the global sea.

On their own initiative but with help from these and other intergovernmental organizations, many States have undertaken regional programs, most recently for the management of marine areas on a more or less comprehensive basis. One very important example is IOMAC (Indian Ocean Marine Affairs Cooperation), which we discuss in Chapter 23 in connection with the emergence of the Indian Ocean as a zone of peace.

Another, also in a region discussed under the same heading, is located in the eastern Caribbean. Here the smaller Commonwealth islands of the subregion are moving in the direction of economic and political unity under the aegis of the Organization of Eastern Caribbean States (OECS), which functions as a subunit of, and in close cooperation with, CARICOM (the Caribbean Community). With the assistance of UNEP, the OECS is gradually creating a subregional program for the comprehensive management of marine resources. This objective, while perfectly logical to an outside observer, has proven elusive. All of the factors that led to the failure of the Federation of the West Indies (covered in Chapters 9 and 10) are still operative here. There is no assurance of success now either, but conditions are more favorable, and it is possible that the sheer necessity of managing the marine environment intelligently and cooperatively will prove to be the *raison d'être* for unity so lacking in the past.

## *The 1994 Agreement on Seabed Mining*

The Reagan administration in Washington and the Thatcher administration in London refused to sign the LOSC largely on ideological grounds, most of which had to do with some provisions of Part XI and certain annexes related to seabed mining. Although the Federal Republic of Germany wanted to sign the Convention, since Hamburg had been designated as the site of the new International Tribunal for the Law of the Sea (which could hardly be located in a country that was not a party to the Convention), it was pressured by its close allies not to do so. Other major industrial countries also expressed displeasure with Part XI but signed the Convention notwithstanding. By the time the Convention entered into force on 16 November 1994, not one major industrial country had ratified it. Clearly, a convention without the active support of such major States

would have much less practical effect than one with their support. Therefore, soon after the LOSC was opened for signature on 10 December 1982, efforts were initiated to find some formula that would bridge the evident gap between the positions of the rich and poor countries on Part XI and some lesser issues.

By mid-1990 these various quiet, informal efforts had not been successful, while slowly the Convention requirement of 60 ratifications for entry into force was being approached. In July 1990 the Secretary-General of the United Nations, Javier Pérez de Cuellar, took the initiative to convene informal discussions aimed at achieving universal participation in the Law of the Sea Convention. Negotiations took place with some frequency and increasing urgency for the next four years. They produced a truly innovative package of devices that effectively amended the Convention without actually doing so and produced a significant gain for the industrial countries. The key to the package is the annex to UN General Assembly resolution 48/263, adopted without opposition on 28 July 1994, titled "Agreement Relating to the Implementation of Part XI of the United Nations Convention on the Law of the Sea of 10 December 1982."

Most of the provisions of this agreement are technical or administrative and of little interest to political geographers. Their net effect, however, is to produce a seabed mining regime based more on "free enterprise" than the original Part XI, one that also discards the erroneous assumptions on which Part XI was based and takes into account all of the political and economic developments in the world between 1982 and 1994. Regardless of how one may feel about individual provisions of the agreement, there is no doubt that it is a splendid example of the value of United Nations good offices, removes nearly all of the rich countries' objections to the 1982 Convention without really damaging the long-term interests of the poor countries significantly, and makes possible the widespread (if not quite universal) acceptance and enforcement of the Convention.

## *Political Geography and the Law of the Sea*

No responsible international lawyer would argue that the 1982 Convention *is* the Law of the Sea, and no responsible political geographer would argue that the Law of the Sea *is* the political geography of the sea, but an understanding of the Convention is essential to the development of both. It deserves much more attention than we have been able to devote to it here.

The Law of the Sea, the political geography of the sea, and marine affairs generally offer a huge and rich lode of fascinating and useful work not only for lawyers, political scientists, economists, and ecologists, but also for geographers—and for a long time to come. Already some geographers are beginning to follow the lead of Lewis Alexander, now retired from the University of Rhode Island, who for more than 30 years has been doing very fine work in the field and has a worldwide reputation for his significant contributions to the development to the new Law of the Sea.

The IGU Commission on Marine Geography has been stimulating research in this area since its organization in 1988. Now, for its 1992–96 program, it is concentrating on global ocean change and on integrated and regional ocean change, mostly the political aspects thereof. Here is an opportunity for young people to get into a field as it is just beginning to develop, and to help guide it into constructive paths leading to wiser and more peaceful and beneficial uses of the sea.*

*Besides the United Nations, the best source of current information on the political geography of the sea is an NGO, the Council on Ocean Law, 2222 King Place, N.W., Washington, D.C. 20007.

# 33

# *Antarctica and the Southern Ocean*

Antarctica is the coldest, windiest, driest, highest, quietest, most remote, and least understood continent on earth. Attached to the continent is a fluctuating apron of shelf and pack ice, a formidable buffer zone making approach to the land far more difficult than to any other continent. Because of its remoteness and inhospitable environment, it was the last continent to be discovered and, except for a few sealers, whalers, and explorers, it was virtually ignored for more than a century after its discovery. Now its importance in the physical geography of the earth—climate, submarine flows of water through the oceans, biomass—has been recognized.

As with any other region, we have a problem defining the limits of Antarctica. More than with any other continent, the surrounding sea is an integral part of the region, and at least a portion of it must be included in our definition. The Antarctic Treaty covers only the region south of latitude 60°S; some observers have suggested latitude 40°S as a boundary. Both of these suggestions have all the advantages and disadvantages of geometric boundaries we have already discussed. The most frequently suggested boundary is the Antarctic Convergence (also called the Polar Frontal Zone), a zone about 50 km (30 miles) wide circling the continent between 55° and 62° South latitude where the cold Antarctic water sinks beneath the warmer waters of the north. Another and perhaps more useful regional delineator is the Subtropical Convergence (also called Sub-Antarctic Convergence), the southernmost reach of the warm, southward moving currents of the Atlantic, Pacific, and Indian oceans. It lies at approximately 40°S, and displays more latitudinal stability than the Antarctic Convergence. Defining the northern boundary of the Southern Ocean is not just an academic exercise. It will be vitally important in establishing a regime (or regimes) for the management of the Antarctic ecosystem.

Why should we now be concerned with a legal regime for an empty, frozen wilderness so far away from the core area of any country? Because at this stage in their political evolution, human beings simply cannot bear to leave any portion of the earth's surface alone. Every square kilometer must be accounted for—claimed, owned, governed, exploited, and perhaps destroyed. The pristine beauty, the inspiring vastness, the tranquilizing solitude of this magnificent region may not last much longer. Already thousands of tourists have intruded into what had been a huge and immensely valuable scientific preserve, and with few legal restrictions on tourism, we can visualize it building to much larger proportions in the near future, bringing to the continent much degradation and little benefit. Even more tempting for commercial exploitation are the mineral deposits of the continent and the super-abundant marine life of the Southern Ocean.

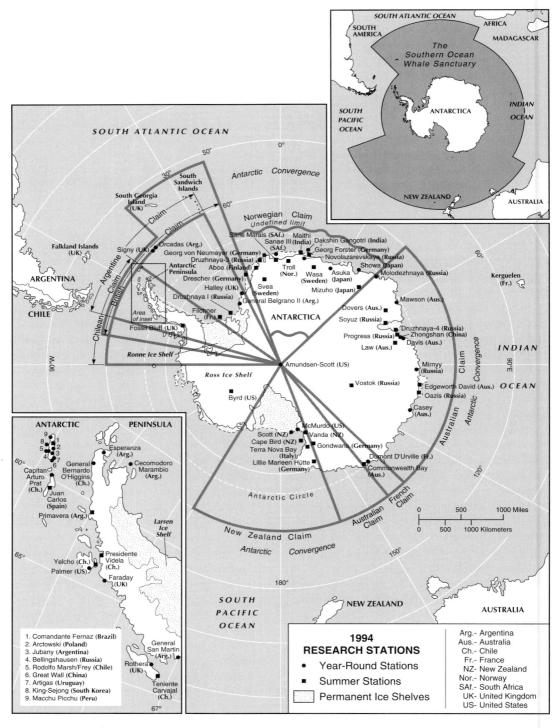

**Antarctica, 1994**

## Territorial Claims and Governance

Although the continent was postulated by the ancient Greeks and probably seen by Polynesian seafarers centuries ago, specific knowledge of Antarctica dates only from the late eighteenth century, when Captain James Cook circumnavigated it (but never saw it) and British and American sealers ventured into Antarctic waters to hunt seals on islands and ice. The continent itself was not spotted until the early nineteenth century, when sealers, whalers, and explorers from several countries discovered the Antarctic Peninsula (called Graham Land by the British, the Palmer Peninsula by the Americans, Tierra O'Higgins by the Chileans, and Península San Martín by the Argentines). The first formal claim to Antarctic territory was made by Britain in 1908. Since then six other countries have made formal claims: New Zealand, 1923; France, 1924; Australia, 1933; Norway, 1939; Chile, 1940; and Argentina, 1943.

Of all the customary bases for claims to territory under international law, which we discuss in Chapter 7, the only one with any validity in the Antarctic is discovery and effective occupation, the basis of most existing claims. Since discovery is difficult to prove and occupation difficult to make "effective," other principles have been created and applied here. Best known is the sector principle. This originated in 1907 when a Canadian senator recommended that Canada declare that it had taken possession of all lands, including islands, lying between its northern coast and the North Pole. Other countries adopted this new concept both in the Arctic (Russia, 1916) and the Antarctic (United Kingdom, first in 1917), but it has not been generally accepted as a valid principle of international law. Other theories are the contiguity doctrine, continuity and hinterland, patrimony and *uti possidetis*, and the region of attraction. All of these have been applied at one time or another, often in combination with the sector principle or discovery and occupation, but none has achieved any significant measure of acceptance by the international community. The matter of sovereignty in Antarctica was so muddled by the 1950s and was becoming so acutely sensitive that there was a need to suspend political competition there.

The opportunity was provided by the International Geophysical Year (IGY), conceived by the International Council of Scientific Unions as a massive, varied, and coordinated study of the earth during a period of great sunspot activity (1957–58). Neither the Cold War nor competing Antarctic claims prevented collaboration among over 10,000 scientists and technicians from 67 countries working at 2500 IGY stations around the world. Included were 50 stations operated by 12 countries in Antarctica.* They did a great deal of scientific exploration, and this international cooperation produced considerable new knowledge about our planet, especially about Antarctica. The international aspect of the operation was epitomized by the exchange of Soviet and American scientists and by the Commonwealth Transantarctic Expedition, which in three months made the first surface crossing of the continent, from Shackleton Base on the Weddell Sea to Scott Base on the Ross Sea via the South Pole. The agreement to allow States working in Antarctica to place scientific stations anywhere on the continent, regardless of prior claims to sovereignty, and the unrestricted exchange of IGY data set precedents for later explorations of the sea and outer space and led directly to the negotiation and success of the Antarctic Treaty.

The Antarctic Treaty was negotiated and signed in Washington in 1959 by the 12 States that participated in IGY activities in Antarctica. Since then 30 States have acceded to the treaty, and 14 have been accepted as Consultative Parties (CPs), equal to the original signatories (see box). The treaty went into effect in 1961 and was reviewable after 30 years, but to date no State has requested a review. It is truly a landmark in international cooperation and in some re-

---

*Argentina, Australia, Belgium, Chile, France, Japan, New Zealand, Norway, South Africa, the United Kingdom, the United States, and the USSR.

## Members of the Antarctic Treaty System (1994)

| Original Consultative Parties (12) | Date of Ratification or Accession | Nonconsultative (16) | Date of Accession |
|---|---|---|---|
| **a. Claimant States (7)** (The date in parentheses is the date of the formal claim.) | | Denmark | 20 May 1965 |
| | | Romania | 15 Sep. 1971 |
| | | Bulgaria | 11 Sep. 1978 |
| United Kingdom (1908) | 31 May 1960 | Papua New Guinea | 16 Mar. 1981 |
| Norway (1939) | 24 Aug. 1960 | Hungary | 27 Jan. 1984 |
| France (1924) | 16 Sep. 1960 | Cuba | 16 Aug. 1984 |
| New Zealand (1923) | 1 Nov. 1960 | Greece | 8 Jan. 1987 |
| Argentina (1943) | 23 June 1961 | Korean Dem. Rep. | 21 Jan. 1987 |
| Australia (1933) | 23 June 1961 | Austria | 25 Aug. 1987 |
| Chile (1940) | 23 June 1961 | Canada | 4 May 1988 |
| **b. Non-claimant States (5)** | | Colombia | 31 Jan. 1989 |
| South Africa | 21 June 1960 | Guatemala | 31 July 1990 |
| Belgium | 26 July 1960 | Switzerland | 15 Nov. 1990 |
| Japan | 4 Aug. 1960 | Ukraine | 28 Oct. 1992 |
| United States | 18 Aug. 1960 | Czech Republic* | 1 Jan. 1993 |
| Russia | 2 Nov. 1960 | Slovak Republic* | 1 Jan. 1993 |
| **Later Consultative Parties (14)** | | | |
| (The date in parentheses is the date of becoming a Consultative Party.) | | * Succeeded Czechoslovakia, which acceded on 14 June 1962. | |
| Poland (29 July 1977) | 8 June 1961 | | |
| Netherlands (19 Nov. 1990) | 30 Nov. 1967 | | |
| Brazil (12 Sep. 1983) | 16 May 1975 | | |
| Germany (3 Mar. 1981) | 5 Feb. 1979 | | |
| Uruguay (7 Oct. 1985) | 11 Jan. 1980 | | |
| Peru (9 Oct. 1989) | 10 Apr. 1981 | | |
| Italy (5 Oct. 1987) | 18 Aug. 1981 | | |
| Spain (21 Sep. 1988) | 31 May 1982 | | |
| PR China (7 Oct. 1985) | 8 June 1983 | | |
| India (12 Sep. 1983) | 19 Aug. 1983 | | |
| Sweden (21 Sep. 1988) | 24 Apr. 1984 | | |
| Finland (9 Oct. 1989) | 15 May 1984 | | |
| Rep. of Korea (9 Oct. 1989) | 28 Nov. 1986 | | |
| Ecuador (19 Nov. 1990) | 15 Sep. 1987 | | |

spects a model for future ventures of this kind in other arenas. It governs the entire continent and has been scrupulously obeyed by all signatories. The spirit of international camaraderie engendered by the IGY has continued to prevail in Antarctica under the treaty. We can hope it will last, but already there are disquieting signs that it may not.

The treaty is short, only 14 articles. To summarize it briefly: Antarctica is to be used for peaceful purposes only; no military ac-tivities of any kind are permitted, though military personnel and equipment may be (and are) used for scientific purposes. Freedom of scientific investigation and cooperation shall continue. Scientific program plans, personnel, observations, and results shall be freely exchanged. No prior territorial claim is recognized, disputed, or established, and no new claims may be made while the treaty is in force. Nuclear explosions and disposal of radioactive waste are prohibited. All land

and ice shelves south of latitude 60°S are covered, but not the high seas of the area. Observers from treaty States have free access to any area and may inspect all stations, installations, and equipment. Treaty States shall meet periodically to exchange information and take measures to further treaty objectives, including the preservation and conservation of "living resources." These consultative meetings shall be open to contracting parties that conduct substantial scientific research in the area. Disputes are to be settled peacefully, ultimately, if necessary, by the International Court of Justice.

Although all the terms of the treaty are being faithfully carried out and we have all benefited from it, some matters the treaty omits have emerged as problems. For one thing, while the treaty is open for accession by nearly all States, only those doing actual and "substantial" scientific work in the area may participate in the consultative meetings that spell out the rules for operations in the Antarctic. This effectively excludes nearly all the developing countries, since they are too poor to bear the heavy costs of Antarctic operations. Some, however, such as Uruguay and Ecuador, already conduct expeditions jointly with richer countries and more are likely to do so in the future.

Although the treaty encourages scientific work and bans military activity in the region, it is silent about commercial activities there. As demand grows for Antarctic resources, especially oil and tourist "destinations," and as technology becomes increasingly able to meet those demands, some regulation (if not outright prohibition) has become essential. The treaty does not come to grips with the question of sovereignty over Antarctic waters: Does ice count as land or water? Are shelf and pack ice to be treated differently? May a State claim territorial waters or other maritime zones around the continent? Even the claims to land territory are only suspended, not resolved.

The sovereignty question may be the most vexing of all. A number of the claimants have organized their claims as colonial territories (Australia, France, UK), complete with their own postage stamps, while others have

incorporated their claims into their national territory (Chile, Argentina), complete with administrative structures, and Norway so adamantly rejects the sector theory that it has refused to set northern or southern limits to its claim. Some countries active in Antarctic exploration and research (South Africa, Japan, Belgium, and, most active of all, the United States and Russia) have made no claims at all and refuse to recognize any existing claims. Japan, in its 1952 peace treaty, renounced its right to make a claim, but the United States and Russia have reserved their right to do so.

It is going to be difficult to get any claimant State to retreat from its position since it has become a matter of prestige for some. New Zealand seems least nationalistic in the matter and may well be willing to transfer its claim to an international body, but Argentina and Chile are different. They have gone to very great lengths indeed to secure their claims, erecting various buildings in unused areas, printing maps showing their claims as part of their territory, and more. President Pinochet of Chile spent a week touring the Chilean-claimed sector in January 1977. That topped the Argentine ploy of having a national cabinet meeting on an Antarctic island in August 1973. But then in the summer of 1978 Argentina sent a pregnant woman to its Esperanza Base, and she gave birth in January to Emilio Marcos Palma, the first child born in Antarctica. In February Argentina staged the first wedding in Antarctica (Sergeant Carlos Alberto Sugliano and Beatriz Buonamio), also at Esperanza.

Both countries have established permanent civilian colonies there, small groups of dependents of military personnel and civil servants serving two-year tours of duty (Argentina on the peninsula in 1977 and Chile on King George Island nearby in 1984). And in October 1984 the Chileans at the latter base (Teniente Marsh) welcomed the first party of tourists to spend the night on shore in Antarctica. Since then, Chile has been accommodating tourists regularly at the base. Continuing this silly and dangerous game of "one-upmanship," Argentina was seriously considering erecting a Holiday Inn at Esperanza! These are only samples of the intensive activities engaged in by these two countries for a purely nationalistic reason: to strengthen their territorial claims in the event that it becomes possible to reassert them, despite the fact that legally, such activities, according to Article IV of the Antarctic Treaty, "cannot constitute a basis for asserting, supporting or denying a claim to territorial sovereignty in Antarctica or create any rights of sovereignty in Antarctica."

## Commercial Activities

We have already referred several times to the likelihood of commercial, as distinguished from scientific, activity in the Antarctic region. The most important has long been, and likely will continue to be, exploitation of the "living resources" of the Southern Ocean.

The Southern Ocean is home to the world's densest population of marine fauna:

**Potential Ecuadorian Antarctica claim.** Although Ecuador has not yet made an Antarctic claim and may not do so, the possibility cannot be ruled out, considering her vigorous pursuit of territorial claims in the Pacific Ocean, the Amazon Basin, and outer space. Such a claim would probably look much like this. This original map appears in an official Ecuadorian government publication. Within the government of Ecuador there is an Antarctica Secretariat, an indication of more than passing interest in the region.

8 species of seal and 12 of whale, some of which have been overharvested almost to extinction; 90 to 100 species of fish, few found elsewhere, but most in small concentrations; large quantities of squid; numerous birds, many found nowhere else; and vast quantities of krill.* Even smaller are the innumerable species of phyto- and zooplankton, produced massively in the mineral-rich Antarctic waters. Of all these animals, the one with the greatest apparent commercial potential is krill, a high-quality food, rich in protein, oil, vitamins, and minerals.

The Soviet Union was harvesting krill at least since 1971, and Japan since the 1972–73 season. Chile, Taiwan, Korea, Spain, Poland, Ukraine and Germany have also engaged in the krill industry, but on a smaller scale. Both the Russians and the Japanese market the krill in various forms, most commonly as a paste used as a food additive and a sandwich spread. Although there are still many problems with the harvesting, processing, and marketing of krill, it would seem that the rapidly growing population of the world, increasing demand for high-quality protein, overfishing of many finfish, exclusion of the major fishing countries from many fisheries by new 200-mile EEZs of coastal States, and other factors will bring more and more krill trawlers to Antarctic waters in the near future. It has been estimated that 50 to 70 million tons of krill could be harvested each year in Antarctic waters without threatening existing stocks. This is about equal to the total catch of all species of fish throughout the world at present.

But this optimistic estimate has already proven false, and most have been far lower. Not only has the harvest never even approached 2 million tons, but it even fell sharply in 1983–85 and has not since returned to previous levels. This is possibly due in part to increased feeding by whales saved from slaughter by tightened rules of

the International Whaling Commission (IWC), but there is new evidence that there simply is not as much krill in the Southern Ocean as had been thought. Not only fish and marine mammals, but even krill seems to have been overfished. The Commission for the Conservation of Antarctic Marine Living Resources, in fact, felt it necessary in 1991 to establish catch limits totaling 1.5 million tons. Despite this restriction, krill catches continue to fall, and the catch from the 1993–94 season may have been as low as one-fifth of the "normal" level.

This problem of overexploitation of the biota of the Southern Ocean led the CPs of the Antarctic Treaty System (ATS) to convene a conference in Canberra in 1980. There they concluded the Convention on the Conservation of Antarctic Marine Living Resources, which entered into force in 1982. Among other things, it provides that:

1. The convention applies for protection of all living organisms in the sea south of the Antarctic Convergence.

2. Harvesting and associated activities are to be carried out in such a manner as to prevent or minimize long-term risk to the marine ecosystem.

3. A Commission for the Conservation of Antarctic Marine Living Resources (CCAMLR) is established to, among other things, facilitate study of the marine ecosystem, publish relevant data, and formulate conservation measures.

4. Observance of provisions of the convention shall be ensured by an observation and inspection system.

The CCAMLR Executive Secretariat is based in Hobart, Tasmania, Australia. It holds regular meetings and has begun publishing some useful materials, but to date its effectiveness has been limited. The same nationalistic and mercenary impulses that have impeded the IWC and other intergovernmental management organizations have been operating here as well. Considering that there have already been violations of the Agreed Measures for the Conservation of Antarctic Fauna and Flora (1964), the first major agreement under the ATS, we may

*Krill is a collective name for many species (11 in Antarctic waters) of large zooplankton—shrimplike crustacea of which *Euphasia superba* is the largest (averaging about 5 cm in length when mature) and most important of the Antarctic species.

well be skeptical about the willingness and/or ability of exploiters of Antarctic resources, whether governments or private parties, to preserve the fragile and valuable ecosystem of the region.

The threat to the Antarctic environment is at least as great from potential exploitation of hydrocarbons and minerals, both on- and offshore. Minerals have been found in great variety in Antarctica, but, so far at least, not in concentrations that can be considered commercially exploitable. Fairly sizable deposits of coal and iron ore exist in the Prince Charles Mountains and the Transantarctic Mountains respectively, but they are inaccessible to commercial exploitation. Sand and gravel are abundant on many raised beaches around the coastline; they are unlikely to be exported, though they can be used for local construction. More than a dozen other minerals occur on the continent, but because of the enormous costs involved, they are likely to remain unattractive for commercial exploitation for some time to come. The narrow and deep continental shelf may contain commercial quantities of petroleum and natural gas, but the difficulties of extracting and delivering them to market will far exceed those of any other offshore deposits in the world or even those of Alaska's Arctic Slope. Extensive areas of the sea floor around the continent have a covering of scattered manganese nodules and related formations, but they tend to be rather lean in the most valuable minerals. There is some but probably not much potential for geothermal energy.

In short, it seems that Antarctica, though rich in varieties of minerals, will probably not be an important commercial producer because of the small concentrations and high costs of exploitation. These factors, however, would not necessarily deter some States from permitting, encouraging, or undertaking mining for what they consider to be security or prestige reasons.

In 1974 the CPs began considering the question of minerals and began in 1979 to try to develop a minerals regime for the region—in great secrecy. Gradually, however, and only under great pressure from non-

governmental environmental groups (NGOs) around the world and from nonparties to the Antarctic Treaty (especially many of the poorer African and Asian countries), the outlines of successive drafts of a minerals treaty became known. The same pressures seem to have forced a gradual shift from an emphasis on exploitation of minerals toward somewhat greater regard for environmental considerations, and from exclusive control by ATS members toward greater participation by nonmember States and by nongovernmental organizations.

In 1988 a Minerals Convention was finally concluded in Wellington, New Zealand. Immediately, opposition to it increased, and in 1990, before it was ratified, Australia and France, then Belgium and Italy, proposed a permanent ban on mining and a new comprehensive environmental protection convention. During the following year, NGOs intensified their opposition to the Minerals Convention and their support for a world park to include all of Antarctica and the Southern Ocean.

This effort bore fruit, as the Minerals Convention was abandoned. In Madrid on 4 October 1991 the ATCPs adopted a Protocol on Environmental Protection to the Antarctic Treaty, with five annexes covering environmental impact assessment, conservation of fauna and flora, waste disposal and waste management, prevention of marine pollution, and protected areas. It includes a qualified 50-year moratorium on minerals mining in the region.

This truly historic event may represent a real reversal of the trend toward erosion of the original commitment to protect Antarctica from the excesses of human greed further north. There is much yet to be done, however, for the protocol provisions are still not adequate, even if they are faithfully executed. Furthermore, to become legally binding, the protocol must still be ratified and implemented by all 26 ATCPs. To date, only 9 ATCPs have ratified the protocol, and only 2 have incorporated the protocol's provisions into domestic law. Other issues critical to environmental protection, such as the completion of a liability annex and the cre-

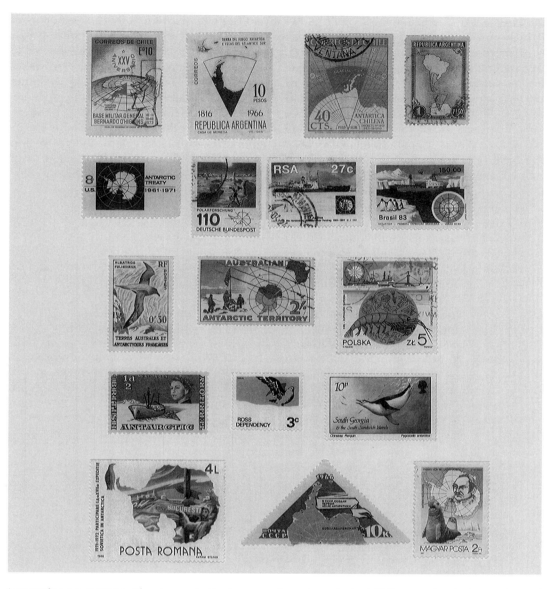

**Antarctica on stamps.** The stamps in the top row show the principal Chilean station in Antarctica and the overlapping Chilean and Argentine claims in the region. The American stamp commemorates the tenth anniversary of the Antarctic Treaty, while the German, South African, and Brazilian stamps depict their countries' research in the region. In the third row are stamps issued for the French and Australian Antarctic claims and one depicting Polish krill fishing in the Southern Ocean. In the fourth are stamps issued for the British and New Zealand Antarctic claims and for South Georgia and the South Sandwich Islands, formerly dependencies of the Falkland Islands. In the bottom row, the first stamp symbolizes Romanian participation in a Soviet expedition, the Soviet stamp highlights the Novolazarevskaya station and shows other research stations and the Hungarian stamp honors Fabian von Bellingshausen, a Russian who explored Antarctic waters in the early nineteenth century. (Martin Glassner)

ation of a secretariat for the ATS, have yet to be resolved.

A more immediate threat to the Antarctic environment is tourism. At present the continent has an estimated population of about 3000 in summer and 1000 in winter; altogether since the beginning of the IGY in 1957 perhaps 28,000 people have spent some time there. Only about half of them so far have been tourists, with a few thousand more overflying or sailing near the continent and its islands without actually landing. But already they have disturbed bird nesting areas, left litter behind, and otherwise sullied the environment. More serious, they are getting themselves trapped in the ice and killed in plane crashes. In 1979 an Air New Zealand plane crashed into the continent's highest peak, Mount Erebus, and all 257 tourists aboard, as well as the crew, were killed. In 1984, on New Year's Eve, eight wealthy American tourists and their two-man

| 1993-94 ANTARCTIC TOURIST ESTIMATES | | |
|---|---|---|
| Vessels | Number of Voyages | Estimated Total Passengers |
| COASTAL—BY SHIP | | |
| World Discoverer | 8 | 919 |
| Explorer | 9 | 649 |
| Hanseatic | 5 | 754 |
| Marco Polo | 4 | 1,823 |
| Columbus Caravelle | 7 | 1,047 |
| Professor Molchanov | 7 | 174 |
| Akademik Vavilov | 4 | 276 |
| Bremen | 4 | 517 |
| Kapitan Khlebnikov | 4 | 373 |
| Ioffe | 12 | 925 |
| Sagafjord | 1 | 500 |
| SUB–TOTALS | 65 | 7,957 |
| INLAND—BY AIRCRAFT | | |
| Company | No. Flights | Estimated Total Passengers |
| Adventure Network | N/A | 59 |
| SUB–TOTALS | N/A | 59 |
| TOTALS | 65 | 8,016 |

*Information supplied by Antarctic Tour Operators to the American Polar Society*

crew were killed when their chartered Chilean plane crashed near Teniente Marsh, to which they had been invited for a party.

Such idiocy is likely to continue since there is as yet no formal mechanism for regulating tourism beyond each country's domestic laws and law enforcement systems. However, at the 1994 Antarctic Treaty Consultative Meeting, the ATCPs adopted recommendations aimed at providing guidance to operators and visitors to Antarctica. These guidelines set minimal standards to protect the Antarctic environment. Even though they are voluntary and have no means for enforcement, the guidelines are a good first step in regulating tourism in the Antarctic. But the cost to science of the diversion of scientists, support personnel, and equipment from their vital work, already tightly constrained by time and money, for search-and-rescue operations is very great.

Tourists and those in the tourist industry have yet to learn that Antarctica is not Honolulu or Cannes or the Seychelles—an exotic, faraway place that simply *must* be visited. Already more than 50 American aircraft have crashed in Antarctica while on legitimate scientific missions. Supply ships have been trapped in ice; scientists and others have been stranded for weeks; fires have seriously damaged and even destroyed bases and field camps; and in 1986 the unoccupied Soviet geophysical base Druzhnaya I on the coast of the Weddell Sea vanished—totally, without a trace. Antarctica, though beautiful, is still a dangerous place. Clearly, the ATS must soon develop and enforce a rational program for regulating Antarctic tourism.

It is possible that other commercial activities will be attempted in the intermediate future, such as towing icebergs northward to provide pure water for arid areas (already done experimentally), operating scheduled airline flights between South America and Australia across the continent, providing supplies and services to people working there, and so on. Each of them poses a host of logistical, environmental, and legal problems—and, of course, political ones. Already (in November 1981) New Zealand customs officials in Christchurch have confiscated

marijuana and other illegal drugs (in 26 parcels) bound for Americans in Antarctica. This is likely to be—to coin a phrase—only the tip of the iceberg. How can commercial or criminal activities be controlled under the current ATS? There is yet no answer.

## Environmental Concerns

We have already mentioned some environmental problems, both actual and potential. But there are more. Under the Antarctic Treaty System, 20 Special Protected Areas have been designated in the region to date, as well as 39 Sites of Special Scientific Interest. But there is no way of legally protecting these areas or of prosecuting violators. The system relies on broad interpretation of the intent of the rules and good-faith compliance, neither of which is likely to survive the growing commercial and strategic interest in the region. The system for protected areas will be improved somewhat with the implementation of the Madrid Protocol, which includes an annex establishing a new and simpler protected areas regime. However, many of the same problems concerning oversight and enforcement will remain.

To date the most blatant violation of all the norms, rules, and promises of environmental protection has been the French airstrip in the Pointe Géologie archipelago near the principal French base, Dumont d'Urville, in Terre Adélie. The site is very close to one of the few—and the most accessible—colonies of Emperor penguins, the least common of all species. From an original number of about 7000 couples before the establishment of the base in the 1950s, the number of couples had declined to an estimated 3000 in 1985. The area is also an important breeding ground for many other species of birds.

Yet in January 1983, the French began blasting, bulldozing, and helicoptering to construct an airfield for the use of French Transall C160 aircraft, even though they could cooperate with other countries operating in Antarctica to supply the base without using an airfield at all. After a brief suspen-

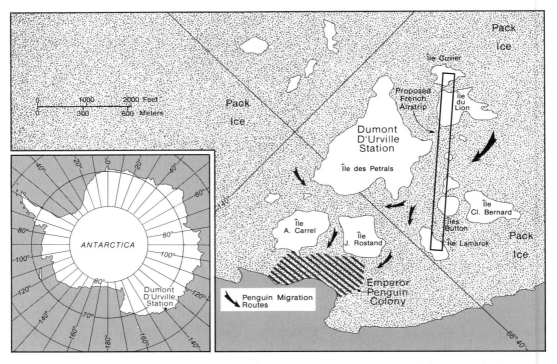

**The controversial French airstrip in Antarctica.** Like the Argentines, Chileans, and others, the French have not been exactly scrupulous in preserving this fragile and irreplaceable ecosystem.

sion because of NGO pressures, construction resumed and the French seemed determined to complete and use this airstrip. Only after it was severely damaged by a storm in early 1994 did the French finally agree to abandon the project. The French have yet to announce plans to clean up the site and are still considering using the remaining portion of the airstrip for small aircraft.

Contrast this with an Australian five-year plan (1985–90) to clean up all Australian stations. Under this plan over 36 tons of discarded machinery and other debris was returned from the Casey Station to Australia in early 1986. Of course, one may wonder why so much junk was allowed to accumulate in the first place.

Another threat to the fragile Antarctic environment is posed by the growing number of vessels of all kinds plying the waters of the Southern Ocean. Even if their passengers and crew scrupulously observe environmental protection rules, accidents can and do

happen. In January 1989, for example, as mentioned in chapter 33, a small Argentine tanker, the *Bahía Paraíso*, ran aground and sank off the west coast of the Antarctic Peninsula. Damage to fish and bird life was considerable, and the last of the ship's remaining cargo of oil was not removed until January 1993.

Despite the best intentions and efforts of most scientists and support personnel, the land and water of Antarctica are threatened. Even the air! Early in the 1980s scientists discovered over Antarctica a hole in the ozone layer, which protects the earth from the ultraviolet rays of the sun. The cause of the hole is known; it results from the release of man-made chemicals, especially chlorofluorocarbons, into the atmosphere. There is now an international agreement to phase out CFCs by early in the twenty-first century, but much damage may already have been done.

If Antarctica is to be spared the less desirable features of modern "civilization," perhaps the best regime for it is some version of a world park. Over the past decade a number of proposals to turn the entire Antarctic and Southern Ocean, or major portions of the region, into a world park of some kind have been offered by various environmental groups, scholars, and others. Some suggest that it could be operated in some way by the United Nations, others that it could be accommodated within the present ATS, others that a wholly new organization needs to be developed for it. Although, as we said earlier, the Antarctic Treaty has been scrupulously observed, its environmental provisions have been far too weak to be meaningful. The Environmental Protocol adopted in Madrid in 1991 may indicate that at last governments are conceding that some kind of world park may eventually be established in the south polar region. Australia and France, in fact, are now vigorously supporting this idea. However, the slow pace of ratification of the protocol has somewhat dampened early hopes for real protection of the Antarctic environment.

## Geostrategy

Article 1 of the Antarctic Treaty states forcefully that "Antarctica shall be used for peaceful purposes only," and apparently it has been. Already, however, there have been disturbing reports of remote sensing devices and satellite tracking facilities functioning within the treaty area. It seems unlikely that nuclear submarines on patrol or engaged in maneuvers have invariably remained north of 60°S latitude, and the Anglo-Argentine war over the Falkland Islands in 1982 generated other questions about military activities in the treaty area. Yet, generally speaking, Antarctica and the Southern Ocean have remained peaceful, probably because the region is so remote from the world's centers of population and of political, economic, and military activity. Not everyone, however, looks at the region quite so benignly, and many South Americans look at it geopolitically.

In Chapter 22 we point out that geopoli-

tics is a lively, respected, and even popular field of study in Latin America, particularly in Brazil and the Southern Cone. It is probably fair to say, in fact, that influential members of the elites, both civilian and military, in this region consider Antarctica and the Southern Ocean to have greater potential geopolitical value than scientific or commercial value. The professional literature in Argentina, Brazil, Chile, and even Peru and Uruguay is laced with discussions of the potential strategic value of the south polar region, especially that portion of it directly south of the South American continent.

The British islands in the South Atlantic—the Falklands and their former dependencies—are seen differently from Buenos Aires than from London: not as remote outposts of empire, useful only for weather stations, some fishing, and a bit of science, but as a screen across the South Atlantic protecting (or controlling) the major approaches to Antarctica from the richest, most industrialized, and most powerful countries in the world. They are also seen differently from Santiago, as the Arc of the Southern Antilles, marking a potential eastern limit of a Chilean claim to Pacific Ocean waters. From Brasilia, they are virtually invisible, having no effect whatever on a potential Brazilian claim to a wedge of Antarctic territory based on *defrontação*, the Brazilian version of the sector theory.

The Brazilian geographer Therezinha de Castro was among the first—in 1958—to urge a Brazilian geopolitical role in the Antarctic. She developed her thesis further in a 1976 book, in which she referred to Antarctica as a future "cornerstone of our destiny . . . a base of warning, interception and departure in . . . the defense of the South Atlantic." Chilean geopoliticians have long argued that Chile should control not only the Straits of Magellan and the Beagle Channel, but also the Drake Passage, which could again become a major world highway if the Panama Canal were interdicted or closed. Argentine geopoliticians argue that Argentine Antarctica, the British-held islands, and all the waters between them ("The Argentine Sea") should be completely incorporated into the country, (the "Tricontinental Argentina concept"), both to protect shipping lanes and to forestall any Brazilian moves southward.

This is only a fraction of Southern Cone geopolitical thinking about the Antarctic. Much more may be found (in English) in the papers and publications of Professor Jack Child of The American University in Washington, and in an article by the author, cited in the bibliography for Part Seven. The important point to be understood, however, is that in the Cono Sur geopolitics is not just abstract, armchair theorizing, but it heavily influences national policymaking. The Argentine invasion of the Falkland Islands and the Argentine-Chilean dispute over three small islands in the Beagle Channel are only two recent examples of this influence. No intelligent discussion of the future of Antarctica can sensibly omit consideration of this viewpoint, regardless of what kind of government each country has at the moment or any other factors. These countries rank the geostrategic value of the Antarctic region much more highly than have either the Europeans or the superpowers. And who knows? They could be right.

## More Vexing Questions

Besides the questions of commercial activities, environmental protection, and geostrategic value, others are unanswered by the present legal documents or are raised by political realities. We cannot go into all of them, but two deserve mention at least briefly.

Earlier in this chapter we refer to a number of legal questions regarding sovereignty over Antarctic waters, the status of ice, and so on. These are all just small parts of the Law of the Sea. Antarctica was deliberately excluded, however, from all discussions in UNCLOS III, and the relationship between the Antarctic Treaty and the United Nations Convention on the Law of the Sea is yet to be worked out. Do any of the LOSC provisions apply to Antarctica and the Southern

Ocean? If so, which ones? To what extent? In a region lacking generally accepted sovereignty, who can accept responsibilities under and derive benefits from the Law of the Sea? The entire matter is incredibly complex and important and will not be settled very soon.

Another question is the relationship between the ATS and the International Whaling Commission. The IWC still has jurisdiction over all whales in the Southern Ocean, a fact specified in both the CCAMLR and the Madrid Protocol. The Southern Ocean has been the primary hunting ground for whales in the twentieth century, and, despite the worldwide IWC moratorium on whaling, dating from 1986, whaling on a limited scale has been continued by Iceland, Norway, and—most vigorously—Japan, all claiming "scientific" reasons for doing so. In response especially to Japan's whaling in the Southern Ocean, the IWC in 1994 voted to declare the entire region a whale sanctuary. Only Japan voted against the sanctuary and soon after filed a formal objection. Russia also filed an objection, presumably to keep its options open. Both Russia and Japan are thus in the contradictory position of supporting protection of the Antarctic environment through the CCAMLR and the Protocol, yet supporting resumption of whaling in the Southern Ocean.

## The Future of Antarctica

In October 1982 the Prime Minister of Malaysia, during the general debate in the UN General Assembly, proposed that the United Nations focus its attention on Antarctica and convene a meeting that would determine the rights of all countries to uninhabited lands, especially Antarctica. Since then, Antarctica has been discussed during each annual session of the General Assembly, and at its request the Secretariat has produced two important reports on Antarctica. At issue is the demand of a number of developing countries, led by Malaysia, that the common heritage principle be applied to the

Antarctic and that in some fashion the developing countries, preferably through the United Nations, be permitted to share in the governance of Antarctica and in its potential "riches." The ATS was viewed as an exclusive club, operating in secrecy for the benefit of a handful of rich (or relatively rich) countries and denying access to a large proportion of the earth's surface to a large proportion of the earth's population. This unfair situation had to be rectified in accordance with the overall objectives of a New International Economic Order.

This viewpoint did not win a great many supporters, either in the ATS, in the UN Secretariat or among the majority of nonparties to the Antarctic Treaty. But the pressure continued. As one means of defusing the potentially explosive situation, the U.S. National Academy of Sciences organized a workshop in Antarctica for a representative group of people. The workshop was held in January 1984 at the Beardmore South Field Camp of the United States at the foot of the Beardmore Glacier in the Transantarctic Mountains. Forty-seven countries were invited and 25 sent representatives, including some from each major grouping of States. In addition, there were two officers of the UN Secretariat and one from the United Nations Environment Programme, representatives of two industries, several university scientists, two re-

porters, and representatives of seven nongovernmental organizations. The participants read formal papers on various aspects of Antarctica, took field trips (including one to the South Pole), and had many informal discussions.

It is impossible to quantify the results of this workshop, but it is noteworthy that since the beginning of the United Nations debate, and especially since the workshop, more countries have acceded to the Antarctic Treaty, including Papua New Guinea, one of the world's poorest countries, and more have been admitted as CPs, including India and China. In addition, the ATS has opened up to more observers (including, for the first time, an NGO, the Antarctic and Southern Ocean Coalition, in Madrid in 1991), has shared more information about its activities, and in general has tried to accommodate some of the demands of Malaysia and others without compromising any of the basic principles and practices of the system. These accommodations have been successful, and at least for a while have prevented broader internationalization of Antarctica under some kind of "common heritage" regime.

The Antarctic Treaty of 1959 represents a very great stride forward in international cooperation. Its provisions for prohibition of military activity, encouragement of science

**Beardmore South Field Camp, Antarctica.** These are the participants in the January 1984 workshop. (Photo courtesy of the U.S. National Academy of Sciences)

and conservation, banning of nuclear explosions and waste dumping, freezing of territorial claims, periodic consultations, and—perhaps most important—formal inspections to ensure compliance with the treaty, were innovative, far-reaching, and inspiring. We can hope that the international community as a whole, not just the parties to the treaty, will resist all attempts at commercial exploitation of the Antarctic region (except for limited and strictly controlled tourism) and preserve it as an international scientific reserve. Only in that way can we bring to life the recognition and the promise of the preamble of the Antarctic Treaty, "that it is in the interest of all mankind that Antarctica shall continue forever to be used exclusively for peaceful purposes and shall not become the scene or object of international discord."*

*Two excellent sources of information about Antarctica are:
International Institute for Environment and Development
3 Endsleigh Street
London WC1H ODD
England
Antarctic and Southern Ocean Coalition
(243 enviromental organizations in 49 countries)
41 Holt Street
Surry Hills, NSW 2010
Australia
or
P.O Box 76920
Washington, DC 20013
USA

# 34

# *Outer Space*

The political geography of outer space? Preposterous! How can there be any kind of geography of outer space when geography by definition is the study of the planet Earth and specifically of the interrelationships between the human inhabitants of the planet and their physical environment? There is no Earth in outer space and no people living there, so geographers can have no professional interest in it. Even if some geographers, such as those specializing in remote sensing, climatology, or transportation, might occasionally glance into outer space in the course of their professional work, why should it be of interest to *political* geographers?

The same arguments were being used as recently as two decades ago with reference to the sea, yet we have seen how the field of marine geography has developed since then—and has even become respectable. While it is certainly premature to attempt to develop a full-blown political geography of outer space, it is not too early to begin laying the foundations, sketching its outlines. And as we do so, we will notice a remarkable resemblance to many aspects of traditional political geography. And why not? Aren't the people who have created political geographical situations on earth and those who have studied them the very people now venturing beyond the atmosphere, even beyond the gravity, of our planet? Ultimately, new terminology, new theories, new approaches, even new topics will emerge as the new field is developed (perhaps by

some of the people reading this book!), but for the present we shall continue with our traditional approach, and merely extend it into a new region.

## *How High Is Up?*

We have pointed out that a State's territory extends, according to traditional international law, from the center of the earth upward "to the heavens." But this raises a serious question that until recently was only a child's riddle: "How high is up?" The answer to the question is that there is no answer, at least, not yet. For more than 60 years we have had regulations on aerial navigation and the use of airspace, in both domestic and international law, and we discuss some of them in Chapter 36, but the important point to make here is that none of this large body of air law includes an upper boundary, a limit to a State's jurisdiction and to the applicability of air law. Since the Soviets launched the first artificial satellite into space in 1957, a body of space law has also developed that covers a variety of subjects, but still not defining the lower limit of outer space. This is not simply an academic matter; at some point the vertical limit of national jurisdiction will have to be fixed, just as horizontal limits are, or we most certainly will have boundary disputes over the interface between air and space.

Over the years, discussions within governments, in the professional literature, and especially in the UN General Assembly's Committee on the Peaceful Uses of Outer Space (COPUOS) have produced a great variety of proposals, not very different from those offered in respect of national jurisdiction in the sea and in the Antarctic region. In fact, we will note many analogies to these other frontier regions in our discussion of the political geography of outer space.

One approach to a definition of the air/space boundary has been the functional approach; that is, some people have argued that the limit should be set according to the uses to be made of both air and space. It has even been suggested that there could be several functional limits that could be invoked as appropriate in individual cases. This approach, however, has not gained a great deal of support, and around the world the trend has generally been toward a spatially measured boundary, either limited or unlimited.

Among the many suggested limits, those mentioned most frequently are 100 nautical miles, 80 kilometers, 100 kilometers and 120–150 kilometers. A number of States and scholars have proposed as the boundary the lowest practical level at which space objects could safely remain in orbit. This level is variable and difficult to measure from the earth, but it roughly approximates 100 kilometers. Since among the qualities sought in any boundary are certainty, recognizability, and ease of measurement, the 100-kilometers proposal seems at present to be gaining support. But both governments and scholars are still divided on the issue, and no one seems to be in any hurry to settle it.

## The Uses of Outer Space

Undoubtedly, the principal users of space in the first half century of the Space Age have been the military establishments of the former Soviet Union and the United States, and secondarily their allies. Space until recently was viewed by them as simply a convenient area in which, and from which, to conduct espionage and communications for military purposes. And, of course, they viewed it as another potential theater of war. Even now, intercontinental ballistic missiles are

launched into space en route to their targets many thousands of kilometers away—so far only in tests—and both countries considered space-based defense systems against such missiles. The entire subject of the military uses of space, in fact, could very well be placed in Part Five of this book, in our discussion of geopolitics and war. As political geographers and not military geographers, however, we are more concerned with the peaceful uses of space.

Already space has many uses and many more potential ones. All geographers and most geography students are by now familiar, at least to some degree, with remote sensing, the observation and recording by various kinds of instruments mounted on orbiting satellites of various features of the earth and its atmosphere. Nearly all of us have experienced directly or indirectly radio, television, telephone, or other kinds of messages relayed from one point on the earth's surface (or below it or above it) to another via satellites. Transportation, astronomy, and on-board scientific experiments have all made headlines and in the future will make even more. Other uses are still in their early stages: space biology and medicine; manufacturing in the weightless vacuum of space of such things as new alloys of metals, pharmaceuticals, glasses, catalysts, and semiconductors; the concentration of solar energy for on-board use and for beaming onward to Earth; mining of the moon, asteroids, and perhaps planets for substances in short supply or even unknown on earth.

Because of the great commercial as well as military value of space, over 16,800 objects were launched into space between 1957 and 1987, and most of them are still orbiting the earth.* While all have been

*Space launches (many of them deploying more than one object), 1957–82:

| | | | |
|---|---|---|---|
| USSR | 1538 | France | 10 |
| United States | 796 | Italy | 4 |
| Japan | 21 | ESA | 3 |
| | | India | 3 |

In 1990 some 7000 orbiting objects were being tracked by the North American Aerospace Defense Command, of which 95 percent were junk and only about 350 were working satellites.

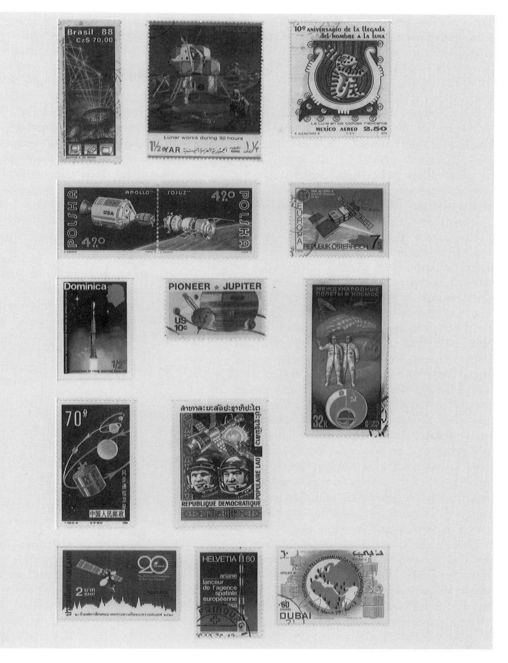

**Outer space on stamps.** Space activities have been one of the most popular subjects on stamps during the past third of a century. This selection illustrates some of the more geographic aspects of the subject. They include manned lunar expeditions, remote sensing of the earth, manned and unmanned space exploration, and telecommunications. In the bottom row, stamps of Thailand, Switzerland and Dubai honor, respectively, INTELSAT, the European Space Agency, and the Outer Space Telecommunications Congress in Paris. (Martin Glassner)

launched by national governments or the European Space Agency (ESA), commercial firms are currently beginning to compete with them, and more countries are also developing space-launch capabilities. India and Israel, for example, have each placed in orbit a number of satellites designed, built, and launched by them. Even countries, companies, institutions, and individuals that cannot launch objects into space have developed satellites to be launched by others or components of satellites or modules for space shuttles or space stations or experiments to be conducted on board.

China, for example, launched its first telecommunications satellite for a foreign client in April 1991, the Asiasat for Hughes Aircraft Company of the United States. Table 34-1 shows payloads launched by Arianespace of France through 1991. All of this activity by so many people in so many countries is bound to generate problems, and some of them are bound to be political and thus of interest to political geographers.

Unlike activities in Antarctica, activities of any kind in space can be perceived as a potential or actual threat to a State, even one some distance away. Satellites designed and placed into orbit for peaceful scientific or commercial purposes could well be used for military purposes with no one on the ground realizing it. Even perfectly innocent information sent to earth by a satellite can be used to prepare an attack or some other military action. Aside from military potential, remote sensing by or through space vehicles can also give the technologically advanced countries exclusive information about potential economic resources of other countries, particularly developing ones, which could then be used to gain commercial advantages for the country possessing the information.

Currently five major politicogeographical problems connected with space activities are engaging lawyers, diplomats, and scholars:

1. **Direct broadcasting from satellites.** One view of this inchoate technology emphasizes the virtues of a free and unrestricted flow of information; another the necessity for screening and regula-

tion of such broadcasting by governments. Some countries object to commercials coming from satellites into societies free of them. Others are apprehensive about being subjected to pornography, excessive violence, or other controversial material. There is also concern about competing religious or political views being beamed into a society unfamiliar with them or about inappropriate educational material or even about simply being drowned in the materialistic culture of the rich Western countries that alone can afford such expensive space operations.

2. **Space transportation systems.** Transportation has come to be regarded as an activity separate and distinct from scientific and technological missions of spacecraft. As such, it raises a host of questions of both a national and an international character. Matters of jurisdiction, scheduling, competing uses of spacecraft and orbits, use of vehicles by nonlaunching States, passage over States during the early stages of launching, and many others are currently receiving the attention of experts. So far the tendency has been to adapt and adjust national legislation to deal with these questions, but very soon it will likely be necessary to develop some new kind of space transportation management and control system.

3. **The environmental impact of space activities.** Human activities have already harmed space, and as they increase in number, variety, scope, and intensity the damage is likely to increase. The atmosphere of the earth has been polluted by the effluents of high-flying aircraft and rocket boosters; near-earth space is littered with man-made debris as indicated by Table 34-2. International space law not only provides for compensation for some kinds of environmental harm, but imposes on States an affirmative duty to avoid it. It remains to be seen, however, how effective current rules and practices will be in the future, even the near future.

**Table 34-1**    *Payloads Launched by Arianespace for the European Space Agency through December 1991*

| EUROPE | | | OUTSIDE EUROPE | | | INTERNATIONAL ORGANIZATIONS | | |
|---|---|---|---|---|---|---|---|---|
| Payload | Ariane | Date | Payload | Ariane | Date | Payload | Ariane | Date |
| FRANCE | | | AUSTRALIA | | | ARABSAT | | |
| Thesee | 1 | 12/81 | Aussat K3 | 3 | 9/87 | Arabsat F1 | 3 | 2/85 |
| Telecom 1A | 3 | 8/84 | | | | | | |
| Telecom 1B | 3 | 5/85 | BRAZIL | | | ESA | | |
| Spot 1 | 1 | 2/86 | Brasilsat F1 | 3 | 2/85 | Meteosat | 1 | 6/81 |
| Telecom 1C | 3 | 3/88 | Brasilsat F2 | 3 | 3/86 | Marecs A | 1 | 12/81 |
| TDF 1 | 2 | 10/88 | | | | Marecs B* | 1 | 9/82 |
| Spot 2 | 4 | 1/90 | CANADA | | | Sirio-2* | 1 | 9/82 |
| TDF 2 | 4 | 7/90 | Anik E2 | 4 | 4/91 | ECS 1 | 1 | 6/83 |
| Sara | 4 | 7/91 | Anik E1 | 4 | 9/91 | ECS 2 | 3 | 8/84 |
| Telecom IIA | 4 | 12/91 | | | | Marecs B2 | 3 | 11/84 |
| | | | INDIA | | | Giotto | 1 | 7/85 |
| GERMANY | | | Apple | 1 | 6/81 | ECS 3* | 3 | 9/85 |
| Firewheel* | 1 | 5/80 | Insat 1C | 3 | 7/88 | ECS 4 | 3 | 9/87 |
| Amsat PIIIA | | | | | | Meteosat P2 | 4 | 6/88 |
| (Oscar 9)* | 1 | 5/80 | JAPAN | | | ECS 5 | 3 | 7/88 |
| Amsat PIIIB | | | JC Sat 1 | 4 | 3/89 | MOP 1 | 4 | 3/89 |
| (Oscar 10) | 1 | 6/83 | Superbird A | 4 | 6/89 | Olympus F1 | 3 | 7/89 |
| TV Sat | 2 | 11/87 | Superbird B* | 4 | 2/90 | Hipparcos | 4 | 8/89 |
| Amsat PIIIC | | | BS-2X | 4 | 2/90 | MOP 2 | 4 | 3/91 |
| (Oscar 13) | 4 | 6/88 | | | | ERS 1 | 4 | 7/91 |
| DFS Kopernicus 1 | 4 | 6/89 | UNITED STATES | | | | | |
| TV Sat 2 | 4 | 8/89 | Spacenet F1 | 1 | 5/84 | EUTELSAT | | |
| DFS Kopernicus 2 | 4 | 7/90 | Spacenet F2 | 3 | 11/84 | Eutelsat II F1 | 4 | 8/90 |
| Tubsat | 4 | 7/91 | G-Star 1 | 3 | 5/85 | Eutelsat II F1 | 4 | 1/91 |
| | | | Spacenet F3* | 3 | 9/85 | | | |
| ITALY | | | G-Star 2 | 3 | 3/86 | INMARSAT | | |
| Italsat 1 | 4 | 11/90 | Spacenet IIIC/ | | | Inmarsat II F3 | 4 | 12/91 |
| | | | Geostar R01 | 3 | 3/88 | | | |
| LUXEMBOURG | | | PanAmerican | 4 | 6/88 | INTELSAT | | |
| Astra 1 | 4 | 12/88 | SBS 5 | 3 | 9/88 | Intelsat V F7 | 1 | 10/83 |
| Astra 1B | 4 | 3/91 | G-Star III/ | | | Intelsat V F8 | 1 | 3/84 |
| | | | Geostar R02 | 3 | 9/88 | Intelsat V F14* | 2 | 5/86 |
| SWEDEN | | | Pascat (Oscar 14) | 4 | 1/90 | Intelsat V F13 | 2 | 5/88 |
| Viking | 1 | 2/86 | Dove (Oscar 15) | 4 | 1/90 | Intelsat V F15 | 2 | 1/89 |
| Tele-x | 2 | 4/89 | WSU (Oscar 16) | 4 | 1/90 | Intelsat VI F2 | 4 | 10/89 |
| | | | Lusat (Oscar 17) | 4 | 1/90 | Intelsat VI F5 | 4 | 8/91 |
| UNITED KINGDOM | | | Amsat NA | 4 | 1/90 | Intelsat VI F1 | 4 | 10/91 |
| Skynet 4B | 4 | 12/88 | SBS 6 | 4 | 10/90 | | | |
| UoSat D | 4 | 1/90 | Galaxy 6 | 4 | 10/90 | * Lost in launch failure | | |
| UoSat E | 4 | 1/90 | G-Star 4 | 4 | 11/90 | | | |
| Skynet 4C | 4 | 8/90 | Satcom C1 | 4 | 11/90 | | | |
| UoSat F | 4 | 7/91 | Orbcom X | 4 | 7/91 | | | |

*Source:*    United Nations: *Space Activities of the United Nations and International Organizations*, New York, 1992, p. 159.

The Ariane rocket, a French development currently in its fourth incarnation, may be the most reliable launch vehicle in the world. Its reliability and relatively low cost make it a commercial, as well as a scientific and technological, success.

***Table 34-2    Estimated Distribution of Man-made Space Debris***

| Object size | Number of Objects | Percentage | Total Mass | Percentage |
|---|---|---|---|---|
| Over 10 cm | 7,000 | 0.2 | 6,611,595.4 lbs | 99.97 |
| 1–10 cm | 17,500 | 0.5 | 2,204.6 lbs | 0.03 |
| Under 1 cm | 3,500,000 | 99.3 | | |
| Total | 3,524,500 | 100.00 | 6,613,800  lbs | 100.00 |

*Source:    Report on Orbital Debris,* Interagency Group (Space), National Security Council, February 1989.

4. **Use of nuclear power systems in space.** Nuclear energy is currently considered the most efficient power source for certain types of space missions, particularly where the operating time of the mission is longer than a few hundred hours. The potential hazards of a malfunctioning reactor, unplanned reentry into the earth's atmosphere, or intentional release of radioactive substances are all being dealt with in a variety of ways through international cooperation, and to date there is no evidence of damage to the earth, its atmosphere, or outer space by any of the hundreds of nuclear power sources that have been used. There is, however, a need to supplement the relevant international space law on the subject, and that may not be easy.

5. **Remote sensing of the earth by satellite.** Remote sensing has long been a useful tool for observing, recording, and interpreting conditions on the surface of the earth, since the first photographs were taken from airborne balloons, in fact. It has achieved a relatively high degree of sophistication in the Space Age, however, and can be used to harm as well as help humanity. We now face such problems as interference with one another's activities by the numerous and multiplying sensors in space, cut-throat competition in the acquisition and dissemination of data collected by satellites, commercial and military spying of all kinds, and monopolization of data collected remotely. In 1987 the UN General Assembly adopted a set of principles on the subject, but much more remains to be done in this area, especially as some countries object to being spied upon from outer space.

### *The Geostationary Orbit*

At present the most important use of outer space is for communications, and most communications satellites are located in geostationary orbital positions. A satellite that orbits the earth in the same direction as the earth rotates and that does so in 24 hours is said to be in geosynchronous, or synchronous, orbit; that is, it completes a revolution about the planet in exactly the same period as the earth rotates once on its axis. Most such orbits are elliptical or inclined with respect to the equator. If, however, such an orbit is directly over the equator, it is said to be geostationary. A satellite in a geostationary orbital position is constantly visible from the surface at the same point in the sky, and it can constantly cover a sizable portion of the earth's surface. A satellite can maintain such a course and speed only at an altitude of ±35,787 kilometers.*

The advantages of such a position for observation and communications are obvious. The number of usable positions within the band is constrained by limitations on the use of the radio-frequency spectrum and by the large portions of the band over essentially empty ocean surfaces. Thus, the geostationary orbit is generally recognized as a valuable resource, and the most desirable portions of it are already getting crowded. Whether it is a "natural" resource, however,

*Actually, because of a number of factors, geostationary satellites move around within a band about 30 kilometers deep extending about 150 kilometers north and south of the equator.

and whether it is truly limited, are highly controversial matters, important to us because of their political implications.

They are likely to become even more important, and more controversial, as very large space stations are introduced into the geostationary orbit to perform functions other than observation and communications. The gathering and transmission of solar energy, for example, can best be done from such a position. This situation has already produced the first attempt to extend contemporary nationalistic impulses and competition among developed and developing countries on earth into outer pace. The result of this attempt is the Bogotá Declaration. It is worth quoting at some length this product of the "First Meeting of Equatorial Countries":

> The undersigned representatives of States traversed by the Equator* met in Bogota, Republic of Colombia, from November 24 through December 3rd, 1976 with the purpose of studying the geostationary orbit that corresponds to their terrestrial, sea and insular territory and considered as a natural resource. After an exchange of information and having studied in detail the different technical, legal and political aspects implied in the exercise of national sovereignty of States adjacent to said orbit, they have reached the following conclusions. . . .
>
> The Equatorial countries declare that the geostationary synchronous orbit is a physical fact linked to the reality of our planet . . . and that is why it must not be considered part of outer space. Therefore, the segments of the geostationary synchronous orbit are part of the territory over which Equatorial States exercise their national sovereignty. The geostationary orbit is a scarce natural resource . . . therefore, the Equatorial countries meeting in Bogota have decided to proclaim and defend on behalf of their peoples, the existence of their sovereignty over this natural resource. . . .

Fortunately, this absurd attempt at neoimperialism has not attracted much support from nonsigners, and it has been generally

*Brazil, Colombia, Congo, Ecuador, Indonesia, Kenya, Uganda, and Zaïre. Nonsigning equatorial countries are Gabon, Kiribati, Nauru, Peru, and Somalia.

ignored by those conducting activities in outer space. It does, however, raise numerous issues that deserve careful attention by political geographers for, while we cannot examine them here, neither the declaration nor the issues it raises will go away.

## Cooperation or Competition in Space?

Not surprisingly, many of the international rivalries and alignments so prominent on earth have been projected out into space—but not all. Although there is a good deal of competition in space, there is also much cooperation. Essentially, there are currently three groupings of States in respect of space activities similar to the groupings surrounding other issues and activities on land and at sea.

First, the United States and its allies and close associates tend to view space not only as a field for new technological breakthroughs, but for new political and legal breakthroughs as well. They see a large role for private enterprise in space, envision some kind of supranational supervision of space activities, and are willing to share their space technology and the results of their exploitation of space with other countries. Their space programs, however, tend to be haphazard, ill-coordinated, shortsighted, and competitive in many respects. The lack of a serious, enduring commitment by the United States to a rational, adequately financed, long-term civilian space program is demonstrated during every annual budget debate in Congress. In 1991, for example, a $2 billion appropriation for a year's work on an eventual space station represented not new money, but money cut out of other NASA projects. By 1994 NASA had adopted a policy of "smaller, faster, cheaper" in response to such developments and was emphasizing civilian application of its activities and projects.

On the other hand, the Soviet Union, before its dissolution, had a long-term and well-financed commitment to space activities of all kinds. It had been concentrating, for example, on gradually building very large, multipurpose, permanently inhabited space

---

### Intergovernmental Organizations

International Telecommunication Satellite Organization (INTELSAT), Washington
  124 member countries in 1992
Council on International Cooperation in the Study and Utilization of Outer Space (INTERCOS-MOS), Moscow
  10 members of the former socialist (Soviet) group; status uncertain in 1992
International Organization of Space Communications (INTERSPUTNIK), Moscow
  12 former socialist countries; in 1992 was in the process of joining INTELSAT
European Space Agency (ESA), Paris
  13 Western European Countries, with Finland and Canada associated
International Maritime Satellite Organization (INMARSAT), London
  64 members in 1992, from all groupings
Arab Satellite Communication Organization (ARABSAT), Riyadh
  All 21 members of the Arab League
European Telecommunications Satellite Organization (EUTELSAT), Paris
  27 European countries in 1992

---

stations rather than the dramatic, one-shot breakthroughs (such as walking on the moon) favored by the United States. They tended, however, to be very conservative in outer space political and legal matters. They did not accept a "common heritage" principle for outer space; they refused to recognize an outer space law independent of general international law; they did not accept individuals or corporations as subjects of law in space; they opposed an outer space agency with supranational functions; and they opposed any compulsion of States operating in outer space to help those unable to reach it.

In the late 1980s, however, the *glasnost* policy of Soviet President Gorbachev was extended to outer space activities, and the former policies were essentially reversed. The Soviet Union began to offer commercial satellite launch services, actively sought Western participation in Soviet space programs, and even offered to launch a U.S. space station into orbit on one of its own giant rockets. It began to participate in international space conferences, signed an agreement with INTELSAT for exchange of information, signed another agreement for an Anglo-Soviet space mission, arranged for exchange visits of cosmonauts and astronauts, began publishing more information about their space activities, and even invited

foreign dignitaries to observe satellite launches from hitherto secret bases. After the Soviet Union dissolved at the end of 1991, the president of Russia, the principal successor State, renewed a long-standing suggestion that that country and the United States cooperate in planning and executing a mission to Mars. Russia is still operating the world's only orbiting space station and looking toward grander space projects.

The developing countries are generally wary of being excluded from outer space altogether merely because they lack the scientific, technological, and financial resources to operate there, just as they fear being excluded from the benefits of development of the sea and Antarctica. They view the rich countries as space-resource States; that is, they have the resources to operate in space. They themselves, then, are non–space-resource States. They tend to favor the extension of the common heritage principle to space, restraints on private enterprise in space, supranational instruments for supervision of activities and resolution of disputes in space, some kind of mandatory transfer of space technology to them, and an equitable distribution of the benefits of space activities. Their primary concern is with the "practical" aspects of space, such as civilian communications and remote sensing of terrestrial features and weather systems. Gradually, as

---

### *Selected Global Nongovernmental Organizations*

International Council of Scientific Unions (ICSU), Paris
  20 unions, including the International Geographical Union
Committee on Space Research (COSPAR), Paris
  34 national academies of science from all groupings, and 12 international scientific unions
International Astronautical Federation (IAF), Paris
  124 member societies and educational and research institutions from 39 countries of all groupings
International Society for Photogrammetry and Remote Sensing (ISPRS), Tokyo
  84 countries and regional organizations

---

some of them develop modern scientific and technological capabilities, they are beginning to engage in space activities themselves.

So far these differing views have not resulted in any serious clashes. The tendency, in fact, has been to compromise, to reconcile differences while dealing with space problems ad hoc, as the need becomes apparent. Pragmatism has generally been winning out over ideology (with the notable exception of the Bogotá Declaration), and nearly all countries are cooperating through the United Nations. The United Nations, in fact, has played a leading role in the development of space law and the coordination of space activities almost from the dawn of the Space Age, largely through COPUOS, the Outer Space Division of the Secretariat, and a number of other organs and specialized agencies. In addition, international cooperation in space activities is carried out through an array of organizations (see box).

Besides its day-to-day work in stimulating, coordinating, and utilizing activities in outer space, the United Nations has fostered international cooperation in space through two other mechanisms. First, it has so far held two conferences on the Exploration and Peaceful Uses of Outer Space, the first in 1968, the second in 1982. These conferences studied contemporary and projected space matters and produced lists of conclusions and recommendations directed to member States and to international agencies and bodies. The final reports of these conferences, together with the documentation they produced, serve as important benchmarks of

the Space Age. In the mid-1990s, spirited debate was taking place over proposals for a third UNISPACE conference. It is possible that one may be held by century's end.

Second, the United Nations has played a vital role in the development of the international law of outer space. To date, it has sponsored five important treaties.

1. Treaty on Principles Governing the Activities of States in the Exploration and Use of Outer Space, Including the Moon and Other Celestial Bodies (the Principles Treaty, or the Outer Space Treaty), 1967.

2. Agreement on the Rescue of Astronauts, the Return of Astronauts, and the Return of Objects Launched into Outer Space, 1968.

3. Convention on International Liability for Damage Caused by Space Objects, 1972.

4. Convention on Registration of Objects Launched into Outer Space, 1976.

5. Agreement Governing the Activities of States on the Moon and Other Celestial Bodies (the Moon Treaty), 1979.

Of these, the Principles Treaty and the Moon Treaty are of greatest interest to political geographers. The first is generally accepted as a pioneering venture. It entered into force on 10 October 1967, just ten years and six days after the Soviet Union launched the world's first artificial satellite (*Sputnik*) into space and only 6 1/2 years after the first manned spaceship (piloted by the Soviet cosmonaut Yuri Gagarin) was launched. There are objections, however, to many of its provisions (such as those by claimants of segments of the geostationary orbit), and others are imperfectly drafted, subject to widely varying interpretations, or out of date. By 1993, however, it had been ratified or acceded to by 91 States, half the States of the world, and it is still recognized as a pillar of space law.

The Moon Treaty, on the other hand, is much more adventurous and may be ahead of its time. It entered into force on 11 July 1984. By 1993, only 14 States had signed it and only 9 had ratified it. Most of the controversy surrounding it centers on paragraph 1 of Article II: "The moon and its natural resources are the common heritage of mankind. . . ." This extension of the common heritage principle from the seabed to outer space, like its extension to Antarctica, is still vigorously opposed and may not be generally accepted for many years. Nevertheless, both treaties are worthy of careful study by

**United Nations leadership in outer space cooperation.** This poster calls attention to the second United Nations Conference on the Exploration and Peaceful Uses of Outer Space, held in Vienna in 1982. (UN)

anyone interested in any aspect of outer space, especially its political geography.

## Beyond Earth Orbits

So far our discussion has concentrated on near-space activities, chiefly those involving orbiting satellites, shuttles, and other objects. But a number of instrumented and a few manned vehicles have already been sent beyond all normal orbital paths around our planet, to the moon and to and beyond other planets of our solar system. These have produced nothing of politicogeographical interest beyond the refusal of the United States, in accordance with the UN Principles Treaty, to claim the moon or any portion of it. But we can easily envision a complex politicogeographical picture emerging in both the near and distant reaches of outer space during coming decades.

There may well be, for example, numerous occupied orbiting space stations, colonies of earthlings living permanently far from their home planet, organized transportation and communications among these space stations and colonies, and trade among them, to say nothing of colonies on the moon, on asteroids, and on other planets and their moons. These will inevitably generate questions of nationality and nationalism and sovereignty, of ownership and use of resources, of the distribution of costs and benefits, of social stratification and cultural differences, of law and loyalties and rivalries and politics, of frontiers and boundaries and power, and perhaps of colonial empires and wars of independence.

The United Nations continues to work diligently to avoid these undesirable consequences of human activities in space. As a device for reinforcing worldwide support for peaceful and cooperative outer-space activities, the General Assembly in 1989 designated 1992 as International Space Year. In addition, the UN has been considering since at least 1988 establishment of a World Space Organization to perform functions similar to those of specialized agencies dealing with earthbound matters.

Political geographers in the future will have plenty to study and analyze and explain and try to improve beyond earth's atmosphere and beyond earth orbits. Meanwhile, there is a great deal to learn about what is being done and what is being planned in space so that we can apply our approach and our techniques to analyze them.*

*Two good sources of information, besides the United Nations and individual governments, are

| Institute and Centre of Air and Space Law McGill University Montreál, P.Q. Canada | International Institute of Air and Space Law Leiden University Leiden The Netherlands |
|---|---|

Although the age-old problem of nationalism leading to territorial claims and boundary disputes is being carried into outer space, we may still be able to contain it. In keeping with Article 3 of the 1967 Principles Treaty, it is still possible that the exploration, use, and scientific investigation of outer space will be carried out "in accordance with international law, including the Charter of the United Nations, in the interest of maintaining international peace and security and promoting international cooperation and understanding."

# BIBLIOGRAPHY FOR PART SEVEN

## *Books and Monographs*

**A**

Ademuni-Odeke, *Protectionism and the Future of International Shipping.* Dordrecht: Nijhoff, 1984.

Ahnish, Faraj Abdulah, *The International Law of Maritime Boundaries and the Practice of States.* Oxford Univ. Press, 1994.

Akaha, Tsuneo, *Japan in Global Ocean Politics.* Honolulu: Univ. of Hawaii Press, 1985.

Alexandersson, Gunnar, *The Baltic Straits.* The Hague: Nijhoff, 1982.

Amin, Sayed Hassan, *Marine Pollution in International and Middle Eastern Law.* Glasgow: Royston, 1986.

Aprieto, Virginia L., *Fishery Management and Extended Maritime Jurisdiction: The Philippine Tuna Fishery Situation.* Honolulu: East-West Center, 1981.

Attard, David Joseph, *The Exclusive Economic Zone in International Law.* Oxford Univ. Press, 1987.

Auburn, F. M., *The Ross Dependency.* The Hague: Nijhoff, 1972.

**B**

Baxter, Richard R., *The Law of International Waterways.* Cambridge, MA: Harvard Univ. Press, 1964.

Beck, Peter J., *The International Politics of Antarctica.* New York: St. Martin's, 1987.

Benko, Marietta and others, *Space Law in the United Nations.* Dordrecht: Nijhoff, 1985.

Birnie, Patricia (ed.), *International Regulation of Whaling: From Conservation of Whaling to Conservation of Whales and Regulation of Whale-Watching,* 2 vols. New York: Oceana, 1985

Blake, Gerald (ed.), *Maritime Boundaries and Ocean Resources.* London: Croom Helm, 1987.

———, *World Boundaries: Maritime Boundaries.* London: Routledge, 1994.

Booner, W. N. and R. I. Lewis Smith, *Conservation Areas in the Antarctic.* Cambridge, UK: Scientific Committee on Antarctic Research, 1985.

Bouchez, Leo J., *The Regime of Bays in International Law.* New York: Oceana, 1978.

Broadus, James M. and Raphael V. Vartanov (eds.), *The Oceans and Environmental Security; Shared U.S. and Russian Perspectives.* Covelo, CA: Island Press, 1994.

Brown, E. D., *Sea-Bed Energy and Minerals: The International Legal Regime* (3 vols.). Kluwer, 1992.

———, *The International Law of the Sea* (2 vols.). Aldershot, UK: Dartmouth, 1994.

Burke, William T., *The New International Law of Fisheries; UNCLOS 1982 and Beyond.* Oxford Univ. Press, 1994.

Bush, W. M., *Antarctica and International Law: A Collection of Inter-state and National Documents.* New York: Oceana, 1988.

Butler, William E., *Northeast Arctic Passage.* Alphen aan den Rijn, Neth.: Sijthoff & Noordhof, 1978.

———, *The Law of the Sea and International Shipping; Anglo-Soviet Post-UNCLOS Perspectives.* New York: Oceana, 1985.

**C**

Carlisle, Rodney P., *Sovereignty for Sale; The Origins and Evolution of the Panamanian and Liberian Flags of Convenience.* Annapolis, MD: Naval Institute Press, 1981.

Chang, Pao-min, *The Sino-Vietnamese Territorial Dispute.* New York: Praeger, 1986.

Charney, Jonathan I. and Lewis M. Alexander (eds.), *International Maritime Boundaries.* Nijhoff, 1992.

Cheng, Chia-jui and others, *The Highways of Air and Outer Space over Asia.* Kluwer, 1992.

Christol, Carl Q., *Space Law: Past, Present and Future.* Kluwer, 1991.

Churchill, Robin R., *E.E.C. Fisheries Law.* Dordrecht: Nijhoff, 1987.

Churchill, Robin R. and Alan V. Lowe, *The Law of the Sea.* 2nd ed. Manchester: Manchester Univ. Press, 1988.

Churchill, Robin R. and Geir Ulfstein, *Marine Management in Disputed Areas.* London: Routledge, 1992.

Cleveland, Harlan, *The Global Commons: Policy for the Planet.* Lanham, MD: Univ. Press of America, 1990.

Cline, Ray S. and William M. Carpenter (eds.), *Secure Passage at Sea.* Lanham, MD: Univ. Press of America, 1993.

Codding, George A., Jr., *The Future of Satellite Communications.* Boulder, CO: Westview, 1990.

Cook, Grahame (ed.), *The Future of Antarctica: Exploitation versus Preservation.* Manchester: Manchester Univ. Press, 1990.

Crawford, James and Donald R. Rothwell (eds.), *The Law of the Sea in the Asian Pacific Region; Developments and Prospects.* Kluwer, 1995.

Cuyvers, Luc, *The Strait of Dover.* Dordrecht: Nijhoff, 1986.

**D**

Dahmani, M., *The Fisheries Regime of the Exclusive Economic Zone.* Dordrecht: Nijhoff, 1987.

Dalhousie Ocean Studies Programme. Halifax, N.S.: Dalhousie Univ. Published studies on various aspects of the Law of the Sea.

David, Steven R. and Peter Digeser, *The United States and the Law of the Sea Treaty.* Lanham, MD: Univ. Press of America, 1990.

DeCesari, P. and others (eds.), *Index of Multilateral Treaties on the Law of the Sea.* Milan: Giuffrè, 1985.

De Yturriaga, Josáe A., *Straits Used for International Navigation; A Spanish Perspective.* Dordrecht: Nijhoff, 1991.

Diedriks-Verschoor, I. H., *An Introduction to Air Law.* Deventer, Neth.: Kluwer, 1985.

**E**

Edeson, William R. and Jean-François Pulvenis, *The Legal Regime of Fisheries in the Caribbean Region.* Berlin: Springer, 1983.

Elferink, Alex G. Oude, *The Law of Maritime Boundary Delimitation: A Case Study of the Russian Federation.* Kluwer, 1994.

El-Hakim, Ali A., *The Middle Eastern States and the Law of the Sea.* Syracuse, NY: Syracuse Univ. Press, 1979.

Evans, Malcolm, *Relevant Circumstances and Maritime Delimitation.* Oxford Univ. Press, 1989.

**F**

Fitzmaurice, M., *International Legal Problems of the Environmental Protection of the Baltic Sea.* Dordrecht: Nijhoff, 1992.

Francalanci, Giampiero and others (eds.), *Atlas of the Straight Baselines.* Milan: Giuffrè, 1986.

Francalanci, Giampiero and Tullio Scovazzi, *Lines in the Sea.* Kluwer, 1994.

Francioni, Francesco (ed.), *International Environmental Law for Antarctica.* Milan: Giuffrè, 1992.

Francioni, Francesco and Tullio Scovazzi (eds.), *International Law for Antarctica/Droit International de L'Antarctique.* Milan: Giuffrè, 1987.

Franckx, Eric, *Maritime Claims in the Arctic: Canadian and Russian Perspectives.* Kluwer, 1993.

Freestone, David and Ton IJlstra (eds.), *The North Sea: Basic Legal Documents on Regional Environmental Co-operation.* Dordrecht: Nijhoff, 1991.

The Fridtjof Nansen Institute of Lysaker, Norway publishes a wide range of useful materials on both polar regions and on marine affairs.

Friedheim, Robert L., *Negotiating the New Ocean Regime.* Columbia: Univ. of South Carolina Press, 1992.

Friedrich, Christoph, *Germany's Antarctic Claim: Secret Nazi Polar Expeditions.* Toronto: Samisdat, 1978.

**G**

Gault, Ian T., *The International Legal Context of Petroleum Operations in Canadian Arctic Waters.* Calgary: Canadian Institute of Resources Law, 1983.

Glassner, Martin Ira, "The Law of the Sea," Special Issue of *Focus* (AGS), March-April, 1978. *Neptune's Domain; A Political Geography of the Sea.* London: Unwin Hyman, 1990.

———, "The Tide Flows On," in *Ocean Yearbook 11.* Chicago: Univ. of Chicago Press, 1994, 9–19.

Gold, Edgar, *Maritime Transport: The Evolution of International Marine Policy and Shipping Law.* Lexington, MA: Heath, 1981.

Gáorbiel, Andrzej, *International Organizations and Outer Space Activities.* Láodáz, Pol.: Redake a Naczelna, Uniwersytet Láodski, 1984.

———, *Outer Space in International Law.* Láodáz: Uniwersytet Láodzki, 1984.

Gorove, Stephen (ed.), *Developments in Space Law; Issues and Policies.* Kluwer, 1991.

Greenfield, Jeanette, *China's Practice in the Law of the Sea.* Oxford Univ. Press, 1992.

Grief, Nicholas, *Public International Law in the Airspace of the High Seas.* Kluwer, 1994.

Griffiths, Franklyn (ed.), *Politics of the Northwest Passage.* Montreal: McGill-Queen's, 1987.

Guerrier, Steven W. and Wayne C. Thompson (eds.), *Space: National Programs and International Cooperation.* Boulder, CO: Westview, 1989.

**H**

Haas, Peter M., *Saving the Mediterranean; The Politics of International Environmental Cooperation.* New York: Columbia Univ. Press, 1990.

Hargrove, Eugene C., *Beyond Spaceship Earth; Environmental Ethics and the Solar System.* San Francisco: Sierra Club, 1987.

Harris, Stuart (ed.), *Australia's Antarctic Policy Options.* CRES Monograph 11. Canberra: Australian National Univ., 1984.

Heine, Irvin Millard, *China's Rise to Commercial Maritime Power*. New York: Greenwood, 1989.

Hey, Ellen, *The Regime for the Exploitation of Transboundary Marine Fisheries Resources*. Dordrecht: Nijhoff, 1989.

Howard, Harry N., *Turkey, the Straits and U.S. Policy*. Baltimore: Johns Hopkins Univ. Press, 1974.

Hudson, Heather, *New Dimensions in Satellite Communications; Challenges for North and South*. Dedham, MA: Artech House, 1985.

Hunt, Constance D., *The Offshore Petroleum Regimes of Canada and Australia*. Calgary: Canadian Institute of Resources Law, 1989.

**I**

IUCN, *A Strategy for Antarctic Conservation*. Cambridge, UK, 1991.

**J**

Jansky, Ronald and others, *Communication Satellites in the Geostationary Orbit*. Dedham, MA: Artech, 1983.

Jasani, Bhupendra and Toshiba Sakata (eds.), *Satellites for Arms Control and Crisis Monitoring*. Oxford Univ. Press, 1987.

Jasentuliyana, Nandasiri (ed.), *Space Law: Development and Scope*. New York: Praeger, 1992.

Jayewardene, Hiran W., *The Regime of Islands in International Law*. Dordrecht: Nijhoff, 1990.

Jelly, Doris, H., *Canada: 25 Years in Space*. Polyscience Pub., 1988.

Jessup, Philip C., *The Law of Territorial Waters and Maritime Jurisdiction*. New York: Jennings, 1927.

Jessup, Philip C. and Howard J. Taubenfeld, *Control for Outer Space and the Antarctic Analogy*. New York: Columbia Univ. Press, 1959.

Johnson, Barbara and Mark W. Zacher (eds.), *Canadian Foreign Policy and the Law of the Sea*. Vancouver: UBC Press, 1977.

Johnston, Douglas M. (ed.), *Canada and the New International Law of the Sea*. Toronto: Univ. of Toronto Press, 1987.

———, *The Theory and History of Ocean Boundary Making*. Montreal: McGill-Queen's, 1988.

Johnston, Douglas M. and P. M. Saunders (eds.), *Ocean Boundary Making; Regional Issues and Developments*. London: Croom Helm, 1987.

Johnston, Douglas M. and Mark J. Valencia, *Pacific Ocean Boundary Problems; Status and Solutions*. Dordrecht: Nijhoff, 1991.

Jorgensen-Dahl, Arnfinn and Willy Ostreng, *The Antarctic Treaty System in World Politics*. Basingstoke, UK: Macmillan, 1991.

Joyner, Christopher C., *Antarctica and the Law of the Sea*. Dordrecht: Nijhoff, 1992.

Joyner, Christopher C. and Sudhir K. Chopra (eds.), *The Antarctic Legal Regime*. Dordrecht: Nijhoff, 1988.

**K**

Keeble, John, *Out of the Channel; the Exxon Valdez Oil Spill in Prince William Sound*. New York: HarperCollins, 1991.

Kimball, Lee A., *Southern Exposure: Deciding Antarctica's Future*. Washington, DC: World Resources Institute, 1990.

Kindt, John Warren, *Marine Pollution and the Law of the Sea*. (6 vols.) Buffalo: Hein, 1986–1988.

Kittichaisaree, Kriangsak, *The Law of the Sea and Maritime Boundary Delimitation in South-East Asia*. Oxford Univ. Press, 1987.

Knight, Gary and Hungdah Chiu, *The International Law of the Sea: Cases, Documents and Readings*. New York: Elsevier, 1991.

Knight, Russell, *Australian Antarctic Bibliography*. Hobart: Univ. of Tasmania, Institute of Antarctic and Southern Ocean Studies, 1987.

Kwiatkowska, Barbara, *The 200 Mile Exclusive Economic Zone in the New Law of the Sea*. Dordrecht: Nijhoff, 1989.

**L**

Lachs, Manfred, *The Law of Outer Space; an Experience in Contemporary Law-making*. Leiden: Sijthoff, 1972.

Lamson, Cynthia (ed.), *The Sea Has Many Voices*. Montreal: McGill-Queens, 1994.

Lamson, Cynthia and David L. Vanderzwaag (eds.), *Transit Management of the Northwest Passage*. Cambridge Univ. Press, 1986.

Lapidoth, Ruth, *Freedom of Navigation with Special Reference to International Waterways in the Middle East*. Jerusalem Papers on Peace Problems No. 13–14. Jerusalem: Hebrew Univ., 1975.

Lapidoth-Eschelbacher, Ruth, *The Red Sea and the Gulf of Aden*. The Hague: Nijhoff, 1982.

Larson, David, *Security Issues and the Law of the Sea*. Washington, DC: World Resources Institute, 1994.

Laursen, Finn (ed.), *Superpower at Sea; U.S. Ocean Policy*. New York: Praeger, 1983.

———, *Small Powers at Sea; Scandinavia and the New International Marine Order*. Kluwer, 1993.

The Law of the Sea Institute at the University of Hawaii publishes its annual conference proceedings, occasional papers, and workshop reports that are most valuable.

Leanza, Umberto and Luigi Sico, *Mediterranean Continental Shelf: Delimitations and Regimes; International and National Legal Sources*. Rome: Univ. of Rome, 1988.

Leifer, Michael, *[Straits of] Malacca, Singapore and Indonesia*. Alphen aan den Rijn, Neth.: Sijthoff & Noordhoff, 1978.

Lewis, John S. and Ruth A. Lewis, *Space Resources: Breaking the Bonds of Earth*. New York: Columbia Univ. Press, 1987.

Lien, Jon and Robert Graham (eds.), *Marine Parks and Conservation; Challenge and Promise*. Toronto: National and Provincial Parks Association of Canada, 1985.

Logan, Rod M., *Canada, the United States, and the Third Law of the Sea Conference*. Montreal and Washington, DC: Canadian-American Committee, 1978.

Lovering, J. F. and J.R.V. Prescott, *Last of Lands—Antarctica*. Melbourne: Melbourne Univ. Press, 1979.

Luther, Sara Fletcher, *The United States and the Direct Broadcast Satellite*. Oxford Univ. Press, 1988.

**M**

Ma Ying-jeou, *Legal Problems of Seabed Boundary Delimitation in the East China Sea*. Occasional Papers/Reprints Series in Contemporary Asian Studies No. 3. Baltimore: Univ. of Maryland School of Law, 1984.

MacDonald, Charles G., *Iran, Saudi Arabia, and the Law of the Sea: Political Interaction and Legal Developments in the Persian Gulf*. Westport, CT: Greenwood, 1980.

Mahmoudi, Said, *The Law of Deep Sea-Bed Mining*. Stockholm: Almqvist & Wiksell, 1987.

Mankabady, Samir, *The International Maritime Organization*. London: Croom Helm, 1984.

Martinez, Larry, *Communications Satellites; Power Politics in Space*. Dedham, MA: Artech, 1985.

Masterson, W. E., *Jurisdiction in Marginal Seas*. New York: Macmillan, 1929.

Matte, Nicholas Mateesco, *Aerospace Law: Telecommunications Satellites*. Toronto: Butterworths, 1982.

McCune, Shannon, *Islands in Conflict in East Asian Waters*. Hong Kong: Asian Research Service, 1984.

McDougal, Myres S. and William T. Burke, *The Public Order of the Oceans; a Contemporary International Law of the Sea*. Dordrecht: Nijhoff, 1987.

McGoodwin, James R., *Crisis in the World's Fisheries; People, Problems, and Policies*. Stanford, CA: Stanford Univ. Press, 1994.

McRae, Donald and Gordon Munro (eds.), *Canadian Oceans Policy; National Strategies and the New Law of the Sea*. Vancouver: UBC Press, 1989.

Meng Qing-nan, *Land-Based Marine Pollution: International Law Development*. Boston: Nijhoff, 1989.

Mericq, Luis H., *Antarctica: Chile's Claim*. Washington, DC: National Defense Univ. Press, 1986.

Mitchell, Barbara and R. Sandbrook, *The Management of the Southern Ocean*. London: International Institute for Environment and Development, 1980.

Morris, Michael A., *International Politics and the Sea; the Case of Brazil*. Boulder, CO: Westview, 1979.

————, *North-South Perspectives on Marine Policy*. Boulder, CO: Westview, 1988.

————, *The Strait of Magellan*. Dordrecht: Nijhoff, 1989.

Munavvar, Mohamed, *Ocean States: Archipelagic Regimes in the Law of the Sea*. Kluwer, 1995.

Myhre, Jeffrey D., *The Antarctic Treaty System; Politics, Law and Diplomacy*. Boulder, CO: Westview, 1986.

**N**

Negrine, Ralph (ed.), *Satellite Broadcasting; The Politics and Implications of the New Media*. New York: Routledge, 1988.

Nelsen, Brent F., *The State Offshore; Petroleum, Politics and State Intervention on the British and Norwegian Continental Shelves*. New York: Praeger, 1991.

The Netherlands Institute for the Law of the Sea (eds.), *International Organizations and the Law of the Sea; Documentary Yearbook*. Kluwer.

Ngantcha, Francis, *The Right of Innocent Passage and the Evolution of the International Law of the Sea*. London: Pinter, 1990.

**O**

Oceans Institute of Canada Compiler, *Maritime Affairs: A World Handbook*. 2nd ed. Detroit: Gale Research, 1991.

O'Connell, D. P., *The International Law of the Sea*. 2 vols. Oxford Univ. Press, 1984.

Oerding, J. B., *Frozen Friction Point: A Geopolitical Analysis of Sovereignty in the Antarctic Peninsula*. Gainesville: Univ. of Florida Press, 1977.

Okolie, Charles C., *International Law of Satellite Remote Sensing and Outer Space.* Dubuque, IA: Kendall/Hunt, 1989.

Orrego Vicuña, Francisco, *Antarctic Mineral Exploitation: The Emerging Legal Framework.* Cambridge Univ. Press, 1988.

——, *The Exclusive Economic Zone; Regime and Legal Nature under International Law.* Cambridge Univ. Press, 1990.

Oxman, Bernard H., *From Cooperation to Conflict: The Soviet Union and the United States at the Third U.N. Conference on Law of the Sea.* Seattle: Univ. of Washington, Washington Sea Grant Project, 1985.


**P**

Pak, Chi Young, *The Korean Straits.* Dordrecht: Nijhoff, 1988.

Papadakis, Nikos, *The International Legal Regime of Artificial Islands.* Leiden: Sijthoff, 1977.

Papadakis, Nikos and Martin Ira Glassner (eds.), *The International Law of the Sea and Marine Affairs: A Bibliography.* The Hague: Nijhoff, 1984.

Parsons, Anthony, *Antarctica.* Cambridge Univ. Press, 1987.

Payoyo, Peter Bautista, *Ocean Governance; Sustainable Development of the Seas.* Tokyo: United Nations Univ. Press, 1994.

Peterson, M. J., *Managing the Frozen South: The Creation and Evolution of the Antarctic Treaty System.* Berkeley: Univ. of California Press, 1988.

Pharand, Donat, *Canada's Arctic Waters in International Law.* Cambridge Univ. Press, 1988.

Pharand, Donat and Umberto Leanzo (eds.), *The Continental Shelf and the Exclusive Economic Zone: Delimitation and Legal Regime.* Kluwer, 1993.

Pharand, Donat, in association with Leonard H. Legault, *The Northwest Passage: Arctic Straits.* Dordrecht: Nijhoff, 1984.

Pinochet de la Barra, Oscar, *Chilean Sovereignty in Antarctica.* Santiago de Chile: Editorial del Pacífico, 1954.

Platzöder, Renate (comp./ed.), *Third United Nations Conference on the Law of the Sea: Documents.* New York: Oceana, 1982 et. seq.

Polar Research Board, National Research Council, *Antarctic Treaty System: An Assessment.* Washington, DC: National Academy Press, 1986. Proceedings of a Workshop held at Beardmore South Field Camp, Antarctica, Jan. 7–13, 1985.

Prescott, John Robert Victor, *Maritime Jurisdiction in Southeast Asia: A Commentary and Map.* Honolulu: East-West Center, 1981.

——, *Australia's Maritime Boundaries.* Canberra: Australian National Univ., Dept. of International Relations, 1985.

——, *The Maritime Political Boundaries of the World.* London: Methuen, 1985.

*Proceedings of Pacific Basin Management of the 200 Nautical Mile Exclusive Economic Zone.* [Honolulu]: Pacific Basin Development Council, 1988.


**Q**

Quartermain, L. B., *New Zealand and the Antarctic.* Wellington: New Zealand Government Printers, 1971.


**R**

Ramazani, Rouhollah K., *The Persian Gulf and the Strait of Hormuz.* Alphen aan den Rijn, Neth.: Sijthoff &: Noordhof, 1979.

Reijnen, Gijsbertha C.M. and William De Graaff, *The Pollution of Outer Space, in Particular of the Geostationary Orbit.* Dordrecht: Nijhoff, 1989.

Rembe, Nasila S., *Africa and the International Law of the Sea.* Alphen aan den Rijn, Neth.: Sijthoff & Noordhoff, 1980.

Research Centre for International Law—Univ. of Cambridge, *International Boundary Cases: The Continental Shelf.* Cambridge: Grotius, 1992.

Reynolds, Glenn H. and Robert P. Nerges, *Outer Space: Problem of Law and Policy.* Boulder, CO: Westview, 1989.

Riddell-Dixon, Elizabeth, *Canada and the International Seabed.* Kingston, Ontario: McGill-Queen's, 1989.

Rothwell, Donald R., *Maritime Boundaries and Resource Development: Options for the Beaufort Sea.* Calgary: Canadian Institute of Resources Law, 1988.

Royal Institute of International Affairs. *Europe's Future in Space.* London: Routledge, 1988.

Rozakis, Christos L. and Petros N. Stagos, *The Turkish Straits.* Dordrecht: Nijhoff, 1987.


**S**

Sahurie, Emilio J. (ed.), *The International Law of Antarctica.* Norwell, MA: Kluwer, 1992.

Sanger, Clyde, *Ordering the Oceans; The Making of the Law of the Sea.* Toronto: Univ. of Toronto Press, 1987.

Schmidhauser, John R. and George O. Totten, *The Whaling Issue in U.S.-Japan Relations.* Boulder, CO: Westview, 1978.

Sebek, Victor, *Eastern European States and the Development of the Law of the Sea*. New York: Oceana, 1979.

Sebenius, James K., *Negotiating the Law of the Sea*. Cambridge, MA: Harvard Univ. Press, 1984.

Shapley, Deborah, *The Seventh Continent: Antarctica in a Resource Age*. Washington, DC: Resources for the Future, 1986.

Singh, Nagendra, *Maritime Flag and International Law*. Leiden: Sijthoff, 1978.

Smith, Hance D. and Adalberto Vallega (eds.), *The Development of Integrated Sea-Use Management*. London: Routledge, 1991.

Symmons, Clive R., *Ireland and the Law of the Sea*. Round Hall Press, 1993.

Székely, Alberto, *Latin America and the Development of the Law of the Sea*. Dobbs Ferry, NY: Oceana, 1976.

**T**

Taishoff, Marika Natasha, *State Responsibility and the Direct Broadcast Satellite*. London: Pinter, 1987.

Tangsubkul, Phiphat, *ASEAN and the Law of the Sea*. Singapore: Institute of Southeast Asian Studies, 1982.

———, *The Southeast Asian Archipelagic States: Concept, Evolution and Current Practice*. Honolulu: East-West Center, 1984.

Truver, Scott C., *The Strait of Gibraltar and the Mediterranean*. Alphen aan den Rijn, Neth.: Sijthoff & Noordhoff, 1980.

**U**

Underdal, Arild, *The Politics of International Fisheries Management: The Case of the Northeast Atlantic*. Oslo: Univ. of Oslo Press, 1980.

**V**

Valencia, Mark J., *International Conference on the Sea of Japan*. Honolulu: East-West Center, 1989.

VanderZwaag, David L., *The Fish Feud: The U.S. and Canadian Boundary Dispute*. Lexington, MA: Heath, 1983.

Van Dyke, Jon M. and others (eds.), *Freedom for the Seas in the 21st Century; Ocean Governance and Environmental Harmony*. Covelo, CA: Island Press, 1993.

Vertzberger, Yaacov Y.I., *Coastal States, Regional Powers, Superpowers and the Malacca-Singapore Straits*. Berkeley: Univ. of California, Institute of East Asian Studies, 1984.

**W**

Watts, Sir Arthur, *International Law and the Antarctic Treaty System*. Cambridge: Grotius, 1992.

Westerman, Gayl L., *The Juridical Bay*. Oxford Univ. Press, 1987.

Westermeyer, William E., *The Politics of Mineral Resources Development in Antarctica*. Boulder, CO: Westview, 1984.

White, Rita and others, *The Law and Regulation of International Space Communications*. Boston: Artech House, 1988.

Wolfrum, R[au]udiger (ed.), *Antarctic Challenge*. Berlin: Duncker & Humblot, 1984 et seq.

———, *The Convention on the Regulation of Antarctic Mineral Resource Activities*. Berlin: Springer, 1991.

**Y**

Young, Oran and Gail Osherenko, *Polar Politics: Creating International Environmental Regimes*, Ithaca, NY: Cornell Univ. Press, 1993.

**Z**

Zhukov, Gennady and Yuri Kolosov, *International Space Law*. New York: Praeger, 1984.

Zumberge, James H., *Possible Environmental Effects of Mineral Exploration and Exploitation*. Cambridge: Scientific Committee on Antarctic Research, 1979.

Zwaan, Tanja L. (ed.), *Space Law: Views of the Future*. Deventer, Neth.: Kluwer, 1988.

# *Periodicals*

**A**

Abrahamsson, Bernhard J., "The Law of the Sea Convention and Shipping," *Political Geography Quarterly*, 5, 1 (Jan. 1986), 13–17.

Alexander, Lewis M., "Geography and the Law of the Sea," *Annals, AAG*, 58, 1 (March 1968), 177–197.

———, "The Delimitation of Maritime Boundaries," *Political Geography Quarterly*, 5, 1 (Jan. 1986), 19–24.

Almond, Harry H., Jr., "Demilitarization and Arms Control: Antarctica," *Case Western Reserve Journal of International Law*, 17, 2 (Spring 1985), 229–285.

*Annals of Air and Space Law.* Montreal: McGill Univ., Institute and Centre of Air and Space Law, published irregularly.

*Antarctic.* Christchurch. New Zealand Antarctic Society. Published quarterly since 1956.

*The Antarctic.* Special issue of *Oceanus* (Woods Hole Oceanographic Institution) 31, 2 (Summer 1988), whole volume.

*Antarctic Bibliography.* U.S. Superintendent of Documents, Washington, DC. Published irregularly since 1965 (Library of Congress).

*Antarctic Journal of the United States.* U.S. Superintendent of Documents, Washington, DC. Published quarterly since 1966.

**B**

Bird, Eric and J.R.V. Prescott, "Rising Global Sea Levels and National Maritime Claims," *Marine Policy Reports*, 1, 3 (1989), 177–196.

Blay, S.K.N., "New Trends in the Protection of the Antarctic Environment: The 1991 Madrid Protocol," *American Journal of International Law*, 86, 2 (April 1992), 377–399.

Bloomfield, Lincoln P., "The Arctic: Last Unmanaged Frontier," *Foreign Affairs*, 60, 1 (Fall 1981), 87–105.

Blum, Yehuda Z., "The Gulf of Sidra Incident," *American Journal of International Law*, 80, 3 (July 1986), 668–678.

Boczek, Boleslaw A., "Global and Regional Approaches to the Protection and Preservation of the Marine Environment," *Case Western Reserve Journal of International Law*, 16 (1984), 39.

———, "The Soviet Union and the Antarctic Regime," *American Journal of International Law*, 78, 4 (Oct. 1984), 834–858.

Boggs, S. Whittemore, "Delimitation of the Territorial Sea," *American Journal of International Law*, 24 (1930), 541.

———, "Problems of Water Boundary Definition: Median Lines and International Boundaries Through Territorial Waters," *Geographical Review*, 27, 3 (July 1937), 445–456.

———, "Delimitation of Seaward Areas under National Jurisdiction," *American Journal of International Law*, 45, 21 (April 1951), 240–266.

———, "National Claims in Adjacent Seas," *Geographical Review*, 41, 21 (April 1951), 185–209.

Bowett, Derek W., "The Second U.N. Conference on the Law of the Sea," *International and Comparative Law Quarterly*, 9 (1960), 415–435.

Boyle, Alan E., "Marine Pollution under the Law of the Sea Convention," *American Journal of International Law*, 79, 2 (April 1985), 347–372.

Burke, William T., "Regulation of Driftnet Fishing on the High Seas and the New International Law of the Sea," *Georgetown International Environmental Law Review*, 3, 2 (Fall 1990).

Burmester, Harry, "The Torres Strait Treaty: Ocean Boundary Delimitation by Agreement," *American Journal of International Law*, 76, 2 (April 1982), 321–349.

**C**

Carnegie, A. R., "The Law of the Sea: Commonwealth Caribbean Perspectives," *Social and Economic Studies*, 36, 3 (Sept. 1987), 99–117.

Chao, K. T., "East China Sea: Boundary Problems Relating to the Tiao-yu-t'ai Islands," *Chinese Yearbook of International Law and Affairs*, 2 (1982), 45–97.

Charney, Jonathan I., "Ocean Boundaries Between Nations: A Theory for Progress," *American Journal of International Law*, 78, 3 (July 1984), 582–606.

———, "Progress in International Maritime Boundary Delimitation Law," *American Journal of International Law*, 88, 2 (April 1994), 227–256.

Child, Jack, "Latin American 'Lebensraum': The Geopolitics of the Ibero-American Antarctica." *Applied Geography*, 10 (1990), 287–306.

Chiu, Hungdah, "Some Problems Concerning the Delimitation of the Maritime Boundary Between the Republic of China and the Philippines," *Chinese Yearbook of International Law and Affairs*, 3 (1983), 1–21.

———, "Political Geography in the Western Pacific after the Adoption of the 1982 United Nations Convention on the Law of the Sea," *Political Geography Quarterly*, 5, 1 (Jan. 1986), 25–32.

Christol, Carl Q., "The Moon Treaty Enters into Force," *American Journal of International Law*, 79, 1 (Jan. 1985), 163–168.

Clingan, Thomas A., Jr., "US-Mexican Maritime Relations in the Aftermath of UNCLOS III," *Political Geography Quarterly*, 5, 1 (Jan. 1986), 57–62.

Cohen, Saul B., "The Oblique Plane Air Boundary," *Professional Geographer*, 10, 6 (1958), 11–15.

Colombos, C. John, "Territorial Waters," *Transactions, Grotius Society*, 9 (1923), 89.

Couper, A. D., "The Marine Boundaries of the United Kingdom and the Law of the Sea," *Geographical Journal*, 151, 2 (July 1985), 228–236.

**D**

D'Amato, Anthony and Sudhir K. Chopra, "Whales: Their Emerging Right to Life," *American Journal of International Law*, 85, 1 (Jan. 1991), 21–62.

Dean, Arthur H., "The Geneva Conference on the Law of the Sea: What Was Accomplished," *American Journal of International Law*, 52, 4 (Oct. 1958), 607–628.

———, "Freedom of the Seas," *Foreign Affairs*, 37, 1 (Oct. 1958), 83–94.

———, "The Second Geneva Conference on the Law of the Sea: The Fight for Freedom of the Seas," *American Journal of International Law*, 54, 4 (Oct. 1960), 751–790.

Deihl, Colin, "Antarctica: An International Laboratory," *Boston College Environmental Affairs Law Review*, 18 (Spring 1991), 423–456.

De Saussure, Hamilton, "The Impact of Manned Space Stations on the Law of Outer Space," *San Diego Law Review*, 21, 5 (Sept. 1984), 985–1014.

*The Development of Marine Regions*. Special issue of *Ocean and Shoreline Management*, 15 (1991), whole volume.

De Vorsey, Louis and Megan C. De Vorsey, "The World Court Decision in the Canada—United States Gulf of Maine Seaward Boundary Dispute: A Perspective from Historical Geography," *Case Western Reserve Journal of International Law*, 18, 3 (1986), 415–442.

de Vries, R.O.G., "The Right of Overflight over Straits and Archepelagic States," *Netherlands Yearbook of International Law*, 61 (1983).

Done, P., "Maritime Boundary Legislation in the Commonwealth Caribbean," *The Hydrographic Journal*, 48 (April 1988), 71–123.

**E**

El Baradei, Mohamed, "The Egyptian-Israeli Peace Treaty and Access to the Gulf of Aqaba: A New Legal Regime," *American Journal of International Law*, 76, 3 (July 1982), 532–554.

Evensen, Jens, "The Anglo-Norwegian Fisheries Case and Its Legal Consequences," *American Journal of International Law*, 46, 4 (Oct. 1952), 609–630.

**F**

Feldman, Mark B., "The Tunisia-Libya Continental Shelf Case: Geographic Justice or Judicial Compromise?" *American Journal of International Law*, 77, 2 (April 1983), 219–238.

Feldman, Mark B. and David Colson, "The Maritime Boundaries of the United States," *American Journal of International Law*, 75, 4 (Oct. 1981), 729–763.

**G**

Georgetown Space Law Group. "The Geostationary Orbit: Legal, Technical and Political Issues Surrounding Its Use in World Telecommunications," *Case Western Reserve Journal of International Law*, 16, 2 (1984), 223–264.

Glassner, Martin Ira, "Israel's Maritime Boundaries," *Ocean Development and International Law*, 1, 4 (Winter 1974), 303–313.

———, "The Illusory Treasure of Davy Jones' Locker," *San Diego Law Review*, 13, 3 (March 1976), 533–551.

———, "The View from the Near North: South Americans View Antarctica and the Southern Ocean Geopolitically," *Political Geography Quarterly*, 5, 1 (Oct. 1985), 329–342.

———, "The New Political Geography of the Sea," *Political Geography Quarterly*, 5, 1 (Jan. 1986), 6–8.

———, "Review Essay: Different Perspectives on the Law of the Sea," *Political Geography Quarterly*, 10, 1 (Jan. 1991), 76–79.

———, "Review Essay: The Frontiers of Earth—and of Political Geography: The Sea, Antarctica and Outer Space," *Political Geography Quarterly*, 10, 4 (Oct. 1991), 422–437.

———, Review Essay: "The Political Geography of the Sea," *Canadian Geographer*, 37, 3 (Fall 1993), 271–279.

———, "Review Essay: Recent Books on Marine Affairs," *Ocean Development and International Law*, 24, 4 (Fall 1993), 431–435.

Gross, Leo, "The Geneva Conference on the Law of the Sea and the Right of Innocent Passage Through the Gulf of Aqaba," *American Journal of International Law*, 53, 3 (1959), 564–594.

Grzybowski, Kazimir A., "The Soviet Doctrine of Mare Clausum and Policies in Black and Baltic Seas," *Journal of Central European Affairs*, 14, 4 (Jan. 1955), 339–353.

**H**

Hage, Robert E., *The Third United Nations Conference on the Law of the Sea: A Canadian Retrospective. Behind the Headlines*, 40, 5 (1983), pamphlet.

Hailbronner, Kay, "Freedom of the Air and the Convention on the Law of the Sea," *American Journal of International Law*, 77, 3 (July 1983), 490–520.

Hannesson, Rögnvaldur, "From Common Fish to Rights Based Fishing; Fisheries Management and the Evolution of Exclusive Rights to Fish," *European Economic Review*, 35 (April 1991), 397–407.

Harry, Ralph L., "The Antarctic Regime and the Law of the Sea Convention: An Australian View," *Virginia Journal of International Law*, 21, 4 (Summer 1981), 727–744.

Herber, Bernard P., "The Common Heritage Principle; Antarctica and the Developing Nations," *American Journal of Economics and Sociology*, 50 (Oct. 1991), 391–406.

Hildreth, R. G., "Managing Ocean Resources: Canada," *International Journal of Estuarine and Coastal Law* (1991), 199–228.

Hodgson, Robert D. and Robert W. Smith, "Boundary Issues Created by Extended National Marine Jurisdiction," *Geographical Review*, 69, 4 (Oct. 1979), 423–433.

Howard, Matthew, "The Convention on the Conservation of Antarctic Marine Living Resources: A Five Year Review," *International and Comparative Law Quarterly*, 38 (1989), 104–149.

Huming, Yu, "Marine Policy in the People's Republic of China," *Marine Policy Reports*, 1, 3 (1989), 237–246.

**J**

Jacobson, Jon L., "International Fisheries Law in the Year 2010," *Louisiana Law Review*, 45 (1985), 1161–1198.

Jasentuliyana, N. and Ralph Chipman, "The Current Legal Regime of the Geostationary Orbit and Prospectives for the Future," *Acta Astronautica*, 17 (1988), 599–606.

Jessup, Philip C. and Howard J. Taubenfeld, "Outer Space, Antarctica, and the United Nations," *International Organization*, 13, 3 (Summer 1959), 363–79.

*Journal of Space Law*. Published semiannually at the Univ. of Mississippi Law Center, Univ. of Mississippi.

Joyner, Christopher C., "Anglo-Argentine Rivalry after the Falklands/Malvinas War: Law, Geopolitics and the Antarctic Connection," *Lawyer of the Americas*, 15, 3 (Winter 1984), 467–502.

——— (ed.), *Polar Politics in the 1980's*. Special issue of *International Studies Notes*, 11, 3 (Spring 1985), six articles.

———, "The Southern Ocean and Marine Pollution: Problems and Prospects," *Case Western Reserve Journal of International Law*, 17, 2 (Spring 1985), 165–194.

———, "1988 Antarctic Minerals Convention," *Marine Policy Reports*, 1, 1 (1989), 69–86.

**K**

Khlestov, O. and Vladimir Golitsyn, "The Antarctic: Arena of Peaceful Cooperation," *International Affairs* (Moscow), (August 1978), 61–65.

Kliot, Nurit, "Cooperation and Conflicts in Maritime Issues in the Mediterranean Basin," *GeoJournal*, 18, 3 (April 1989), 263–272.

Koh, Tommy T.B., "The Third United Nations Conference on the Law of the Sea; What Was Accomplished?" *Law and Contemporary Problems*, 46, 2 (Spring 1983), 5–9.

———, "The Exclusive Economic Zone," *Malaya Law Review*, 30, 1 (July 1988), 1–33.

Kriwoken, Lorne K. and Marcus Haward, "Marine and Estuarine Protected Areas in Tasmania, Australia; The Complexities of Policy Development," *Ocean and Shoreline Management*, 15, 2 (1991), 143–163.

**L**

Lamontagne, Michele R., "United States Commercial Space Policy: Impact on International and Domestic Law," *Syracuse Journal of International Law and Commerce*, 13, 1 (Fall 1986), 129–154.

Lapidoth, Ruth, "The Strait of Tiran, the Gulf of Aqaba, and the 1979 Treaty of Peace Between Egypt and Israel," *American Journal of International Law*, 77, 1 (Jan. 1983), 84–108.

Leanza, Umberto, "Marine Delimitation in the Persian Gulf and Right of Passage in the Strait of Hormuz," *Marine Policy Reports*, 1, 3 (1989), 217–236.

Lee Yong Leng, "Malacca Strait, Kra Canal, and International Navigation," *Pacific Viewpoint*, 19, 1 (1978), 65–74.

Legault, L. H. and Blair Hankey, "From Sea to Seabed: The Single Maritime Boundary in the Gulf of Maine Case," *American Journal of International Law*, 79, 4 (Oct. 1985), 961–991.

Lundquist, Thomas R., "The Iceberg Cometh?: International Law Relating to Antarctic Iceberg Exploitation," *Natural Resources Journal*, 17 (Jan. 1977), 1–41.

**M**

Malone, James L., "UN Law of the Sea Convention and U.S. Ocean Policy," *Marine Policy Reports,* 1, 3 (1989), 169–176.

*Marine Policy.* London: Butterworths, published quarterly. since 1976.

*Maritime Policy and Management.* London: Taylor & Francis, published quarterly since 1973.

Marks, Beth C. and James N. Barnes, "The Future of Antarctica under the Environmental Protocol," *Journal of Environment and Development,* 2, 2 (Summer 1993), 169–174.

McDorman, Ted L., "International Fishery Relations in the Gulf of Thailand," *Contemporary Southeast Asia,* 12, 1 (June 1990), 40–54.

———, "The Canada—France Maritime Boundary Case: Drawing a Line Around St. Pierre and Miquelon," *American Journal of International Law,* 84, 1 (Jan. 1990), 157–189.

McGowan, Glenn, "Geographical Disadvantages as a Basis for Marine Resource Sharing Between States," *Monash Univ. Law Review,* 13, 4 (Dec. 1987), 200–229.

McLaughlin, Richard J., "UNCLOS and the Demise of the United States' Use of Trade Sanctions to Protect Dolphins, Sea Turtles, Whales and Other International Marine Living Resources," *Ecology Law Quarterly,* 21 (1994), 1–78.

McRae, Donald M., "Delimitation of the Continental Shelf Between the United Kingdom and France: The Channel Arbitration," *Canadian Yearbook of International Law,* 15 (1977), 173–197.

Melamid, Alexander, "The Division of Narrow Seas," *Political Geography Quarterly,* 5, 1 (Jan. 1986), 39–42.

Molde, Jörgen, "The Status of Ice in International Law," *Nordisk Tidsskrift for International Ret,* 51, 3–4 (1982), 164–178.

Moneta, Carlos J. "Antarctica, Latin America and the International System in the 1980's," *Journal of Interamerican Studies and World Affairs,* 23 (Feb. 1981), 29–68.

Moore, John Norton, "The Regime of Straits and the Third United Nations Conference on the Law of the Sea," *American Journal of International Law,* 74, 1 (Jan. 1980), 77–121.

**N**

Nanda, Ved P., "The Exclusive Economic Zone," *Political Geography Quarterly,* 5, 1 (Jan. 1986), 9–11.

Nandan, Satya, "A Constitution for the Ocean: The 1982 UN Law of the Sea Convention," *Marine Policy Reports,* 1, 1 (1989), 1–12.

Nash, Marian L., "U.S. Maritime Boundaries with Mexico, Cuba, and Venezuela," *American Journal of International Law,* 75, 1 (Jan. 1981), 161–162.

Nelson, L.D.M., "The Roles of Equity in the Delimitation of Maritime Boundaries," *American Journal of International Law,* 84, 4 (Oct. 1990), 837–858.

**O**

*Ocean Development and International Law.* London: Taylor & Francis, published quarterly since 1973.

*Ocean Management.* Amsterdam: Elsevier Scientific Publishing Co., published quarterly since 1974.

*Ocean Yearbook.* Chicago: Univ. of Chicago Press; International Ocean Institute of Malta.

Oda, Shigeru, "Fisheries under the United Nations Convention on the Law of the Sea," *American Journal of International Law,* 77, 4 (Oct. 1983), 739–755.

Oxman, Bernard H., "The Third United Nations Conference on the Law of the Sea: The Eighth Session (1979)," *American Journal of International Law,* 74, 1 (Jan. 1980), 1–47.

———, "The Third United Nations Conference on the Law of the Sea: The Ninth Session (1980)," *American Journal of International Law,* 75, 2 (April 1981), 211–256.

———, "The Third United Nations Conference on the Law of the Sea: The Tenth Session (1981)," *American Journal of International Law,* 76, 1 (Jan. 1982), 1–23.

**P**

Panel discussion on the new political geography of the sea, *Political Geography Quarterly,* 5, 1 (Jan. 1986), 33–38.

Pardo, Arvid, "Who Will Control the Seabed?" *Foreign Affairs,* 47, 1 (Oct. 1968), 123–137.

———, "Sovereignty Under the Sea: The Threat of National Occupation," *Round Table,* 232 (1968), 341–355.

———, "Before and After," *Law and Contemporary Problems,* 46, 2 (Spring 1983), 95–105.

Pardo, Arvid and Richard Young, "The Legal Regime of the Deep Sea Floor," *American Journal of International Law,* 62, 3 (July 1968), 641–653.

Payne, Richard J., "Southern Africa and the Law of the Sea: Economic and Political Implications," *Journal of Southern African Affairs,* 4, 2 (April 1979), 175–186.

Pearcy, G. Etzel, "Geographical Aspects of the Law of the Sea," *Annals, AAG*, 49, 1 (March 1959), 1–23.

———, "Hawaii's Territorial Sea," *Professional Geographer*, 11, 6 (Nov. 1959), 2–6.

———, "The Continental Shelf: Physical Versus Legal Definition," *Canadian Geographer*, 5, 3 (1961), 26–29.

*Polar Regions*. Special issue of *Ambio* (Sweden) 18, 1 (1989), whole volume.

Pounds, Norman J.G., "The Political Geography of the Straits of Gibraltar," *Journal of Geography*, 51, 4 (April 1952), 165–170.

Prescott, John Robert Victor, "Australia's Maritime Claims and the Great Barrier Reef," *Australian Geographical Studies*, 19, 1 (April 1981), 99–106.

———, "Maritime Boundary Agreements: Australia-Indonesia and Australia-Solomon Islands," *Marine Policy Reports*, 1, 1 (1989), 37–46.

**R**

Reisman, W. Michael, "The Regime of Straits and National Security: An Appraisal of International Lawmaking," *American Journal of International Law*, 74, 1 (Jan. 1980), 48–76.

Rhee, Sang-Myon, "Equitable Solutions to the Maritime Boundary Dispute Between the United States and Canada in the Gulf of Maine," *American Journal of International Law*, 75, 3 (July 1981), 590–628.

———, "Sea Boundary Delimitations Between States Before World War II," *American Journal of International Law*, 76, 3 (July 1982), 555–588.

Ricketts, P., "Geography and International Law: The Case of the 1984 Gulf of Maine Boundary Dispute," *Canadian Geographer*, 30 (1986), 194–205.

Rothblatt, Martin A., "Satellite Communication and Spectrum Allocation," *American Journal of International Law*, 76, 1 (Jan. 1982), 56–77.

Roucek, Joseph S., "The Geopolitics of Antarctica and the Falkland Islands," *World Affairs Interpreter*, 22 (April 1951), 44–56.

Ryan, K. W. and M.W.D. White, "The Torres Strait Treaty," *Australian Yearbook of International Law*, 7 (1981).

**S**

*San Diego Law Review*. Annual issues on the Law of the Sea, 1969–1989.

*Sea Changes; the Developing Râegime of the Sea in Law and State Practice*. Cape Town: Univ. of Cape Town, Institute of Marine Law, published semiannually.

Silverstein, Andrew L., "Okinotorishima: Artificial Preservation or a Speck of Sovereignty," *Brooklyn Journal of International Law*, 16, 2 (1990), 409–432.

Smith, Brian, "Innocent Passage as a Rule of Decision: Navigation v. Environmental Protection," *Columbia Journal of Transnational Law*, 21, 1 (1982), 49–102.

Smith, Hance O., "The Role of the Sea in the Political Geography of Scotland," *Scottish Geographical Magazine*, 100, 3 (Dec. 1984), 138–150.

Song, Yann-Huei, "The British 150-Mile Fishery Conservation and Management Zone Around the Falkland (Malvinas) Islands," *Political Geography Quarterly*, 7, 2 (April 1988), 183–196.

Sorensen, Christina, "Drug Trafficking on the High Seas," *Emory International Law Review*, 4 (1990), 207.

Soroos, Marvin S., "The International Commons: A Historical Perspective," *Environmental Review*, 12, 1 (Spring 1988), 1–22.

*Space Policy*, London: Butterworths. Published quarterly since 1985.

Stevenson, John R. and Bernard H. Oxman, "The Future of the United Nations Convention on the Law of the Sea," *American Journal of International Law*, 88, 3 (July 1994), 488–499.

Symmons, Clive R., "Maritime Boundary Disputes in the Irish Sea and Northeast Atlantic Ocean," *Marine Policy Reports* (Univ. of Delaware), 9, 1 (Sept. 1986), 1–7.

Symonides, Janusz, "Geographically Disadvantaged States and the New Law of the Sea," *Polish Yearbook of International Law*, 8 (1976), 55–73.

———, "Delimitation of Maritime Areas Between States with Opposite or Adjacent Coasts," *Polish Yearbook of International Law*, 13, (1984), 19–46.

Symposium: "The International Legal Regime for Antarctica," *Cornell International Law Journal*, 19, 2 (Summer 1986), eight articles.

**T**

Taijudo, Kanae, "Japan and the Problems of Sovereignty over the Polar Regions," *Japanese Annual of International Law*, 3 (1959), 12–17.

Taitt, Branford M., "The Exclusive Economic Zone: A Caribbean Perspective," *West Indian Law Journal*, 7 (May 1983), 36–55 and 8 (May 1984), 26–44.

Tangsubkul, Phiphat and Frances Lai Fung-Wai, "The New Law of the Sea and Development in Southeast Asia," *Asian Survey*, 23, 7 (July 1983), 858–878.

Toma, Peter A., "The Soviet Attitude Toward the Acquisition of Territorial Sovereignty in Antarctica," *American Journal of International Law*, 50 (July 1956), 611–626.

Treves, Tullio, "Military Installations, Structures, and Devices on the Seabed," *American Journal of International Law*, 74, 4 (Oct. 1980), 808–857.

Triggs, Gillian, "The Antarctic Treaty Regime: A Workable Comprise or a 'Purgatory of Ambiguity'?" *Case Western Reserve Journal of International Law*, 17, 2 (Spring 1985), 195–228.

**V**

Van Dyke, Jon and Susan Heftel, "Tuna Management in the Pacific: An Analysis of the South Pacific Forum Fisheries Agency," *Univ. of Hawaii Law Review*, 3, 1 (1981), 1–65.

Vidas, Davor, "Antarctic Tourism: A Challenge to the Legitimacy of the Antarctic Treaty System?," *German Yearbook of International Law*, 36 (1993), 187–224.

**W**

Wang, Erik B., *Canada-United States Fisheries and Maritime Boundary Negotiations: Diplomacy in Deep Waters. Behind the Headlines* (Toronto), 38, 6 (1981); 39, 1 (1981).

Ward, Veronica, "Regime Norms as 'Implicit' Third Parties; Explaining the Anglo-Argentine Relationship," *Review of International Studies*, 17 (April 1991), 167–192.

Winseck, Dwayne and Marlene Cuthbert, "Space WARC: A New Regulatory Environment for Communication Satellites?" *Gazette*, 47, 3 (1991), 195–203.

**Y**

Yanagida, Joy A., "The Pacific Salmon Treaty," *American Journal of International Law*, 81, 3 (July 1987), 577–592.

**Z**

Zumberge, James H., "Mineral Resources and Geopolitics in Antarctica," *American Scientist*, 67 (Jan. 1979), 68–77.

# PART EIGHT

## *The Political Geography of Everyday Life*

The Friendship Bridge connecting Thailand and Laos.

# 35

# *The Politics of Religion, Language, and Ethnic Diversity*

In our discussion of the term *nation* we observed that it properly refers to a group with a common culture, of which two major components are religion and language. Much has and will continue to be written about these two cultural elements, but we are concerned here with their political aspects and relationships and in particular those that are linked with geography. One problem we face is trying to separate the two. There is certainly no necessary correlation or linkage between religion and language, yet many nations (or peoples or ethnic groups) have both distinctive religions *and* languages. Thus an ethnic minority in a country may demand cultural autonomy that would include both or may simply be distinguishable from other groups because of both. The Québécois, for example, are distinguished from their fellow Canadians not only by being predominantly French-speaking, but also by being overwhelmingly Roman Catholic in an otherwise largely Protestant country, although language has been the basis of the separation movement. Thus, in this chapter, as we refer to an ethnic group (or people or nation), we may be referring to linguistic or religious factors or both.*

## *Religion*

No longer, as in centuries past, is religion a dominant force in the determination of State boundaries or even in the creation of States. In the twentieth century the partitions of Ireland, Syria, India, and Palestine were all designed to create separate political units for Catholics and Protestants, Christians and Muslims, Muslims and Hindus, and Jews and Arabs, respectively. In each case, however, while the political map of the world was altered and the course of history changed accordingly, minority groups were included in the new units, and some of them have continued to pose problems of integration into the national system. These partitions, furthermore, led to the creation of only four new States out of more than a hundred born in this century. And religion was an insignificant factor in the breakup of the USSR and Yugoslavia and the reunification of Germany. It was, however, as we point out in Chapter 17, an instrument of Tsarist Russian imperialism, and missionaries were vitally important in establishing all of the great overseas empires.

No longer are new religions sweeping across the face of the earth, gaining adher-

---

*The term ethnic is used loosely to include inherited physical features that some people imply in the term "race." While physical characteristics may be important in the identification of some groups, including self-identification, there is no necessary connection between race and either language or religion. We therefore ignore "racial" factors here. The species *homo sapiens,* in fact, constitutes a single biological race and the common concept of "race" is purely cultural and not biological.

**Baha'i in Nicaragua.** One of the world's fastest-growing religions, Baha'i flourished in Nicaragua even under the leftist government of the Sandinistas (1979–1990) and symbolizes the erosion of Catholic domination in Latin America. This billboard, photographed outside Managua in 1988, features a quotation of Bahâ'u'llâh, the great 19th century leader of the faith. (Martin Glassner)

ents by the millions, often under threat of the sword. The only religions gaining significant numbers of new adherents today are Islam, primarily in the northern parts of East and West Africa, and Christianity, in much of Africa. There is also a steady growth of a number of Protestant and other denominations in nominally Roman Catholic Latin America and of the Baha'i faith worldwide. Generally speaking, however, we are in a period of relative religious stability, in which the major changes are taking place *within* religions.

Although the religious wars that characterized Europe for centuries are fortunately behind us and we are living in an era of religious toleration, there has been no shortage of local religious wars and domestic conflicts in which religion has played a significant role. In Northern Ireland, for example, the conflict is not clearly between Catholics and Protestants; that is, it is not a war over religion but rather of different positions and orientations of the various communities within the province. Many members of the large Catholic minority want the province to be reunited with the dominantly Catholic Republic of Ireland, while most of the Protestants want to remain united with Great Britain. Catholics over the centuries have been discriminated against in employment, and even some Protestants feel discriminated against by English control of the Ulster

economy. But while the religious lines in Ulster are not rigidly drawn and economic and ideological factors are important components of the conflict, it is the religious component that attracts attention and tends to polarize people not only in Ulster but also in Britain, Ireland, and elsewhere.

The Nigerian civil war, which lasted for 30 months in the late 1960s, was particularly bloody. Again, the war had many causes, but because the Ibo nation, largely Roman Catholic, felt that it was not being, and could not be, treated fairly in largely Muslim and animist Nigeria, it broke away to form the State of Biafra. The Ibos lost the war and Biafra was obliterated, but it seems that the Ibo nation is being treated more fairly now in reorganized Nigeria.

A similar uprising, but one more drawn out and less well organized, is the continuing one of the Moro National Liberation Front in Mindanao and the Sulu Archipelago against the government of the Philippines. This group represents many of the Muslims who form a majority in five of the 13 provinces of these islands, and they have been fighting for autonomy in a chiefly Christian country.

Almost lost in the avalanche of news and opinion about the war in Vietnam and neighboring countries from the early 1960s to 1975 was the fact that two indigenous religious groups, the Hoa Hao and the Cao Dai, controlled several important provinces. They fig-

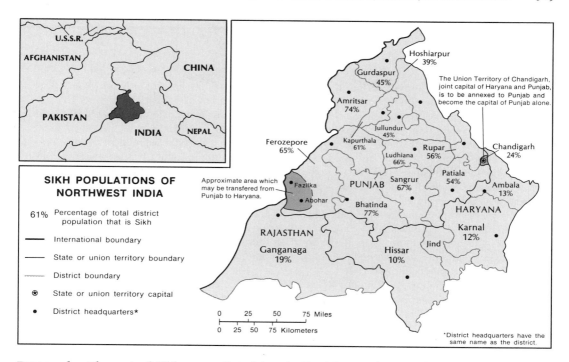

**Proposed settlement of Sikh separatist claims in Punjab.** In July 1985 then Prime Minister Rajiv Gandhi and Sikh leader Sant Harchand Singh Longowal agreed on territorial transfers to settle Sikh claims for a state of their own. However, Longowal and Gandhi were later assassinated, opposition to the agreement is strong among both Hindu and Sikh extremists, and implementation of the agreement has been postponed indefinitely.

ured importantly in the domestic political maneuvering, and they maintained their own armies that at various times fought the Japanese, French, Saigon government, and Viet Cong. While armed and militant religious groups of this type are no longer very common in the world, their quarrels can and sometimes do cause great hardship and suffering. The clearest recent example is Lebanon—once a peaceful, prosperous, cosmopolitan, and reasonably democratic country, but now devastated by years of civil war, fragmented into bitterly hostile factions, and occupied by foreign troops.

In the United States, religious groups are not armed and militant, and they do not oppose one another politically, but there is no doubt that religion influences politics in many areas of the country. The Mormon influence in Utah and neighboring states, the Catholic majorities or near-majorities in southern New England and the Middle Atlantic states, the "Bible Belt" of fundamen-

talist Protestants in the South, the concentration of Jews in the major cities, especially in the Northeast and Midwest—these are just a few religious factors to consider when trying to explain or predict electoral behavior in the United States. Religious doctrines, though no longer as important as they once were in determining candidates for public office and the outcome of elections, are still influential in referenda and in legislative action on certain issues.

Religious minorities often play roles, make contributions, and exercise influence in societies far out of proportion to their numbers. This includes both indigenous and immigrant minorities. Among them are Jews, Parsees, Mennonites, Catholic Ibos, Mormons, Quakers, Armenians, Sikhs, and others. While they tend to focus on the economic sphere and are sometimes excluded from politics, these groups are always political factors, especially in areas where they are concentrated. In some countries, espe-

## CASE STUDY–LEBANON ˙χ

Although its roots go back to ancient Phoenicia, Lebanon under the Ottoman Empire was simply Mount Lebanon in Syria. When the French gained control of Syria after World War I, they added to Mount Lebanon the coastal strip and the Bekaa Valley, and on 1 September 1920, proclaimed there the State of Greater Lebanon—still under French mandate. The people of Mount Lebanon were predominantly Christian (mostly Maronites) and pro-French; the other areas, mainly populated by Muslims, were included within Lebanon to enhance the new country's political and economic position, and the total population had only a slight Christian majority. Tensions between Christians and Muslims grew until 1943 when they formulated the ``National Pact.'' This provided for proportional representation by religious community in the Chamber of Deputies, as shown in Table 35-1. It was also agreed to continue the tradition of a Maronite president, Sunni prime minister, and Shiite speaker of the Chamber of Deputies. Informally, cabinet posts, army and civil service jobs, and other positions were allocated according to religious affiliation, but not necessarily proportionately. This precarious arrangement was upset in 1958 by a number of internal and external factors, and it completely disintegrated after 1975, when civil war broke out and lasted until 1991.

During these 16 years, fighting raged among numerous private militias maintained by religious and political groups, and assorted local warlords and clan chiefs. Most of the fighting was among various factions of Christians, Muslims, and Druzes, with alliances changing kaleidoscopically. The war was also notable for the introduction of two Western multinational forces and an Arab

**Table 35–1**   *Lebanon: Estimated Proportions of Religious Groups in Total Population and Allocation of Seats in the Chamber of Deputies*

| Religious Groups | Percentage (1970)* | Deputies (1970) | Estimated Percentage of Population (1980) |
|---|---|---|---|
| Christians (54%) | | | Christians (40%) |
|   Catholics | 38% | 38 | |
|     Maronites | 75% | 30 | 23 |
|     Greek Catholics | 20 | 6 | 5 |
|     Armenian Catholics | 2.0 | 1 | |
|     Roman Catholics | 1.5 | 1 | |
|     Syriacs | 1.0 | 0 | |
|     Chaldeans | 0.5 | 0 | |
|   Greek Orthodox | 11 | 11 | 7 |
|   Armenian Orthodox | 4 | 4 | All other |
|   Protestants | 1 | 1 | Christians 5 |
| Muslims (39%) | | | Muslims (53%) |
|   Sunni Muslims | 20 | 20 | 26 |
|   Shiite Muslims | 19 | 19 | 27 |
| Others (7%) | | | Others (7%) |
|   Druzes | 6.0 | 6 | 7 |
|   Jews | 0.5 | 0 | |
|   'Alawites | 0.5 | 0 | |
| | 100% | 99 deputies | |

*Embedded within these figures are perhaps 1 percent of the total population who are Ismaili Muslims, Syrian Orthodox, Baha'i, and other groups too small to be represented as such in the Chamber of Deputies. Druzes and 'Alawites are sometimes considered as Muslims whether or not they are aligned politically with Muslims. There has been no census in Lebanon since 1932, and all population figures, including those of the 1932 census, are very likely wrong, perhaps by a considerable margin. Nevertheless, the political system was based on this enumeration until June 1991, when a new provisional constitution went into effect.

League force composed almost entirely of Syrian troops, an Israeli incursion in 1982 that went all the way to Beirut and precipitated the expulsion of the Palestine Liberation Organization (PLO), the creation of an Israeli "security zone" along the southern border protected by a largely Christian militia armed and trained by Israel, the rise to prominence of at least two fundamentalist Shiite factions (complete with their own militias), the return of the PLO, the occupation of nearly all of the country by Syria, the insertion of a United Nations peace-keeping force (UNIFIL) into the southern border region, large-scale flight of Lebanese to safer countries, intervals of relative calm, hostage-taking by many of the factions (including some Western victims), massive physical destruction and human suffering, and many other events and situations, each of which could be the subject of a book. Many are of politicogeographical interest and are discussed in earlier chapters of this book. Now we are concerned with the outcome of the war.

In the autumn of 1989 Lebanese deputies representing most factions met in Taif, Saudi Arabia, and hammered out the National Reconciliation Charter. This document became effective in June 1991 and serves as a temporary constitution, replacing the National Pact of 1943. It is a lengthy document, covering political reforms, disbanding of the militias and reconstitution of the national armed forces, repositioning of Syrian troops and Lebanese-Syrian relations, and "the liberation of Lebanon from Israeli occupation." Briefly, the main provisions for our purposes are:

1. As an interim measure, "until Parliament enacts an election law which is not based on religious affiliation records," seats in Parliament will be allocated much as before except that the Muslims get 9 additional seats for a total of 54 (including Druzes), to equal the number of Christian seats. *proportional represent.*

2. The president will continue to be a Maronite and the prime minister a Sunni, at least *ad interim,* but the powers of the president are reduced while those of the prime minister are increased.

3. "Political deconfessionalization is a principal national objective. . . . "

4. Some authority is devolved upon the administrative regions, but the country remains "one unified State with a powerful centralized authority."

Although the wording of the document is frequently very vague and even hortatory, it does express the national aspiration for enduring peace based on elimination of religious (or confessional) affiliation from the political system. We can hope that this will be accomplished soon; if not, resumption of tribal warfare is all but inevitable.

cially in Europe, Asia, and Israel, there are political parties formed by, and representing, particular religious groups. The Christian Democratic parties of Europe and Latin America are generally associated, however loosely, with the Roman Catholic Church, although the influence of the Vatican on them is often exaggerated by their opponents. The National Religious (Jewish) Party in Israel has been a coalition partner in every government since the founding of the State in 1948, and there are other Jewish and Muslim parties as well. The Komeito in Japan and the Jan Sangh in India are similarly parties representing the orthodox wings of major religions, Buddhism and Hinduism, respectively.

Although religious groups are active in politics around the world, there is probably no true theocracy. The last one was eliminated when the Chinese occupied Tibet in the 1950s, drove the Dalai Lama (the spiritual and temporal ruler) into exile, and destroyed the Lamaist Buddhist monasteries that were an integral part of the governing system. The nearest thing to a theocracy today is probably Iran, in which not only does the Shari'a (Islamic law) dominate the society, but all major government posts are held by clerics. In Saudi Arabia, the puritanical Wahhabi sect is still influential and the government takes seriously its role as protector of the holiest places of Islam, but it is still a secular State. Similarly, Libya, Sudan and Pakistan—other countries in which Islamic law has been incorporated into the civil and criminal codes—continue to be governed by laymen and remain secular States.* Other countries have established churches that often receive some degree of financial and other support from the government, but the churches do not control the governments, and generally religious freedom is granted to other denominations. In Japan the official religion until the end of World War II was Shinto, a nationalistic, militaristic indigenous religion, but it was disestablished under the American occupation after the war and the emperor renounced the doctrine of his divine descent.

Finally, we should note the numerous religious communal settlements around the world. By their very nature, they are generally uninvolved in politics, sometimes even refusing to acknowledge citizenship in the host State. Nevertheless questions of land ownership, water rights, education, taxation, military service, and so on often involve them in politics willy-nilly. Cultural geographers have long been interested in these settlements, but they deserve the attention of political geographers as well.

What pattern emerges from all these observations about the politics of religion today? Generally, it is a picture of a world in which religion no longer dominates the lives of people as it once did. Despite survivals of past orthodoxies and occasional revivals of religious feeling and activity (such as in Islam today), we are living in a secularizing world in which religion remains important to many individuals and is locally important politically, but is no longer a force that shapes history. Religious wars and massacres still break out from time to time in some countries, but seldom do these spill over into other countries and not for a long time have they generated massive crusades. Nationalism, it would seem, is a more powerful influence on most people in the world than religion, though it will not necessarily remain so. In fact, there are already growing

strength and passions among militant fundamentalist groups within several religions.

## Language

Throughout this book we have repeatedly referred to the strength of nationalism in the world today. One of the most important components of nationalism has long been language. The European nation-state evolved out of a desire of people to be ruled by people who spoke their language; that is, who were of the same nation. Other cultural components, of course, go into the blend of symbols and feelings and attitudes that bind together a nation, but historically, few even approach language as the most important single element.

Ethnic minorities within many countries cling to their native languages in the face of cultural imperialism of the majority or even the obsolescence of the language in the face of technological change. It is a badge of identification, of belonging to a distinctive group with a proud tradition of its own. Cornish has died out, but Gaelic (both Scottish and Irish varieties), Welsh, and Breton of the old Celtic languages are preserved still and used by ardent nationalists as touchstones of their movements for greater autonomy or even independence. In France, Italy, The Netherlands and other European countries, formerly dying and even proscribed regional languages, such as Occitan (Provençal), Friulian, and Frisian, are being revived, officially encouraged, and even financed by a bureau of the European Union. The political significance of this trend is immense: As Europe slowly unites and English spreads as its *lingua franca*, State governments have come to understand that local cultural diversity can be valuable in the face of economic and eventual political unity.

Of all the ancient languages, however, only Hebrew has been revived in modern times, updated, and firmly established as the national language of a modern State, in everyday use by the great majority of the population of Israel. One reason for the survival of Hebrew through nearly 2000 years of the dispersion of its users throughout the

---

*Vatican City may be considered a theocracy since the Pope is both spiritual and temporal ruler and all officials are clergymen, but this is a very special case.

**Religious minorities far from their places of origin.** Top: These Hutterite women hoeing a field in Alberta are among the some 14,000 members of the Hutterian Brethren living in about 115 communal farm settlements in the Canadian prairie provinces, the Dakotas, Montana, and Washington. This Mennonite sect originated in Austria in 1528, fled to Germany after persecution in Switzerland, then to Bohemia, to Hungary, and eventually to Russia in the eighteenth century. About 400 members of the sect came from Russia to South Dakota in 1874–77 to begin the North American settlements. (Photo courtesy of the National Film Board of Canada) Bottom: Rihaniya is a Circassian village in Upper Galilee, Israel, one of two in the country. These people are Sunni Muslims from the Caucasus who fled Russian persecution in the nineteenth century to settle in Turkish Palestine. They have nearly abandoned their own language in favor of Arabic, but retain their religion and pastoral farming economy. (Martin Glassner) Both of these minority peoples live in peace within the dominant cultures of their refuges, but others around the world are not so fortunate.

world was its retention as a liturgical language. Other liturgical languages, such as Latin, have also served as unifying elements among peoples who share a religion but little else. No other, however, has retained the almost mystical loyalty of a people for so long as Hebrew among the Jewish people. It never really died out as an everyday language in its homeland either, as clusters of Jews have always lived there, using the ancient language regularly. When they were joined by East European Zionists beginning in the nineteenth century, the newcomers adopted the traditional tongue. By 1948, when the State of Israel was born, the debate over a national language among the 600,000 Jews of Palestine had been settled: Hebrew had won out over Yiddish, English, and other candidates. It was a symbol of the rebirth of the ancient Jewish State, a matter of enormous political significance.

The case of Namibia, another former League of Nations mandate, is different. There, the leaders decided to abandon Afrikaans as the official language when the country became independent in 1990, and rejected all other native languages as a substitute. They chose English, mother tongue of only 5 percent of the population, as the new official language. Afrikaans, however, is likely to continue to be the *lingua franca* of the country for a time, and German will continue to be used as well as the local languages. In South Africa, a very different policy was adopted in May 1994, when a new democratic government took office under a new constitution. The former official languages, Afrikaans and English, are now only 2 of 11 official languages, though, of course, still the most widely used.

The multitude of languages in the world and the tendency of all languages to change through time have created a need for *linguae francae*, languages used in trade and general communication among peoples who speak other languages of their own. Some of these evolved naturally and have been accepted by the people as, at least in part, their own. Examples are Swahili in East Africa, Hausa in West Africa, Pidgin in the South Pacific, and Tupí-Guaraní in Brazil

and nearby areas. Others, however, have been introduced by conquerors and either adopted willingly by natives wishing to advance under the new rulers or imposed by the rulers as instruments of cultural imperialism. Quechua, Arabic, French, Latin, Spanish, and English are only a few of the languages spread far and wide in this manner. By now there are literally hundreds of pidgin and creole languages, derivations from or amalgamations of established languages, and many more have become extinct.

Since World War II English has replaced French as the nearest approximation of a worldwide *lingua franca,* not because of any inherent virtues it might have, but because the United States emerged from the war as the world's most powerful country, with its troops, businesspeople, teachers, scholars, scientists, and tourists penetrating the farthest reaches of the globe—and often unwilling to learn the local language. American influence, added to British influence within the Commonwealth and beyond, has made English the world's most important language—for the present. Already Spanish has replaced French as the most popular foreign language for Americans to learn, and the Spanish-speaking population of the United States is growing rapidly. Some observers estimate that early in the twenty-first century the United States will be a bilingual country, which should not be a disturbing prospect at all. It could lead, in fact, to closer relations with the Spanish-speaking world.*

Esperanto, the most successful of many invented languages, was devised by Lazarus Ludwig Zamenhof (1859–1917), a Jewish physician and linguist of Russian Poland, in the hope that it would become a worldwide, nonpolitical, value-free *lingua franca.* Only 20 years after he published the first tract on his new language, two "reformed" versions of it were developed in France. Since then it has indeed spread around the world, and

*Politically motivated or fear-induced movements in the United States favoring "Black English" and "English first" and their variations have fortunately not won widespread support for they are truly divisive. Bilingualism, on the other hand, is a unifying force.

**Survival of an ancient language.** Around the world many linguistic minorities are trying to preserve their languages in the face of enormous pressures to abandon them in favor of the dominant languages of their countries. Use of the Manx language, for example, has declined drastically in the past half century but survives as a badge of distinctiveness on the Isle of Man, a dependency of the British crown. This bilingual sign is an example. (Photo courtesy of the British Tourist Authority)

there are numerous publications, including periodicals, in the language. But it has never attracted more than perhaps 3 million speakers altogether, and today there are probably no more than half a million, mostly intellectuals. In the United States it was for a time identified by ultraconservatives as an instrument of "one worldism" propagated by communists, and therefore anathema. Politics aside, it has no chance of replacing a major world language, such as English, as a preferred second language or *lingua franca.*

A *bilingual* country is one in which two languages are dominant rather than one. Examples are Belgium (French and Flemish), Canada (English and French), Sri Lanka (Sinhalese and Tamil), and Paraguay (Spanish and Guaraní). In a recent case close to home, the governor of Puerto Rico on 5 April 1991 signed into law a bill that disestablished English, which had been imposed as an official language in 1902, leaving Spanish as the sole official language of the commonwealth. This was by no means an anti-American act, however, only a recognition of an existing fact, and English would continue to be widely used with no penalties. Puerto Rico remains, moreover, an integral part of the United States. Pro-statehood forces, in fact, were successful in 1993 in having English reestablished as an official language, and in a 1993 plebiscite,

only 4.4% of the votes were cast for independence.

In nearly all bilingual countries, other languages are also spoken, but generally by small numbers of people. In a great many countries, the overwhelming majority, in fact, many languages are spoken by the native peoples. In most of these *multilingual* countries, one or two languages have achieved dominance, either because they are spoken by the largest groups of the population or because they are the languages of the ruling elites. Sometimes these dominant languages obscure the basic multilingual character of the country. Mexico, Nigeria, South Africa, India, China, and Russia are only a few of the more prominent multilingual States.

In some bilingual and multilingual States, attempts to introduce (or impose) an official "national" language or to suppress a local language have led to vigorous protests, bloody riots, and even full-scale rebellion or attempted secession, as in India, Sri Lanka, Belgium, Spain, and South Africa. In most of them, however, a *modus vivendi* has been reached among the various language groups, and the selection of a national language or languages has been accomplished peacefully.

In some countries, such as Norway and Britain, social groups are identified by the

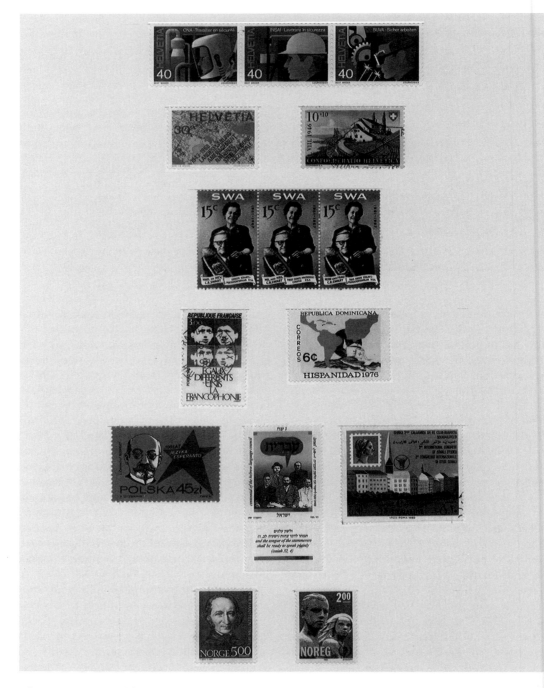

**The importance of language in both domestic and international politics.**    Many countries issue stamps with inscriptions in two languages, since many are bilingual. Switzerland, however, has issued sets of three stamps in three of the four official languages, all three languages on one stamp, and the country's official name in Latin. More recently, Swiss stamps handle the language problem by just saying "Helvetia." Southwest Africa (now Namibia) issued two *se tenant* sets of stamps in 1967 in Afrikanns, English, and German. In the fourth row are stamps touting the international binding force of the French and Spanish languages. In the next row a Polish stamp honors the Polish inventor of Esperanto, an Israeli stamp commemorates the centennial of the Hebrew Language Council (all Israeli stamps have the country's dialect of the national language they speak. In others, regional dialects label people and sometimes restrict their participation in political life. In still other societies secret or ritual languages confer power or prestige on the few initiates. In others, mastery of a classical or literary language or dialect is essential for advancement in society.

On the other hand, Jean-Bertrand Aristide was elected President of Haiti in January 1991 in the first democratic elections in the country's history, supervised by the United Nations, largely because he campaigned among the masses in Creole, their native language, while the other candidates used only French, the language of the country's tiny elite. And, in Taiwan, where the Nationalist government that fled the mainland of China in 1949 had imposed Mandarin as the official language and banned all use of Taiwanese, the local language has not only become respectable again, but it is now seen as essential for success in the country's politics. This is a reflection of both the democratization of Taiwan and the growing movement for official and permanent, not just *de facto*, independence form China.

Nearly everywhere in the world language has political as well as cultural significance. But, if language is such an important component of nationalism and if nationalism is as powerful a force in the world today as we have indicated earlier, why do the political boundaries of the world not conform to linguistic boundaries? Part of the explanation lies in the mobility of people. Throughout history peoples have migrated, often over considerable distances, carrying their cultural baggage with them. Although they may have returned to their homelands or amalgamated with the local people, their languages—or traces of them—remain to distinguish regions that otherwise may not

differ greatly. Language then becomes only one element, and sometimes a subordinate element, in the culture. Even culture as a whole may not determine boundaries. As we have seen, military, economic, dynastic, physiographic, and other factors are often determinative in boundary making, and language may be ignored. In twentieth-century Asia and Africa, anticolonialism has often been the strongest component of nationalism, a rallying cry that unites most people in a colony, regardless of the language they speak. And, as we have also seen, most colonial boundaries have remained intact after independence as a matter of policy, even though they frequently cut across linguistic lines.*

### Ethnic Minorities and Polycultural States

There are few true nation-states in the world today. Most countries are home for two, three, or numerous cultural or ethnic groups. It is not just the largest countries, such as the former Soviet Union or China, that are polycultural or plural societies, but even many of the smallest, such as Mauritius, Fiji, Comoros, and Trinidad and Tobago. Cultural pluralism, in fact, is not only worldwide but also a prominent element in political unrest nearly everywhere.

Nigeria and Lebanon are not the only countries in which a census is more than just a useful source of routine statistical information but a sensitive political issue as well. In 1976, for example, the Slovenes of Carinthia

---

*In the late 1980s the Organization of African Unity was actively promoting use of indigenous languages in place of colonial ones, under the rubric of "the linguistic liberation and unity of Africa."

---

name in Hebrew, English and Arabic), and a Somali stamp notes—in four languages—the Second International Congress of Somali Studies. Finally, Norway has traditionally used the country's name in the Riksmal (Bokmal) version of Norwegian, which is largely Danish and spoken in the urban centers and in the eastern part of the country. Recently, as on the stamp on the right, the Landsmal (Nynorsk) version, invented in 1850 and used in Western Norway and in rural regions, has been used on some stamps. (Martin Glassner)

Legend:

- (P) Polish communities demanding the creation of nationality-based administrative units in Belorussian SSR and Lithuanian SSR
- (G) Communities of deported Germans seeking to re-establish the former Volga German ASSR
- (T) Communities of deported Crimean Tatars seeking to re-establish the former Crimean Tatar ASSR
- (M) Communities of deported Meshketian Turks seeking to return to Georgia SSR

Area where claims require adjustment in borders only

Approximate areas in dispute as reported in the Soviet press before March 1990

Map labels: Claim from Karelian ASSR for portion of Murmansk Oblast (Kola Peninsula); Murmansk Oblast; Karel'skaya ASSR; Claim from Lithuanian SSR for portion of Kaliningrad Oblast; KALININGRAD OBLAST (RSFSR); Claim from Belorussian SSR for southern portion of Lithuanian SSR; Claim by Lithuania for portion of Belorussian SSR; Claim from Ukrainian SSR for northern portion of Moldavian SSR; Demand from Gagauz communities for creation of autonomous region in southern Moldavian SSR; Slavic majority northeast of the Dniestr River demands session from Moldova; Former Crimean Tatar ASSR; Crimea demands secession from Ukraine; Former Volga German ASSR; Rostov Oblast; RUSSIAN SOVIET FEDERATIVE SOCIALIST REPUBLIC (RSFSR) (RUSSIAN FEDERATED REPUBLIC); Chuvashkaya ASSR; Mordovskaya ASSR; Tatarskaya ASSR; Claim from Baskir ASSR for Tatar ASSR; Claim from Tatar ASSR for Bashkir ASSR; Bashkirskaya ASSR; Claims from Kalmyk ASSR for portions of Rostov Oblast and Stavropol Kray; Kalmytskaya ASSR; Stavropol'skiy Kray; See Inset map; Claim from Kazakh SSR for Karakalpak ASSR; Karakalpakskaya ASSR; Claim from Tajik SSR for southern portion of Uzbek SSR; TURKMEN SSR (TURKMENISTAN); UZBEK SSR (UZBEKISTAN); KAZAKH SSR (KAZAKHSTAN); Lake Balkhash; Claims from Tuvin ASSR for portions of Gorno-Altay AO and Krasnoyarsk Kray; Gorno-Altayskaya AO; Claim from Kazakh SSR for undetermined area of Kirghiz SSR; KIRGHIZ SSR (KYRGYZSTAN); TAJIK SSR (TAJIKISTAN); AFGHANISTAN; CHINA

Boundaries:
- International
- Union republic (SSR)
- Autonomous republic (ASSR), oblast, or kray
- Autonomous oblast (AO) or autonomous okrug (AOk)

**SOURCE:** Adapted from map by Department of State, 1990

**Ethnicity and political boundaries in the former Soviet Union.** Although each of the 15 constituent republics of the former Union of Soviet Socialist Republics is now an independent State, there is no reason to believe that the fragmentation of this huge and ethnically complex territory has ended. This map shows only the more important of the numerous claims and demands for territorial changes on ethnic grounds.

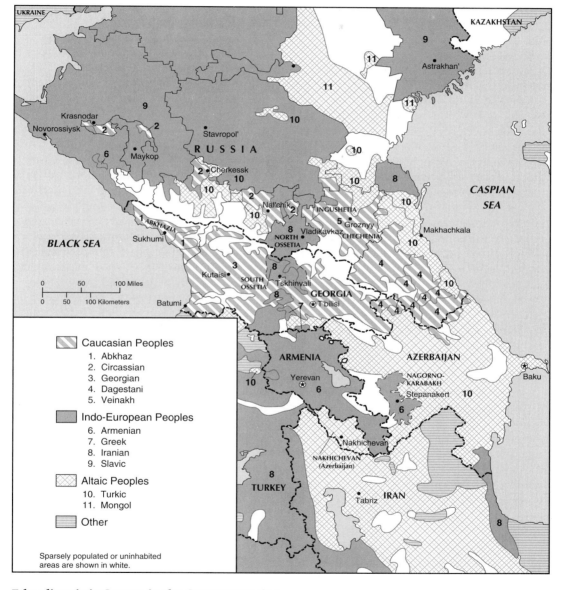

**Ethnolinguistic Groups in the Caucasus Region.**

in Austria, supported by Yugoslavia and opposed by German-speaking Austrian nationalists, vigorously protested a proposed ethnic census in their province. And Saudi Arabia has yet to conduct a formal census and publish the figures, at least in part because it would inevitably show that the native population is quite small and the number of foreigners, mostly Palestinians and other Arabs, is large.

But cultural pluralism need not lead to political instability. In few countries, in fact, are the ethnic groups as bitterly hostile as in Cyprus or Lebanon or Sudan. In most, ethnic differences have been at least temporarily sublimated or suppressed by nationalist or repressive governments. Where the harmony is only superficial and ethnic animosities simmer close to the surface, we may witness in the near future more outbreaks of communal fighting based on racial, religious, linguistic, tribal, or other cultural rivalries. But if such outbreaks can be postponed long enough for a true sense of national identity to develop, for people to be identified and judged as individuals and as members of a larger society instead of as members of a particular ethnic group, we can find a gradual reduction in intercommunal tensions and thus in civil and international strife.

As pointed out in Chapter 18, however, tribal warfare broke out in the former Soviet Union and in Yugoslavia as repression in these countries, particularly the former, was relaxed and rickety economies collapsed during the 1989–92 period. Similar conditions prevail throughout Central and Eastern Europe, Southeast Asia (including the islands), and the Middle East. All of these areas face similar ethnic explosions unless political, economic, and social democracy can be established to an acceptable degree before the pressures build up to explosive force. Otherwise, all of these regions will continue to be shatterbelts, and the Balkans may once again become "the tinderbox of Europe."

Perhaps the world's best current example of a plural society that has welded itself into a nation-state is Switzerland. With four major languages and two major religions, it could be a land of strife, but it is not. There are a number of reasons for this.

First, there is no correlation between language, religion, place of residence, and socioeconomic status. German- and French-speaking Catholics, for example, live in both cities and countryside and are found at all socioeconomic levels. Thus, people identify with one another in a number of ways depending on the situation, but always as Swiss, clearly distinct from those across the borders who might share their language or religion.

Second, the peoples of Switzerland came together (usually voluntarily) over a long period of time, from 1291 to 1815, partly to protect themselves against more powerful countries surrounding them and partly to share the advantages of a central location in Europe. In order to preserve these advantages, and to avoid taking sides in outside conflicts that might prove internally divisive, Switzerland has since 1815 maintained a stance of permanent, armed neutrality.

Third, the country is not just a federation, but a confederation, in which the cantons retain a large measure of authority, both by exercising primary responsibility in certain areas and by executing many federal laws. Since religion and language often cut across cantonal lines, many cantons have to reach compromises in most political matters. Finally, Switzerland, more than any other country in the world, practices direct democracy, involving frequent use of the initiative and referendum (even permitting referenda on certain types of treaties), so the likelihood of any ethnic group imposing anything on any other is very small.*

We are not implying here that Switzerland is a model to be emulated by other plural societies. Switzerland is the product of unique circumstances and of 700 years of

*Although the Swiss are generally quite conservative, their system does have some flexibility. In 1975, the French-speaking population of predominantly German Berne canton won the right to secede from Berne, and in 1978 the new canton of Jura was admitted into the confederation.

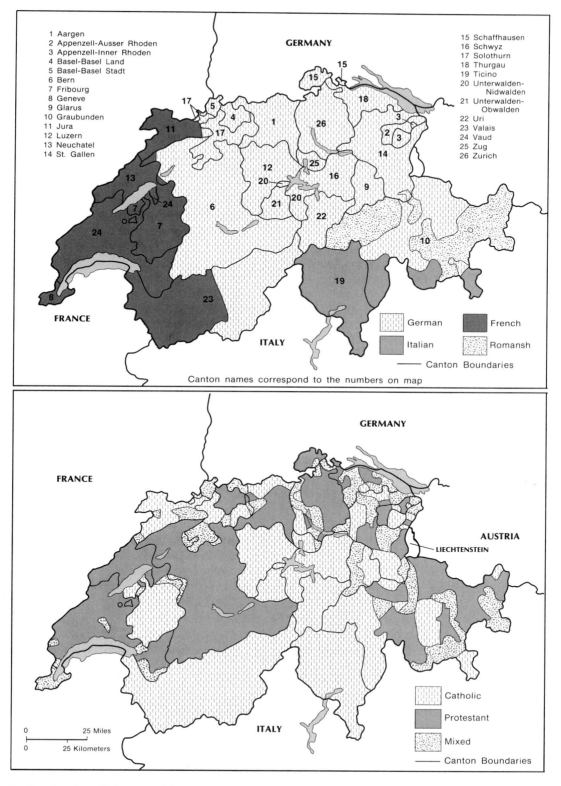

1 Aargen
2 Appenzell-Ausser Rhoden
3 Appenzell-Inner Rhoden
4 Basel-Basel Land
5 Basel-Basel Stadt
6 Bern
7 Fribourg
8 Geneve
9 Glarus
10 Graubunden
11 Jura
12 Luzern
13 Neuchatel
14 St. Gallen

15 Schaffhausen
16 Schwyz
17 Solothurn
18 Thurgau
19 Ticino
20 Unterwalden-
  Nidwalden
21 Unterwalden-
  Obwalden
22 Uri
23 Valais
24 Vaud
25 Zug
26 Zurich

GERMANY

FRANCE

ITALY

German          French
Italian         Romansh
— Canton Boundaries

Canton names correspond to the numbers on map

GERMANY

FRANCE

AUSTRIA

LIECHTENSTEIN

ITALY

0    25 Miles
0    25 Kilometers

Catholic

Protestant

Mixed

— Canton Boundaries

**Switzerland—religions and languages**

**A symbol of Swiss unity.** One of the centripetal forces binding together the people of Switzerland is compulsory military service for males beginning with basic training at age 20, followed by service in the active reserves until age 50 (55 for officers). This billboard in Geneva carries announcements of the reserve units being called up for training and other duties for specified periods, as well as other notices concerning military service. The Swiss are rightfully proud of their "armed neutrality," regardless of their language or religion. (Martin Glassner)

trial and error. Nevertheless, it is a living example of a country whose people live in harmony despite differences of language and religion. These differences can be tolerated in order to attain mutually beneficial objectives, not only here, but elsewhere as well.

Considering that the politics of religion and language touch the lives of most people in the world at some time or another, it is remarkable how little serious work had been done in this area by political geographers until recently. We could use many more studies, especially for areas outside Europe.

# 36

# *The Politics of Transportation and Communications*

In our discussion of the State as a political entity, we pointed out that one of its characteristics is a circulation system, a system of transportation and communications that permits a flow of goods, people, and ideas within the State and—equally essential today—between the State and other parts of the world. While such a system has, of course, its social, economic, and technological aspects, even its very existence may be political in nature. The great highway systems of the Incas and the Romans were designed and maintained primarily to facilitate administration of their vast empires, while the rulers of early nineteenth century Japan and Paraguay (among many others) deliberately kept their internal circulation systems primitive and their countries isolated from the rest of the world for similar reasons. A decision to build a road from one place to another or to develop a national merchant fleet or restrict the ownership of shortwave radios is frequently a political decision. Circulation, therefore, is of great interest to the political geographer.

## *Transportation*

Simple paths or tracks served as the principal transportation arteries for the great majority of the world's people until quite recently. They still lace together farms, homes, and settlements in much of the world. Gradually, however, they are being replaced by streets, roads, and highways. The result is increased mobility and expanded horizons for individuals and greater economic, cultural, and political unity for societies. Even warfare has changed as new vehicles traveling over relatively flexible road systems have given armies greater mobility and adaptability than railroads did earlier.

## *Roads*

Nearly everywhere road building and maintenance are the responsibility of government, generally with various levels of government responsible for specific types of roads. Every phase of the process, from the decision to build a road or a network of roads to the allocation of funds for repairing and maintaining roads, has a political component. New superhighways built in urban areas of the United States, for example, are generally routed through slums or working-class areas, disrupting the communities and displacing many people, rather than through the much less densely populated wealthy suburbs, largely because of the distribution of political power. In hundreds of towns in the southeastern United States, streets in sections inhabited by blacks or what used to be called "poor white trash" are still unpaved and essentially unserviced. The Interstate and Defense Highway System, begun in the 1950s and only completed in 1991, is, like the *autobahnen* of Hitler's Germany, designed for tanks, heavy artillery, and other military traffic, though of course civilians benefit from it also.

Roads through mountain passes have special significance around the world. Sometimes they open up new areas for settlement or connect important population centers or provide invasion routes. Khyber Pass, Brenner Pass, Cumberland Gap, Mitla Pass, Dzungarian Gates, Cilician Gates, South Pass—the names ring through history, recalling epic struggles between people and nature and among peoples. A highway connecting China and Pakistan through some of the most rugged and forbidding terrain in the world, opened in 1971 after six years of construction, closely follows an ancient caravan route, but its strategic significance is quite modern.

Roads of continental dimensions have also had enormous political significance ever since the old Silk Road connected the Chinese and Roman empires. Today the Asian Highway reaches from Istanbul to Singapore and Ho Chi Minh City. Even longer is the Pan-American Highway, linking Fairbanks, Alaska with Puerto Montt, Chile, and Buenos Aires, Argentina.* Another international highway is the Carretera Marginal Bolivariana de la Selva (commonly known as "La Marginal") running along the eastern foothills of the Andes through Colombia, Peru, and Bolivia. Designed to link the rivers and roads running east and west through the Andes and to stimulate settlement of the region, its construction slowed considerably after the overthrow by the army of its principal promoter, President Fernando Belaunde Terry of Peru. There is no comparable highway yet in Africa, reflecting the

political fragmentation of the continent as well as physiographic, demographic, and economic factors. Some are being planned and constructed, however, coordinated by the staff of the UN Second Transportation and Communications Decade for Africa (1991–2000). As mentioned in Chapter 29, the newly independent former Soviet republics of the Caucasus and Central Asia are now attempting to develop new economic, diplomatic and strategic links to the south and west. Key to all these objectives is improved transportation and communications, chiefly roads and pipelines. The European Union is assisting them through development of TRACECA (the Transport Corridor Europe–Caucasia–Asia) and the Transport Corridor Europe–Asia.

### *Railroads*

Possibly even more dramatic and long-lasting has been the political significance of railroads, even though they are only a century and a half old. Railroads are far more efficient (thus cheaper) than roads for carrying low-value bulk products long distances. They are also very efficient for carrying large numbers of people short distances through densely settled country. Consequently, railroads have been instrumental in opening up vast new areas of the world for agriculture, mining, and forestry as well as in stimulating and servicing the urban sprawl so typical of most industrial societies. The political significance of both of these movements is clear.

Most of the world's railroads are government owned, and nearly all privately owned railroads are government regulated. Even if privately built, railroads are subject to political as well as economic influences. The land-grant program of the U.S. government to spur railroad construction in the West in the mid-nineteenth century is only the most obvious of these influences. Towns all through the country were born, thrived, shriveled, or died according to whether the railroad came through them or passed them by, and many sprang up along the railroad as the track was laid. Even in Russia, the growth

---

*Latin Americans insist on calling it the Inter-American Highway, believing that "Pan-American" smacks too much of Yankee imperialism. When a bridge replaced a ferry across the Panama Canal in 1962, linking major segments of the highway, Latin Americans stridently protested naming it the Thatcher Ferry Bridge. Reluctantly, the United States agreed to call it the Bridge of the Americas. And the Darien Gap in the highway has still not been closed, at least in part because of cool relations between Panama and Colombia.

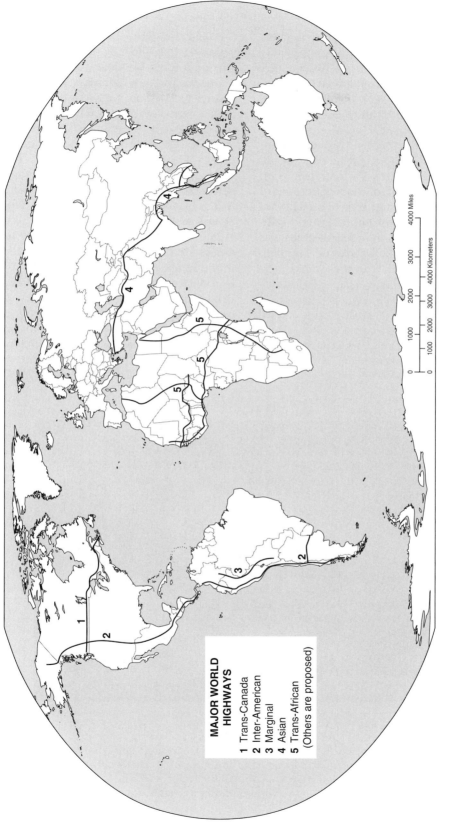

MAJOR WORLD
HIGHWAYS

1 Trans-Canada
2 Inter-American
3 Marginal
4 Asian
5 Trans-African
(Others are proposed)

Major international highways in Asia, Africa, and the Americas.

of Tomsk was stunted when the Trans-Siberian Railroad was built far to the south, while in Africa, a small camp used by workers building the Uganda Railway became Nairobi, Kenya, now the transportation and communications hub of all of eastern Africa.

In the former Soviet Union even today, railroads are given priority by the governments over roads, and truck haulage is still rudimentary. The Trans-Siberian has been double-tracked, rerouted north of, instead of through, Manchuria, and improved in other ways. Perhaps the most important and difficult railroad construction in recent times was the 4300-kilometer Baikal–Amur Mainline Railroad (BAM), paralleling the Trans-Siberian in Yakutia but much more comfortably north of the Chinese border. It was laid in some of the coldest, bleakest, remotest, and most rugged areas on earth at the rate of some 300 miles per year, for reasons partly economic and partly strategic. Another railroad laid largely for strategic reasons was

the Tanzam Railway (TAZARA), built by China to link the rail nets of Zambia and Tanzania after Rhodesia declared its independence in 1965. Zambia was placed in the politically uncomfortable position of continuing to use the Rhodesian Railways for its own vital copper exports or of boycotting the Rhodesian Railways and suffering the inevitable economic consequences. The decision was to sever links with southern Africa and establish new ones with East Africa by means of a railroad and pipeline to Dar es Salaam.

The Tanzam Railway, though new, is not a new idea. It was included by Cecil Rhodes in the nineteenth century as part of his projected Cape-to-Cairo Railway, a dream of British empire builders through much of the scramble for Africa. The dream was never realized, however, as Germany managed to hold on to German East Africa (Tanganyika), thus blocking the most feasible routes. After Britain obtained a mandate over Tan-

**The Tanzam Railway terminal in Dar es Salaam.** The railway and its facilities are operated by TAZARA—the Tanzania-Zambia Railway Authority. Its fortunes wax and wane depending on the political situation in southern Africa. Now that South Africa has joined SADC and most of Zambia's transit traffic is passing through South African ports, the railway must find a new *raison d'etre* or become simply a local line serving southwest Tanzania. (Martin Glassner)

ganyika, it welded together an efficient and practical rail system with more than 3000 miles (4830 kilometers) of track in its territories of Tanganyika, Uganda, and Kenya. This system was operated by all three territories through the East African Railways and Harbours (EAR&H). It continued functioning after independence as a service of the East African Community. But the Community collapsed in late 1976, and EAR&H broke up in February 1977. Then each of the former partners had to reorganize its share of the rails and equipment and try to operate its own national rail system.

Another uncompleted transcontinental railroad project has also been the victim of politics. A railroad linking Arica, Chile, with Santos, Brazil, was proposed in 1928, and over the years various segments of it have been put into service. There is still a gap, however, in the mountainous area between Cochabamba and Santa Cruz, Bolivia. Bolivia has accorded this stretch a low priority, and construction proceeds very slowly. Part of Bolivia's reluctance to complete the line may be attributed to Brazil's interest in its completion; Bolivia, with considerable historical justification, is suspicious of Brazil's motives.

The role of railways as political instruments cannot be overemphasized. Canada came into existence in 1867 only because the British acceded to the demand of the Maritimes and built the Intercolonial Railway linking St. John, New Brunswick, with Montréal. The line was uneconomic, but its purpose was political and it was successful. Later, British Columbia demanded to be linked with Canada by rail before it would join the confederation, and did so in 1871 after the Canadian Pacific Railway was built. Similar examples of the unifying effect of railways can be found on every continent. So can examples of railways as instruments of colonial policy, of political jealousies and suspicions inhibiting the development of railroads or the standardization of gauges and equipment, of political factors in setting freight rates, of railroads being built primarily for military use, and of political battles between rail and road for the carriage of

urban and commuting passengers. Indeed, it would be difficult to exaggerate the political factors inherent in rail transport.

*Waterways*

Inland waterways similarly have been subject to political influence and have had their effects on politics. Rivers have been great natural highways for millennia and have linked peoples more often than they have divided them. Still, as we have seen, international rivers have frequently been foci of contention among not only riparian States, but other users as well. More than one country has felt it necessary to use force to open up a river for international navigation. And unlike highways and railways, rivers have other uses besides transportation. In fact, the nonnavigational uses of rivers are growing in importance, and the various uses are frequently in conflict with one another. As mentioned earlier, the International Law Commission has been working on this problem.

Artificial waterways give rise to similar problems. There are numerous tales that can be told of canals proposed and constructed; of rivers straightened, deepened, and canalized, all embroiled in political controversy both domestic and international. The United States abounds in them: the Houston Ship Channel, the Cross-Florida Barge Canal, the Tennessee–Tombigbee Waterway (Tenn–Tom), and the proposal to open a route to Oklahoma City for oceangoing ships, to mention only a few recent examples. But the all-time classic story has to be that of the St. Lawrence Seaway, opened in 1959 after half a century of complex and bitter battles.

There were rivalries between Montréal and New York for seaborne trade, between Canada and the United States, between railroad and shipping interests, between hydropower and navigation needs, between advocates of state and federal action, between Canadian prime ministers and provincial premiers, and on and on. In the end, the United States concurred in its construction only after Canada threatened to build a seaway entirely on its own territory and charge

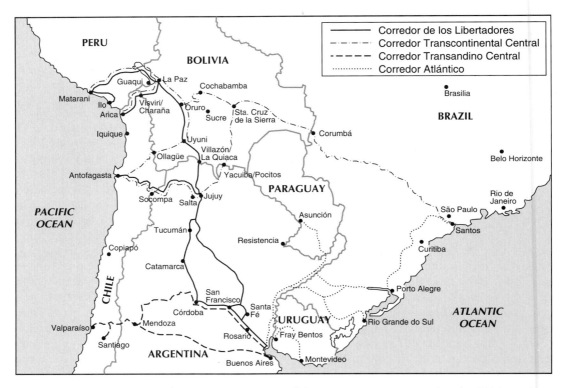

**Integration of railways in the Cono Sur.** As one of the activities commemorating the 500th anniversary of Columbus' voyages, the major railways of Spain and of the Southern Cone of South America undertook Proyecto Libertadores. This major project involves not only heavy investment in railway infrastructure, but also improvements in telecommunications, customs cooperation, containerization, and linkages among rail, road, and water transport in the four "corridors" shown on this map. It is only the latest effort along these lines, expanding and tying together earlier projects sponsored by the Latin American Railways Association (ALAF), the Plata Basin countries, the Latin American Integration Association (ALADI), the United Nations Economic Commission for Latin America and the Caribbean (CEPAL), and other agencies. If successful, it should vastly improve transport within the region, including the transit transport of the land-locked countries of Bolivia and Paraguay. Note that most of the Corredor Transcontinental Central includes the old but still incomplete Santos-Arica line.

American ships very high tolls to use it! Two considerations made the seaway seem desirable to the U.S. government: the discovery of iron ore on the Ungava Peninsula near the Québec–Labrador boundary that could replace the depleted iron ranges of Minnesota in feeding Midwestern steel mills, and the belief that the seaway would make a material contribution to the defense of the continent. For many years after its opening it was profitable and almost too successful, cited as a model of practicality, of wise multipurpose development of an important resource, and of international cooperation!

Since the mid-1980s, however, both traffic and revenues have declined, largely because of changes in the economies of both Canada and the United States. In order to stimulate business, the United States no longer charges tolls on its portion of the seaway, and Canada in late 1994 was also considering dropping tolls altogether on its section.

### Pipelines

Pipelines in their modern form are relatively new, though their ancestors are the numerous aqueducts of antiquity. Now every con-

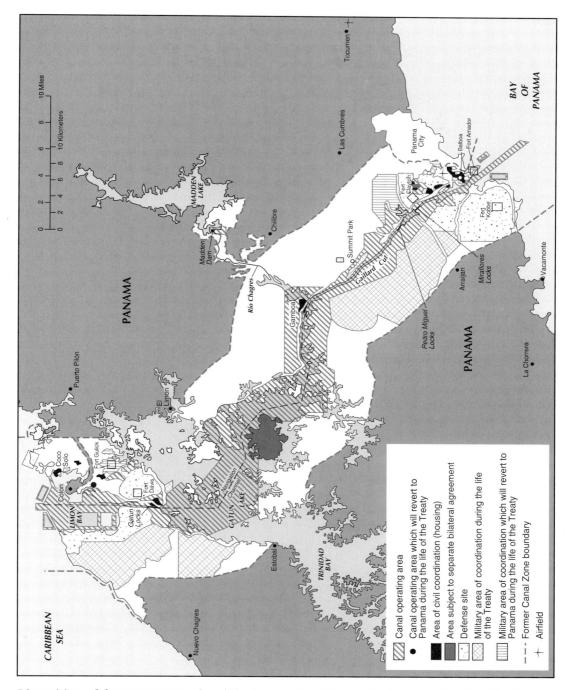

**Disposition of the Panama Canal and the former Canal Zone.** The three treaties signed by Panama and the United States in 1978 provide for a gradual and orderly transfer of the canal and related facilities to Panama by the end of 1999, and for protection of the canal and its employees during this period. This is in sharp contrast to the way the Suez Canal was abruptly nationalized by Egypt in July 1956, provoking an attempt by Britain and France in October of that year to try to seize it back by force. It is also an example of peaceful decolonization while maintaining intact a vital international waterway. Since 1990 the Panama Canal Commission, which replaced the Panama Canal Company in 1979, has been headed by a Panamanian. Imagine! A United States Government agency headed by a non-American! To date, this revolutionary development has not led to the collapse of the United States or even closure of the Panama Canal.

tinent is webbed with pipelines carrying all manner of petroleum and petroleum products, natural gas, water, wastes, and even coal slurry. Some pipelines have been of heroic proportions, both in length and in difficulties overcome. The Trans-Arabia Pipeline (Tapline) from the Persian Gulf oilfields of Saudi Arabia to the Mediterranean in Lebanon; the Friendship Pipeline carrying petroleum products from the Volga-Urals fields of the USSR to Poland, Czechoslovakia, and East Germany; and the Alyeska Pipeline from Prudhoe Bay on the Arctic Ocean to Valdez on the Pacific coast of Alaska are only a few of the better-known ones. Each has its obvious strategic advantages, but each is uniquely vulnerable to attack by saboteurs or aircraft.

The construction and routing of pipelines are no less political than for any other mode of transport. Witness the Sumed Pipeline built by Egypt in the late 1970s from Suez to the Mediterranean in direct competition with the lines built by Israel from Eilat to the Mediterranean a decade earlier. And the drawn-out debate over the best route for a gas pipeline from Prudhoe Bay to Calgary and the midwestern United States, any of which would require settlement of Canadian native land claims.

The Soviet Union's 1981–85 economic plan included a provision for six new pipelines to carry natural gas from Western Siberia to Western Europe. They sought to purchase some Western equipment and technology for one of them. The United States strenuously opposed these sales on political and security grounds and instituted economic sanctions against all potential suppliers. This policy was unsuccessful. The equipment was supplied, the pipeline opened ahead of schedule in late 1983, and Western Civilization has somehow managed to survive! But we may expect more such rows over pipelines in the future. Now undersea pipelines are becoming common, generally linking oilfields or storage tanks with offshore tanker terminals, but they are destined to become longer and probably more politically—as well as environmentally—controversial.

## Bridges and Tunnels

The past two decades have seen the construction of major bridges and tunnels around the world, many of them crossing strategic waterways and many having major geostrategic as well as economic importance. To name only a few: the bridge across the Bosporus at Istanbul to link Europe with Asia, the bridge-tunnel connecting Honshu and Kyushu in Japan; the bridge from Ciudad del Este (formerly Puerto Presidente Stroessner), Paraguay across the Río Paraguay to Brazil; the Channel Tunnel (the "Chunnel"), a rail link under the English Channel connecting Britain and France; and the Friendship Bridge across the Mekong River between Thailand and Laos.

The United Nations is studying construction of a "Europe–Africa permanent link through the Strait of Gibraltar," probably to be a bridge-tunnel combination; the Gulf Cooperation Council is considering an eventual bridge-tunnel across the Persian Gulf; Italy has been probing a possible bridge across the Strait of Messina between Sicily and the mainland; and a rail tunnel under Bering Strait connecting Alaska and Siberia has been discussed by officials of the two countries. Eventually, some version of all these links will probably be built, tying together even more tightly islands and continents and the peoples who inhabit them. They should help, among other things, to break down the insularity of many countries and peoples and thus reinforce even more the world's interdependence.

## Maritime Transport

We have already discussed many features of maritime transport in our chapters on geopolitics, international trade, the sea, and outlaws and merchants of death. More needs to be said, however, as we are becoming more, not less, dependent on seaborne trade with the passage of time. The increase in number and size of vessels sailing the world's sea lanes has necessitated more stringent rules of the road, tighter controls on ship design and construction, establish-

**The Channel Tunnel.** This historic fixed link between Great Britain and the Eurasian mainland entered into service in 1994. Otherwise known as the Eurotunnel or The Chunnel, its long-term political significance is likely to be considerable—provided it remains open and functions properly. These highway signs at the French end refer to La Manche, the French name for the English Channel. (Patrice Thomas/Photo Researchers)

ment of one-way lanes and traffic control systems in many straits, and other reductions in the traditional freedom of the seas. All this is quite apart from the extension of internal waters and territorial seas farther out to sea and the creation of archipelagic waters.

The major maritime powers made it quite clear from the outset of the Third UN Conference on the Law of the Sea (UNCLOS III) that they would tolerate no unjustified interference with freedom of navigation. As a result of this position and the realization by most other States that they too benefit from a free flow of international commerce, agreement was reached early in the Conference on regimes of transit through straits, archipelagos, territorial waters, and exclusive economic zones. Other agreements on restrictive measures necessary to prevent collisions at sea, marine pollution, and other hazards have been adopted not only in UNCLOS III, but also in IMO, the International Maritime Organization. International cooperation is essential and is generally maintained

in international shipping, but this does not mean that there are or will be no problems.

Changing technology of maritime transport, changes in the types of goods carried, entry of new countries into the shipping industry, development of new seaports, new sources of fuels and raw materials, and differential rates of economic growth are some factors that are leading to changes in shipping routes and hence the political significance of certain waterways and nearby coasts and ports. Despite the reopening of the Suez Canal after the signing of the Egypt–Israel Peace Treaty in 1979, for example, traffic around the Cape of Good Hope did not diminish as much as had been expected and may increase again despite the hazards presented by formidable storms and currents.

Some waterways are governed by special regimes. These include the Panama, Suez, and Kiel canals and the Turkish Straits (the Dardanelles and the Bosporus). International trade is protected in all of them, but under different conditions, and the rules are sometimes interpreted by the State controlling the waterway to suit itself. The 1888 Convention of Constantinople, for example, states explicitly that the Suez Canal is to be free and open to the ships of all countries without distinction both in peace and war. Yet Egypt prohibited passage of Israeli ships and of all ships bound to or from Israel through the canal for a quarter of a century, until late in the 1970s. And under the Montreux Convention of 1936 that governs the Turkish Straits, aircraft carriers are forbidden passage in either direction, yet in July 1976 Turkey permitted the *Kiev*, a new Soviet aircraft carrier, to pass from the Black Sea into the Mediterranean, accepting at face value the Soviet claim that the ship was really a cruiser!

The shipping industry itself is probably the most complex in the world, rivaled only by the oil industry. Much of the complexity, including the highly controversial issue of *flags of convenience* (called by the users, however, "flags of necessity"), is due to economic factors, but some of it is political also. Some countries, for example, require that privately owned merchant and passenger vessels registered under their flags be available to the government in times of emergency, as during the Falkland Islands War in 1982 when even the *Queen Elizabeth II* was pressed into service. Some maintain "mothballed" or reserve fleets of inactive ships that can be reactivated as necessary. The U.S. National Defense Reserve Fleet currently consists of some 300 vessels berthed in Virginia, Texas, and California, of which 86 are in the Ready Reserve Force, capable of going into action in 5 to 20 days. It was used most recently during the 1990–92 Gulf War and related operations, and in the 1994 Haitian crisis. U.S. flag ships in service in 1993 numbered only 551, but many of the other 80,000 ships in the world are owned in whole or in part by Americans.

Many countries have cargo reservation laws, providing that some or all government cargoes be carried only on national-flag-ships. Many have cargo preference systems, allocating all export cargoes among national,

regional, and worldwide carriers. An increasing number of developing countries have followed the lead of Panama, Liberia, and Honduras and allow foreign-owned and foreign-operated vessels to be registered there under very liberal conditions. The world's ten largest merchant fleets in 1993 included those flying the flags of Panama, Liberia, Cyprus, The Bahamas, and Malta, none of which is among the world's largest trading countries. Saint Vincent and The Grenadines, a tiny Caribbean island State with a population of only 115,000, ranks nineteenth in the world, just behind China, in the size of its merchant fleet. Of course, most of its ships consist of recently reflagged vessels making one last voyage before heading for the scrap yard, but still—there it is, another example (if another is needed) of how statistics must be viewed most warily. During the Iran–Iraq war of 1980–88, a number of tankers owned by friendly Persian Gulf States were transferred to United States registry to protect them from attack by the combatants, a political move of questionable validity under international law.*

Many of the historic practices in the shipping industry have had the effect, if not the intention, of making it very difficult for developing countries to obtain efficient, economical maritime transport service. These countries have had to resort to cargo preferences and other devices to try to reduce their high costs of trade. UNCTAD has been working for decades to improve the situation and has had some success in the areas of insurance, regional shipping lines, and training of crews. It has developed a Code of

---

*In 1950 U.S. flag ships carried 42.6 percent of world seaborne trade, which in that year amounted to 117 million tons; in 1992 U.S. flag ships carried 3.9 percent of the 852 million tons of seaborne trade. The U.S. merchant fleet in 1993 ranked eleventh in the world and was the third oldest, behind Greece and Saint Vincent, with ships averaging 21 years. In early 1994, only two commercial vessels were on order in U.S. shipyards. (figures from U.S. Maritime Administration, 1994, and *Lloyd's Register World Fleet Statistics*, December 1993.) *All* shipping statistics are subject to interpretation, and none should be accepted unquestioningly. Nevertheless, the overall patterns and relative orders of magnitude are clear and generally accepted.

Conduct for merchant ships which requires, among other things, a "genuine link" between a ship and its country of registry. The net result of all this activity, however, is unimpressive. Poor countries in general still pay more for shipping than rich ones, an important element in their competitive disadvantage in the global market.

Before leaving the subject of maritime transport, it would be well to reiterate something of the maritime problems of landlocked countries. Lacking seaports, those marvelous windows on the world, these countries are handicapped initially by being isolated, to a greater or lesser extent, from the swirls and eddies of human intercourse along the coasts and across the seas. They are further handicapped in participating in the shipping industry by their lack of home ports and of opportunities to train seamen and masters. Switzerland has been operating a merchant fleet for some time now, using Basel on the Rhine River and the Italian port of Genoa as its home ports, and the Czech and Slovak Republics and Hungary also have oceangoing ships. Bolivia again has a merchant fleet, composed of one ship serving Europe and South America. And the Royal Nepal Shipping Corporation operated one chartered ship on one voyage from Calcutta to Bremen in 1972. Land-locked States, especially the poorer ones, are even more at the mercy of shipping companies than coastal States without fleets of their own. This is one more reason that they need special help.

## Air Transport

Air transport is very likely the most politicized of all modes of transport for a number of reasons. First, airspace, unlike ocean space, is not free for anyone's use. It belongs to the State under it, and no commercial or military aircraft has a right of innocent passage. States have absolute jurisdiction over use of their airspace and ground facilities. Second, air travel is relatively new, developing only after the modern State system was well established, and flowering during the period of decolonization. As a result, it

has developed not through custom and practice over a long period, but very quickly and by the agreement of a relative handful of States. Third, the technology of aviation is now so costly and so attractive that States are willing to go into debt in order to acquire and maintain an air fleet both for legitimate security reasons and purely for prestige. Indeed, a national airline, like a seat in the United Nations or a steel mill, is today part of the iconography of new States, as we saw in Chapter 19.

As a result, such matters in international air transport as routes flown, cities served, frequency of flights, and types of aircraft used are thrashed out by government representatives primarily on the basis of reciprocity and spelled out carefully in bilateral agreements. There is no most-favored-nation clause relating to air transport and few blanket agreements covering multiple countries and arrangements. Even the friendliest of States, such as the United States and the United Kingdom, may engage in arduous and protracted negotiations, perhaps offering to swap an extra flight a week into Hong Kong for an American carrier for the right of a British carrier to fly from London to Denver nonstop. Package deals are sometimes worked out that contain provisions unrelated to air transport. This helps to account for the seemingly irrational practices so common in the international air transport industry.

Since overflight rights must be specifically granted by States, the lack of such rights may force planes to travel extraordinarily long distances at great cost to avoid overflying particular countries. Only in the early 1980s did the Soviet Union and China begin allowing overflights and even landing by selected noncommunist aircraft; most other countries still forbid overflights by aircraft of countries deemed unfriendly. The map of world airline routes shows large areas with a very low density of international flights. On the other hand, Aeroflot, airline of the former Soviet State, provided extensive international service nearly everywhere, except Australia and parts of South America. This service was maintained chiefly to obtain convertible currencies and to carry the So-

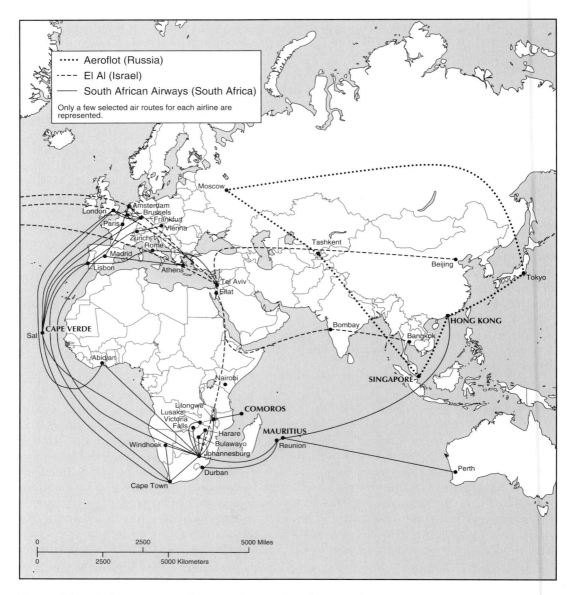

**The politics of air transport.** This map shows selected routes of three major international air carriers. Note how El Al and SAA had to fly costly circuitous routes in order to avoid overflying hostile territory. The Aeroflot route was proposed during the 1960s, when Sino-Soviet tension was at its peak. (Manila was considered a possible substitute for Hong Kong.) It never actually went into effect. Today Aeroflot has routes directly across China between Moscow and Beijing and Pyongyang, North Korea, and the other two airlines also have greater flexibility because of improved political relations with nearby countries.

viet presence as far as feasible. The actual operations of Aeroflot abroad, such as marketing, public relations, and flight service, were allegedly marked by inefficiency, unethical sales practices, and a general lack of professionalism. This was probably an ex- tension of the Soviet domestic system, which stressed political and strategic goals over commercial operation, and its network could not be justified by the amount of traffic carried. It was, nevertheless, generally accepted by many Western governments as being similar to other airlines, but they did not insist on the customary reciprocity—probably for political reasons.

Other airline routes and service patterns are distorted by such factors as some countries' desire to serve their colonies and former colonies, even if there is little local demand for air transport; a desire to preempt carriers of other countries; and hostility toward certain countries that are denied overflight for their carriers. Vietnam, on the other hand, earns considerable foreign exchange from overflight charges, even to aircraft of former enemies. Many countries prohibit flights by foreign aircraft over particular areas or even permit overflights only in certain designated corridors. The reasons are not limited to the safety of air navigation or the possibility of disturbing people on the ground. Even though the Cold War is over, the United States still requires prior approval of each routing of charter flights of all CIS countries into its airspace and the first routing of scheduled flights.

Another vital factor in the airline industry is the predominance of government ownership and subsidy. There are many reasons for this, such as the desire to attract tourist trade and foreign investment, national security, the need to conserve foreign exchange, the need to serve small or remote places that generate little revenue, and so on. Although these reasons individually might be defensible, they contribute to the confusion.

Fares and standards of service were until the early 1980s controlled for its members by the International Air Transport Association (IATA), a nongovernmental industry association whose members include many government-owned airlines and whose rulings require government approval. In order to restrict competition among international carriers, IATA even regulated the charges for in-flight films and the types of meals that could be served as part of the plane fare, down to the size and content of sandwiches! IATA's stranglehold over fares was broken, in part, by nonmembers, such as Ecuatoriana, Loftleidir, and Aeroflot; the increasingly popular low-fare charter flights; and such innovations as very low fare–no frills scheduled

transatlantic flights and low-fare–standby–noreservation systems.

International nonmilitary air transport is largely governed by the International Civil Aviation Organization (ICAO), a specialized agency of the United Nations with headquarters in Montréal. All of the bilateral arrangements and restrictions described above have developed within the overall framework of the Chicago Convention of 1944, which is still the basic document in the field and is administered by ICAO. Its primary function is to promote safe and efficient international air transport, and its activities are chiefly technical. It works closely with the International Telecommunication Union and other organs and agencies of the United Nations system. It cannot avoid politics, however, and it has been embroiled in the disputes over the membership and role of such countries as Franco Spain, apartheid South Africa, and Communist China. Many other global, regional, and bilateral disputes have filtered into ICAO and, however subtly, influenced its functioning.

One of ICAO's perennial problems has been the large and shifting area of restricted or closed airspace over such trouble spots as the Aegean Sea, the Persian Gulf, Indochina, the Indian subcontinent, Cyprus, the Falkland Islands, and North Korea. Another is frequent conflicts between the needs of civil aviation and those of military operations and interdiction of airborne drug trafficking. The end of the Cold War and generally improving international security (except, of course, for numerous "brushfire wars") should reduce those conflicts and lead to improved civil aviation generally.

The powerful ICAO Council is still dominated by "the States of chief importance in air transport [and] the States not otherwise included which make the largest contribution to the provision of facilities for international civil air navigation," in the words of the Chicago Convention. The numerous developing countries are thus left to compete for the few remaining seats allocated on a geographical basis. This and other structural and procedural features have not helped the struggling, marginal airlines of

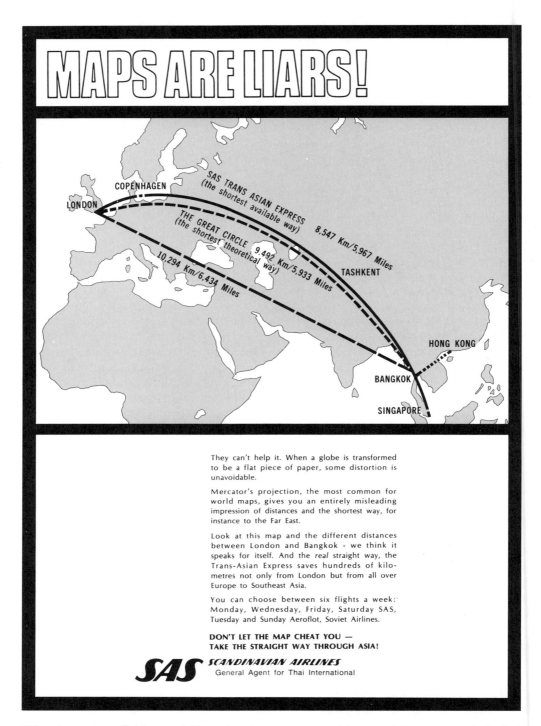

**"The shortest available way."** This advertisement appeared in several European periodicals in late 1971 and early 1972. It shows not only how most maps distort distances, shapes, and directions, so the charted route is seldom what it appears to be, but how it is nearly impossible to fly a true great circle route for very long because of both technical and political considerations. It also shows how even large, sophisticated corporations can err in their advertising since, according to Professor Joseph Schwartzberg (emeritus, University of Minnesota), the true metric distance via SAS from London to Bangkok is 9,603 km. The points about map distortion and great circle routes are still valid, however. (Used by permission of SAS.)

developing countries. Nevertheless, some of them have been able to grow and become first-class airlines respected for safe, efficient, and extensive service. The national carriers of Singapore and Ethiopia are examples. ICAO has provided valuable technical assistance to developing country airlines and has aided in their economic development, but they are still at a disadvantage in bilateral negotiations for routes, schedules, landing rights, and so on. The organization has, however, been creative in adjusting to new conditions, and the overall situation is improving.

Airline deregulation, begun in the United States in the late 1970s, in Canada in 1986, and later in Europe, is complicating, not simplifying, this picture. We shall see whether it results in safe, efficient, and affordable air travel for the average person.

## Communications

Like it or not, perhaps even without realizing it, everyone in the world is enmeshed in an intricate web of messages emanating from innumerable sources and carried by many means, from the silent gesture to the most sophisticated devices of modern technology. We are all inevitably, unavoidably influenced by what we see, hear, and feel. Sometimes the influence is benign, even helpful, and sometimes quite the opposite, but even if it is not quantifiable, it is always significant. We have seen examples of this influence throughout this book, but more needs to be said, not about the messages but about the media that carry them.

Taken together, telecommunications—telephone, telegraph, radio, and television—have had an incalculable impact on people everywhere. Modern transportation would be impossible without them. So would democratic elections, efficient government, and destructive wars. Radio has been a particularly effective means of entertaining, informing, educating, and persuading even illiterate people in all but the remotest of locations, for by now the loudspeaker and the transistor radio are ubiquitous. In most

of the world telecommunications are wholly or partially government owned and everywhere they are at least regulated by government. Some governments are quite responsible in exercising their power in the public interest, whereas others are much less so.

Even more than transportation, telecommunications have an international character and impact that are unmistakable. Radio and television signals leap international boundaries without permission or interference, and jamming them is very costly and only partially effective. Radio Moscow, the Voice of America, and the highly respected World Service of the British Broadcasting Corporation broadcast by shortwave around the world, supplemented by such largely propaganda stations as Radio Free Europe, Radio Martí, and Radio Liberty. Radio Beijing and the Voice of the Arabs (from Cairo) are two other powerful shortwave stations beaming heavy doses of propaganda far afield. Even the United Nations Radio Service in 1978 began producing 15-minute programs in four languages denouncing apartheid and supporting self-determination for the native African peoples, programs broadcast through the facilities of member States. Now it beams 37 programs in 19 languages to 165 States and territories.

But Radio RSA, the South African government's external shortwave service, responded in kind with 208 hours a week of broadcasts in English, French, German, Portuguese, Spanish, Dutch, Afrikaans, Swahili, Chichewa, Lozi, and Tsonga plus another 20 hours a day in six languages designed exclusively for listeners in Africa. Another politically important shortwave service is the External Services Division of All-India Radio, which broadcasts over 74 hours worth of programming daily in English, 16 other foreign languages and eight Indian languages, reaching most of the world except the eastern Pacific and the Americas.

News and entertainment programs can be influential without any deliberate attempt to persuade. American music, fashions, humor, and attitudes infiltrate even the most closed societies on earth, inadvertently influencing attitudes of others toward them. The process works in reverse also: Radio broadcasts from London during the Battle of Britain in World War II and on-the-scene television coverage of the Vietnam War demonstrably influenced American attitudes toward those conflicts. Round-the-clock, round-the-world, on-the-spot coverage of the Gulf War in 1991 established the American Cable News Network not only as a commercial success but also as a potent political force, even influencing in many ways the course of the war as well as attitudes toward it. Television also played an important role in the separation of Eastern Europe from the Soviet Union and in the reunification of Germany. Even political enemies listen to one another's broadcasts and watch one another's telecasts: Jordanians and Israelis, for example, enjoyed one another's television programs designed primarily for domestic consumption, well before they made peace in 1994.

New technology is constantly intensifying the telecommunications mesh. Undersea cables that carry hundreds of messages simultaneously, teletype and telex, messages carried by fiber optic cables and laser beams, and programs of all kinds relayed by orbiting satellites and even broadcast directly from satellites—all these and others are not futuristic fantasies but are here now, in current use. Even the relatively simple facsimile (fax) machine fed accurate information almost instantly from Hong Kong and Western countries to the students and others demonstrating against dictatorship and corruption in China in 1989. Many facsimile pages from foreign newspapers were posted in public places, and students explained them to workers and peasants gathered around. These new technologies, however, raise a host of political, legal, and ethical questions we can only mention here.

"Pirate," or unauthorized, broadcasting beamed at a country from ships in international waters, a practice begun in the 1950s in the North Sea but later brought under international control, made the world debate proposals to regulate the content as well as the frequencies of radio and television broadcasts that cross State boundaries. Now the debate focuses on communications satel-

lites. The United Nations has been working since 1977 on "an international convention on principles governing the use by States of artificial earth satellites for direct television broadcasting." At issue, as noted in Chapter 34, are not only blatant propaganda messages, but even programs tending to condone or encourage racial bigotry, religious programs received in countries where atheism is the official religion, even commercials received in countries where noncommercial broadcasting is the norm. These are not easy problems to solve, and they are unlikely to be solved soon.

The print media—newspapers, magazines, books, pamphlets and broadsides, trade and professional journals—are also widespread and influential, in some countries even more so than the broadcast media. They also are subject to government ownership or at least control. And to an even greater extent than broadcast media, they serve as mouthpieces of special interest groups of all kinds: labor unions, political parties, churches, ethnic groups, manufacturers, farmers, veterans, retirees, and so on. Even in illiterate societies, the written word may be a powerful force, if only because it may be endowed with mystical qualities, and those who can read are automatically in an advantageous position.

Government censorship, both before and after publication, is widespread in the world; a free press is quite rare, in fact. Even where the press is not overtly owned, controlled, or censored by government, it may be cowed into docility by harassment, threats, revokable licenses, control of paper supplies, and still more subtle devices.

Why do governments go to such lengths to control the printed word? Partly because, sometimes for the best of motives, they wish to channel people's thinking along particular lines, to guide the political and social development of their countries as well as economic development, and partly because such control has been repeatedly demonstrated to be both feasible and effective. Not completely effective, of course—witness the *samizdat*, or mimeographed newsletters clandestinely produced and circulated among the intelligentsia in the USSR, and the forbidden books smuggled into South Africa. Nevertheless, it does work, and people's thinking is influenced by it.

Many agencies, both public and private, are working around the world to encourage and maintain greater freedom in the print media. The Inter-American Press Association, for example, annually rates the countries of the Western Hemisphere according to the degree of press freedom enjoyed there. The Pan-American Institute of Geography and History (PAIGH) and the United Nations Educational, Scientific and Cultural Organization (UNESCO) are sponsoring long-range international programs for writing "objective" histories of parts of the world. And the American Civil Liberties Union stoutly defends the First Amendment to the U.S. Constitution guaranteeing freedom of the press, religion, and assembly.

### External Linkages

Two related points need to be made here, and both are elements of topics we have already covered. In Chapter 19 we discussed the movement for a New International Economic Order (NIEO). In the mid-1970s, shortly after this movement was born, another emerged, also from the ranks of the developing countries, chiefly ex-colonies in Africa and Asia. This was called the *New World Information and Communication Order* (NWICO).

Its basic premise was that the flow of information in the world was largely controlled by a few powerful and technologically developed countries, and that the flow of information was therefore both in one direction only (from the rich countries to the poor countries) and culturally and ideologically biased. The developing countries, therefore, were victims of an oligopoly in both the generation of information and its transmission. This had two basic effects: The wire services and the various electronic media were saturating the South with the cultures and the ideological perspectives of the rich countries, smothering their indige-

nous cultures and ideologies, and these media transmitted back to the North highly selective and grossly distorted information about the South. This was an important component of the cultural imperialism discussed in Chapter 17. They therefore sought to rectify the situation in five overlapping and interconnected areas:

1. **Access to telecommunications facilities and services.** This basic matter is immensely complicated, since the disparity between North and South is so great in all of its aspects. It ranges from vast differences in simple equipment (e.g., the city of Tokyo has more telephones than all of Africa) to control over electronic pathways (e.g., the radio spectrum and the geostationary orbit). The claim by some equatorial countries to portions of the geostationary orbit may therefore be seen as a component of the drive for an NWICO.

2. **Political concerns.** Since the rich countries control the messengers, they can also dictate the message. Countries that have recently emerged from colonial domination are especially concerned about developing their own identities, and to do this they need some measure of control over the information received by their people. The gap between the information-poor and the information-rich may, in this "Information Age," be even more important than the gaps between the oil-poor and the oil-rich, or the technology-poor and the technology-rich.

3. **Economic concerns.** Since communications are a vital part of economic development, they should receive higher priority in aid programs, trade arrangements, and other economic activities than they do. This, then, is a direct link to the NIEO concept. It raises the same questions of State control versus free enterprise, monopoly versus competition. It also calls into question the allocation of costs and benefits of INTELSAT,

**International transportation and communication on stamps.** Although most countries routinely pay tribute to elements of their circulation systems on stamps, far fewer recognize international circulation. These stamps commemorate the St. Lawrence Seaway, Bridge of the Americas over the Panama Canal, British Broadcasting Corporation's World Service, the reopening of the Suez Canal in 1957 after its closure by Egypt in 1956, the Addis Ababa-Djibouti railway, The Andean Marginal Highway, World Communication Year, East African Airways (defunct), Kiel Canal, ASEAN Submarine Cable Network, Asia Pacific Broadcasting Network, Radio Moscow, international shipping on the Rhine River, Asian-Oceanic Postal Union, the Saimaa Canal, Universal Postal Union, Tanzam Railway, Nairobi-Addis Ababa Highway, reopening of the Strait of Tiran in 1956 and opening of the Rhine-Main-Danube Canal connections in 1992. (Martin Glassner)

which has tended to serve only the most profitable routes.

4. **Technological questions.** There must be an adequate infrastructure in the form of facilities and channel/spectrum space to serve the poor as well as the rich, and technical standards must be set to ensure the interconnectivity of all relevant equipment. And there must be some way to assist the poor countries not only to catch up technologically, but to manage the costs of keeping up with rapidly changing technology.

5. **International institutions.** The battle for an NWICO has been waged in UNESCO, the ITU (International Telecommunication Union), and other entities within the United Nations system. The introduction of essentially political matters into such technical or cultural fora raised the ire of the United States, the United Kingdom and other rich countries, only partly because this threatened their traditional control of these agencies.

There is much more to be said about all this, of course, but we can only point out here two of the many reasons why the campaign for an NWICO faltered and faded. For one thing, the attempt by the developing countries to start their own international wire service, to provide the North with "factual" information about the South, ran aground when battered by the forces listed above: technology, economics, insufficient trained personnel, politics. Second, some of the most ardent advocates of an NWICO were, in fact, representatives of some of the world's most brutal and corrupt dictators, who greatly restricted the flow of information at home. The purity of their motives was questionable at best.

Apart from, but linked with, the NWICO, was the Uruguay Round of trade negotiations under GATT, discussed in Chapter 25. International communications (transborder data flows, in UN parlance) is a major component of the trade in services that proved so difficult to fit into a system essentially designed to regulate trade in goods. Not only

telecommunications, but even computer software and computer services of all kinds came under this heading. If the gap between rich and poor in the world is ever to narrow, then these services must be made available to the poor on reasonable terms, while at the same time inventors and investors, innovators and technicians, continue to be suitably rewarded for their contributions.

This leads to the protection of intellectual property—patents, licenses, copyrights, trademarks (relatively few of which as yet originate in the South)—which was a separate perplexing problem on the agenda of the Uruguay Round. There are inventors, innovators, investors, and intellectuals in the South as well as the North, but generally the risks and hurdles they face are far greater than those in the North, and their rewards are far smaller. This accounts in large part for the infamous "brain drain," discussed in more detail in Chapter 37.

The NWICO was, as a practical matter interred by UNESCO in 1989, when its members reaffirmed its constitutional mandate to promote the "free flow of information" and explicitly endorsed "freedom of the press" as an international goal. But the problems of improving the flows of information of all kinds, both within and among the poor countries, and between them and the rich countries, are likely to become both more severe and more difficult to resolve in the coming decades. At the same time, our ability to resolve the problems is likely to grow also, perhaps even more rapidly. The real question is whether the political will to do so will grow as well.

### Circulation

In 1977 the General Assembly of the United Nations endorsed the recommendation made earlier by the Economic Commission for Africa and proclaimed 1978–88 as the United Nations Decade on Transport and Communications in Africa. We are now into the second such decade. Both authorized extensive programs designed to develop these facilities in Africa and to mobilize the technical and financial resources required for that purpose. This is fitting, for Africa is the continent with, overall, the poorest circulation system. In part, this is a result of the practices of the colonial powers, which developed multiple and unconnected circulation systems designed to tap the human and natural resources of the areas under their control, to facilitate public administration, and to maintain contact with the home countries. When the colonies became independent, they found these facilities useful, of course, but quite inadequate for the development of modern States and modern economies. They were even less suited for the development of regional and subregional cooperation and normal intercourse. Even now there is still no continuous overland route connecting the countries of West Africa from Senegal to Cameroon. Between Sierra Leone and Somalia, one still flies via Rome or Harare. And until recently, in order to place a telephone call between Accra in the former British Gold Coast (now Ghana) and Lomé in formerly French Togo, only 125 km apart, it was necessary to route it through Paris and London!

Circulation, including an efficient and inexpensive postal system, is part of the infrastructure that is absolutely essential to any economic development program, and it is being strongly emphasized in Latin America and Asia as well as Africa. In fact, the UN also sponsored the Transport and Communications Decade for Asia and the Pacific (1985–94), whose goals and activities were similar to those in Africa. But as circulation improves in these regions and worldwide, time and distance will shrink still more rapidly every aspect of our lives, not least the political aspects.

Who can predict the long-term implications of a world so seemingly shrunken by technology that almost literally everyone will be everyone else's neighbor? Can our present multitude of "sovereign" States survive? Or our multitude of languages? Or belief systems? Or political systems? Will the human species become homogenized culturally and intellectually as well as physically? If so, what are the implications of these changes? If not, what will prevent them? And what will be the implications of a failure of the species to adjust its thinking to its technology? Would it be better to adjust our technology to our cultural and ethical values? We cannot answer these questions, only ask them. But we can say with absolute assurance that inevitably we are all being drawn closer together, we are becoming steadily more interdependent, and we must begin thinking of ourselves as members of one species and citizens of one world.

# 37

# *The Politics of Population, Migration, and Food*

Ever since Thomas Robert Malthus published his celebrated essay on population in 1798, population has been of greater interest, if not concern, to people in many walks of life around the world. Population geography has been a recognized specialization within our discipline for some time, and the newer field of demography is even more specialized. Population is important in political geography also, and political geographers must be concerned with matters in addition to population density and distribution. Such demographic factors as literacy and education, health and age structure, skills and abilities contribute to people's mobility, awareness, and political consciousness.

## *Population*

### *Population Quality*

Illiteracy has been a major problem facing the emergent ex-colonial States, for it is impossible to transfer allegiances from tribe and region to nation and State without the tools of understanding. Thus, many of these emergent States are spending much of their revenue on education. Unfortunately, the consequences are not always salutary; increased education in the village often causes emigration from rural areas to the urban centers, where thousands of unemployed must be accommodated, for jobs are few. These unemployed form a major political threat, and several countries have taken steps to limit migration to the cities.

Of equal or even greater importance is the health of the population, and this question is related directly to that of food supply. Comparisons between Western European States and some Asian countries in terms of life expectancy, daily food calories available, dietary balance, and incidence of disease quickly indicate why available energies differ in quantity and are expended in different directions. A population that is largely engaged in day-to-day survival under a subsistence form of agriculture cannot be expected to make any great contribution to the State, economically or otherwise. The struggle for food is a major factor in political geography, affecting as it does the internal condition of the State as well as its international relationships.

The ideal or model nation-state must possess a people of sufficient size and quality who consider themselves to be a nation. Quality includes factors such as health and well-being, education, and skills. But other qualities have much to do with the forging (or failing) of a nation. In North America, and especially in the United States, we need only look around to see the detrimental impact of racial and other inequities. Millions of people who have come from Eastern and Western Europe have become U.S. citizens in the real sense of the word, emotionally as well as legally. But many people who came from other continents, especially Africa, have been accepted far less completely and effectively. This situation fosters what might be called "retribalization," and the danger to any State inherent in such a situation is apparent. In addition, as we pointed out in Chapter 20, raw population numbers lose much of their impact in any responsible power inventory if much of the population is prohibited from making its maximum contribution to the society because of discrimination on the basis of "race," social class, national or tribal origin, gender, sexual orientation, mental or physical health, or any other artificial distinction.

Another important characteristic of the population of any State can be illustrated quickly by means of the age/sex pyramid. This indicates the number of persons within various age groups for both sexes living in the country at a given time so that the ratios can be recognized at a glance. Although these pyramids do not reveal anything regarding the spatial distribution of the people involved, they do indicate much concerning longevity, changing male/female ratios (which can be severely disturbed by war or migration), and perhaps the effects of epidemics. They are one of the tools used by the political geographer who wishes to gain an insight into the problems of any State as a whole.

## Population Growth

World population, which presently totals over 5.6 billion, is undergoing a rapid expansion in which several distinct stages can be recognized. Various States are at different stages in this process, known as the *demographic transition*, which are closely related to the introduction of modern amenities in such fields as health and education, and are reflected by the State's degree of urbanization, industrial development, literacy rates, and so forth.

The demographic transition functions essentially in this manner: During the first stage (in which most of the world found itself prior to the Industrial Revolution), both birth rates and death rates are high, resulting in a fluctuating population that increases, but slowly, in response to famine

and epidemics and later recovery. During the second stage, as various aspects of development take effect, death rates are reduced. Sanitary standards are raised, diseases are fought successfully, food production and distribution improve. The result is an ever-increasing excess of births over deaths (the *demographic gap*) and an "exploding" population. This is largely true because while health conditions get better, socioeconomic conditions remain virtually unchanged. It is still considered desirable to have a large family to help with the farm work and to contribute income. In many colonial areas, the Western world brought better hygienic conditions, but did not stimulate social change. The third stage, as exemplified by most of Western Europe and North America, is one of renewed stability, and "leveling off" of the rate of increase. As people become educated and urbanization and industrialization take effect, there are social and financial restraints to having a large family. Thus both birth and death rates are very low, and population growth is again very slow.

Although it is difficult to classify the world's States according to the growth stage in which they find themselves—and various regions of States may be at different stages in the transition—the matter is, nevertheless, crucial in political geography. It is probably true that every group of people who became united in a State—either in recent centuries in Western Europe or thousands of years ago in the Middle East—underwent changes in this respect. Probably the most dramatic changes ever to occur took place as a result of the technological revolutions of Europe, but rapid technological progress also took place in all of the world's eight culture hearths, and they probably also experienced a mushrooming of population.

The model of the demographic transition is based largely on European experience. Is it likely to be replicated in today's poorer countries? Those States that are today still in the first stage of the demographic transition (some African and Asian countries) are in many ways unlike preindustrial Europe. Population densities are higher today than

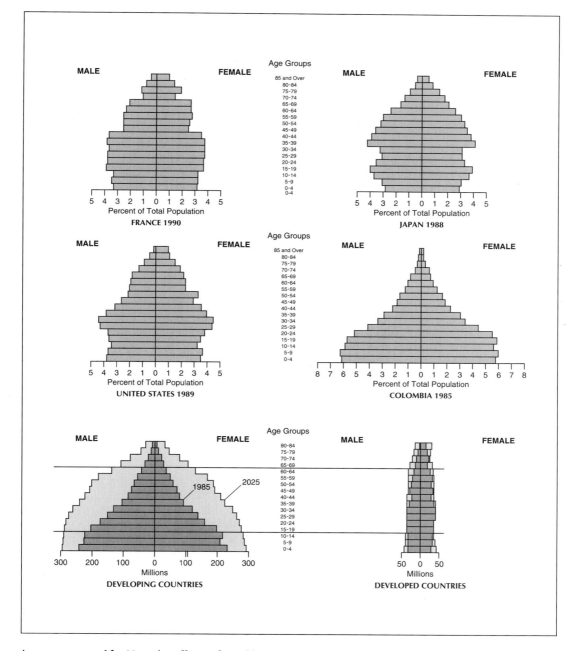

**Age-sex pyramids.** Note the effects of World War II on the population growth rates, the relatively large number of older people in France, and the "population explosion" in Colombia. In the bottom row are composite pyramids for the developing and developed countries prepared by the United Nations for 1985 and projected to 2025 on the basis of a mid-range estimate between the low and high estimates derived from various possible developments. Note the very great differences between the two pyramids, both in the current situation and the one projected for 2025. These differences, of course, have profound political implications, some of which are indicated in this book. In the words of the UN report, "In 1985, 37 percent of the total population of the developing world consisted of children below the age of 15 and only 4 percent were 65 years or older. By contrast, the developed countries' population of children under 15 was 22 percent of the total, with 11 percent 65 years or older." *Source*: United Nations Population Fund (UNFPA), *Population and the Environment: The Challenges Ahead*, August 1991.

they were three centuries ago however, so that these States have a different point of departure. Europe's mortality rates declined slowly as the changes caused by the Industrial Revolution took effect; the European colonies experienced far more sudden declines in death rates. Thus, the rates of population increase (the "explosion") in developing, urbanizing, industrializing Western Europe in its second stage were probably less than they are and will be in many developing countries now experiencing these conditions. Thus the argument that the current population explosion will soon subside as it did in Europe and Europeanized countries that industrialized as much as a century ago may well be invalid.

## Population Policies

Many States have population policies that reflect their internal demographic conditions. Those that have expansive population policies desire an increase in the growth rate, and reward large families by tax reduction and even elimination, "baby bounties," and even public praise. The Soviet Union, which desired growth in view of the vastness of its territory and the need for eastward spread of effective occupancy, was an example. Most of the former communist countries of Eastern and Central Europe have long been hampered by labor shortages and have adopted a number of policies designed both to bring more women into the labor force and to encourage them to bear more children. Many excellent and cheap nursery school/day-care centers, generous maternity leaves at full pay, and a system of baby bounties are among the major features of these efforts. Canada, Israel, Brazil, and France are other countries that have been pursuing "pro-natalist" policies.

### Restricting Growth

While some countries pursue expansive population policies, encouraging births and immigration, others pursue *restrictive* policies, encouraging limitation of births and

even in a few cases, such as that of the exceedingly densely populated Netherlands, encouraging emigration. Unfortunately, there is not yet sufficient correspondence between the countries with restrictive policies and those that need them. A cursory glance at current population figures and trends is sufficient to reveal that population growth is most dramatic—and most alarming—in the poorer countries of the world and the poorer segments of the populations of the richer countries. That is, the population is growing most rapidly in those societies that can least afford to support more people. It is as true now as it ever was that, in the words of the old song, "the rich get richer and the poor have babies."

During the period 1970–76 the population of the world grew at an average annual rate of 1.9 percent. By 1991 this rate had fallen, but only to 1.8 percent. As usual, the average masks the range of the figures—from a low of 0.3 percent in Northern Europe to a high of 2.9 percent in Africa. A growth rate of 1.8 percent is serious enough, but a rate of 2.9 percent is staggering, and even this average hides local growth rates of as high as 3.9 percent per year. When we consider that some States and parts of others are still in the first stage of the demographic transition—that is, where the death rates are still very high (on the order of 3.0 percent)—we face the frightening prospect of enormous populations when the death rate drops to levels approaching those of the industrialized countries (on the order of 0.9 percent).*

A most unfortunate coincidence in this period of history occurs between skin color and wealth. With the major exception of Japan and a few minor exceptions, the societies with the highest per capita incomes and the lowest population growth rates are predominantly light-skinned, whereas the rapidly growing poor populations are predominantly dark-skinned. This led many poor countries to reject the advice of the developed countries that they take urgent

*As a point of reference, we might keep in mind that a population growth rate of 3.0 percent per year will result in a 19-fold increase in a century.

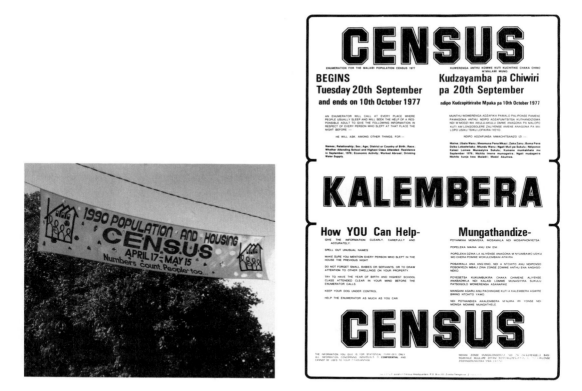

**Censuses of population and housing are vital.** All countries need accurate and up-to-date census figures to assist in planning of all kinds, and most exhort their residents to cooperate with their censuses. On the left, a banner over Frederick Street near the central business district of Port of Spain, Trinidad, contains just such a message. (Martin Glassner) On the right, a poster in English and Chichewa gives detailed instructions for assisting the census enumerator and assures confidentiality. (Government of Malawi, collected by Martin Glassner)

**Emotional resistance to family planning.** A slogan painted on the wall around a derelict house in Kingston, Jamaica reads "Birth control; Plan to kill blacks." The picture was taken in 1974. Thirteen years earlier, the author saw a slogan nearby that read, "Birth control is a plot to kill Negroes." The fear persists despite semantic changes, but in Jamaica, as in most developing countries, family planning is gradually being accepted by large segments of the population as both necessary and desirable. (Martin Glassner)

and effective measures to limit births, viewing this as demographic warfare, even genocide. There has been a tendency to view population growth as an asset in the struggle to redress the balance between rich and poor and to overcome the disabilities deriving from colonialism.

Fortunately, this attitude has changed. Perhaps it was the example of Japan that stimulated the change when that country demonstrated incontrovertibly the spectacular gains to be achieved from vigorous action to limit population growth. Perhaps it was the sudden realization in such pro-natalist countries as Brazil and China that the growth of population was not only gobbling up the economic gains achieved through great effort and sacrifice, but it was even threatening to reduce an already low level of living. Perhaps it was the erosion around the world of religion's hold on the minds and habits of people. Very likely it was a combination of all these factors and others that resulted in a slowing of the world's population growth rate in 1977 for the first time in memory. Yet even here, the slight drop was due largely to the reversal of China's population policy, and in 1978 China experienced setbacks in its new campaign to reduce the birth rate. Since then, the rate of growth worldwide has been dropping very slowly.

It is politically as well as sociologically significant that most populations that have recently experienced sharp and perhaps permanent drops in their birth rates are both poor and dark-skinned: Mauritius, Singapore, Hong Kong, Taiwan, Barbados, Trinidad and Tobago, and Puerto Rico are outstanding in this respect, along with Chile, Argentina, Costa Rica, and Uruguay, all of which are less poor and less dark. Other countries that have made notable progress in this area are Sri Lanka, South Korea, and Cuba. All this is encouraging, but it is sobering to observe that these are all small countries whose combined population is less than 170 million. We cannot be sanguine about the population problem until we can see comparable—and sustained—progress in such countries as China, India, Indonesia, Pakistan, Brazil, Mexico, Nigeria, Kenya, and Egypt.

Nor are population problems confined to the poor countries. Even rich and middle-class countries contribute to the excessive demands we are placing on our abundant but finite store of "natural resources." After all, poor people, for all their numbers, do not consume very much, while the United States, with some 5 percent of the world's population, consumes about 50 percent of the world's goods and services. This situation raises moral questions as well as political ones. It cannot be sustained forever, and possibly not for very much longer.

Following up on its World Population Conference in Bucharest in 1974 and its International Conference on Population in Mexico City in 1984, the United Nations sponsored the International Conference on Population and Development in Cairo in September 1994. This latest conference was very different and even more important than its predecessors. The very fact that it was held in Egypt was remarkable, for in Mexico City Egypt was one of the last few holdouts against extensive family planning policies, stressing religious and political objections. Yet in mid-1994, on the eve of the Cairo conference, Egyptian President Hosni Mubarak was awarded the UN's Population Prize. In his acceptance speech in Geneva, President Mubarak noted that, although his country's population growth rate had been rising for more than half a century, it had been dropping for the past decade, "thereby confirming that a real change has taken place in Egyptian society." He attributed the drop from a 2.8 percent rate in 1980 to 2.2 percent in 1994 to a basic change in attitudes among the Egyptian people, as a result of active government encouragement and its cooperation with various United Nations agencies.

Much of what transpired at the conference in Cairo lies in the realm of sociology, but some aspects of it are relevant to our context as well. First, of course, is the emphatic linkage between population control and economic and social development, a connection that had been gingerly avoided at the previous two conferences. By now it was so obvious that there were scarcely any naysayers. Even pro-natalist Brazil agreed with the

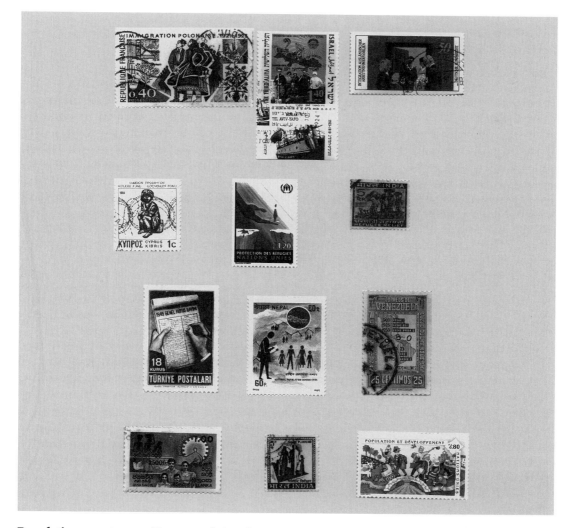

**Population on stamps.** Many population themes appear on stamps from time to time. In this selection, the top row displays stamps from France, Israel and Germany celebrating immigration from many countries and cultures. In the second row, stamps from Cyprus, the United Nations and India urge assistance to refugees. Then, Turkey, Nepal and Venezuela urge cooperation in the census, and in the last row, Sri Lanka, India and The United Nations link population and development. (Martin Glassner)

basic promise. From all over the developing world—Colombia, Bangladesh, Botswana, Iran, and others—came reports of success in reducing birth rates. Even the Catholic Church chose to withdraw from its forward position opposing any controls on population growth at the beginning of the conference and join the rest of the delegates in debating methods of limiting population growth while improving the quality of life and protecting the human rights of those people who were born, especially women. As in many other major conferences in recent decades, NGOs from around the world were exceptionally active and effective.

In the end, the conference adopted a 16-chapter, 20-year Program of Action. It laid out a new strategy for addressing population issues, emphasizing the numerous links between population and development, and focusing on meeting the needs of individuals rather than on achieving demographic targets. The key to the new program is empowering women, especially by giving them greater access to education, training, and health services. It advocates making family planning universally available by 2015 at the latest. It also addresses issues related to the environment, consumption patterns, family life, migration, the AIDS pandemic, information, technology, and others. The program was so carefully prepared and thoroughly debated that even the Holy See surprised many delegates and observers by supporting its two basic themes: economic development and empowerment of women. The 1994 International Conference on Population and Development can be expected to have profound influences on the course of world history for at least the next several decades.

### International Migration

A few countries are stimulating population growth by encouraging immigration. Today such countries as Brazil, Canada, Israel, and Australia are actually subsidizing immigrants who are expected to become permanent residents. Of course, such immigration is highly selective. No country except Israel (and certainly not the United States) as a matter of policy is willing any longer to welcome "your tired, your poor, your huddled masses yearning to breathe free, the wretched refuse of your teeming shore," as Emma Lazarus so poignantly expressed it in her sonnet inscribed on the base of the Statue of Liberty. The immigrants most desired today are those who are educated, healthy, skilled, financially independent, and easily assimilable; that is, precisely the ones most needed in their own countries.

Most countries, as we have seen, have heterogeneous populations, and typically the various ethnic groups within the population display differential rates of growth. Not infrequently, governing elites will attempt to improve their demographic position by for-

**Family planning campaign in a developing country.** This billboard outside Port-au-Prince, Haiti, is typical of government efforts throughout Africa, Asia, and Latin America to encourage couples to have no more than two children. These campaigns have had increasing success. (Martin Glassner)

bidding intermarriage among groups, by discrimination of various kinds against the less favored groups, ethnically selective immigration policies, and even, in extreme but by no means rare cases, genocide. The concerns of the rulers about being outnumbered by rival ethnic groups are based on solid demographic, cultural, and historical facts. Flemings have multiplied faster than Walloons in Belgium, for example, and now outnumber them; Muslims are no longer a minority in Christian-dominated Lebanon; and the Slavic rulers of the Soviet Union were perplexed by the rapid growth of the Muslim population of Soviet Central Asia.

Both the Rhodesian and South African governments, representing very small minorities of their populations, attempted to strengthen their positions by encouraging immigration of white Europeans. In both cases the effort failed, in part because of potential immigrants' reluctance to enter a potentially dangerous situation. The South African government virtually abandoned its policy after it discovered that the overwhelming majority of new immigrants were joining the English-speaking rather than the Afrikaans-speaking sector of the population and held distinctly more liberal views than those of the governing elite.

A policy adopted by some countries to ease what appear to be temporary or spot labor shortages is that of importing temporary or specialized workers. This situation has been most prominent in the past few decades in Western Europe, the Arab oil States on the Persian Gulf, and southern Africa, where many millions of people are involved. The economic recession beginning in the late 1970s in Europe, however, prompted a recession of the tide of migration, back to Turkey, Greece, Yugoslavia, Italy, Spain, Portugal; back in many cases to joblessness and despair. Smaller movements of this type in this century include those of Basque sheepherders and Mexican farm laborers into the United States, Jamaican laborers to Panama to build the canal, Pacific islanders to Australia to harvest sugar cane, and Egyptian teachers to the Persian Gulf States. This type of movement was much more common in the nineteenth century when, for example, Italian *golondrinas* (swallows) migrated seasonally between Argentina and Italy to work on farms during the Southern and Northern Hemisphere growing seasons.

Selected immigrant groups are also invited to settle in sparsely populated areas of a country, to help expand the ecumene and contribute to the economy. Such movements, which are generally very small, are typified by Okinawans and Japanese settling in the eastern lowlands of Bolivia and Mennonites in the Gran Chaco of Paraguay and the Mexican state of Chihuahua.

In the past 30 years or so, while international migration for farm labor and agricultural colonization has remained stable or even declined, migration across international boundaries for temporary work in factories, mines, and other industrial settings and in other nonagricultural occupations has vastly increased. And it has become more politicized. We have already mentioned the huge temporary worker programs in Europe and southern Africa, but they deserve more attention.

In the prosperous countries of postwar Northwestern Europe, labor shortages developed quickly following reconstruction. Individually, countries began developing programs for bringing unemployed or underemployed workers from the poorer areas of Southern Europe, North Africa, and Turkey to as far north as Sweden. Some of the workers became modern *golondrinas*, migrating northward for short periods of work and returning home seasonally or when the crisis in the factory had passed or when they were displaced by machines or when they had earned and saved enough to meet their immediate needs—and doing this repeatedly. Others, however, particularly Arabs in France and Turks in Germany, settled down for longer periods, perhaps permanently, sending for their families or starting families in their new homelands. These are the people who were hurt most by the economic hard times of the 1980s, and these are the ones who have had the greatest cultural, as well as economic, impact on the landscape.

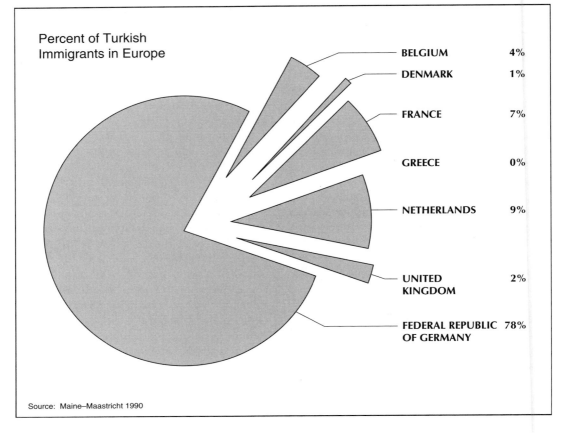

Percent of Turkish
Immigrants in Europe

| | |
|---|---|
| BELGIUM | 4% |
| DENMARK | 1% |
| FRANCE | 7% |
| GREECE | 0% |
| NETHERLANDS | 9% |
| UNITED KINGDOM | 2% |
| FEDERAL REPUBLIC OF GERMANY | 78% |

Source: Maine–Maastricht 1990

**Table 37-1   Foreigners in West Germany at the beginning of 1990**

| | | | |
|---|---|---|---|
| Turks | 1,612,600 | Moroccans | 61,800 |
| Yugoslavs | 610,500 | Vietnamese | 33,400 |
| Italians | 519,500 | Sri Lankans | 32,700 |
| Greeks | 293,600 | Czechs | 31,700 |
| Poles | 220,400 | Hungarians | 31,600 |
| Austrians | 171,100 | Lebanese | 30,100 |
| Spaniards | 127,000 | Tunisians | 24,300 |
| Dutch | 101,200 | Indians | 23,900 |
| Britons | 85,700 | Afghans | 22,500 |
| Americans (US) | 85,700 | Romanians | 21,100 |
| Iranians | 81,300 | Japanese | 20,100 |
| French | 77,600 | Total | 4,845,900 |
| Portuguese | 74,900 | | |

*Source of Figures:   The Courier* (ACP-EC) No. 129, Sep.–Oct. 1991, p. 62, from Federal Statistics Office.

These two figures illustrate some of the characteristics of the new demographic situation in Western Europe in 1990. First, a very high proportion of "guest workers" from North Africa, Southern Europe, and Turkey have become permanent residents, thereby creating a more heterogeneous population than ever before, in a region in which there have long been problems with ethnic minorities and discrimination against them. Second, some immigrant nationalities have tended to concentrate in certain countries: Algerians in France and Turks in Germany, for example. Third, each European country, even each member of the European Union, maintains its own immigration policies. Those of Germany are probably the most liberal; those of Switzerland probably the most restrictive. The EC will eventually have a uniform immigration code, or at least some harmonization or coordination of immigration policies after 1996. Fourth, these figures represent the situation *before* the large wave of immigration began arriving from Eastern Europe, including new refugees from Yugoslavia. The figures also do not reveal illegal immigrants, who probably total in the hundreds of thousands, or the hundreds of thousands of ethnic Germans from the former USSR and Central and Eastern Europe, who are admitted freely into Germany and not listed as foreigners. All of this also helps to explain why already densely populated Europe is not accepting large numbers of refugees from developing countries.

Mosques have been built in Germany, whole neighborhoods in major cities now house Muslim workers and their families, and some social frictions have developed. Table 37-1 gives a rough picture of the nature and scope of these changes. In France, Switzerland, Germany and other countries, antimigrant demagogues with unconcealed racist attitudes have been urging expulsion of the swarthy foreigners, and these proposals have become major political issues. One wonders what will happen in these countries when their economies again expand rapidly and again there is a need for swarthy labor.

In southern Africa the situation is similar but different in important ways. The Republic of South Africa is the giant magnet attracting desperate workers from all of the poverty-stricken countries of the region. In mid-1985 nearly 400,000 of them were officially registered, with possibly four times that number in the country. Unlike Europe, South Africa at least tries very hard to control this migration strictly. Migrants have been forbidden to bring their families and have had to live in barracks provided for them, work at specific jobs for specified periods, and have no hope of settling permanently, of entering the mainstream of society, or even of getting any education or advanced training. Little or nothing has been done to reduce tribal, ethnic, or national frictions among the migrants. But the money they send home is not just beneficial, it is essential to their families and their countries' economies as well. The threat of expulsion of these foreign workers in the event of imposition of sanctions was a powerful weapon in the hands of the government. Of course, South Africa itself would inevitably suffer from the sudden departure of most of the people who do the dirty, hard, and dangerous work, but the cost-benefit ratio of such expulsion is almost impossible to calculate.

In both Europe and South Africa there have been profound changes in the migration picture recently. In Western Europe many of the "guest workers" of the 1960s and 1970s have remained there rather than returning to their native countries. Now many of them have children born in Europe. Nevertheless, depending on the nationality laws in each country, they or their children may or may not be eligible for permanent residence or even citizenship. This problem is now compounded by a veritable flood of migrants from the East, chiefly into Germany, unleashed by the final crumbling of the remaining political barriers to movement as communist governments were replaced by more democratic ones. Europe has to deal with this exceedingly difficult problem. In South Africa also, liberalization of government policies and economic recession have resulted in a return of many migrants voluntarily to their home countries, and in far fewer restrictions and better working and living conditions for those who remain. Both groups may well play important roles in the possible economic and political integration of southern Africa.

Currently, there are other significant international labor migration flows. Among them are those into Argentina from all of its neighbors, into the United States from Mexico and the Caribbean, and into Senegal from Mali and Mauritania. Just as the rise in oil prices in the 1970s triggered a recession in Europe that generated an outflow of temporary workers, so the collapse of oil prices in the 1980s triggered recessions in many of the oil-exporting countries that forced migrants back from Venezuela to Colombia, from Nigeria to a number of countries to the north and west, and from Saudi Arabia and the Gulf States to many other Arab countries, Korea, the Philippines, and South Asia. Hundreds of thousands more returned home as a consequence of the Iraqi occupation of Kuwait and the subsequent Gulf War in 1990–91. When oil prices rise again, will the flow again be reversed? Is this the best way to provide needed workers? Is economic efficiency worth the political and social costs?

Not all migrants from poor countries to rich ones, however, are uneducated, unskilled workers. The "brain drain" has been a major concern of many developing countries since the late 1940s, and it is constantly growing. Britain, France, Germany, Canada, Australia, and especially the United States

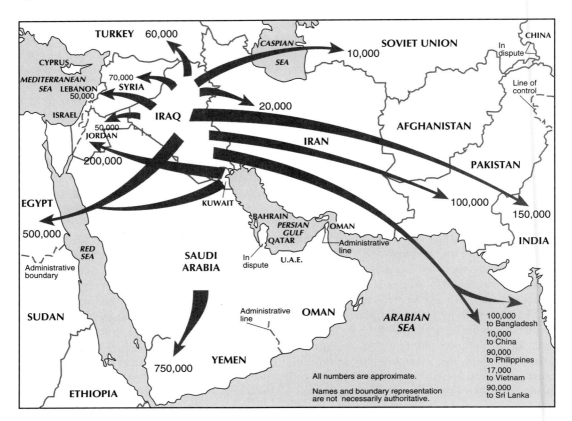

**Foreign Workers Flee the Gulf.** As a result of the Iraqi invasion and annexation of Kuwait in August 1990 and the economic sanctions and war against Iraq in 1990–91, more than 2 million people fled these two countries. The first wave was of foreign workers in Kuwait, who began leaving as soon as the Iraqis arrived. The map indicates that most of them went to Jordan and Egypt, but additional thousands went to other countries. When the coalition campaign against Iraq began, hundreds of thousands fled that country, as the map shows, while Yemenis were encouraged to leave Saudi Arabia because Yemen supported Iraq, and Saudi Arabia was a leader of the anti-Iraq coalition. Most of those who left Iraq and Kuwait for Jordan, Syria, and Lebanon were not nationals of those countries, but rather considered themselves Palestinians. In a third mass migration after the war (in late winter 1991), about 1.4 million Iraqis went to Iran and another 400,000 went to Turkey, all fleeing the ruthless army suppression of a Shiite revolt in the south and a Kurdish revolt in the north. By the end of 1991 nearly all of these people had returned to Iraq, but few of the guest workers had returned. In early 1992 there were still about 35,000 refugees in Saudi Arabia and nearly 900 in Kuwait, all being cared for by the Office of the United Nations High Commissioner for Refugees. That agency also did heroic work with the Kurds and Shiites, aided by the coalition forces. *Source of map:* US Department of State, *Geographic Notes* No. 14, 1 October 1991, p. 15.

are attracting—and welcoming—educated and trained people from developing countries who are seeking higher salaries, better working conditions, better research facilities, more opportunities for advancement, and perhaps more political freedom. Even the character of labor migration from Puerto Rico and Mexico to the United States has changed as more managers, skilled technicians, and professionals migrate northward and blend readily into the general population. Recent immigrants from many Asian, Latin American, and even African countries now occupy important positions in North American and European laboratories, hospitals, computer centers, universities, and corporations. Other significant "brain drains" in recent years include those from South Pacific islands to Australia and New Zealand and the mass exodus since the mid-1980s of scientists, managers, and other skilled professionals from Hong Kong, anticipating its return to China in 1997. They are going chiefly to Singapore, Australia, and Canada.

This form of "foreign aid" to the United States and other rich countries from some of the poorest countries in the world is rarely discussed in the public media or the professional geographic literature. It has been of concern, however, to the United Nations, which has created an Inter-Agency Group on Reverse Transfer of Technology to consider its nature, scale, and effects and to recommend ways of dealing with it. We would do well to do likewise.

### Refugees

Not all migrants who cross international boundaries do so voluntarily. The history of the world from biblical times on is repeatedly punctuated by horrific tales of mass expulsions of peoples, of people forced to flee from famine and pestilence and natural disasters. Whole populations have been uprooted by war, revolution, and boundary changes, fleeing or being driven to foreign and sometimes hostile lands. Never before in history, however, have we experienced such a lengthy period of massive and sustained refugee flows as we have in the twentieth century. To recall but a few of them: the partition of India; the breakup of the Ottoman Empire; the creation of Israel and Bangladesh; the civil wars in Sudan and Nigeria; the independence of Algeria; the westward migration of Poland after World War II; Stalin's eastward evacuation of ethnic minorities; the outpouring of political refugees from Hungary in 1956, Cuba in the 1960s, and Indochina in the 1970s; Paraguayans settling in Argentina, Ugandans in Kenya, Jews from Arab countries in Israel; the millions of people forced back and forth across Europe by the Nazis and the dazed survivors who were called "displaced persons." The list can go on and on, and is likely to extend far into the future.

Not only is the magnitude of refugee flows far greater than ever before in history, but they are no longer simply incidental to the climactic political events or natural disasters that generate them. Refugees have always evoked some measure of sympathy and support, and, more recently, organized humanitarian aid, but what is needed now is detailed study of the geographic, political, economic, and social causes of refugee movements, their nature and characteristics, and their effects on the sending and receiving countries and on the global community as well. Here we can only give a brief introduction to the subject.

The first problem we face is a definition of the term *refugee*. According to the 1951 United Nations Convention Relating to the Status of Refugees and its 1967 protocol, a refugee is

> any person who, owing to a well-founded fear of being persecuted for reasons of race, religion, nationality, membership of a particular social group or political opinion, is outside the country of his nationality and is unable or, owing to such fear, is unwilling to avail himself of the protection of that country, or who, not having a nationality and being outside the country of his former habitual residence, is unable or, owing to such fear, is unwilling to return to it.

This definition, with some variations, has been adopted by a number of countries, including the United States, but it is far from satisfactory. How does one, for example, define *and* prove "well-founded fear" or "persecuted?" Is it really necessary for every single member of a group to prove that he or she is in fact a "refugee"? Much more important, however, this definition does not include people escaping war and civil strife or those who do not cross international boundaries. The Organization of African Unity adopted in 1969 a Convention Governing the Specific Aspects of Refugee Problems in Africa that includes the first category but not the second; even in Africa, for the most part, a person does not become a refugee until he or she crosses a border. There are many other definitional problems, but the important point is that *all* such definitions are inherently political.

**Tibetan refugee camp in Patan, Nepal.** Many Tibetans fled the Chinese occupation of their country in 1950, mostly to India, and another 100,000 fled with the Dalai Lama after an uprising in 1959 was crushed by Chinese troops. Several thousand escaped to Nepal, where they were given asylum. Many have integrated into Nepali society, but in 1989, when this photograph was taken, many others were still living in camps in or near major population centers where they live rather normal lives, though far from home. Between December 1990 and the end of 1994, Nepal received another wave of refugees, some 86,000 Bhutanese of Nepali origin expelled from Bhutan in a local version of "ethnic cleansing." Negotiations for their return had not been successful by mid–1995. (Martin Glassner)

Even if we confine our consideration to genuine political refugees, ignoring those who flee economic deprivation and natural calamities, we face other problems. A refugee loses this status, for example, when he or she is given citizenship in another country or returns to the home country. More important is the fundamental UN principle of *nonrefoulement,* the right of a refugee not to be forcibly repatriated to the country of origin. States of asylum are often all too anxious, often for domestic or external political reasons, to get rid of refugees as rapidly as possible and ignore this right. Then there are the questions arising when persons flee *anticipated* persecution or violence without having actually suffered it, when refugees are interdicted en route to a country of refuge and turned away, or when refugees are interned for political reasons or to discourage further influxes of refugees.

To illustrate some of these points: The world was horrified in the early 1990s by the forcible repatriation of Burmese and Cambodians from Thailand, Vietnamese from Hong Kong, Albanians from Italy, and Haitians from the United States. Even if, as the unwilling hosts contended, none of these people met the strictest definition of "refugee," surely some more humane way of managing the situations could have been found—if there had been a political will to do so.

This stands in stark contrast to Operation Moses, the clandestine evacuation in 1984 of several thousand Ethiopian Jews to Israel by Israeli military and civilian personnel, who rescued them from Sudanese refugee camps and brought them to Israel. This remarkable feat was surpassed in May 1991, when almost the entire remaining Jewish community in Ethiopia, 14,400 souls, were airlifted from refugee camps and slums in Addis Ababa to Israel in just over 24 hours! This evacuation, meticulously planned and flawlessly exe-

**Mozambican refugees in Malaŵi.** Many thousands of Mozambicans fled into neighboring countries to escape famine and civil war. Many have been sheltered in camps in Malaŵi, which receives some help from other countries and from international agencies for their care and maintenance. Although many have returned to their homes since the advent of peace in Mozambique in the mid-1990s, the remainder are still a heavy burden on one of the world's poorest countries. This camp is south of Dedza and very close to the Mozambique border. (Martin Glassner)

cuted as Eritrean rebel armies were investing Addis and about to take the city, will stand for generations as a true epic of man's ability to achieve the impossible—given the will to do so. Regrettably, this episode is likely to remain unique. It is difficult to imagine any other country taking such risks and expending so much money and energy to rescue from poverty and persecution anyone else's tired, poor, huddled masses yearning to breathe free.

Probably the first major refugee flows in modern times resulted from the expulsion in the late fifteenth and early sixteenth centuries of the Jews and Muslims who had been living in Spain for centuries. This was followed by the expulsion of Protestants from what is now Belgium, of Scots and Irish and heretics from the British Isles, of Huguenots from France, and many more. More recently, wars of independence inspired by the nation-state ideal have gener-

ated numerous outflows of people, before, during, and after the wars. After the success of the American War of Independence, for example, thousands of British Loyalists fled to Canada, the Bahamas, the West Indies, and elsewhere to continue living under the British flag. In our century, similar and more massive flows followed the creation of independent Indonesia, Algeria, Kenya, Vietnam, Mozambique, and Israel, among others. Although decolonization has largely been completed, its legacy lingers on, and parts of Africa and Asia are still reeling from the aftershocks.

Whether in Central America, the Horn of Africa, Pakistan's North-West Frontier, or anywhere else, the problem of the disposition of refugees is difficult, largely because of political considerations. Basically, refugees can be removed from that status by any of three means: voluntary repatriation to the country of origin, integration into the country of refuge, or resettlement in a third country. All of these depend on political decisions by governments, with many domestic and external considerations influencing them. Meanwhile, the refugees need help. The limited help offered refugees by the United States is illustrated by Table 37-2.

The two intergovernmental organizations with primary responsibility for refugees are the Office of the United Nations High Commissioner for Refugees (UNHCR) and the Intergovernmental Organization for Migration (IOM), both with headquarters in Geneva.* Their mandates are different and complementary, and they generally work as a team. The UNHCR is chiefly responsible for protecting the legal rights of refugees and providing them with physical safety and security, assistance with documentation and supervision, and coordination of emergency assistance, including education and training programs, but it is not itself an operational agency. Instead it relies on other agencies to plan and implement programs. These in-

*Until 1989 the IOM was known as the Intergovernmental Committee for Migration.

clude many of the organs and specialized agencies of the United Nations, the OAU, and the OAS as well as numerous private voluntary organizations, of which the most important are the International Committee of the Red Cross and Red Crescent and the League of Red Cross Societies.

The UNHCR continues its heroic work, often under extraordinarily difficult and dangerous conditions, principally, in the mid-1990s, in Southeast Asia, the Horn of Africa, Mozambique and neighboring countries, Pakistan, Afghanistan, Rwanda and neighboring countries, Bosnia and Herzegovina, and West Africa, especially Liberia. Its operations, however, are nearly worldwide.

The IOM implements programs of voluntary repatriation and third-country resettlement. Its mandate is broader than that of the UNHCR in that it may assist persons not classified as refugees under UNHCR definitions. Thus, it aids asylum seekers, economic migrants, labor migration, family reunion, return of individual self-exiles, and transfer of highly qualified personnel among its Latin American member countries. In 1974 it also launched a "return of talent" program to help

**Table 37-2    Refugees Resettled and Persons Granted Asylum in the United States**

| Origin | 1983 | 1984 | 1988 | 1993 |
|---|---|---|---|---|
| Africa | 2,490 | 3,396 | 1,500 | 6,839 |
| Ethiopia | 2,274 | 3,037 | 1,454 | 2,722 |
| East Asia | 39,853 | 53,148 | 35,144 | 38,198 |
| Vietnam | 21,465 | 25,132 | 17,594 | 30,920 |
| Cambodia | 13,163 | 20,551 | 2,900 | 156 |
| Eastern Europe and USSR | 13,605 | 12,391 | 28,268 | 58,611 |
| Latin America | 1,496 | 1,602 | 3,000 | 4,557 |
| Near East and South Asia | 12,296 | 13,611 | 8,400 | 7,399 |
| Iran | 7,922 | 10,470 | 6,309 | 1,302 |
| Other | 63 | 6 | 79 | 59 |
| Total | 69,803 | 84,154 | 76,391 | 113,152 |

*Sources:* 1983 and 1984: U.S. Department of State, Bureau for Refugee Programs, *World Refugee Report*, September 1985. 1988: U.S. Department of Health and Human Services, Office of Refugee Resettlement, *Report to the Congress*, 31 January 1989. 1993: U.S. Immigration and Naturalization Service, statistical report, Table 26, p. 81, 1994.

These figures are very revealing. Although the United States consistently accepts for asylum and/or resettlement the most refugees and asylum-seekers in absolute numbers, as a percentage of the country's own population, the United States ranks well below Sweden, Australia, Canada, and Denmark. Furthermore, the number admitted annually is a minuscule fraction of the world refugee population, estimated at over 19 million in 1993. Note also the origins of the refugees accepted: 44 percent in 1993 were from Eastern Europe and the Soviet Union, but only .02 percent came from Africa, whereas in the world's refugee population as a whole, the proportions are roughly the opposite. Clearly, humanitarianism is not the controlling factor in U.S. refugee policy.

reverse the brain drain. In carrying out its work, it cooperates with some three dozen nongovernmental organizations (NGOs), even more than the UNHCR. Between 1952 and 1989 this little-known organization resettled more than 4 million people in 126 countries. Like the UNHCR, its work will, sadly, have to continue for a very long time.

International agencies, however, cannot solve refugee problems; only governments can. They can do this in two ways: first, by helping refugees *in situ* and by terminating their refugee status by one of the three methods described above; second, and far more important than refugee relief, by eliminating the causes of refugee flows in the first place. This is far more difficult and costly in political terms than sending food and medicine to refugees or even inviting them in as permanent residents. Not only are governments failing to consider potential refugee problems when they undertake military action, for example, whether within or across their boundaries, but they often deliberately uproot innocent civilians and evacuate them or simply drive them out of their homes, as the Serbs did on a large scale in Croatia and Bosnia in the early 1990s. Even domestic security laws, conscription, persecution of minorities, and suppression of rebellions cause peaceful people to become refugees, and these conditions can only be rectified by governments. To date there is no evidence of a firm commitment by States to prevent refugee problems, and there are more refugees and other displaced persons in the world today than ever before. Many of them are refugees from famine.

## *The Problem of Food*

There are many people around the world who argue, sometimes very cogently and forcefully, that there *is* no population problem. Many deny that there is a food problem. They base their arguments on interpretations of available statistics (which frequently are incomplete or unreliable) if they are inclined toward quantification, or on simple but firm faith, whether the faith be in God, Marxism, Fate, or Science. Since it is most unwise to

rely on either statistics or faith alone, we must try to understand at least the political aspects of population and food, using a combination of approaches. This kind of analysis leads us to believe that there is a world population problem and that it is both a cause and a result of a world food problem; in fact, both are really aspects of a single problem. But let us not be misled; the problem is *not* one of too many people or too little food in the world. It is far more complex than that, and here we can only point out some of its more important political aspects.

The United Nations recognized the linkage between the two when it convened in 1974 both the World Population Conference in Bucharest and the World Food Conference in Rome. In both these conferences political factors were prominent, and subsequent UN attempts to follow up these conferences with action have been limited or even frustrated by politics. We can perhaps understand the political aspects of population just described, but the politics of food are less well known and perhaps more difficult to comprehend.

### *World Food Supply*

The 1974 World Food Conference took place in an atmosphere of grave concern. Only months before, the OPEC oil embargo and quadrupling of crude petroleum prices had generated a near-disaster in agriculture as tractors ran out of gas, the prices of petroleum-based fertilizers and pesticides increased alarmingly, and many other products made from petroleum for use in agriculture became either too costly for farmers to buy or unavailable altogether. Combined with the lingering Sahel drought and adverse weather conditions elsewhere, the embargo and price rise led to localized famines and threats of much greater ones. The specter of Thomas Malthus was ominously present at the conference.

By 1976, however, the world food situation had improved, responding to the drawing down of food stocks for emergency aid, better weather, the UN's program of assis-

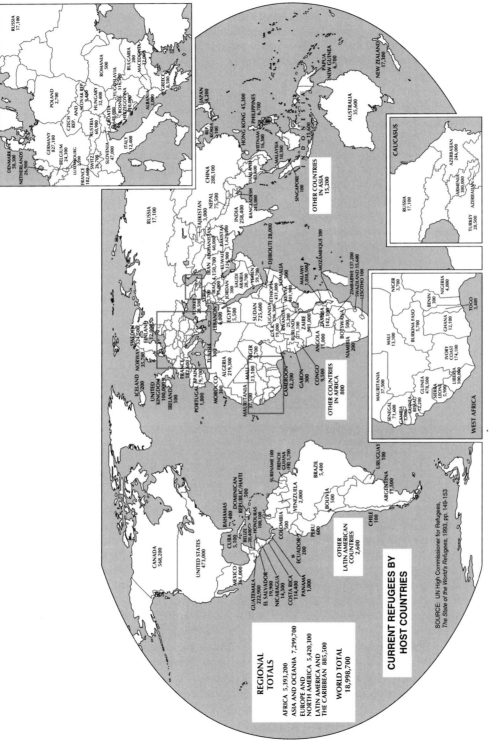

**Current refugee movements.** Since World War II hundreds of millions of people have been forced to leave their homes and seek refuge in another country. A very high proportion of them have been Africans, though their plight is seldom publicized in Europe or North America. The map shows only refugees so classified and registered by the UNHCR. Several million more are being assisted by UNRWA, the IOM and national programs. Most figures are provided by governments based on their own records and methods of estimation, and some include people considered as being in "refugee-like situations."

SOURCE: UN High Commissioner for Refugees,
*The State of the World's Refugees*, 1993, pp. 149-153

**CURRENT REFUGEES BY HOST COUNTRIES**

**REGIONAL TOTALS**

AFRICA 5,393,200
ASIA AND OCEANIA 7,299,700
EUROPE AND NORTH AMERICA 5,420,300
LATIN AMERICA AND THE CARIBBEAN 885,500

**WORLD TOTAL 18,998,700**

tance to the countries most seriously affected by the oil price rise, and other factors. Malthus was forgotten by many. Governments, international agencies, academics, and even Lester R. Brown, one of the most perceptive, prolific, and dynamic observers of world ecological problems and long-time president of Worldwatch Institute, were optimistic about our chances of increasing agricultural production greatly enough and rapidly enough to keep up with, or even get ahead of, population growth. But in 1978 the World Food Council reported that "beyond the foreseeable recovery of food production in 1976 and 1977, little real progress has taken place toward achieving the [World Food] conference objectives." Then in 1979 Worldwatch Institute reported that worldwide per capita production of fish, wool, beef, oil, mutton, wood, and cereals "has leveled off or begun to decline" since 1960, and Lester Brown, who wrote the UN-financed report, argued for concerted efforts to curb population growth.

Two decades after the World Food Conference the pattern was repeated. In November 1989 a group of government planners, world hunger scholars, opinion leaders, and scientists from 14 countries, including the United States, the USSR, Sri Lanka, Japan, China, and Mexico, as well as the United Nations, met in Bellagio, Italy. For four days they discussed the problem of food and concluded that in the 1980s, "the lost '80s," the world actually lost momentum in the race to keep up with rapidly rising populations. Their Bellagio Declaration was cautiously optimistic, suggesting four "achievable" goals for the 1990s, which would cut world hunger at least in half.

Their optimism seemed justified in early 1991 when the Food and Agriculture Organization of the United Nations (FAO) reported that in the year just past, world food production exceeded consumption for the first time since 1986–87, and food prices had declined. They projected another 8 percent increase in food stocks for the coming year. But, ominously, most of the production increase had been in the OECD countries and a few in Asia. Production had actually fallen

in Africa and Latin America. Then in December 1991 FAO reported a drastic drop in food production worldwide, led by a 23 percent decline of wheat production in the USSR. Once again, severe food shortages loomed.

Optimism and pessimism about food production have alternated and intermingled just like good harvests and bad harvests, like favorable and unfavorable weather. The reasons are not hard to find. First, we really know little about such things as food production totals, human nutritional needs, food consumption, food preferences and taboos, and many other factors essential in any analysis or projection. Second, even what we do know is interpreted in widely different ways according to the background, attitudes, and motives of the interpreters. Third, projections of food sufficiency or scarcity are based on a variety of projections, for example, of population growth, technological development, and climatic change—all of which are decidedly imprecise. Finally, few analyses or projections give sufficient weight to political factors. Before reviewing some of these political factors, it would be well to make a few very brief and very general observations about the present world food situation.

1. Through nearly the entire twentieth century, total world food production has kept ahead of population increase—but only just barely.

2. This pattern has left well over a billion people in the world chronically hungry, and about 700–800 million suffer chronically from malnutrition.

3. The carrying capacity of the planet is still very much greater than current production; that is, food production can still be increased substantially.

4. Experience has demonstrated that increasing food production substantially and permanently is a very costly and long-range process.

5. Even when a remarkable breakthrough is made, such as the development of high-yielding varieties of wheat in

Mexico and of rice in the Philippines, which introduced the "Green Revolution" in 1965, its impact is unlikely to be felt very widely, very soon, or for very long.*

6. The most dramatic increases in food production have been mostly in the industrialized countries, and rarely in the largely agricultural developing countries; during the 1970s per capita food production actually *declined* in 69 (more than half) of the developing countries, and the trend continued in the 1980s.

7. The proportion of foods exported by developed countries to other developed countries has been steadily increasing, so that now it accounts for more than half of all world food trade; poor countries are getting less and less of the food surpluses of the rich countries.

8. Extraordinary quantities—perhaps one quarter to one third in many poorer countries—of all food produced never gets to the people for whom it is intended because of spillage, spoilage, fouling and consumption by insects and rodents, fire and flood damage, diversion into black markets, and other factors.

9. Despite UN urging and efforts, there is still no worldwide program for the production, storage, and distribution of food, not even a world food reserve.

10. Food aid has never been adequate in either quantity or type, and the situation is worsening. Food aid, moreover, is all too frequently used as a tool or a weapon for political purposes.

11. There are grounds for hope that we

*In his acceptance speech on receiving the Nobel Peace Prize in 1970 for his work in initiating the Green Revolution, Norman Borlaug, an American agronomist, emphasized that all he and his colleagues had done was "to buy a little time" and urged population control as the only real solution to the problem.

can solve the food problem. Some of them follow.

(a) The seriousness of the situation is at last beginning to be understood.

(b) Some developing countries have dramatically increased their food output recently.

(c) There is now an array of experienced agencies providing technical and financial assistance to agriculture.

(d) More and better fertilizers have become available recently.

(e) For the first time in history serious attention is being given to improving productivity of subsistence crops in the humid tropics.

(f) There have recently been great advances in the production of chemical and biological controls for some major diseases and insect pests.

(g) Recent improvements in transport and communications can be applied to the improvement of food production and distribution.

(h) There is increasing awareness of the need for integrated rural development rather than simply increases of food production.

12. The hope is yet to be realized and will not be realized until there is a firm and permanent *commitment* on the part of governments everywhere to reorder their priorities and cooperate to assure everyone on earth an adequate diet.

## Food Policies

The last point is perhaps the most important, for it means far more than simply tinkering with gadgets and chemicals and budgets. It means profound changes in the political concepts and practices that may inhibit better food production and distribution, even more than a shortage of water, soil erosion, biological limits to the yield of cows and soybeans, or any of the other very real ecological constraints.

It means, for example, genuine land reform as part of a genuine program of agrarian reform. The deficiencies of the *latifundia* and *minifundia* systems that still characterize much of the agricultural world are well known. But land ownership in many countries is the basis of both economic and political power, and those who possess such power do not give it up easily. We discuss land reform and agrarian reform in more detail in Chapter 38; here we need only point out that such reforms not only increase food production, but also slow the flow of people to the cities, expand the market for locally manufactured goods, and improve the chances of developing a democratic society.

It means shifting much land, labor, and capital from the production on plantations of industrial crops and luxury foods destined for sale in the rich countries to the production of staple foods for the local people. But the plantation system is a vital and persistent element of colonialism that has brought wealth and power not only to the owners (usually foreign), but also to the local elites with whom they are allied.

It means shifting to less complex and more appropriate technology in areas unsuited for giant machines and potent chemicals. But this means less profit and power for the (usually foreign) manufacturers and (often local) distributors and service personnel.

It means reducing the number of middlemen and the amount of packaging and processing, all of which add substantially to the cost of food to the consumer without necessarily helping the farmer. But this again would weaken the positions of those who profit from high-priced food.

It means changing the systems of agricultural subsidies that prevail in the industrialized countries to new ones that would encourage more production rather than simply maintain high incomes for farmers. But this would shift political power and disturb entrenched bureaucracies.

It means developing a rational system for utilizing living marine resources so as to protect the ecosystem of the sea while still maximizing the production of high-quality protein for human consumption instead of for animal feed. But we have already seen that this is unlikely to happen very soon, if ever, as nationalism and exploitation continue to dominate our relations with the sea.

It means shifting financial, material, and human resources away from the production of luxury manufactures into research and development of food products, especially high-protein and tropical foods. But are the American people, for example, willing to give up their electric carving knives and their motorboats and their hair dryers?

It means organizing food aid so that the right kinds of food go to the right people in the right quantities in emergencies, with more emphasis on incentives for local food production where appropriate. But which countries are willing to cease using food aid programs as a means to dump surplus agricultural products and obtain political rewards in return?*

It means shifting the emphasis in the developing countries from rapid industrialization to improving food production. But how many corrupt and authoritarian leaders are willing to give up either the modern status symbols that swell their egos or the deals with (usually foreign) manufacturers that swell their bank accounts?

Perhaps the most essential and potentially most effective method of increasing the food supply and the overall quality of life for all human beings would be to rechannel

---

*Perhaps the most successful food relief project in history was the response of the international community to a devastating drought in southern Africa during 1991–92. A partnership was formed between the Southern African Development Community and the World Food Programme of the United Nations, with the cooperation of apartheid South Africa, to ship some 12 million metric tons of food into the region through a dozen ports and over many railways, including those of South Africa, to destinations in the region. Despite its remarkable success over a year and a half in averting a calamity that might have gotten a few minutes of air time on Western television programs, the project did experience delays in the distribution of food imports and higher than normal costs, and it still has not been followed up by a world food reserve, a long-range plan to improve agriculture in the poorest countries or any permanent arrangement for the efficient and economical movement of food in emergencies.

**Drought relief for Kenya.** Kenya, normally self-sufficient in basic foodstuffs, has had to import large quantities of food during the recurrent African droughts of the 1970s–1990s. The problem is exacerbated in Kenya by rapid population growth. In this photo, maize is being unloaded in Kilindini Harbour, Mombasa for transport to western Kenya. (Martin Glassner)

**The World Food Programme in action.** The United Nations agency operates a food depot outside Kampala, Uganda. When photographed in August 1994, it was very busy collecting and dispatching relief supplies for Rwanda, wracked by civil war and frightful massacres. (Martin Glassner)

money, natural resources, time, and talent currently expended on military equipment and weapons into investments in *Homo sapiens*. We referred to this in Chapters 19 and 23, but it bears repeating here.

Finally, governments must somehow be educated or compelled to treat their own people decently during times of stress brought on by food shortages. The behavior of the Ethiopian government during the drought and famine that afflicted the Sahel and most of Eastern and Southern Africa during the 1980s was less than admirable and must not be repeated anywhere, ever!

One might be tempted to look for models for expanded food production and more equitable food distribution to the "socialist" countries, those with centrally planned economies. But the first of them—the Soviet Union—was not able to achieve what we are suggesting in its 74-year existence. Marxism simply does not seem to work in agriculture, whatever virtues it might have elsewhere. Those socialist countries sometimes cited as models because they were able to feed their people adequately were precisely the ones that did the greatest violence to Marxist principles. Cuba continues to export tobacco, sugar, and citrus while importing nearly half of its rice, its basic staple food. China has since 1949 emphasized food production and

a tight organization for allocating the scarce resources necessary for high agricultural productivity. And in Hungary nearly all the agricultural land was privately owned, and there was nearly a free market economy in agricultural goods even before the abandonment of socialism there. Finally, in 1988 the Warsaw Pact countries alone accounted for 35 percent of world military expenditures and half of all arms transfers, mostly to developing countries.*

This is not to suggest that socialism was a total failure in agriculture. Certainly, the socialist countries were able, on the basis of their own experience, to contribute some useful ideas and techniques to the pool that also draws from experts of all kinds in the industrialized market-economy countries, from leaders in the developing countries, and even, perhaps especially, from the dirt farmers themselves, including the peasant farmers of the poorest countries. Nothing less than a rational and sustained cooperative effort can bring population and food once more into balance so we will never again have refugees from famine. Despite the very real obstacles and handicaps, it can be done.

*U.S. Arms Control and Disarmament Agency. *World Military Expenditures and Arms Transfers 1989*. Washington, 1990, Tables I and II.

# 38

# *The Politics of Ecology, Energy, and Land Use*

There is little need to elaborate on the quite well-known fact that our physical environment around the world is being rapidly and perhaps irreparably damaged by intense and growing human depredations. This degradation of the environment is the product of four factors we have already discussed: rapid population growth, increasing wealth leading to increased per capita consumption, ever more sophisticated technology, and accelerating urbanization. These trends are all interrelated and in varying degree interdependent. They are all, furthermore, influenced and shaped in some degree by politics—by positions, policies, and decisions based on political considerations.

Concern with the environment and its relation to politics is not new. We have discussed, for example, concepts of environmental determinism, whose origins go back at least to Hippocrates and Aristotle. Ibn Khaldun, as we pointed out in Chapter 1, theorized about the effect of the environment on the political units of his time and the life cycle of the State. Montesquieu, Ratzel, Kjellén, and Huntington are only a few others who have followed in this tradition. Toynbee, Wittfogel, and the Sprouts have more recently developed less deterministic concepts of "the impress of the environment on politics," as pointed out by

Kasperson and Minghi.* Study of "the impress of politics upon the environment" does not reach back nearly so far, but Ratzel and Huntington both recognized that it was important, and half a century ago Whittlesey examined it in some detail. Kasperson and Minghi suggest that a study of this kind of relationship might be organized into four major components: "political goals, agents of impress, processes and effects." They also consider a third type of environment–politics relationship, "the public management of the environment." Here they review some of the work on this subject by Barrows, Colby, White, Burton, and themselves, all during the twentieth century. They organize their discussion under three main headings—environmental policy and planning, resource allocations, and spatial linkages and area repercussion. These are most useful concepts and can help us understand the three types of relationships as we examine three aspects of the environment–politics linkage.

## *Ecology*\*\*

Before the mid-1960s, "ecology" was a word used commonly by geographers, biologists,

*Roger E. Kasperson and Julian V. Minghi, eds., *The Structure of Political Geography*, Chicago: Aldine, 1969, pp. 423–435.

\*\*We prefer the term *ecology* to *environment* because ecology is the study of the mutual relations between organisms and their environment. When it becomes a public issue, however, or when linked with human activity, environment serves just as well, and we therefore use the terms more or less interchangeably.

and other scientists but seldom heard or seen by the general public. Since then it has been adopted by a new generation concerned about rapid worldwide destruction of our planet's environment. They have broadened and fortified the conservation movement and impressed the public with both the urgency and the practicality of protecting our physical environment from further destruction and actually reversing the trend and restoring the environment if possible to its original state. The euphoria and élan of the early days have largely disappeared, however, and the ecology movement has settled down into a persistent, dogged, and frequently successful battle against entrenched interests, rigid thinking, and obsolete laws. Heartening also has been the spread of the ecology movement from the United States to other countries.

In only one generation Americans have lived through an economy of scarcity during the Great Depression, an economy of abundance in the post–World War II period, and an economy of waste at present. Through it all they have never lost faith in the eternal bounty of nature and the virtue of exploiting it enthusiastically. Now we must shift gears and reverse all this, but every stage, every step involves a political struggle. In less than half a century we have experienced titanic battles in the Congress and the press over the Glen Canyon Dam (Arizona), the Dickey-Lincoln project (Maine), the Cross-Florida Barge Canal, the Central Arizona Project, the Alaska Pipeline, and dozens of other proposals, large and small, for modifying our physical environment, allegedly in order to improve our economic and perhaps social environment. We have come to expect government at all levels (but particularly the federal government) to subsidize projects of this nature, but we are still unwilling to accept their regulation and control by government. Worse, we have yet to develop a national consensus on environmental matters that can be expressed in a national plan or at least guidelines.

The Environmental Policy Act of 1970 was a landmark, a giant step in the right direction. It spelled out goals and policies to guide all federal actions that would have an impact on the quality of our environment. It made a concern for environmental amenities and values a part of the mandate of every federal agency. It established the Council on Environmental Quality to identify the policy issues and alternatives for environmental administration. Finally, it required an annual report on the quality of the environment. This act was followed by other, more specialized legislation, such as the Clean Water Act and the Clean Air Act, all the products of diligent and persistent efforts of citizens individually and collectively working with— and on—legislators.

These have helped rectify two of the three basic problems that had prevented the federal government from playing an effective role in long-term environmental planning. When long-range planning has been undertaken, it has generally been intended to deal with problems posed by projected trends rather than to achieve desirable goals, and public policies too often have been defined and carried out in fragmented, narrow programs by mission-oriented agencies. Now there is somewhat more order in both setting and reaching goals, but there is still considerable scope for improvement, both in the planning and in the execution of plans. Both are made more difficult than necessary, however, by the failure so far to deal with a third basic problem: the fact that public administration in general is geared to annual appropriations that tend to favor short-term considerations.

It would be difficult indeed to find someone who would speak forcefully in favor of deliberately destroying our environment. Yet many argue in favor of postponing decisions on ecological problems or insist that they are not really so serious or that other problems should be attended to first or that we really cannot afford to protect wildlife or clean up streams or restore land eroded by careless farming practices or reduce noise pollution. In fact, all these arguments have some merit and it is unwise to ignore them. Since we cannot do everything first or well or at all, we must make choices. These choices are, in part, moral ones—but they are largely political. Decision making in these cases is always difficult.

The types of decisions to be made fall into three categories.

1. **Priorities of resource allocation.** There is so much to be done. As abundant as our resources are, they are limited and must be distributed among the various tasks to be accomplished. Money, energy, talent, and time must be allocated on some basis other than simply greasing the squeaky wheels, yet we have not yet devised such a system.

2. **Distribution of costs.** We all recognize by now that nothing is free; everything costs, even clean water and air. But who is to pay for achieving and maintaining a livable environment? Ultimately, of course, we all pay for everything, but the real questions are whether we pay now or later; through higher taxes or higher prices; in cash or in kind, or simply by foregoing luxuries and reducing our consumption of material things; and, of course, who is to bear what proportion of the costs?

3. **Distribution of benefits.** Should the benefits of an improved environment be distributed evenly throughout the entire country, through all socioeconomic levels and among all ethnic groups? That would seem to be very democratic, but we may question whether it is practical or desirable. Should not special attention be given to the physically and mentally handicapped, to deprived minorities, to low-paid workers, to the unemployed? Should the wealthy get subsidized marinas and rural people beautiful parks?

Clearly, there are no easy answers to any of these questions. Two contemporary but long-standing problems may serve to illustrate these points.

### Strip Mining

There are at present more than 3000 surface mines in the United States, spread widely across the midsection of the country from the Appalachians to the Rockies, most of them coal mines and most of them relatively small. Until adoption of the federal Strip Mining Control Act of 1977, regulation of strip mining was largely left to the states. Long after the severe ecological damage of strip mining had been amply documented, its control was spotty and inadequate. Kentucky, the country's leading producer of strip-mined coal, provides a good example.*

> Kentucky's efforts to control strip mining have followed a most peculiar course. The period 1947–1967 was marked by a continual increase in the strength and scope of the regulatory effort which culminated in a tremendous burst of initiative in the years 1965 and 1966. In that short span of time, the control program was vastly expanded and improved. Stringent and detailed administrative regulations were promulgated. A new strip mine control law, considered to be the toughest in the nation, was passed. Aggressive enforcement of both the regulations and the statute was undertaken. The available geological data for those years indicate that the improved control program actually achieved a palpable reduction in strip mine-related ecological damage.
>
> The strip mine control program reached a peak of effectiveness in 1967. Then, just as the rest of the country was commencing a period of unprecedented concern for environmental quality, it entered a period of decline.

In this case, apparently neither pressure on the Kentucky political system by outside forces nor scandal or other crisis mobilizing public opinion was significant in achieving stricter controls. Rather, "the crucial political actor in the initiation of the program, as well as in its passage by the legislature and its successful implementation, was the Governor, Edward T. Breathitt." But his achievement did not survive his term of office: "The deterioration [in strip mine control] was the

---

*The material on Kentucky is based on Marc Karnis Landy, *The Politics of Environmental Reform*, Washington D.C.: Resources for the Future, 1976. All quotations are from Chapter 1, pp. 1–16.

result of the loss of political autonomy which resulted from electoral defeat suffered by the administration faction in 1967 and its subsequent failure to return to power in the gubernatorial election in 1971."

Because of the tenuous and uncertain nature of state control of strip mining, the Congress was finally persuaded, after decades of effort by ecologists and others, to pass federal legislation. There have been many reports of abuse and evasion of the 1977 Strip Mining Control Act, notably illegal strip mining on federal land. But rigorous enforcement of legislation requires not only money but also sufficient evidence to permit successful prosecution when necessary, and this is often difficult to obtain in regions where strip mining plays such an important role in the lives of people. The problems are likely to increase, not decrease, if demand for coal rises to compensate for shortfalls in petroleum supplies or problems with nuclear energy.

## Water Projects

One of the traditions of American politics since the founding of the republic has been the regular congressional appropriations to support what were once termed "internal improvements," more recently referred to as "rivers and harbors projects" and informally known as "pork-barrel" legislation. In April 1977 President Carter, less than three months in office, decided to break with tradition and slashed more than $7 billion worth of water projects from the budget. Despite the ample justification for most of these cuts on grounds of both economy and ecology, congressmen and senators from nearly all sections of the country raised a storm of protest. Their pet projects, those designed to benefit their constituents and win votes, were threatened, and they fought to protect them. In the end there was a compromise, and the president signed a bill that still provided over $10 billion for public works, including some big and expensive projects he had opposed. The president and the public learned how hard it is to overcome local demands for federally financed "improvements."

## A Brief Catalogue of Ecopolitical Woes

We do not have the space here to discuss in detail the manifold ecological problems of the planet, or even of the United States, that are caused or aggravated by politics. The problems are so numerous, complex, widespread, and intertwined that volumes would be necessary even to outline them properly. So here is a brief, random list of some current issues that are not particularly well publicized. They are presented in no particular order.

1. **Irrigation:** Irrigation agriculture has sustained civilization for thousands of years and is still vitally important for the subsistence of millions of people around the world. Yet in the United States it is not a matter of civilization or even of subsistence; it is a matter of luxury. We discussed some elements of this issue under the heading of public lands in Chapter 14, but the point warrants expansion here. Irrigation of rice fields in California and sugar cane fields in Florida is destroying the natural environment in order to provide crops heavily subsidized by the taxpayers that could be more cheaply imported to meet domestic needs. But Florida and California have large and growing congressional delegations that seem more interested in protecting the huge agribusiness complexes in their states than in protecting the consumer, the small farmer, the taxpayer, or the environment.

2. **Ocean fisheries:** During the 1970s some American commercial fishermen, chiefly those in New England and to a lesser extent in the Pacific Northwest, lobbied hard for a 200-mile exclusive fishing zone in order to be able to exclude foreign fishing fleets that they claimed were seriously depleting fish stocks. They won the legal battle but are losing the ecological and economic ones. Only 15 years after the president's proclamation of a 200-mile exclusive fisheries zone, several fish stocks were seriously depleted, many fishermen

were unable or unwilling to modernize their vessels and their techniques, and the United States had signed agreements with a number of countries allowing their fishermen to harvest stocks American fishermen cannot or do not wish to harvest.

3. **Waste disposal:** The disposal of the refuse of an economy based on waste has emerged as a major environmental, political, and social problem. We now have the spectacle of household garbage being trucked across the country or barged to Europe and Africa in a search for cheap, legal, and perhaps secret dump sites. Ocean dumping has been restricted for sound ecological reasons, and shooting garbage into outer space is not yet feasible, so we must find some terrestrial repositories. Toxic and radioactive wastes are even more difficult to manage. The most logical and the ecologically soundest method of dealing with the problem—produce less waste by reducing consumption—is considered un-American and is unlikely to be adopted very soon.

4. **Wetlands:** The biologically richest and most essential physical environments in the world are coastal wetlands, followed closely by inland wetlands. Yet the American propensity for living, working, and playing on or near water has drastically depleted the wetlands, perhaps past the point of no return. State and federal legislation to protect these fragile areas from the ravages of "developers" is hard won and easily lost. President Bush's solution to the problem of competition between commercial and environmental concerns was announced on 9 August 1991: redefine "wetlands" so as to remove huge areas from potential federal protection and make them available for "development."

5. **Wildlife:** The Alaska Native Claims Settlement Act of 1971 allocated to new native corporations large portions of the state. One of them, Old Harbor Native Corporation, now faces destitution unless it can sell or lease its land to commercial interests to develop for logging camps and sawmills, hunting lodges, canneries, fishing camps, airstrips, and other activities. The people would like to sell their land, or most of it, back to the federal government, which does not want it. The problem is that their land is within the Kodiak National Wildlife Refuge, established in 1941 to protect the Kodiak bear, the largest land predator on earth, of which there are currently 2500 to 3000, as well as bald eagles and many other wild animals. This is a classic case of conflict between preservation of the environment and its destruction through commercial development, only here the Aleuts are caught in the middle. Similar problems are faced by indigenous peoples throughout North America and elsewhere.

6. **Feral animals:** In the western United States, especially in the Great Basin, large herds of wild horses and donkeys roam freely, protected since 1971 as "living symbols of the historic and pioneer spirit of the West." The legislation was a product of a huge outcry of many conservation and animal-rights groups as well as the general public, who protested the mass slaughter of these feral animals whose numbers were far in excess of the carrying capacity of their environment. The result of the protection is that the number of horses alone has grown from about 35,000 to between 50,000 and 75,000. The "adopt-a-horse" and "adopt-a-donkey" programs haven't worked, zoos don't want the animals, the law protects them from the pet-food producers, and the Bureau of Land Management does not have the funds to manage them properly. So the feral animals drive away the native wildlife in the competition for scarce water and forage; even water just below the surface is being depleted, and erosion is being accelerated.

7. **Water:** Not only is improper and wasteful irrigation causing waterlogging and

salinization of good agricultural land, but the runoff water is contaminated with chemical fertilizers, pesticides, and fungicides. The per capita consumption of water in the United States is the highest in the world, and consumption is increasing with increasing population, wealth, and water-using technology, at the same time as we are losing potable water to pollution from many sources. Again, the politically acceptable approach to the problem has been to spend more money on water treatment, recycling of wastewater, and other palliatives rather than to reduce per capita consumption of water in the first place. Logical, of course, but politically unacceptable at present.

The overall problem of managing the environment wisely is compounded in the United States by its federal system. Since the early 1980s the federal government has abdicated its leadership role in this area, and some states are beginning to step in as understudies. States, counties, and municipalities are regulating the emission of toxic chemicals into the air, requiring and regulating recycling programs, regulating the use of cancer-causing substances, strengthening liability rules for oil spills, controlling automobile emissions, modifying allocations of irrigation water, and so on. Some of these programs are innovative, and some are ef-

fective locally. Since the problems are regional, national, even global, however, even the best local programs in the long run will simply not be good enough. Political geographers could contribute substantially to the effort at all levels of government and among the general population as well as within the academy. But few have done so at this point.

## Ecology in Other Countries

Concern with environmental matters has been manifested mostly among the educated and well-to-do. Poor people have other and higher priorities. The same is true of States. Generally, as we have pointed out in our discussion of marine pollution, the poorer countries of the world have only recently begun to realize that their environments are also in danger and that ecological damage may well cost more than industrial development will earn. All over the world, major development projects are being reexamined in light of new understanding of ecological principles. The Jonglei Canal project in Sudan and the Trans-Amazonian Highway in Brazil are examples. Even Egypt's pride, the Aswan High Dam, completed only in 1970, has already caused so much ecological damage in the Nile Valley and the eastern Mediterranean that there is talk of dismantling it. But the destruction goes on: The large wild animals of East and Central Africa are rapidly being exterminated,

**Ecology consciousness spreads to China.** Even China is now ignoring Marxist doctrine to some extent and making some efforts to reduce environmental degradation. This billboard in a public park in Zhengzhou, Henan, reads: "Theme of World Environment Day 5 June 1989—Warning: The Globe is Getting Warmer." It was erected by the Zhengzhou Environmental Protection Bureau. (Martin Glassner)

the forests of Southeast Asia are disappearing at a frightening rate, overirrigation continues to destroy cropland through salinization and waterlogging in North Africa and Southwest Asia, and soil erosion continues almost unchecked in the highlands of Latin America. It will be difficult to overcome the suspicion born of colonialism and foreign exploitation and convince the people in these regions that conservation is in their interest.

Even the Soviet Union began in the late 1970s to take action on environmental matters, turning slightly from its customary emphasis on heavy industry and massive development projects. In early 1979, the main governmental agency on the environment was upgraded to the status of a State committee and intensified a six-year-old program of cooperation with the United States in 11 major environmental areas, including air and water pollution and earthquake prediction. They issued a *Red Data Book* listing endangered species of plants and animals and began reducing chemical pollution, moving factories out of the central areas of major cities, removing sulfur emissions of hydroelectric plants, and in general demonstrating a serious commitment to ecology. Although we have no detailed information, we might suppose that the political infighting that led to this commitment was at least as brisk as in the United States.

The frightful devastation of the environment throughout Central and Eastern Europe and the Soviet Union became clear, however, only as their communist governments were crumbling. Newly organized nongovernmental organizations, newly liberated mass media, and newly unshackled scientists and other intellectuals began voicing loudly and documenting what hitherto they had only been able to conjecture and grumble about *sotto voce*. Now all of them are faced with daunting problems of restoring a reasonably acceptable degree of ecological sanity along with economic, social, and political sanity. We can hope that they will be successful—with considerable outside help—but we cannot be sure that any of them will not revert to dictatorship, despair, and decay.

Canada's Green Plan for a Healthy Environment was unveiled in December 1990. It contains more than 100 new proposals, policies, programs, and standards to clean up, protect, and enhance Canada's land, water, air, renewable resources, the Arctic, parks, and wildlife, and to reduce waste generation and energy use. This may be the world's most comprehensive national environmental plan. Much of it is based on the United Nations concept of sustainable development. Since the plan is the product of democratic give and take, however, it is full of compromises and really pleases no one. On the other hand, it is far more progressive than anything produced south of the border and has few determined opponents. We shall see how vigorously and effectively it is implemented.

One hopeful sign may be the fate of the projected Grande Baleine (Great Whale) Complex, phase II of the massive James Bay Project. This scheme to dam the La Grande and other rivers flowing through northern Québec into southeastern Hudson Bay, including James Bay, was initiated by Hydro-Québec, one of Canada's largest corporations. In 1971 the Québec government created the Société d'Energie de la Baie James, and construction got under way with little fanfare or opposition. Gradually, however, opposition began to grow, led by environmentalists and Cree Indian leaders. From the beginning, however, the project had the single-minded support of Robert Bourassa, Premier of Québec, and it became a major political issue in several elections. The opposition focused on the undoubted serious and varied environmental damage done by the dams, dikes, reservoirs, roads, airfields, and other facilities, and on the disruption of the traditional way of life of the Crees, Naskapis, and Inuit in the area. The supporters focused on the jobs created (mostly for Caucasians coming in from outside the area) and on the revenue from the sale of electricity (mostly to New York and New England). Politics won and James Bay I was completed by 1995.

Meanwhile, planning was going ahead for James Bay II, the Grand Baleine Complex to the north of the original project. In 1988, when the Québec government announced that it would proceed with the project, it looked unstoppable. Hydro-Québec had learned from experience that it had to be much more sophisticated in its public relations campaign, but so had the opposition. Hydro-Québec hoped that its initial support for the 1975 James Bay and Northern Québec Agreement and the subsequent Northeastern Québec Agreement (both discussed in Chapter 16) would calm the native peoples of the region and that the environmental impact sections of their own $400 million feasibility study would calm everyone else. By this time, however, Canadians

generally and Québécois in particular had swung toward support of the concepts incorporated into the 1990 Green Plan, the Crees had become more united and politically powerful, and the Parti Québécois government of Premier Jacques Parizeau was not nearly as committed to James Bay as the Liberal government of Bourassa had been. On 18 November 1994 Premier Parizeau announced that the $13 billion project was not a priority for his government and that it would not be constructed in the foreseeable future.

India has been building huge dams almost since independence, and generally they have been sources of pride. They have made

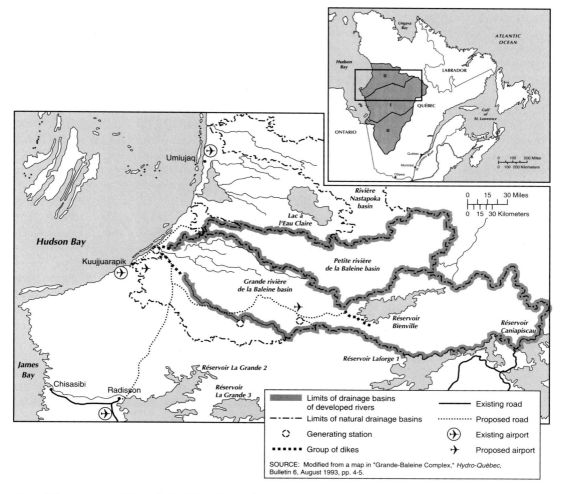

**Canada's proposed Grande Baleine Complex.**

available immense quantities of electric power and irrigation water, and have helped India's economic development keep pace with its rapidly growing population. Only recently have the environmental costs of these economic benefits begun to be considered. In 1989 there were popular protests against the Narmada Basin Plan to build 30 large, 135 medium, and more than 3000 small dams on the Narmada River and its tributaries. One of the outcomes of this project would be the world's largest man-made lake behind the planned Sardar Sarovar Dam. The lake would displace some 70,000 people in all, including many tribal people. The project, in Madhya Pradesh, is designed to benefit many people in Gujarat, Rajasthan, and Maharashtra as well. Opposition has been rising, however, as some experts are predicting "a major ecological calamity."

Africa's largest water diversion project and one of the largest public-works projects anywhere is under way now in Lesotho, the country that perforates South Africa. The Lesotho Highlands Water Project, initiated in 1986, involves building five major dams, hundreds of kilometers of highways and tunnels, and an underground hydroelectric power plant over a 30-year period. The purpose of this gigantic project is to provide massive amounts of electricity and irrigation water—largely for South Africa. Although it is being financed in part by the World Bank, the European Union, and the African Development Bank, much of the capital is coming from the South African government and private banks. The South African role generates controversy, of course, but so do the impending displacement of 20,000 people, potential erosion and sedimentation problems, endangered plants and wildlife, and the destruction of valuable archaeological sites. Although the project is being intensely monitored by ecologists and great efforts are being made to meet every objection, there is no assurance that the project will result in a net long-term gain for the people of Lesotho.

In Chile, the military dictatorship (1973–89) headed by General Augusto Pinochet essentially ignored environmental issues. After the democratic government of Patricio Aylwin took over in 1990, a national debate began over environmental policy. As in most of Latin America, the debate swirled around two competing conceptualizations of "sustainable development," the concept formulated in 1987 by the Brundtland Commission (explained in more detail later in this chapter) and now underpinning all development efforts of the United Nations and of many other intergovernmental organizations, governments, and voluntary groups. Sustainable development requires programs designed to stimulate economic growth, promote social equity, and protect the environment.

The debate in Chile was about how to achieve these three goals. One of the two competing views held that rapid economic growth based on free market economic restructuring would in the nature of things eventually result in improved housing, health, education, and other basic needs that constitute social equity, and would also allow people, as they got richer, the luxury of concern about the environment. The alternative approach to sustainable development does not subordinate social equity and environmental quality to market-oriented economic growth. Rather, it considers that market-based growth historically has not, especially in developing countries, led to the other goals and that it is therefore necessary for government to assist the process through grass-roots development projects and local control over resources. It is a more holistic approach to development and emphasizes the linkages among all aspects of all three goals.

The debate was strongly influenced by a variety of internal and external factors, including the return to democracy, the entrenched power of the traditional elites, pressure from the United States and the World Bank to adopt environmental protection policies, the nature and role of the opposition during Chile's gradual transition to democracy (1983–89), the strengths and weaknesses of the responsible technocrats in the government, changing world market conditions and the need to attract investors. In the end, the comprehensive environmental law enacted in March 1994 facilitates the market-oriented approach to sustainable development.

This does not mean, however, either that Chile is totally controlled by international capitalism or that it is headed for ecological disaster. It means that Chile has chosen, for the present at least, to build its development strategy on traditional foundations with less government involvement in the development process than the supporters of the alternative view had wanted. In addition, since Chile is a democracy, the alternative concept will still be expounded and will be able to influence decisions in many individual cases. The first effect of the legislation, however, was to cool the enthusiasm of the United States to have Chile join the North American Free Trade Area as the next step in creating a hemisphere-wide free trade area.

## *Energy*

The "oil crisis" of 1973–74 suddenly and painfully brought to the attention of the American public what scholars, government officials, ecologists, and others had been saying for many years: that the supply of hydrocarbon fuels, although very large and still unknown, is limited, whereas demand generally rises at an accelerated rate; that the United States and other countries have become too dependent on such fuels; and that petroleum can be used as a political weapon. The spate of books and articles on the politics of oil that appeared during the 1950s and 1960s was suddenly engulfed by a flood of new analyses and exhortations. The result has been a new line of thinking, not only in the industrialized countries but around the world, and new plans of action to forestall another such "crisis."

For the first time Americans (and others) began thinking seriously about the total energy picture instead of isolated portions of it, about the folly of building a society based on the expectation of unlimited supplies of cheap energy, about the ease with which mighty States can be held hostage by a few countries poor in technology but rich in fuel.

The fall of the Shah of Iran early in 1979 generated another "oil shock" of falling stocks and rising prices. Suddenly, for the first time in memory, the U.S. government and even private industry were preaching energy conservation, investing heavily in alternative energy research, building up strategic petroleum reserves, and intensifying the search for new sources of petroleum within the United States, including the continental shelf—all in the name of "energy independence."

But then, as happens so often with commodities, the price of oil dropped on the world market, down to about half of the peak prices. Almost immediately, research on alternative energy sources was shut down, the search for new oil reserves was slowed almost to a halt, offshore oil rigs were laid up, promotion of conservation was left to the NGOs and to private firms that stood to gain from it, and Americans resumed building and buying larger automobiles and other motor vehicles. "Project Independence" went a-glimmering. American petroleum imports, which had fallen from a high of 46.5 percent of total supplies in 1977 to a low of 28.1 percent in 1982, began rising again until they reached 42.2 percent of total supplies in 1991 (a much larger total than in 1973, of course), and were still rising. A quarter of U.S. oil supplies, in fact, came from OPEC countries in 1990, compared with 17.3 percent in 1973.

The bar graph shows the energy picture for the United States early in the 1990s. The most striking change from the 1973 picture is the considerable rise in the proportion of total U.S. energy derived from coal and nuclear fission, especially for the generation of electricity. Another is the actual drop in the proportion of energy derived from nonnuclear and nonfossil fuels. Geothermal energy, tidal power, wind power, wave power, ocean thermal energy conversion (OTEC), solar power—where are they? Hydroelectric power has dropped somewhat percentage-wise, and the "nonconventional" sources of energy have nearly vanished. So much for the American commitment to clean, renewable energy.

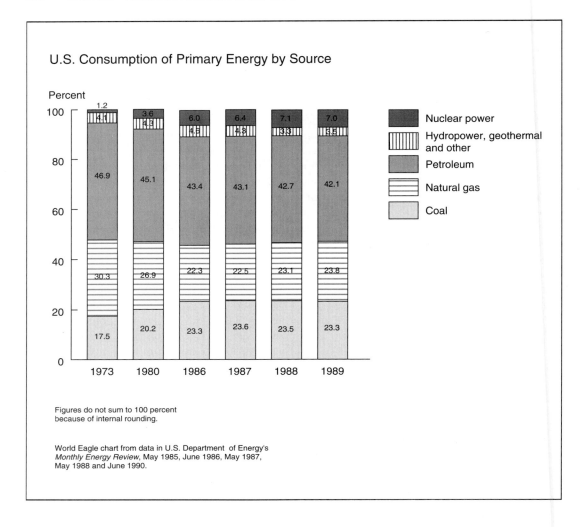

## U.S. Consumption of Primary Energy by Source

Figures do not sum to 100 percent
because of internal rounding.

World Eagle chart from data in U.S. Department of Energy's
*Monthly Energy Review*, May 1985, June 1986, May 1987,
May 1988 and June 1990.

In the United States more than 90 percent of the natural gas and nearly 75 percent of the crude oil produced domestically come from only five states—Texas, Louisiana, Oklahoma, New Mexico, and Kansas. Despite the undoubted wealth and vaunted political power of the "oil lobby," the fact remains that Americans still enjoy huge amounts of relatively cheap energy. It is clear, however, that this situation cannot last long; it is much too fragile. Difficult energy situations may well recur more frequently and more seriously until the country adopts and maintains a policy based on conservation of energy. We need an appropriate mix of fossil fuels, nuclear power, and such "unconventional"

sources of energy as the sun, wind, tides, geothermal steam, alcohol, and thermal layers in the sea, and we must have a rational means of paying for energy.

In addition, we must drastically reduce our overall per capita consumption of energy to something close to the world average. Considering our frightfully wasteful agricultural, manufacturing, transportation, and household uses of energy, this could be done sensibly without causing undue hardship for anyone. As for "energy independence," it is a chimera. Autarky in energy is as foolish and impractical as it is in automobiles, textiles, consumer electronics, or nearly anything else. Real energy security

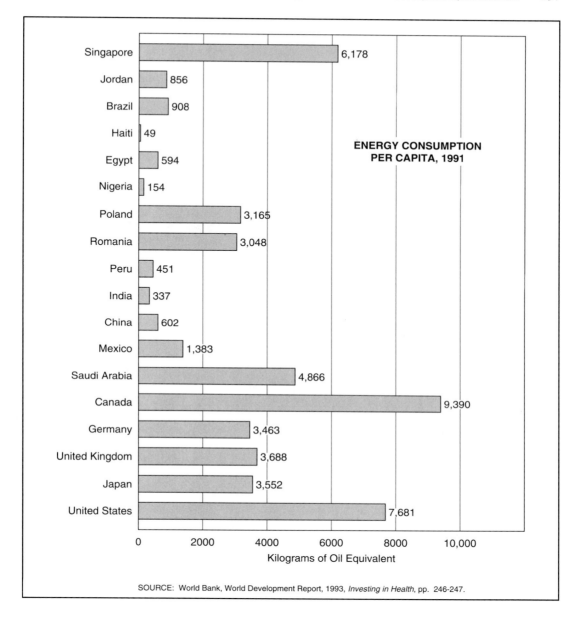

**ENERGY CONSUMPTION PER CAPITA, 1991**

| Country | Kilograms of Oil Equivalent |
|---|---|
| Singapore | 6,178 |
| Jordan | 856 |
| Brazil | 908 |
| Haiti | 49 |
| Egypt | 594 |
| Nigeria | 154 |
| Poland | 3,165 |
| Romania | 3,048 |
| Peru | 451 |
| India | 337 |
| China | 602 |
| Mexico | 1,383 |
| Saudi Arabia | 4,866 |
| Canada | 9,390 |
| Germany | 3,463 |
| United Kingdom | 3,688 |
| Japan | 3,552 |
| United States | 7,681 |

Kilograms of Oil Equivalent

SOURCE: World Bank, World Development Report, 1993, *Investing in Health*, pp. 246-247.

may well lie in international cooperation rather than competition. But other countries' energy pictures are not the same as ours, and both their goals and their policies will differ accordingly.

Western Europe, where the Industrial Revolution began, using first water power and then coal, is today much more dependent on external sources for its energy than the United States. Petroleum and natural gas under and around the North Sea have re-

lieved this situation somewhat, and still newer reserves in the form of tar sands in France and heavy crude under the Adriatic Sea may in the future be important, but the basic energy dependence of Western Europe remains unchanged. Consequently, nuclear energy is proportionately more important there than anywhere else in the world, nearly all the hydroelectric potential has been developed, and the world's only large-scale tidal power project functions in the es-

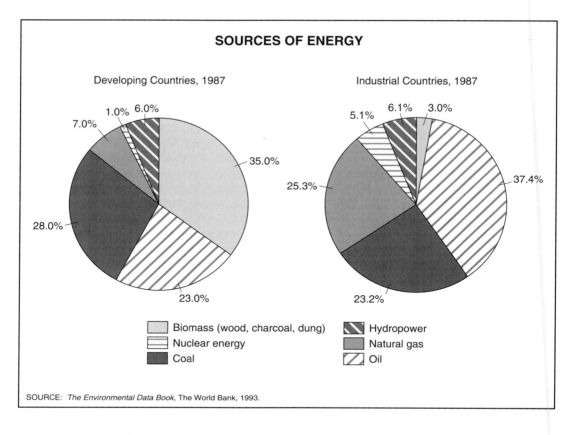

## SOURCES OF ENERGY

Developing Countries, 1987

Industrial Countries, 1987

SOURCE: *The Environmental Data Book*, The World Bank, 1993.

tuary of the Rance River in France. Nevertheless, Western Europe is still vulnerable to oil blackmail, and the OPEC oil embargo of 1973–74 did influence the policies of most States in the region (with the notable exception of The Netherlands) toward the Arab-Israel dispute, though not necessarily to the extent hoped for by the Arabs.

Japan is even more vulnerable, having a much smaller domestic energy stock and no friendly neighbors on whom to call for assistance (nothing like the electricity intertie system binding most of the continental Western European countries, for example), and being located at the end of a very long oil supply line vulnerable itself to interdiction at a number of strategic points. Thus Japan, despite its close links with the United States, is forced to pay some deference to the political demands of its principal suppliers of petroleum while trying to develop new sources of supply outside the Middle East.

The developing countries present a more complex picture. Some are exporters of petroleum and members of OPEC, and therefore do not have to worry about sources of energy, at least for the short term. But as they invest their oil revenues in industrialization and modernization in general, they become more dependent on imports of all manner of consumer and capital goods, technology, and skills obtainable only in the developed countries. At the same time their domestic energy needs grow, making a smaller proportion of production available for export (barring a sharp increase in production, which is unlikely because of the desire to conserve finite reserves). The result is likely to be decreased bargaining power and a restoration of something approaching a community of interests.

The non–oil-exporting developing countries still present a varied picture. Some—the least developed countries—consume so little inanimate energy that for the present, at least, local supplies of wood, cow dung, agricultural wastes, coal, peat, and other traditional organic fuels are adequate to meet present needs. But these materials could also be used for fertilizers, chemicals, and industrial raw materials if other fuels were reliably available at reasonable prices, and cutting of trees for firewood and charcoal contributes substantially to the devastation of forests causing so much damage to the planet's environment.

Other countries are well along the road to industrialization and thus more energy-dependent than the poorest countries. Others have attained middle-class status, largely on the basis of the export of agricultural and nonfuel mineral commodities. Both of these groups of countries are hard-hit when oil prices rise steeply, but can get along because they have more financial resources of their own, better credit ratings, and more bargaining power than the poorest countries. Those most seriously affected are generally those whose infant industries, budding transport systems, and nascent urbanization are heavily dependent on foreign energy sources but whose economies cannot yet bear the new costs. These are the countries receiving the most aid from the United Nations, the OECD, and even some of the OPEC countries.*

A factor that must be considered, although it cannot yet be evaluated, is the entry of new suppliers of petroleum and natural gas into the market, none of which has joined OPEC. Mexico is already self-sufficient, has become an important exporter, and may have one of the world's largest oil reserves. Brazil, Vietnam, Egypt, and other developing countries may become important producers in the future, with inevitable though unpredictable political consequences.

*Consumer countries have formed organizations to counterbalance OPEC. The International Energy Agency, composed of 20 major consumers, mostly OECD countries, and the Latin American Energy Organization are trying to work out programs to conserve petroleum, use petroleum more efficiently, and develop alternative energy sources. Their prospects for success are still uncertain.

We emphasize petroleum here because worldwide it is, and is likely to remain for a long time, the world's most important fuel and the most important fuel in international trade. Coal and natural gas seem unlikely to increase substantially in international trade, although of course locally they may well become more important. Nuclear energy, despite the reevaluation undertaken around the world following the nuclear power plant accident at Chernobyl in the Soviet Ukraine in April 1986, will very likely play a greater role in the world's energy picture than it does now—though probably less of a role than had been predicted two decades ago. One reason for the caution, of course, is concern over the safety of nuclear energy; another is the enormous increase in the capital costs of building nuclear facilities. But caution is also warranted by the uncertainties of nuclear power. Neither commercial-grade uranium nor advanced nuclear technology is as abundant or widespread as fossil fuels and the technology necessary to utilize them. The possibilities of nuclear blackmail are therefore at least as great for the suppliers of either the fuel or the technology.

Here is still another example of the intricate interdependence of all the States and peoples of the world and of the need for worldwide cooperation rather than competition. The bar graph also shows the great disparities in consumption of energy in the world, another situation calling for cooperation.

## Land Use

Increasing population is putting severe pressure on another finite resource: land. Land ownership and use have been matters of concern and study in many parts of the world for a long time, but have never been more important than right now. At all levels, from the individual house lot to whole countries and regions, they present problems so difficult that at times one despairs of trying to solve them. They can be solved, probably, but not very soon and probably not very satisfactorily.

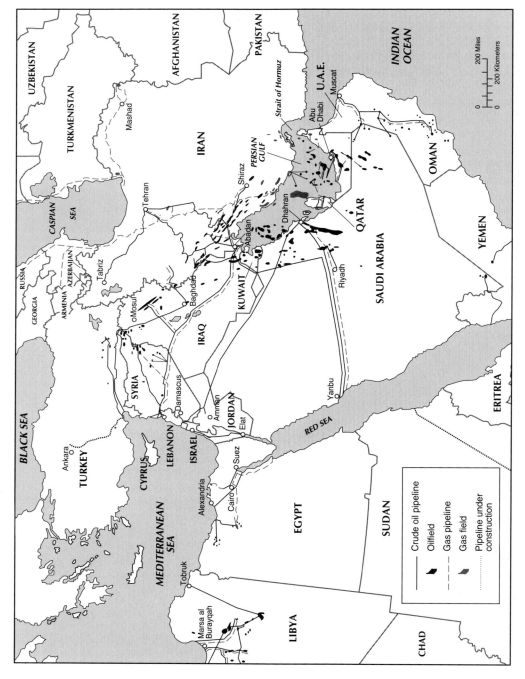

**Petroleum and natural gas in the Middle East.** The world's largest reserves of oil and gas are shown on this map. Their political and economic importance in the contemporary world can hardly be exaggerated. The Arab oil embargo and OPEC's quadrupling of the price of oil in 1973–74, a second "oil crisis" following the Iranian Revolution in 1979, the Iran–Iraq War (1980–1988), and the Iraqi invasion and occupation of Kuwait in 1990 followed by the Gulf War between Iraq and a large coalition of disparate countries in 1991 are only the most dramatic illustrations of the reserves' geopolitical importance, and of their vulnerability to political instability.

Legend:
— Crude oil pipeline
Oilfield
- - - Gas pipeline
Gas field
····· Pipeline under construction

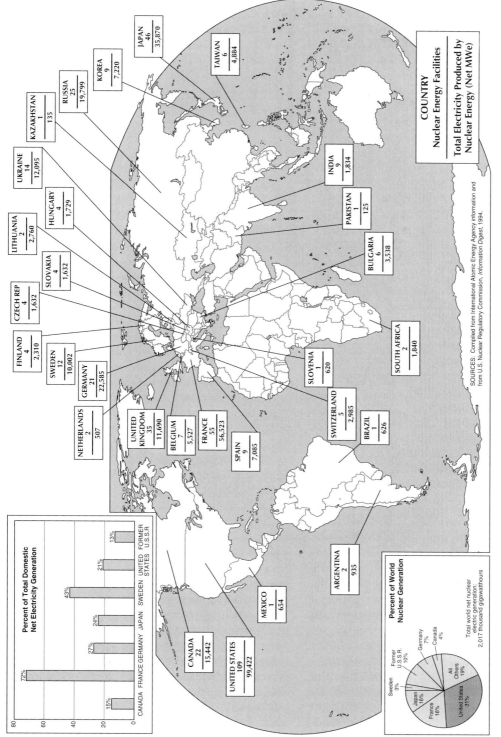

**Worldwide nuclear generation of electricity.** Despite the accidents at Three Mile Island and Chernobyl, the world is not likely to abandon the use of nuclear fission to produce electricity until a more satisfactory method is developed, probably not until well into the twenty-first century. In 1993, there were 419 nuclear energy plants in use around the world. The United States generated about 31 percent of the world's total nuclear energy, and France about 16 percent. Since 1983, the average annual gross percentage factor has increased 17 percent in the United States and 7 percent in France.

## Zoning

One of the most interesting examples of conflicts over land use, and one attracting the attention of some geographers, is that of local zoning. Systematic land zoning within urban areas of the United States dates only from 1916, and even today many urbanizing areas have only rudimentary zoning systems or none at all. In some areas where urban sprawl has become a serious problem, the old zoning regulations have proven inadequate to cope with new situations. Suburbanization has tended to rob central cities of their centrality, thus undercutting the fundamental premise on which zoning was originally based. Now suburbs are themselves becoming central cities, surrounded not by other suburbs but by other central cities. The long-term implications of the suburbanization of the hinterland and the urbanization of suburbia are unforeseeable, though we have already referred to some of them in our chapters on civil divisions and special purpose districts. Here we can only introduce some specific problems relating to zoning in urban and suburban areas.

The concept of zoning is based on the notion that society in general and most individuals benefit from the separation of various types of land use, generally divided into industrial, business, and residential, and subdivided further according to local perceptions. Those who have lived in countries where zoning is unknown or rare, however, and have been able to live in quiet, charming neighborhoods within easy walking distance of employment, shopping, and services, all of which are sprinkled throughout the city rather than confined to particular zones, may question the validity of this basic concept.

More serious from a political standpoint are the social effects of zoning. One of the popular ideas of the 1920s, when zoning spread rapidly across the country, was that cities should be surrounded by "garden cities," spacious, quiet, neighborhoods, green with vegetation surrounding single-family homes on large lots facing broad and sometimes deliberately curved streets. This "ideal" suburban community has had distinct advantages for those who live in them and has provided some needed relief from the "asphalt jungle" of the city. But it has created some problems as well. For one thing, as whites became more affluent and moved into these suburbs, they tended to be replaced by blacks, Chicanos, Puerto Ricans, and other minorities, leading to patterns of black cities surrounded by white suburbs. Zoning that includes minimum lot sizes of 1 or 2 acres, prohibits multifamily dwellings, severely restricts the location of businesses providing goods and services, and provides other benefits for the well-to-do family with more automobiles than children has had the effect of keeping the minorities bottled up in the cities with little opportunity to follow the whites out to the suburbs, even when they acquire the inclination and means to do so. Environmental zoning can become, by design or chance, exclusionary zoning.

Another effect of the insistence on low-density suburbs has been the acceleration of urban sprawl, with huge areas of often productive farmland given over to subdivisions. The ecological effects of urban sprawl are at least as serious as the social and political ones. Most states by now have adopted some laws protecting open spaces around cities from the ravages of the "developers," but such laws drive up the price of developable land, deprive municipalities of additional revenue-producing land, intensify traffic congestion in the urban areas, and have other undesirable side effects. This is not to say that it is wrong to insist on green belts around cities, only that better planning is necessary to permit them to perform effectively the tasks for which they are designed.

Another type of zoning that is beginning to be taken more seriously is hazard zoning, that is, prohibition or regulation of land use in areas subject to frequent natural hazards. So far this has been used mainly to exclude residences, businesses, and most industries from floodplains (clearly a public good), but there are still many problems with such zoning even where it is in force. We still have, moreover, virtually no zoning for areas subjected to other natural hazards. An example of the need for such zoning is provided by the Los Angeles area. It has become fashion-

**The politics of land use.** This dramatic satellite photograph of the border area of Alberta (top) and Montana shows clearly the effects of an international boundary on land use in an environmentally homogeneous region. Because of different governmental agricultural policies, grazing is the dominant economic activity on the Canadian side and wheat farming on the American side of the 49th parallel. (Courtesy Geometrics Canada)

able since World War II to build homes on the slopes of the canyons in the Santa Monica Mountains and the Hollywood Hills in the northern part of the city and in other nearby areas. Nearly every summer and fall the drought dries up the grass and brush cover that is then set afire by natural or human action. Powerful winds spread the fires through the canyons, endangering and even destroying houses. Then, with the vegetation cover removed, the slopes generate massive floods and mudslides during the winter and spring rains, endangering and even destroying still more houses. The hazards are well known, yet the taxpayers are annually called upon to provide emergency services for people who insist on living in these hazardous areas. Similar examples abound in various physical environments around the country.

A still larger problem is one that has been scarcely, if ever, mentioned publicly. It can be bluntly summarized in a single question: How long will it be before we begin to re-

strict the occupation of desert areas that require enormous expenditures of public funds for water, roads, power, and other facilities to make them habitable, to say nothing of the effects of large-scale human habitation and public works on fragile desert ecosystems? Zoning of this type on a national scale might seem unthinkable now, but in fact there is ample precedent for it.

## Land Reform

Contrary to the situation in the United States, it is private, not public, land ownership that is controversial in many other countries. During the eighteenth and nineteenth centuries and well into the twentieth, the dominant form of land tenure throughout Latin America, North Africa, Southern and Eastern Europe, and most of Asia was the *latifundio*, the large family-owned estate known by a variety of local names, of which the most common is *hacienda.* Under the *latifundia* system, a large proportion of the land is owned by a very small percentage of the people. Before the 1789 revolution in France, for example, 40 percent of the land was owned by only 3 percent of the people. In Bolivia 10 percent of the farmers owned nearly 95 percent of the land in 1950. In industrial, urbanized societies, these figures might not be alarming, but in a traditional society in which land plays such a great role in the lives of people, they are serious indeed.

Land reform has been an essential prelude to, or component of, every important economic and social revolution in recent history. Land reform in Japan before industrialization began generated a doubling of agricultural production between 1870 and 1914. Similar increases in both agricultural production and farmers' incomes have followed (after an interval of disorganization and uncertainty) the revolutions of 1910 in Mexico and 1952 in Bolivia. Land reform has been fundamental in the programs of all communist governments and of many democratic governments as well. In fact, it has been amply demonstrated that *latifundismo* retards economic and social progress, while more equitable distribution of land encourages them.

Maldistribution of land also inhibits political democracy. Quite commonly, the large landowners are aligned with the local and national military and religious authorities to form a triumvirate that controls the State. Each reinforces and protects the other two. (See Chapter 20.) This is the major reason why meaningful land reform is so difficult to initiate and sustain without a violent or at least radical revolution. Where it has been attempted peacefully, as in Venezuela, the process does not seem to be so effective. In Venezuela, from the initiation of serious land reform in 1960 to 1973, 75 percent of the land redistributed came from the public domain; only about 12 percent of the country's private estate land had been affected. But land reform began in Venezuela when economic development, fueled by oil money, had already begun and only a third of the labor force was in agriculture. Expropriated lands were paid for generously, again out of

oil revenues, and after a brief early period of expropriation of private land, the reform evolved into what is essentially a colonization program. These three factors—redistribution of public lands, small agricultural space, and the colonization program—together permitted an increase in agricultural production as part of overall economic development under a stable democratic system, but these three factors are seldom present together. Furthermore, there are still grave imbalances and inequities in the country's socioeconomic picture. The country's rate of economic development slowed considerably after the drop in oil prices, and its former political stability has been seriously disrupted. Perhaps real land reform could have mitigated the damage of falling oil prices and helped to maintain political stability.

Land reform is not enough, however. Agricultural production cannot be increased and social inequities redressed without comprehensive programs of *agrarian reform* and *rural development.* These include such things as access to, and better utilization of, land, water, forests, and other natural resources; the development of a rural infrastructure, including electric power, farm-to-market roads, storage and transport facilities, irrigation and drainage projects, pure water, schools, and medical facilities; provision of agricultural inputs (such as seeds, fertilizers, pesticides, and machinery), crop insurance, generous credit, and agricultural education and extension training; development of nonagricultural activities in rural areas; and greater participation of rural people, especially women, in rural development. All this costs money, takes time, and requires a reordering of political priorities. And it inevitably has profound and often unpredictable political consequences. Yet in the long run, it is not only desirable but essential to have such agrarian reform and rural development if even the roughly 1 billion rural people in the world today living below an absolute poverty line of $200 per capita income per year are to experience any betterment of their condition (to say nothing of the hundreds of millions who will be joining them in the next generation).

## International Environmental Problems

The most important lesson to be learned from our experience with ecology, energy, and land use has yet to be learned by mankind as a whole: that the planet Earth and its atmosphere constitutes one unified ecosystem and that damage to any part of it inevitably results in damage to the whole. Everything is connected to everything else. Statesmen rarely give much thought to the long-range environmental consequences of their decisions and actions, and politicians do so even more rarely. Even though, as we pointed out earlier, an environmental consciousness has begun to spread around the world, it has not spread far enough or fast enough. Pollution, for example, knows no boundaries, yet there is still no adequate mechanism for preventing, containing, or reversing pollution at the international level.

The nearest thing to a global environmental agency is the United Nations Environment Programme (UNEP), with headquarters in Nairobi. Its Oceans and Coastal Areas Programme has generated and coordinated environmental protection and cleanup activities in designated portions of the global sea. Its Mediterranean Action Plan has had considerable success in cleaning up the Mediterranean Sea, one of the most polluted large bodies of water in the world. Progress has also been made in the Baltic, where the problem is at least as serious. Other international and regional organizations have expressed interest in environmental matters, but seldom do they rank very high on their priority lists. Even the World Bank has come under heavy fire for ignoring ecological considerations when funding huge infrastructure projects in developing countries, and it is beginning to reorder its priorities as a result. Some measure of the Bank's conversion to environmental protection is its expenditure of $1.6 billion on it in fiscal 1991, compared with $404 million the year before and almost nothing in previous years.

Meanwhile, agricultural pesticides applied in the rich countries of the Northern Hemisphere wash down rivers into the global sea

and eventually end up in the bodies of penguins in Antarctica. Acid precipitation caused by the mixing of atmospheric water with airborne industrial and automotive pollutants is devastating lakes and forests in North America and northern and central Europe and is threatening farms and monuments in India, Mexico, Indonesia, Zambia, Brazil, and other developing countries.

Industrialized countries in Western Europe were sending 10,000 to 20,000 shipments of hazardous wastes every year to Eastern Europe for disposal and were seeking disposal sites in developing countries until such practices came under the strict control provisions of the 1989 Basel Convention on the Control of Transboundary Movement of Hazardous Wastes and their Disposal. In 1994, 64 parties to this convention agreed that such shipments from OECD members to nonmembers will be completely banned by the end of 1997; however, the practice continues on a reduced scale.

The threat of major worldwide climate change, usually expressed as "global warming," is growing, with all of its ramifications, such as desiccation of agricultural areas and rising sea level as glaciers and icecaps melt. The depletion of the ozone layer in the upper atmosphere, which protects the planet from excessive ultraviolet radiation from outer space, is another global problem. So is deforestation, the rapid disappearance of the forests that once blanketed much of the earth and that provide much of our atmospheric oxygen as well as wildlife habitat, watershed protection, sustenance for soil, and regenerative forest products. This is one of the factors contributing to desertification, the spreading of deserts, particularly in Africa, over land that had been at least steppe and even savanna.*

---

*Although the nature, causes, and even existence of desertification are challenged by some scientists, most governments consider it a genuine problem. In 1977 the UN Conference on Desertification was held in Nairobi, and the UN Conference on Environment and Development, (Rio de Janeiro, 1992) called for a UN Convention on Desertification which was negotiated subsequently and opened for signature in October 1994.

All of these intensify the pressure on finite global supplies of fresh water, which is rapidly becoming so scarce a commodity in much of the world that serious consideration is being given to towing icebergs from the Southern Ocean to coastal deserts, as has already been done experimentally. Another result of both natural forces and especially man's activities is the rapid extinction of species of plants and animals. Who can tell how many potential sources of food, medicine, pest control, beauty, and spiritual joy have been lost in this way?

The political and social consequences of this degradation of the global environment are incalculable. We are already experiencing growing, not diminishing, hunger and malnutrition in the world and huge waves of environmental refugees fleeing drought, floods, famine, destruction of the land, and other ecological problems. The possibility looms of resource wars, water wars, food wars, even land wars, just as we used to have before ideology became the chief rationalization for war.

All of this, of course, both stems from and reinforces the maldistribution of both wealth and power in the world, generated in part by the Industrial Revolution and in part by the staggering growth of the world's population that results in part from the Industrial Revolution. All these points are discussed in greater detail elsewhere in this book, but here they are seen in a broader context. Another point that bears repetition: Although at the present stage of world history populations are growing most rapidly in the poorest countries, it is likely that the greatest environmental damage is caused by the rich countries and their agents in the poor countries. Furthermore, it is in the rich countries that industrial and urban pollution is greatest.

Western Europe, for example, rich, sophisticated, and proud, was caught completely off guard in November 1986 when a series of accidental spills of toxic chemicals in Switzerland flowed down the Rhine River, killing hundreds of thousands of fish, contaminating drinking water for millions of people, and otherwise damaging the environment of four countries. This calamity called to mind the words of the English Romantic poet Samuel Taylor Coleridge:

The river Rhine, it is well known,
Doth wash your city of Cologne;
But tell me, nymphs, what power divine
Shall henceforce wash the river Rhine?

What power indeed? Perhaps some global plan of action—backed by a worldwide consensus, determination, sustained effort, and plenty of cash—will emerge as a result of the 1987 report of the World Commission on Environment and Development. This commission, composed of 23 people from all parts of the world and headed by Gro Harlem Brundtland, prime minister of Norway, was created by the United Nations but was independent of it. It labored for three years before issuing its massive report on an enormous range of contemporary world problems. Its recommendations, if followed urgently and faithfully, would certainly result in a better world for all of us.

Since release of the Brundtland report, a number of specialized international conferences have been held to discuss some of the global environmental problems listed above. Each has generated serious activity, including international conventions that bind their parties to specific actions to help reverse the destruction of our environment. It is doubtful, however, if, even after the end of the Cold War, there is yet the political will outside a few small countries to take the difficult and costly actions necessary to achieve the goal laid out by the Brundtland Commission: *sustainable* economic and social progress in an environmentally healthy world.

There are some signs of hope, other than the governmental efforts described already. One sign is the growing effectiveness of environmental NGOs around the world. The moratorium on whaling, the ban on large-scale driftnet fishing, the outlawing of chloroflurocarbons discharged into the atmosphere, and many other small victories can be credited at least in part to their efforts. A creative and potentially important arrangement advocated by environmental NGOs is that of forgiving portions of a developing country's foreign debt if it dedicates areas in the country as biosphere reserves or other types of protected parks or reserves. This "debt-for-nature swap" has great potential to accomplish a number of environmental and developmental objectives at once—if it is administered wisely, on a large enough scale, and over a long enough period of time.

Early experience with this arrangement, however, is not encouraging. In the words of a recent GAO report,

> From 1987 through 1990, 13 countries completed 26 debt swaps. These swaps retired debts totaling about $126 million (less than one-twentieth of 1 percent of the countries' external debt and less than one-fifth of one percent of their commercial debt.) Of the $126 million, $86.4 million was exchanged by Costa Rica. Fifteen swaps for nature accounted for nearly 90 percent of the $126 million, while 11 swaps for development accounted for about 10 percent.*

Another sign of hope is the "greening" of politics at the local and national level, which may eventually spread to the international level. "Green" parties were organized in a number of European countries in the 1970s and 1980s to contest elections on environmentalist platforms. They have had some success in some countries, but so far have not really caught on elsewhere. Single-issue parties such as this can only be successful in electoral systems based on proportional representation, and even there the system places limits on what they can accomplish. If they try to broaden their appeal and win more votes, either by incorporating other planks in their platform or forging alliances with other parties with different programs, then they automatically dilute both their message and their energy, possibly to the detriment of both the parties and the environmental movement. It is a classic dilemma in politics, and it has no simple solution.

*United States General Accounting Office, *Developing Country Debt; Debt Swaps for Development and Nature Provide Little Debt Relief.* Washington, D.C., December 1991, pp. 1–2.

Nevertheless, formal, legal political activity has both publicized and legitimated the notion of environmental protection, and this has demonstrably influenced some national policies.

Perhaps the crucial test not only of the concept of linking economic development and environmental protection but also of international action as a means to bring it about is the United Nations Conference on Environment and Development (Rio de Janeiro, 1–12 June 1992). This conference, scheduled just 20 years after the United Nations Conference on the Human Environment in Stockholm, was voted by the General Assembly in 1989 and preparations for it began almost immediately. Countries were represented by heads of State or government; it was the largest "summit" meeting held to date. Another unique feature of the conference was the full and active participation of both nongovernmental organizations and private-sector interests. The scope, magnitude, and pioneering nature of this conference were breathtaking. It was certainly the most important conference since UNCLOS III and perhaps the most important since the San Francisco conference in 1945 that established the United Nations. Its true significance, however, will not be measurable for years—perhaps even decades—for the real measure was not what was said or done in Rio but how the resolutions, declarations, plans, treaties, and other documents that emanated from it are interpreted and implemented afterward. We shall see.

## Conflicts in Ecology, Energy, and Land Use

In this chapter we have pointed out a number of problems relating to environmental politics from the urban neighborhood to the global one. Each is difficult enough in itself, but their interrelatedness compounds the complexity. Here we suggest a few interrelationships only to illustrate the point.

1.  In the rush to "energy independence," we are considering a return to coal as a primary fuel because it is domesti-

cally abundant, but we forget the reasons that we converted from coal to oil and gas in the first place: Coal is bulky and dirty, and, except for expensive anthracite, not very efficient. Underground coal mining, moreover, is hazardous and unhealthful, while surface mining is ecologically ruinous.

2.  Expansion of the land under cultivation to increase food production means in most of the world using marginal or submarginal land, which would require a great deal of energy and can have most unfortunate ecological consequences.

3.  Dispersal of industry to improve the urban environment and provide rural employment can simply accelerate urban sprawl, increase transport costs and energy consumption, and transfer the less desirable features of industry to rural areas, making these areas less attractive than they are now.

4.  An increase in energy consumption means additional pipelines through fragile environments, oil spills, and blowouts; unsightly and perhaps dangerous electricity transmission lines; more rail, road, and water transport, requiring more energy and raw materials to produce and operate; uncertainties about nuclear energy; and changes in the composition of the earth's atmosphere.

**A small portion of a large petroleum refinery.** This is the desulfurization unit of the huge Texaco refinery in Convent, Louisiana. The immense complexity and capital cost of a large modern refinery preclude such installations in poor countries without external assistance in the form of capital, technology, equipment, management, or some combination of these elements provided by transnational corporations, international organizations, governments of rich countries or some combination of these sources. Petroleum refineries generally employ relatively few workers, since in order to be efficient and profitable, they must be highly automated. They pose serious environmental problems in the form of emission of noxious gases and toxic effluents, besides the ever-present dangers of oil spills and fires. They provide raw materials for a great variety of petrochemicals which have become important to both producers and consumers in modern industrial societies and in poor agricultural societies alike. They are therefore good candidates for vertical integration as the core of a modern sector of a traditional country. (Photo courtesy of Texaco and the American Petroleum Institute)

5. Increased raising of livestock for food means increased competition with wildlife for forage and destruction of their habitats.

6. Landfills in urban areas or even for offshore facilities, such as airports and oil terminals, can have adverse effects on the ecology of the coastal zone.

7. Economic development, considered a desirable political goal, inevitably requires intensified land use, ever-increasing energy supplies, and ecological damage.

8. Hydroelectric dams, which generate nonpolluting energy at relatively low cost, can interfere with the migrations of anadromous and catadramous fish, cause siltation behind the dam and desiccate wetlands below it, and devastate the land and culture of nearby indigenous peoples.

9. The readily accessible fossil fuels have already been found; new deposits are likely to be found in environmentally fragile areas where the risks and costs are far greater than in the older ones.

10. Orderly economic growth today requires broad policies and plans for land use, energy, and ecology that can be developed and administered by government only at the expense of some limitations on free enterprise, private ownership of land, and individual behavior.

11. Successes already achieved in environmental protection are threatened by reversal as economic interests organize to achieve "wise use" or "balanced use" or "multiple use" of land and resources, thereby obliterating wilderness altogether.

12. A growing population would require economic growth merely to maintain the present unsatisfactory levels of consumption of goods and services for the great majority of the people of the

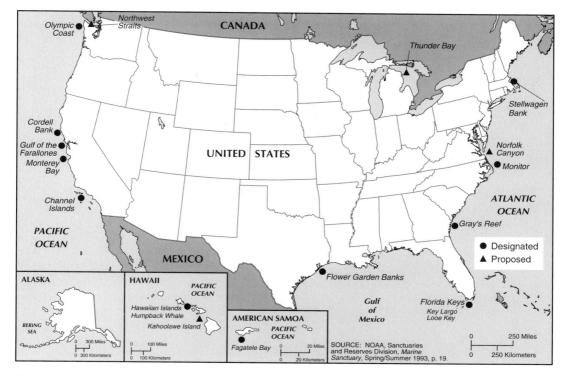

**Marine Sanctuaries of the United States.**

world, so that cutbacks in production can only make a bad situation worse.

13. A growing population growing wealthier would also require more recreational facilities on land and offshore in areas not yet urbanized, thereby placing even greater pressure on our remaining rural and wilderness areas and in the most biologically productive area of the sea.

14. Establishment of marine parks, sanctuaries, and reserves to protect rare, scenic, or scientifically important portions of the sea can conflict with marine transportation and fisheries needed by a growing population.

15. Political democracy can best be achieved and maintained in a society that is reasonably prosperous and in which wealth is reasonably equitably distributed, but these conditions can only be developed over a period of time at some environmental cost.

In these 15 points alone, without even considering other summary points that could be made or our more elaborate discussion that led up to them, we can see all the components suggested by Kasperson and Minghi: "political goals, agents of impress, processes and effects." But using these components in a linear fashion to analyze environment–politics relationships could quickly lead to an analytical dead end. They themselves are interrelated in complex ways. The agents of impress, for example, may select the political goals they wish to reach but find that they have a very limited range of processes from which to choose, while the effects of their choices are felt by people who played no role in the selection of goals, agents, or processes and may object to all of them.

The political geography of everyday life, as we have seen, affects every single individual on the planet, in multiple ways, everywhere one goes, awake or asleep. Students of the subject understand the factors we have discussed and others we have had to omit; they see the world through different eyes than other people, and that should enable them to make wiser decisions. They should be able to help make the world just a little bit better.

# BIBLIOGRAPHY FOR PART EIGHT

## *Books and Monographs*

**A**

Abu-Amr, Ziad, *Islamic Fundamentalism in the West Bank and Gaza—Muslim Brotherhood and Islamic Jihad.* Bloomington: Indiana Univ. Press, 1994.

Adams, William M., *Green Development: Environment and Sustainability in the Third World.* New York: Routledge, 1992.

Addleton, Jonathan Stuart, *Undermining the Centre; the Gulf Migration and Pakistan.* Oxford Univ. Press, 1992.

Adelman, Howard and John Sorenson, *African Refugees.* Boulder, CO: Westview, 1994.

Afendras, Evangelos A. and Eddie C.Y. Kuo (eds.), *Language and Society in Singapore.* Singapore: Singapore Univ. Press, 1980.

Aird, John S., *Slaughter of the Innocents: Coercive Birth Control in China.* Lanham, MD: Univ. Press of America, 1990.

Alexiev, Alexander R. and S. Enders Wimbush (eds.), *Ethnic Minorities in the Red Army.* Boulder, CO: Westview, 1988.

Altstadt, Audrey L., *The Azerbaijani Turks.* Stanford, CA: Hoover Institution Press, 1992.

Amy, Douglas J., *The Politics of Environmental Mediation.* New York: Columbia Univ. Press, 1987.

Andersen, Svein S., *The Struggle over North Sea Oil and Gas; Government Strategies in Denmark, Britain and Norway.* Oxford Univ. Press, 1993.

Anderson, David and Richard H. Grove (eds.), *Conservation in Africa; Peoples, Policies and Practice.* Cambridge Univ. Press, 1987.

Antoun, Richard T. and Mary Elaine Hegland (eds.), *Religious Resurgence; Contemporary Cases in Islam, Christianity and Judaism.* Syracuse: Syracuse Univ. Press, 1987.

Arizpe, Lourdes and others, *Population and the Environment; Rethinking the Debate.* Boulder, CO: Westview, 1994.

Arkin, William M. and others, *On Impact; Modern Warfare and the Environment: A Case Study of the Gulf War.* Washington, DC: Greenpeace, 1991.

Arnold, Guy, *Strategic Highways of Africa.* New York: St. Martin's, 1977.

Ashworth, G. (ed.), *World Minorities,* 2 vols. London: Quartermaine House, 1977, 1978.

Augelli, John P., *The Panama Canal Area in Transition Part I: The Treaties and the Zonians.* AUFS Reports North America Series 1981/No. 3.

———, *Panama Canal Area in Transition Part II: The Challenge of Integration & Development.* AUFS Reports North America Series 1981/No. 4.

Austin, James E. and Gustavo Esteva (eds.), *Food Policy in Mexico: The Search for Self-Sufficiency.* Ithaca, NY: Cornell Univ. Press, 1987.

Ayubi, Nazih, *Political Islam; Religion and Politics in the Arab World.* New York: Routledge, 1994.

**B**

Baldwin-Edwards, Martin and Martin Schain (eds.), *The Politics of Immigration in Western Europe.* London: Frank Cass, 1994.

Banac, Ivo, *The National Question in Yugoslavia: Origins, History, Politics.* Ithaca, NY: Cornell Univ. Press, 1984.

Banister, Judith, *China's Changing Population.* Stanford, CA: Stanford Univ. Press, 1991.

Bankes, Nigel (ed.), *Public Disposition of Natural Resources.* Toronto: Carswell, 1984.

Banks, John P. and others, *Nuclear Power: The Promise of New Technologies.* Washington, DC: CSIS, 1991.

Banuazizi, Ali and Myron Weiner (eds.), *The State, Religion and Ethnic Politics; Afghanistan, Iran and Pakistan.* Syracuse, NY: Syracuse Univ. Press, 1986.

Barber, Chip, *Cutting Our Losses; Policy Reform to Sustain Tropical Forest Resources.* Washington, DC: World Resources Institute, 1991.

Barry, Tom, *Roots of Rebellion; Land and Hunger in Central America.* Boston: South End Press, 1987.

Barzetti, Valerie and Yanina Rovinski (eds.), *Toward a Green Central America; Integrating Conservation and Development.* West Hartford, CT: Kumarian Press, 1992.

Bean, Frank D. and others (eds.), *Mexican and Central American Population and U.S. Immigration Policy.* Austin: Univ. of Texas Press, 1989.

Beatley, Timothy, *Ethical Land Use; Principles of Policy and Use.* Baltimore: Johns Hopkins Univ. Press, 1994.

Becker, Jorgen (ed.), *Information Technology and a New International Order.* Melbourne, FL: Krieger, 1984.

Becker, Jorgen and T. Szecskö (eds.), *Europe Speaks to Europe; International Information Flows Between Eastern and Western Europe.* Oxford: Pergamon, 1989.

Beer, William R. and James E. Jacob (eds.), *Lan-*

*guage Policy and National Unity.* Rowman & Allanheld, 1985.

Beissinger, Mark and Lubomyr Hajda (eds.), *The Nationalities Factor in Soviet Politics and Society.* Boulder, CO: Westview, 1990.

Bell, David and John Gaffney (eds.), *European Immigration Policy.* Oxford: Pergamon, 1990.

———, *Politics and Religion.* Oxford: Pergamon, 1990.

Benard, Cheryl and Zalmay Khalilzad, *"The Government of God.": Iran's Islamic Republic.* New York: Columbia Univ. Press, 1984.

Benedick, Richard Eliott, *Ozone Diplomacy; New Directions in Safeguarding the Planet.* Cambridge, MA: Harvard Univ. Press, 1991.

Bennett, Olivia (ed.), *Greenwar; Environment and Conflict.* London: Panos, 1991.

Benson, Linda K., *The Ili Rebellion; The Moslem Challenge to Chinese Authority in Xinjiang, 1944–1949.* Armonk, NY: Sharpe, 1989.

Berkhout, Frans, *Radioactive Waste; Politics and Technology.* New York: Routledge, 1991.

Bernstein, Henry and others (eds.), *The Food Question; Profits Versus People?.* New York: Monthly Review Foundation, 1993.

Berry, R. T. (ed.), *Environmental Dilemmas.* New York: Routledge, 1993.

Besmeres, John, *Socialist Population Politics: The Political Implications of Demographic Trends in the USSR and Eastern Europe.* Armonk, NY: Sharpe, 1980.

Birnie, Patricia W. and Alan E. Boyle, *International Law and the Environment.* Oxford Univ. Press, 1993.

Black, Richard and Vaughan Robinson (eds.), *Geography and Refugees: Patterns and Processes of Change.* New York: Wiley, 1993.

Blackburn, John O., *The Renewable Energy Alternative.* Durham, NC: Duke Univ. Press, 1987.

Blatherwick, David E.S., *The International Politics of Telecommunications.* Berkeley: Univ. of California, Institute of International Studies, 1987.

Blaut, James M., *The National Question; Decolonizing the Theory of Nationalism.* Atlantic Highlands, NJ: Humanities, 1987.

Blinken, Anthony, *Ally Verses Ally: America, Europe and the Siberian Pipeline Crisis.* New York: Praeger, 1987.

Blinn, Keith W. and others, *International Petroleum Exploration and Exploitation Agreements: Legal, Economic and Policy Aspects.* London: Euromoney Publications, 1986.

Boal, Frederick W. and J.N.H. Douglas (eds.), *Integration and Division: Geographical Perspectives on the Northern Ireland Problem.* London: Academic Press, 1982.

Boardman, Robert, *Global Regimes and Nation-States; Environmental Issues in Australian Politics.* Ottawa: Carleton Univ. Press, 1990.

Boehmer-Christiansen, J. and J. Skea, *Acid Politics; Environmental and Energy Policies in Britain and Germany.* London: Pinter, 1991.

Böhning, W. R., *Studies in International Labor Migration.* New York: St. Martin's, 1985.

Bothe, Michael and others (eds.), *Amazonia and Siberia; Legal Aspects of the Preservation of the Environment and Development in the Last Open Spaces.* London: Graham & Trotman, 1993.

Boué, Juan Carlos, *Venezuela; the Political Economy of Oil.* Oxford Univ. Press, 1994.

Bourdeaux, Michael, *The International Politics of Eurasia.* Vol. 3: The Influence of Religion. Armonk, NY: Sharpe, 1994.

Bourgault, Louise M., *Mass Media in Sub-Saharan Africa.* Bloomington: Indiana Univ. Press, 1995.

Bovard, James, *The Farm Fiasco.* San Francisco: ICS Press, 1990.

Braude, Benjamin and Bernard Lewis (eds.), *Christians and Jews in the Ottoman Empire.* New York: Holmes & Meier, 1991.

Brockett, Charles D., *Land, Power, and Poverty; Agrarian Transformation and Political Conflict in Central America.* Boulder, CO: Westview, 1990.

Brower, David J. and Daniel S. Carol, *Managing Land Use Conflicts.* Durham, NC: Duke Univ. Press, 1987.

Brown, Janet W. and Gareth Porter, *Global Environmental Politics.* Boulder, CO: Westview, 1991.

Brown, Lester R. and others, *State of the World.* New York: Norton, published annually.

Brown, Peter G. and Henry Shue, *The Border That Joins: Mexican Migrants and U.S. Responsibility.* Totawa, NJ: Rowman & Littlefield, 1983.

Browne, William P. and others, *Sacred Cows and Hot Potatoes; Agrarian Myths in Agricultural Policy.* Boulder, CO: Westview, 1992.

Brubaker, William Rogers (ed.), *Immigration and the Politics of Citizenship in Europe and North America.* Lanham, MD: Univ. Press of America, 1989.

Brunn, Stanley D. and Thomas R. Leinbach (eds.), *Collapsing Space and Time: Geographic Aspects of Communication and Information.* New York: HarperCollins, 1991.

Brunnée, Jutta, *Acid Rain and Ozone Layer Depletion: International Law and Regulation.* Dobbs Ferry, NY: Transnational, 1988.

Buck, Stephen W. and Charles F. Doran, *The Gulf, Energy, and Global Security: Political and Economic Issues.* Boulder, CO: Lynne Rienner, 1991.

Buckley, Helen, *From Wooden Ploughs to Welfare*. Montreal: McGill-Queen's, 1992.

Bugajski, Janusz, *Ethnic Politics in Eastern Europe: A Guide to Nationality Policies, Organizations, and Parties*. Armonk, NY: Sharpe, 1994.

Burg, Steven, *Ethnic Conflict and International Intervention, The Crisis in Bosnia-Herzegovina 1990–1993*. Armonk, NY: Sharpe, 1994.

Burgat, Francois and Wm. Dowell, *The Islamic Movement in North America*. Austin: Univ. of Texas Press, 1992.

Burr, J. Millard and Robert O. Collins, *Requiem for the Sudan; War, Drought and Disaster Relief on the Nile*. Boulder, CO: Westview, 1994.


**C**

Cairns, A. and C. Williams, *The Politics of Gender, Ethnicity, and Language in Canada*. Toronto: Univ. of Toronto Press, 1987.

Calavita, Kitty, *Inside the State; The Bracero Program, Illegal Immigration, and the I.N.S.* New York: Routledge, 1992.

Campiglio, Luigi and others (eds.), *The Environment after Rio; International Law and Economics*. Kluwer, 1993.

Card, Brigham Y. and others (eds.), *The Mormon Presence in Canada*. Edmonton: Univ. of Alberta Press, 1990.

Carroll, John E. (ed.), *International Environmental Diplomacy; The Management and Resolution of Transfrontier Environmental Problems*. Cambridge Univ. Press, 1990.

Castles, Stephen and Mark J. Miller, *The Age of Migration; International Population Movements in the Modern World*. New York: Guilford, 1993.

Caves, Roger W., *Land Use Planning; The Ballot Box Revolution*. Newbury Park, CA: Sage, 1992.

Center for International Development and Environment (ed.), *Religious Beliefs and Environmental Protection: The Malshegu Sacred Grove in Northern Ghana*. Washington, DC: World Resources Institute, 1991.

Chami, Joseph G., *Days of Tragedy: Lebanon 1975–76*. New Brunswick, NJ: Transaction, 1980.

———, *Days of Wrath: Lebanon 1977–1982*. New Brunswick, NJ: Transaction, 1983.

Chaney, Rick, *Regional Emigration and Remittances in Developing Countries: The Portuguese Experience*. New York: Praeger, 1986.

Chazan, Naomi and Timothy M. Shaw, *Coping with Africa's Food Crisis*. London: Pinter, 1987.

Chehabi, H. E., *Iranian Politics and Religious Modernism*. London: Tauris, 1990.

Chibnik, Michael, *Risky Rivers; The Economics and Politics of Floodplain Farming in Arizona*. Tucson: Univ. of Arizona Press, 1994.

Chiswick, Barry R. (ed.), *Immigration, Language and Ethnicity; Canada and the United States*. LaVergne, TN: AEI, 1992.

Choucri, Nazli (ed.), *Global Accord; Environmental Challenges and International Responses*. Cambridge, MA: MIT Press, 1993.

Cioffi-Revilla, Claudio and others (eds.), *Communications and Interaction in Global Politics*. Newbury Park, CA: Sage, 1987.

Clark, Robert P., *The Basques; The Franco Years and Beyond*. Reno: Univ. of Nevada Press, 1991.

———, *Negotiating with ETA; Obstacles to Peace in the Basque Country 1975–1988*. Reno: Univ. of Nevada Press, 1991.

Clarke, Colin and others (eds.), *Geography and Ethnic Pluralism*. London: Allen & Unwin, 1984.

———, *South Asians Overseas; Migration and Ethnicity*. Cambridge Univ. Press, 1990.

Clarke, John I. and Leszek A. Kosinski, *Redistribution of Population in Africa*. Portsmouth, NH: Heinemann, 1982.

Clay, Jason W., *Indigenous Peoples and Tropical Forests: Models of Land Use and Management from Latin America*. Cambridge, MA: Cultural Survival, Inc., 1988.

Clay, Jason W. and Bonnie K. Holcomb, *Politics and the Ethiopian Famine 1984–1985*. Cambridge, MA: Cultural Survival, 1986.

Coakley, John (ed.), *Politics, the Territorial Management of Ethnic Conflict*. London: Frank Cass, 1993.

Cockburn, Alexander and Susanna Hecht, *The Fate of the Forest; Developers, Destroyers and Defenders of the Amazon*. New York: Routledge, 1989.

Codding, George A., Jr. and Anthony M. Rutkowski, *The International Telecommunications Union in a Changing World*. Dedham, MA: Artech House, 1982.

Cohen, Shaul Ephraim, *The Politics of Planting; Israeli-Palestinian Competition for Control of Land in the Jerusalem Periphery*. Chicago: Univ. of Chicago Press, 1993.

Cohn, Theodore H., *Canadian Food Aid: Domestic and Foreign Policy Implications*. Denver: Univ. of Denver, 1979.

Cole, W. Owen and Piara Singh Samblai, *The Sikhs*. New York: Routledge, 1978.

Coleman, William, *The Independence Movement in Quebec 1945–1980*. Toronto: Univ. of Toronto Press, 1984.

Committee on Western Water Management, National Research Council, *Water Transfers in the West*. Cambridge Univ. Press, 1992.

Conway, Dennis, *Caribbean Migrants; Opportunistic and Individualistic Sojourners*. UFSI Reports Latin America Series 1986/No. 24.

Croll, Elisabeth and others (eds.), *China's One-Child Family Policy*. New York: St. Martin's, 1985.

Crowe, David and John Kolsti (eds.), *The Gypsies of Eastern Europe*. Armonk, NY: Sharpe, 1991.

Cullingworth, J. Barry, *Energy, Land, and Public Policy*. New Brunswick, NJ: Transaction, 1990.

Currey, Bruce and Graeme Hugo (eds.), *Famine as a Geographical Phenomenon*. Dordrecht: Reidel, 1984.

Cuthbertson, Ian M. and Jane Leibowitz, *Minorities: The New Europe's Old Issue*. Boulder, CO: Westview, 1994.


**D**

Dando, William A., *The Geography of Famine*. New York: Wiley, 1980.

Darmstadter, Joel (ed.), *Reconciling Global Development and the Environment; Perspectives on Sustainability*. Baltimore: Johns Hopkins Univ. Press, 1992.

Das Gupta, Jyotirindra, *Language Conflict and National Development; Group Politics and National Language Policy in India*. Berkeley: Univ. of California Press, 1970.

Davis, Clarence B. and Kenneth E. Wilburn (eds.), *Railway Imperialism*. Westport, CT: Greenwood, 1990.

de Alcántara, Cynthia Hewitt (ed.), *Real Markets; Social and Political Issues of Food Policy Reform*. London: Frank Cass, 1993.

Dean, Robert W., *Nationalism and Political Change in Eastern Europe: The Slovak Question and the Czechoslovak Reform Movement*. Denver: Univ. of Denver, 1973.

Debardeleben, Joan (ed.), *To Breathe Free; Eastern Europe's Environmental Crisis*. Baltimore: Johns Hopkins Univ. Press, 1991.

de Castro, Josué, *The Geography of Hunger*. Boston: Little, Brown, 1952.

Dejene, Alemneh, *Environment, Famine, and Politics in Ethiopia: A View from the Village*. Boulder, CO: Lynne Rienner, 1990.

de Leon, Pablo M. (ed.), *Air Transport Law and Policy in the 1990s; Controlling the Boom*. Norwell, MA: Kluwer, 1991.

Dempsey, Paul Stephen, *Law and Foreign Policy in International Aviation*. Ardsley-on-Hudson, NY: Transnational, 1987.

Denber, Rachel (ed.), *The Soviet Nationality Reader; The Crisis in Context*. Boulder, CO: Westview, 1992.

Deng, Francis M., *Protecting the Dispossessed; A Challenge for the International Community*. Washington, DC: Brookings, 1993.

Deng, Francis M. and Larry Minear, *The Challenges of Famine Relief; Emergency Operations in the Sudan*. Washington, DC: Brookings, 1992.

Dew, Edward, *The Difficult Flowering of Surinam; Ethnicity and Politics in a Plural Society*. The Hague: Nijhoff, 1978.

Diani, Mario, *Green Networks; A Structural Analysis of the Italian Environmental Movement*. Edinburgh: Edinburgh Univ. Press, 1994.

Díaz-Briquets, Sergio and Sidney Weintraub (eds.), *Determinants of Emigration from Mexico, Central America, and the Caribbean*. Boulder, CO: Westview, 1991.

————, *The Effects of Receiving Country Policies on Migration Flows*. Boulder, CO: Westview, 1991.

————, *Migration Impacts of Trade and Foreign Investment; Mexico and Caribbean Basin Countries*. Boulder, CO: Westview, 1991.

Dirks, Gerald E., *Canada's Refugee Policy*. Montreal: McGill-Queen's, 1978.

Dobson, Alan P., *Peaceful Air Warfare; The United States, Britain, and the Politics of International Aviation*. Oxford Univ. Press, 1991.

Dominian, Leon, *The Frontiers of Language and Nationality in Europe*. New York: Holt, 1917.

Doran, Charles F. and Stephen W. Buck (eds.), *The Gulf, Energy, and Global Security: Political and Economic Issues*. Boulder, CO: Lynne Rienner, 1991.

Dowty, Alan, *Closed Borders*. New Haven, CT: Yale Univ. Press, 1987.

Driedger, Leo and others (eds.), *Ethnic Demography; Canadian Immigrant, Racial and Cultural Variations*. Ottawa: Carleton Univ. Press, 1990.

Drüke, Luise, *Preventive Action for Refugee Producing Situations*. 2nd ed. Peter Lang, 1993.

Drummond, Phillip and others (eds.), *National Identity and Europe*. Bloomington: Indiana Univ. Press, 1993.

Drury, Bruce and others (eds.), *Agrarian Reform in Reverse*. Boulder, CO: Westview, 1987.

Dundas, Paul, *The Jains*. New York: Routledge, 1992.

Dunn, Charles W. (ed.), *Religion in American Politics*. Washington, DC: Congressional Quarterly, 1988.

Dunn, Dennis J. (ed.), *Religion and Nationalism*

*in Eastern Europe and the Soviet Union.* Boulder, CO: Lynn Rienner, 1987.

Durning, Alan Thein, *Guardians of the Land: Indigenous Peoples and the Health of the Earth.* Washington, DC: Worldwatch Institute, 1992.

Dyson, Kenneth (ed.), *Politics of the Communications Revolution in Western Europe.* London: Frank Cass, 1986.

### E

Eastman, Clyde, *Immigration Reform and New Mexico Agriculture.* El Paso: Univ. of Texas, 1984.

Ebel, Robert E., *Energy Choices in the Near Abroad; The Haves and Have-Nots Face the Future.* Boulder, CO: Westview, 1995.

Echeverria, John D., *Rivers at Risk: The Concerned Citizen's Guide to Hydropower.* Washington, DC: Island Press, 1989.

Edelman, Marc, *The Logic of the Latifundio.* Stanford, CA: Stanford Univ. Press, 1992.

Eelens, F. and others, *Labour Migration to the Middle East.* London: Routledge, 1992.

Eggert, Roderick G. (ed.), *Mining and the Environment; International Perspectives on Public Policy.* Baltimore: Johns Hopkins Univ. Press, 1994.

El-Ghonemy and M. Riad, *Land, Food and Rural Development in North Africa.* Boulder, CO: Westview, 1992.

Esman, Milton J. and Itamar Rabinovich (eds.), *Ethnicity, Pluralism, and the State in the Middle East.* Ithaca, NY: Cornell Univ. Press, 1988.

Esposito, John L. *Islam and Politics.* 3rd ed. Syracuse, NY: Syracuse Univ. Press, 1991.

———, *The Islamic Threat; Myth or Reality?* Oxford Univ. Press, 1992.

### F

Faber, Daniel, *Environment under Fire.* New York: Monthly Review Foundation, 1993.

Fabian, Johannes, *Language and Colonial Power; The Appropriation of Swahili in the Former Belgian Congo 1880–1938.* Cambridge Univ. Press, 1986.

Faeth, Paul (ed.), *Agriculture Policy and Sustainability: Case Studies from India, Chile, the Philippines and the U.S.* Washington, DC: World Resources Institute, 1993.

Falla, Jonathan, *True Love and Bartholomew; Rebels on the Burmese Border.* Cambridge Univ. Press, 1991.

Fardon, Richard and Graham Furniss (eds.), *African Languages, Development and the State.* New York: Routledge, 1994.

Fawcett, James, *The International Protection of Minorities.* Report No. 41. London: Minority Rights Group, 1979.

Feldman, Elliot J. and Michael A. Goldberg (eds.), *Land Rites and Wrongs; the Management, Regulation, and Use of Land in Canada and the United States.* Cambridge, MA: Lincoln Institute Publications, 1987.

Ferreira, Eduardo de Sousa and Guy Clausse (eds.), *Closing the Migratory Cycle: The Case of Portugal.* Breitenbach, 1986.

Ferris, Elizabeth G. (ed.), *Refugees and World Politics.* New York: Praeger, 1985.

———, *The Central American Refugees.* New York: Praeger, 1987.

Field, John Osgood (ed.), *The Challenge of Famine.* West Hartford, CT: Kumarian Press, 1993.

Finkle, Jason L. and Alison Mcintosh (eds.), *The New Politics of Population; Conflict and Consensus in Family Planning.* Oxford Univ. Press, 1994.

Finkle, Peter and Douglas M. Johnston, *Acid Precipitation in North America: The Case for Transboundary Cooperation.* Calgary: Canadian Institute of Resources Law, 1983.

Finn, James (ed.), *Ethiopia: The Politics of Famine.* Lanham, MD: Univ. Press of America, 1990.

Fletcher, Lehman B., *World Food Trade and Aid in the 1990s.* Boulder, CO: Westview, 1992.

Flinterman, Cees and others, *Transboundary Air Pollution.* Dordrecht: Kluwer, 1986.

Flynn, James and others, *One Hundred Centuries of Solitude: Redirecting America's High-Level Nuclear Waste Policy.* Boulder, CO: Westview, 1995.

Foon, Chew Sock, *Ethnicity and Nationality in Singapore.* Athens: Ohio Univ. Press, 1987.

Foresta, Ronald A., *Amazon Conservation in the Age of Development,* Gainesville: Univ. of Florida Press, 1991.

Fowler, Cary and Pat Mooney, *Shattering; Food, Politics, and the Loss of Genetic Diversity.* Tucson: Univ. of Arizona Press, 1990.

Francioni, Francesco and Tullio Scovazzi (eds.), *International Responsibility for Environmental Harm.* Norwell, MA: Kluwer, 1991.

Frankel, Edith Rogovin, *The Soviet Germans.* New York: St. Martin's, 1986.

Fraser, Angus, *The Gypsies.* Oxford: Blackwell, 1992.

Friedman, Francine, *The Bosnian Muslims.* Boulder, CO: Westview, 1995.

Fukuri, Katsuyoshi and John Maskakis (eds.), *Ethnicity and Conflict in the Horn of Africa*. Athens: Ohio Univ. Press, 1994.

Fuller, Graham E. and Ian O. Leser, *A Sense of Siege; The Geopolitics of Islam and the West*. Boulder, CO: Westview, 1994.

**G**

Galtung, Johan and Richard C. Vincent, *Global Glasnost; Toward a New World Information and Communications Order?* Cresskill, NJ: Hampton Press, 1992.

Gardiner, C. Harvey, *Pawns in a Triangle of Hate: The Peruvian Japanese and the United States*. Seattle: Univ. of Washington Press, 1981.

Gedicks, Al, *The New Resource Wars; Native and Environmental Struggles Against Multinational Corporations*. Boston: South End, 1993.

Geisler, Charles C. and Frank J. Popper (eds.), *Land Reform, American Style*. Totowa, NJ: Rowman & Allanheld, 1984.

Genizi, Haim, *America's Fair Share; the Administration and Resettlement of Displaced Persons, 1945–1952*. Detroit: Wayne State Univ. Press, 1994.

Gerrard, Michael B., *Whose Backyard, Whose Risk; Fear and Fairness in Toxic and Nuclear Waste Siting*. Cambridge, MA: MIT Press, 1994.

Ghai, Dharam and Lawrence D. Smith, *Agricultural Prices, Policy and Equity in Sub-Saharan Africa*. London: Pinter, 1987.

Ghose, Ajit Kumar, *Agrarian Reform in Contemporary Developing Countries*. New York: St. Martin's, 1983.

Gibb, Richard (ed.), *The Geography of the Channel Tunnel*. New York: Wiley, 1994.

Gibbon, Peter and others, *A Blighted Harvest; The World Bank and African Agriculture in the 1980s*. Trenton, NJ: African World, 1993.

Giblin, James, *The Politics of Environmental Control in Northeastern Tanzania, 1840–1940*. Philadelphia: Univ. of Pennsylvania Press, 1993.

Gibney, Mark (ed.), *Open Borders? Closed Societies? The Ethical and Political Issues*. New York: Greenwood, 1988.

Gillespie, Andrew and others (eds.), *Transport and Communication Innovation in Europe*. London: Belhaven, 1993.

Gillroy, John Martin (ed.), *Environmental Risk, Environmental Values and Political Choices; Beyond Efficiency Tradeoffs in Public Policy Analysis*. Boulder, CO: Westview, 1993.

Glantz, Michael H., *Drought and Hunger in Africa*. New York: Cambridge, 1987.

Glaser, William A., *The Brain Drain; Emigration and Return*. New York: Pergamon, 1978.

Glebe, Günther and John O'Loughlin (eds.), *Foreign Minorities in Continental European Cities*. Stuttgart: Frantz Steiner, 1987.

Goldwin, Robert A. and others (eds.), *Forging Unity Out of Diversity; The Approaches of Eight Nations*. Lanham, MD: AEI Press, 1989.

Gonzalez, Nancie L. and Carolyn S. McCommon (eds.), *Conflict, Migration, and the Expression of Ethnicity*. Boulder, CO: Westview, 1989.

Goran, Rystad (ed.), *The Uprooted*. Melbourne, FL: Krieger, 1990.

Gordenker, Leon, *Refugees in International Politics*. New York: Columbia Univ. Press, 1987.

Gorman, Robert F., *Mitigating Misery; An Inquiry into the Political and Humanitarian Aspects of U.S. and Global Refugee Policy*. Lanham, MD: Univ. Press of America, 1993.

Goulbourne, Harry, *Ethnicity and Nationalism in Post-Imperial Britain*. Cambridge Univ. Press, 1991.

Gregory, Robert G., *South Asians in East Africa*. Boulder, CO: Westview, 1992.

Griffiths, Stephen Iwan, *Nationalism and Ethnic Conflict; Threats to European Security*. Oxford Univ. Press, 1993.

Grigg, David, *World Food Problem*. Oxford: Blackwell, 1993.

Grubb, Michael and others, *The Earth Summit Agreements; A Guide and Assessment*. Washington, DC: Brookings, 1993.

Guidieri, Remo and others (eds.), *Ethnicities and Nations; Processes of Interethnic Relations in Latin America, S.E. Asia, and the Pacific*. Austin: Univ. of Texas Press, 1988.

Guimarães, Roberto P., *The Ecopolitics of Development in the Third World: Politics and Environment in Brazil*. Boulder, CO: Lynne Rienner, 1991.

Gunatilleke, Godfrey, *Migration to the Arab World*. Tokyo: United Nations Univ. Press, 1991.

Gurr, Ted Robert, *Minorities at Risk: A Global View of Ethnopolitical Conflicts*. Washington, DC: USIP Press, 1993.

Gurr, Ted Robert and Barbara Harff, *Ethnic Conflict in World Politics*. Boulder, CO: Westview, 1994.

**H**

Hadari, Ze'ev Venia, *Second Exodus: The Full Story of Jewish Illegal Immigration to Palestine, 1945–1948*. London: Frank Cass, 1991.

Hall, Richard and Hugh Peyman, *The Great Uhuru Railway: China's Showpiece in Africa.* London: Gollancz, 1976.

Hamelink, Cees J., *The Politics of World Communication.* Newbury Park, CA: Sage, 1994.

Hamilton, Kimberly A., *Migration and the New Europe.* Washington, DC: CSIS, 1994.

Hamlin, Christopher and Philip T. Shepard, *Deep Disagreement in U.S. Agriculture.* Boulder, CO: Westview, 1992.

Hanna, Willard A., *The Kra Isthmus Canal.* American Universities Field Staff Reports, Southeast Asia Series, vol. 15, No. 12, 1967.

Hannum, Hurst (ed.), *Documents on Autonomy and Minority Rights.* Kluwer, 1993.

Harrell-Bond, Barbara E., *Imposing Aid: Emergency Assistance to Refugees.* Oxford Univ. Press, 1986.

Harris, Joseph E. (ed.), *Global Dimensions of the African Diaspora.* Washington, DC: Howard Univ. Press, 1994.

Harrison, Paul, *The Third Revolution; Environment, Population and Sustainable World.* London: Tauris, 1992.

Harrison, Selig S., *In Afghanistan's Shadow: Baluch Nationalism and Soviet Temptations.* New York: Carnegie Endowment, 1981.

Hartmann, Betsy, *Reproductive Rights and Wrongs; The Global Politics of Population Control.* Boston: South End, 1994.

Hatch, Michael T., *Politics and Nuclear Power: Energy Policy in Western Europe.* Lexington: Univ. Press of Kentucky, 1986.

Hathaway, James C., *The Law of Refugee Status.* Toronto: Butterworths, 1991.

Hawkins, Freda, *Critical Years in Immigration; Canada and Australia Compared,* 2nd ed. Montreal: McGill-Queen's, 1991.

Haynes, Jeffrey, *Religion in Third World Politics.* Boulder, CO: Lynne Rienner, 1993.

Haywood, Keith, *International Collaboration in Civil Aerospace.* London: Pinter, 1986.

Hecht, Susanna and Alexander Cockburn, *The Fate of the Forest: Developers, Destroyers and Defenders of the Amazon.* London: Verso, 1989.

Hechter, Michael, *Internal Colonialism: The Celtic Fringe in British National Development, 1536–1966.* London: Routledge, 1975.

Hein, Jeremy, *States and Political Migrants.* Boulder, CO: Westview, 1992.

Heraclides, Alexis, *The Self-determination of Minorities in International Politics.* London: Frank Cass, 1991.

Herberer, Thomas, *China and Its National Minorities; Autonomy or Assimilation?* Armonk, NY: Sharpe, 1990.

Hindley, Reg, *The Death of the Irish Language; A Qualified Obituary.* New York: Routledge, 1991.

Hiro, Dilip, *Holy Wars; The Rise of Islamic Fundamentalism.* New York: Routledge, 1989.

H[au]oll, Otmar (ed.), *Environmental Cooperation in Europe; The Political Dimension.* Boulder, CO: Westview, 1994.

Holliday, Ian and others, *The Channel Tunnel; Public Policy, Regional Development and European Integration.* London: Pinter, 1991.

Hollist, W. Ladd and F. LaMond Tullis, *Pursuing Food Security: Strategies and Obstacles in Africa, Asia, Latin America, and the Middle East.* Boulder, CO: Lynne Rienner, 1987.

Holzberg, Carol S., *Minorities and Power in a Black Society; The Jewish Community of Jamaica.* Lanham, MD: North-South, 1987.

Horwich, George and David L. Weimer (eds.), *Responding to International Oil Crises.* Lanham, MD: AEI Press, 1988.

Houts, Peter S. and others, *The Three Mile Island Crisis: Psychological, Social, and Economic Impacts on the Surrounding Population.* University Park: Penn State Press, 1988.

Hovannisian, Richard (ed.), *The Armenian Genocide.* New York: St. Martin's, 1991.

Howard, Michael C., *Fiji: Race and Politics in an Island State.* Vancouver: UBC Press, 1991.

Hull, Terence H. and Wayne Jiye (eds.), *Population and Development Planning in China.* Sydney: Allen & Unwin, 1991.

Hummel, Monte (ed.), *Endangered Spaces.* Toronto: Key Porter Books, 1989.

Hurrell, Andrew and Benedict Kingsbury (eds.), *The International Politics of the Environment; Actors, Interests, and Institutions.* Oxford Univ. Press, 1992.

Hurst, Philip, *Rainforest Politics; Ecological Destruction in South-East Asia.* Washington, DC: Humanities Press, 1990.

Hussain, Arthur and others (eds.), *The Political Economy of Hunger.* Oxford Univ. Press, 1994.

Huttenbach, Henry R. (ed.), *Soviet Nationality Policies.* Mansell, 1990.

———, *Nationalities in the Soviet Union.* Melbourne, FL: Krieger, 1993.

**I**

Illyes, Elemer, *National Minorities in Romania.* New York: Columbia Univ. Press, 1982.

Ismael, Tareq Y. and Jacqueline S. Ismael (eds.), *Government and Politics in Islam.* New York: St. Martin's, 1985.

Ispahani, Mahnaz Z., *Roads and Rivals; The Polit-*

*ical Uses of Access in the Borderlands of Asia.* Ithaca: Cornell Univ. Press, 1989.

**J**

Jacob, Gerald, *Site Unseen: The Politics of Siting a Nuclear Waste Depository.* Pittsburgh: Univ. of Pittsburgh Press, 1990.

Jacob, James E., *Hills of Conflict; Basque Nationalism in France.* Reno: Univ. of Nevada Press, 1993.

James, Daniel, *Illegal Immigration: An Unfolding Crisis.* Lanham, MD: Univ. Press of America, 1991.

Jancar-Webster, Barbara (ed.), *Environmental Action in Eastern Europe; Response to Crisis.* Armonk, NY: Sharpe, 1993.

Jentleson, Bruce W., *Pipeline Politics: The Complex Political Economy of East-West Energy Trade.* Ithaca, NY: Cornell Univ. Press, 1986.

Jesudason, James V., *Ethnicity and the Economy; The State, Chinese Business and Multinationals in Malaysia.* Oxford Univ. Press, 1989.

Johnson, Stanley P., *World Population—Turning the Tide; Three Decades of Progress.* Norwell, MA: Kluwer, 1994.

Johnston, Douglas M. and Cynthia Sampson, *Religion: the Missing Dimension of Statecraft.* Washington, DC: CSIS, 1994.

Joly, Daniele and others, *Refugee Asylum in Europe.* Boulder, CO: Westview, 1992.

Jönsson, Christer, *International Aviation and the Politics of Regime Change.* New York: St. Martin's, 1987.

Jordan, Wayne R. (ed.), *Water and Water Policy in World Food Supplies.* College Station: Texas A&M Univ. Press, 1987.

Judge, David (ed.), *Politics; A Green Dimension for the European Community.* London: Frank Cass, 1993.

**K**

Kalt, Joseph P. and Frank C. Schuller (eds.), *Drawing the Line on Natural Gas Regulation.* Westport, CT: Greenwood, 1987.

Kapar, *Sikh Separatism.* London: Allen & Unwin, 1985.

Katzenstein, Mary F., *Ethnicity and Equality: The Shiv Sena Party and Preferential Policies in Bombay.* Ithaca, NY: Cornell Univ. Press, 1979.

Keeler, John T.S., *Politics of Neocorporatism in France: Farmers, the State, and Agricultural Policy-Making in the Fifth Republic.* Oxford Univ. Press, 1987.

Keely, Charles B. and others, *Immigration and U.S. Foreign Policy.* Boulder, CO: Westview, 1990.

Keen, David, *Refugees: Rationing the Right to Life; the Crisis in Emergency Relief.* London: Zed Books, 1992.

Kellerman, Aharon, *Society and Settlement: Jewish Land of Israel in the Twentieth Century.* Albany, NY: SUNY Press, 1993.

————, *Telecommunications and Geography.* London: Belhaven, 1993.

Kellner, Douglas, *The Persian Gulf TV War.* Boulder, CO: Westview, 1992.

Khalaf, Samir, *Lebanon's Predicament.* New York: Columbia Univ. Press, 1987.

Kimball, Lee A., *Forging International Agreements: The Role of Institutions in Environment and Development.* Washington, DC: World Resources Institute, 1992.

King, Russell, *Mass Migrations in Europe; the Legacy and the Future.* London: Belhaven, 1993.

Kirk, John and others, *Studies in Linguistic Geography.* London: Longwood, 1985.

Kirkwood, Michael (ed.), *Language Planning in the Soviet Union.* New York: St. Martin's, 1990.

Kiss, Alexandre and Dinah Shelton, *International Environmental Law.* Norwell, MA: Kluwer, 1991.

Kliot, Nurit, *Water Resources and Conflict in the Middle East.* New York: Routledge, 1994.

Kliot, Nurit and Stanley Waterman (eds.), *Pluralism and Political Geography; People, Territory and State.* London: Croom Helm, 1983.

Koehn, Peter H., *Refugees from Revolution; U.S. Policy and Third-World Migration.* Boulder, CO: Westview, 1991.

Kofele-Kale, Ndiva, *Tribesmen and Patriots: Political Culture in a Poly-ethnic African State.* Lanham, MD: Univ. Press of America, 1981.

Kozloff, Keith L. and Roger C. Dower, *A New Power Base: Renewable Energy Policies for the Nineties and Beyond.* Washington, DC: World Resources Institute, 1993.

Kramer, L., *EEC Treaty and Environmental Protection.* England: Sweet and Maxwell, 1992.

Krannich, Ronald L. and Caryl Rae Krannich, *The Politics of Family Planning Policy: Thailand—A Case of Successful Implementation.* Lanham, MD: Univ. Press of America, 1983.

Kraybill, Donald B., *The Amish and the State.* Baltimore: Johns Hopkins Univ. Press, 1993.

Kuhn, Raymond, *The Politics of Broadcasting.* New York: St. Martin's, 1985.

**L**

LaFeber, Walter, *The Panama Canal; the Crisis in Historical Perspective.* Oxford Univ. Press, 1978.

Landau, Jacob M., *The Politics of Pan-Islam, Ide-*

*ology and Organization.* Oxford Univ. Press, 1990.

Lang, Winfried and others (eds.), *Environmental Protection and International Law.* Norwell, MA: Kluwer, 1991.

Laponce, J. A., *Languages and Their Territories.* Toronto: Univ. of Toronto Press, 1987.

Lawless, Richard and Laila Monahan (eds.), *War and Refugees; The Western Sahara Conflict.* London: Pinter, 1987.

Layard, Richard and others, *East-West Migration.* Tokyo: United Nations Univ. Press, 1992.

Lee, Changsoo and George de Vos, *Koreans in Japan.* Berkeley: Univ. of California Press, 1982.

Leinbach, Thomas R. and Chiu Lin Sien, *South-East Asian Transport: Issue in Development.* Oxford Univ. Press, 1989.

Leiss, William (ed.), *Ecology Versus Politics in Canada.* Toronto: Univ. of Toronto Press, 1979.

Leith, James A. and others (eds.), *Planet Earth.* Montreal: McGill-Queens, 1995.

Le Marchand, René, *Burundi: Ethnocide as Discourse and Practice.* Cambridge Univ. Press, 1993.

LeMay, Michael C., *From Open Door to Dutch Door: An Analysis of U.S. Immigration Policy Since 1820.* New York: Praeger, 1987.

Lemco, Jonathan, *Tensions at the Border: Energy and Environment Concerns in Canada and the United States.* New York: Praeger, 1992.

Leo, Christopher, *Land and Class in Kenya.* Toronto: Univ. of Toronto Press, 1984.

Levine, Barry B. (ed.), *The Caribbean Exodus.* New York: Praeger, 1986.

Levine, Marc V., *The Reconquest of Montreal; Language Policy and Social Change in a Bilingual City.* Philadelphia: Temple Univ. Press, 1990.

Lewin-Epstein, Noah and Moshe Semyonov, *The Arab Minority in Israel's Economy; Patterns of Ethnic Inequality.* Boulder, CO: Westview, 1993.

Liebman, Charles S. and Eliezar Don-Yehiya, *Religion and Politics in Israel.* Bloomington: Indiana Univ. Press, 1984.

Linder, Wolf, *Swiss Democracy; Possible Solutions to Conflict in Multicultural Societies.* New York: St. Martin's, 1994.

Litfin, Karen T., *Ozone Discourse: Science and Politics in Global Environmental Cooperation.* Irvington, NY: Columbia Univ. Press, 1994.

Loescher, Gil, *Beyond Charity: International Cooperation and the Global Refugee Crisis.* Oxford Univ. Press, 1993.

Loescher, Gil and Laila Monahan (eds.), *Refugees and International Relations.* Oxford Univ. Press, 1989.

Loescher, Gilbert D. and John A. Scanlon, *Calculated Kindness; Refugees and America's Half-opened Door, 1945 to the Present.* New York: Free Press, 1986.

Long, James W., *From Privileged to Dispossessed: The Volga Germans, 1860–1917.* Lincoln: Univ. of Nebraska Press, 1988.

Lowi, Miriam R., *Water and Power; the Politics of a Scarce Resource in the Jordan River Basin.* Cambridge Univ. Press, 1993.

Lustick, Ian S., *For the Land and the Lord; Jewish Fundamentalism in Israel.* New York: Council on Foreign Relations, 1988.

Luter, James P. and Ann Bowman (eds.), *The Politics of Hazardous Waste Management.* Durham, NC: Duke Univ. Press, 1984.

Lynge, Finn, *Arctic Wars, Animal Rights, Endangered Peoples.* Univ. Press of New England, 1992.

Lyster, Simon, *International Wildlife Law.* Cambridge: Grotius, 1985.

## M

Maasdorp, Gavin G., *Current Political and Economic Factors in Transportation in Southern Africa.* Braamfontein: South African Institute of International Affairs, February 1988.

MacDonnell, Lawrence J. and Sarah F. Bates, *Natural Resources Policy and Law; Trends and Directions.* Washington, DC: Island Press, 1993.

Mackenas, Colin, *China's Minorities; Integration and Modernization in the Twentieth Century.* Oxford Univ. Press, 1994.

Maingot, Anthony P. (ed.), *Small Country Development and International Labor Flows; Experiences in the Caribbean.* Boulder, CO: Westview, 1991.

Major, John, *Prize Possession; The United States and The Panama Canal, 1903–1979.* Cambridge Univ. Press, 1993.

Manogaran, Chelvadurai, *Ethnic Conflict and Reconciliation in Sri Lanka.* Honolulu: Univ. of Hawaii Press, 1987.

Mansell, Gerard, *Let the Truth Be Told: 50 Years of BBC External Broadcasting.* London: Weidenfeld & Nicolson, 1982.

Marcus, Alfred Allen, *Controversial Issues in Energy Policy.* Vol. 2. Newbury Park, CA: Sage, 1992.

Markiewicz, Dana, *The Mexican Revolution and the Limits of Agrarian Reform, 1915–1946.* Boulder, CO: Lynne Rienner, 1992.

Marks, Steven Gary, *Road to Power: The Trans-*

*Siberian Railroad and the Colonization of Asian Russia, 1850–1917*. Ithaca, NY: Cornell Univ. Press, 1991.

Marsh, John, *The Changing Role of Common Agricultural Policy*. London: Belhaven, 1991.

Marshall, Dawn I., *The Haitian Problem: Illegal Migration to the Bahamas*. Kingston, Jamaica: Institute of Social and Economic Research, Univ. of the West Indies, 1979.

Martin, David, *Tongues of Fire; The Explosion of Protestantism in Latin America*. Oxford: Blackwell, 1990.

Martin, David A. (ed.), *The New Asylum Seekers: Refugee Law in the 1980's*. London: Nijhoff, 1988.

Martz, John D., *Regime, Politics, and Petroleum*. New Brunswick, NJ: Transaction, 1987.

Mason, Robert J., *Contested Lands: Conflict and Compromise in New Jersey's Pine Barrens*. Philadelphia: Temple Univ. Press, 1992.

Masud-Piloto, Felix Roberto, *With Open Arms: The Political Dynamics of the Migration From Revolutionary Cuba*. Totowa, NJ: Rowman & Littlefield, 1987.

Mazrui, Alamin and Ibrahim Noor Shariff, *The Swahili; Idiom and Identity of an African People*. Trenton, NJ: African World, 1994.

Mbuyi, Benjamin Mulamba (ed.), *Refugees and International Law*. Toronto: Carswell, 1994.

McFarland, Andrew S., *Cooperative Pluralism; the National Coal Policy Experiment*. Lawrence: Univ. Press of Kansas, 1994.

McGarry, John and Brendan O'Leary (eds.), *The Politics of Ethnic Conflict Regulation*. New York: Routledge, 1993.

McGuire, Thomas R., *Politics and Ethnicity on the Río Yaqui; Potam Revisited*. Tucson: Univ. of Arizona Press, 1986.

McIntosh, C. Alison, *Population Policy in Western Europe: Responses to Low Fertility in France, Sweden, and West Germany*. Armonk, NY: Sharpe, 1983.

McNeill, William H., *Polyethnicity and National Unity in World History*. Toronto: Univ. of Toronto Press, 1985.

McPhail, Thomas L., *Electronic Colonialism: The Future of International Broadcasting and Communication*. 2nd ed. Newbury Park, CA: Sage, 1987.

McRae, Kenneth D., *Conflict and Compromise in Multilingual Societies*. Vol. 1 Switzerland; Vol. 2 Belgium. Waterloo, Ont.: Wilfrid Laurier Univ. Press, 1983 and 1992.

Mehmet, Ozay, *Islamic Identity and Development; Studies of the Islamic Periphery*. New York: Routledge, 1991.

Menashri, David (ed.), *Central Asia Meets the Middle East*. London: Frank Cass, 1994.

Mendes, Chico with Tony Gross, *Fight for the Forest*. New York: Monthly Review Foundation, 1993.

Meyer, Carrie A. and others, *Population Growth, Poverty, and Environmental Stress: Frontier Migration in the Philippines and Costa Rica*. Washington, DC: World Resources Institute, 1992.

Miall, Hugh (ed.), *Minority Rights in Europe; Prospects for a Transnational Regime*. New York: Council on Foreign Relations, 1995.

Mickiewicz, Ellen, *Split Signals; Television and Politics in the Soviet Union*. Oxford Univ. Press, 1988.

Middleton, Niel and others, *The Tears of the Crocodile; From Rio to Reality in the Developing World*. Boulder, CO: Westview, 1993.

Mikesell, Raymond F., *Economic Development and the Environment*. Mansell, 1992.

Miller, Marian A.L., *The Third World in Global Environmental Politics*. Boulder, CO: Lynne Rienner, 1995.

Minault, Gail, *The Khilafat Movement: Religious Symbolism and Political Mobilization in India*. New York: Columbia Univ. Press, 1982.

Minority Rights Group. *World Directory of Minorities*. Harlow, UK: Longman, 1991.

Mitchell, Ronald B., *Intentional Oil Pollution at Sea; Environmental Policy and Treaty Compliance*. Cambridge, MA: MIT, 1994.

Moch, Leslie Page, *Moving Europeans; Migration in Western Europe Since 1650*. Urbana: Indiana Univ. Press, 1992.

Molvaer, Reidulf K., *Environmental Cooperation and Confidence Building in the Horn of Africa*. Newbury Park, CA: Sage, 1993.

Moosa, Matti, *Extremist Shiites; The Ghulat Sects*. Syracuse, NY: Syracuse Univ. Press, 1993.

———, *The Maronites in History*. Syracuse, NY: Syracuse Univ. Press, 1993.

Moriyama, Alan T., *Imingaisha*. Honolulu: Univ. of Hawaii Press, 1985.

Mosely, George (ed.), *The Party and the National Question in China*. Cambridge, MA: MIT Press, 1966.

Motyl, Alexander J., *Sovietology, Rationality, Nationality; Coming to Grips with Nationalism in the USSR*. New York: Columbia Univ. Press, 1990.

Mounfield, Peter R., *World Nuclear Power; A Geographical Appraisal*. New York: Routledge, 1991.

Mowlana, Hamid and others (eds.), *Triumph of the Image; the Media's War in the Persian*

*Gulf—A Global Perspective.* Boulder, CO: Westview, 1992.

Moynihan, Daniel Patrick, *Pandaemonium; Ethnicity in International Politics.* Oxford Univ. Press, 1993.

Moyser, George (ed.), *Politics and Religion in the Modern World.* New York: Routledge, 1991.

Muntarbhorn, Vitit, *The Status of Refugees in Asia.* Oxford Univ. Press, 1992.

Murphy, Alexander B., *The Regional Dynamics of Language Differentiation in Belgium: A Study in Cultural-Political Geography.* Univ. of Chicago, Dept. of Geography Research Paper No. 227, 1988.

Murphy, Brian, *International Politics of New Information Technology.* New York: St. Martin's, 1986.

Mutalib, Hussin, *Islam and Ethnicity in Malay Politics.* Oxford Univ. Press, 1990.

Mutalib, Hussin and Taj Ul-Blam Hashmi, *Islam, Muslims and the Modern State.* New York: St. Martin's, 1993.

Mutukwa, Kasuka Simwinji, *Politics of the Tanzania-Zambia Railway Project.* Lanham, MD: Univ. Press of America, 1979.


**N**

Nanda, Ved P. (ed.), *World Climate Change; the Role of International Law and Institutions.* Boulder, CO: Westview, 1983.

———, *Refugee Law and Policy; International and U.S. Responses.* Westport, CT: Greenwood, 1989.

Narayana, G. and John F. Kantner, *Doing the Needful; the Dilemma of India's Population Policy.* Boulder, CO: Westview, 1992.

Ngwenya, Sindiso (ed.), *The Transport and Communications Sector in Southern Africa.* Harare: SAPES Books, 1993.

Nielsen, Jargen, *Muslims in Western Europe.* Edinburgh: Edinburgh Univ. Press, 1992.

Nivola, Pietro S., *The Politics of Energy Conservation.* Washington, DC: Brookings, 1986.

Nollkaemper, Andre, *The Legal Regime for Transboundary Water Pollution: Between Discretion and Constraint.* Kluwer, 1993.

Norton, Robert, *Race and Politics in Fiji.* New York: St. Martin's, 1978.


**O**

Ofer, Dalia, *Escaping the Holocaust; Illegal Immigration to the Land of Israel, 1939-1944.* Oxford Univ. Press, 1991.

Olzak, Susan, *The Dynamics of Ethnic Competition and Conflict.* Stanford, CA: Stanford Univ. Press, 1992.

Omran, Abdel R. and Farzaneh Roudi, *The Middle East Population Puzzle.* Washington, DC: Population Reference Bureau, 1993.

Opalski, Magdalena M. and Israel Bartel, *Poles and Jews; A Failed Brotherhood.* Univ. Press of New England, 1992.

Opie, John, *The Law of the Land; Two Hundred Years of American Farmland Policy.* Lincoln: Univ. of Nebraska Press, 1987.

Owen, Roger, *Migrant Workers in the Gulf.* Cambridge, MA: Cultural Survival, 1985.


**P**

Papademetriou, Demetrios G., *At the Precipice? Europe and Migration.* Washington, DC: Brookings, 1993.

Parnwell, Mike, *Population Movements and the Third World.* New York: Routledge, 1993.

Payne, Stanley G., *Basque Nationalism.* Reno: Univ. of Nevada Press, 1991.

Pearson, Scott and others, *Rice Policy in Indonesia.* New York: Cornell Univ. Press, 1991.

Percy, David R., *The Framework of Water Rights Legislation in Canada.* Calgary: Canadian Institute of Resources Law, 1988.

Peterson, D. J., *Troubled Lands; The Legacy of Soviet Environmental Destruction.* Boulder, CO: Westview, 1992.

Pilat, J. F., *Ecological Politics; the Rise of the Green Movement.* Lanham, MD: Univ. Press of America, 1980.

Pipa, Arshi, *Politics of Language in Socialist Albania.* New York: Columbia Univ. Press, 1989.

Pirages, Dennis, *Global Ecopolitics: The New Context for International Relations.* North Scituate, MA: Duxbury, 1978.

———, *Global Technopolitics: The International Politics of Technology and Resources.* Pacific Grove, CA: Brooks/Cole, 1989.

Pisani, Edgard and others (eds.), *European Minorities.* New York: Pergamon, 1992.

Platt, Rutherford H., *Land Use Control: Geography, Law and Public Policy.* Englewood Cliffs, NJ: Prentice Hall, 1991.

Plender, Richard (ed.), *Basic Documents on International Migration Law.* Dordrecht: Nijhoff, 1988.

Polansky, Antony and Norman Darris (eds.), *Jews in Eastern Poland and USSR, 1939–46.* New York: St. Martin's, 1991.

Porter, Gareth and Janet W. Brown, *Global Environmental Politics.* 2nd ed. Boulder, CO: Westview, 1995.

Portney, Kent E., *Controversial Issues in Environmental Policy: Science vs. Economics vs. Politics.* Newbury Park, CA: Sage, 1992.

Prendiville, Brendan, *Environmental Politics in France*. Boulder, CO: Westview, 1994.

Price, Terence, *Political Electricity; What Future for Nuclear Energy?* Oxford Univ. Press, 1990.

Prindle, David F., *Petroleum Politics and the Texas Railroad Commission*. Austin: Univ. of Texas Press, 1981.

Pringle, D. G., *One Island, Two Nations? A Political Geographical Analysis of the National Conflict in Ireland*. New York: Wiley, 1985.

Pryde, Philip R., *Environmental Management in the Soviet Union*. Cambridge Univ. Press, 1991.

**Q**

Quester, George H., T*he International Politics of Television*. Lexington, MA: Heath, 1990.

**R**

Raikes, Philip, *Modernising Hunger; Famine, Food Surplus and Farm Policy in the E.E.C. and Africa*. Heinemann, 1988.

Rao, M.S.A., *Studies in Migration: Internal and International Migration in India*. Delhi: Manohar, 1986.

Raphael, Ray, *More Tree Talk; the People, Politics and Economics of Timber*. Covelo, CA: Island Press, 1994.

Rapp, David, *How the U.S. Got into Agriculture; And Why It Can't Get Out*. Washington, DC: Congressional Quarterly, 1988.

Read, Jan, *The Catalans*. London: Faber & Faber, 1978.

Reeves, Geoffrey W., *Communications and the "Third World."* New York: Routledge, 1993.

Reich, Michael R., *Toxic Politics; Responding to Chemical Disasters*. New York: Cornell Univ. Press, 1991.

Reichley, A. James, *Religion in American Public Life*. Washington, DC: Brookings, 1985.

Repetto, Robert and Malcolm Gillis (eds.), *Public Policies and the Misuse of Forest Resources*. Washington, DC: World Resources Institute, 1988.

*Resources for the Future, Inc.* in Washington, DC. Publishes many studies on resource and environmental problems.

Reynell, Josephine, *Political Pawns; Refugees on the Thai-Kampuchean Border*. Oxford: Refugee Studies Programme, 1989.

Ribeiro, Gustavo Lins, *Transnational Capitalism and Hydropolitics in Argentina: The Yacyretá High Dam*. Gainesville: Univ. Press of Florida, 1994.

Rivera, Mario A., *Decision and Structure: U.S. Refugee Policy and the Mariel Crisis*. Lanham, MD: Univ. Press of America, 1991.

R'oi, Yaacov, *The Struggle for Soviet Jewish Emigration, 1948–1967*. Cambridge Univ. Press, 1990.

Romaine, Suzanne, *Pidgin and Creole Languages*. London: Longman, 1988.

Rorlich, Azade-Ayse, *The Volga Tatars; a Profile in National Resistance*. Stanford, CA: Hoover Institution Press, 1986.

Rose, Laurel L., *The Politics of Harmony; Land Dispute Strategies in Swaziland*. Cambridge Univ. Press, 1991.

Rosenbaum, Walter A., *Environmental Politics and Policy*. 3rd ed. Washington, DC: Congressional Quarterly, 1994.

Rosenfeld, Stanley B., *The Regulation of International Commercial Aviation*. Dobbs Ferry, NY: Oceana, 1984.

Ross, Lester and Mitchell A. Silk, *Environmental Law and Policy in the People's Republic of China*. Westport, CT: Quorum Books, 1987.

Ross, Monique and J. Owen Saunders (eds.), *Growing Demands on a Shrinking Heritage: Managing Resource-Use Conflicts*. Calgary: Univ. of Calgary, 1992.

Rothchild, Donald and Victor A. Olorunsola (eds.), *State Versus Ethnic Claims; African Policy Dilemmas*. Boulder, CO: Westview, 1983.

Rowles, James P., *Law and Agrarian Reform in Costa Rica*. Boulder, CO: Westview, 1985.

Rüdig, Wolfgang, *Green Politics III*. Edinburgh: Edinburgh Univ. Press, 1994.

Rudolph, Joseph R., Jr. and Robert J. Thompson (eds.), *Ethnoterritorial Politics, Policy, and the Western World*. Boulder, CO: Lynne Rienner, 1989.

Rusinow, Dennison I., *The Other Albania: Kosovo 1979 Part I: Problems and Prospects*. AUFS Reports Europe Series 1980/No. 5.

———, *Part II: The Village, the Factory, and the Kosovars*. AUFS Reports Europe Series 1980/No. 6.

———, *Unfinished Business: The Yugoslav "National Question."* AUFS Reports Europe Series 1981/No. 35.

———, *Yugoslavia's Muslim Nation*. UFSI Reports Europe Series 1982/No. 8.

Ruttan, Vernon W. (ed.), *Why Food Aid?* Baltimore: Johns Hopkins Univ. Press, 1993.

Rwelamira, Medard, *Refugees in a Chess Game: Reflections on Botswana, Lesotho and Swaziland Refugee Policies*. Uppsala: Scandinavian Institute of African Studies, 1990.

Ryan, Stephen, *Ethnic Conflict and International Relations*. 2nd ed. U.K.: Dartmouth, 1995.

Rywkin, Michael, *Moscow's Muslim Challenge: Soviet Central Asia*. Armonk, NY: Sharpe, 1990.

**S**

Salitan, Laurie P., *Politics and Nationality in Contemporary Soviet Jewish Emigration, 1968–89.* New York: St. Martin's, 1991.

Salter, M. J., *Studies in the Immigration of the Highly Skilled.* Canberra: Australian National Univ. Press, 1978.

Salvatore, Dominick (ed.), *World Population Trends and Their Impact on Economic Development.* Westport, CT: Greenwood, 1988.

Sand, Peter H. (ed.), *The Effectiveness of International Environmental Agreements, A Survey of Existing Legal Instruments.* Cambridge: Grotius, 1992.

Saunders, J. Owen, *The Legal Challenge of Sustainable Development.* Calgary: Canadian Institute of Resources Law, 1989.

Savage, James G., *The Politics of International Telecommunications Regulation.* Boulder, CO: Westview, 1989.

Schirazi, Asghar, *Islamic Development Policy: The Agrarian Question in Iran.* Boulder, CO: Lynne Rienner, 1993.

Schloss, Aran, *The Politics of Development: Transportation Policy in Nepal.* Lanham, MD: Univ. Press of America, 1983.

Schuler, G. Henry M., *The Venezuelan-U.S. Petroleum Relationship: Past, Present, and Future.* Washington, DC: CSIS, 1991.

Sherington, Geoffrey, *Australia's Immigrants 1788–1988.* 2nd ed. Allen & Unwin, 1991.

Shiva, Vandana, *Ecology and the Politics of Survival; Conflicts Over Natural Resources in India.* Newbury Park, CA: Sage, 1991.

———, *The Violence of the Green Revolution: Third World Agriculture, Ecology and Politics.* Washington, DC: Zed Books, 1991.

Shlapentokh, Vladimir and others (eds.), *The New Russian Diaspora; Russian Minorities in the Former Soviet Republics.* Armonk, NY: Sharpe, 1994.

Short, K. R., *Western Broadcasting over the Iron Curtain.* New York: St. Martin's, 1986.

Sicular, Terry (ed.), *Food Price Policy in Asia: A Comparative Study.* Ithaca, NY: Cornell Univ. Press, 1989.

Simon, Gerhard, *Nationalism and Policy Toward the Nationalities in the Soviet Union.* Boulder, CO: Westview, 1991.

Simon, Julian L., *Population and Development in Poor Countries.* Princeton, NJ: Princeton Univ. Press, 1992.

Simon, Rita J. and others, *International Migration; the Female Experience.* New York: Barnes & Noble, 1985.

Sisk, Timothy D., *Islam and Democracy: Religion, Politics, and Power in the Middle East.* Washington, DC: USIP Press, 1992.

Sjöstedt, Gunnar, *International Environmental Negotiation.* Newbury Park, CA: Sage, 1992.

Skeldon, Ronald, *Population Mobility in Developing Countries; A Reinterpretation.* London: Pinter, 1990.

Skidmore, Thomas (ed.), *Television, Politics, and the Transition to Democracy in Latin America.* Baltimore: Johns Hopkins Univ. Press, 1993.

Smith, Anthony, *Geopolitics of Information: How Western Culture Dominates the World.* Oxford Univ. Press, 1981.

Smits, Jan, *Legal Aspects of Implementing International Telecommunication Links: Institutions, Regulations, and Instruments.* Dordrecht: Nijhoff, 1991.

Sochor, Eugene, *The Politics of International Aviation.* Iowa City: Univ. of Iowa Press, 1991.

Sohn, Louis B. and Thomas Buergenthal (eds.), *The Movement of Persons Across Borders.* Washington, DC: American Society of International Law, 1992.

Soley, Lawrence C. and John C. Nichols, *Clandestine Radio Broadcasting.* New York: Praeger, 1986.

Solomos, John and John Wrench (eds.), *Racism and Migration in Western Europe.* Herndon, VA: Berg, 1993.

Spickard, Paul (ed.), *Pacific Island Peoples in Hawaii.* Honolulu: Univ. of Hawaii Press, 1994.

Stamp, L. Dudley, *Land for Tomorrow: The Underdeveloped World.* New York: American Geographical Society, 1952.

Starke, Linda and Gro Harlem Brundtland, *Signs of Hope; Working Towards Our Common Future.* Oxford Univ. Press, 1990.

Starkie, D.N.M., *The Motorway Age: Road and Traffic Politics in Britain 1896–1970.* Oxford: Pergamon, 1982.

Stavis, Benedict, *The Politics of Agricultural Mechanization in China.* Ithaca, NY: Cornell Univ. Press, 1978.

Steenland, Kyle, *Agrarian Reform under Allende; Peasant Revolt in the South.* Albuquerque: Univ. of New Mexico Press, 1977.

Stevens, Paul, *International Gas.* New York: St. Martin's, 1986.

Stevenson, Garth, *The Politics of Canada's Airlines from Diefenbaker to Mulroney.* Toronto: Univ. of Toronto Press, 1987.

Stonich, Susan C., *"I Am Destroying the Land!" The Political Ecology of Poverty and Environmental Destruction in Honduras.* Boulder, CO: Westview, 1993.

Studies of Nationalities in the USSR. Series of

books produced in Stanford, CA by the Hoover Institution Press beginning in 1978. In print to date: *The Crimean Tatars, The Kazakhs, The Volga Tatars, The Modern Uzbeks, Estonia and the Estonians, Azerbaijani Turks, The Ukraine, The Moldavian Republic, The Last Empire; Nationality and the Soviet Future* and *The Making of the Georgian Nation* (available from Indiana Univ. Press).

Suksamran, Somboon, *Political Buddhism in Southeast Asia.* New York: St. Martin's, 1977.

Super, John C. and Thomas C. Wright (eds.), *Food, Politics, and Society in Latin America.* Lincoln: Univ. of Nebraska Press, 1985.

Susskind, Lawrence E., *Environmental Diplomacy; Negotiating More Effective Global Agreements.* Oxford Univ. Press, 1994.

**T**

Tapper, Richard, *The Conflict of Tribe and State in Iran and Afghanistan.* New York: St. Martin's, 1983.

Teich, Mikulas and Roy Porter (eds.), *The National Question in Europe in Historical Context.* Cambridge Univ. Press, 1993.

Terrie, Philip G., *Forever Wild: Environmental Aesthetics and the Adirondacks Forest Preserve.* Philadelphia: Temple Univ. Press, 1985.

Thiesenhusen, William C., *Latin American Agriculture: Structure and Reform.* London: Allen & Unwin, 1987.

Thomas, Caroline, *The Environment in International Relations.* London: RIIA, 1992.

———, *Rio: Unravelling the Consequences.* London: Frank Cass, 1994.

Thomas, David S.G. and Nicholas J. Middleton, *Desertification: Exploding the Myth.* New York: Wiley, 1994.

Thompson, Carol, *Harvests under Fire; Regional Co-operation for Food Security in Southern Africa.* London: Zed Books, 1991.

Thompson, Paul B., *The Ethics of Aid and Trade; U.S. Food Policy, Foreign Competition, and the Social Contract.* Cambridge Univ. Press, 1992.

Thornberry, Patrick, *International Law and the Rights of Minorities.* Oxford Univ. Press, 1993.

Thorsell, J. (ed.), *Parks on the Borderline; Experience in Transfrontier Conservation.* Cambridge: IUCN Publications, 1990.

Thränhardt, Dietrich, *Europe—A New Immigration Continent: Policies and Politics in Comparable Perspective.* Boulder, CO: Westview, 1994.

Thukral-Ganguly, Enakshi (ed.), *Big Dams, Displaced People.* Newbury Park, CA: Sage, 1992.

Tickell, Crispin, *Climatic Change and World Affairs.* Lanham, MD: Univ. Press of America, 1986.

Tobin, Richard J., *Expendable Future: U.S. Politics and the Protection of Biological Diversity.* Durham, NC: Duke Univ. Press, 1990.

Toke, David, *Energy and Environment, The Political and Economic Debate.* Boulder, CO: Westview, 1995.

Toland, Judith D. (ed.), *Ethnicity and the States; Political and Legal Anthropology.* Vol. 10. New Brunswick, NJ: Transaction, 1992.

Tolba, Mostafa Kamal, *Saving Our Planet; Challenges and Hopes.* Chapman and Hall, 1992.

Tolba, Mostafa Kamal and Asit K. Biswas (eds.), *Earth and Us; Population, Resources, Environment, Development.* Letchworth, UK: Butterworth-Heinemann, 1991.

Tolley, Rodney and Brian Turton, *Transport Systems, Policy and Planning: A Geographical Approach.* New York: Wiley, 1995.

Tyson, James L., *U.S. International Broadcasting and National Security.* New York: Ramapo Press, 1983.

**U**

United Nations High Commission for Refugees. *The State of the World's Refugees 1993; The Challenge of Protection.* New York: Penguin, 1993.

Uvin, Peter, *The International Organization of Hunger.* Irvington, NY: Columbia Univ. Press, 1994.

**V**

Van Der Pugt, Joop, *Nuclear Energy and the Public.* Oxford: Blackwell, 1992.

Van Der Veer, Peter (ed.), *Nation and Migration: The Politics of Space in the South Asian Diaspora.* Philadelphia: Univ. of Pennsylvania Press, 1995.

Varnis, Steven L., *Reluctant Aid or Aiding the Reluctant?; U.S. Food Aid Policy and Ethiopian Famine Relief.* New Brunswick, NJ: Transaction, 1990.

Vellinga, Pier and Michael Grubb (eds.), *Climate Change Policy in the European Community; A Workshop Report.* Washington, DC: Brookings, 1993.

Vernez, Georges (ed.), *Immigration and International Relations; Proceedings of a Conference on the International Effects of the 1986 Immigration Reform and Control Act (IRCA).* Lanham, MD: Univ. Press of America, 1990.

Vietor, Richard H.K., *Environmental Politics and*

*the Coal Coalition.* College Station: Texas A&M Univ. Press, 1986.

Vig, Norman J. and Michael E. Kraft, *Environmental Policy in the 1990's.* 2nd ed. Washington, DC: Congressional Quarterly, 1993.

**W**

Wagaw, Teshome G., *For Our Soul; Ethiopian Jews in Israel.* Detroit: Wayne State Univ. Press, 1994.

Wald, Kenneth, *Religion and Politics in the United States.* 2nd ed. Washington, DC: Congressional Quarterly, 1991.

Watson, Ian, *Fighting over the Forests.* Allen & Unwin, 1991.

Webb, Patrick and Joachim von Braun, *Famine and Food Security in Ethiopia: Lessons for Africa.* New York: Wiley, 1994.

Weinberg, William J., *War on the Land; Ecology and Politics in Central America.* London: Zed Books, 1991.

Weiss, Anita M. (ed.), *Islamic Reassertion in Pakistan; The Application of Islamic Laws in a Modern State.* Syracuse, NY: Syracuse Univ. Press, 1986.

Weiss, Edith Brown (ed.), *Environmental Change and International Law.* Tokyo: United Nations Univ. Press, 1992.

Whalley, John, *Trading for the Environment.* Washington, DC: Institute for International Economics, 1991.

Whiteford, Scott and Anne E. Ferguson (eds.), *Harvest of Want: Hunger and Food Security in Central America and Mexico.* Boulder, CO: Westview, 1991.

Whiteside, James, *Regulating Danger: The Struggle for Mine Safety in the Rocky Mountain Coal Industry.* Lincoln: Univ. of Nebraska Press, 1990.

Wilhelm, Donald, *Global Communications and Political Power.* New Brunswick, NJ: Transaction, 1990.

Williams, Colin H., *Language in Geographic Context.* Clevedon, U.K.: Multilingual Matters, Ltd., 1988.

Wilson, A. Jeyaratnam, *The Break-up of Sri Lanka; The Sinhalese-Tamil Conflict.* Honolulu: Univ. of Hawaii Press, 1988.

Wionczek Guzman, Miguel and others (eds.), *Energy Policy in Mexico: Problems and Prospects for the Future.* Boulder, CO: Westview, 1985.

Wirth, John C. (ed.), *Latin American Oil Companies and the Politics of Energy.* Lincoln: Univ. of Nebraska Press, 1985.

Wistrich, Enid, *The Politics of Transportation.* London: Longman, 1983.

Wolfson, Richard, *Nuclear Choices A Citizens Guide to Nuclear Technology.* (revised edition), Cambridge, MA: MIT, 1993.

*Worldwatch Papers.* Washington, DC: Worldwatch Institute. Monographs; topics include environment, energy, food, population, national security, health, migration, and related topics.

Wrench, John (ed.), *Racism and Migration in Western Europe.* Providence, RI: Berg, 1993.

**Y**

Yaffee, Steven Lewis, *The Wisdom of the Spotted Owl; Policy Lessons for a New Century.* Covelo, CA: Island Press, 1994.

Yans-McLaughlin, Virginia (ed.), *Immigration Reconsidered; History, Sociology, and Politics.* Oxford Univ. Press, 1990.

Yarnold, Barbara M., *Refugees Without Refuge: Formation and Failed Implementation of U.S. Political Asylum Policy in the 1980's.* Lanham, MD: Univ. Press of America, 1990.

Young, Oran R., *International Cooperation: Building Regimes for Natural Resources and the Environment.* Ithaca, NY: Cornell Univ. Press, 1989.

**Z**

Zachariah, K.C. and Julien Condé, *Migration in West Africa; Demographic Aspects.* Oxford Univ. Press, 1981.

Zamosc, Leon, *The Agrarian Question and the Peasant Movement in Colombia.* Cambridge Univ. Press, 1986.

Zegeye, A. and S. Ishemo (eds.), *Forced Labor Migration: Patterns of Movement Within Africa.* London: Hans Zell, 1989.

Zetterqvist, Jenny, *Refugees in Botswana in the Light of International Law.* Uppsala: Scandanavian Institute of African Studies, 1990.

Zolberg, Aristide R. and others, *Escape from Violence; Conflict and the Refugee Crisis in the Developing World.* Oxford Univ. Press, 1989.

Zucker, Norman L. and Naomi Flink Zucker, *The Guarded Gate; the Reality of American Refugee Policy.* New York: Harcourt Brace Jovanovich, 1987.

Zylice, Marek, *International Air Transport Law.* Norwell, MA: Kluwer, 1992.

# *Periodicals*

**A**

Armstrong, Graham, "Canadian Energy Policy in the 1990s," *National Economic Review* (Feb. 1991), 13–20.

Auer, Matthew R., "Prospects for Environmental Cooperation in the Yellow Sea," *Emory International Law Review*, 5, 1 (Spring 1991).

**B**

Barker, Mary L., "National Parks, Conservation, and Agrarian Reform in Peru," *Geographical Review*, 70, 1 (Jan. 1980), 1–18.

Bierman, Don E. and W. Rydzkowski, "Regional Politics in Public Works Projects: The Tennessee-Tombigbee Waterway," *Transportation Quarterly*, 45, 2 (April 1991), 169–180.

Birnie, Patricia W. (ed.), *The International Law of Migratory Species.* Special issue of *Natural Resources Journal*, 29, 4 (Fall 1989), whole volume.

Black, Richard, "Refugees and Displaced Persons: Geographical Perspectives and Research Directions," *Progress in Human Geography*, 15, 3 (1991), 281–298.

Blowers, Andrew, "The Triumph of Material Interests—Geography, Pollution and the Environment," *Political Geography Quarterly*, 3, 1 (Jan. 1984), 49–68.

Bole, Janice J., "Feast or Famine: Do Ethiopians Have a Choice?" *Dickinson Journal of International Law*, 5, 1 (Fall 1986), 103–131.

Bonanno, Alessandro, "The Agro-Food Sector and the Transnational State," *Political Geography*, 12, 4 (July 1993), 341–360.

Bryant, Raymond, "Political Ecology: An Emerging Research Agenda in Third-World Studies," *Political Geography*, 11, 1 (Jan. 1992), 12–36.

Bunge, William, "Comment: Racial Continents," *Political Geography Quarterly*, 9, 1 (Jan. 1990), 5–8.

**C**

*Canadian Ethnic Studies.* Published thrice a year by Canadian Ethnic Studies Association, Dept. of Sociology, Univ. of Saskatchewan, Saskatoon.

Cluver, August D. de V., "Namibia's New Language Policy," *Africa Insight*, 20, 3 (1990), 161–168.

Coughlin, Robert E., "Formulating and Evaluating Agricultural Zoning Programs," *American Planning Association Journal*, 57 (Spring 1991), 183–192.

**D**

Dalby, Simon, "Ecopolitical Discourse: 'Environmental Security' and Political Geography," *Progress in Human Geography*, 16, 4 (Dec. 1992), 503–522.

De Glopper, Donald R., "Chinese Nationality and 'The Tibetan Question'," *Problems of Communism*, 39 (Nov. 1990), 81–89.

Drjkink, Gertjan and Herman van der Wusten, "Green Politics in Europe: The Issues and the Voters," *Political Geography*, 11, 1 (Jan. 1992), 7–11.

Douglas, Neville, Review Essay: "Historical Myth and Materialist Reality: Nationalist and Structuralist Perspectives on the Conflict in Ireland," *Political Geography Quarterly*, 6, 1 (Jan. 1987), 89–101.

DuMars, Charles T. (ed.), "New Challenges to Western Water Law." Special issue of *Natural Resources Journal*, 29, 2 (Spring 1989), whole volume.

**E**

Emel, Jody and others, "Ideology, Property, and Groundwater Resources: An Exploration of Relations," *Political Geography*, 11, 1 (Jan. 1992), 37–54.

*Energy Policy.* Published 10 times a year by Butterworth-Heinemann, Letchworth, U.K.

*The Environmental Consequences of Nuclear War.* Special issue of *Environment*, 30 (June 1988).

*Environmental Politics.* Published quarterly since 1992 by Frank Cass in London.

*Environmental Policy Review.* Published semiannually by the Faculty of Social Sciences, Hebrew Univ., Jerusalem.

*Ethnic and Racial Studies.* Routledge, published quarterly since 1977.

*Ethnic Conflict and International Security.* Special issue of *Survival*, 35, 1 (Spring 1993), whole volume.

**F**

Feeley, Michael Scott, "Reclaiming the Beautiful Island: Taiwan's Emerging Environmental Regulation." *San Diego Law Review*, 27, 4 (1990), 907–936.

Feiler, Gil, "Migration and Recession; Arab Labor Mobility in the Middle East, 1982–89," *Population and Development Review*, 17 (March 1991), 134–155.

*Food Policy*. London: Butterworths, published quarterly since 1975.

French, Hilary F., "Green Revolutions: Environmental Reconstruction in Eastern Europe and the Soviet Union," *Columbia Journal of World Business*, 26 (Spring 1991), 28–51.

**G**

Garber, Larry and Courtney M. O'Connor, "The 1984 UN Sub-Commission on Prevention of Discrimination and Protection of Minorities," *American Journal of International Law*, 79, 1 (Jan. 1985), 168–180.

Gibb, Richard A., "Imposing Dependence: South Africa's Manipulations of Regional Railways," *Transport Reviews*, 11, 1 (Jan. 1991), 19–39.

*Global Environmental Problems: A Legal Perspective*. Annual Symposium, *Syracuse Journal of International Law and Commerce*.

*The Global Political Economy of Food*. Special issue of *International Organization*, 32, 3 (Summer 1978).

*The Greening of World Politics*. Special issue of *International Journal* (Toronto), 45, 1 (Winter 1989–90).

Griffiths, Ieuan L., "The Tazama Oil Pipeline," *Geography*, 54, 2 (April 1969), 214–217.

Gupta, Dipankar, "The Indispensable Centre: Ethnicity and Politics in the Indian Nation State," *Journal of Contemporary Asia*, 20, 4 (1990), 521–539.

Gurr, Ted Robert, "Peoples Against States: Ethnopolitical Conflict and the Changing World System," *International Studies Quarterly*, 38, 3 (Sept. 1994), 347–377.

Gurtov, Mel, "Open Borders: A Global-Humanist Approach to the Refugee Crisis," *World Development*, 19 (May 1991), 485–496.

**H**

Heisler, Martin O. and Barbara Schmitter Heisler (eds.), *From Foreign Workers to Settlers? Transnational Migration and the Emergence of New Minorities. Annals, Am. Acad. of Polit. and Soc. Sci.*, 485 (May 1986), whole volume.

Hilz, Christoph and John R. Ehrenfeld, "Transboundary Movements of Hazardous Wastes," *International Environmental Affairs*, 3 (Winter 1991), 26–63.

**I**

*Immigrants and Minorities*. Published thrice a year since 1981 by Frank Cass in London.

*International Environmental Affairs*. Published quarterly by Univ. Press of New England, Hanover, NH.

*International Journal of Refugee Law*. Published quarterly since 1988 by Oxford Univ. Press.

*International Migration Review*. Published quarterly by the Center for Migration Studies of New York, Staten Island, NY.

**J**

Jones, P. N. and M. T. Wild, "Western Germany's 'Third Wave' of Migrants: The Arrival of the *Aussiedler.*" *Geoforum*, 23, 1 (Feb. 1992), 1–11.

*Journal of Ethnic Studies*. Published quarterly since 1973 at Western Washington Univ., Bellingham.

*Journal of Refugee Studies*. Published quarterly since 1987 by Oxford Univ. Press.

*Journal of Soviet Nationalities*. Published quarterly at Duke Univ., Durham, NC.

*Journal of Transport Geography*. Published quarterly since 1993 in Oxford by Butterworth-Heinemann.

Juaregui, G., "Nationalism in the Basque Country," *European Journal of Political Research*, 16 (1987), 587–605.

**K**

Kak, Subhash C., "Religion and Politics in East Punjab," *Journal of Social, Political and Economic Studies*, 15 (Winter 1990), 435–456.

Kaplan, David H., "Population and Politics in a Plural Society: The Changing Geography of Canada's Linguistic Groups." *Annals, AAG*, 84, 1 (1994), 46–67.

Katz, Yossi, "Transfer of Population as a Solution to International Disputes: Population Exchanges Between Greece and Turkey as a Model for Plans to Solve the Jewish-Arab Dispute in Palestine During the 1930s," *Political Geography*, 11, 1 (Jan. 1992), 55–72.

Kenzer, Martin S. (ed.), *Global Refugee Issues at the Beginning of the 1990s*. Special issue of The *Canadian Geographer*, 35 (Summer 1991).

Kimerling, Judith, "Disregarding Environmental Law: Petroleum Development in Protected Natural Areas and Indigenous Homelands in the Ecuadorean Amazon," *Hastings International and Comparative Law Review*, 14, 4 (1991).

Kliot, Nurit, "The Collapse of the Lebanese State," *Middle Eastern Studies*, 23, 1 (Jan. 1987), 54–74.

———, "Mediterranean Potential for Ethnic Conflict: Some Generalizations," *Tidjschrift voor Economische en Sociale Geografie*, 80 (1989), 147–163.

Kodras, Janet E., "Shifting Global Strategies of U.S. Foreign Food Aid, 1955–90," *Political Geography*, 12, 3 (May 1993), 232–246.

Kosinski, L. A., "The International Geographical Union and Global Change Programmes," *Canadian Geographer*, 35 (Spring 1991), 90–92.

**L**

Lake, R. W. and R. A. Johns, "Legitimation Conflicts: The Politics of Hazardous Waste Siting Law," *Urban Geography*, 11 (1990), 488–508.

*Land Use Policy.* Published quarterly by Butterworth-Heinemann, Letchworth, U.K.

Laponce, J. A., "The French Language in Canada: Tensions Between Geography and Politics," *Political Geography Quarterly*, 3, 2 (April 1984), 91–104.

———, "More about Languages and Their Territories: A Reply to Pattanayak and Bayer," *Political Geography Quarterly*, 6, 3 (July 1987), 265–267.

Lavrov, S. B., "Geoecology: Theory and Some Practical Issues," *Soviet Geography*, 30 (1989), 670–679.

Lee, Hochul, "Political Economy of Land Reforms in Korea and Bolivia; State and Class in Rural Structure," *Asian Perspective*, 15 (Spring/Summer 1991), 215–229.

Lee, Luke T., "The UN Group of Governmental Experts on International Co-operation to Avert New Flows of Refugees," *American Journal of International Law*, 78, 2 (April 1984), 480–484.

Lee Yong Leng, "Language and National Cohesion in Southeast Asia," *Contemporary Southeast Asia*, 2, 3 (Dec. 1980), 226–240.

Leitner, Helga, "International Migration and the Politics of Admission and Exclusion in Postwar Europe," *Political Geography*, 14, 3 (April 1995), 259–278.

Lipshitz, Gabriel, "Ethnic Differences in Migration Patterns—Disparities Among Arabs and Jews in the Peripheral Regions of Israel," *Professional Geographer*, 43, 4 (Nov. 1991), 445–455.

Logan, Bernard I., "An Assessment of the Environmental and Economic Implications of Toxic-waste Disposal in Sub-Saharan Africa," *Journal of World Trade*, 25 (Feb. 1991), 61–76.

Loya, A., "Radio Propaganda of the United Arab Republic: An Analysis," *Middle East Journal*, (1962), 98–110.

**M**

McCarthy, J. J. and M. Swilling, "South Africa's Emerging Politics of Bus Transportation," *Polit-*

ical Geography Quarterly*, 4, 3 (July 1985), 235–249.

Miller, Jake, C., "The Haitian Refugees and the Closed 'Golden Door,'" *Transafrica Forum*, 2, 3 (Fall 1984), 49–64.

Miller, Mark J. "Strategies for Immigration Control; An International Comparison." *Annals, Am. Acad. of Polit. and Soc. Sci.*, 534 (July 1994), whole volume.

Moline, Molly J., "Debt-for-Nature Exchanges; Attempting to Deal Simultaneously with Two Global Problems," *Law and Policy in International Business*, 22, 1 (1991), 133–158.

Moore, Richard H., "Resistance of Japanese Rice Policy; A Case-Study of the Hachirogota Model Farm Project," *Political Geography*, 12, 3 (May 1993), 278–296.

Murphy, Alexander B., "Territorial Policies in Multiethnic States," *Geographical Review*, 79, 4 (Oct. 1989), 410–421.

Mwase, Ngila, "Reflections on the Proposed Botswana-Namibia Trans-Kalahari Railway," *Eastern Africa Economic Review*, 3, 1 (1987), 65–75.

Myers, Robert J. (ed.), *Religion and the State: The Struggle for Legitimacy and Power. Annals, Am. Acad. of Polit. and Soc. Sci.*, 483 (Jan. 1986), whole volume.

**N**

*Nationalism and Ethnic Politics.* Published quarterly since 1994 in London by Frank Cass.

*Nationalities Papers.* Published semiannually by the Dept. of Sociology, Univ. of Nebraska, Omaha.

*The Nations Within: Ethnic Group Demands in a Changing World.* Symposium, *Cornell International Law Journal*, 25, 3 (1992), 481–623.

Newman, David, "The Development of the Yishuv Kehillati: Political Process and Settlement Form," *Tijdschrift voor Economische en Sociale Geografie*, 75 (1984).

———, "Ideological and Political Influences on Israeli Rurban Colonization: The West Bank and Galilee Mountains," *Canadian Geographer*, 28, 2 (Summer 1984).

**O**

Ofuatey-Kodjoe, W., "The Political Economy of Food Production in Africa," *International Journal of World Peace*, 7 (Dec. 1990), 33–53.

Oommen, T. K., "State and Religion in Multireligious Nation-states: The Case of South Asia," *South Asia Journal*, 4, 1 (July 1990), 17–33.

Osman, Abdillahi Said Osman, "International Cooperation to Avert the Flows of Refugees in Africa," *Nordisk Tidsskrift for International Ret,* 51, 3–4 (1982), 179–188.

Overton, J. and R. G. Ward, "The Coups in Retrospect: The New Political Geography of Fiji," *Pacific Viewpoint,* 30 (1989), 207–230.

**P**

Park, Chris, "Transfrontier Air Pollution; Some Geographical Issues," *Geography,* 76 (Jan. 1991), 21–35.

Paterson, Christopher, "Television News from the Frontline States [Southern Africa]," *Transafrica Forum,* 8 (Spring 1991), 59–75.

Pattanayak, D. P. and J. M. Bayer, "Laponce's 'The French Language in Canada: Tensions Between Geography and Politics'—A Rejoinder." *Political Geography Quarterly,* 6, 3 (July 1987), 261–263.

Patterson, John G. and Nanda R. Shrestha, "Population Growth and Development in the 'Third World': The Neocolonial Context," *Studies in Comparative International Development,* 23, 3 (Fall 1988).

Perumal, C. A. and R. Thandavan, "Ethnic Violence in Sri Lanka: Cause and Consequences," *Indian Journal of Political Science,* 50, 1 (Jan. 1989), 1–17.

Pirie, Gordon H., "The Decivilizing Rails: Railways and Underdevelopment in Southern Africa," *Tijdschrift voor Economische en Sociale Geografie,* 73, 4 (1982), 221–228.

———, "Race Zoning in South Africa: Board, Court, Parliament, Public," *Political Geography Quarterly,* 3, 3 (July 1984), 207–221.

———, "Transport, Food Insecurity and Food Aid in Sub-Saharan Africa," *Journal of Transport Geography,* 1, 1 (March 1993), 12–19.

*The Politics of Conservation in Southern Africa.* Special issue of *Journal of Southern African Studies,* 15, 2 (Jan. 1989), whole volume.

*The Politics of International Telecommunications.* Special issue of *International Journal* (Toronto), 42, 2 (Spring 1987), whole number.

*Population and Development Review.* Published quarterly by the Population Council, New York.

*Population and Environment.* Published quarterly by Human Sciences Press, New York.

Premdas, Ralph R., "Fiji under a New Political Order: Ethnicity and Indigenous Rights," *Asian Survey,* 31, 6 (June 1991), 540–558.

**R**

Rabe, Barry G., "Exporting Hazardous Waste in North America," *International Environmental Affairs,* 3 (Spring 1991), 108–123.

Raffestin, Claude and Roderick Lawrence, "Comment: Human Ecology and Environmental Policies: Prospects for Politics and Planning," *Political Geography Quarterly,* 9, 2 (April 1990), 103–107.

*Refuge; Canada's Periodical on Refugees.* Published quarterly by the Centre for Refugee Studies, York Univ., North York, Ontario.

Reichardt, Markus and David Duncan, "Rail Transport and the Political Economy of Southern Africa, 1965–1980," *Africa Insight,* 20, 2 (1990), 100–110.

Romsa, G. and others, "From the Economic to the Political: Regional Planning in West Germany," *Canadian Geographer,* 33 (1989), 47–52.

Rosebrock, Jens and Harald Sondhof, "Debt-for-nature Swaps; A Review of the First Experiences," *Intereconomics,* 26 (March 1991), 82–87.

**S**

Savage, Christopher, "Middle East Water," *Asian Affairs* (London), 22 (Feb. 1991), 3–10.

Schoenbaum, Thomas J., "Agora: Trade and Environment—Free International Trade and Protection of the Environment: Irreconcilable Conflict?," *American Journal of International Law,* 86, 4 (Oct. 1992), 700–727.

Schoolmaster, F. Andrew, "Water Marketing and Water Rights Transfers in the Lower Rio Grande Valley, Texas," *Professional Geographer,* 43, 3 (August 1991), 292–303.

Schöpflin, George, "Nationalism and National Minorities in East and Central Europe," *Journal of International Affairs,* 45 (Summer 1991), 51–65.

Schultheis, Michael J., "The Geopolitics of Forced Displacement," *African Studies Review,* 32, 1 (April 1989), 3–29.

Schultz, Cynthia B. and Tamara Raye Crockett, "Economic Development, Democratization and Environmental Protection in Eastern Europe," *Boston College Environmental Affairs Law Review,* 18 (Fall 1990), 53–84.

Segal, Brian, "Geopolitics of Broadcasting," *Washington Quarterly,* 6, 2, 140–148.

Shrestha, Nanda R., "The Political Economy of Underdevelopment and External Migration in Nepal," *Political Geography Quarterly,* 4, 4 (Oct. 1985), 149–161.

Simon, Rita J. (ed.), *Immigration and American Public Policy. Annals, Am. Acad. of Polit. and Soc. Sci.,* 487 (Sept. 1986), whole volume.

Singh, Chanra Pal, "NAGI Commission on Politics and the Environment," *Political Geography,* 11, 4 (July 1992), 413–414.

Skeldon, Ronald, "Emigration and the Future of Hong Kong," *Pacific Affairs*, 63 (Winter 1990-91), 500–523.

Smith, Graham, "Gorbachev's Greatest Challenge: Perestroika and the National Question," *Political Geography Quarterly*, 8, 1 (Jan. 1989), 7–20.

Stevis, Dimitris and Stephen P. Mumme, "Nuclear Power, Technical Autonomy, and the State in Mexico," *Latin American Research Review*, 26, 3 (1991), 55–82.

Strong, Maurice F., "ECO '92: Critical Challenges and Global Solutions," *Journal of International Affairs*, 44 (Winter 1991), 287–300.

Stuyt, Jan, "The Comprehensive Plan of Action for Indochinese Refugees—A NGO Approach," *Chinese Yearbook of International Law and Affairs*, 11 (1991–92), 34–49.

Sutherland, William "Nationalism, Racism and the State; Class Rule and the Paradox of Race Relations in Fiji," *Pacific Viewpoint*, 31 (Oct. 1990), 60–72.

**T**

Teclaff, Ludwik A. and Albert E. Utton (eds.), *The International Law of the Hydrologic Cycle*. Special issue of *Natural Resources Journal*, 31 (Winter 1991), whole volume.

*Telecommunications Policy*. Published nine times a year since 1976 by Butterworth-Heinemann.

Tickner, Vincent, "Southern Africa: The Politics of Railways," *New African Yearbook* (London), (1980), 41–45.

**U**

Ufkes, Frances M., "The Globalization of Agriculture," *Political Geography*, 12, 3 (May 1993), 194–197.

———, "Trade Liberalization, Agrofood Politics and the Globalization of Agriculture," *Political Geography*, 12, 3 (May 1993), 215–231.

**V**

von Moltke, Konrad, "Debt-for-Nature: The Second Generation," *Hastings International and Comparative Law Review*, 14, 4 (1991).

**W**

Warf, Barney, "Telecommunications and the Globalization of Financial Services," *Professional Geographer*, 41, 3 (August 1989), 257–271.

Watts, Michael J., "Politics, the State and Agrarian Development: A Comparative Study of Nigeria and the Ivory Coast," *Political Geography Quarterly*, 5, 2 (April 1986), 103–125.

Weiss, Edith Brown, "Agora: Trade and Environment—Environment and Trade as Partners in Sustainable Development: A Commentary," *American Journal of International Law*, 86, 4 (Oct. 1992), 728–735.

Widgren, Jonas, "International Migration and Regional Stability," *International Affairs*, 66 (Oct. 1990), 749–766.

Williams, Colin H., "Ideology and the Interpretation of Minority Cultures," *Political Geography Quarterly*, 3, 2 (April 1984), 105–125.

Wood, William B., "The Political Geography of Asylum: Two Models and a Case Study," *Political Geography Quarterly*, 8, 2 (April 1989), 181–196.

———, "Long Time Coming: The Repatriation of Afghan Refugees," *Annals, AAG*, 79, 3 (Sept. 1989), 345–369.

*World Refugee Survey*. Washington, DC: U.S. Committee for Refugees, Annual.

Wright, Theodore P., Jr., "Center-periphery Relations and Ethnic Conflict in Pakistan: Sindhis, Muhajirs and Punjabis," *Comparative Politics*, 23 (April 1991), 299–312.

# PART NINE

# *Looking Ahead*

Managua after the earthquake—trigger of the Sandinista rebellion.

# 39

# *Political Geography for the future*

Throughout this book, as we have explored most of the highways and some of the byways of political geography, we have had glimpses of topics new and undeveloped, of ideas that need investigation, of problems that political geographers have little studied but that could be moved toward resolution by their contributions. We have called attention to some of these opportunities as we have gone along. Now it seems appropriate to peer into the future and suggest some other subjects that are virtually virgin territory for political geographers. Some have already been mentioned but require additional discussion; others we have saved until now. All are offered here to suggest how much there is left to do and how political geographers can help to do it. These suggestions are by no means exhaustive.

## *Integration and Disintegration*

In a number of places in this book, we have referred to centripetal and centrifugal forces operating at all levels of political organization from tribal to global. These forces lead respectively to integration and disintegration of political units of the same or different levels. Today, as perhaps never before, we see both processes operating concurrently around the globe. How can we explain the breakup of the USSR, Yugoslavia, Czechoslovakia, and Ethiopia at the same time as the European Union, the North American Free Trade Association, and the Economic

Cooperation Organization were being created or expanded? Is it a contrast between economic unity and political disunity? Or ethnic fragmentation and class coalescence? Or something else? Is there any pattern at all or simply a concurrence of random events? Can we predict what will happen next and where?

## *Very Small Places*

We have devoted two chapters to discussions of minor civil divisions, municipalities, and special-purpose districts. Elsewhere throughout the book we have referred to the territories of street gangs, exclaves of States, neutral zones, international straits, capital cities, and other very small places, most of which have considerable politicogeographical importance. But other very small places in the world deserve analysis by political geographers. We offer a few suggestions here.

In our discussion of decolonization we raised the question of independence for very small fragments of colonial territory, and in our discussion of the size of States we pointed out that there is no real correlation between size and success. Small can be not only beautiful but practical as well. But is there no lower limit to the size of States? Must every remaining scrap of colonial empire become a separate State? And what about tiny States that are already independent? If they cannot bear the costs of independence, must they forever be maintained

**A foreign economic enclave in a developing country.** Chuquicamata, in the Atacama Desert in northern Chile, is the site of the world 's largest open-pit copper mine. Until it was nationalized in the late 1960s, it was owned and operated by the Chile Exploration Company (Chilex), a subsidiary of Anaconda, a major transnational corporation based in the United States. Only people working in the mine or smelters (one of which appears in the middle of the picture) were permitted to live in the company-owned townsite. One section of the town housed the *obreros*, or wage-earning workers (mostly Chilean) in barrack-like buildings, while the *empleados*, or salaried white-collar workers (mostly foreigners) lived in the American-style suburban community shown here. Given the limitations of a remote desert environment, life here closely resembled that in the United States rather than in Chile. (Martin Glassner)

as international charity cases? Is there some realistic and acceptable alternative to statehood for the people who live in very small places?

A widespread and growing practice in the world is the establishment of economic enclaves within countries. These include foreign-owned mines and plantations and their "company towns" as well as free ports and free trade zones—all of which have been mentioned before in other contexts. Free zones have become especially important. At the turn of the century there were only 11 around the world; in 1985 there were more than 400 to be found on every continent except Antarctica. Now there are probably a thousand or more. Of particular importance in political geography are export-processing zones and transit zones for land-locked countries. In terms of numbers, distribution, and impact on the host countries, they may well be extremely important.

**United States foreign trade zones.** These are called free trade zones or free ports in other countries. They are secured areas legally outside the customs territory of the country. Their purpose is to attract and promote international trade, and they are normally located in or near customs ports of entry. In the United States they are operated as public utilities by states, political subdivisions, or qualified corporations under Customs supervision. Foreign and domestic merchandise may be moved into the zones for storage, exhibition, assembly, disassembly, packing, repacking, sorting, grading, and cleaning, and may be manufactured or otherwise processed without payment of customs duties unless and until the goods enter the customs territory for domestic consumption. Such zones are very popular around the world and are still proliferating; they are making a great contribution to the increase and free flow of trade.

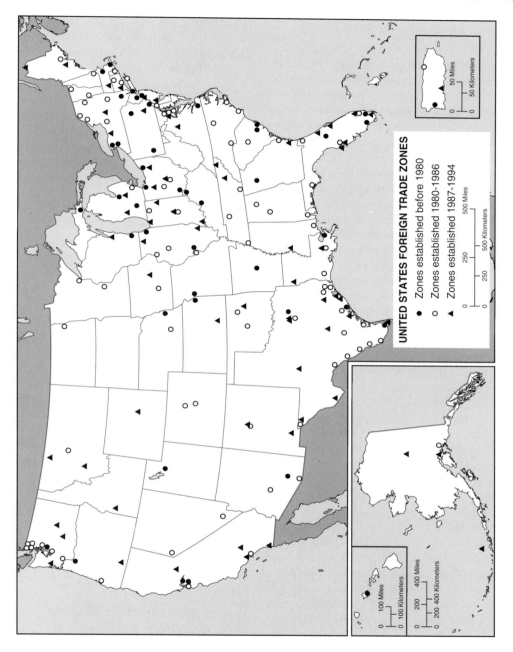

UNITED STATES FOREIGN TRADE ZONES

● Zones established before 1980
○ Zones established 1980-1986
▲ Zones established 1987-1994

## Nongovernmental Organizations

In several chapters we have called attention to the role played by international non-governmental organizations (INGOs) in many matters of concern to geographers, especially political geographers. In individual countries, particularly in North America, Western Europe, Israel, Australia, and New Zealand, nongovernmental organizations (NGOs) are enormously important in coordinating activities; transmitting information to their members and the general public; setting professional, technical, and ethical standards; providing development aid to poor countries and to poor sectors of rich countries; influencing policies of governments; and forecasting economic, political, social, demographic, and ecological trends.

Here is an important subject only skimmed by political scientists (principally in studies of "pressure groups") and virtually ignored by geographers, even political geographers interested in decision-making and politicogeographical behavior. Much can be revealed about the importance of NGOs and INGOs in matters of concern to political geographers by studies of some of the several thousand such groups in dozens of countries and their roles in particular events and processes. Cooperation of political geographers with political scientists, sociologists, and psychologists could be most rewarding in understanding the interactions of these groups and the attitudes and behavior of their members and nonmembers. Individuals, acting through such groups, may be more influential than we realize. One place to start is with the Oslo Declaration and Plan of Action, adopted by the global NGO and UNHCR Conference in June 1994 on cooperation among official anad voluntary organizations engaged in helping refugees.

## The AIDS Pandemic

Apparently, AIDS originated in Africa and has spread rapidly around the world in a surprisingly short time, despite strenuous efforts in many countries to contain it. The World Health Organization estimated in July 1990 that there were approximately 8 million HIV-infected adults worldwide; only a year and a half later the estimate was 12 million, of whom 1 million were children not yet sexually active. WHO estimates that by the end of the century the figure will reach 30 million; other experts expect closer to 40 million. If sheer numbers or the characteristics of the infection itself were the only factor, it would not be of interest to *political* geographers. But it is of interest, for two reasons. First, even now, AIDS is being treated as a political problem, with various groups squabbling over causes, competing for public funds to prevent or treat it, and supporting or opposing candidates on the basis of their perceived positions on AIDS. Second, AIDS is decimating the ranks of people who are politically, economically, or culturally active in many places, especially in Africa, with unforeseeable consequences. We are losing a lot of talent in places where there is little to spare.

## Gender Issues

As mentioned in Chapter 1, political geographers have jumped on the bandwagon of "gender issues," "feminist perspectives," and the like. A survey of the literature, however, reveals little that is both geographic *and* political. Most should probably be classified as sociology, social theory, political theory, or political science; some may be social or cultural geography. If there really is a legitimate "feminist" or "gender-based" political geography, we should be the beneficiaries of serious, scholarly studies in the field.

## States with Centrally Planned and Transition Economies

Until recently there were 16 countries in Europe, Asia, and Latin America frequently referred to as "communist" but more accurately described as having centrally planned economies. Perhaps another score, mostly in Africa, have practiced some variety of socialism. They have scarcely been considered by political geographers except in terms of boundaries, ethnic minorities, military alliances, and other traditional studies. Perhaps if good studies had been made of such

**Life in countries with centrally planned economies.** Marxist countries constantly use various devices to urge their people to produce more. Two of them are illustrated by these photographs. Top: An oxygen plant near the airport of Managua, Nicaragua, is emblazoned with this inscription in lights: "We hail the 25th Anniversary of the FSLN [ruling Marxist party]. Producing oxygen for the revolution." (Martin Glassner) Bottom: "Glory to Work" reads this imposing metal sign near the infamous ravine of Babi Yar in Kiev, Ukraine. Taken in July 1993, this photograph symbolizes the difficulty of purging former Communist societies of their less admirable features and of successfully assimilating the better features of capitalism. (Martin Glassner) Since these pictures were taken, both countries (and many others) have abandoned central planning and adopted more conventional monetary incentives to increase production.

matters as their political economy, foreign relations, ecological problems, trade groupings, populations, and centrifugal and centripetal forces, we would have been better prepared for the collapse of their economies. Even now, though, there are a few of these communist countries remaining, as well as a number of others, chiefly in Africa and around the Persian Gulf, that also practice central planning to a large extent. It might be instructive to analyze why central planning has failed in most places where it has been tried, why it persists after being abandoned where it originated, and whether it has any value at all. And why have a number of former communist countries returned former communists to high office in democratic elections? Is there a geographic factor operating here?

## Individual Countries and Regions

A staple of political geography during the first half of this century was the study of the political geography of an individual country or of a region or subregion. Indeed, most political geography textbooks and anthologies were organized regionally rather than systematically, as this one is. While the world is changing too rapidly to make regional political geography practical for texts now, there is a real need to return to the production of articles and monographs that will help us understand contemporary events. Would not a serious political geography of China be useful? Or of Nicaragua or the Arctic or the Middle East or Africa? In their quest for ever-new behavioral, statistical, or theoretical subjects, many contemporary political geographers seem to be forgetting their roots. If they are, indeed, geographers—not social historians, political scientists, economists, or sociologists—then they must be concerned with *places*, where things are happening. Those concerned with political units smaller than a State would perform a useful service if they would produce political geographies of Punjab, of the Midlands, of the Swiss cantons, of New York City, and of some of the component units of the United Arab Emirates. Naturally, to be scientific and practical, they would have to go beyond description and use the best tools of modern political geography for analysis. This would help us all to understand better the stories behind the headlines.

## International Financial Flows

The world is now bound together by a tight mesh of interconnections of all kinds. We have already discussed a number of them:

migration and resultant interpersonal relationships, circulation systems, international trade, international organizations and law, cultural ties, and so on. Now there is a need for studies of the flows of money (or its symbolic representations) across national boundaries. Normal commercial transactions are studied by some economic geographers, but we need good analyses of those that generally have a political component, such as the flows of public and private military and economic development aid to developing countries; the remittances sent home by migrants, legal and illegal, temporary and permanent; and the financial flows to, from, and among the outlaws and merchants of death that we discussed in Chapter 30. The financial operations of the BCCI (Bank of Credit and Commerce International) alone provide fertile fields for political geographers to plow because of the bank's alleged links with many governments and its apparent influence on both domestic and foreign policies.

## U.S.–Canada, U.S.–Mexico, and Canada–Mexico Relations

In Chapter 26 we presented some basic information about the North American Free Trade Area, which includes Canada, the United States, and Mexico. In order to study the pros and cons of this issue and its possible ramifications, however, we should know something of how each pair of countries relate to one another. One pair—Canada and Mexico—have interacted relatively little, and the other two linkages are more important at present, but this is likely to change. A few examples of bilateral issues will illustrate the point.

**U.S.–Canada.** Chiefly environmental problems, such as pollution of boundary waters, dangers of oil spills along the coasts, flooding of transborder lakes and streams, protection of wilderness areas and migratory birds, and "acid rain," which was seriously addressed by the Canada–United States Air Quality Accord of March 1991. Others include fluctuating levels of the Great Lakes; maritime boundaries in the Beaufort Sea, Dixon Entrance, and Strait of Juan de Fuca; defense of North

**Surgical mask sewing operation in Ciudad Juáfez, Mexico.** Of the hundreds of free zones around the world, the zone along the ·U.S. border in Mexico is among the most unusual. In 1965 the Border Industrial Program was inaugurated to promote border industrialization, stimulate Mexican supporting industries, and reduce unemployment along the border. In 1986, there were 806 textile and assembly plants called *maquiladoras* within the 20 km (12.2 mi) deep zone. They are mostly owned by U.S. and Japanese firms and employ more than 236,000 people, mostly women. There were also 59 *maquiladoras* employing nearly 14,000 people located in other free zones in the interior of Mexico. (Photo courtesy of El Paso Industrial Development Corportion)

America; U.S. "cultural imperialism" in Canada; and illegal immigration of third-country nationals through Canada into the United States.

**U.S.–Mexico.** Illegal movements of drugs and migrants from Mexico into the United States, poor wages and working conditions in northern Mexico, insufficient and polluted irrigation water permitted to flow from United States into Mexico, a disputed maritime boundary segment in the Gulf of Mexico, diseased cattle and beef moving from Mexico to United States, inadequate environmental protection legislation and enforcement in Mexico.

## The Political Effects of Natural Events

The reaction of governments and political leaders to the disastrous effects of natural events on their constituents has aided or destroyed many political careers at all levels from the municipal to the national. Everyone can recall mayors or governors who have been defeated for reelection because they failed to act sufficiently to mitigate the

effects of snowstorms or hurricanes. But many such events have also led to changes in national government and even to the creation of new countries. Some examples are the earthquake that destroyed Managua, Nicaragua and led to the overthrow of the Somoza dictatorship; the cyclones that devastated East Bengal and led to the creation of Bangladesh; the volcanic eruption that rendered U.S. bases in the Philippines inoperable and ended the political debate over their future; and the droughts that devastated Ethiopia and Somalia and contrubuted to the overthrow of their governments and the independence of Eritrea.

## The Role of Political Geographers in the Future

In this book we have suggested many topics that need to be studied by political geographers. The clear implication has been that if political geographers investigate these and similar topics, analyze them using all available tools and methodologies and devise new ones where necessary, and make their conclusions known through theses, dissertations, lectures, and publications, the studies would be not only interesting, but useful to those exposed to them.

Political geographers should and will continue in their primary roles as objective observers, students, and critics, remaining apart from events and contributing their insights to the world at large. They have proved their value in such roles in the past and will continue to do so. We suggest, however, that political geographers can also make valuable contributions to society by entering the fields of political combat, testing their theories in the crucible of hard experience, offering practical solutions to real and immediate problems.

This suggestion for the practical application of the concepts of political geography, even ignoring geopolitics, is neither new nor unique. Several distinguished political geographers have served as The Geographer in the U.S. Department of State, including S. Whittemore Boggs, G. Etzel Pearcy, Robert Hodgson, Lewis Alexander, George Demko, and the incumbent, William Wood. All have been advisors to the government as well as collectors, organizers, and analysts of politicogeographical information. Other political geographers have served the government in various capacities. Isaiah Bowman, one of our most outstanding political geographers, was an advisor to the American Delegation to Negotiate Peace at the Paris Peace Conference in 1919 and later became an advisor and confidant of President Franklin D. Roosevelt. More recently, one political geographer was deeply involved in the dispute between Maine and New Hampshire over their maritime boundary; another was very active in redrawing congressional and other political district boundaries in the State of Washington; another was advisor to Nepal in its negotiations with India for a new transit treaty.

Political geographers can serve as advisors, technicians, and expert witnesses for governments, international agencies, transnational corporations, courts, even political parties. They can advise on drawing boundaries of electoral districts and planning regions, on locations of seats of government, on mapping of political phenomena, on regulations for international rivers, and myriad other matters within their purview.

In applying their approaches, knowledge, techniques, and insights, however, political geographers, like other scholars, must avoid being too closely identified with particular policies, thus losing their status as experts and becoming simply partisans. They must also recognize that their contributions are only being added to the contributions of other experts in various fields, including the art of politics, and are not likely to be decisive. Finally, they must not get so involved in applications that they neglect study and research; that is, they can be scholar-practitioners but not solely practitioners. Within these limits, political geographers can make real contributions to the formulation and execution of policy at every level—local, national, and international. It is an opportunity and a challenge they should not ignore.

## *Books and Monographs*

**A**

Adam, Jan (ed.), *Economic Reforms and Welfare Systems in The USSR, Poland and Hungary.* New York: St. Martin's, 1991.

**B**

Barnes, James F., *Gabon.* Westview, 1992.

Barnett, Tony and Piers Blaikie, *AIDS in Africa; Its Present and Future Impact.* New York: Guilford, 1992.

Barry, Tom and Beth Sims, *The Challenge of Cross-Border Environmentalism: The U.S.–Mexico Case.* Albuquerque, N. Mex.: Resource Center, 1994.

Barry, Tom and others, *Crossing the Line; Immigrants, Economic Integration, and Drug Enforcement on the U.S.–Mexico Border.* Albuquerque, N. Mex.: Resource Center, 1994.

Bebbington, Anthony and Graham Thiele (eds.), *Non-Governmental Organisations and the State in Latin America.* New York: Routledge, 1994.

Berberoglu, Berch, *The Internationalization of Capital.* New York: Praeger, 1987.

Bilateral Commission on the Future of United States-Mexican Relations. *The Challenge of Interdependence.* Lanham, MD: Univ. Press of America, 1989.

Bloomfield, Lincoln M. and Gerald F. Fitzgerald, *Boundary Waters Problems of Canada and the United States.* Toronto: Carswell, 1958.

Boateng, E.A., *A Political Geography of Africa.* Cambridge Univ. Press, 1978.

Bradshaw, Michael J. (ed.), *The Soviet Union; A New Regional Geography?.* London: Belhaven, 1993.

Bryson, Phillip J. and Manfred Melzer, *The End of the East German Economy.* New York: St. Martin's, 1991.

Bustamante, George A. and others (eds.), *U.S.-Mexico Relations—Labor Market Interdependence.* Stanford Univ. Press, 1992.

**C**

Carroll, Thomas F., *Intermediary NGOs; The Supporting Link in Grassroots Development.* West Hartford, CT: Kumarian Press, 1992.

Cavanagh, John and others (eds.), *Beyond Bretton Woods; Alternatives to The Global Economic Order.* London: Pluto Press, 1994.

Caviedes, Cáesar, *The Politics of Chile: A Socio-Geographical Assessment.* Boulder, CO: Westview, 1979.

Clark, John, *Democratizing Development; The Role of Voluntary Organizations.* West Hartford, CT: Kumarian Press, 1991.

Comisso, Ellen and Laura Tyson (eds.), *Power, Purpose, and Collective Choice: Economic Strategy in Socialist States.* Ithaca, NY: Cornell Univ. Press, 1986.

**D**

Dale, Richard, *The Regulation of International Banking.* Cambridge: Woodhead-Faulkner, 1986.

Deener, David R. (ed.), *Canada-United States Treaty Relations.* Durham, NC: Duke Univ. Press, 1963.

Diamond, Walter H. and Dorothy B. Diamond, *Tax-Free Trade Zones of the World.* New York: Matthew Bender. Loose-leaf; updates issued frequently.

Drysdale, Alasdair and Gerald H. Blake, *The Middle East and North Africa.* Oxford Univ. Press, 1985.

**E**

Eaton, David J. and John M. Andersen, *The State of the Rio Grande/Río Bravo; A Study of Water Resource Issues Along the Texas/Mexico Border.* Tucson: Univ. of Arizona Press, 1987.

Edwards, Michael and David Hulme (eds.), *Making a Difference: NGOs and Development in a Changing World.* London: Earthscan, 1992.

Efrat, Elisha, *Geography and Politics in Israel Since 1967.* London: Frank Cass, 1988.

**F**

Farrington, John and David J. Lewis (eds.), *Non-Governmental Organisations and the State in Asia.* New York: Routledge, 1994.

Farrington, John and others, *Reluctant Partners? Non-Governmental Organizations, the State and Sustainable Agricultural Development.* New York: Routledge, 1993.

Fields, Rona M., *Northern Ireland.* New Brunswick, NJ: Transaction, 1981.

Forbes, Dean, *The Geography of Underdevelopment.* London: Croom Helm, 1984.

Fowler, Alan and others, *Institutional Development and NGOs in Africa: Policy Perspectives for European Development Agencies.* Oxford: INTRAC, 1992.

**G**

George, Susan and Fabrizio Sabelli, *Faith and Credit; The World Bank's Secular Empire.* Boulder, CO: Westview, 1994.

Gilmore, W.C. (ed.), *International Efforts to Combat Money Laundering.* Cambridge: Grotius, 1992.

*Global HIV/AIDS: A Strategy for U.S. Leadership.* Washington, DC: CSIS, 1994.

Gould, Peter, *The Slow Plague; A Geography of the AIDS Pandemic,* Oxford: Blackwell, 1993.

**H**

Haglund, David G. and Joel J. Sokolsky (eds.), *The U.S.-Canada Security Relationship: The Politics, Strategy, and Technology of Defense.* Boulder, CO: Westview, 1989.

Harkavy, R.E., *Bases Abroad: The Global Foreign Military Presence.* Oxford Univ. Press, 1989.

Herzog, Lawrence A. (ed.), *Planning the International Border Metropolis: Trans-boundary Policy Options in the San Diego/Tijuana Region.* Center for U.S.-Mexican Studies, 1986.

———, *Where North Meets South; Cities, Space, and Politics on the United States-Mexico Border.* Austin: Univ. of Texas Press, 1990.

Howlett, Michael and M. Ramesh, *Political Economy of Canada.* London: UCL Press, 1992.

**J**

Jamail, Milton H. and Margo Gutiáerrez, *The Border Guide; Institutions and Organizations of the United States-Mexico Borderlands,* 2nd ed. Austin: Univ. of Texas Press, 1991.

Jones, Paul, *The Structure and Reform of Centrally Planned Economic Systems.* New York: Columbia Univ. Press, 1990.

**K**

Karan, Pradumna P. and Hiroshi Ishii, *Nepal: Development and Change in a Landlocked Himalayan Kingdom.* Tokyo: Tokyo Univ. of Foreign Studies, 1994.

Kim, Byoung-Lo P., *Two Koreas in Development; A Comparative Study of Principles and Strategies of Capitalist and Communist Third World Development.* New Brunswick, NJ: Transaction, 1991.

Kohler, Miles (ed.), *The Politics of International Debt.* Ithaca, NY: Cornell Univ. Press, 1986.

Köves, Andráas and Paul Marer (eds.), *Central and East European Economies in Transition.* Boulder, CO: Westview, 1991.

**L**

Lee Yong Leng, *Southeast Asia: Essays in Political Geography.* Singapore: Singapore Univ. Press, 1982.

Lefever, Ernest W., *Amsterdam to Nairobi; The World Council of Churches and the Third World.* Lanham, MD: Univ. Press of America, 1979.

———, *Nairobi to Vancouver; The World Council of Churches and the World, 1975-87.* Univ. Press of America, 1987.

Lemon, A. (ed.), *Geography of Change in South Africa.* Wiley, 1995.

Lessard, Donald and John Williamson (eds.), *Capital Flight and Third World Debt.* Washington: Institute for International Economics, 1987.

Lipset, Seymour Martin, *Continental Divide; The Values and Institutions of the United States and Canada.* Routledge, 1991.

**M**

Mahant, Edelgard E., *American-Canadian Relations.* Melbourne, FL: Krieger, 1993.

Martellaro, Joseph A., *Economic Reform in China, Hungary, and the USSR.* Hong Kong: Asian Research Service, 1989.

Martínez, Legorreta, Omar (ed.), *Relations Between Mexico and Canada.* Mexico City: Centro de Estudios Internacionales, El Colegio de México, 1990.

Martinez, Oscar J., *Troublesome Border.* Tucson: Univ. of Arizona, 1988.

McDonald, John W. Jr. and Diane B. Bendahmane (eds.), *U.S. Bases Overseas: Negotiations with Spain, Greece, and the Philippines.* Boulder, CO: Westview, 1990.

Mendez, Ruben P., *International Public Finance; A New Perspective on Global Relations.* Oxford Univ. Press, 1992.

Mohan, J. (ed.), *The Political Geography of Contemporary Britain.* London: Macmillan, 1989.

Murphy, Alexander B., *Western Investment in East-Central Europe: Emerging Patterns and Implications for State Stability.* Oxford: Blackwell, 1992.

**N**

Nissman, David B., *The Soviet Union and Iranian Azerbaijan.* Boulder, CO: Westview, 1987.

**O**

O'Loughlin, John and Herman van der Wusten (eds.), *The New Political Geography of Eastern Europe.* London: Belhaven, 1993.

Onwuka, Ralph I. and Amadu Sesay, *The Future*

*of Regionalism in Africa*. New York: St. Martin's, 1985.

**P**

Peet, Richard (ed.), *International Capitalism and Industrial Restructuring*. Winchester, MA: Allen & Unwin, 1987.

Picardi, Elizabeth, *Lebanon: A Divided Land*. New York: Holmes and Meier, 1991.

Piper, Don Courtney, *The International Law of the Great Lakes; A Study of Canadian-United States Co-operation*. Durham, NC: Duke Univ. Press, 1967.

**R**

Reich, Bernard and Gershan R. Kieval, *Israel*. Boulder, CO: Westview, 1993.

Robins-Mowry, Dorothy (ed.), *Canada-U.S. Relations: Perceptions and Misperceptions*. Lanham, MD: Univ. Press of America, 1988.

Rolef, S.H., *The Political Geography of Palestine*. New York: Middle East Review Special Studies No. 3 (1983).

**S**

Schmandt, Jurgen and Hillard Roderick (eds.), *Acid Rain and Friendly Neighbours: the Policy Dispute Between Canada and the U.S.* Durham, NC: Duke Univ. Press, 1985.

Schmitz, Gerald, *Chiapas and After: The Mexican Crisis and Implications for Canada*. Background Paper BP-384E. Ottawa: Research Branch, Library of Parliament, Feb. 1994.

Shannon, Gary W. and others, *The Geography of AIDS*. New York: Guilford, 1991.

Shaw, Denis J.B. (ed.), *The Post-Soviet Republics: A Systematic Geography*. New York: Wiley, 1995.

Simon, Reeva S., *Iraq Between the Two World Wars*. New York: Columbia Univ. Press, 1986.

Sivetz, Laurie and others, *Doing Good; The Australian NGO Community*. London: UCL Press, 1992.

Sjoberg, Orjan and others (eds.), *Economic Crisis and Reform in the Balkans*. New York: St. Martin's, 1991.

Sklair, Leslie, *Assembling for Development: The Maquila Industry in Mexico and the United States*. Boston: Unwin Hyman, 1989.

Smith, Graham, *Planned Development in the Socialist World*. Cambridge Univ. Press, 1989.

Staar, Richard F. (ed.), *East-Central Europe and the USSR*. New York: St. Martin's, 1991.

Stairs, Dennis and Gilbert R. Winham, *Canada and the International Political/Economic Environment*. Univ. of Toronto Press, 1987.

————, *Politics of Canada's Economic Relationship With the U.S.* Toronto: Univ. of Toronto Press, 1987.

Stewart, Gordon T., *The American Response to Canada Since 1776*. Michigan State Univ. Press, 1992.

Stoianovich, Traian, *Balkan Worlds; The First and Last Europe*. Armonk, NY: Sharpe, 1994.

Story, Jonathan (ed.), *The New Europe*. Oxford: Blackwell, 1992.

Sylvester, Christine, *Zimbabwe*. Boulder, CO: Westview, 1991.

Szporluk, Roman (ed.), *The International Politics of Eurasia*, Vol. 2: The Influence of National Identity. Armonk, NY: Sharpe, 1994.

**T**

Taagepera, Rein, *Estonia*. Boulder, CO: Westview, 1993.

Theunis, Sjef, *Non-Governmental Development Organizations of Developing Countries: And the South Smiles*. Norwell, MA: Kluwer, 1992.

Thoman, Richard S., *Free Ports and Foreign Trade Zones*. Cambridge, MD: Cornell Maritime Press, 1956.

Trifunovska, Snezana (ed.), *Yugoslavia Through Documents; From Its Creation to Its Dissolution*. Norwell, MA: Kluwer, 1994.

Truell, Peter and Larry Gurwin, *False Profits; The Inside Story of BCCI, The World's Most Corrupt Financial Empire*. New York: Houghton Mifflin, 1992.

Tsoukalis, Loukas, *The Political Economy of International Money*. Newbury Park, CA: Sage, 1985.

**V**

Vizulis, I. Joseph, *Nations Under Duress: The Baltic States*. Port Washington, NY: Associated Faculty Press, 1985.

**W**

Waggener, Thomas R., *Forests, Timber and Trade; Emerging Canadian and U.S. Relations Under the Free Trade Agreement*. Canadian-American Public Policy Paper No. 4. Orono: Univ. of Maine Press, December 1990.

Wallace, Cynthia Day and John M. Kline, *EC 92 and Changing Global Investment Patterns*. Boulder, CO: Westview, 1992.

Weisman, Alan, *La Frontera; The United States Border With Mexico*. Tucson: Univ. of Arizona Press, 1991.

Wellard, Kate and James G. Copestake (eds.), *Non-Governmental Organisations and the State in Africa*. New York: Routledge, 1994.

Whalley, John, *Domestic Policies and the International Economic Environment*. Toronto: Univ. of Toronto Press, 1987.

Wilson, A. Jeyaratnam and Dennis Dalton (eds.), *The States of South Asia*. Honolulu: Univ. of Hawaii Press, 1983.

Wilson, Patricia A., *Exports and Local Development: Mexico's New Maquiladoras*, Austin: Univ. of Texas Press, 1992.

Wolch, Jennifer R., *The Shadow State; Govern-ment and the Voluntary Sector in Transition*. New York: Foundation Center, 1990.

**Z**

Zaprundnik, Ia, *Belarus: At a Crossroads in History*. Boulder, CO: Westview, 1993.

Zimbalist, Andrew and Claes Brandenius, *The Cuban Economy; Measurement and Analysis of Socialist Performance*. Baltimore: Johns Hopkins Univ. Press, 1989.

## *Periodicals*

**A**

Agnew, John A., "Beyond Core and Periphery: The Myth of Regional Political-Economic Restructuring and Sectionalism in Contemporary American Politics," *Political Geography Quarterly*, 7, 2 (April 1988), 127–240.

———, "Better Thieves Than Reds'? The Nationalization Thesis and the Possibility of a Geography of Italian Politics," *Political Geography Quarterly*, 7, 4 (Oct. 1988), 307–322.

**B**

Badcock, Blair, "Removing the Spatial Bias from State Housing Provision in Australian Cities," *Political Geography Quarterly*, 1, 2 (April 1982), 137–157.

Banaszak, Lee Ann and Jan E. Leighley, "How Employment Affects Women's Gender Attitudes: The Workplace as a Locus of Contextual Effects," *Political Geography Quarterly*, 10, 2 (April 1991), 174–185.

Bassett, Keith, "Labour in the Sunbelt: The Politics of Local Economic Development Strategy in an 'M4-Corridor' Town," *Political Geography Quarterly*, 9, 1 (Jan. 1990), 67–84.

Bassett, Keith and Anthony Hoare, "Bristol and the Saga of Royal Portbury: A Case Study in Local Politics and Municipal Enterprise," *Political Geography Quarterly*, 3, 3 (July 1984), 223–250.

Bennett, Sari and Carville Earle, "Socialism in America: A Geographical Interpretation of Its Failure," *Political Geographical Quarterly*, 2, 1 (Jan. 1983), 31–55.

Blackman, Tim, "The Politics of Full Employment," *Political Geography Quarterly*, 6, 4 (Oct. 1987), 313–333.

Blomley, Nicholas K., "Mobility, Empowerment and the Rights Revolution," *Political Geography*, 13, 5 (Sept. 1994), 407–422.

Bunge, William, "The Cave of Coulibistrie,"
*Political Geography Quarterly*, 2, 1 (Jan. 1983), 57–70.

**C**

*Canada and the U.S. in a Changing Global Environment*. Special issue of *International Journal* (Toronto), 46, 1 (Winter 1990–91), whole volume.

*Canada—U.S. Outlook*. Published quarterly by National Planning Association, Washington.

Chouinard, Vera, "State Formation and the Politics of Place: The Case of Community Legal Aid Clinics," *Political Geography Quarterly*, 9, 1 (Jan. 1990), 23–38.

Clapham, C., "The Political Economy of Conflict: The Horn of Africa," *Survival*, 32 (1990), 403–420.

Clark, Gordon L., "Urban Impact Analysis: A New Tool for Monitoring the Geographical Effects of Federal Policies," *Professional Geographer*, 32, 1 (Feb. 1980), 82–85.

———, "A Question of Integrity: The National Labor Relations Board, Collective Bargaining and the Relocation of Work," *Political Geography Quarterly*, 7, 3 (July 1988), 209–228.

Corbridge, Stuart, "Crisis, What Crisis?" *Political Geography Quarterly*, 3, 4 (Oct. 1984), 331–345.

———, "The Asymmetry of Interdependence: The United States and the Geopolitics of International Financial Relations," *Studies in Comparative International Development*, 23, 2 (Summer 1988), 3–54.

Cox, B. A. and C. M. Rogerson, "The Corporate Power Elite in South Africa: Interlocking Directorships among Large Enterprises," *Political Geography Quarterly*, 4, 3 (July 1985), 219–234.

Cox, Kevin R., "Residential Mobility, Neighborhood Activism and Neighborhood Problems," *Political Geography Quarterly*, 2, 2 (April 1983), 99–117.

Crane, Melissa, "Diminishing Water Resources

and International Law; U.S.–Mexico, A Case Study," *Cornell International Law Journal*, 24, 2 (Spring 1991).

Crouch, Colin and David Marquand (eds.), *Towards Greater Europe? A Continent Without an Iron Curtain*. Special edition of *The Political Quarterly*. Oxford: Blackwell, 1992.

Crump, Jeff R. and J. Clark Archer, "Spatial and Temporal Variability in the Geography of American Defense Outlays," *Political Geography*, 12, 1 (Jan. 1993), 38–63.

**D**

Da Ponte, John J., Jr., "United States Foreign-Trade Zones: Adapting to Time and Space," *The Maritime Lawyer*, 5, 2 (Fall 1980), 197–217.

De Bres, Karen and Patty Ernst, "Insiders and Outsiders, or Core and Periphery in the Novels of Kenneth Roberts," *Political Geography*, 11, 5 (Sept. 1992), 449–460.

De Oliver, Miguel, "The Hegemonic Cycle and Free Trade: The U.S. and Mexico," *Political Geography*, 12, 5 (Sept. 1993), 457–472.

Duncan, S. S. and M. Goodwin, "The Local State: Functionalism, Autonomy and Class Relations in Cockburn and Saunders," *Political Geography Quarterly*, 1, 1 (Jan. 1982), 77–96.

Dunleavy, Patrick, Comments: "Class, Consumption and Radical Explanations in Urban Politics: A Rejoinder to Hooper, *Political Geography Quarterly*, 1, 2 (April 1982), 187–192.

**E**

*Environment and Planning C: Government and Policy*. London: Pion, quarterly journal published since 1982.

**F**

Filipp, Karlheinz, "Facing the Political Map of Germany," *Political Geography Quarterly*, 3, 3 (July 1984), 251–258.

Fincher, Ruth, "The State Apparatus and the Commodification of Quebec's Housing Cooperatives," *Political Geography Quarterly*, 3, 2 (April 1984), 127–143.

Fryer, Donald W., "The Political Geography of International Lending by Private Banks," *Transactions, Institute of British Geographers NS*, 12 (1987), 413–432.

Fuller, Gary and Forrest R. Pitts, "Youth Cohorts and Political Unrest in South Korea," *Political Geography Quarterly*, 9, 1 (Jan. 1990), 9–22.

**G**

Gares, Paul A., "Geographers and Policy-Making: Lessons Learned from the Failure of the New Jersey Dune Management Plan," *Professional Geographer*, 41, 1 (Feb. 1989), 20–28.

**H**

Hall, Peter, "The New Political Geography: Seven Years On," *Political Geography Quarterly*, 1, 1 (Jan. 1982), 65–76.

Hessenius, Charles, "Explaining Who Writes to Congressmen: A Contextual Analysis," *Political Geography Quarterly*, 10, 2 (April 1991), 149–161.

Hoare, Anthony G., "Dividing the Pork Barrel: Britain's Enterprise Zone Experience," *Political Geography Quarterly*, 4, 1 (Jan. 1985), 29–46.

Hoggart, Keith, "Political Parties and Local Authority Capital Investments in English Cities," *Political Geography Quarterly*, 3, 1 (Jan. 1984), 5–32.

——, "Geography, Political Control and Local Government Policy Outputs," *Progress in Human Geography*, 10, 1 (March 1986), 1–23.

Hundley, Norris, Jr., "The Colorado Waters Dispute," *Foreign Affairs*, 42, 3 (April 1964), 495–500.

——, "The Politics of Water and Geography: California and the Mexican-American Treaty of 1944," *Pacific Historical Review*, 36, 2 (1967), 209–226.

**J**

Jackson, Peter, "The Politics of the Streets: A Geography of Caribana," *Political Geography*, 11, 2 (March 1992), 130–151.

Janda, Kenneth and Robin Gillies, "How Well Does 'Region' Explain Political Party Characteristics?" *Political Geography Quarterly*, 2, 3 (July 1983), 179–203.

Johnson, R. W., "The Canada–United States Controversy over the Columbia River," *Univ. of Washington Law Review*, 41 (1966), 676–763.

Jones, Bryan D., Review Essay: "Government and Business: The Automobile Industry and the Public Sector in Michigan," *Political Geography Quarterly*, 5, 4 (Oct. 1986), 369–384.

**K**

Kirby, Andrew, Review Essay: "A Tribute to John House," *Political Geography Quarterly*, 3, 4 (Oct. 1984), 347–348.

Kliot, Nurit and Stanley Waterman, "The Political Impact on Writing the Geography of Palestine/Israel," *Progress in Human Geography*, 14 (1990), 237–260.

Knapp, Lawrence, "Some Theoretical Implications of Gay Involvement in an Urban Land Market," *Political Geography Quarterly*, 9, 4 (Oct. 1990), 337–352.

**L**

Lattimore, Owen, "The New Political Geography of Inner Asia," *Geographical Journal*, 119, 1 (March 1953), 17–32.

Lutz, James M., "The Spatial and Temporal Diffusion of Selected Licensing Laws in the United States," *Political Geography Quarterly*, 5, 2 (April 1986), 141–159.

**M**

Mair, Andrew, "The Homeless and the Post-Industrial City," *Political Geography Quarterly*, 5, 4 (Oct. 1986), 351–368.

Malecki, Edward J., "Government-Funded R&D: Some Regional Implications," *Professional Geographer*, 33, 1 (Feb. 1981), 72–82.

Marston, Sallie A., "Public Rituals and Community Power: St. Patrick's Day Parades in Lowell, Massachusetts, 1841–1874," *Political Geography Quarterly*, 8, 3 (July 1989), 255–270.

Martis, Kenneth C., "Sectionalism and the United States Congress," *Political Geography Quarterly*, 7, 2 (April 1988), 99–110.

McCalla, Robert J., "The Geographical Spread of Free Zones Associated with Ports," *Geoforum*, 21, 1 (1990), 121–134.

McIntyre, Alistair, "Caribbean: Free Ports Among the Islands," *Ceres*, 1, 6 (Nov. 1968), 38–41.

Michalak, Wieslaw Z. and Richard A. Gibb, "A Debt to the West: Recent Developments in the International Financial Situation of East-Central Europe," *Professional Geographer*, 44, 3 (August 1992), 260–271.

Mingst, Karen A., "The Ivory Coast at the Semi-Periphery of the World Economy," *International Studies Quarterly*, 32, 3 (Sept. 1988), 259–274.

Mitchell, Don, "Iconography and Locational Conflict from the Underside: Free Speech, People's Park, and the Politics of Homelessness in Berkeley, California," *Political Geography*, 11, 2 (March 1992), 152–169.

Mohan, John, "State Policies and the Development of the Hospital Service of Northeast England, 1948–1982," *Political Geography Quarterly*, 3, 4 (Oct. 1984), 275–295.

Murrell, Peter and Mancur Olson, "The Devolution of Centrally Planned Economies," *Journal of Comparative Economics*, 15 (June 1991), 239–265.

**N**

*Nongovernmental Organizations, the United Nations and Global Government*. Special issue of *Third World Quarterly*, 16, 3 (Fall 1995), whole volume.

**O**

O'hUalracháin, Breandán and Neil Reid, "Source Country Differences in the Spatial Distribution of Foreign Direct Investment in the United States," *Professional Geographer*, 44, 3 (August 1992), 272–285.

O'Loughlin, John and Herman van der Wusten, "The Political Geography of Panregions," *Geographical Review*, 80 (1990), 1–20.

ÓTuathail, Gearòid and Timothy W. Luke, "Present at the (Dis)integration: Deterritorialization and Reterritorialization in the New Wor(l)d Order," *Annals, AAG*, 84, 3 (Sept. 1994), 381–398.

**P**

Palan, R., "The European Miracle of Capital Accumulation," *Political Geography*, 11, 4 (July 1992), 401–406.

Pannell, Clifton W., "Economic Reforms and Readjustment in the People's Republic of China and Some Geographic Consequences," *Studies in Comparative International Development*, 22, 4 (Winter 1987–88), 54–73.

Peck, Jamie A., "TECs and the Local Politics of Training," *Political Geography*, 11, 4 (July 1992), 335–354.

Páerez-Láopez, Jorge F., "Swimming Against the Tide: Implications for Cuba of Soviet and Eastern European Reforms in Foreign Economic Relations," *Journal of Interamerican Relations and World Affairs*, 33 (Summer 1991), 81–139.

Pfister, Ulrich and Christian Suter, "International Financial Relations as Part of the World-System," *International Studies Quarterly*, 31, 3 (Sept. 1987), 239–272.

**R**

Raghuram, Parvati, "Comment: The New World Order from Below: The Role of Women in Transforming Indian Politics," *Political Geography*, 11, 4 (July 1992), 331–334.

Reynolds, David R. and Fred M. Shelley, "Procedural Justice and Local Democracy," *Political Geography Quarterly*, 4, 4 (Oct. 1985), 267–288.

Rich, Jan Gilbreath, "Bordering on Trouble," [maquiladoras] *Environmental Forum*, 8 (May 1991), 26–33.

Rose, Damaris, "'Collective Consumption' Revisited: Analysing Modes of Provision and Access to Childcare Services in Montréal, Quebec," *Political Geography Quarterly*, 3, 4 (Oct. 1990), 353–380.

Routledge, Paul, "Putting Politics in Its Place: Baliapal, India, as a Terrain of Resistance," *Political Geography*, 11, 6 (Nov. 1992), 588–611.

Rowley, Gwyn, "Local Government or Central Government Agency?—the British Case," *Political Geography Quarterly*, 3, 3 (July 1984), 265–268.

**S**

Sadler, David, "Works Closure at British Steel and the Nature of the State," *Political Geography Quarterly*, 3, 4 (Oct. 1984), 297–311.

Saunders, P., Comments: "Urban Politics," *Political Geography Quarterly*, 1, 2 (April 1982), 181–186.

Scramstad, Barbara, "Transboundary Movement of Hazardous Waste from the United States to Mexico," *The Transnational Lawyer*, 4 (Spring 1991).

Seth, Vijay K., "State and Spatial Aspects of Industrialization in Post-Independence India," *Political Geography Quarterly*, 5, 4 (Oct. 1986), 331–350.

South, Robert B., "Transnational 'Maquiladora' Location," *Annals, AAG*, 80, 4 (Dec. 1990), 549–571.

Staeheli, Lynn A., "Empowering Political Struggle: Spaces and Scales of Resistance," *Political Geography*, 13, 5 (Sept. 1994), 387–391.

———, "Empowering Women's Citizenship," *Political Geography*, 13, 5 (Sept. 1994), 443–460.

**T**

Taylor, Peter J., "Political Geography—Research Agendas for the Nineteen Eighties," *Political Geography*, 1, 1 (Jan. 1982), 1–17.

———, "Research Agendas for the Nineteen Eighties: Comments, Additions, and Critiques," *Political Geography Quarterly*, 1, 2 (April 1982), 167–180.

———, Review Essay: "Chaotic Conceptions, Antinomies, Dilemmas and Dialectics: Who's Afraid of the Capitalist World-Economy," *Political Geography Quarterly*, 3, 4 (Oct. 1984), 87–93.

Thrall, Grant Ian, "Three Pure Planning Scenarios and the Consumption Theory of Land Rent," *Political Geography Quarterly*, 2, 3 (July 1983), 219–231.

Thrift, Nigel and Andrew Leyshon, "A Phantom State? The Detraditionalization of Money, the International Financial System and International Financial Centres," *Political Geography*, 13, 4 (July 1994), 299–327.

Trubowitz, Peter, "Political Conflict and Foreign Policy in the United States; A Geographical Interpretation," *Political Geography*, 12, 2 (March 1993), 121–135.

**V**

Valauskas, Charles C., "China's Special Economic Zones in Perspective: A Contextual Discussion with Emphasis on the Shekou Industrial Zone," *Hastings International and Comparative Law Review*, 9, 2 (Winter 1986), 149–234.

van Brabant, Jozef M., "Renewal of Cooperation and Economic Transition in Eastern Europe," *Studies in Comparative Communism*, 24 (June 1991), 151–172.

**W**

Wade, Larry L. and John B. Gates, "A New Tariff Map of the United States (House of Representatives)," *Political Geography Quarterly*, 9, 3 (July 1990), 284–304.

Whitelegg, John, "Transport Policy, Fiscal Discrimination and the Role of the State," *Political Geography Quarterly*, 3, 4 (Oct. 1984), 313–329.

Wilson, David, "Toward a Revised Urban Managerialism: Local Managers and Community Development Block Grants," *Political Geography Quarterly*, 8, 1 (Jan. 1989), 21–42.

# Index of Persons

The index pages are designed to supplement the Table of Contents and List of Maps and do not repeat the information contained therein. Boldface page numbers indicate photographs; italic numbers indicate maps, graphs, and other illustrations.

# Subject Index

The index pages are designed to supplement the Table of Contents and List of Maps and do not repeat the information contained therein. Boldface page numbers indicate photographs; italic numbers indicate maps, graphs, and other illustrations.